Contents

The Structure and Dynamics of Networks

Mark Newman
Albert-László Barabási
Duncan J. Watts

Editors

Princeton University Press
Princeton and Oxford

Published by Princeton University Press, 41 William Street,
Princeton, New Jersey 08540
In the United Kingdom: Princeton University Press, 3 Market Place,
Woodstock, Oxfordshire OX20 1SY

Library of Congress Control Number: 2005921569

ISBN-13: 978-0-691-11356-2 (cl. alk. paper)
ISBN-10: 0-691-11356-4 (cl. alk. paper)

ISBN-13: 978-0-691-11357-9 (paper alk. paper)
ISBN-10: 0-691-11357-2 (paper alk. paper)

British Library Cataloging-in-Publication Data is available

The publisher would like to acknowledge the editors of this volume for providing, other than the previously published material, the camera-ready copy from which this book was printed.

Printed on acid-free paper.

pup.princeton.edu

Printed in the United States of America

10 9 8 7 6 5

Preface

Networks such as the Internet, the World Wide Web, and social and biological networks of various kinds have been the subject of intense study in recent years. From physics and computer science to biology and the social sciences, researchers have found that a great variety of systems can be represented as networks, and that there is much to be learned by studying those networks. The study of the web, for instance, has led to the creation of new and powerful web search engines that greatly outperform their predecessors. The study of social networks has led to new insights about the spread of diseases and techniques for controlling them. The study of metabolic networks has taught us about the fundamental building blocks of life and provided new tools for the analysis of the huge volumes of biochemical data that are being produced by gene sequencing, microarray experiments, and other techniques.

In this book we have gathered together a selection of research papers covering what we believe are the most important aspects of this new branch of science. The papers are drawn from a variety of fields, from many different journals, and cover both empirical and theoretical aspects of the study of networks. Along with the papers themselves we have included some commentary on their contents, in which we have tried to highlight what we believe to be the most important findings of each of the papers and offer pointers to other related literature. (Note that within the text of our commentary we have for convenience marked in **bold text** citations to papers that themselves are reproduced within this book; we hope this will save the reader some unnecessary trips to the library.)

After a short introduction (Chapter 1), the book opens with a collection of historical papers (Chapter 2) that predate the current burst of interest in networks, but that lay important foundations for the later work. Chapter 3 reproduces a selection of papers on empirical studies of networks in various fields, the raw experimental data on which many theoretical developments build. Then in Chapter 4, which occupies the largest portion of the book, we look at models of networks, focusing particularly on random graph models, small-world models, and models of scale-free networks. Chapter 5 deals with applications of network ideas to particular real-world problems, such as epidemiology, network robustness, and search algorithms. Finally, in Chapter 6 we give a short discussion of the most recent developments and where we see the field going in the next few years. There will of course be many developments that we cannot anticipate at present, and we look forward with excitement to the new ideas researchers come up with as we move into the 21st century.

This field is growing at a tremendous pace, with many new papers appearing every day, so there is no hope of making a compilation such as this exhaustive; inevitably many important and deserving papers have been left out. Nonetheless, we hope that by collecting a representative selection of papers together in one volume, this book will prove useful to students and researchers alike in the field of networks.

A number of people deserve our thanks for their help with the creation of this book. First, our thanks must go to our editor Vickie Kearn and everyone else at Princeton University Press for taking on this project and helping to make it a success. Many thanks also to Ádám Makkai, who translated from the original Hungarian the remarkable short story *Chains* that forms the first article reproduced in the book. And of course we have benefited enormously from conversations with our many erudite colleagues in the field. It is their work that forms the bulk of the material in this volume, and we are delighted to be a part of such an active and inspiring community of scientists.

Mark Newman
Albert-László Barabási
Duncan Watts

Chapter One
Introduction

Networks are everywhere. From the Internet and its close cousin the World Wide Web to networks in economics, networks of disease transmission, and even terrorist networks, the imagery of the network pervades modern culture.

What exactly do we mean by a network? What different kinds of networks are there? And how does their presence affect the way that events play out? In the past few years, a diverse group of scientists, including mathematicians, physicists, computer scientists, sociologists, and biologists, have been actively pursuing these questions and building in the process the new research field of network theory, or the "science of networks" (Barabási 2002; Buchanan 2002; Watts 2003).

Although it is still in a period of rapid development and papers are appearing daily, a significant literature has already accumulated in this new field, and it therefore seems appropriate to summarize it in a way that is accessible to researchers unfamiliar with the topic. That is the purpose of this book. We begin by sketching in this introductory chapter a brief history of the study of networks, whose beginnings lie in mathematics and more recently sociology. We then place the "new" science of networks in context by describing a number of features that distinguish it from what has gone before, and explain why these features are important. At the end of the chapter we give a short outline of the remainder of the book.

1.1 A BRIEF HISTORY OF THE STUDY OF NETWORKS

The study of networks has had a long history in mathematics and the sciences. In 1736, the great mathematician Leonard Euler became interested in a mathematical riddle called the Königsberg Bridge Problem. The city of Königsberg was built on the banks of the Pregel River in what was then Prussia,[1] and on two islands that lie in midstream. Seven bridges connected the land masses, as shown in Figure 1.1. (There are many more than that today.) A popular brain-teaser of the time asked, "Does there exist any single path that crosses all seven bridges exactly once each?" Legend has it that the people of Königsberg spent many fruitless hours trying to find such a path before Euler proved the impossibility of its existence. The proof, which perhaps seems rather trivial to us now, but which apparently wasn't obvious in 1736, makes use of a *graph*—a mathematical object consisting of points, also called *vertices* or *nodes*, and lines, also called *edges* or *links*, which abstracts away

[1] Today Königsberg lies in Russia and is called Kaliningrad.

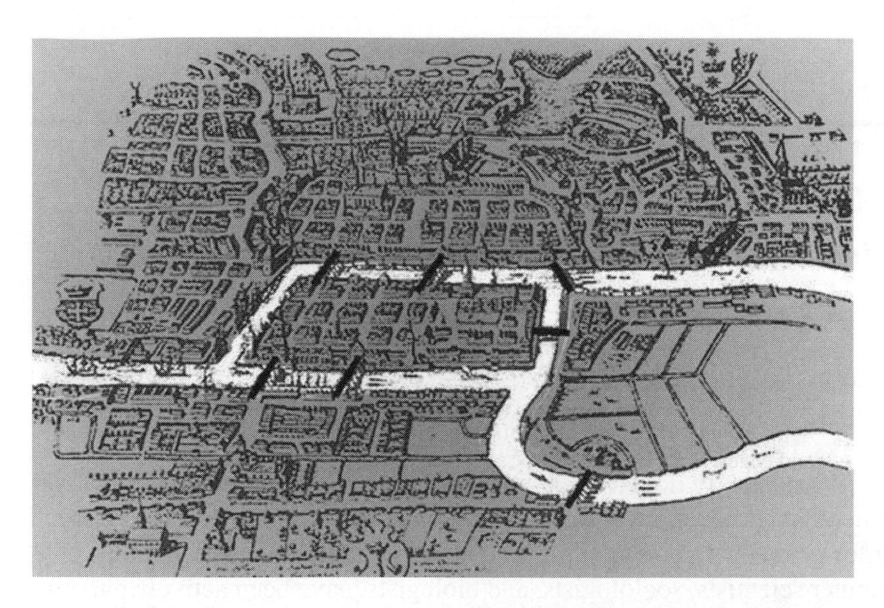

FIGURE 1.1 A map of eighteenth century Königsberg, with its seven bridges highlighted.

all the details of the original problem except for its connectivity. In this graph there are four vertices representing the four land masses and seven edges joining them in the pattern of the Königsberg bridges (Figure 1.2). Then the bridge problem can be rephrased in mathematical language as the question of whether there exists any *Eulerian path* on the network. An Eulerian path is precisely a path that traverses each edge exactly once. Euler proved that there is not, by observing that, since any such path must both enter and leave every vertex it passes through, except the first and last, there can at most be two vertices in the network with an odd number of edges attached. In the language of graph theory, we say that there can at most be two vertices with odd *degree*, the degree of a vertex being the number of edges attached to it.[2] Since all four vertices in the Königsberg graph have odd degree, the bridge problem necessarily has no solution. The problem of the existence of Eulerian paths on networks, as well as the related problem of *Hamiltonian paths* (paths that visit each vertex exactly once), is still of great interest to mathematicians, with new results being discovered all the time.

Many consider Euler's proof to be the first theorem in the now highly developed field of discrete mathematics known as *graph theory*, which in the past three centuries has become the principal mathematical language for describing the properties of networks (Harary 1995; West 1996). In its simplest form, a network is nothing more than a set of discrete elements (the vertices), and a set of connections (the edges) that link the elements, typically in a pairwise fashion. The elements

[2]Within physics some authors have referred to this quantity as the "connectivity" of a vertex, and the reader will see this usage in some of the papers reproduced in this book. As the word connectivity already has another meaning in graph theory, however, this choice of nomenclature has given rise to some confusion. To avoid such confusion, we will stick to standard usage in this book and refer to the degree of a vertex.

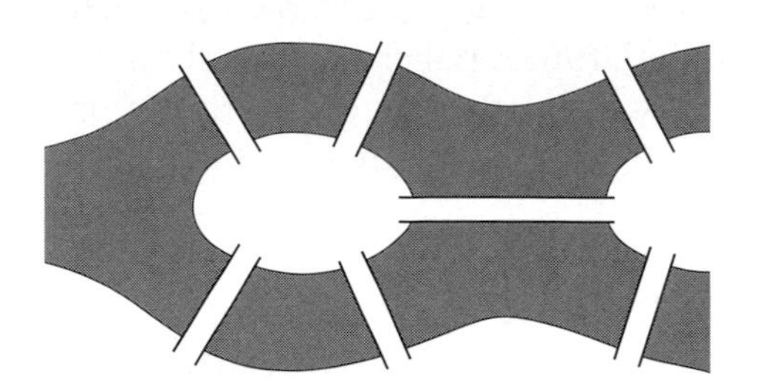

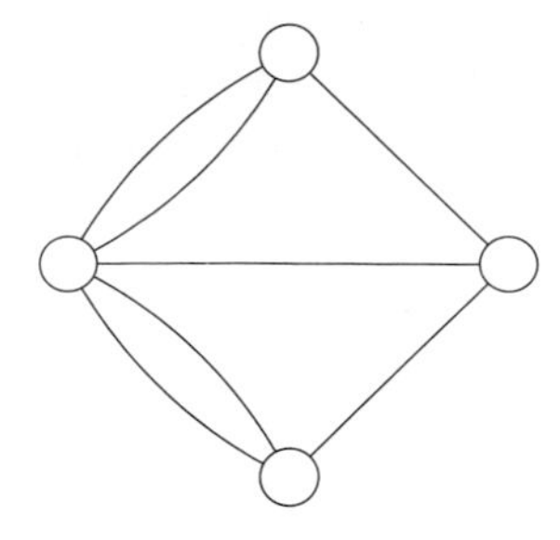

FIGURE 1.2 Left: a simplified depiction of the pattern of the rivers and bridges in the Königsberg bridge problem. Right: the corresponding network of vertices and edges.

and their connections can be almost anything—people and friendships (Rapoport and Horvath 1961), computers and communication lines **(Faloutsos *et al.* 1999)**, chemicals and reactions **(Jeong *et al.* 2000; Wagner and Fell 2001)**, scientific papers and citations (**Price 1965**; Redner 1998)[3]—causing some to wonder how so broad a definition could generate anything of substantive interest. But its breadth is precisely why graph theory is so powerful. By abstracting away the details of a problem, graph theory is capable of describing the important topological features with a clarity that would be impossible were all the details retained. As a consequence, graph theory has spread well beyond its original domain of pure mathematics, especially in the past few decades, to applications in engineering (Ahuja *et al.* 1993), operations research (Nagurney 1993), and computer science (Lynch 1996). Nowhere, however, has graph theory found a more welcome home than in sociology.

Starting in the 1950s, in response to a growing interest in quantitative methods in sociology and anthropology, the mathematical language of graph theory was coopted by social scientists to help understand data from ethnographic studies (Wasserman and Faust 1994; Degenne and Forsé 1999; Scott 2000). Much of the terminology of social network analysis—actor centrality, path lengths, cliques, connected components, and so forth—was either borrowed directly from graph theory or else adapted from it, to address questions of status, influence, cohesiveness, social roles, and identities in social networks. Thus, in addition to its role as a language for describing abstract models, graph theory became a practical tool for the analysis of empirical data. Also starting in the 1950s, mathematicians began to think of graphs as the medium through which various modes of influence—information and disease in particular—could propagate (**Solomonoff and Rapoport 1951**; **Erdős and Rényi 1960**). Thus the structural properties of networks, especially their connectedness, became linked with behavioral characteristics like the expected size of an epidemic or the possibility of global information transmission. Associated

[3]Throughout this book citations highlighted in **bold text** refer to papers that are reproduced within this book.

with this trend was the notion that graphs are properly regarded as stochastic objects (**Erdős and Rényi 1960**; Rapoport 1963), rather than purely deterministic ones, and therefore that graph properties can be thought of in terms of probability distributions—an approach that has been developed a great deal in recent years.

1.2 THE "NEW" SCIENCE OF NETWORKS

So what is there to add? If graph theory is such a powerful and general language and if so much beautiful and elegant work has already been done, what room is there for a new science of networks? We argue that the science of networks that has been taking shape over the last few years is distinguished from preceding work on networks in three important ways: (1) by focusing on the properties of real-world networks, it is concerned with empirical as well as theoretical questions; (2) it frequently takes the view that networks are not static, but evolve in time according to various dynamical rules; and (3) it aims, ultimately at least, to understand networks not just as topological objects, but also as the framework upon which distributed dynamical systems are built. As we will see in Chapter 3, elements of all these themes predate the recent explosion of interest in networks, but their synthesis into a coherent research agenda is new.

Modeling real-world networks

The first difference between the old science of networks and the new is that, social network analysis aside, traditional theories of networks have not been much concerned with the structure of naturally occurring networks. Much of graph theory qualifies as pure mathematics, and as such is concerned principally with the combinatorial properties of artificial constructs. Pure graph theory is elegant and deep, but it is not especially relevant to networks arising in the real world. Applied graph theory, as its name suggests, is more concerned with real-world network problems, but its approach is oriented toward design and engineering. By contrast, the recent work that is the topic of this book is focused on networks as they arise naturally, evolving in a manner that is typically unplanned and decentralized. Social networks and biological networks are naturally occurring networks of this kind, as are networks of information like citation networks and the World Wide Web. But the category is even broader, including networks—like transportation networks, power grids, and the physical Internet—-that are intended to serve a single, coordinated purpose (transportation, power delivery, communications), but which are built over long periods of time by many independent agents and authorities. Social network analysis, for its part, is strongly empirical, but tends to be descriptive rather than constructive in nature. With the possible exception of certain types of random graph models (Holland and Leinhardt 1981; Strauss 1986; Anderson *et al.* 1999), network analysis in the social sciences has largely avoided modeling, preferring simply to describe the properties of networks as observed in collected data.

In contrast to traditional graph theory on the one hand, and social network analysis on the other, the work described in this book takes a view that is both theoretical and empirical. In order to develop new graph-theoretic models that can account for the structural features of real-world networks, we must first be able to say what those features are and hence empirical data are essential. But adequate theoretical models are equally essential if the significance of any particular empirical finding is to be correctly understood. Just as in traditional science, where theory and experiment continually stimulate one another, the science of networks is being built on the twin foundations of empirical observation and modeling.

That such an obvious requirement for scientific validity should have made its first appearance in the field so recently seems surprising at first, but is understandable given the historical difficulty of obtaining high quality, large-scale network data. For most of the past fifty years, the collection of network data has been confined to the field of social network analysis, in which data have to be collected through survey instruments that not only are onerous to administer, but also suffer from the inaccurate or subjective responses of subjects. People, it turns out, are not good at remembering who their friends are, and the definition of a "friend" is often quite ambiguous in the first place.

For example, the General Social Survey[4] requests respondents to name up to six individuals with whom they discuss "important matters." The assumption is that people discuss matters that are important to them with people who are important to them, and hence that questions of this kind—so-called "name generators"—are a reliable means of identifying strong social ties. However, a recent study by Bearman and Parigi (2004) shows that when people are asked about the so-called "important matters" they are discussing, they respond with just about every topic imaginable, including many that most of us wouldn't consider important at all. Even worse, some topics are discussed with family members, some with close friends, some with coworkers, and others with complete strangers. Thus, very little can be inferred about the network ties of respondents simply by looking at the names generated by the questions in the General Social Survey. Bearman and Parigi also find that some 20% of respondents name no one at all. One might assume that these individuals are "social isolates"—people with no one to talk to—yet nearly 40% of these isolates are married! It is possible that these findings reveal significant patterns of behavior in contemporary social life—perhaps many people, even married people, really do not have anyone to talk to, or anything important to talk about. But apparently the respondent data are so contaminated by diverse interpretations of the survey instrument, along with variable recollection and even laziness, that any inferences about the corresponding social network must be regarded with skepticism.

The example of the General Social Survey is instructive because it typifies the uncertainties associated with traditional, survey-based collection of network data. If people have difficulty identifying even their closest confidants, how can one expect to extract reliable information concerning more subtle relations? And if, in response to this obstacle, survey instruments become more elaborate and spe-

[4]See http://www.norc.uchicago.edu/projects/gensoc.asp.

cific, then as the size of the surveyed population increases, the work required of the researcher to analyze and understand the resulting volume of raw data becomes prohibitive. A better approach would be to record the activities and interactions of subjects directly, thus avoiding recall problems and allowing us to apply consistent criteria to define relationships. In the absence of accurate recording technologies, however, such direct observation methods are even more onerous than the administration of surveys.

Because of the effort involved in compiling them, social network datasets rarely document populations of more than a hundred people and almost never more than a thousand. And although other kinds of (nonsocial) networks have not suffered from the same difficulties, empirical examples prior to the last decade have been few—probably because other network-oriented disciplines have lacked the empirical focus of sociology. The lack of high quality, large-scale network data has, in turn, delayed the development of the kind of statistical models with which much of the work in this book is concerned. Such models, as we will see, can be very successful and informative when applied to large networks, but tend to break down, or simply don't address the right questions, when applied to small ones. As an example, networks of contacts between terrorists have been studied recently by, for instance, Krebs (2002), but they are poor candidates for statistical modeling because the questions of interest in these networks are not statistical in nature, focusing more on the roles of individuals and small groups within the network as a whole. The traditional tools of social network analysis—centrality indices, structural measures, and measures of social capital—are more useful in such cases.

Recent years, however, have witnessed a dramatic increase in the availability of network datasets that comprise many thousands and sometimes even millions of vertices—a consequence of the widespread availability of electronic databases and, even more important, the Internet. Not only has the Internet focused popular and scientific attention alike on the topic of networks and networked systems, but it has led to data collection methods for social and other networks that avoid many of the difficulties of traditional sociometry. Networks of scientific collaborations, for example, can now be recorded in real time through electronic databases like Medline and the Science Citation Index (**Newman 2001a**; Barabási *et al.* 2002), and even more promising sources of network data, such as email logs (Ebel *et al.* 2002; Guimerà *et al.* 2003; Tyler *et al.* 2003) and instant messaging services (Smith 2002; Holme *et al.* 2004), await further exploration. Being far larger than the datasets of traditional social network analysis, these networks are more amenable to the kinds of statistical techniques with which physicists and mathematicians are familiar. As the papers in Chapter 3 of this volume demonstrate, real networks, from citation networks and the World Wide Web to networks of biochemical reactions, display properties—like local clustering and skewed degree distributions—that were not anticipated by the idealized models of graph theory, and that have forced the development of new modeling approaches, some of which are introduced in Chapter 4.

Networks as evolving structures

A second distinguishing feature of the work described in this book is that, whereas in the past both graph theory and social network analysis have tended to treat networks as static structures, recent work has recognized that networks evolve over time (**Barabási and Albert 1999**; Watts 1999). Many networks are the product of dynamical processes that add or remove vertices or edges. For instance, a social network of friendships changes as individuals make and break ties with others. An individual with many acquaintances might, by virtue of being better connected or better known, be more likely to make new friends than someone else who is less well connected. Or individuals seeking friends might be more likely to meet people with whom they share a common acquaintance. The ties people make affect the form of the network, and the form of the network affects the ties people make. Social network structure therefore evolves in a historically dependent manner, in which the role of the participants and the patterns of behavior they follow cannot be ignored.

Similar statements apply to other kinds of networks as well: processes operating at the local level both constrain and are constrained by the network structure. A principal objective of the new science of networks (as dealt with by a number of papers in Chapter 4), is an understanding of how structure at the global scale (say, the connectivity of the network as a whole) depends on dynamical processes that operate at the local scale (for example, rules governing the appearance and connections of new vertices).

Networks as dynamical systems

The final feature that distinguishes the research described in this book from previous work is that traditional approaches to networks have tended to overlook or oversimplify the relationship between the structural properties of a networked system and its behavior. A lot of the recent work on networks, by contrast, takes a dynamical systems view according to which the vertices of a graph represent discrete dynamical entities, with their own rules of behavior, and the edges represent couplings between the entities. Thus a network of interacting individuals, or a computer network in which a virus is spreading, not only has topological properties, but has dynamical properties as well. Interacting individuals, for instance, might affect one another's opinions in reaching some collective decision (voting in a general election, for example), while an outbreak of a computer virus may or may not become an epidemic depending on the patterns of connections between machines. Which outcomes occur, how frequently they occur, and with what consequences, are all questions that can only be resolved by thinking jointly about structure and dynamics, and the relationship between the two.

Questions of this nature are not easily tackled, however; dynamical problems lie at the forefront of network research, where there are many unanswered questions. One class of problems on which some progress has been made, and which is addressed in Section 5.1, is that of contagion dynamics. Whether we are interested in the spread of a disease or the diffusion of a technological innovation, it is frequently the case that contagion occurs over a network. Not only physical but also social contacts can significantly influence the probability that a particular disease or piece of information will be transmitted, and also what effect it will have. In traditional mathematical epidemiology, as well as research on the diffusion of information, it is usually assumed that all members of the population have equal likelihood of interaction with all others. Clearly this assumption requires modification once we take network structure into account. As the papers in Section 5.1 demonstrate, the particular structure of the network through which a contagious agent is transmitted can have a dramatic impact on outcomes at the level of entire populations.

1.3 OVERVIEW OF THE VOLUME

Mirroring the themes introduced above, this volume is divided into a number of parts, each of which is preceded by an introduction that outlines the general theme and summarizes the contributions of the papers in that part. Chapter 2 sets the stage by presenting a selection of papers that we feel are important historical antecedents to contemporary research. Although recent work on networks takes a distinctly different approach from traditional network studies, a careful reading of Chapter 2 reveals that many of the basic themes were anticipated by mathematicians and social scientists years or even decades earlier. Given their age, some of these contributions seem remarkably familiar and modern, occasionally to the extent that recent papers almost exactly replicate previous results. Power-law distributions, random networks with local clustering, the notion of long-range shortcuts, and the small-world phenomenon were all explored and analyzed well before the new science of networks reconstituted the same ideas in the language of mathematical physics.

Chapter 3 emphasizes the empirical side of the new science of networks, and Chapter 4 presents some of the foundational modeling ideas that have generated a great deal of subsequent interest and activity. By exploring some tentative applications of the ideas introduced in Chapters 3 and 4, Chapter 5 takes the reader to the cutting edge of network science, the relationship between network structure and system dynamics. From disease spreading and network robustness to search algorithms, Chapter 5 is a potpourri of topics at this poorly understood but rapidly expanding frontier. Finally, Chapter 6 provides a short discussion of what we see as some of the most interesting directions for future research. We hope the reader will be encouraged to strike out from where the papers in this volume leave off, adding his or her own ideas and results to this exciting and fast-developing field.

Chapter Two
Historical developments

The study of networks has had a long history in mathematics and the sciences, stretching back at least as far as Leonhard Euler's 1736 solution of the Königsberg Bridge Problem discussed in Chapter 1. In this chapter we present a selection of historical publications on the subject of networks of various kinds. Of particular interest to us are papers from mathematical graph theory and from the literature on social networks. For example, the classic model of a network that we know of as the *random graph*, and which is discussed in greater detail in Section 4.1, was first described by the Russian mathematician and biologist Anatol Rapoport in the early 1950s, before being rediscovered and analyzed extensively by Paul Erdős and Alfréd Rényi in a series of papers in the late 1950s and early 1960s. Around the same time, a social scientist and a mathematician, Ithiel de Sola Pool and Manfred Kochen, in collaboration gave a beautiful and influential discussion of the "small-world effect" in an early preprint on social networks.

Thus, while much of this book is devoted to recent work on networks in the physics and applied mathematics communities, many of the crucial ideas that have motivated that work were well known, at least to some, many decades earlier. The articles reproduced in this chapter provide an overview of some of the original work on these topics and set the scene for the material that appears in the following chapters.

Karinthy (1929)

The first publication reproduced in this chapter is in fact not a scientific paper at all, but a translation of a short story, a work of fiction, originally published in Hungarian in 1929. Certainly this is an unusual way to start a volume of scientific reprints, but, as the reader will see, this brief story, published more than seventy years ago, describes beautifully one of the fundamental truths about network structure that has driven scientific research in the field for the last few decades, the concept known today as the "small-world effect," or "six degrees of separation."

The writer Frigyes Karinthy (1887–1938) became an overnight sensation in Hungary following the publication in 1912 of his first book, a volume of literary caricature, which is required reading in Hungarian schools even today. Karinthy's 1929 volume of short stories, entitled *Everything is Different,* did not receive the same warm welcome from the literary establishment. Friends and critics alike believed the book to be little more than a scheme for making some quick money by stringing together a set of short pieces with scant respect for coherence or flow. Among the pieces in this collection, however, is one gem of a story, entitled "Chains," in which the writer raised in a fictional context some of the questions that network theory would be struggling with for much of the rest of the century. Without any pretensions to scientific rigor or proof, Karinthy tackled and suggested answers to one of the deep problems in the theory of networks.

In "Chains" Karinthy argues, as Jules Verne did fifty years earlier, that the world is getting smaller. Unlike Verne in "Around the World in Eighty Days," however, Karinthy proposes to demonstrate his thesis not by physical means—circumnavigating the globe—but by a *social* argument. He claims that people are increasingly connected to each other via their acquaintances, and that the dense web of friendship surrounding each person leads to an interconnected world in which everyone on Earth is at most *five acquaintances away from anyone else.*

To back up this remarkable claim, Karinthy demonstrates that it is possible to connect a Nobel prize winner to himself via a chain of just five acquaintances. He also points out, however, that this may not be an entirely fair example, because famous people with many social connections can be more easily connected to others, an insight whose relevance in the study of networks has only been fully appreciated quite recently. To show that his "five degrees of separation" claim also applies to less prominent people than Nobel laureates, he connects a worker in a Ford factory to himself, again via five acquaintances. Finally, he argues that the changing nature of human acquaintance patterns is a consequence of human exploration, of the demolition of geographical boundaries, and of new technologies that allow us to stay in touch even when we are thousands of miles apart.

The idea of chains of acquaintances linking distant individuals has been revisited many times in the decades since Karinthy's story. Jane Jacobs in her influential 1961 book *The Death and Life of Great American Cities* recalls:

> When my sister and I first came to New York from a small city, we used to amuse ourselves with a game we called Messages. The idea was to pick two wildly dissimilar individuals—say a head hunter in the Solomon Islands and a cobbler in Rock Island, Illinois—and assume that one had to get a message to the other by word of mouth; then we would each silently figure out a plausible, or at least possible, chain of persons through which the message could go. The one who could make the shortest plausible chain of messengers won. The head hunter would speak to the head man of his village, who would speak to the trader who came to buy cobra, who would speak to the Australian patrol officer when he came through, who would tell the man who was next slated to go to Melbourne on leave, etc. Down at the other end the cobbler would hear from his priest, who got it from the mayor, who got it from a state senator, who got it from the governor, etc. We soon had these close-to-home messengers down to a routine for almost everybody we could conjure up.
>
> — Jacobs (1961), pp. 134–135

Jacobs settles on an unusually long chain, however; the path in her example is at least nine links long, not counting the links that are presumably missing between the man heading to Melbourne and the governor.

Solomonoff and Rapoport (1951)

Scientific interest in the structure of networks began to develop in earnest in the 1940s and 1950s. Perhaps the most profound thinker in the field during this period was Anatol Rapoport, a Russian immigrant to the United States who worked not in sociology but in mathematical biology. Trained first as a pianist in Vienna, Rapoport turned to mathematics after realizing that a successful career as a concert performer would require the support of a wealthy patron, which he didn't have (Spencer 2002). He was unusual in developing an interest for mathematical biology at a time when mathematicians and biologists hardly spoke to each other, and he developed startling and prescient views about many topics that fall into the area we now call complex systems. In particular, he was decades ahead of his time in his views on the properties and importance of networks, developing methods that concentrated, as we often do today, on general statistical properties of networks, rather than individual properties of network nodes or edges. In a 1961 paper with William H. Horvath, he wrote,

> The theoretician's interest, however, is seldom focused on a particular large sociogram [i.e., network]. Rather, the interesting features of large sociograms are revealed in their gross, typical properties. Thus one seeks to define classes of sociograms, or else describe them by a few well-chosen parameters. It is perhaps natural to consider statistical parameters, since one is interested in trends or averages, or distributions rather than particulars.
>
> — Rapoport and Horvath (1961)

This remarkable statement could easily serve as a manifesto for the revolution in the study of networks that has recently taken place, four decades later, in the physics and mathematics communities.

In this chapter, we reproduce the important 1951 paper by Rapoport and Ray Solomonoff which presents the first systematic study of what we would now call a *random graph*. The paper is important both because it introduces the random graph for the first time and because it demonstrates one of the most crucial properties of the model: as the ratio of the number of edges to vertices in the graph is increased, the network reaches a point at which it undergoes an abrupt change from a collection of disconnected vertices to a connected state in which, in modern parlance, the graph contains a *giant component*.

The paper starts by considering a graph composed of a collection of vertices randomly connected to one another by edges (or axons, to use the paper's neurologically inspired terminology). The authors discuss three natural systems in which such networks might appear: neural networks, the social networks of physical contacts that are responsible for the spread of epidemic disease, and a network problem rooted in genetics.

Solomonoff and Rapoport define a quantity called the *weak connectivity*, which is the expected number of vertices reachable through the network from a randomly chosen vertex. In the modern terminology of networks, the weak connectivity is the average component size in the network. Solomonoff and Rapoport then derive an iteration relation for the weak connectivity by reasoning about the behavior of a simple component-finding algorithm which is equivalent to what we would today call a burning algorithm or breadth-first search. This result leads them to conclude that the average component size depends crucially on the mean degree a, where the degree again is the number of edges connected to a vertex. They show that for $a < 1$ the network is broken into many small isolated islands, but that when the mean degree exceeds $a = 1$ a giant component forms that contains a finite fraction of all the vertices in the network. Thus, although they did not use this language, Solomonoff and Rapoport predicted[1] in 1951 the existence of a phase transition from a fragmented network for $a < 1$ to one dominated by a giant component for $a > 1$.

Erdős and Rényi (1960)

Despite the early contributions of Solomonoff and Rapoport, random graph theory did not really take off until the late 1950s and early 1960s, when several important papers on the subject appeared almost simultaneously (Ford and Uhlenbeck 1957; Erdős and Rényi 1959, 1960; Gilbert 1959). Among these, the most influential, and the most relevant to current work, were the papers by Paul Erdős and Alfréd Rényi, who are considered the fathers of the modern theory of random graphs. Between 1959 and 1968 Erdős and Rényi published eight papers on random graphs that set the tone for network research for many decades to come. The next paper reproduced in this section **(Erdős and Rényi 1960)** is probably the most important of these. It deals with the evolution of the structure of random graphs as the mean degree is increased.

In this paper, the authors showed that many properties of random graphs emerge not gradually but suddenly, when enough edges are added to the graph. They made use of the following definition: if the probability of a graph having property Q approaches 1 as the size of the graph $N \to \infty$, then we say that *almost every* graph of N vertices has the property Q. They studied the behavior of a variety of different properties as a function of the probability p of the existence of an edge between any two vertices, and showed that for many properties there is a critical probability $p_c(N)$ such that if $p(N)$ grows more slowly than $p_c(N)$ as $N \to \infty$ then almost every graph with connection probability $p(N)$ fails to have the property Q. Conversely, if $p(N)$ grows faster than $p_c(N)$ then almost every graph has the property Q. Thus the probability that a graph with N nodes and

[1] This result, and indeed the invention of the random graph itself, is usually attributed to Erdős and Rényi (1959), but it is clear that Solomonoff and Rapoport had many of the crucial results almost a decade earlier. Erdős and Rényi appear not to have been aware of Solomonoff and Rapoport's work, and rediscovered their results independently. Erdős and Rényi's work also went much farther than that of Solomonoff and Rapoport and maintained a substantially higher level of rigor.

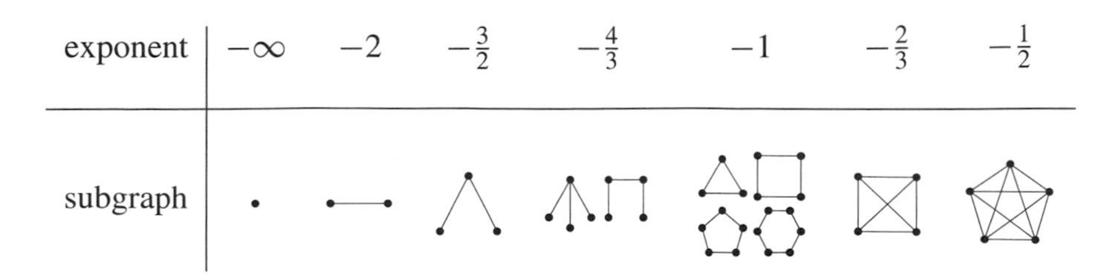

TABLE 2.1 The threshold probabilities at which different subgraphs appear in a random graph. For $pN^{3/2} \to 0$ the graph consists only of isolated nodes or pairs connected by edges. When $p \sim N^{-3/2}$ trees with 3 edges appear and at $p \sim N^{-4/3}$ trees with 4. If $p \sim N^{-1}$ trees of all sizes are present, as well as cycles of all lengths. When $p \sim N^{-2/3}$ the graph contains complete subgraphs of 4 vertices and for $p \sim N^{-1/2}$ there are complete subgraphs of 5 vertices. As the exponent approaches 0, the graph contains complete subgraphs of increasing order.

connection probability $p = p(N)$ has property Q satisfies

$$\lim_{N\to\infty} P_{N,p}(Q) = \begin{cases} 0 & \text{if } p(N)/p_c(N) \to 0, \\ 1 & \text{if } p(N)/p_c(N) \to \infty. \end{cases} \tag{2.1}$$

As an example, let us consider one of the first cases discussed by Erdős and Rényi, namely the appearance of a given subgraph within a random graph. For low values of the edge probability p, the graph is very sparse and the likelihood of finding, for example, a single vertex connected to two others is very low. One might imagine that in general the probability of there being such a structure somewhere on the graph would increase slowly with increasing p, but Erdős and Rényi prove that this is not the case. Instead, the probability of finding a connected trio of vertices is negligible if $p < cN^{-1/2}$ for some constant c, but tends to 1 as N becomes large if $p > cN^{-1/2}$. In other words, almost all graphs contain a connected trio of vertices if the number of links is greater than a constant times $N^{1/2}$, but almost none of them do if the number of links is less than this.

Erdős and Rényi generalized this result further to show that the probability of occurrence of a tree of k vertices in the graph (i.e., a connected set of k vertices containing no loops) tends to 1 on large graphs with more than a constant times $N^{(k-2)/(k-1)}$ edges. They also extended their method to cycles, that is, closed loops of vertices in which every two consecutive vertices, and only these, are joined by an edge. Cycles also show a threshold behavior—they are present with probability 1 above some critical value of p as the graph becomes large. In Table 2.1 we summarize some of the thresholds found in the evolution of random graphs.

An area of study closely related to random graphs is *percolation theory* (Stauffer and Aharony 1992; Bunde and Havlin 1994, 1996), which has been the object of attention within the physics community for many years, since the introduction in the 1950s of the original percolation model by Hammersley and others (Broadbent and Hammersley 1957; Hammersley 1957). In bond percolation models, one studies the properties of the system in which the bonds on a lattice or network are either occupied or not with some occupation probability p, asking questions such as what the mean sizes are of the clusters of lattice sites connected together by occupied bonds, and whether or not there exists a "spanning cluster" in the limit of large system size (i.e., a cluster that connects opposite sides of the lattice via a path of occupied bonds). It is clear that the random graph model is equivalent to bond percolation on a complete graph (i.e., a graph in which every vertex is connected to every other), and hence the methods developed for studying percolation can be applied to random graphs also. In particular, there has been much effort devoted to the study of the behavior of percolation models close to the phase transition at which a spanning cluster forms, which in random graph language is the point at which a *giant component* appears. It is known, for example, that many properties display *universal* behavior close to the phase transition, behavior that is dependent on the system dimension but not on the details of the lattice. And even the dimension dependence vanishes above the so-called *upper critical dimension*, giving way to generic behavior that can be extracted using simple mean-field theories. Since a complete graph is a formally infinite-dimensional object in the limit of large system size, the behavior of the random graph near the phase transition therefore falls into this *mean-field universality class*, and many results for the random graph, such as values for critical exponents can then be extracted from mean-field theory.

Pool and Kochen (1978)

In the late 1950s, around the same time that Erdős and Rényi were beginning their work on random graphs, the sociological community started developing an interest in applications of graph theory. The next paper reproduced in this section is the influential article on patterns of social contacts by the political scientist Ithiel de Sola Pool and the mathematician Manfred Kochen (**Pool and Kochen 1978**). This paper was actually written in 1958, and circulated for many years in preprint form. In it Pool and Kochen addressed for the first time many of the questions that the field would be struggling with for the next few decades, and yet they felt that they had not dealt satisfactorily with the issues and so didn't submit their work for publication in a journal. It was not until twenty years after its first appearance that the authors consented to the publication of this important work in the new journal *Social Networks*, on page 1 of volume 1. Pool and Kochen's work provided the inspiration for, among other things, the famous "small-world" experiments conducted in the 1960s by Stanley Milgram (Milgram 1967; **Travers and Milgram 1969**), which are the subject of the following paper in this book. Hence it is appropriate that we here reproduce Pool and Kochen's work ahead of Milgram's, even though Milgram's bears the earlier date of publication.

In the introduction to their paper, Pool and Kochen formulate some of the questions that have come to define the field of social networks:

i) How many other people does each individual in a network know? In other words what is the person's degree in the network? (Pool and Kochen refer to this quantity as the "acquaintance volume.")

ii) What is the distribution of the degrees? What are their mean and their largest and smallest values?

iii) What kinds of people have large numbers of contacts? Are these the most influential people in the network?

iv) How exactly are the contacts organized? What is the structure of the network?

In addition to these general questions about individuals and about the network as a whole, Pool and Kochen looked also at questions about paths between pairs of individuals:

i) What is the probability that two people chosen at random from the population will know each other?

ii) What is the chance that they have a friend in common?

iii) What is the chance that the shortest chain between them requires two intermediates? Or more than two?

Pool and Kochen start by discussing the difficulty of determining the number of social contacts people have. There are two primary problems: ambiguity about what exactly constitutes a social contact, and the fact that people are not very good at estimating the number of their acquaintances even if the definition of an acquaintance is clear. Typically most people underestimate their acquaintance volume.

Given the limited and unreliable nature of network data, Pool and Kochen resort to mathematical models. Inspired by Rapoport's work **(Solomonoff and Rapoport 1951)**, they base their work on the random graph, using this simple model to make conjectures about the characteristics of social networks.

This paper discusses for the first time in scientific terms the phenomenon we now call the small-world effect. Starting with the assumption that each person has about 1,000 acquaintances, they predict that most pairs of people on Earth can be connected via a path that goes through just two intermediate acquaintances. They give arguments reminiscent of those found in Karinthy's 1929 short story, reproduced in this chapter, to make this counterintuitive claim more plausible. They also consider the possibility that community groupings and social stratification within the network would affect their conclusions. But, after some laborious calculation, they conclude, apparently somewhat to their own surprise, that social strata have only a small effect on the average distance between individuals.

Travers and Milgram (1969)

Although network ideas were already becoming popular among sociologists in the 1950s and 1960s, it was an experimentalist, Stanley Milgram, who propelled the field into the public consciousness in the late 1960s with his famous small-world experiment. Milgram, at that time working at Harvard and influenced by the thinking of Harrison White and Ithiel Pool, both also in the Boston area, was inspired to devise an experiment that could test Pool and Kochen's surprising conjectures about path lengths between individuals in social networks.

Milgram published several papers about his small-world experiments. The earliest and best known is a 1967 piece that he wrote for the popular newsstand magazine *Psychology Today* (Milgram 1967). Although entertaining and thought provoking, this is not a rigorous piece of scientific writing, and many of the details of his work are left out of the discussion. After the first set of experiments, Milgram started collaborating with Jeffrey Travers, and repeated the experiments with new subjects and more detailed quantitative analyses. The 1969 article Milgram coauthored with Travers contains a clear and thorough explanation of these new experiments, and it is this second paper that we reproduce here.

Milgram's experiments started by selecting a target individual and a group of starting individuals. A package was mailed to each of the starters containing a small booklet or "passport" in which participants were asked to record some information about themselves. Then the participants were to try and get their passport to the specified target person by passing it on to someone they knew on a first-name basis who they believed either would know the target, or might know somebody who did. These acquaintances were then asked to do the same, repeating the process until, with luck, the passport reached the designated target. At each step participants were also asked to send a postcard to Travers and Milgram, allowing the researchers to reconstruct the path taken by the passport, should it get lost before it reached the target. Travers and Milgram recruited 296 starting individuals, 196 from Omaha, Nebraska and the other 100 from Boston. The target was a stockbroker who lived in Sharon, Massachusetts, a small town outside Boston.

In the end, 64 of the 296 chains reached the target, 29% of those that started out. The number of intermediate acquaintances between source and target varied from 1 to 11, the median being 5.2. Five intermediate acquaintances means that there were six steps along the chain, a result that has passed into popular myth in the phrase "six degrees of separation," which was the title of a 1990 Broadway play by John Guare in which one of the characters discusses the small-world effect.

To what degree can we trust the results of Milgram's experiments? Are we indeed just six steps from anyone else on average, or could the real result be closer to three as predicted by Pool and Kochen? Or perhaps the average separation is larger than six? This question is discussed in some detail by Travers and Milgram. The letters were more likely to get lost if they took a longer path from source to target, and hence the completed chains that Travers and Milgram used to estimate the average chain length are probably biased toward the shorter lengths. As Travers and Milgram describe in a footnote, however, White (1970) calculated a correction to the raw results to allow for this effect and found that the change in the figures was not large: the correction increases the average separation from 6 to 8. But there are other effects acting in the opposite direction also, potentially making the mean separation shorter than six. In particular, there is no guarantee that Travers and Milgram's subjects would have found the shortest path through the network to the target person. They forwarded the letter to the person of their acquaintance who they *thought* was closest to the target person, but they could easily have had another acquaintance who—unknown to them—was acquainted directly with the target. Thus the real separation between participants could be much shorter than that recorded by the experiment.

Price (1965)

At about the same time that Milgram was developing his first small-world experiment, the empirical study of networks was being taken up in another very different branch of the scientific community, information science. Derek de Solla Price's 1965 article "Networks of Scientific Papers," which appeared in the journal *Science*, is a hidden treasure, largely unknown within the mathematics and physics communities. In this paper, Price studies one of the oldest of information networks, the network of citations between scientific papers, in which each vertex represents a paper and a directed edge from one paper to another indicates that the first paper cites the second in its bibliography. Price indeed appears to have been one of the first to suggest that we view the pattern of citations as a network at all, and to present detailed statistical analyses of this network, for which he made use of the databases of citations that were just starting to become available, thanks to the work of Eugene Garfield and others.

Since citation networks are directed, each paper in such a network has both an out-degree (the numbers of papers that it cites) and an in-degree (the number of papers in which it is cited). Price studied the distributions of both in- and out-degrees and found that both have power-law tails, with exponents of about -2 and -3, respectively. Networks with power-law degree distributions are now known to occur in a number of different settings and are often called "scale-free networks" (see Chapter 3 and Section 4.3).

The quality of citation data has improved markedly in the years since Price's pioneering work, and particularly since the advent of computer tabulation of data, and a number of more recent studies have improved upon Price's results. Of particular interest is the paper by Redner (1998), in which the author independently discovered Price's power law using two large databases of citations of physics papers. Redner investigated the citation frequency of 783 339 papers published in 1981 and cited over 6 million times between 1981 and 1997, using data collected by the Institute for Scientific Information, the commercial enterprise that grew out of Garfield's early work on citation. The careful analysis presented in the paper shows that the in-degree of the citation network does indeed have a power-law tail, with an exponent roughly equal to -3. A second data set compiled from the bibliographies of 24 296 papers published in the journal *Physical Review D* between 1975 and 1994 shows similar results.

The paper reproduced here is not Derek Price's only contribution to the study of citation networks. A decade later he published a second remarkable paper in which he proposed a possible mechanism for the generation of the power laws seen in the citation distribution (Price 1976). Building on previous work by Simon (1955), he proposed that papers that have many citations receive further citations in proportion to the number they already have. He called this process "cumulative advantage," and gave a mathematical model of it, which he solved to demonstrate that it does indeed give rise to power-law distributions as observed in the data. The cumulative advantage process is more commonly known today under the name "preferential attachment," and is widely accepted as the explanation for the occurrence of power-law degree distributions in networks as diverse as the World Wide Web, social networks, and biological networks.

De Castro and Grossman (1999)

Our final paper in this chapter deals with another network formed by the patterns of scientific publication, and while it is a relatively recent work, having been published in 1999, we feel it belongs here in this historical section, as it summarizes an idea that has been current in the mathematics community for some decades but has rarely been studied formally.

Paul Erdős was a stunningly prolific mathematician who lived from 1913 to 1996. During his long life, he authored over 1500 papers with more than 500 coauthors, including his papers on random graphs with fellow Hungarian Alfréd Rényi, which are discussed earlier in this section. His staggering output, together with his pivotal role in the development of the theory of networks, prompted some of his colleagues to see him as a central node of the worldwide collaboration network of mathematicians and other scientific researchers.

Consider the network whose vertices are mathematicians and scientists, with an edge between any two vertices if the researchers they represent have coauthored one or more papers together. For each vertex we define the *Erdős number* to be the length of the shortest path from that vertex to Paul Erdős along the edges of the network. As de Castro and Grossman describe it,

> Paul Erdős himself has Erdős number 0, and his co-authors have Erdős number 1. People not having Erdős number 0 or 1 who have published with someone with Erdős number 1 have Erdős number 2, and so on. Those who are not linked in this way to Paul Erdős have Erdős number ∞.

For many years now, it has been a popular cocktail-party pursuit among mathematicians to calculate their Erdős number, or more strictly an upper bound on their Erdős number, since it is rarely possible to be certain one has considered all possible paths through the network. Most mathematicians, and many in other subjects as well, have no difficulty establishing a fairly low upper bound on their Erdős number. As de Castro and Grossman conjecture, "Most mathematical researchers of the twentieth century have a finite (and rather small) Erdős number."

In the paper, de Castro and Grossman argue in favor of this conjecture by charting paths through the collaboration network to Erdős from a wide variety of starting individuals. As well as mathematicians, they derive upper bounds on the Erdős numbers of Nobel Prize winners in physics, economics, biology, and chemistry. And since it seems likely that most scientists in those fields could be connected to the corresponding Nobel laureates in a small number of steps, it is reasonable to suppose that most scientists have small Erdős numbers.

This exercise of course constitutes another demonstration of the small-world effect, this time in the context of the scientific community. With the recent increase in interest in networks, the Erdős number has been elevated from mathematical anecdote to the subject of serious (if playful) scientific inquiry. **Newman** (**2001a**, 2001b, 2001c), for instance, has studied in detail collaboration networks from a variety of subjects, including networks of biologists, physicists, and computer scientists, while Barabási and coworkers (2002) have focused on understanding the time evolution of collaboration networks, using data for publications in mathematics and neuroscience.

CHAIN-LINKS

by
Frigyes Karinthy

We were arguing energetically about whether the world is actually evolving, headed in a particular direction, or whether the entire universe is just a returning rhythm's game, a renewal of eternity. "There has to be something of crucial importance," I said in the middle of debate. "I just don't quite know how to express it in a new way; I hate repeating myself."

Let me put it this way: Planet Earth has never been as *tiny* as it is now. It shrunk – relatively speaking of course – due to the quickening pulse of both physical and verbal communication. This topic has come up before, but we had never framed it quite this way. We never talked about the fact that anyone on Earth, at my or anyone's will, can now learn in just a few minutes what I think or do, and what I want or what I would like to do. If I wanted to convince myself of the above fact: in couple of days I could be — *Hocus pocus*! — where I want to be.

Now we live in fairyland. The only slightly disappointing thing about this land is that it is smaller than the real world has ever been.

Chesterton praised a tiny and intimate, small universe and found it obtuse to portray the Cosmos as something *very big*. I think this idea is peculiar to our age of transportation. While Chesterton rejected technology and evolution, he was finally forced to admit that the fairyland he dreamed of could only come about through the scientific revolution he so vehemently opposed.

Everything returns and renews itself. The difference now is that the *rate* of these returns has increased, in both space and time, in an unheard-of fashion. Now my thoughts can circle the globe in minutes. Entire passages of world history are played out in a couple of years.

Something must result from this chain of thoughts. If only I knew what! (I feel as if I knew the answer to all this, but I've forgotten what it was or was overcome with doubt. Maybe I was *too close* to the truth. Near the North Pole, they say, the needle of a compass goes haywire, turning around in circles. It seems as if the same thing happens to our beliefs when we get too close to God.)

A fascinating game grew out of this discussion. One of us suggested performing the following experiment to prove that the population of the Earth is closer together now than they have ever been before. We should select any person from the 1.5 billion inhabitants of the Earth – anyone, anywhere at all. He bet us that, using no more than *five* individuals, one of whom is a personal acquaintance, he could contact the selected individual using nothing except the network of personal acquaintances. For example, "Look, you know Mr. X.Y., please ask him to contact his friend Mr. Q.Z., whom he knows, and so forth."

"An interesting idea!" — someone said — "Let's give it a try. How would you contact Selma Lagerlöf?"[1]

---

[1] Swedish novelist Selma Lagerlöf (1858–1940), who received the Nobel Prize for literature in 1909, was a champion of the return of Swedish romanticism with a mystical overtone. She also wrote novels for children.

"Well now, Selma Lagerlöf," the proponent of the game replied, "Nothing could be easier." And he reeled off a solution in two seconds: "Selma Lagerlöf just won the Nobel Prize for Literature, so she's bound to know King Gustav of Sweden, since, by rule, he's the one who would have handed her the Prize. And it's well known that King Gustav loves to play tennis and participates in international tennis tournaments. He has played Mr. Kehrling,[2] so they must be acquainted. And as it happens I myself also know Mr. Kehrling quite well." (The proponent was himself a good tennis player.) "All we needed this time was two out of five links. That's not surprising since it's always easier to find someone who knows a famous or popular figure than some run-of-the-mill, insignificant person. Come on, give me a harder one to solve!"

I proposed a more difficult problem: to find a chain of contacts linking myself with an anonymous riveter at the Ford Motor Company — and I accomplished it in four steps. The worker knows his foreman, who knows Mr. Ford himself, who, in turn, is on good terms with the director general of the Hearst publishing empire. I had a close friend, Mr. Árpád Pásztor, who had recently struck up an acquaintance with the director of Hearst publishing. It would take but one word to my friend to send a cable to the general director of Hearst asking him to contact Ford who could in turn contact the foreman, who could then contact the riveter, who could then assemble a new automobile for me, should I need one.

And so the game went on. Our friend was absolutely correct: nobody from the group needed more than five links in the chain to reach, just by using the method of acquaintance, any inhabitant of our Planet.

---

[2] Béla Kehrling, (1891–1937) was a noted Hungarian sportsman, soccer, ping-pong and tennis player. In tennis, he emerged victorious in 1923 in Gothenberg, Sweden, both indoors and in the open; he placed third in the Wimbledon doubles. He also played soccer and ice hockey.

And this leads us to another question: Was there ever a time in human history when this would have been impossible? Julius Caesar, for instance, was a popular man, but if he had got it into his head to try and contact a priest from one of the Mayan or Aztec tribes that lived in the Americas at that time, he could not have succeeded — not in five steps, not even in three hundred. Europeans in those days knew less about America and its inhabitants than we now know about Mars and its inhabitants.

So something is going on here, a process of contraction and expansion which is beyond rhythms and waves. Something coalesces, shrinks in size, while something else flows outward and grows. How is it possible that all this expansion and material growth can have started with a tiny, glittering speck that flared up millions of years ago in the mass of nerves in a primitive human's head? And how is it possible that by now, this continuous growth has the inundating ability to reduce the entire physical world to ashes? Is it possible that power can conquer matter, that the soul makes a mightier truth than the body, that life has a meaning that survives life itself, that good survives evil as life survives death, that God, after all, is more powerful than the Devil?

I am embarrassed to admit — since it would look foolish — that I often catch myself playing our well-connected game not only with human beings, but with objects as well. I have become very good at it. It's a useless game, of course, but I think I'm addicted to it, like a gambler who, having lost all of his money, plays for dried beans without any hope of real gain — just to see the four colors of the cards. The strange mind-game that clatters in me all the time goes like this: how can I link, with three, four, or at most five links of the chain, trivial, everyday things of life. How can I link one phenomenon to another? How can I join the relative and the ephemeral with steady, permanent things — how can I tie up the part with the whole?

It would be nice to just live, have fun, and take notice *only* of the utility of things: how much pleasure or pain they cause me. Alas, it's not possible. I hope that this game will help me find something else in the eyes that smile at me or the fist that strikes me, something beyond the urge to draw near to the former and shy away from the latter. One person loves me, another hates me. Why? Why the love and the hatred?

There are two people who do not understand one another, but I'm supposed to understand both. How? Someone is selling grapes in the street while my young son is crying in the other room. An acquaintance's wife has cheated on him while a crowd of hundred and fifty thousand watches the Dempsey match, Romain Roland's[3] last novel bombed while my friend Q changes his mind about Mr. Y. Ring-a-ring o' roses, a pocketful of posies. How can one possibly construct any chain of connections between these random things, without filling thirty volumes of philosophy, making only reasonable suppositions. The chain starts with the matter, *and its last link leads to me,* as the source of everything.

Well, just like this gentleman, who stepped up to my table in the café where I am now writing. He walked up to me and interrupted my thoughts with some trifling, insignificant problem and made me forget what I was going to say. Why did he come here and disturb me? The first link: he doesn't think much of people he finds scribbling. The second link: this world doesn't value scribbling nearly as much as it used to just a quarter of a century ago. The famous worldviews and thoughts that marked the end of the 19th century are to no avail today. Now we disdain the intellect. The third link: this disdain is the source of the hysteria of fear and terror that grips Europe today. And so to the fourth link: the order of the world has been destroyed.

Well, then, let a New World Order appear! Let the new Messiah of the world come! Let the God of the universe show himself once more through the burning rosehip-bush! Let there be peace, let there be war, let there be revolutions, so that — and here is

---

[3] Romain Roland, the noted French novelist, lived from 1866 until 1944. He was awarded the Nobel Prize for literature in 1915. Nearly all of his works were translated into Hungarian, just as in the case of Selma Lagerlöf.

the fifth link — it cannot happen again that someone should dare disturb me when I am at play, when I set free the phantoms of my imagination, when I think!

Translated from Hungarian and annotated by
Adam Makkai
Edited by ***Enikö Jankó***

BULLETIN OF
MATHEMATICAL BIOPHYSICS
VOLUME 13, 1951

CONNECTIVITY OF RANDOM NETS

RAY SOLOMONOFF AND ANATOL RAPOPORT
DEPARTMENT OF PHYSICS AND COMMITTEE ON MATHEMATICAL BIOLOGY
THE UNIVERSITY OF CHICAGO

The weak connectivity γ of a random net is defined and computed by an approximation method as a function of a, the axone density. It is shown that γ rises rapidly with a, attaining 0.8 of its asymptotic value (unity) for $a = 2$, where the number of neurons in the net is arbitrarily large. The significance of this parameter is interpreted also in terms of the maximum expected spread of an epidemic under certain conditions.

Numerous problems in various branches of mathematical biology lead to the consideration of certain structures which we shall call "random nets." Consider an aggregate of points, from each of which issues some number of outwardly directed lines (axones). Each axone terminates upon some point of the aggregate, and the probability that an axone from one point terminates on another point is the same for every pair of points in the aggregate. The resulting configuration constitutes a *random net.*

The existence of a *path* in a random net from a point A to a point B implies the possibility of tracing directed lines from A through any number of intermediate points, on which these lines terminate, to B.

We shall say that B is t axones removed from A, if t is the smallest number of axones contained in any of the paths from A to B. Point A itself is zero axones removed from A. All the other points upon which the axones of A terminate are one axone removed. The points upon which the axones from these latter points terminate, and which are not one or zero axones removed, are two axones removed, etc.

The notion of a random net may be generalized, if it is not assumed that the probability of direct connection between every pair of points in the net is the same. In that case it is necessary to define this probability for every pair of points. This can be done, for example, in terms of the distance between them or in some other way.

If the connections are not equiprobable, we shall speak of a net with a bias.

The following examples illustrate problems in which the concept of a net, defined by the probability of the connections among its points, seems useful.

1. *A problem in the theory of neural nets.* Suppose the points of a net are neurons. What is the probability that there exists a path between an arbitrary pair of neurons in the net? If the net has bias, what is the probability that there exists a path between a specified pair? In particular, what is the probability that a neuron is a member of a cycle (i.e., there exists a path from the neuron to itself through any positive number of internuncials)? Or, one may ask, what is the probability that there exists a path from a given neuron to every other neuron in the net?

2. *A problem in the theory of epidemics.* Suppose a number of individuals in a closed population contract a contagious disease, which lasts a finite time and then either kills them or makes them immune. If the probability of transmission is defined for each pair of individuals, what is the expected number of individuals which will contract the disease at a specified time? In particular, what is the

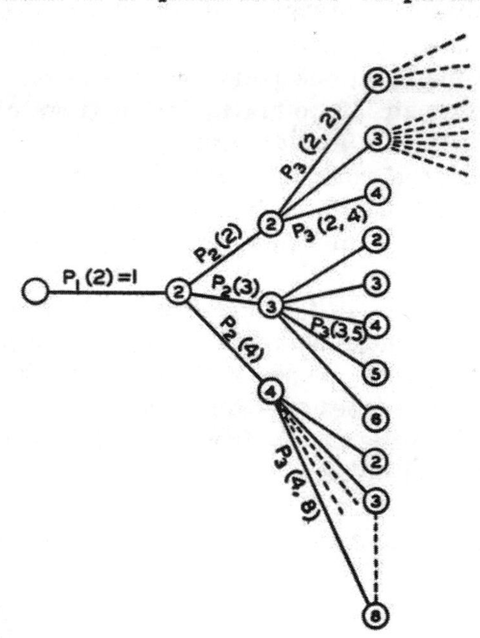

FIGURE 1. The probability tree for the number of ancestors of a single individual.

expected number of individuals which will eventually (after an infinite time) contract the disease? Or else, what is the probability that the entire population will succumb? Note that if the probability of transmission is the same for each pair of individuals, we are dealing with a random net.

3. *A problem in mathematical genetics.* Given the probability of mating between each pair of individuals in a population (as a function of their distance, or kinship, or the like), what is the expected number of ancestors of a given order for each individual? Clearly, the less the expected number of ancestors, the greater the genetic homogeneity of the population.

Each of these problems can be formalized by constructing a "probability tree." As an example, a tree for the genetic problem is illustrated in Figure 1.

We note that the tree consists of "nodes" connected by lines. The nodes can be designated by "first order," "second order," etc., depending on their distance from the "root." The number at the node indicates a possible number of ancestors of a given order. The lines connecting the nodes are labeled with the corresponding probabilities. Thus $p_1(2) = 1$, since it is certain that an individual has exactly two ancestors of the first order (parents). However, the parents may have been siblings or half-siblings. Therefore it is possible that the number of grandparents is 2, 3, or 4. The corresponding probabilities are $p_2(2)$, $p_2(3)$, and $p_2(4)$. The probability of having a certain number of great-grandparents depends on how many grandparents one has had. Consequently, those probabilities must be designated by $p_3(i,j)$ where $i = 2, \cdots 4$ and $j = 2, \cdots 8$. In general, the probability of having a certain number of ancestors of order k will depend on how many ancestors of each of the smaller orders one has had. If, however, we simplify the problem by supposing that the probability of having a certain number of ancestors of the kth order depends only on how many ancestors of the $(k-1)$th order one has, then the probability that an individual has exactly n ancestors of the mth order will be given by

$$P_m(n) = \sum_{r=2}^{2^m} \cdots \sum_{j=2}^{8} \sum_{i=2}^{4} p_2(2,i) p_3(i,j) p_4(j,k) \cdots p_m(r,n). \qquad (1)$$

The expected number of ancestors of the mth order will then be

$$E(m) = \sum_{n=2}^{2^m} nP(n). \qquad (2)$$

Clearly, a similar tree can be constructed for the neural net problem. Here the numbers at the nodes of the kth order would designate the possible number of neurons k axones removed from a given neuron. The p's would designate the corresponding transition probabilities from a certain number of neurons $(k-1)$ axones removed to a certain number k axones removed, etc. If N is the number of neurons in the aggregate, clearly, a neuron B is at most N axones removed from a neuron A, or else there exists no path from A to B. Hence $E(N)$ represents the expected number of neurons in the aggregate to which there exist paths from an arbitrary neuron, if the neurons are not in any way distinguished from each other. This expected number we shall call the *weak connectivity* of a random net and will designate it by γ.

The contagion problem could be formulated in similar terms. Here weak connectivity would represent the expected number of individuals which will contract the disease eventually. If we define Γ, the *strong connectivity* as the probability that from an arbitrary point in a random net there exist paths to every other point, then Γ will represent the probability that the entire population will succumb in the epidemic described above. In this case, the number of "axones" represents the number of individuals infected by a carrier before he recovers or dies.

The weak connectivity of a random net. We shall compute the weak connectivity of a neural net in terms of certain approximations whose justification will be given in subsequent papers. It will be assumed that:

1. The number of axones per neuron a is constant throughout the net. This constant (the axone density) need not be an integer, since it may equally well be taken as the average number of axones per neuron.

2. Connections are equiprobable, i.e., an axone synapses upon one or another neuron in the aggregate with equal probability.

A. Shimbel (1950) has formulated the problem in terms of the following differential-difference equation

$$dx/dt = [N - x(t)][x(t) - x(t - \tau)]. \quad (3)$$

Here $x(t)$ is a function related to the expected number of neurons t axones removed from an arbitrary neuron, and τ is related to the axone density. Then the problem of finding γ is equivalent to the problem of finding $x(\infty)$. A somewhat generalized form of equation (3) is given also by M. Puma (1939). The solution of the equation is, however, not given.

An approximate expression for γ where N is large was derived by one of the authors (Rapoport, 1948) where the number of axones per neuron is exactly one. This case will be generalized here to a axones per neuron, which are supposed constant through out the aggregate.

The axone-tracing procedure. Let us start with an arbitrarily selected neuron A and consider the set of all neurons removed by not more than t axones from A. Let x be the expected number of these neurons. Then evidently $x = x(N, a, t)$ depends on the total number of neurons in the net, on the axone density, and on t. Moreover, the weak connectivity of the net can be expressed as

$$\gamma(N, a) = x(N, a, N)/N. \tag{4}$$

Since N and a are fixed, we shall refer to the expected number of points removed from A by not more than t axones by $x(t)$. Note that t is a positive integer.

We seek a recursion formula for $x(t)$ which will give us an approximate determination of that function. To give a rigorous treatment of the problem, one would need to deal with distribution functions instead of expected values. For example, $p(i,t)$, denoting the probability that there are *exactly* i neurons not more than t axones removed from A, would determine the distribution for t. Successive distributions (for $t + 1$, etc.) would then depend on previous *distributions*, instead of merely upon the first moments of these distributions (expected values). The "probability tree" method does take these relations into account. An "exact" approach to the problem will be given in a subsequent paper. Meanwhile, however, we shall develop an approximation method in which it will be assumed that the expected value $x(t)$ depends only upon previous expected values, and, of course, upon the parameters of the net.

The recursion formula. We now seek an expression for $x(t + 1) - x(t)$. This is evidently the expected number of neurons *exactly* $(t + 1)$ axones removed from A. We shall make use of the following formula, which may be readily verified. Let s marbles be placed independently and at random into N boxes. Then the expected number of boxes occupied by one or more marbles will be given by

$$N[1 - (1 - 1/N)^s]. \tag{5}$$

In our axone-tracing procedure there are $a[x(t) - x(t-1)]$ axones of the *newly* contacted neurons to be traced on each step. Then the total number of neurons contacted on the $(t+1)$th tracing will be, according to formula (5),

$$N[1-(1-1/N)^{a[x(t)-x(t-1)]}]. \tag{6}$$

But of these neurons the fraction $x(t)/N$ has already been contacted. Hence the expected number of newly contacted neurons will be given by

$$x(t+1)-x(t)=[N-x(t)][1-(1-1/N)^{a[x(t)-x(t-1)]}], \tag{7}$$

which is our desired recursion formula.

Determination of γ. Let us set

$$y(t)=N-x(t). \tag{8}$$

Then equation (7) may be written as

$$y(t+1)=y(t)(1-1/N)^{a[y(t-1)-y(t)]}, \tag{9}$$

or

$$y(t+1)(1-1/N)^{ay(t)}=y(t)(1-1/N)^{ay(t-1)}. \tag{10}$$

Hence

$$y(t+1)(1-1/N)^{ay(t)}=\text{constant}=K. \tag{11}$$

We proceed to evaluate K. We have

$$y(t+1)=K(1-1/N)^{-ay(t)}. \tag{12}$$

But $y(t)$ represents the expected number of uncontacted points in the tth step. Since before the tracing began one point constituted the set of contacted points, therefore we have

$$y(0)=N-1, \tag{13}$$

and using formula (5),

$$y(1)=(N-1)^{a+1}N^{-a}. \tag{14}$$

Letting $t=0$ in (12), we obtain

$$K=N^{-aN}(N-1)^{aN+1}. \tag{15}$$

Furthermore, since $y(1) \leqq y(0)$ and $(1-1/N)^{-a} > 1$, we have $y(2) \leqq y(1)$, etc., so that $y(t)$ is a non-increasing function of t (this is also intuitively evident from the definition of y). Since $y \geq 0$ for all t, $y(t)$ must approach a limit as t grows without bound. Hence

$$\lim_{t\to\infty} y(t+1) = \lim_{t\to\infty} y(t) = Y. \tag{16}$$

Note that $\gamma = x(N)$ may also be considered as $\lim_{t\to\infty} x(t)/N$. This is so since contacting no new neurons on any tracing implies that no new neurons will be contacted on any subsequent tracings. If we continue to carry out tracings "symbolically," it is evident that at some tracing not greater than the Nth no new neurons will be contacted, and all subsequent tracings will be "dummy" tracings.

Using equations (12) and (15), we see that Y satisfies the transcendental equation

$$Y = (N-1)(1-1/N)^{a(N-Y)}. \tag{17}$$

For large N, this can be approximated by

$$Y \sim N \operatorname{Exp}\{a(Y/N-1)\}. \tag{18}$$

Hence, for large N,

$$Y/N \sim \operatorname{Exp}\{a(Y/N-1)\}. \tag{19}$$

But $\gamma = x(\infty)/N = 1 - Y/N$. Substituting this value into (19), we obtain the transcendental equation which defines γ implicitly as a function of a, namely,

$$\gamma = 1 - e^{-a\gamma}. \tag{20}$$

We note that for $\gamma = 0$, every a is a solution of (20). If $\gamma \neq 0$, then equation (20) can be solved explicitly for a giving

$$a = \frac{-\log(1-\gamma)}{\gamma}. \tag{21}$$

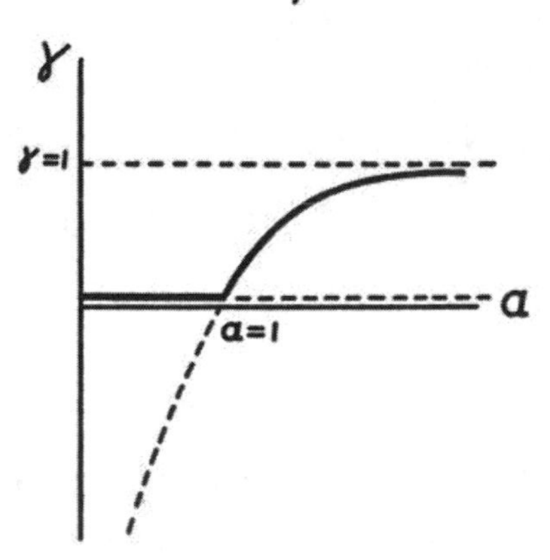

FIGURE 2. Weak connectivity as a function of axone density.

The right side of (21) is analytic in every neighborhood of the origin and tends to unity as γ approaches 0. Expanding that function in powers of γ, we have

$$a = 1 + \gamma/2 + \gamma^2/3 \cdots, \tag{22}$$

which allows us to plot a against γ (cf. Fig. 2). This graph consists of two branches, namely, the entire a-axis and the function (21). Negative values of γ, being physically meaningless, must be discarded. Thus in the region $0 \leqq a \leqq 1$, we have $\gamma \equiv 0$, as is intuitively evident. We must show, however, that for $a > 1$, γ follows the nonzero branch of the graph, otherwise we get the unlikely result that for sufficiently large N the fraction of individuals eventually infected in an epidemic will be negligible, regardless of the number of individuals infected by each carrier of the disease. Actually, the solution $\gamma \equiv 0$ is extraneous for $a > 1$ and appears in our equation because we have let N increase without bound *before* determining the relation between a and γ. In any physical situation N is finite. Hence a physically meaningful procedure is to determine γ as a function of a and N and *then* allow N to increase without bound. Such a function is given by equation (17). Proceeding from that equation we obtain

$$Y/(N-1) = (1-1/N)^{a(N-Y)}, \tag{23}$$

$$\log Y - \log(N-1) = a(N-Y)\log(1-1/N), \tag{24}$$

$$a = \frac{\log Y - \log(N-1)}{(N-Y)\left[\log(N-1) - \log(N)\right]}. \tag{25}$$

Let us write $Y = N - \phi(N) = N[1 - \phi(N)/N]$. Then equation (25) may be written as

$$\begin{aligned} &\log N - \log(N-1) + \log[1-\phi(N)/N] \\ &\quad = a\,\phi(N)\left[\log(N-1) - \log N\right]. \end{aligned} \tag{26}$$

Since $\phi(N) < N$ for all N, we may expand the last term of the left side of (26) and obtain

$$\begin{aligned} &\log N - \log(N-1) - \phi(N)/N - \tfrac{1}{2}[\phi(N)/N]^2 \\ &\quad - 1/3[\phi(N)/N]^3 \cdots = a\,\phi(N)\left[\log(N-1) - \log N\right]. \end{aligned} \tag{27}$$

We now expand $\log(N-1) - \log N$ which appears in the right side of (27) and after rearrangements obtain

$$\log N - \log(N-1)$$
$$= \frac{\phi(N)}{N}\left[1 - a + \frac{\phi(N) - a}{2N} + \frac{[\phi(N)]^2 - a}{3N^2} + \cdots\right] \quad (28)$$
$$< \frac{\phi(N)}{N}\left[1 - a + (1 - \phi(N)/N)^{-1}\right].$$

Now if a is fixed and greater than unity, the limit of $\phi(N)/N$ cannot be zero as N increases without bound, because otherwise for N sufficiently large the right side of (28) becomes negative, while the left side is always positive, a contradiction of inequality (28). Therefore, the limit of Y/N, as N increases without bound, cannot be unity for $a > 1$. But this means that $\gamma \neq 0$ if $a > 1$. Hence, for $a > 1$, the non-zero branch of our curve is the only meaningful one.

An examination of the meaningful part of the graph of equation (20) shows that as long as the axone density does not exceed one axone per neuron, $\gamma = 0$, i.e., for very large N, the number of neurons to which there exist paths from an arbitrary neuron is negligible compared with the total number of neurons in the net. On the other hand, as the axone density increases from unity, γ increases rather rapidly, starting with slope 2. Already for $a = 2$, γ reaches about 0.8 of its asymptotic value (unity) and is within a fraction of one per cent of unity for quite moderate a (say > 6). This means that no matter how large the net is, it is practically certain that there will exist a path between two neurons picked at random, provided only the axone density is a few times greater than unity. The interpretation in terms of an epidemic with equiprobable contacts is entirely analogous.

The case $a = 1$. This case was treated by one of the authors (Rapoport, 1948) by a different method. It was shown that for large N, the probability that a neuron was member of a cycle was given by $\sqrt{\pi/2N}$. This gives the probability of the existence of a path from a neuron over any number of internuncials greater than one to itself. But under the assumption of equiprobable connections, this may well represent the probability of the existence of a path from the given neuron to any *other* neuron in the net. Therefore we should have for large N, in the case $a = 1$,

$$\gamma \sim \sqrt{\pi/2N}. \quad (29)$$

For $N = \infty$, γ reduces to zero, as it should according to equation (20). We shall, however, examine the asymptotic behavior of γ for large N deduced from our approximate method, in order to compare it with the asymptotic behavior (29) deduced from an exact treatment of the special case. Dividing both sides of (17) by N, we may write for $a=1$

$$Y/N=[(N-1)/N]^{N-Y+1}, \tag{30}$$

whence, since $Y/N=1-\gamma$,

$$\begin{aligned} 1-\gamma &= [(N-1)/N]^{N\gamma+1} \\ &= \operatorname{Exp}\{\ln(1-1/N)+N\gamma\ln(1-1/N)\}. \end{aligned} \tag{31}$$

We let $z=N^{-1}$ and examine the behavior of γ for small values of z. Expanding the right side of (31) by power series and retaining only terms of the second order (note that z and γ vanish together), we obtain

$$\begin{aligned} 1-\gamma = 1 &+ [-z-z^2/2\cdots\cdot]+[-\gamma-\gamma z/2-\cdots\cdot] \\ &+ z^2/2+\gamma^2/2+\gamma z+\cdots\cdot. \end{aligned} \tag{32}$$

Hence,

$$0=-z+\gamma^2/2+\gamma z/2+\cdots\cdot. \tag{33}$$

Differentiating with respect to γ, we get

$$dz/d\gamma=\gamma+\gamma/2\cdot dz/d\gamma+z/2+\cdots\cdot, \tag{34}$$

$$dz/d\gamma\sim(\gamma+z/2)/(1-\gamma/2). \tag{35}$$

Therefore $dz/d\gamma$ vanishes at $z=0$, $\gamma=0$. Differentiating once again with respect to γ, we obtain

$$\left.\frac{d^2z}{d\gamma^2}\right|_{\substack{z=0\\ \gamma=0}}=1. \tag{36}$$

Hence the power series representing z as a function of γ begins as follows:

$$z=\gamma^2/2+\cdots\cdot. \tag{37}$$

Thus

$$\gamma^2\sim 2z=2/N, \tag{38}$$

$$\gamma\sim\sqrt{2/N}\cong 1.41\sqrt{N}. \tag{39}$$

The "exact" result as expressed by (22) gives

$$\gamma \sim 1.2/\sqrt{N}.$$

Thus the approximate method applied to the case $a = 1$ implies an asymptotic behavior of γ for large N which does not depart too sharply from that deduced by the exact method. The limiting value for γ is zero in both cases. The question of how well the limiting values of γ are approached by the approximate method for $a > 1$ remains open.

This investigation is part of the work done under Contract No. AF 19(122)-161 between the U. S. Air Force Cambridge Research Laboratories and The University of Chicago.

LITERATURE

Puma, Marcello. 1939. *Elementi per una teoria matematica del contagio.* Rome: Editoriale Aeronautica.

Rapoport, Anatol. 1948. "Cycle Distributions in Random Nets." *Bull. Math. Biophysics*, 10, 145–57.

Shimbel, Alfonso. 1950. "Contributions to the Mathematical Biophysics of the Central Nervous System with Special Reference to Learning." *Bull. Math. Biophysics*, 12, 241–75.

ON THE EVOLUTION OF RANDOM GRAPHS

by
P. ERDŐS and A. RÉNYI

Dedicated to Professor P. Turán at his 50th birthday.

Introduction

Our aim is to study the probable structure of a random graph $\Gamma_{n,N}$ which has n given labelled vertices $P_1, P_2, \ldots, P_n$ and N edges; we suppose that these N edges are chosen at random among the $\binom{n}{2}$ possible edges, so that all $\binom{\binom{n}{2}}{N} = C_{n,N}$ possible choices are supposed to be equiprobable. Thus if $G_{n,N}$ denotes any one of the $C_{n,N}$ graphs formed from n given labelled points and having N edges, the probability that the random graph $\Gamma_{n,N}$ is identical with $G_{n,N}$ is $\frac{1}{C_{n,N}}$. If A is a property which a graph may or may not possess, we denote by $\mathbf{P}_{n,N}(A)$ the probability that the random graph $\Gamma_{n,N}$ possesses the property A, i. e. we put $\mathbf{P}_{n,N}(A) = \frac{A_{n,N}}{C_{n,N}}$ where $A_{n,N}$ denotes the number of those $G_{n,N}$ which have the property A.

An other equivalent formulation is the following: Let us suppose that n labelled vertices $P_1, P_2, \ldots, P_n$ are given. Let us choose at random an edge among the $\binom{n}{2}$ possible edges, so that all these edges are equiprobable. After this let us choose an other edge among the remaining $\binom{n}{2} - 1$ edges, and continue this process so that if already k edges are fixed, any of the remaining $\binom{n}{2} - k$ edges have equal probabilities to be chosen as the next one. We shall study the "evolution" of such a random graph if N is increased. In this investigation we endeavour to find what is the "typical" structure at a given stage of evolution (i. e. if N is equal, or asymptotically equal, to a given function $N(n)$ of n). By a "typical" structure we mean such a structure the probability of which tends to 1 if $n \to +\infty$ when $N = N(n)$. If A is such a property that $\lim_{n \to +\infty} \mathbf{P}_{n,N(n)}(A) = 1$, we shall say that „almost all" graphs $G_{n,N(n)}$ possess this property.

The study of the evolution of graphs leads to rather surprising results. For a number of fundamental structural properties A there exists a function $A(n)$ tending monotonically to $+\infty$ for $n \to +\infty$ such that

$$\lim_{n\to+\infty} \mathbf{P}_{n,N(n)}(A) = \begin{cases} 0 & \text{if} \quad \lim\limits_{n\to+\infty} \dfrac{N(n)}{A(n)} = 0 \\ 1 & \text{if} \quad \lim\limits_{n\to+\infty} \dfrac{N(n)}{A(n)} = +\infty . \end{cases} \tag{1}$$

If such a function $A(n)$ exists we shall call it a *"threshold function"* of the property A.

In many cases besides (1) it is also true that there exists a probability distribution function $F(x)$ so that if $0 < x < +\infty$ and x is a point of continuity of $F(x)$ then

$$\lim_{n\to+\infty} \mathbf{P}_{n,N(n)}(A) = F(x) \quad \text{if} \quad \lim_{n\to+\infty} \frac{N(n)}{A(n)} = x . \tag{2}$$

If (2) holds we shall say that $A(n)$ is a *„regular threshold function"* for the property A and call the function $F(x)$ the *threshold distribution function* of the property A.

For certain properties A there exist two functions $A_1(n)$ and $A_2(n)$ both tending monotonically to $+\infty$ for $n \to +\infty$, and satisfying $\lim\limits_{n\to+\infty} \dfrac{A_2(n)}{A_1(n)} = 0$, such that

$$\lim_{n\to+\infty} \mathbf{P}_{n,N(n)}(A) = \begin{cases} 0 & \text{if} \quad \lim\limits_{n\to+\infty} \dfrac{N(n) - A_1(n)}{A_2(n)} = -\infty \\ 1 & \text{if} \quad \lim\limits_{n\to+\infty} \dfrac{N(n) - A_1(n)}{A_2(n)} = +\infty . \end{cases} \tag{3}$$

Clearly (3) implies that

$$\lim_{n\to+\infty} \mathbf{P}_{n,N(n)}(A) = \begin{cases} 0 & \text{if} \quad \limsup\limits_{n\to+\infty} \dfrac{N(n)}{A_1(n)} < 1 \\ 1 & \text{if} \quad \liminf\limits_{n\to+\infty} \dfrac{N(n)}{A_1(n)} > 1 . \end{cases} \tag{4}$$

If (3) holds we call the pair $(A_1(n), A_2(n))$ a pair of *"sharp threshold"*-functions of the property A. It follows from (4) that if $(A_1(n), A_2(n))$ is a pair of sharp threshold functions for the property A then $A_1(n)$ is an (ordinary) threshold function for the property A and the threshold distribution function figuring in (2) is the degenerated distribution function

$$F_1(x) = \begin{cases} 0 & \text{for} \quad x \leq 1 \\ 1 & \text{for} \quad x > 1 \end{cases}$$

and convergence in (2) takes place for every $x \neq 1$. In some cases besides (3) it is also true that there exists a probability distribution function $G(y)$ defined for $-\infty < y < +\infty$ such that if y is a point of continuity of $G(y)$ then

$$\lim_{n\to+\infty} \mathbf{P}_{n,N(n)}(A) = G(y) \quad \text{if} \quad \lim_{n\to+\infty} \frac{N(n) - A_1(n)}{A_2(n)} = y. \tag{5}$$

If (5) holds we shall say that we have a *regular sharp threshold* and shall call $G(y)$ the *sharp-threshold distribution function of the property* A.

One of our chief aims will be to determine the threshold respectively sharp threshold functions, and the corresponding distribution functions for the most obvious structural-properties, e. g. the presence in $\Gamma_{n,N}$ of subgraphs of a given type (trees, cycles of given order, complete subgraphs etc.) further for certain global properties of the graph (connectedness, total number of connected components, etc.).

In a previous paper [7] we have considered a special problem of this type; we have shown that denoting by C the property that the graph is connected, the pair $C_1(n) = \frac{1}{2} n \log n$, $C_2(n) = n$ is a pair of strong threshold functions for the property C, and the corresponding sharp-threshold distribution function is $e^{-e^{-2y}}$; thus we have proved[1] that putting $N(n) = \frac{1}{2} n \log n + yn + o(n)$ we have

$$\lim_{n\to+\infty} \mathbf{P}_{n,N(n)}(C) = e^{-e^{-2y}} \qquad (-\infty < y < +\infty). \tag{6}$$

In the present paper we consider the evolution of a random graph in a more systematic manner and try to describe the gradual development and step-by-step unravelling of the complex structure of the graph $\Gamma_{n,N}$ when N increases while n is a given large number.

We succeeded in revealing the emergence of certain structural properties of $\Gamma_{n,N}$. However a great deal remains to be done in this field. We shall call in § 10. the attention of the reader to certain unsolved problems. It seems to us further that it would be worth while to consider besides graphs also more complex structures from the same point of view, i. e. to investigate the laws governing their evolution in a similar spirit. This may be interesting not only from a purely mathematical point of view. In fact, the evolution of graphs may be considered as a rather simplified model of the evolution of certain communication nets (railway, road or electric network systems, etc.) of a country or some other unit. (Of course, if one aims at describing such a real situation, one should replace the hypothesis of equiprobability of all connections by some more realistic hypothesis.) It seems plausible that by considering the random growth of more complicated structures (e. g. structures consisting of different sorts of "points" and connections of different types) one could obtain fairly reasonable models of more complex real growth processes (e. g.

[1] Partial result on this problem has been obtained already in 1939 by P. ERDŐS and H. WHITNEY but their results have not been published.

the growth of a complex communication net consisting of different types of connections, and even of organic structures of living matter, etc.).

§§ 1—3. contain the discussion of the presence of certain components in a random graph, while §§ 4—9. investigate certain global properties of a random graph. Most of our investigations deal with the case when $N(n) \sim cn$ with $c > 0$. In fact our results give a clear picture of the evolution of $\Gamma_{n,N(n)}$ when $c = \frac{N(n)}{n}$ (which plays in a certain sense the role of time) increases. In § 10. we make some further remarks and mention some unsolved problems.

Our investigation belongs to the combinatorical theory of graphs, which has a fairly large literature. The first who enumerated the number of possible graphs with a given structure was A. Cayley [1]. Next the important paper [2] of G. Pólya has to be mentioned, the starting point of which were some chemical problems. Among more recent results we mention the papers of G. E. Uhlenbeck and G. W. Ford [5] and E. N. Gilbert [6]. A fairly complete bibliography will be given in a paper of F. Harary [8]. In these papers the probabilistic point of view was not explicitly emphasized. This has been done in the paper [9] of one of the authors, but the aim of the probabilistic treatment was there different: the existence of certain types of graphs has been shown by proving that their probability is positive. Random trees have been considered in [14].

In a recent paper [10] T. L. Austin, R. E. Fagen, W. F. Penney and J. Riordan deal with random graphs from a point of view similar to ours. The difference between the definition of a random graph in [10] and in the present paper consists in that in [10] it is admitted that two points should be connected by more than one edge ("parallel" edges). Thus in [10] it is supposed that after a certain number of edges have already been selected, the next edge to be selected may be any of the possible $\binom{n}{2}$ edges between the n given points (including the edges already selected). Let us denote such a random graph by $\Gamma^*_{n,N}$. The difference between the probable properties of $\Gamma_{n,N}$ resp. $\Gamma^*_{n,N}$ are in most (but not in all) cases negligible. The corresponding probabilities are in general (if the number N of edges is not too large) asymptotically equal. There is a third possible point of view which is in most cases almost equivalent with these two; we may suppose that for each pair of n given points it is determined by a chance process whether the edge connecting the two points should be selected or not, the probability for selecting any given edge being equal to the same number $p > 0$, and the decisions concerning the different edges being completely independent. In this case of course the number of edges is a random variable, having the expectation $\binom{n}{2}p$; thus if we want to obtain by this method a random graph having in the mean N edges we have to choose the value of p equal to $\frac{N}{\binom{n}{2}}$. We shall denote such a random graph by $\Gamma^{**}_{n,N}$. In many (though not all) of the problems treated in the present paper it does not cause any essential difference if we consider instead of $\Gamma_{n,N}$ the random graph $\Gamma^{**}_{n,N}$.

Comparing the method of the present paper with that of [10] it should be pointed out that our aim is to obtain threshold functions resp. distributions, and thus we are interested in asymptotic formulae for the probabilities considered. Exact formulae are of interest to us only so far as they help in determining the asymptotic behaviour of the probabilities considered (which is rarely the case in this field, as the exact formulae are in most cases too complicated). On the other hand in [10] the emphasis is on exact formulae resp. on generating functions. The only exception is the average number of connected components, for the asymptotic evaluation of which a way is indicated in § 5. of [10]; this question is however more fully discussed in the present paper and our results go beyond that of [10]. Moreover, we consider not only the number but also the character of the components. Thus for instance we point out the remarkable change occuring at $N \sim \frac{n}{2}$. If $N \sim nc$ with $c < 1/2$ then with probability tending to 1 for $n \to +\infty$ all points except a bounded number of points of $\Gamma_{n,N}$ belong to components which are trees, while for $N \sim nc$ with $c > \frac{1}{2}$ this is no longer the case. Further for a fixed value of n the average number of components of $\Gamma_{n,N}$ decreases asymptotically in a linear manner with N, when $N \leqq \frac{n}{2}$, while for $N > \frac{n}{2}$ the formula giving the average number of components is not linear in N.

In what follows we shall make use of the sysmbols O and o. As usually $a(n) = o\left(b(n)\right)$ (where $b(n) > 0$ for $n = 1, 2, \ldots$) means that $\lim\limits_{n \to +\infty} \frac{|a(n)|}{b(n)} = 0$, while $a(n) = O\left(b(n)\right)$ means that $\frac{|a(n)|}{b(n)}$ is bounded. The parameters on which the bound of $\frac{|a(n)|}{b(n)}$ may depend will be indicated if it is necessary; sometimes we will indicate it by an index. Thus $a(n) = O_\varepsilon\left(b(n)\right)$ means that $\frac{|a(n)|}{b(n)} \leqq K(\varepsilon)$ where $K(\varepsilon)$ is a positive constant depending on ε. We write $a(n) \sim b(n)$ to denote that $\lim\limits_{n \to +\infty} \frac{a(n)}{b(n)} = 1$.

We shall use the following definitions from the theory of graphs. (For the general theory see [3] and [4].)

A finite non-empty set V of labelled points $P_1, P_2, \ldots, P_n$ and a set E of different unordered pairs (P_i, P_j) with $P_i \in V$, $P_j \in V$, $i \neq j$ is called a *graph*; we denote it sometimes by $G = \{V, E\}$; the number n is called the *order* (or *size*) of the graph; the points $P_1, P_2, \ldots, P_n$ are called the *vertices* and the pairs (P_i, P_j) the *edges* of the graph. Thus we consider *non-oriented finite graphs without parallel edges and without slings.* The set E may be empty, thus a collection of points (especially a single point) is also a graph.

A graph $G_2 = \{V_2, E_2\}$ is called a *subgraph* of a graph $G_1 = \{V_1, E_1\}$ if the set of vertices V_2 of G_2 is a subset of the set of vertices V_1 of G_1 and the set E_2 of edges of G_2 is a subset of the set E_1 of edges of G_1.

A sequence of k edges of a graph such that every two consecutive edges and only these have a vertex in common is called a *path* of order k.

A cyclic sequence of k edges of a graph such that every two consecutive edges and only these have a common vertex is called a *cycle* of order k.

A graph G is called *connected* if any two of its points belong to a path which is a subgraph of G.

A graph is called a *tree* of order (or size) k if it has k vertices, is connected and if none of its subgraphs is a cycle. A tree of order k has evidently $k - 1$ edges.

A graph is called a *complete graph* of order $\binom{k}{2}$ if it has k vertices and $\binom{k}{2}$ edges. Thus in a complete graph of order k any two points are connected by an edge.

A subgraph G' of a graph G will be called an *isolated subgraph* if all edges of G one or both endpoints of which belong to G', belong to G'. A connected isolated subgraph G' of a graph G is called a *component* of G. The number of points belonging to a component G' of a graph G will be called the *size* of G'.

Two graphs shall be called *isomorphic*, if there exists a one-to-one mapping of the vertices carrying over these graphs into another.

The graph $\overline{G}$ shall be called *complementary graph* of G if $\overline{G}$ consists of the same vertices $P_1, P_2, \ldots, P_n$ as G and of those and only those edges (P_i, P_j) which do not occur in G.

The number of edges starting from the point P of a graph G will be called the *degree* of P in G.

A graph G is called a *saturated even graph of type* (a, b) if it consists of $a + b$ points and its points can be split in two subsets V_1 and V_2 consisting of a resp. b points, such that G contains any edge (P, Q) with $P \in V_1$ and $Q \in V_2$ and no other edge.

A graph is called *planar*, if it can be drawn on the plane so that no two of its edges intersect.

We introduce further the following definitions: If a graph G has n vertices and N edges, we call the number $\frac{2N}{n}$ the "*degree*" of the graph. (As a matter of fact $\frac{2N}{n}$ is the average degree of the vertices of G.) If a graph G has the property that G has no subgraph having a larger degree than G itself, we call G a *balanced* graph.

We denote by $\mathbf{P}(\ldots)$ the probability of the event in the brackets, by $\mathbf{M}(\xi)$ resp. $\mathbf{D}^2(\xi)$ the mean value resp. variance of the random variable ξ. In cases when it is not clear from the context in which probability space the probabilities or respectively the mean values and variances are to be understood, this will be explicitly indicated. Especially $\mathbf{M}_{n,N}$ resp. $\mathbf{D}^2_{n,N}$ will denote the mean value resp. variance calculated with respect to the probabilities $\mathbf{P}_{n,N}$.

We shall often use the following elementary asymptotic formula:

$$\binom{n}{k} \sim \frac{n^k e^{-\frac{k^2}{2n}-\frac{k^3}{6n^2}}}{k!} \quad \text{valid for } k = o(n^{3/4}). \tag{7}$$

Our thanks are due to T. Gallai for his valuable remarks.

§ 1. Thresholds for subgraphs of given type

If N is very small compared with n, namely if $N = o(\sqrt{n})$ then it is very probable that $\Gamma_{n,N}$ is a collection of isolated points and isolated edges, i. e. that no two edges of $\Gamma_{n,N}$ have a point in common. As a matter of fact the probability that at least two edges of $\Gamma_{n,N}$ shall have a point in common is by (7) clearly

$$1 - \frac{\binom{n}{2N}(2N)!}{2^N N! \binom{\binom{n}{2}}{N}} = O\left(\frac{N^2}{n}\right).$$

If however $N \sim c\sqrt{n}$ where $c > 0$ is a constant not depending on n, then the appearance of trees of order 3 will have a probability which tends to a positive limit for $n \to +\infty$, but the appearance of a connected component consisting of more than 3 points will be still very improbable. If N is increased while n is fixed, the situation will change only if N reaches the order of magnitude of $n^{2/3}$. Then trees of order 4 (but not of higher order) will appear with a probability not tending to 0. In general, the threshold function for the presence of trees of order k is $n^{\frac{k-2}{k-1}}$ $(k = 3, 4, \ldots)$. This result is contained in the following

Theorem 1. *Let $k \geq 2$ and l $\left(k-1 \leq l \leq \binom{k}{2}\right)$ be positive integers. Let $\mathcal{B}_{k,l}$ denote an arbitrary not empty class of connected balanced graphs consisting of k points and l edges. The threshold function for the property that the random graph considered should contain at least one subgraph isomorphic with some element of $\mathcal{B}_{k,l}$ is $n^{2-\frac{k}{l}}$.*

The following special cases are worth mentioning

Corollary 1. *The threshold function for the property that the random graph contains a subgraph which is a tree of order k is $n^{\frac{k-2}{k-1}}$ $(k = 3, 4, \ldots)$.*

Corollary 2. *The threshold function for the property that a graph contains a connected subgraph consisting of $k \geq 3$ points and k edges (i. e. containing exactly one cycle) is n, for each value of k.*

Corollary 3. *The threshold function for the property that a graph contains a cycle of order k is n, for each value of $k \geq 3$.*

Corollary 4. *The threshold function for the property that a graph contains a complete subgraph of order* $k \geqq 3$ *is* $n^{2\left(1-\frac{1}{k-1}\right)}$.

Corollary 5. *The threshold function for the property that a graph contains a saturated even subgraph of type* (a, b) (*i. e. a subgraph* consisting of $a+b$ points $P_1, \ldots, P_a, Q_1, \ldots Q_b$ and of the ab edges (P_i, Q_j) *is* $n^{2-\frac{a+b}{ab}}$.

To deduce these Corollaries one has only to verify that all 5 types of graphs figuring in Corollaries 1—5. are balanced, which is easily seen.

Proof of Theorem 1. Let $B_{k,l} \geqq 1$ denote the number of graphs belonging to the class $\mathscr{B}_{k,l}$ which can be formed from k given labelled points. Clearly if $P_{n,N}(\mathscr{B}_{k,l})$ denotes the probability that the random graph $\Gamma_{n,N}$ contains at least one subgraph isomorphic with some element of the class $\mathscr{B}_{k,l}$, then

$$\mathbf{P}_{n,N}(\mathscr{B}_{k,l}) \leq \binom{n}{k} B_{k,l} \cdot \frac{\binom{\binom{n}{2}-l}{N-l}}{\binom{\binom{n}{2}}{N}} = O\left(\frac{N^l}{n^{2l-k}}\right). \tag{1.1}$$

As a matter of fact if we select k points (which can be done in $\binom{n}{2}$ different ways) and form from them a graph isomorphic with some element of the class $\mathscr{B}_{k,l}$ (which can be done in $B_{k,l}$ different ways) then the number of graphs $G_{n,N}$ which contain the selected graph as a subgraph is equal to the number of ways the remaining $N-l$ edges can be selected from the $\binom{n}{2}-l$ other possible edges. (Of course those graphs, which contain more subgraphs isomorphic with some element of $\mathscr{B}_{k,l}$ are counted more than once.)

Now clearly if $N = o(n^{2-\frac{k}{l}})$ then by

$$\mathbf{P}_{n,N}(\mathscr{B}_{k,l}) = o(1)$$

which proves the first part of the assertion of Theorem 1. To prove the second part of the theorem let $\mathscr{B}_{k,l}^{(n)}$ denote the set of all subgraphs of the complete graph consisting of n points, isomorphic with some element of $\mathscr{B}_{k,l}$. To any $S \in \mathscr{B}_{k,l}^{(n)}$ let us associate a random variable $\varepsilon(S)$ such that $\varepsilon(S)=1$ or $\varepsilon(S)=0$ according to whether S is a subgraph of $\Gamma_{n,N}$ or not. Then clearly (we write in what follows for the sake of brevity $\mathbf{M}$ instead of $\mathbf{M}_{n,N}$)

$$\mathbf{M}\left(\sum_{S \in \mathscr{B}_{k,l}^{(n)}} \varepsilon(S)\right) = \sum_{S \in \mathscr{B}_{k,l}^{(n)}} \mathbf{M}(\varepsilon(S)) = \binom{n}{k} B_{k,l} \frac{\binom{\binom{n}{2}-l}{N-l}}{\binom{\binom{n}{2}}{N}} \sim \frac{B_{k,l}}{k!} \frac{(2N)^l}{n^{2l-k}}. \tag{1.2}$$

On the other hand if S_1 and S_2 are two elements of $\mathcal{S}_{k,l}^{(n)}$ and if S_1 and S_2 do not contain a common edge then

$$\mathsf{M}(\varepsilon(S_1)\,\varepsilon(S_2)) = \frac{\binom{\binom{n}{2}-2l}{N-2l}}{\binom{\binom{n}{2}}{N}}.$$

If S_1 and S_2 contain exactly s common points and r common edges $(1 \leqq r \leqq l-1)$ we have

$$\mathsf{M}(\varepsilon(S_1)\,\varepsilon(S_2)) = \frac{\binom{\binom{n}{2}-2l+r}{N-2l+r}}{\binom{\binom{n}{2}}{N}} = O\left(\frac{N^{2l-r}}{n^{4l-2r}}\right).$$

On the other hand the intersection of S_1 and S_2 being a subgraph of S_1 (and S_2) by our supposition that each S is balanced, we obtain $\frac{r}{s} \leqq \frac{l}{k}$ i. e. $s \geqq \frac{rk}{l}$ and thus the number of such pairs of subgraphs S_1 and S_2 does not exceed

$$B_{k,l}^2 \sum_{j \geq \frac{rk}{l}}^{k} \binom{n}{k}\binom{k}{j}\binom{n-k}{k-j} = O\left(n^{2k-\frac{rk}{l}}\right).$$

Thus we obtain

$$\mathsf{M}\left(\left(\sum_{S\in\mathcal{S}_{k,l}^{(n)}} \varepsilon(S)\right)^2\right) =$$

(1.3)

$$= \sum_{S\in\mathcal{S}_{k,l}^{(n)}} \mathsf{M}(\varepsilon(S)) + \frac{n!\,B_{k,l}^2}{k!^2(n-2k)!}\,\frac{\binom{\binom{n}{2}-2l}{N-2l}}{\binom{\binom{n}{2}}{N}} + O\left(\left(\frac{N^l}{n^{2l-k}}\right)^2 \sum_{r=1}^{l}\left(\frac{n^{2-\frac{k}{l}}}{N}\right)^r\right).$$

Now clearly

$$\frac{n!}{k!^2(n-2k)!}\,\frac{\binom{\binom{n}{2}-2l}{N-2l}}{\binom{\binom{n}{2}}{N}} \leqq \binom{n}{k}^2 \frac{\binom{\binom{n}{2}-l}{N-l}^2}{\binom{\binom{n}{2}}{N}^2}.$$

If we suppose that

$$\frac{N}{n^{2-\frac{k}{l}}} = \omega \to +\infty ,$$

it follows that we have

$$(1.4) \qquad \mathbf{D}^2\Big(\sum_{S\in\mathscr{B}_{k,l}^{(n)}} \varepsilon(S)\Big) = O\left(\frac{\big(\sum_{S\in\mathscr{B}_{k,l}^{(n)}} \mathbf{M}(\varepsilon(S))\big)^2}{\omega}\right).$$

It follows by the inequality of *Chebysheff* that

$$\mathbf{P}_{n,N}\left(\Big|\sum_{S\in\mathscr{B}_{k,l}^{(n)}} \varepsilon(S) - \sum_{S\in\mathscr{B}_{k,l}^{(n)}} \mathbf{M}(\varepsilon(S))\Big| > \frac{1}{2} \sum_{S\in\mathscr{B}_{k,l}^{(n)}} \mathbf{M}(\varepsilon(S))\right) = O\left(\frac{1}{\omega}\right)$$

and thus

$$(1.5) \qquad \mathbf{P}_{n,N}\left(\sum_{S\in\mathscr{B}_{k,l}^{(n)}} \varepsilon(S) \le \frac{1}{2} \sum_{S\in\mathscr{B}_{k,l}^{(n)}} \mathbf{M}(\varepsilon(S))\right) = O\left(\frac{1}{\omega}\right).$$

As clearly by (1.2) if $\omega \to +\infty$ then $\sum_{S\in\mathscr{B}_{k,l}^{(n)}} \mathbf{M}(\varepsilon(S)) \to +\infty$ it follows not only that the probability that $\Gamma_{n,N}$ contains at least one subgraph isomorphic with an element of $\mathscr{B}_{k,l}$ tends to 1, but also that with probability tending to 1 the number of subgraphs of $\Gamma_{n,N}$ isomorphic to some element of $\mathscr{B}_{k,l}$ will tend to $+\infty$ with the same order of magnitude as ω^l.

Thus Theorem 1 is proved.

It is interesting to compare the thresholds for the appearance of a subgraph of a certain type in the above sense with probability near to 1, with the number of edges which is needed in order that the graph should have *necessarily* a subgraph of the given type. Such "compulsory" thresholds have been considered by P. TURÁN [11] (see also [12]) and later by P. ERDŐS and A. H. STONE [17]). For instance for a tree of order k clearly the compulsory threshold is $\left[\frac{n(k-2)}{2}\right] + 1$; for the presence of at least one cycle the compulsory threshold is n while according to a theorem of P. TURÁN [11] for complete subgraphs of order k the compulsory threshold is $\frac{(k-2)}{2(k-1)}(n^2 - r^2) + \binom{r}{2}$ where $r = n - (k-1)\left[\frac{n}{k-1}\right]$. In the paper [13] of T. KŐVÁRI, V. T. SÓS and P. TURÁN it has been shown that the compulsory threshold for the presence of a saturated even subgraph of type (a, a) is of order of magnitude not greater than $n^{2-\frac{1}{a}}$. In all cases the "compulsory" thresholds in TURÁN's sense are of greater order of magnitude as our "probable" thresholds.

§ 2. Trees

Now let us turn to the determination of threshold distribution functions for trees of a given order. We shall prove somewhat more, namely that if $N \sim \varrho\, n^{\frac{k-2}{k-1}}$ where $\varrho > 0$, then the number of trees of order k contained in $\Gamma_{n,N}$ has in the limit for $n \to +\infty$ a Poisson distribution with mean value $\lambda = \frac{(2\varrho)^{k-1} k^{k-2}}{k!}$. This implies that the threshold distribution function for trees of order k is $1 - e^{-\lambda}$.

In proving this we shall count only *isolated trees* of order k in $\Gamma_{n,N}$, i. e. trees of order k which are isolated subgraphs of $\Gamma_{n,N}$. According to Theorem 1. this makes no essential difference, because if there would be a tree of order k which is a subgraph but not an isolated subgraph of $\Gamma_{n,N}$, then $\Gamma_{n,N}$ would have a connected subgraph consisting of $k+1$ points and the probability of this is tending to 0 if $N = o\left(n^{\frac{k-1}{k}}\right)$ which condition is fulfilled in our case as we suppose $N \sim \varrho n^{\frac{k-2}{k-1}}$.

Thus we prove

Theorem 2a. *If* $\lim\limits_{n\to+\infty} \frac{N(n)}{n^{\frac{k-2}{k-1}}} = \varrho > 0$ *and* τ_k *denotes the number of isolated trees of order* k *in* $\Gamma_{n,N(n)}$ *then*

$$\lim_{n\to+\infty} \mathbf{P}_{n,N(n)}(\tau_k = j) = \frac{\lambda^j e^{-\lambda}}{j!} \tag{2.1}$$

or $j = 0, 1, \ldots,$ *where*

$$\lambda = \frac{(2\varrho)^{k-1} k^{k-2}}{k!}. \tag{2.2}$$

For the proof we need the following

Lemma 1. *Let* $\varepsilon_{n1}, \varepsilon_{n2}, \ldots, \varepsilon_{nl_n}$ *be sets of random variables on some probability space; suppose that* $\varepsilon_{ni} (1 \leq i \leq l_n)$ *takes on only the values* 1 *and* 0. *If*

$$\lim_{n\to+\infty} \sum_{1\leq i_1<i_2<\ldots<i_r\leq l_n} \mathbf{M}(\varepsilon_{ni_1} \varepsilon_{ni_2} \ldots \varepsilon_{ni_r}) = \frac{\lambda^r}{r!} \tag{2.3}$$

uniformly in r *for* $r = 1, 2, \ldots,$ *where* $\lambda > 0$ *and the summation is extended over all combinations* $(i_1, i_2, \ldots, i_r)$ *of order* r *of the integers* $1, 2, \ldots, l_n$, *then*

$$\lim_{n\to+\infty} \mathbf{P}\left(\sum_{i=1}^{l_n} \varepsilon_{n_i} = j\right) = \frac{\lambda^j e^{-\lambda}}{j!} \qquad (j = 0, 1, \ldots) \tag{2.4}$$

i. e. the distribution of the sum $\sum_{i=1}^{l_n} \varepsilon_{ni}$ *tends for* $n \to +\infty$ *to the Poisson-distribution with mean value* λ.

Proof of Lemma 1. Let us put

$$P_n(j) = \mathbf{P}\left(\sum_{i=1}^{l_n} \varepsilon_{n_i} = j\right). \tag{2.5}$$

Clearly

$$\sum_{1 \le i_1 < i_2 < \ldots < i_r \le l_n} \mathbf{M}(\varepsilon_{ni_1} \varepsilon_{ni_2} \ldots \varepsilon_{ni_r}) = \sum_{j=r}^{+\infty} \binom{j}{r} P_n(j) \tag{2.6}$$

thus it follows from (2.3) that

$$\lim_{n \to +\infty} \sum_{j=r}^{+\infty} P_n(j) \binom{j}{r} = \frac{\lambda^r}{r!} \qquad (r = 1, 2, \ldots) \tag{2.7}$$

uniformly in r.

It follows that for any z with $|z| < 1$

$$\lim_{n \to +\infty} \sum_{r=1}^{\infty} \left(\sum_{j=r}^{+\infty} P_n(j) \binom{j}{r}\right) z^r = \sum_{r=1}^{\infty} \frac{(\lambda z)^r}{r!} = e^{\lambda z} - 1. \tag{2.8}$$

But

$$\sum_{r=1}^{\infty} \left(\sum_{j=r}^{+\infty} P_n(j) \binom{j}{r}\right) z^r = \sum_{j=0}^{+\infty} P_n(j) (1 + z)^j - 1. \tag{2.9}$$

Thus choosing $z = x - 1$ with $0 < x \le 1$ it follows that

$$\lim_{n \to +\infty} \sum_{j=0}^{+\infty} P_n(j) x^j = e^{\lambda(x-1)} \qquad \text{for } 0 < x \le 1. \tag{2.10}$$

It follows easily that (2.10) holds for $x = 0$ too. As a matter of fact putting $G_n(x) = \sum_{j=0}^{+\infty} P_n(j) x^j$, we have for $0 < x \le 1$

$$|P_n(0) - e^{-\lambda}| \le |G_n(x) - e^{\lambda(x-1)}| + |G_n(x) - P_n(0)| + |e^{\lambda(x-1)} - e^{-\lambda}|.$$

As however

$$|G_n(x) - P_n(0)| \le x \sum_{j=1}^{+\infty} P_n(j) \le x$$

and similarly

$$|e^{\lambda(x-1)} - e^{-\lambda}| \le x$$

it follows that

$$|P_n(0) - e^{-\lambda}| \le |G_n(x) - e^{\lambda(x-1)}| + 2x.$$

Thus we have

$$\limsup_{n \to +\infty} |P_n(0) - e^{-\lambda}| \le 2x;$$

as however $x > 0$ may be chosen arbitrarily small it follows that

$$\lim_{n \to +\infty} P_n(0) = e^{-\lambda}.$$

i. e. that (2.10) holds for $x = 0$ too. It follows by a well-known argument that

$$\lim_{n \to +\infty} P_n(j) = \frac{\lambda^j e^{-\lambda}}{j!} \qquad (j = 0, 1, \ldots). \tag{2.11}$$

As a matter of fact, as (2.10) is valid for $x = 0$, (2.11) holds for $j = 0$. If (2.11) is already proved for $j \leq s - 1$ then it follows from (2.10) that

$$\lim_{n \to +\infty} \sum_{j=s}^{+\infty} P_n(j)\, x^{j-s} = \sum_{j=s}^{+\infty} \frac{\lambda^j e^{-\lambda}}{j!} x^{j-s} \qquad \text{for } 0 < x \leq 1. \tag{2.12}$$

By the same argument as used in connection with (2.10) we obtain that (2.12) holds for $x = 0$ too. Substituting $x = 0$ into (2.12) we obtain that (2.11) holds for $j = s$ too. Thus (2.11) is proved by induction and the assertion of Lemma 1 follows.

Proof of Theorem 2a. Let $T_k^{(n)}$ denote the set of all trees of order k which are subgraphs of the complete graph having the vertices $P_1, P_2, \ldots, P_n$. If $S \in T_k^{(n)}$ let the random variable $\varepsilon(S)$ be equal to 1 if S is an *isolated* subgraph of $\Gamma_{n,N}$; otherwise $\varepsilon(S)$ shall be equal to 0. We shall show that the conditions of Lemma 1 are satisfied for the sum $\sum_{S \in T_k^{(n)}} \varepsilon(S)$ provided that $N = N(n) \sim \sim \varrho n^{\frac{k-2}{k-1}}$ and λ is defined by (2.2). As a matter of fact we have for any $S \in T_k^{(n)}$.

$$\mathsf{M}(\varepsilon(S)) = \frac{\binom{\binom{n-k}{2}}{N-k+1}}{\binom{\binom{n}{2}}{N}} = \left(\frac{2N}{n^2}\right)^{k-1} e^{-\frac{2Nk}{n}} \left(1 + O\left(\frac{N}{n^2}\right)\right). \tag{2.13}$$

More generally if $S_1, S_2, \ldots, S_r$ ($S_j \in T_k^{(n)}$) have pairwise no point in common then clearly we have for each fixed $k \geq 1$ and $r \geq 1$ provided that $n \to +\infty$, $N \to +\infty$

$$\mathsf{M}(\varepsilon(S_1)\, \varepsilon(S_2) \ldots \varepsilon(S_r)) = \frac{\binom{\binom{n-rk}{2}}{N-r(k-1)}}{\binom{\binom{n}{2}}{N}} = \left(\frac{2N}{n^2}\right)^{(k-1)r} e^{-\frac{2Nrk}{n}} \left(1 + O\left(\frac{r^2 N}{n^2}\right)\right) \tag{2.14}$$

where the bound of the O term depends only on k. If however the S_j ($j = = 1, 2, \ldots, r$) are not pairwise disjoint, we have

$$\mathsf{M}(\varepsilon(S_1)\, \varepsilon(S_2) \ldots \varepsilon(S_r)) = 0. \tag{2.15}$$

Taking into account that according to a classical formula of CAYLEY [1] the number of different trees which can be formed from k labelled points is equal to k^{k-2}, it follows that

$$\sum \mathsf{M}(\varepsilon(S_1)\,\varepsilon(S_2)\ldots\varepsilon(S_r)) = \left(\frac{k^{k-2}}{k!}\right)^r \frac{n^{kr}}{r!}\left(\frac{2N}{n^2}\right)^{rk-1} e^{-\frac{2Nrk}{n}}\left(1+O\left(\frac{r^2N}{n^2}\right)\right) \tag{2.16}$$

where the summation on the left hand side is extended over all r-tuples of trees belonging to the set $T_k^{(n)}$ and the bound of the O-term depends only on k. Note that (2.16) is valid independently of how N is tending to $+\infty$. This will be needed in the proof of Theorem 3.

Thus we have, uniformly in r

$$\lim_{\frac{N(n)}{n^{\frac{k-2}{k-1}}}\to\varrho} \sum \mathsf{M}(\varepsilon(S_1)\,\varepsilon(S_2)\ldots\varepsilon(S_r)) = \frac{\lambda^r}{r!} \qquad \text{for } r = 1, 2, \ldots \tag{2.17}$$

where λ is defined by (2.2).

Thus our Lemma 1 can be applied; as $\tau_k = \sum_{S\in T_k^{(n)}} \varepsilon(S)$ Theorem 2 is proved.

We add some remarks on the formula, resulting from (2.16) for $r = 1$

$$\mathsf{M}(\tau_k) = \frac{n^2}{2N}\,\frac{\left(\frac{2N}{n}e^{-\frac{2N}{n}}\right)^k k^{k-2}}{k!}\left(1+O\left(\frac{N}{n^2}\right)\right). \tag{2.18}$$

Let us investigate the functions $m_k(t) = \dfrac{k^{k-2}\,t^{k-1}\,e^{-kt}}{k!}$ $(k = 1, 2, \ldots)$. According to (2.18) $nm_k\left(\dfrac{2N}{n}\right)$ is asymptotically equal to the average number of trees of order k in $\Gamma_{n,N}$. For a fixed value of k, considered as a function of t, the value of $m_k(t)$ increases for $t < \dfrac{k-1}{k}$ and decreases for $t > \dfrac{k-1}{k}$; thus for a fixed value of n the average number of trees of order k reaches its maximum for $N \sim \dfrac{n}{2}\left(1-\dfrac{1}{k}\right)$; the value of this maximum is

$$M_k^* \sim n\,\frac{\left(1-\frac{1}{k}\right)^{k-1} e^{-(k-1)}\,k^{k-2}}{k!}.$$

For large values of k we have evidently

$$M_k^* \sim \frac{n}{\sqrt{2\pi}\,k^{5/2}}.$$

It is easy to see that for any $t > 0$ we have

$$m_k(t) \geqq m_{k+1}(t) \qquad (k = 1, 2, \ldots).$$

The functions $y = m_k(t)$ are shown on Fig. 1.

It is natural to ask what will happen with the number τ_k of isolated trees of order k contained in $\Gamma_{n,N}$ if $\frac{N(n)}{n^{\frac{k-2}{k-1}}} \to +\infty$. As the Poisson distribution $\left\{\frac{\lambda^j e^{-\lambda}}{j!}\right\}$ is approaching the normal distribution if $\lambda \to +\infty$, one can guess that τ_k will be approximately normally distributed. This is in fact true, and is expressed by

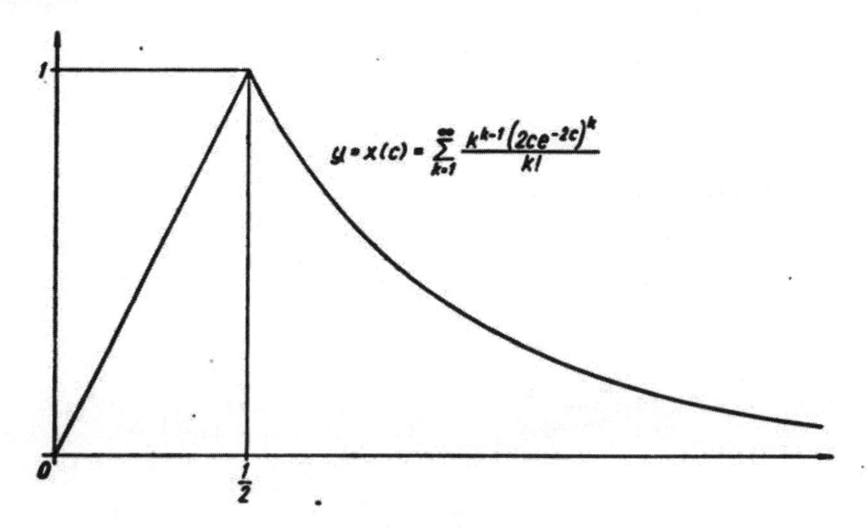

Figure 1a.

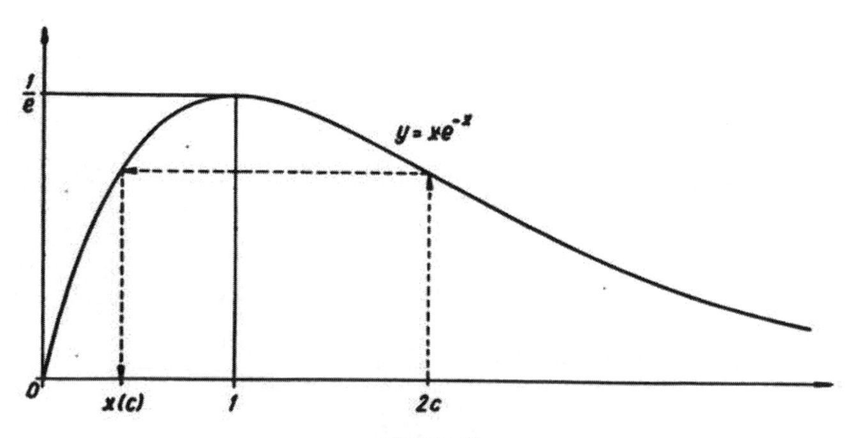

Figure 1b.

Theorem 2b. *If*

$$\frac{N(n)}{n^{\frac{k-2}{k-1}}} \to +\infty \tag{2.19}$$

but at the same time

$$\lim_{n\to+\infty} \frac{N(n) - \frac{1}{2k} n \log n - \frac{k-1}{2k} n \operatorname{loglog} n}{n} = -\infty, \tag{2.20}$$

then denoting by τ_k the number of disjoint trees of order k contained as subgraphs in $\Gamma_{n,N(n)}$ ($k = 1, 2, \ldots$), we have for $-\infty < x < +\infty$

$$\lim_{n\to+\infty} \mathbf{P}_{n,N(n)}\left(\frac{\tau_k - M_{n,N(n)}}{\sqrt{M_{n,N(n)}}} < x\right) = \Phi(x) \tag{2.21}$$

where

$$M_{n,N} = n\frac{k^{k-2}}{k!}\left(\frac{2N}{n}\right)^{k-1} e^{-\frac{2kN}{n}} \tag{2.22}$$

and

$$\Phi(x) = \frac{1}{\sqrt{2\pi}} \int_{-\infty}^{x} e^{-\frac{u^2}{2}} du. \tag{2.23}$$

Proof of Theorem 2b. Note first that the two conditions (2.19) and (2.20) are equivalent to the single condition $\lim_{n\to+\infty} M_{n,N(n)} = +\infty$, and as $\mathbf{M}(\tau_k) \sim M_{n,N}$ this means that the assertion of Theorem 2b can be expressed by saying that the number of isolated trees of order k is asymptotically normally distributed always if n and N tend to $+\infty$ so, that the average number of such trees is also tending to $+\infty$. Let us consider

$$\mathbf{M}(\tau_k^r) = \mathbf{M}((\sum_{S\in T_k^{(n)}} \varepsilon(S))^r).$$

Now we have evidently, using (2.16)

$$\mathbf{M}(\tau_k^r) = \left(1 + O\left(\frac{r^2 N}{n^2}\right)\right) \sum_{j=1}^{r} \left(\sum_{\sum_{i=1}^{j} h_i = r,\, h_i \geq 1} \frac{r!}{h_1!\, h_2! \ldots h_j!} \right) \frac{M_{n,N}^j}{j!}$$

where $M_{n,N}$ is defined by (2.22). Now as well known (see [16], p. 176)

$$\frac{1}{j!} \sum_{\sum_{i=1}^{j} h_i = r,\, h_i \geq 1} \frac{r!}{h_1!\, h_2! \ldots h_j!} = \sigma_r^{(j)} \tag{2.24}$$

where $\sigma_r^{(j)}$ are the Stirling numbers of the second kind (see e. g. [16], p. 168) defined by

$$x^r = \sum_{j=1}^{r} \sigma_r^{(j)} x(x-1)\dots(x-j+1). \tag{2.25}$$

Thus we obtain

$$\mathbf{M}(\tau_k^r) = \left(1 + O\left(\frac{r^2 N}{n^2}\right)\right) \cdot \sum_{j=1}^{r} \sigma_r^{(j)} M_{n,N}^j. \tag{2.26}$$

Now as well known (see e. g. [16], p. 202)

$$e^{\lambda(e^x-1)} - 1 = \sum_{j=1}^{+\infty} \sum_{r=j}^{+\infty} \sigma_r^{(j)} \frac{x^r \lambda^j}{r!} = \sum_{r=1}^{\infty} \frac{x^r}{r!} \left(\sum_{j=1}^{r} \sigma_r^{(j)} \lambda^j\right). \tag{2.27}$$

Thus it follows that

$$\sum_{j=1}^{r} \sigma_r^{(j)} \lambda^j = \left[\frac{d^r}{dx^r} e^{\lambda(e^x-1)}\right]_{x=0} = \sum_{k=0}^{+\infty} \frac{\lambda^k}{k!} e^{-\lambda} k^r. \tag{2.28}$$

We obtain therefrom

$$\mathbf{M}\left(\left(\frac{\tau_k - M_{n,N}}{\sqrt{M_{n,N}}}\right)^r\right) = \left(\frac{1}{M_{n,N}^{1/2}} \sum_{k=0}^{+\infty} \frac{M_{n,N}^k}{k!} e^{-M_{n,N}} (k - M_{n,N})^r\right)\left(1 + O\left(\frac{r^2 N}{n^2}\right)\right). \tag{2.29}$$

Now evidently $\sum_{k=0}^{+\infty} \frac{\lambda^k}{k!} e^{-\lambda} (k-\lambda)^r$ is the r-th central moment of the Poisson distribution with mean value λ. It can be however easily verified that the moments of the Poisson distribution appropriately normalized tend to the corresponding moments of the normal distribution, i. e. we have for $r = 1, 2, \dots$

$$\lim_{\lambda \to +\infty} \frac{1}{\lambda^{\frac{r}{2}}} \left(\sum_{k=1}^{+\infty} \frac{\lambda^k e^{-\lambda}}{k!} (k-\lambda)^r\right) = \frac{1}{\sqrt{2\pi}} \int_{-\infty}^{+\infty} x^r e^{-\frac{x^2}{2}} dx. \tag{2.30}$$

In view of (2.29) this implies the assertion of Theorem 2b.

In the case $N(n) = \frac{1}{2k} n \log n + \frac{k-1}{2k} n \operatorname{loglog} n + yn + o(n)$ when the average number of isolated trees of order k in $\Gamma_{n,N(n)}$ is again finite, the following theorem is valid.

Theorem 2c. *Let τ_k denote the number of isolated trees of order k in $\Gamma_{n,N}$ $(k = 1, 2, \dots)$. Then if*

$$N(n) = \frac{1}{2k} n \log n + \frac{k-1}{2k} n \operatorname{loglog} n + yn + o(n) \tag{2.31}$$

where $-\infty < y < +\infty$, we have

$$\lim_{n \to +\infty} \mathbf{P}_{n,N(n)}(\tau_k = j) = \frac{\lambda^j e^{-\lambda}}{j!} \qquad (j = 0, 1, \dots) \tag{2.32}$$

where

$$\lambda = \frac{e^{-2ky}}{k \cdot k!}. \tag{2.33}$$

Proof of Theorem 2c. It is easily seen that under the conditions of Theorem 2c

$$\lim_{n\to+\infty} \mathbf{M}_{n,N(n)}(\tau_k) = \lambda .$$

Similarly from (2.16) it follows that for $r = 1, 2, \ldots$

$$\lim_{n\to+\infty} \sum_{S_j \in T_k^{(n)}} \mathbf{M}_{n,N(n)}(\varepsilon(S_1)\,\varepsilon(S_2)\ldots\varepsilon(S_r)) = \frac{\lambda^r}{r!}$$

and the proof of Theorem 2c is completed by the use of our Lemma 1 exactly as in the proof of Theorem 2a.

Note that Theorem 2c generalizes the results of the paper [7], where only the case $k = 1$ is considered.

§ 3. Cycles

Let us consider now the threshold function of cycles of a given order. The situation is described by the following

Theorem 3a. *Suppose that*

$$N(n) \sim cn \text{ where } c > 0 . \tag{3.1}$$

Let γ_k denote the number of cycles of order k contained in $\Gamma_{n,N}$ $(k = 3, 4, \ldots)$. Then we have

$$\lim_{n\to+\infty} \mathbf{P}_{n,N(n)}(\gamma_k = j) = \frac{\lambda^j e^{-\lambda}}{j!} \qquad (j = 0, 1, \ldots) \tag{3.2}$$

where

$$\lambda = \frac{(2c)^k}{2k} . \tag{3.3}$$

Thus the threshold distribution corresponding to the threshold function $A(n) = n$ for the property that the graph contains a cycle of order k is $1 - e^{-\frac{1}{2k}(2c)^k}$.

It is interesting to compare Theorem 3a with the following two theorems:

Theorem 3b. *Suppose again that* (3.1) *holds. Let γ_k^* denote the number of isolated cycles of order k contained in $\Gamma_{n,N}$ $(k = 3, 4, \ldots)$. Then we have*

$$\lim_{n\to+\infty} \mathbf{P}_{n,N(n)}(\gamma_k^* = j) = \frac{\mu^j e^{-\mu}}{j!} \qquad (j = 0, 1, \ldots) \tag{3.4}$$

where

$$\mu = \frac{(2c\,e^{-2c})^k}{2k} . \tag{3.5}$$

Remark. Note that according to Theorem 3b for isolated cycles there does not exist a threshold in the ordinary sense, as $1 - e^{-\mu}$ reaches its maximum $1 - e^{-\frac{1}{2ke^k}}$ for $c = \frac{1}{2}$ $\left(\text{i. e. for } N(n) \sim \frac{n}{2}\right)$ and then again decreases;

thus the probability that $\Gamma_{n,N}$ contains an *isolated* cycle of order k never approaches 1.

Theorem 3c. *Let δ_k denote the number of components of $\Gamma_{n,N}$ consisting of $k \geq 3$ points and k edges. If* (3.1) *holds then we have*

$$\lim_{n\to+\infty} \mathbf{P}_{n,N(n)}(\delta_k = j) = \frac{\omega^j e^{-\omega}}{j!} \qquad (j = 0, 1, \ldots) \tag{3.6}$$

where

$$\omega = \frac{(2ce^{-2c})^k}{2k}\left(1 + k + \frac{k^2}{2!} + \ldots + \frac{k^{k-3}}{(k-3)!}\right). \tag{3.7}$$

Proof of Theorems 3a., 3b. and 3c. As from k given points one can form $\frac{1}{2}(k-1)!$ cycles of order k we have evidently for fixed k and for $N = O(n)$

$$\mathbf{M}(\gamma_k) = \frac{1}{2}\binom{n}{k}(k-1)!\frac{\binom{\binom{n}{2}-k}{N-k}}{\binom{\binom{n}{2}}{N}} \sim \frac{\left(\frac{2N}{n}\right)^k}{2k} \tag{3.8}$$

while

$$\mathbf{M}(\gamma_k^*) = \frac{1}{2}\binom{n}{k}(k-1)!\frac{\binom{\binom{n-k}{2}}{N-k}}{\binom{\binom{n}{2}}{N}} \sim \frac{\left(\frac{2N}{n}e^{-\frac{2N}{n}}\right)^k}{2k}. \tag{3.9}$$

As regards Theorem 3c it is known (see [10] and [15]) that the number of connected graphs $G_{k,k}$ (i. e. the number of connected graphs consisting of k labelled vertices and k edges) is exactly

$$\Theta_k = \frac{1}{2}(k-1)!\left(1 + k + \frac{k^2}{2} + \ldots + \frac{k^{k-3}}{(k-3)!}\right). \tag{3.10}$$

Now we have clearly

$$\mathbf{M}(\delta_k) = \binom{n}{k}\Theta_k\frac{\binom{\binom{n-k}{2}}{N-k}}{\binom{\binom{n}{k}}{N}} \sim \frac{\left(\frac{2N}{n}e^{-\frac{2N}{n}}\right)^k}{2k}\left(1 + k + \frac{k^2}{2!} + \ldots + \frac{k^{k-3}}{(k-3)!}\right). \tag{3.11}$$

For large values of k we have (see [15])

$$\Theta_k \sim \sqrt{\frac{\pi}{8}}\, k^{k-1/2} \tag{3.12}$$

and thus

$$\mathbf{M}(\delta_k) \sim \frac{\left(\frac{2N}{n} e^{1-\frac{2N}{n}}\right)^k}{4k}. \tag{3.13}$$

For $N \sim \frac{n}{2}$ we obtain by some elementary computation using (7) that for large values of k (such that $k = o(n^{3/4})$.

$$\mathbf{M}(\delta_k) \sim \frac{e^{-\frac{k^3}{n^2}}}{4k}. \tag{3.14}$$

Using (3.8), (3.9) and (3.11) the proofs of Theorems 3a, 3b and 3c follow the same lines as that of Theorem 2a, using Lemma 1. The details may be left to the reader.

Similar results can be proved for other types of subgraphs, e. g. complete subgraphs of a given order. As however these results and their proofs have the same pattern as those given above we do not dwell on the subject any longer and pass to investigate *global properties of the random graph* $\Gamma_{n,N}$.

§ 4. The total number of points belonging to trees

We begin by proving

Theorem 4a. *If* $N = o(n)$ *the graph* $\Gamma_{n,N}$ *is, with probability tending to* 1 *for* $n \to +\infty$, *the union of disjoint trees.*

Proof of Theorem 4a. A graph consists of disjoint trees if and only if there are no cycles in the graph. The number of graphs $G_{n,N}$ which contain at least one cycle can be enumerated as was shown in § 1 for each value k of the length of this cycle. In this way, denoting by T the property that the graph is a union of disjoint trees, and by $\overline{T}$ the opposite of this property, i. e. that the graph contains at least one cycle, we have

$$\mathbf{P}_{n,N}(\overline{T}) \leq \sum_{k=3}^{n} \binom{n}{k} (k-1)!\, \frac{\binom{\binom{n}{2}-k}{N-k}}{\binom{\binom{n}{2}}{N}} = O\left(\frac{N}{n}\right). \tag{4.1}$$

It follows that if $N = o(n)$ we have $\lim_{n \to +\infty} \mathbf{P}_{n,N}(T) = 1$ which proves Theorem 4a.

If N is of the same order of magnitude as n i. e. $N \sim cn$ with $c > 0$, then the assertion of Theorem 4a is no longer true. Nevertheless if $c < 1/2$, still almost all points (in fact $n - O(1)$ points) of $\Gamma_{n,N}$ belong to isolated trees. There is however a surprisingly abrupt change in the structure of $\Gamma_{n,N}$ with $N \sim cn$ when c surpasses the value $\frac{1}{2}$. If $c > 1/2$ in the average only a positive fraction of all points of $\Gamma_{n,N}$ belong to isolated trees, and the value of this fraction tends to 0 for $c \to +\infty$.

Thus we shall prove

Theorem 4b. *Let $V_{n,N}$ denote the number of those points of $\Gamma_{n,N}$ which belong to an isolated tree contained in $\Gamma_{n,N}$. Let us suppose that*

$$\lim_{n\to+\infty} \frac{N(n)}{n} = c > 0 . \tag{4.2}$$

Then we have

$$\lim_{n\to+\infty} \frac{\mathbf{M}(V_{n,N(n)})}{n} = \begin{cases} 1 & \text{for } c \leq 1/2 \\ \dfrac{x(c)}{2c} & \text{for } c > \dfrac{1}{2} \end{cases} \tag{4.3}$$

where $x = x(c)$ is the only root satisfying $0 < x < 1$ of the equation

$$x e^{-x} = 2ce^{-2c} , \tag{4.4}$$

which can also be obtained as the sum of a series as follows:

$$x(c) = \sum_{k=1}^{\infty} \frac{k^{k-1}}{k!} (2ce^{-2c})^k . \tag{4.5}$$

Proof of Theorem 4b. We shall need the well known fact that the inverse function of the function

$$y = x e^{-x} \qquad (0 \leq x \leq 1) \tag{4.6}$$

has the power series expansion, convergent for $0 \leq y \leq \frac{1}{e}$

$$x = \sum_{k=1}^{+\infty} \frac{k^{k-1} y^k}{k!} . \tag{4.7}$$

Let τ_k denote the number of isolated trees of order k contained in $\Gamma_{n,N}$. Then clearly

$$V_{n,N} = \sum_{k=1}^{n} k \tau_k \tag{4.8}$$

and thus

$$\mathbf{M}(V_{n,N}) = \sum_{k=1}^{n} k\, \mathbf{M}(\tau_k) . \tag{4.9}$$

By (2.18), if (4.2) holds, we have

$$\lim_{n\to+\infty} \frac{1}{n} \mathbf{M}(\tau_k) = \frac{1}{2c} \frac{k^{k-2}}{k!} (2ce^{-2c})^k . \tag{4.10}$$

Thus we obtain from (4.10) that for $c \leq 1/2$

$$\liminf_{n\to+\infty} \frac{\mathbf{M}(V_{n,N(n)})}{n} \geq \frac{1}{2c}\sum_{k=1}^{s} \frac{k^{k-1}(2ce^{-2c})^k}{k!} \text{ for any } s \geq 1. \tag{4.11}$$

As (4.11) holds for any $s \geq 1$ we obtain

$$\liminf_{n\to+\infty} \frac{\mathbf{M}(V_{n,N(n)})}{n} \geq \frac{1}{2c}\sum_{k=1}^{\infty} \frac{k^{k-1}(2ce^{-2c})^k}{k!}. \tag{4.12}$$

But according to (4.7) for $c \leq 1/2$ we have

$$\sum_{k=1}^{\infty} \frac{k^{k-1}(2ce^{-2c})^k}{k!} = 2c.$$

Thus it follows from (4.12) that for $c \leq 1/2$

$$\liminf_{n\to+\infty} \frac{\mathbf{M}(V_{n,N(n)})}{n} \geq 1. \tag{4.13}$$

As however $V_{n,N(n)} \leq n$ and thus $\limsup_{n\to+\infty} \frac{M(V_{n,N(n)})}{n} \leq 1$ it follows that if (4.2) holds and $c \leq 1/2$ we have

$$\lim_{n\to+\infty} \frac{\mathbf{M}(V_{n,N(n)})}{n} = 1. \tag{4.14}$$

Now let us consider the case $c > \frac{1}{2}$. It follows from (2.18) that if (4.2) holds with $c > 1/2$ we obtain

$$\mathbf{M}(V_{n,N(n)}) = \frac{n^2}{2N}\sum_{k=1}^{n} \frac{k^{k-1}}{k!}\left(\frac{2N(n)}{n}e^{-\frac{2N(n)}{n}}\right)^k + O(1) \tag{4.15}$$

where the bound of the term $O(1)$ depends only on c. As however for $N(n) \sim nc$ with $c > 1/2$

$$\sum_{k=n+1}^{\infty} \frac{k^{k-1}}{k!}\left(\frac{2N(n)}{n}e^{-\frac{2N(n)}{n}}\right)^k = O\left(\frac{1}{n^{3/2}}\right)$$

it follows that

$$\mathbf{M}(V_{n,N(n)}) = \frac{n^2}{2N(n)}\, x\left(\frac{N(n)}{n}\right) + O(1) \tag{4.16}$$

where $x = x\left(\frac{N(n)}{n}\right)$ is the only solution with $o < x < 1$ of the equation $xe^{-x} = \frac{2N(n)}{n}e^{-\frac{2N(n)}{n}}$. Thus it follows that if (4.2) holds with $c > 1/2$ we have

$$\lim_{n\to+\infty} \frac{\mathbf{M}(V_{n,N(n)})}{n} = \frac{x(c)}{2c} \tag{4.17}$$

where $x(c)$ is defined by (4.5).

The graph of the function $x(c)$ is shown on Fig. 1a; its meaning is shown by Fig. 1b. The function

$$y = \begin{cases} 1 & \text{for } c \leq 1/2 \\ \dfrac{x(c)}{2c} & \text{for } c > 1/2 \end{cases}$$

is shown on Fig. 2a.

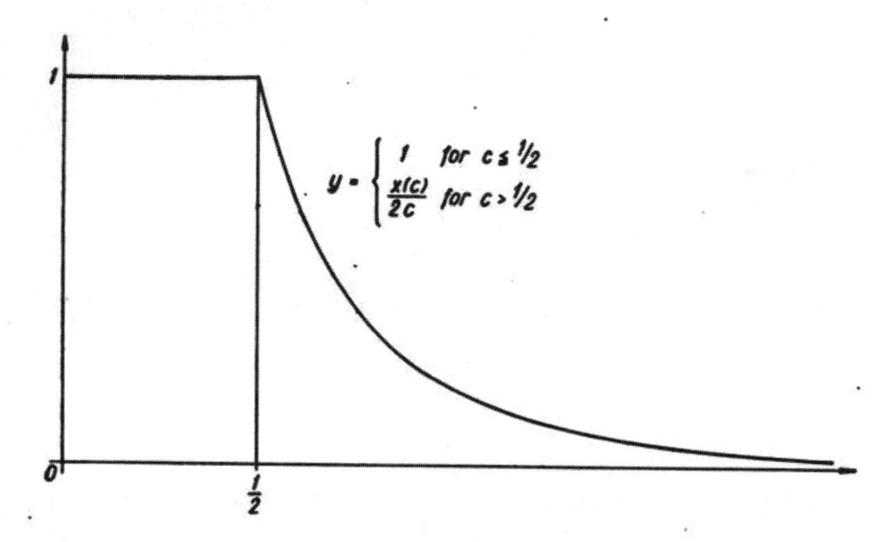

Figure 2a.

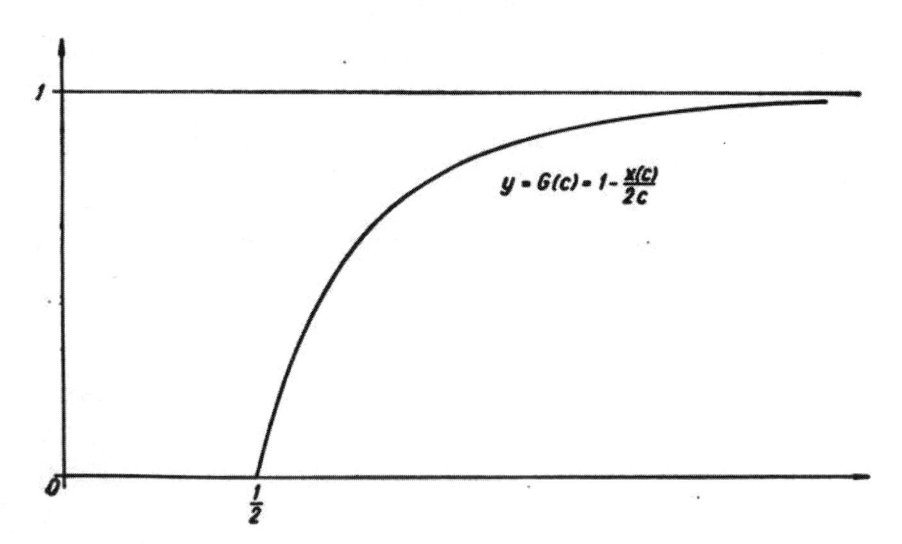

Figure 2b.

Thus the proof of Theorem 4b is complete. Let us remark that in the same way as we obtained (4.16) we get that if (4.2) holds with $c < 1/2$ we have

$$\mathbf{M}(V_{n,N(n)}) = n - O(1) \tag{4.18}$$

where the bound of the $O(1)$ term depends only on c. (However (4.18) is not true for $c = \frac{1}{2}$ as will be shown below.)

It follows by the well known inequality of Markov

$$\mathbf{P}(\xi > a) \leq \frac{1}{a}\mathbf{M}(\xi) \tag{4.19}$$

valid for any nonnegative random variable ξ and any $a > \mathbf{M}(\xi)$, that the following theorem holds:

Theorem 4c. *Let $V_{n,N}$ denote the number of those points of $\Gamma_{n,N}$ which belong to isolated trees contained in $\Gamma_{n,N}$. Then if ω_n tends arbitrarily slowly to $+\infty$ for $n \to +\infty$ and if* (4.2) *holds with $c < 1/2$ we have*

$$\lim_{n \to +\infty} \mathbf{P}(V_{n,N(n)} \geqq n - \omega_n) = 1\,. \tag{4.20}$$

The case $c > 1/2$ is somewhat more involved. We prove

Theorem 4d. *Let $V_{n,N}$ denote the number of those points of $\Gamma_{n,N}$ which belong to an isolated tree contained in $\Gamma_{n,N}$. Let us suppose that* (4.2) *holds with $c > 1/2$. It follows that if ω_n tends arbitrarily slowly to $+\infty$, we have*

$$\lim_{n \to +\infty} \mathbf{P}\left(\left|V_{n,N(n)} - \frac{n^2}{2\,N(n)}\,x\left(\frac{N(n)}{n}\right)\right| > \sqrt{n}\,\omega_n\right) = 0 \tag{4.21}$$

where $x = x\left(\frac{N(n)}{n}\right)$ is the only solution with $0 < x < 1$ of the equation

$$xe^{-x} = \frac{2\,N(n)}{n}\,e^{-\frac{2N(n)}{n}}\,.$$

Proof. We have clearly, as the series $\sum_{k=1}^{\infty} \frac{k^k}{k!}(2\,ce^{-2c})^k$ is convergent, $\mathbf{D}^2(V_{n,N(n)}) = O(n)$. Thus (4.21) follows by the inequality of *Chebyshev.*

Remark. It follows from (4.21) that we have for any $c > 1/2$ and any $\varepsilon > 0$

$$\lim_{n \to +\infty} \mathbf{P}\left(\left|\frac{V_{n,N(n)}}{n} - \frac{x(c)}{2\,c}\right| < \varepsilon\right) = 1 \tag{4.22}$$

where $x(c)$ is defined by (4.5).

As regards the case $c = {}^1/_2$ we formulate the theorem which will be needed latter.

Theorem 4e. *Let $V_{n,N}(r)$ denote the number of those points of $\Gamma_{n,N}$ which belong to isolated trees of order $\geq r$ and $\tau_{n,N}(r)$ the number of isolated trees of order $\geq r$ contained in $\Gamma_{n,N}$. If $N(n) \sim \frac{n}{2}$ we have for any $\delta > o$*

$$\lim_{n \to +\infty} \mathbf{P}\left(\left|\frac{V_{n,N(n)}(r)}{n} - \sum_{k=r}^{+\infty} \frac{k^{k-1}}{k!} e^{-k}\right| < \delta\right) = 1 \tag{4.23}$$

and

$$\lim_{n \to +\infty} \mathbf{P}\left(\left|\frac{\tau_{n,N(n)}(r)}{n} - \sum_{k=r}^{+\infty} \frac{k^{k-2}}{k!} e^{-k}\right| < \delta\right) = 1. \tag{4.24}$$

The proof follows the same lines as those of the preceding theorems.

§ 5. The total number of points belonging to cycles

Let us determine first the average number of all cycles in $\Gamma_{n,N}$. We prove that this number remains bounded if $N(n) \sim cn$ and $c < {}^1/_2$ but not if $c = {}^1/_2$.

Theorem 5a. *Let $H_{n,N}$ denote the number of all cycles contained in $\Gamma_{n,N}$. Then we have if $N(n) \sim cn$ holds with $c < \frac{1}{2}$*

$$\lim_{n \to +\infty} \mathbf{M}(H_{n,N(n)}) = \frac{1}{2} \log \frac{1}{1-2c} - c - c^2 \tag{5.1}$$

while we have for $c = \frac{1}{2}$

$$\mathbf{M}(H_{n,N(n)}) \sim \frac{1}{4} \log n. \tag{5.2}$$

Proof. Clearly if γ_k is the number of all cycles of order k contained in $\Gamma_{n,N}$ we have

$$H_{n,N} = \sum_{k=1}^{n} \gamma_k.$$

Now (5.1) follows easily, taking into account that (see (3.8))

$$\mathbf{M}(\gamma_k) = \frac{1}{2}\binom{n}{k}(k-1)!\frac{\binom{\binom{n}{2}-k}{N-k}}{\binom{\binom{n}{2}}{N}} = \frac{\left(\frac{2N}{n}\right)^k}{2k}\left(1 + O\left(\frac{k^2}{n}\right)\right). \tag{5.3}$$

If $c = {}^1/_2$ we have by (3.8)

$$\mathbf{M}(\gamma_k) \sim \frac{1}{2k} e^{-\frac{3k^2}{2n}}. \tag{5.4}$$

As $\sum_{k=3}^{n} \frac{1}{2k} e^{-\frac{3k^2}{2n}} \sim \frac{1}{4} \log n$, it follows that (5.2) holds. Thus Theorem 5a is proved.

Let us remark that it follows from (5.2) that (4.18) is not true for $c = {}^1/_2$.

Similarly as before we can prove corresponding results concerning the random variable $H_{n,N}$ itself.

We have for instance in the case $c = {}^1/_2$ for any $\varepsilon > o$

$$\lim_{n \to +\infty} \mathbf{P}\left(\left|\frac{H_{n,N(n)}}{\log n} - \frac{1}{4}\right| < \varepsilon\right) = 1. \tag{5.5}$$

This can be proved by the same method as used above: estimating the variance and using the inequality of Chebyshev.

An other related result, throwing more light on the appearance of cycles in $\Gamma_{n,N}$ runs as follows.

Theorem 5b. *Let K denote the property that a graph contains at least one cycle. Then we have if $N(n) \sim nc$ holds with $c \leq {}^1/_2$*

$$\lim_{n \to +\infty} \mathbf{P}_{n,N(n)}(K) = 1 - \sqrt{1 - 2c}\, e^{c + c^2}. \tag{5.6}$$

Thus for $c = \frac{1}{2}$ it is „almost sure" that $\Gamma_{n,N(n)}$ contains at least one cycle, while for $c < \frac{1}{2}$ the limit for $n \to +\infty$ of the probability of this is less than 1.

Proof. Let us suppose first $c < \frac{1}{2}$. By an obvious sieve (taking into account that according to Theorem 1 the probability that there will be in $\Gamma_{n,N(n)}$ with $N(n) \sim nc$ ($c < {}^1/_2$) two circles having a point in common is negligibly small) we obtain

$$\lim_{n \to +\infty} \mathbf{P}_{n,N(n)}(\overline{K}) = e^{-\lim_{n \to +\infty} \mathbf{M}(H_{n,N(n)})} = \sqrt{1 - 2c}\, e^{c + c^2}. \tag{5.7}$$

Thus (5.6) follows for $c < {}^1/_2$. As for $c \to {}^1/_2$ the function on the right of (5.6) tends to 1, it follows that (5.6) holds for $c = {}^1/_2$ too. The function $y = 1 - \sqrt{1 - 2c}\, e^{c + c^2}$ is shown on Fig. 3.

We prove now the following

Theorem 5c. *Let $H^*_{n,N}$ denote the total number of points of $\Gamma_{n,N}$ which belong to some cycle. Then we have for $N = N(n) \sim cn$ with $0 < c < {}^1/_2$*

$$\lim_{n \to +\infty} \mathbf{M}(H^*_{n,N(n)}) = \frac{4c^3}{1 - 2c}. \tag{5.8}$$

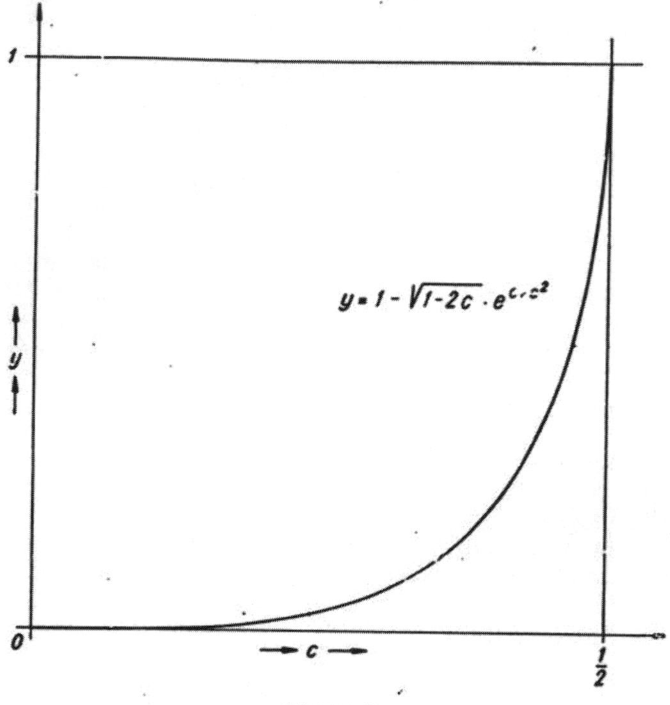

Figure 3.

Proof of Theorem 5c. As according to Theorem 1 the probability that two cycles should have a point in common is negligibly small, we have by (5.3)

$$\mathbf{M}(H^*_{n,N(n)}) \sim \sum_{k=1}^{n} k\gamma_k \sim \frac{(2c)^3}{2(1-2c)} = \frac{4c^3}{1-2c}.$$

The size of that part of $\Gamma_{n,N}$ which does not consist of trees is still more clearly shown by the following

Theorem 5d. *Let $\vartheta_{n,N}$ denote the number of those points of $\Gamma_{n,N}$ which belong to components containing exactly one cycle. Then we have for $N = N(n) \sim \sim cn$ in case $c \neq 1/2$*

$$\lim_{n\to+\infty} \mathbf{M}(\vartheta_{n,N(n)}) = \frac{1}{2}\sum_{k=3}^{+\infty}(2ce^{-2c})^k\left(1+\frac{k}{1!}+\frac{k^2}{2!}+\dots+\frac{k^{k-3}}{(k-3)!}\right) \tag{5.9}$$

while for $c = 1/2$ we have

$$\mathbf{M}(\vartheta_{n,N(n)}) \sim \frac{\Gamma\left(\frac{1}{3}\right)}{12} n^{2/3} \tag{5.10}$$

where $\Gamma(x)$ denotes the gamma-function $\Gamma(x) = \int_0^\infty t^{x-1}e^{-t}\,dt$ for $x > 0$.

Proof of Theorem 5d. (5.9) follows immediately from (3.11); for $c=1/2$ we have by (3.14)

$$\mathbf{M}(\vartheta_{n,N(n)}) \sim \frac{1}{4}\sum_{k=3}^{n} e^{-\frac{k^3}{n^2}} \sim \frac{\Gamma\left(\frac{1}{3}\right)}{12} n^{2/3}.$$

Remark. Note that for $c \to 1/2$

$$\frac{1}{2}\sum_{k=3}^{\infty}(2ce^{-2c})^k\left(1+\frac{k}{1!}+\ldots+\frac{k^{k-3}}{(k-3)!}\right) \sim \frac{1}{4(1-2c)^2}.$$

Thus the average number of points belonging to components containing exactly one cycle tends to $+\infty$ as $\frac{1}{4(1-2c)^2}$ for $c \to 1/2$.

We now prove

Theorem 5e. *For $N(n) \sim cn$ with $0 < c < 1/2$ all components of $\Gamma_{n,N(n)}$ are with probability tending to 1 for $n \to +\infty$, either trees or components containing exactly one cycle.*

Proof. Let $\psi_{n,N}$ denote the number of points of $\Gamma_{n,N}$ belonging to components which contain more edges than vertices and the number of vertices of which is less than $\sqrt{\log n}$. We have clearly for $N(n) \sim cn$ with $c < 1/2$

$$\mathbf{M}(\psi_{n,N(n)}) \leq \sum_{k=4}^{[\sqrt{\log n}]} k\binom{n}{k}2^{\binom{k}{2}}\frac{\binom{\binom{n-k}{2}}{N-k-1}}{\binom{\binom{n}{2}}{N}} = O\left(n^{\frac{\log 2}{2}-1}\right).$$

Thus

$$\mathbf{P}(\psi_{n,N(n)} \geq 1) = O\left(\frac{1}{n^{1-\frac{\log 2}{2}}}\right).$$

On the other hand by Theorem 4c the probability that a component consisting of more than $\sqrt{\log n}$ points should not be a tree tends to 0. Thus the assertion of Theorem 5e follows.

§ 6. The number of components

Let us turn now to the investigation of the average number of components of $\Gamma_{n,N}$. It will be seen that the above discussion contains a fairly complete solution of this question. We prove the following

Theorem 6. *If $\zeta_{n,N}$ denotes the number of components of $\Gamma_{n,N}$ then we have if $N(n) \sim cn$ holds with $0 < c < \frac{1}{2}$*

$$\mathbf{M}(\zeta_{n,N(n)}) = n - N(n) + O(1) \tag{6.1}$$

where the bound of the O-term depends only on c. If $N(n) \sim \frac{n}{2}$ we have

$$\mathbf{M}(\zeta_{n,N(n)}) = n - N(n) + O(\log n). \tag{6.2}$$

If $N(n) \sim cn$ holds with $c > \frac{1}{2}$ we have

$$\lim_{n\to+\infty} \frac{\mathbf{M}(\zeta_{n,N(n)})}{n} = \frac{1}{2c}\left(x(c) - \frac{x^2(c)}{2}\right) \tag{6.3}$$

where $x = x(c)$ is the only solution satisfying $0 < x < 1$ of the equation $xe^{-x} = 2ce^{-2c}$, i. e.

$$x(c) = \sum_{k=1}^{\infty} \frac{k^{k-1}}{k!}(2ce^{-2c})^k. \tag{6.4}$$

Proof of Theorem 6. Let us consider first the case $c < \frac{1}{2}$. Clearly if we add a new edge to a graph, then either this edge connects two points belonging to different components, in which case the number of components is decreased by 1, or it connects two points belonging to the same component in which case the number of components does not change but at least one new cycle is created. Thus[2]

$$\zeta_{n,N} - (n - N) \leq H_{n,N} \tag{6.5}$$

where $H_{n,N}$ is the total number of cycles in $\Gamma_{n,N}$. Thus by Theorem 5a it follows that (6.1) holds.

Similarly (6.2) follows also from Theorem 5a. Now we consider the case $c > \frac{1}{2}$.

It is easy to see that for $o \leq y \leq \frac{1}{e}$ we have (see e. g. [14])

$$\sum_{k=1}^{+\infty} \frac{k^{k-2} y^k}{k!} = x - \frac{x^2}{2} \tag{6.6}$$

where

$$x = \sum_{k=1}^{+\infty} \frac{k^{k-1} y^k}{k!}. \tag{6.7}$$

[2] In fact according to a well known theorem of the theory of graphs (see [4], p. 29) being a generalization of Euler's theorem on polyhedra we have $N - n + \zeta_{n,N} = \varkappa^*_{n,N}$, where $\varkappa^*_{n,N}$ — the „cyclomatic number" of the graph $\Gamma_{n,N}$ — is equal to the maximal number of independent cycles, in $\Gamma_{n,N}$ (For a definition of independent cycles see [4] p. 28).

x can be characterized also as the only solution satisfying $0 < x \leq 1$ of the equation $xe^{-x} = y$.

It follows that if $N(n) \sim nc$ holds with $c < {}^1/_2$ we have

$$\mathbf{M}(\zeta_{n,N(n)}) = \frac{n^2}{2\,N(n)}\left(\frac{2\,N(n)}{n} - \frac{4\,N^2(n)}{2\,n^2}\right) + O(1) = n - N(n) + O(1) \tag{6.8}$$

which leads to a second proof of the first part of Theorem 6.

To prove the second part, let us remark first that the number of components of order greater than A is clearly $\leq \frac{n}{A}$. Thus if $\zeta_{n,N}(A)$ denotes the number of components of order $\leq A$ of $\Gamma_{n,N}$ we have clearly

$$\mathbf{M}(\zeta_{n,N}) = \mathbf{M}(\zeta_{n,N}(A)) + O\left(\frac{n}{A}\right). \tag{6.9}$$

The average number of components of fixed order k which contain at least k edges will be clearly according to Theorem 1 of order $\left(\frac{N}{n}\right)^k$, i. e. bounded for each fixed value of k. As A can be chosen arbitrarily large we obtain from (6.9) that

$$\mathbf{M}(\zeta_{n,N}) \sim \sum_{k=1}^{n} \mathbf{M}(\tau_k). \tag{6.10}$$

According to (2.18) it follows that

$$\mathbf{M}(\zeta_{n,N}) \sim \frac{n^2}{2\,N} \sum_{k=1}^{+\infty} \frac{k^{k-2}}{k!}\left(\frac{2\,N}{n}e^{-\frac{2N}{n}}\right)^k \tag{6.11}$$

and thus, according to (6.6) if $N(n) \sim cn$ holds with $c > {}^1/_2$ we have

$$\lim_{n \to +\infty} \frac{\mathbf{M}(\zeta_{n,N(n)})}{n} = \frac{1}{2\,c}\left(x(c) - \frac{x^2(c)}{2}\right) \tag{6.12}$$

where $x(c)$ is defined by (6.4). Thus Theorem 6 is completely proved.

Let us add some remarks. Theorem 6 illustrates also the fundamental change in the structure of $\Gamma_{n,N}$ which takes place if N passes $\frac{n}{2}$. While the average number of components of $\Gamma_{n,N}$ (as a function of N with n fixed) decreases linearly if $N \leq \frac{n}{2}$ this is no longer true for $N > \frac{n}{2}$; the average number of components decreases from this point onward more and more slowly. The graph of

$$z(c) = \lim_{\frac{N(n)}{n} \to c} \frac{\mathbf{M}(\zeta_{n,N(n)})}{n} = \begin{cases} 1 - c & \text{for} \quad 0 \leq c \leq \frac{1}{2} \\ \frac{1}{2\,c}\left(x(c) - \frac{x^2(c)}{2}\right) & \text{for} \quad c > {}^1/_2 \end{cases} \tag{6.13}$$

as a function of c is shown by Fig. 4.

From Theorem 6 one can deduce easily that in case $N(n) \sim cn$ with $c < {}^1/_2$ we have for any sequence ω_n tending arbitrarily slowly to infinity

$$\text{(6.14)} \qquad \lim_{n\to+\infty} \mathbf{P}(|\zeta_{n,N(n)} - n + N(n)| < \omega_n) = 1$$

(6.14) follows easily by remarking that clearly $\zeta_{n,N} \geq n - N$.

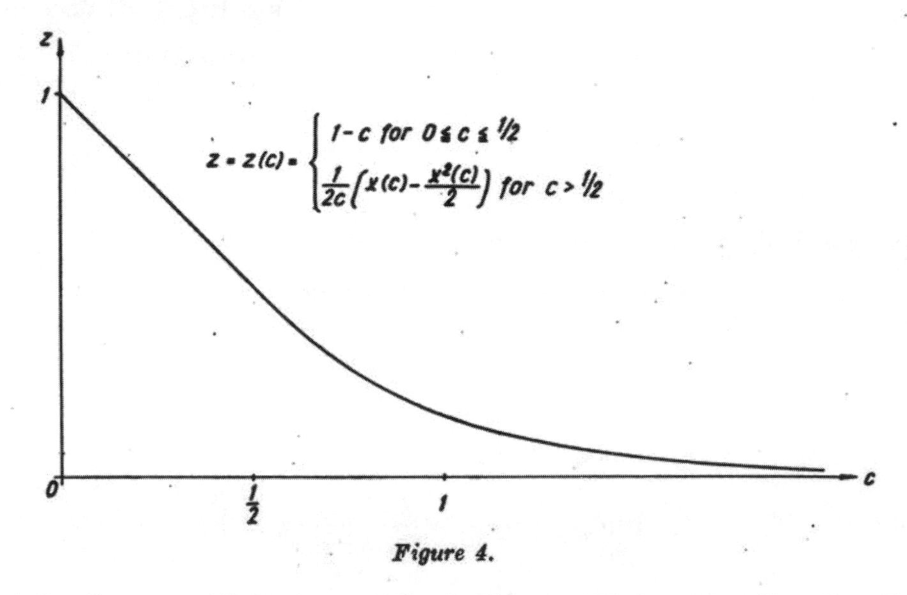

Figure 4.

For the case $N(n) \sim cn$ with $c \geq {}^1/_2$ one obtains by estimating the variance of $\zeta_{n,N(n)}$ and using the inequality of Chebyshev that for any $\varepsilon > 0$

$$\text{(6.15)} \qquad \lim_{n\to+\infty} \mathbf{P}\left(\left|\frac{\zeta_{n,N(n)}}{n} - \frac{1}{2c}\left(x(c) - \frac{x^2(c)}{2}\right)\right| < \varepsilon\right) = 1 .$$

The proof is similar to that of (4.21) and therefore we do not go into details.

§ 7. The size of the greatest tree

If $N \sim cn$ with $c < {}^1/_2$ then as we have seen in § 6 all but a finite number of points of $\Gamma_{n,N}$ belong to components which are trees. Thus in this case the problem of determining the size of the largest component of $\Gamma_{n,N}$ reduces to the easier question of determining the greatest tree in $\Gamma_{n,N}$. This question is answered by the following.

Theorem 7a. *Let $\Delta_{n,N}$ denote the number of points of the greatest tree which is a component of $\Gamma_{n,N}$. Suppose $N = N(n) \sim cn$ with $c \neq {}^1/_2$. Let ω_n be a sequence* *tending arbitrarily slowly to* $+\infty$. *Then we have*

$$\lim_{n\to+\infty} \mathbf{P}\left(\Delta_{n,N(n)} \geqq \frac{1}{\alpha}\left(\log n - \frac{5}{2}\log\log n\right) + \omega_n\right) = 0 \tag{7.1}$$

and

$$\lim_{n\to+\infty} \mathbf{P}\left(\Delta_{n,N(n)} \geqq \frac{1}{\alpha}\left(\log n - \frac{5}{2}\log\log n\right) - \omega_n\right) = 1 \tag{7.2}$$

where

$$e^{-\alpha} = 2\,ce^{1-2c} \qquad (\text{i. e. } \alpha = 2c - 1 - \log 2c \tag{7.3}$$

and thus $\alpha > 0$.)

Proof of Theorem 7a. We have clearly

$$\mathbf{P}(\Delta_{n,N(n)} \geqq z) = \mathbf{P}\left(\sum_{k\geqq z} \tau_k \geqq 1\right) \leqq \sum_{k\geqq z} \mathbf{M}(\tau_k) \tag{7.4}$$

and thus by (2.18)

$$\mathbf{P}(\Delta_{n,N(n)} \geqq z) = O\left(\frac{ne^{-\alpha z}}{z^{5/2}}\right). \tag{7.5}$$

It follows that if $z_1 = \frac{1}{\alpha}\left(\log n - \frac{5}{2}\log\log n\right) + \omega_n$ we have

$$\mathbf{P}(\Delta_{n,N(n)} \geqq z_1) = O(e^{-\alpha\omega_n}). \tag{7.6}$$

This proves (7.1). To prove (7.2) we have to estimate the mean and variance of τ_{z_2} where $z_2 = \frac{1}{\alpha}\left(\log n - \frac{5}{2}\log\log n\right) - \omega_n$. We have by (2.18)

$$\mathbf{M}(\tau_{z_2}) \sim \frac{\alpha^{5/2}}{2c\sqrt{2\pi}} e^{\alpha\omega_n} \tag{7.7}$$

and

$$\mathbf{D}^2(\tau_{z_2}) = O(\mathbf{M}(\tau_{z_2})). \tag{7.8}$$

Clearly

$$\mathbf{P}(\Delta_{n,N(n)} \geqq z_2) \geqq \mathbf{P}(\tau_{z_2} \geqq 1) = 1 - \mathbf{P}(\tau_{z_2} = 0)$$

and it follows from (7.7) and (7.8) by the inequality of Chebyshev that

$$\mathbf{P}(\tau_{z_2} = 0) = O(e^{-\alpha x_n}). \tag{7.9}$$

Thus we obtain

$$\mathbf{P}(\Delta_{n,N(n)} \geqq z_2) \geqq 1 - O(e^{-\alpha\omega_n}). \tag{7.10}$$

Thus (7.2) is also proved.

Remark. If $c < \frac{1}{2}$ the greatest tree which is a component of $\Gamma_{n,N}$ with $N \sim cn$ is — as mentioned above — at the same time the greatest component of $\Gamma_{n,N}$, as $\Gamma_{n,N}$ contains with probability tending to 1 besides trees only components containing a single circle and being of moderate size. This follows evidently from Theorem 4c. As will be seen in what follows (see § 9) for $c > \frac{1}{2}$ the situation is completely different, as in this case $\Gamma_{n,N}$ contains a very large component (in fact of size $G(c)n$ with $G(c) > 0$) which is not a tree. Note that if we put $c = \frac{1}{2k} \log n$ we have $\alpha = \frac{1}{k} \log n$ and $\frac{1}{\alpha} \log n \sim k$ in conformity with Theorem 2c.

We can prove also the following

Theorem 7b. *If $N \sim cn$, where $c \neq \frac{1}{2}$ and $e^{-\alpha} = 2ce^{1-2c}$ then the number of isolated trees of order $h = \frac{1}{\alpha}\left(\log n - \frac{5}{2} \log\log n\right) + l$ resp. of order $\geq h$ (where l is an arbitrary real number such that h is a positive integer) contained in $\Gamma_{n,N}$ has for large n approximately a Poisson distribution with the mean value* $\lambda = \dfrac{\alpha^{5/2} e^{-\alpha l}}{2c\sqrt{2\pi}}$ *resp.* $\mu = \dfrac{\alpha^{5/2} e^{-\alpha l}}{2c\sqrt{2\pi}(1 - e^{-\alpha})}$.

Corollary. The probability that $\Gamma_{n,N(n)}$ with $N(n) \sim nc$ where $c \neq \frac{1}{2}$ does not contain a tree of order $\geq \frac{1}{\alpha}\left(\log n - \frac{5}{2} \log\log n\right) + l$ tends to $\exp\left(-\dfrac{\alpha^{5/2} e^{-\alpha l}}{2c\sqrt{2\pi}(1 - e^{-\alpha})}\right)$ for $n \to +\infty$, where $\alpha = 2c - 1 - \log 2c$.

The size of the greatest tree which is a component of $\Gamma_{n,N}$ is fairly large if $N \sim \frac{n}{2}$. This could be guessed from the fact that the constant factor in the expression $\frac{1}{\alpha}\left(\log n - \frac{5}{2} \log\log n\right)$ of the „probable size" of the greatest component of $\Gamma_{n,N}$ figuring in Theorem 7a becomes infinitely large if $c = \frac{1}{2}$.

For the size of the greatest tree in $\Gamma_{n,N}$ with $N \sim \frac{n}{2}$ the following result is valid:

Theorem 7c. *If $N \sim \frac{n}{2}$ and $\Delta_{n,N}$ denotes again the number of points of the greatest tree contained in $\Gamma_{n,N}$, we have for any sequence ω_n tending to $+\infty$ for $n \to +\infty$*

$$\lim_{n \to +\infty} \mathbf{P}(\Delta_{n,N} \geq n^{2/3} \omega_n) = 0 \tag{7.11}$$

and

$$\lim_{n \to +\infty} \mathbf{P}\left(\Delta_{n,N} \geq \frac{n^{2/3}}{\omega_n}\right) = 1. \tag{7.12}$$

Proof of Theorem 7c. We have by some simple computation using (7)

$$\mathsf{M}(\tau_k) = \frac{\binom{n}{k} k^{k-2} \binom{\binom{n-k}{2}}{N-k+1}}{\binom{\binom{n}{2}}{N}} \sim \frac{nk^{k-2}e^{-k}}{k!} e^{-\frac{k^3}{6n^2}}. \tag{7.13}$$

Thus it follows that

$$\mathsf{P}(\Delta_{n,N} \geq n^{2/3}\omega_n) \leq \sum_{k \geq n^{2/3}\omega_n} \mathsf{M}(\tau_k) = O\left(\frac{1}{\sqrt{\omega_n}}\right) \tag{7.14}$$

which proves (7.11).

On the other hand, considering the mean and variance of $\tau^* = \sum_{k \geq \frac{n^{2/3}}{\omega_n}} \tau_k$, it follows that

$$\mathsf{M}(\tau^*) \geq A\,\omega_n^{3/2} \text{ where } A > 0 \text{ and } \mathsf{D}^2(\tau^*) = O(\omega_n^{3/2})$$

and (7.12) follows by using again the inequality of *Chebyshev*. Thus Theorem 7c is proved.

The following theorem can be proved by developing further the above argument and using Lemma 1.

Theorem 7d. *Let $\tau(\mu)$ denote the number of trees of order $\geq \mu n^{2/3}$ contained in $\Gamma_{n,N(n)}$ where $0 < \mu < +\infty$ and $N(n) \sim \frac{n}{2}$. Then we have*

$$\lim_{n \to +\infty} \mathsf{P}_{n,N(n)}(\tau(\mu) = j) = \frac{\lambda^j e^{-\lambda}}{j!} \qquad (j = 0, 1, \ldots) \tag{7.15}$$

where

$$\lambda = \frac{1}{\sqrt{12\pi}} \int_{\frac{1}{6}\mu^3}^{+\infty} \frac{e^{-x}\,dx}{x^{3/2}}. \tag{7.16}$$

§ 8. When is $\Gamma_{n,N}$ a planar graph?

We have seen that the threshold for a subgraph containing k points and $k + d$ edges is $n^{2-\frac{k}{k+d}}$; thus if $N \sim cn$ the probability of the presence of a subgraph having k points and $k + d$ edges in $\Gamma_{n,N}$ tends to 0 for $n \to +\infty$, for each particular pair of numbers $k \geq 4$, $d \geq 1$. This however does not imply that the probability of the presence of a graph of arbitrary order having more edges than vertices in $\Gamma_{n,N}$ with $N \sim nc$ tends also to 0 for $n \to +\infty$. In fact this is not true for $c \geq 1/2$ as is shown by the following

Theorem 8a. *Let $\chi_{n,N}(d)$ denote the number of cycles of $G_{n,N}$ of arbitrary order which are such that exactly d diagonals of the cycle belong also to $\Gamma_{n,N}$. Then if* $N(n) = \frac{n+\lambda\sqrt{n}}{2} + o(\sqrt{n})$ where $-\infty < \lambda < +\infty$, *we have*

$$\lim_{n\to+\infty} \mathbf{P}(\chi_{n,N(n)}(d) = j) = \frac{\varrho^j e^{-\varrho}}{j!} \qquad (j = 0, 1, \ldots) \tag{8.1}$$

where

$$\varrho = \frac{1}{2\cdot 6^d \cdot d!} \int_0^{+\infty} y^{2d-1} e^{\frac{\lambda y}{\sqrt{3}}} \cdot e^{-\frac{y^2}{2}} dy. \tag{8.2}$$

Proof of Theorem 8a. We have clearly as the number of diagonals of a k – gon is equal to $\frac{k(k-3)}{2}$

$$\mathbf{M}(\chi_{n,N}(d)) = \sum_{k=4}^{n} \frac{1}{2}\binom{n}{k}(k-1)!\binom{\frac{k(k-3)}{2}}{d}\frac{\binom{\binom{n}{2}-\binom{k}{2}}{N-k-d}}{\binom{\binom{n}{2}}{N}} \tag{8.3}$$

and thus if $N(n) = \frac{n+\lambda\sqrt{n}}{2} + o(\sqrt{n})$

$$\mathbf{M}(\chi_{n,N(n)}(d)) \sim \frac{1}{2^{d+1}\cdot d!\, n^d} \sum_{k=4}^{n} k^{2d-1}\left(1 + \frac{\lambda}{\sqrt{n}}\right)^k e^{-\frac{3k^2}{2n}}. \tag{8.4}$$

It follows from (8.4) that

$$\lim_{n\to+\infty} \mathbf{M}(\chi_{n,N(n)}(d)) = \frac{1}{2\cdot 6^d\, d!} \int_0^{\infty} y^{2d-1} e^{\frac{\lambda y}{\sqrt{3}} - \frac{y^2}{2}} dy. \tag{8.5}$$

The proof can be finished by the same method as used in proving Theorem 2a.

Remark. Note that Theorem 8a implies that if $N(n) = \frac{n}{2} + \omega_n\sqrt{n}$ with $\omega_n \to +\infty$ then the probability that $\Gamma_{n,N(n)}$ contains cycles with any prescribed number of diagonals tends to 1, while if $N(n) = \frac{n}{2} - \omega_n\sqrt{n}$ the same probability tends to 0. This shows again the fundamental difference in the structure of $\Gamma_{n,N}$ between the cases $N < \frac{n}{2}$ and $N > \frac{n}{2}$. This difference can be expressed also in the form of the following

Theorem 8b. *Let us suppose that* $N(n) \sim nc$. *If* $c < \frac{1}{2}$ *the probability*

that the graph $\Gamma_{n,N(n)}$ is planar is tending to 1 while for $c > \frac{1}{2}$ this probability tends to 0.

Proof of Theorem 8b. As well known trees and connected graphs containing exactly one cycle are planar. Thus the first part of Theorem 8b follows from Theorem 5e. On the other hand if a graph contains a cycle with 3 diagonals such that if these diagonals connect the pairs of points (P_i, P_i') $(i = 1, 2, 3)$ the cyclic order of these points in the cycle is such that each pair (P_i, P_i') dissects the cycle into two paths which both contain two of the other points then the graph is not planar. Now it is easy to see that among the $\binom{\frac{k(k-3)}{2}}{3}$ triples of 3 diameters of a given cycle of order k there are at least $\binom{k}{6}$ triples which have the mentioned property and thus for large values of k approximately one out of 15 choices of the 3 diagonals will have the mentioned property. It follows that if $N(n) = \frac{n}{2} + \omega_n \sqrt{n}$ with $\omega_n \to +\infty$, the probability that $\Gamma_{n,N(n)}$ is not planar tends to 1 for $n \to +\infty$. This proves Theorem 8b. We can show that for $N(n) = \frac{n}{2} + \lambda\sqrt{n}$ with any real λ the probability of $\Gamma_{n,N(n)}$ not being planar has a positive lower limit, but we cannot calculate ts value. It may even be 1, though this seems unlikely.

§ 9. On the growth of the greatest component

We prove in this § (see Theorem 9b) that the size of the greatest component of $\Gamma_{n,N(n)}$ is for $N(n) \sim cn$ with $c > 1/2$ with probability tending to 1 approximately $G(c)n$ where

$$G(c) = 1 - \frac{x(c)}{2c} \tag{9.1}$$

and $x(c)$ is defined by (6.4). (The curve $y = G(c)$ is shown on Fig. 2b).

Thus by Theorem 6 for $N(n) \sim cn$ with $c > 1/2$ almost all points of $\Gamma_{n,N(n)}$ (i. e. all but $o(n)$ points) belong either to some small component which is a tree (of size at most $1/\alpha$ $(\log n - \frac{5}{2}\log\log n) + O(1)$ where $\alpha = 2c - 1 - \log 2c$ by Theorem 7a) or to the single "giant" component of the size $\sim G(c)n$.

Thus the situation can be summarized as follows: the largest component of $\Gamma_{n,N(n)}$ is of order $\log n$ for $\frac{N(n)}{n} \sim c < 1/2$, of order $n^{2/3}$ for $\frac{N(n)}{n} \sim \frac{1}{2}$ and of order n for $\frac{N(n)}{n} \sim c > 1/2$. This double "jump" of the size of the largest component when $\frac{N(n)}{n}$ passes the value $1/2$ is one of the most striking facts concerning random graphs. We prove first the following

Theorem 9a. *Let $\mathscr{H}_{n,N}(A)$ denote the set of those points of $\Gamma_{n,N}$ which belong to components of size $> A$, and let $H_{n,N}(A)$ denote the number of elements of the set $\mathscr{H}_{n,N}(A)$. If $N_1(n) \sim (c - \varepsilon)\, n$ where $\varepsilon > 0$, $c - \varepsilon \geq 1/2$ and $N_2(n) \sim cn$ then with probability tending to 1 for $n \to +\infty$ from the $H_{n,N_1(n)}(A)$ points belonging to $\mathscr{H}_{n,N_1(n)}(A)$ more than $(1 - \delta)\, H_{n,N_1(n)}(A)$ points will be contained in the same component of $\Gamma_{n,N_2(n)}$ for any δ with $0 < \delta < 1$ provided that*

$$A \geq \frac{50}{\varepsilon^2 \delta^2}. \tag{9.2}$$

Proof of Theorem 9a. According to Theorem 2b the number of points belonging to trees of order $\leq A$ is with probability tending to 1 for $n \to +\infty$ equal to

$$n \left(\sum_{k=1}^{A} \frac{k^{k-1}}{k!} [2(c - \varepsilon)]^{k-1} e^{-2(c-\varepsilon)} \right) + o(n) .$$

On the other hand, the number of points of $\Gamma_{n,N_1(n)}$ belonging to components of size $\leq A$ and containing exactly one cycle is according to Theorem 3c $o(n)$ for $c - \varepsilon \geq 1/2$ (with probability tending to 1), while it is easy to see, that the number of points of $\Gamma_{n,N_1(n)}$ belonging to components of size $\leq A$ and containing more than one cycle is also bounded with probability tending to 1.)

Our last statement follows by using the inequality (4.19) from the fact that the average number of components of the mentioned type is, as a simple calculation similar to those carried out in previous §§, shows, of order $O\left(\frac{1}{n}\right)$.

Let $E_n^{(1)}$ denote the event that

$$|H_{n,N_1(n)}(A) - nf(A, c - \varepsilon)| < \tau nf(A, c - \varepsilon) \tag{9.3}$$

where $\tau > 0$ is an arbitrary small positive number which will be chosen later and

$$f(A, c) = 1 - \frac{1}{2c} \sum_{k=1}^{A} \frac{k^{k-1}}{k!} (2\, ce^{-2c}) > 0 \tag{9.4}$$

and let $\overline{E}_n^{(1)}$ denote the contrary event. It follows from what has been said that

$$\lim_{n \to +\infty} \mathbf{P}(\overline{E}_n^{(1)}) = 0 . \tag{9.5}$$

We consider only such $\Gamma_{n,N_1(n)}$ for which (9.3) holds.

Now clearly $\Gamma_{n,N_2(n)}$ is obtained from $\Gamma_{n,N_1(n)}$ by adding $N_2(n) - N_1(n) \sim n\varepsilon$ new edges at random to $\Gamma_{n,N_1(n)}$. The probability that such a new edge should connect two points belonging to $\mathscr{H}_{n,N_1(n)}(A)$, is at least $\dfrac{\binom{H_{n,N_1(n)}(A)}{2} - N_2(n)}{\binom{n}{2}}$,

and thus by (9.3) is not less than $(1 - 2\tau) f^2 (A, c - \varepsilon)$, if n is sufficiently large and τ sufficiently small.

As these edges are chosen independently from each other, it follows by the law of large numbers that denoting by ν_n the number of those of the $N_2(n) - N_1(n)$ new edges which connect two points of $\mathscr{H}_{n,N_1(n)}$ and by $E_n^{(2)}$ the event that

$$\nu_n \geqq \varepsilon(1 - 3\tau) f^2(A, c - \varepsilon)\, n \tag{9.6}$$

and by $\overline{E}_n^{(2)}$ the contrary event, we have

$$\lim_{n \to +\infty} \mathbf{P}(\overline{E}_n^{(2)}) = 0\,. \tag{9.7}$$

We consider now only such $\Gamma_{n,N_2(n)}$ for which $E_n^{(2)}$ takes place. Now let us consider the subgraph $\Gamma^*_{n,N_2(n)}$ of $\Gamma_{n,N_2(n)}$ formed by the points of the set $\mathscr{H}_{n,N_1(n)}(A)$ and only of those edges of $\Gamma_{n,N_2(n)}$ which connect two such points.

We shall need now the following elementary

Lemma 2. *Let $a_1, a_2, \ldots, a_r$ be positive numbers, $\sum_{j=1}^{r} a_j = 1$. If $\max_{1 \leq j \leq r} a_j \leq \alpha$ then there can be found a value $k\,(1 \leq k \leq r - 1)$ such that*

$$\frac{1-\alpha}{2} \leq \sum_{j=1}^{k} a_j \leq \frac{1+\alpha}{2}$$

(9.8) *and*

$$\frac{1-\alpha}{2} \leq \sum_{j=k+1}^{n} a_j \leq \frac{1+\alpha}{2}\,.$$

Proof of Lemma 2. Put $S_j = \sum_{i=1}^{j} a_i \; (j = 1, 2, \ldots, r)$. Let j_0 denote the least integer, for which $S_j > {}^1\!/_2$. In case $S_{j_0} - {}^1\!/_2 > {}^1\!/_2 - S_{j_0-1}$ choose $k = j_0 - 1$, while in case $S_{j_0} - {}^1\!/_2 \leq {}^1\!/_2 - S_{j_0-1}$ choose $k = j_0$. In both cases we have $| S_k - {}^1\!/_2 | \leq \frac{a_{j_0}}{2} \leq \frac{\alpha}{2}$ which proves our Lemma.

Let the sizes of the components of $\Gamma^*_{n,N_2(n)}$ be denoted by $b_1, b_2, \ldots, b_r$. Let $E_n^{(3)}$ denote the event

$$\max b_j > H_{n,N_1(n)}(A)\,(1 - \delta) \tag{9.9}$$

and $\overline{E}_n^{(3)}$ the contrary event. Applying our Lemma with $\alpha = 1 - \delta$ to the numbers $a_j = \frac{b_j}{H_{n,N_1(n)}(A)}$ it follows that if the event $\overline{E}_n^{(3)}$ takes place, the set $\mathscr{H}_{n,N_1(n)}(A)$ can be split in two subsets $\mathscr{H}_n'$ and $\mathscr{H}_n''$ containing H_n' and H_n'' points such that $H_n' + H_n'' = H_{n,N_1(n)}(A)$ and

$$H_{n,N_1(n)}(A)\frac{\delta}{2} \leq \min(H_n', H_n'') \leq \max(H_n', H_n'') \leq H_{n,N_1(n)}(A)\left(1 - \frac{\delta}{2}\right) \tag{9.10}$$

further no point of $\mathscr{H}_n'$ is connected with a point of $\mathscr{H}_n''$ in $\Gamma^*_{n,N_2(n)}$.

It follows that if a point P of the set $\mathscr{H}_{n,N_1(n)}(A)$ belongs to $\mathscr{H}_n'$ (resp. $\mathscr{H}_n''$) then all other points of the component of $\Gamma_{n,N_1(n)}$ to which P belongs are also contained in $\mathscr{H}'_n$ (resp. $\mathscr{H}''_n$). As the number of components of size $> A$ of $\Gamma_{n,N_1(n)}$ is clearly $< \frac{H_{n,N_1(n)}(A)}{A}$ the number of such divisions of the set $\mathscr{H}_{n,N_1(n)}(A)$ does not exceed $2^{\frac{1}{A} H_{n,N_1(n)}(A)}$.

If further $\bar{E}_n^{(3)}$ takes place then every one of the ν_n new edges connecting points of $\mathscr{H}_{n,N_1(n)}(A)$ connects either two points of $\mathscr{H}'_n$ or two points of $\mathscr{H}''_n$. The possible number of such choices of these edges is clearly

$$\binom{\binom{H'_n}{2} + \binom{H''_n}{2}}{\nu_n}.$$

As by (9.10)

$$\frac{\binom{H'_n}{2} + \binom{H''_n}{2}}{\binom{H_n}{2}} \leq \frac{\delta^2}{4} + \left(1 - \frac{\delta}{2}\right)^2 = 1 - \delta + \frac{\delta^2}{2} \leq 1 - \frac{\delta}{2} \tag{9.11}$$

it follows that

$$\mathbf{P}(\bar{E}_n^{(3)}) \leq 2^{\frac{1}{A} H_{n,N_1(n)}(A)} \left(1 - \frac{\delta}{2}\right)^{\varepsilon(1-3\tau) f(A,c-\varepsilon) n} \tag{9.12}$$

and thus by (9.3) and (9.6)

$$\mathbf{P}(\bar{E}_n^{(3)}) \leq \exp\left[n f(A, c - \varepsilon) \left(\frac{(1 + \tau) \log 2}{A} - \frac{\varepsilon(1 - 3\tau) f(A, c - \varepsilon) \delta}{2}\right)\right]. \tag{9.13}$$

Thus if

$$A \varepsilon\delta(1 - 3\tau) f(A, c - \varepsilon) > (1 + \tau) \log 4 \tag{9.14}$$

then

$$\lim_{n \to +\infty} \mathbf{P}(\bar{E}_n^{(3)}) = 0. \tag{9.15}$$

As however in case $c - \varepsilon > 1/2$ we have $f(A, c - \varepsilon) \geq G(c - \varepsilon) > 0$ for any A, while in case $c - \varepsilon = 1/2$

$$f\left(A, \frac{1}{2}\right) = 1 - \sum_{k=1}^{A} \frac{k^{k-1}}{k!\, e^k} = \sum_{k=A+1}^{\infty} \frac{k^{k-1}}{k!\, e^k} \geq \frac{1}{2\sqrt{A}} \quad \text{if } A \geq A_0. \tag{9.15a}$$

the inequality (9.13) will be satisfied provided that $\tau < \frac{1}{10}$ and $A > \frac{50}{\varepsilon^2 \delta^2}$. Thus Theorem 9a is proved.

Clearly the "giant" component of $\Gamma_{n,N_2(n)}$ the existence of which (with probability tending to 1) has been now proved, contains more than

$$(1 - \tau)(1 - \delta)\, n f(A, c - \varepsilon)$$

points. By choosing ε, τ and δ sufficiently small and A sufficiently large, $(1 - \tau)(1 - \delta) f(A, c - \varepsilon)$ can be brought as near to $G(c)$ as we want. Thus we have incidentally proved also the following

Theorem 9b. *Let $\varrho_{n,N}$ denote the size of the greatest component of $\Gamma_{n,N}$. If $N(n) \sim cn$ where $c > 1/2$ we have for any $\eta > 0$*

$$\lim_{n \to +\infty} \mathbf{P}\left(\left|\frac{\varrho_{n,N(n)}}{n} - G(c)\right| < \eta\right) = 1 \tag{9.16}$$

where $G(c) = 1 - \frac{x(c)}{2c}$ and $x(c) = \sum_{k=1}^{\infty} \frac{k^{k-1}}{k!} (2c\, e^{-2c})^k$ *is the solution satisfying* $0 < x(c) < 1$ *of the equation* $x(c)\, e^{-x(c)} = 2ce^{-2c}$.

Remark. As $G(c) \to 1$ for $c \to +\infty$ it follows as a corollary from Theorem 9b that the size of the largest component will exceed $(1 - \alpha)n$ if c is sufficiently large where $\alpha > 0$ is arbitrarily small. This of course could be proved directly. As a matter of fact, if the greatest component of $\Gamma_{n,N(n)}$ with $N(n) \sim nc$ would not exceed $(1 - \alpha)n$ (we denote this event by $B_n(\alpha, c)$) one could by Lemma 2 divide the set V of the n points $P_1, \ldots, P_n$ in two subsets V' resp. V'' consisting of n' resp. n'' points so that no two points belonging to different subsets are connected and

$$\frac{\alpha n}{2} \leq \min(n', n'') \leq \max(n', n'') \leq \left(1 - \frac{\alpha}{2}\right) n. \tag{9.17}$$

But the number of such divisions does not exceed 2^n, and if the n points are divided in this way, the number of ways N edges can be chosen so that only points belonging to the same subset V' resp. V'' are connected, is

$$\binom{\binom{n'}{2} + \binom{n''}{2}}{N}.$$

As $\binom{n'}{2} + \binom{n''}{2} \leq \frac{n^2}{2}\left(1 - \frac{\alpha}{2}\right)$, it follows

$$\mathbf{P}(B_n(\alpha, c)) \leq 2^n \left(1 - \frac{\alpha}{2}\right)^{N(n)} \leq 2^n e^{-\frac{N(n)\alpha}{2}}. \tag{9.18}$$

Thus if $\alpha c > \log 4$, then

$$\lim_{n \to +\infty} \mathbf{P}(B_n(\alpha, c)) = 0 \tag{9.19}$$

which implies that for $c > \frac{\log 4}{\alpha}$ and $N(n) \sim cn$ we have

$$\lim_{n \to +\infty} \mathbf{P}(\varrho_{n,N(n)} \geq (1 - \alpha)\, n) = 1. \tag{9.20}$$

We have seen that for $N(n) \sim cn$ with $c > 1/2$ the random graph $\Gamma_{n,N(n)}$ consists with probability tending to 1, neglecting $o(n)$ points, only of isolated trees (there being approximately $\frac{n}{2c}\frac{k^{k-2}}{k!}(2c\,e^{-2c})^k$ trees of order k) and of a single giant component of size $\sim G(c)n$.

Clearly the isolated trees melt one after another into the giant component, the "danger" of being absorbed by the "giant" being greater for larger components. As shown by Theorem 2c for $N(n) \sim \frac{1}{2k} n \log n$ only isolated trees of order $\leq k$ survive, while for $\frac{N(n) - 1/2\, n \log n}{n} \to +\infty$ the whole graph will with probability tending to 1 be connected.

An interesting question is: what is the "life-time" distribution of an isolated tree of order k which is present for $N(n) \sim cn$? This question is answered by the following

Theorem 9c. *The probability that an isolated tree of order k which is present in $\Gamma_{n,N_1(n)}$ where $N_1(n) \sim cn$ and $c > 1/2$ should still remain an isolated tree in $\Gamma_{n,N_2(n)}$ where $N_2(n) \sim (c + t)\, n$ $(t > 0)$ is approximately e^{-2kt}; thus the „life-time" of a tree of order k has approximately an exponential distribution with mean value $\frac{n}{2k}$ and is independent of the "age" of the tree.*

Proof. The probability that no point of the tree in question will be connected with any other point is

$$\prod_{j=N_1(n)+1}^{N_2(n)} \left(\frac{\binom{n-k}{2} - j + k}{\binom{n}{2} - j} \right) \sim e^{-2kt}.$$

This proves Theorem 9c.

§ 10. Remarks and some unsolved problems

We studied in detail the evolution of $\Gamma_{n,N}$ only till N reaches the order of magnitude $n \log n$. (Only Theorem 1 embraces some problems concerning the range $N(n) \sim n^\alpha$ with $1 < \alpha < 2$.) We want to deal with the structure of $\Gamma_{n,N(n)}$ for $N(n) \sim cn^\alpha$ with $\alpha > 1$ in greater detail in a fortcoming paper; here we make in this direction only a few remarks.

First it is easy to see that $\Gamma_{n,\binom{n}{2} - N(n)}$ is really nothing else, than the complementary graph of $\Gamma_{n,N(n)}$. Thus each of our results can be reformulated to give a result on the probable structure of $\Gamma_{n,N}$ with N being not much less than $\binom{n}{2}$. For instance, the structure of $\Gamma_{n,N}$ will have a second abrupt change when N passes the value $\binom{n}{2} - \frac{n}{2}$; if $N < \binom{n}{2} - cn$ with $c > 1/2$ then the complementary graph of $\Gamma_{n,N}$ will contain a connected graph of order $f(c)n$, while for $c < 1/2$ this (missing) "giant" will disappear.

To show a less obvious example of this principle of getting result for N near to $\binom{n}{2}$, let us consider the maximal number of pairwise independent points in $\Gamma_{n,N}$. (The vertices P and Q of the graph Γ are called *independent* if they are not connected by an edge).

Evidently if a set of k points is independent in $\Gamma_{n,N(n)}$ then the same points form a complete subgraph in the complementary graph $\overline{\Gamma}_{n,N(n)}$. As however $\overline{\Gamma}_{n,N(n)}$ has the same structure as $\Gamma_{n,\binom{n}{2}-N(n)}$ it follows by Theorem 1, that there will be in $\Gamma_{n,N(n)}$ almost surely no k independent points if $\binom{n}{2} - N(n) = o\left(n^{2\left(1-\frac{1}{k-1}\right)}\right)$ i. e. if $N(n) = \binom{n}{2} - o\left(n^{2\left(1-\frac{1}{k-1}\right)}\right)$ but there will be in $\Gamma_{n,N(n)}$ almost surely k independent points if $N(n) = \binom{n}{2} - \omega_n n^{2\left(1-\frac{1}{k-1}\right)}$ where ω_n tends arbitrarily slowly to $+\infty$. An other interesting question is: what can be said about the degrees of the vertices of $\Gamma_{n,N}$. We prove in this direction the following

Theorem 10. *Let $D_{n,N(n)}(P_k)$ denote the degree of the point P_k in $\Gamma_{n,N(n)}$ (i. e. the number of points of $\Gamma_{n,N(n)}$ which are connected with P_k by an edge). Put*

$$\underline{D}_n = \min_{1 \le k \le n} D_{n,N(n)}(P_k) \quad \text{and} \quad \overline{D}_n = \max_{1 \le k \le n} D_{n,N(n)}(P_k) .$$

Suppose that

$$\lim_{n \to +\infty} \frac{N(n)}{n \log n} = +\infty . \tag{10.1}$$

Then we have for any $\varepsilon > o$

$$\lim_{n \to +\infty} \mathbf{P}\left(\left|\frac{\overline{D}_n}{\underline{D}_n} - 1\right| < \varepsilon\right) = 1 . \tag{10.2}$$

We have further for $N(n) \sim cn$ for any k

$$\lim_{n \to +\infty} \mathbf{P}(D_{n,N(n)}(P_k) = j) = \frac{(2c)^j e^{-2c}}{j!} \qquad (j = 0, 1, \ldots). \tag{10.3}$$

Proof. The probability that a given vertex P_k shall be connected by exactly r others in $\Gamma_{n,N}$ is

$$\frac{\binom{n-1}{r}\binom{\binom{n-1}{2}}{N-r}}{\binom{\binom{n}{2}}{N}} \sim \frac{\left(\frac{2N}{n}\right)^r e^{-\frac{2N}{n}}}{r!}$$

thus if $N(n) \sim cn$ the degree of a given point has approximately a Poisson distribution with mean value $2c$. The number of points having the degree r is thus in this case approximately

$$n\frac{(2c)^r e^{-2c}}{r!} \qquad (r = 0,1,\ldots).$$

If $N(n) = (n \log n)\,\omega_n$ with $\omega_n \to +\infty$ then the probability that the degree of a point will be outside the interval $\frac{2N(n)}{n}(1-\varepsilon)$ and $\frac{2N(n)}{n}(1+\varepsilon)$ is approximately

$$\sum_{|k-2\log n\cdot\omega_n|>\varepsilon\cdot 2\log n\cdot\omega_n} \frac{(2\omega_n\cdot\log n)^k e^{-2\omega_n\log n}}{k!} = O\left(\frac{1}{n^{\varepsilon^2\omega_n}}\right)$$

and thus this probability is $o\left(\frac{1}{n}\right)$, for any $\varepsilon > 0$.

Thus the probability that the degrees of not all n points will be between the limit $(1 \pm \varepsilon)\,2\omega_n \log n$ will be tending to 0. Thus the assertion of Theorem 10 follows.

An interesting question is: what will be the chromatic number of $\Gamma_{n,N}$? (The *chromatic number* $Ch(\Gamma)$ of a graph Γ is the least positive integer h such that the vertices of the graph can be coloured by h colours so that no two vertices which are connected by an edge should have the same colour.)

Clearly every tree can be coloured by 2 colours, and thus by Theorem 4a almost surely $Ch(\Gamma_{n,N}) = 2$ if $N = o(n)$. As however the chromatic number of a graph having an equal number of vertices and edges is equal to 2 or 3 according to whether the only cycle contained in such a graph is of even or odd order, it follows from Theorem 5e that almost surely $Ch(\Gamma_{n,N}) \leq 3$ for $N(n) \sim nc$ with $c < 1/2$.

For $N(n) \sim \frac{n}{2}$ we have almost surely $Ch(\Gamma_{n,N(n)}) \geq 3$.

As a matter of fact, in the same way, as we proved Theorem 5b, one can prove that $\Gamma_{n,N(n)}$ contains for $N(n) \sim \frac{n}{2}$ almost surely a cycle of odd order. It is an open problem how large $Ch(\Gamma_{n,N(n)})$ is for $N(n) \sim cn$ with $c > 1/2$!

A further result on the chromatic number can be deduced from our above remark on independent vertices. If a graph Γ has the chromatic number h, then its points can be divided into h classes, so that no two points of the same class are connected by an edge; as the largest class has at least $\frac{n}{h}$ points it follows that if f is the maximal number of independent vertices of Γ we have $f \geq \frac{n}{h}$. Now we have seen that for $N(n) = \binom{n}{2} - o\left(n^{2\left(1-\frac{1}{k}\right)}\right)$ almost surely $f \leq k$; it follows that for $N(n) = \binom{n}{2} - o\left(n^{2\left(1-\frac{1}{k}\right)}\right)$ almost surely $Ch(\Gamma_{n,N(n)}) > \frac{n}{k}$.

Other open problems are the following: for what order of magnitude of $N(n)$ has $\Gamma_{n,N(n)}$ with probability tending to 1 a *Hamilton-line* (i.e. a path which passes through all vertices) resp. in case n is even *a factor of degree* 1 (i.e. a set of disjoint edges which contain all vertices).

An other interesting question is: what is the threshold for the appearance of a "topological complete graph of order k" i.e. of k points such that any two of them can be connected by a path and these paths do not intersect. For $k > 4$ we do not know the solution of this question. For $k = 4$ it follows from Theorem 8a that the threshold is $\frac{n}{2}$. It is interesting to compare this with an (unpublished) result of G. Dirac according to which if $N \geqq 2n - 2$ then $G_{n,N}$ contains certainly a topological complete graph of order 4.

We hope to return to the above mentioned unsolved questions in an other paper.

Remark added on May 16, 1960. It should be mentioned that N. V. Smirnov (see e. g. *Математический Сборник* **6** (1939) p. 6) has proved a lemma which is similar to our Lemma 1.

(Received December 28, 1959.)

REFERENCES

[1] Cayley, A.: *Collected Mathematical Papers.* Cambridge, 1889—1897.

[2] Pólya, G.: "Kombinatorische Anzahlbestimmungen für Gruppen, Graphen und chemische Verbindungen". *Acta Mathematica* **68** (1937) 145—254.

[3] König, D.: *Theorie der endlichen und unendlichen Graphen.* Leipzig, 1936.

[4] Berge, C.: *Théorie des graphes et ses applications.* Paris, Dunod, 1958.

[5] Ford, G. W.—Uhlenbeck, G. E.: "Combinatorial problems in the theory of graphs, I." *Proc. Nat. Acad. Sci.* **42** (1956) USA 122—128.

[6] Gilbert, E. N.: "Enumeration of labelled graphs". *Canadian Journal of Math.* **8** (1957) 405—411.

[7] Erdős, P.—Rényi, A.: "On random graphs, I". *Publicationes Mathematicae (Debrecen)* **6** (1959) 290—297.

[8] Harary, F.: "Unsolved problems in the enumeration of graphs" In this issue, p. 63.

[9] Erdős, P.: „Graph theory and probability." *Canadian Journal of Math.* **11** (1959) 34—38.

[10] Austin, T. L.—Fagen, R. E.—Penney, W. F.—Riordan, J.: "The number of components in random linear graphs". *Annals of Math. Statistics* **30** (1959) 747—754.

[11] Turán P.: "Egy gráfelméleti szélsőértékfeladatról". *Matematikai és Fizikai Lapok* **48** (1941) 436—452.

[12] Turán, P.: "On the theory of graphs", *Colloquium Mathematicum* **3** (1954) 19—30.

[13] Kővári, T.—Sós, V. T.—Turán, P.: "On a problem of K. Zarenkiewicz". *Colloquium Mathematicum* **3** (1954) 50—57.

[14] Rényi, A.: "Some remarks on the theory of trees". *Publications of the Math. Inst. of the Hung. Acad. of Sci.* **4** (1959) 73—85.

[15] Rényi, A.: "On connected graphs, I.". *Publications of the Math. Inst. of the Hung. Acad. of Sci.* **4** (1959) 385—387.

[16] Jordan, Ch.: *Calculus of finite differences.* Budapest, 1939.

[17] Erdős, P.—Stone, A. H.: "On the structure of linear graphs". *Bull. Amer. Math. Soc.* **52** (1946) 1087—1091.

О РАЗВЁРТЫВАНИЕ СЛУЧАЙНЫХ ГРАФОВ

P. ERDŐS и A. RÉNYI

Резюме

Пусть даны n точки $P_1, P_2, \ldots, P_n$, и выбираем случайно друг за другом N из возможных $\binom{n}{2}$ ребер (P_i, P_j) так что после того что выбраны k ребра каждый из других $\binom{n}{2} - k$ ребер имеет одинаковую вероятность быть выбранным как следующии. Работа занимается вероятной структурой так получаемого случайного графа $\Gamma_{n,N}$ при условии, что $N = N(n)$ известная функция от n и n очень большое число. Особенно исследуется изменение этой структуры если N нарастает при данном очень большом n. Случайно развёртывающий граф может быть рассмотрен как упрощенный модель роста реальных сетей (например сетей связы).

Social Networks, 1 (1978/79)5–51 5

Contacts and Influence

Ithiel de Sola Pool
*Massachusetts Institute of Technology**

Manfred Kochen
*University of Michigan***

This essay raises more questions than it answers. In first draft, which we have only moderately revised, it was written about two decades ago and has been circulating in manuscript since then. (References to recent literature have, however, been added.) It was not published previously because we raised so many questions that we did not know how to answer; we hoped to eventually solve the problems and publish. The time has come to cut bait. With the publication of a new journal of human network studies, we offer our initial soundings and unsolved questions to the community of researchers which is now forming in this field. While a great deal of work has been done on some of these questions during the past 20 years, we do not feel that the basic problems have been adequately resolved.

1. Introduction

Let us start with familiar observations: the "small world" phenomenon, and the use of friends in high places to gain favors. It is almost too banal to cite one's favorite unlikely discovery of a shared acquaintance, which usually ends with the exclamation "My, it's a small world!". The senior author's favorite tale happened in a hospital in a small town in Illinois where he heard one patient, a telephone lineman, say to a Chinese patient in the next bed: "You know, I've only known one Chinese before in my life. He was —— from Shanghai." "Why that's my uncle," said his neighbor. The statistical chances of an Illinois lineman knowing a close relative of one of (then) 600 000 000 Chinese are minuscule; yet that sort of event happens.

The patient was, of course, not one out of 600 000 000 random Chinese, but one out of the few hundred thousand wealthy Chinese of Westernized families who lived in the port cities and moved abroad. Add the fact that the Chinese patient was an engineering student, and so his uncle may well have been an engineer too – perhaps a telecommunications engineer. Also there were perhaps some geographic lines of contact which drew the members of one family to a common area for travel and study. Far from surprising, the encounter seems almost natural. The chance meetings that we have are a clue to social structure, and their frequency an index of stratification.

*MIT, Center for International Studies, 30 Wadsworth Street, Cambridge, Mass. 02139, U.S.A.
**Mental Health Research Institute, The University of Michigan, Ann Arbor, Mich. 48104, U.S.A.

Less accidental than such inadvertent meetings are the planned contacts sought with those in high places. To get a job one finds a friend to put in a good word with his friend. To persuade a congressman one seeks a mutual friend to state the case. This influence is peddled for 5%. Cocktail parties and conventions institutionalize the search for contacts. This is indeed the very stuff of politics. Influence is in large part the ability to reach the crucial man through the right channels, and the more channels one has in reserve, the better. Prominent politicians count their acquaintances by the thousands. They run into people they know everywhere they go. The experience of casual contact and the practice of influence are not unrelated. A common theory of human contact nets might help clarify them both.

No such theory exists at present. Sociologists talk of social stratification; political scientists of influence. These quantitative concepts ought to lend themselves to a rigorous metric based upon the elementary social events of man-to-man contact. "Stratification" expresses the probability of two people in the same stratum meeting and the improbability of two people from different strata meeting. Political access may be expressed as the probability that there exists an easy chain of contacts leading to the power holder. Yet such measures of stratification and influence as functions of contacts do not exist.

What is it that we should like to know about human contact nets?

– For any *individual* we should like to know how many other people he knows, *i.e.* his acquaintance volume.

– For a *population* we want to know the distribution of acquaintance volumes, the mean and the range between the extremes.

– We want to know what kinds of people they are who have many contacts and whether those people are also the influentials.

– We want to know how the lines of contact are stratified; what is the structure of the network?

If we know the answers to these questions about individuals and about the whole population, we can pose questions about the implications for *paths* between pairs of individuals.

– How great is the probability that two persons chosen at random from the population will know each other?

– How great is the chance that they will have a friend in common?

– How great is the chance that the shortest chain between them requires two intermediaries; *i.e.*, a friend of a friend?

The mere existence of such a minimum chain does not mean, however, that people will become aware of it. The surprised exclamation "It's a small world" reflects the shock of discovery of a chain that existed all along.[1] So another question is:

---

[1]In the years since this essay was first written, Stanley Milgram and his collaborators (Milgram 1967; Travers and Milgram 1969; Korte and Milgram 1970) have done significant experiments on the difficulty or ease of finding contact chains. It often proves very difficult indeed.

– How far are people aware of the available lines of contact? A friend of a friend is useful only if one is aware of the connection. Also a channel is useful only if one knows how to use it. So the final question is, what sorts of people, and how many, try to exert influence on the persons with whom they are in contact: what sorts of persons and how many are opinion leaders, manipulators, politicists (de Grazia 1952; Boissevain 1974; Erickson and Kringas 1975)?

These questions may be answered at a highly general level for human behavior as a whole, and in more detail for particular societies. At the more general level there are probably some things we can say about acquaintanceship volume based on the nature of the human organism and psyche. The day has 24 hours and memory has its limits. There is a finite number of persons that any one brain can keep straight and with whom any one body can visit. More important, perhaps, there is a very finite number of persons with whom any one psyche can have much cathexis.

There are probably some fundamental psychological facts to be learned about the possible range of identifications and concerns of which a person is capable (Miller 1956).

These psychic and biological limits are broad, however. The distribution of acquaintanceship volumes can be quite variable between societies or social roles. The telephone makes a difference, for example. The contact pattern for an Indian villager *sans* radio, telephone, or road to his village is of a very different order from that of a Rotarian automobile dealer.

There is but little social science literature on the questions that we have just posed.[2] Even on the simplest question of the size of typical acquaintanceship volumes there are few data (Hammer, n.d.; Boissevain 1967). Some are found in anecdotal descriptions of political machines. In the old days there was many a precinct captain who claimed to know personally every inhabitant of his area. While sometimes a boastful exaggeration, there is no doubt that the precinct worker's success derived, among other things, from knowing 300 - 500 inhabitants of his neighborhood by their first names and family connections (Kurtzman 1935). At a more exalted level too, the art of knowing the right people is one of the great secrets of political success; James Farley claimed 10 000 contacts. Yet no past social science study has tested how many persons or what persons any politician knows. The estimates remain guesswork.

There exists a set of studies concerning acquaintanceship volume of delinquent girls in an institutional environment: J. L. Moreno and Helen Jennings asked girls in a reform school (with 467 girls in cottages of 23 or 24 apiece) to enumerate all other girls with whom they were acquainted (Jennings 1937). It was assumed they knew all the girls in their own cottage.

---

[2]In the last few years, however, the literature on human networks has started proliferating. There are articles dealing with information and help-seeking networks in such fields as mental health (Saunders and Reppucci 1977; Horowitz 1977; McKinlay 1973). There is also some anthropological literature on networks in different societies (Nutini and White 1977; Mitchell 1969; Jacobson 1970).

Computed that way, the median number of acquaintances was approximately 65. However, the range was tremendous. One girl apparently knew 175 of her fellow students, while a dozen (presumably with low I.Q.s) could list only four or fewer girls outside of their own cottage.

These figures have little relevance to normal political situations; but the study is valuable since it also tested the hypothesis that the extent of contact is related to influence. The girls were given sociometric tests to measure their influence. In each of two separate samples, a positive correlation (0.4 and 0.3) was found between contact range and influence.

One reason why better statistics do not exist on acquaintanceship volume is that they are hard to collect. People make fantastically poor estimates of the number of their own acquaintances (Killworth and Russell 1976). Before reading further, the reader should try to make an estimate for himself. Define an acquaintance as someone whom you would recognize and could address by name if you met him. Restrict the definition further to require that the acquaintance would also recognize you and know your name. (That excludes entertainment stars, public figures, *etc.*) With this criterion of acquaintance, how many people do you know?

The senior author tried this question on some 30 colleagues, assistants, secretaries and others around his office. The largest answer was 10 000; the smallest was 50. The median answer was 522. What is more, there seemed to be no relationship between the guesses and reality. Older or gregarious persons claimed no higher figures than young or relatively reclusive ones. Most of the answers were much too low. Except for the one guess of 10 000 and two of 2000 each, they were all probably low. We don't know that, of course, but whenever we have tried sitting down with a person and enumerating circles of acquaintances it has not taken long before he has raised his original estimate as more and more circles have come to mind: relatives, old school friends, merchants, job colleagues, colleagues on former jobs, vacation friends, club members, neighbors, *etc.* Most of us grossly underestimate the numer of people we know for they are tucked in the recesses of our minds, ready to be recalled when occasion demands.

Perhaps a notion of the order of magnitude of acquaintanceship volume can be approached by a *gedankenexperiment* with Jennings' data on the reform school. The inmates were young girls who had not seen much of the world; they had but modest I.Q.s and memories; they had come from limited backgrounds; and in the recent past they had been thoroughly closed off from the world. We know that the average one knew 65 inmates. Is it fair to assume that we may add at least 20 teachers, guards, and other staff members known on the average? Somewhere the girls had been in school before their internment. Perhaps each knew 40 students and 10 teachers from there. These girls were all delinquents. They were usually part of a delinquent gang or subculture. Perhaps an average of 30 young people were part of it. They had been arrested, so they knew some people from the world of lawyers, judges, policemen, and social workers. Perhaps there were 20 of them. We have not yet mentioned families and relatives; shall we say another 30? Then there were neighbors in the place they had lived before, perhaps adding up to 35. We have already reached 250 acquaintances which an average girl might have, based solely on the typical life history of an inmate. We have not yet included friends made in club or church, nor merchants, nor accidental contacts. These might add another 50. Nor have we allowed for the girls who had moved around – who had been in more than one school or neighborhood or prison. Perhaps 400 acquaintances is not a bad guess of the average for these highly constricted, relatively inexperienced young girls. Should we not suspect that the average for a mature, white collar worker is at least double that?

Perhaps it is, but of course we don't know. All we have been doing so far is trying to guess orders of magnitude with somewhat more deliberation than was possible by the respondents to whom we popped the question "How many people do you know?". There has been no real research done to test such estimates.

It could be done by a technique analogous to that used for estimating a person's vocabulary. In any given time period during which we observe, a person uses only some of the words he knows and similarly has contact with only some of the people he knows. How can we estimate from this limited sample how many others are known to him? In each case (words and friends) we can do it by keeping track of the proportion of new ones which enter the record in each given time period. Suppose we count 100 running words. These may contain perhaps 60 different words, with some words repeated as many as 6 or 7 times, but most words appearing once. Adding a second 100 running words may add 30 new ones to the vocabulary. A third hundred may add 25 new ones, and so on down. If we extrapolate the curve we reach a point where new words appear only every few thousand running words, and if we extrapolate to infinity we have an estimate of the person's total vocabulary. In the same way, on the first day one may meet 30 people. On the second day one may meet another 30 but perhaps only 15 of them are new, the other 15 being repeaters. On the third day perhaps the non-repeaters may be down to 10, and so on. Again by extrapolating to infinity an estimate of the universe of acquaintances may be made.

Extrapolation to infinity requires strong assumptions about the number of very rarely seen acquaintances. If there are very many who are seen but once in a decade, then a much longer period of observation is required. If the number of people seen once in two decades is not significantly smaller than the number seen in a shorter period, then there are methodological difficulties in estimation.

Two further cautions are necessary. It turns out that the lumpiness in the schedules of our lives makes this technique unusable except over long periods. Perhaps we start on Thursday and go to work. Friday we go to work and see almost the same people. Saturday we go to the beach and have an entirely new set of contacts. Then Monday, perhaps, we are sent on a trip to another office. In short, the curves are highly irregular. Long and patient observation is called for.

Also note that at the end of a lengthy experiment (say after one year), it is necessary to check back over the early lists to determine who are forgotten and no longer acquaintances. Just as new persons enter the acquaintanceship sphere, old ones drop out of it. In one record, for example, a subject recorded 156 contacts in five successive days, with 117 different persons whom he then knew. Two years and ten months later, though still working in the same place, he could no longer recall or recognize 31 of these; *i.e.*, 86 (or 74%) were still acquaintances.

It is important to collect more such empirical information. Section 2 of this paper describes some empirical findings that we have obtained. But before we can decide what to collect we need to think through the logical model of how a human contact net works. We shall do that roughly and non-mathematically in this introduction. Section 3 of the paper deals with it more formally.

One question that quite properly is raised by readers is what do we mean by acquaintanceship, or friendship, or contact. For the mathematical model, the precise definition of "knowing" is quite irrelevant. What the mathematical model gives us is a set of points each of which is connected with some of the other points. As we look away from our model to the world for which it stands, we understand that each point somehow represents a person, and each connection an act of knowing. The model is indifferent to this, however. The points could stand for atoms, or neurons, or telephones, or nations, or corporations. The connections could consist of collisions, or electric charges, or letters written, or hearing about, or acquaintanceship, or friendship, or marriage. To use the model (and satisfy ourselves that it is appropriate) we shall have to pick definitions of person (*i.e.*, point) and knowing (*i.e.*, connectedness) related to the problem at hand. But we start with a model that is quite general. We do indeed impose some constraints on the points and on their connections. These constraints are the substance of our theory about the nature of human contacts.

One simplification we make in our model is to assume that the act of knowing is an all-or-none relationship. That is clearly not true and it is not assumed by Hammer (n.d.), Gurevich (1961) and Schulman (1976). There are in reality degrees of connectedness between persons. There are degrees of awareness which persons have of each other, and there are varied strengths of cathexis. But we cannot yet deal with these degrees. For the moment we want to say of any person, A, that he either does or does not know any given other person, B.

The criterion of human acquaintanceship might be that when A sees B he recognizes him, knows a name by which to address him, and would ordinarily feel it appropriate that he should greet him. That definition excludes, as we have noted, recognition of famous persons, since as strangers we do not feel free to greet them. It excludes also persons whom we see often but whose names we have never learned; *e.g.*, the policeman on the corner. It is, however, a useful operational definition for purposes of contact net studies, because without knowing a name it is hard to keep a record.

Alternatively, the criterion might be a relationship which creates a claim on assistance. In politics, that is often the important kind of knowing. One might well find that a better predictor of who got a job was a man's position in the network of connections defined by obligation than his position in the network of mere acquaintance connections.

For some anthropological studies the connection with which we are concerned might be kinship. As many societies operate, the most important fact in the dealings of two persons is whether they are kin or not. Kinship creates obligations and thus provides a protection of each by the other. Blood kinship is a matter of degree fading off imperceptibly; we are all ultimately related to everyone else. But society defines the limit of who are recognized as kin and who are unrelated. This varies from society to society, and sometimes is hard to establish. In many societies, Brazil and India for example, the first gambit of new acquaintances is to talk about relatives to see if a connection can be established. For such societies kinship is clearly an important criterion of connectedness.

Another criterion of connectedness, of considerable relevance in the United States, is the first-name index. This makes a sharp distinction between levels of knowing, just as does *Sie* and *du* in German or *vous* and *tu* in French.

Whatever definition of knowing we choose to use, our model proceeds by treating connectedness as an all-or-none matter. In short, we are trying to develop not a psychological model of *the* knowing relationship, but a model for treating data about knowing relationship*s* (however defined) which can be applied using whatever knowing relationship happens to be of interest.

The political scientist, using an appropriate definition, may use a contact net model to study influence (Gurevich and Weingrod 1976; n.d.). He asks the number of "connections" of a political kind a person has. The sociologist or anthropologist, using an appropriate definition, may use such a model to study social structure. He asks what kinds of persons are likely to be in contact with each other. The communications researcher may use such a model to study the channels for the flow of messages. Psychologists may use it to examine interrelationships within groups.

So far we have imposed only one restriction on the knowing relationship in our model, namely, that it be all-or-none. There are a few further things we can say about it. When a mathematician describes a relationship he is apt to ask three questions about it: Is it reflexive? Is it symmetric? Is it transitive? The "equals" (=) relationship is reflexive, symmetric, and transitive.

The knowing relationship about which we are talking is clearly not an equality relationship. Anything equals itself; *i.e.,* the equals relation is reflexive. Acquaintanceship is reflexive or not as one chooses to define it. The issue is a trivial one. One could say that by definition everyone knows himself, or one could say that by definition the circle of acquaintances does not include oneself. (We have chosen in our examples below to do that latter and so to define the knowing relation as nonreflexive.)

There is no reason why the knowing relation has to be symmetric. Many more people knew the film star Marilyn Monroe than she knew. If we use the definition of putting a face together with a name then, clearly, persons with good memories know persons with bad memories who do not know them. Similarly, it has been found in some studies that persons are more apt to know the names of persons with higher than lower social status. Thus, privates know each others' names *and* the names of their officers. Officers know each others' names and the names of those they serve, but not necessarily those of privates. Those served may only know servants categorically as, for example, "the tall blond waitress". All in all, to define any knowing relationship as a symmetric one is a great constraint on reality, but it is one which simplifies analysis enormously. It helps so much that for the most part we are going to make that assumption in the discussion below. And, for many purposes, it is largely correct. A kinship relationship is clearly symmetric; if A is a kin to B, B is a kin to A. Also the recognition relationship is mostly symmetric. Most of the time if A can recognize and greet B, B can recognize and greet A. It is generally convenient in our model to define away the minority of cases where this does not hold.

On the other hand, the assumption of transitivity is one that we cannot usefully make. If A knows B, and B knows C, it does not follow that A knows C. If it did follow, then all of society would decompose into a set of one or more cliques completely unconnected with each other. It would mean that everyone you knew would know everyone else you knew, and it follows that you could not know anyone who was outside the clique (*i.e.*, not known to all your friends).[3] Clustering into cliques does occur to some extent and is one of the things we want to study. We want to measure the extent to which these clusters are self-contained, but they are not that by definition.

Thus one useful model of a contact network consists of a set of individuals each of whom has some knowing relationships with others in the set of a kind which we have now defined: all-or-none, irreflexive, symmetric, not necessarily transitive.

We would like to be able to describe such a network as relatively unstructured or as highly structured. Intuitively that is a meaningful distinction, but it covers a considerable variety of strictly defined concepts. Figure 1 describes three hypothetical groups of eight people each, in which each individual has three friends. In the first there are no cliques, in the third there are two completely disjoint cliques, and the second group is intermediate. In the first any two people can be connected by at most one intermediary; in the second some pairs (*e.g.*, A and E) require two intermediaries to be connected; in the third some individuals cannot be connected at all. We are inclined to describe the third group as the most stratified or structured and the first as least so,

[3]Most sociometric literature deals with "liking" rather than "knowing". Preference relationships do tend to be transitive (Hallinan and Felmlee 1975).

and in some senses that is true. But, of course, the first graph is also a rigid structure in the sense that all individuals are alike. In general, however, when we talk of a network as showing more social stratification or clustering in this paper, we mean that it departs further from a random process in which each individual is alike except for the randomness of the variables. The clustering in a society is one of the things which affects who will meet whom and who can reach whom.[4] Any congressman knows more congressmen than average for the general populace; any musician knows more musicians.

Figure 1. *Networks of different structuredness.*

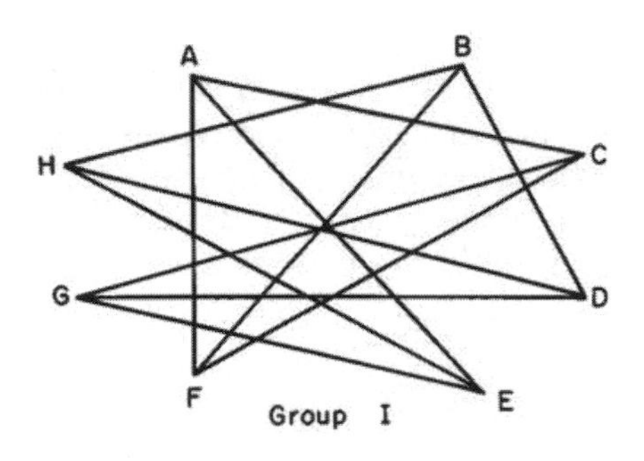

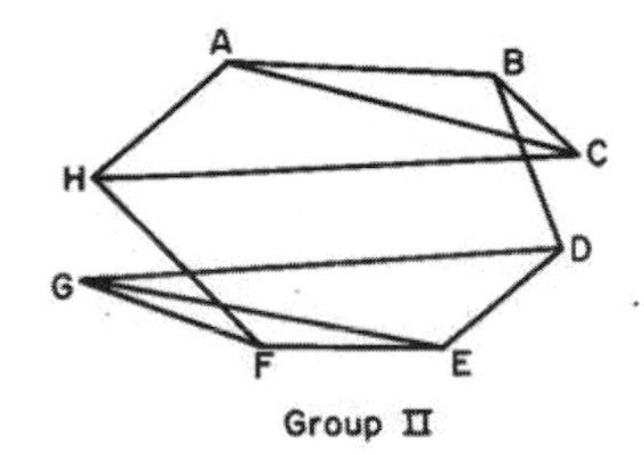

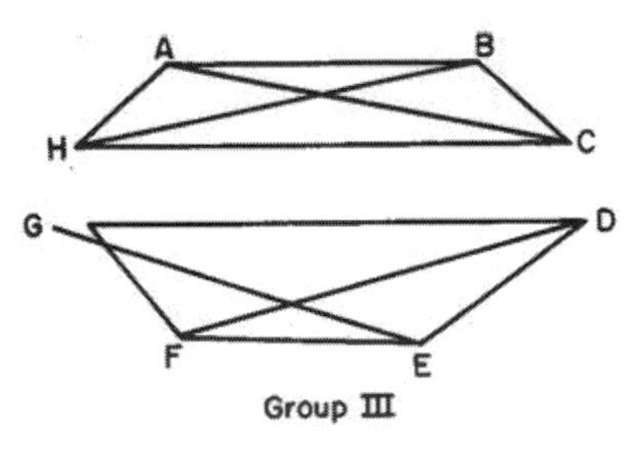

[4]A growing literature exists on structures in large networks (Boorman and White 1976; Lorrain 1976; Lorrain and White 1971; Rapoport and Horvath 1961; Foster *et al.* 1963; Foster and Horvath 1971; Wolfe 1970; McLaughlin 1975; Lundberg 1975; Alba and Kadushin 1976).

The simplest assumption, and one perhaps to start with in modelling a large contact net, is that the number of acquaintances of each person in the population is a constant. We start then with a set of N persons each of whom knows n persons from among the N in the universe; n is the same for all N persons.

If in such a population we pick two persons at random and ask what is the probability that they know each other, the answer can quickly be given from knowing N and n (or, if n is a random variable, the mean n). We know nothing about A and B except that they are persons from a population of size N each of whom on the average knows n other persons in that population. The probability that B is one of the n persons in the circle of acquaintances of A is clearly n/N. If we were talking of a population of 160 000 000 adults and each of them knew, on the average, 800 persons, the chances of two picked at random knowing each other would be one in 200 000.

Suppose we pick A and B who do not know each other, what is the probability of their having an acquaintance in common? The answer to that question, even with random choice of A and B, no longer depends just on n and N. The results now depend also on the characteristic *structure* of interpersonal contacts in the society, as well as on the size of the population and the number of acquaintances each person has. To see the reason why, we turn to an example which we outline diagrammatically in Fig. 2. This Figure represents parts of two networks in which $n = 5$; *i.e.*, each person knows five others in the population. We start with A; he knows B, C, D, E, and F; this is his circle of acquaintances. Next we turn to B; he also knows five people. One of these, by the assumption of symmetry, is A. So, as the acquaintanceship tree fans out, four persons are added at each node.

Figure 2. *Structure in a population.*

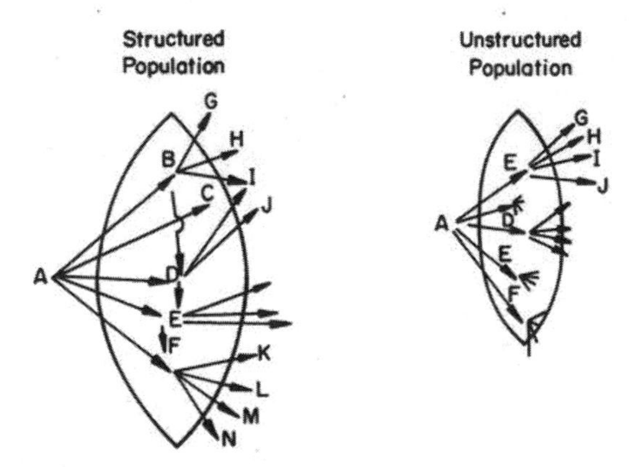

However, here we note a difference between the structured and the unstructured population. In a large population without structure the chance of any of A's acquaintances knowing each other is very small (one in 200 000 for the U.S.A. figures used above). So, for a while at least, if there is no structure the tree fans out adding four entirely new persons at each node: A knows five people; he has 20 friends of friends, and 80 friends of friends of friends, or a total of 125 people reachable with at most two intermediaries. That unstructured situation is, however, quite unrealistic. In reality, people who have a friend in common are likely to know each other (Hammer, n.d.). That is the situation shown in the slightly structured network on the left side of Fig. 2. In that example one of D's acquaintances is B and another is E. The effect of these intersecting acquaintanceships is to reduce the total of different people reached at any given number of steps away from A. In the left-hand network A has five friends, but even with the same n only 11 friends of friends.

So we see, the more cliquishness there is, the more structure there is to the society, the longer (we conjecture) the chains needed on the average to link any pair of persons chosen at random. The less the acquaintanceship structure of a society departs from a purely random process of interactions, in which any two persons have an equal chance of meeting, the shorter will be the average minimum path between pairs of persons.[5] Consider the implications, in a random network, of assuming that n, the mean number of acquaintances of each person, is 1000. Disregarding duplications, one would have 1000 friends, a million (1000^2) friends-of-friends, a billion (1000^3) persons at the end of chains with two intermediaries, and a trillion (1000^4) with three. In such a random network two strangers finding an acquaintance in common (*i.e.*, experiencing the small-world phenomenon) would still be enjoying a relatively rare event; the chance is one million out of 100 or 200 million. But two intermediaries would be all it would normally take to link two people; only a small minority of pairs would not be linked by one of those billion chains.

Thus, in a country the size of the United States, if acquaintanceship were random and the mean acquaintance volume were 1000, the mean length of minimum chain between pairs of persons would be well under two intermediaries. How much longer it is in reality because of the presence of considerable social structure in the society we do not know (nor is it necessarily longer for all social structures). Those are among the critical problems that remain unresolved.

Indeed, if we knew how to answer such questions we would have a good quantitative measure of social structure. Such an index would operationalize the common sociological statement that one society is more structured than

---

[5] Let us state this more carefully for a network of n nodes and m links, in which $n! \gg m$, but all nodes are reachable from all nodes. In that case, m pairs know each other. The question is what structure will minimize the average number of steps between the $n! - m$ remaining pairs. Whenever the m pairs who know each other are also linked at two steps, then the two-step connection is wasted. The same is true for pairs linked by more than one two-step route. Such wastage occurs often when there are dense clusters of closely related nodes in a highly structured network. It happens rarely (because $n! \gg m$) in a random network structure – but it does happen. The minimum average chain would occur not in a random structure, but in one designed to minimize wasted links. However, when $n! \gg m$, the random structure will depart from that situation only to a small extent.

another. The extent to which the mean minimum chain of contacts departs from that which would be found in a random network could be a convenient index of structuredness.

There are all sorts of rules for the topology of a network that can make its graph depart from random linkages. Perhaps the simplest and most important structure is that of triangular links among a given person's friends. If two persons both know person A, the odds are much better than otherwise that they will know each other; if they do know each other the acquaintanceship links form a triangle. For an example see Fig. 3. Disregarding the symmetric

Figure 3. *Effect of structure.*

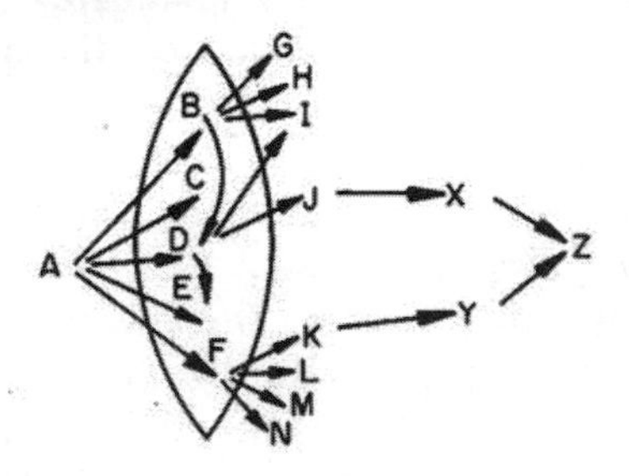

path (*i.e.*, A knows B so B knows A), let us ask ourselves how many links it takes to go from A out to each of his acquaintances and back to A *via* the shortest path. If we start out on the path from A to B, we can clearly return to A *via* a triangle, A,B,D,A. We can also return by a triangle if we go from A to D or A to E. On the other hand, there is no triangle which will take one back if one starts on the path from A to F. Sooner or later there will be a path back, in this instance a path of eight links. (The only instance in which there would be no path back would be if the society were broken into two cliques linked at no point (see Fig. 1), or at only one point.) Clearly, the number of triangles among all the minimum circular chains is a good index of the tightness of the structure, and one that is empirically usable. It is perfectly possible to sample and poll the acquaintances of A to estimate how many of them know each other. That figure (which measures the number of triangles) then provides a parameter of the kind for which we are looking (Hammer, n.d.; Wasserman 1977).

The fact that two persons have an acquaintance in common means that to some extent they probably move in the same circles. They may live in the same part of the country, work in the same company or profession, go to the same church or school, or be related. These institutions provide a nucleus of contacts so that one acquaintance in common is likely to lead to more. One way to describe that situation can be explained if we turn back to Fig. 3. Suppose we inquire of a person whether he knows A. If the answer is yes, then the chances of his knowing B are better than they would otherwise have been. Conversely if the answer is no, that reduces the chances of his knowing B. If he has told us that he does not know either A or B the chances of his knowing C are still further reduced. And so on down the list. This fact suggests that a second measure of structuredness would be the degree to which the chance of knowing a subsequent person on the list of acquaintances of A is reduced by the information that a person does not know the previous person on the list. In a society that is highly segmented, if two persons have any acquaintances in common they will have many, and so each report of nonacquaintanceship reduces more markedly than it otherwise would the chances of finding one common acquaintance on the list.

We require a measure, such as one of those two we have just been discussing, of the degree of clusteredness in a society, to deal with the question with which we started a few pages back, namely, the distribution of length of minimum contact chains: how many pairs of persons in the population can be joined by a single common acquaintance, how many by a chain of two persons, how many by a chain of three, *etc.*?

The answer depends on three values: N, n, and a parameter measuring structuredness. Increased social stratification reduces the length of chains between persons in the same stratum and at the same time lengthens the chains across strata lines. Thus, for example, two physicians or two persons from the same town are more likely to have an acquaintance in common than persons who do not share such a common characteristic. While some chains are thus shortened and others are lengthened by the existence of clusters within a society, it seems plausible to conjecture that the mean chain averaged over pairs of persons in the population as a whole is lengthened. Two persons chosen at random would find each other more quickly in an unstructured society than in a structured one, for most of the time (given realistic values of N, n, and clustering) persons chosen at random will not turn out to be in the same strata.

We might conjecture, for example, that if we had time series data of this kind running over the past couple of decades, we would find a decline in structuredness based on geography. The increased use of the long-distance telephone (and in the future of computer networks), and also of travel, probably has made acquaintanceship less dependent on geographic location than ever in the past.

In the final section of this paper we turn to an exploration of some of the alternative ways of modelling a network of the kind just described. The central problem that prevents an entirely satisfactory model is that we do not know how to deal with the structuredness of the population. Because of its lovely mathematical simplicity, there is an almost irresistable tendency to want to assume that whenever we do not know how the probability of acquaintanceship within two pairs of persons differs, we should treat it as equal; but it is almost never equal (Hunter and Shotland 1974; White 1970a). The real-world population lives in an n-dimensional space distributed at varying social distances from each other. But it is not a Euclidean space. Person A may be very close to both B and C and therefore very likely to know them both, but B and C may be very far from each other.

In the hope of getting some clues as to the shape of the distribution of closeness among pairs in real-world populations, we undertook some research on the actual contact networks of some 27 individuals. These data we shall describe in Part 2 of this paper. While we learned a lot from that exercise, it failed to answer the most crucial questions because the most important links in establishing the connectedness of a graph may often be not the densely travelled ones in the immediate environment from which the path starts, but sparse ones off in the distance. How to go between two points on opposite sides of a river may depend far more critically on where the bridge is than on the roads near one's origin or destination. The point will become clear as we examine the data.

2. Empirical estimates of acquaintanceship parameters

One is awed by the way in which a network multiplies as links are added. Even making all allowances for social structure, it seems probable that those whose personal acquaintances range around 1000, or only about 1/100 000 of the U.S. adult population, can presumably be linked to another person chosen at random by two or three intermediaries on the average, and almost with certainty by four.

We have tried various approaches to estimating such data. We start with *gedankenexperiments,* but also have developed a couple of techniques for measuring acquaintance volume and network structure.

Consider first a rather fanciful extreme case. Let us suppose that we had located those two individuals in the U.S. between whom the minimum chain of contacts was the longest one for any pair of persons in the country. Let us suppose that one of these turned out to be a hermit in the Okefenokee Swamps, and the other a hermit in the Northwest woods. How many intermediaries do we need to link these two?

Each hermit certainly knows a merchant. Even a hermit needs to buy coffee, bread, and salt. Deep in the backwood, the storekeeper might never have met his congressman, but among the many wholesalers, lawyers, inspectors, and customers with whom he must deal, there will be at least one who is acquainted with his representative. Thus each of the hermits, with two intermediaries reaches his congressman. These may not know each other, though more likely they do, but in any case they know a congressman in common. Thus the maximum plausible minimum chain between any two persons in the United States requires no more than seven intermediaries.

This amusing example is not without significance. Viewed this way, we see Congress in a novel but important aspect, that of a communication node. The Congress is usually viewed as a policy choosing, decision-making instrument, which selects among pre-existing public opinions which are somehow already diffused across the country. Its more important function, however, is that of a forum to which private messages come from all corners, and within which a public opinion is created in this process of confrontation of attitudes and information. Congress is the place which is quickly reached by messages conveying the feelings and moods of citizens in all walks of life. These feelings themselves are not yet public opinion for they are not crystallized into policy stands; they are millions of detailed worries concerning jobs, family, education, old age, *etc.* It is in the Congress that these messages are quickly heard and are revised and packaged into slogans, bills, and other policy formulations. It is these expressions of otherwise inchoate impulses that are reported in the press, and which become the issues of public opinion. Thus the really important function of the Congress, distinguishing it from an executive branch policy making body, is as a national communication center where public reactions are transformed into public opinion. Its size and geographically representative character puts it normally at two easily found links from everyone in the country. Its members, meeting with each other, formulate policies which express the impulses reaching them from outside. Through this communication node men from as far apart as the Okefenokee Swamps and the north woods can be put in touch with the common threads of each other's feelings expressed in a plank of policy. A body of 500 can help to weld a body of 100 000 000 adults into a nation.

While thinking about such matters has its value, it is no substitute for trying to collect hard data.

Empirical collection of contact data is possible but not easy:

First of all, people are not willing to reveal some or all of their contacts.

Second, it is hard to keep track of such massive and sequential data.

Third, because contacts run in clusters and are not statistically independent events, the statistical treatment of contact data is apt to be hard.

Reticence is probably the least serious of the difficulties. It is certainly no more of a problem for studies of contacts than for Kinsey-type research or for research on incomes or voting behavior, all of which have been successfully conducted, though with inevitable margins of error. As in these other areas of research, skill in framing questions, patience, proper safeguards of confidence, and other similar requirements will determine success, but there is nothing new or different about the difficulties in this field. Reticence is less of an obstacle to obtaining valid information about contacts than are the tricks played by our minds upon attempts at recall.

Indeed it is usually quite impossible for persons to answer questions accurately about their contacts. We noted above the bewilderment which respondents felt when asked how many people they knew, and how most gave fantastic underestimates. Over one day, or even a few hours, recall of contacts is bad. Given more than a very few contacts, people find it hard to recall whom they have seen or conversed with recently. They remember the lengthy or emotionally significant contacts, but not the others. The person who has been to the doctor will recall the doctor, but may neglect to mention the receptionist. The person who has been to lunch with friends may forget about contact with the waiter. In general, contacts which are recalled are demonstrably a highly selected group.

Most importantly, they are selected for prestige. A number of studies have revealed a systematic suppression of reports of contacts down the social hierarchy in favor of contacts up it (Warner 1963; Festinger *et al.* 1950; Katz and Lazarsfeld 1955). If one throws together a group of high status and low status persons and later asks each for the names of the persons in the group to whom he talked, the bias in the outcome is predictably upward. Unaided recall is not an adequate instrument for collecting contact data except where the problem requires recording only of emotionally meaningful contacts. If we wish to record those, and only those, we can use the fact of recall as our operational test of meaningfulness. Otherwise, however, we need to supplement unaided recall.

Some records of contacts exist already and need only be systematically noted. Noninterview sources of contact information include appointment books, committee memberships, and telephone switchboard data. The presidential appointment book is a fascinating subject for study.

Telephone switchboard data could be systematically studied by automatic counting devices without raising any issues of confidence. The techniques are already available and are analogous to those used for making load estimates. They could have great social science value too. A study, for example, of the ecology of long-distance telephone contacts over the face of the country would tell us a great deal about regionalism and national unity. A similar study of the origin and destination of calls by exchange could tell us a great deal about neighborhoods, suburbanism, and urbanism in a metropolitan region. This would be particularly interesting if business and residential phones could be segregated. The pattern of interpersonal contact could be studied by counting calls originating on any sample of telephones. (What proportion of all calls from any one phone are to the most frequently called other phone? What proportion to the 10 most frequently called others?) How many different numbers are called in a month or a year? Would the results on such matters differ for upper and lower income homes, urban and rural, *etc.*?

In similar ways mail flows can tell us a good deal (Deutsch 1956, 1966). The post office data are generally inadequate, even for international flows, and even more for domestic flows. Yet sample counts of geographic origins and destinations are sometimes made, and their potential use is clear.

Not all the information we want exists in available records. For some purposes interviews are needed for collection of data. Various devices suggest themselves for getting at the range of a person's contacts. One such device is to use the telephone book as an *aide-memoire*. We take a very large book, say the Chicago or Manhattan book. We open it to a page selected by a table of random numbers. We then ask our respondent to go through the names on that page to see if they know anyone with a name that appears there or a name that would appear there if it happened to be in that book. Repeat the operation for a sample of pages. One can either require the subject to think of all the persons he knows with such names, which is both tedious and, therefore, unreliable, or assume that the probability of a second, third, or fourth known person appearing on a single page is independent of the previous appearance of a known name on the page. Since that is a poor assumption we are in a dilemma. Depending on the national origins of our respondent, he is apt to know more persons of certain names; he may know more Ryans, or Cohens, or Swansons according to what he is. Nationality is a distorting factor in the book, too. The Chicago phone book will contain a disproportionate number of Polish names, the Manhattan phone book a disproportionate number of Jewish ones. Also if the subject knows a family well he will know several relatives of the same name. In short, neither the tedious method of trying to make him list all known persons of the name, nor the technique in which one simply counts the proportion of pages on which no known name occurs (and uses that for p, $1 - p = q$, and then expands the binomial), gives a very satisfactory result. Yet with all those qualifications, this technique of checking memory against the phone book gives us a better estimate of approximate numbers of acquaintances than we now have.

One of the authors tried this technique on himself using a sample of 30 pages of the Chicago phone book and 30 pages of the Manhattan phone book. The Chicago phone book brought back names of acquaintances on 60% of the pages, yielding an estimate that he knows 3100 persons. The Manhattan phone book, with 70% of the pages having familiar names, yielded an estimate of 4250 acquaintances. The considerations raised above suggested that the estimate from the Manhattan phone book should be higher, for the author is Jewish and grew up in Manhattan. Still the discrepancy in estimates is large. It perhaps brings us closer to a proper order of magnitude, but this technique is still far from a solution to our problem.

To meet some of these problems we developed a somewhat better method which involves keeping a personal log of all contacts of any sort for a number of sample days. Each day the subject keeps a list (on a pad he carries with him) of all persons whom he meets and knows. The successive lists increasingly repeat names which have already appeared. By projecting the curve one hopes to be able to make estimates of the total size of the acquaintanceship volume, and from the lists of names to learn something of the character of the acquaintances.

The rules of inclusion and exclusion were as follows:

(1) A person was not listed unless he was already known to the subject. That is to say, the first time he was introduced he was not listed; if he was met again on a later day in the 100 he was. The rationale for this is that we meet many people whom we fail to learn to recognize and know.

(2) Knowing was defined as facial recognition and knowing the person's name – any useful name, even a nickname. The latter requirement was convenient since it is hard to list on a written record persons for whom we have no name.

(3) Persons were only listed on a given day if when the subject saw them he addressed them, if only for a greeting. This eliminated persons seen at a distance, and persons who the subject recognized but did not feel closely enough related to, to feel it proper to address.

Table 1. *100-day contacts of respondents*

Sex	Job	Age	*(a)* No. of different persons seen in 100 days	*(b)* No. of contact events	Ratio *b/a*
Blue collar					
M	Porter	50 - 60	83	2946	35.5
M	Factory labor	40 - 50	96	2369	24.7
M	Dept. store receiving	20 - 30	137	1689	12.3
M	Factory labor	60 - 70	376	7645	20.3
M	Foreman	30 - 40	510	6371	12.5
F	Factory labor and unemployed	30 - 40	146	1222	8.4
White collar					
F	Technician	30 - 40	276	2207	8.0
F	Secretary	40 - 50	318	1963	6.2
M	Buyer	20 - 30	390	2756	7.1
M	Buyer	20 - 30	474	4090	8.6
M	Sales	30 - 40	505	3098	6.1
F	Secretary	50 - 60	596	5705	9.5
Professional					
M	Factory engineer	30 - 40	235	3142	13.5
F	T.V.	40 - 50	533	1681	3.2
M	Adult educator	30 - 40	541	2282	4.2
M	Professor	40 - 50	570	2175	3.8
M	Professor	40 - 50	685	2142	3.1
M	Lawyer-politician	30 - 40	1043	3159	3.0
M	Student	20 - 30	338	1471	4.4
M	Photographer	30 - 40	523	1967	4.8
M	President*	50 - 60	1404**	4340**	3.1**
Housewives					
F	–	30 - 40	72	377	5.2
F	–	20 - 30	255	1111	4.4
F	–	20 - 30	280	1135	4.0
F	–	30 - 40	363	1593	4.4
F	–	30 - 40	309	1034	3.3
F	–	50 - 60	361	1032	2.9
Adolescent					
M	Student	10 - 20	464	4416	9.5

*Data estimated from Hyde Park records.
**Record for 85 days.

(4) Telephone contacts were included. So were letters written but not letters received. The rationale for the latter is that receiving a letter and replying to it is a single two-way communication such as occurs simultaneously in a face-to-face contact. To avoid double counting, we counted a reply as only half the act. Of course, we counted only letters written to people already known by the above criterion.

(5) A person was only listed once on a given day no matter how often he was seen. This eliminated, for example, the problem of how many times to count one's secretary as she walked in and out of the office.

The task of recording these contacts is not an easy one. It soon becomes a tedious bore. Without either strong motivation or constant checking it is easy to become forgetful and sloppy. But it is far from impossible; properly controlled and motivated subjects will do it.

The data on 27 persons were collected mostly by Dr. Michael Gurevich (1961) as part of a Ph.D. dissertation which explored, along with the acquaintanceship information itself, its relation to a number of dependent variables. As Table 1 shows us, the respondents, though not a sample of any defined universe, covered a range of types including blue collar, white collar, professional, and housewives.

Among the most important figures in the Table are those found in the right-hand column. It is the ratio between the number of different persons met and the number of meetings. It is what psychologists call the type–token ratio. It is socially very indicative, and is distinctive for different classes of persons.

Blue collar workers and housewives had the smallest number of different contacts over the 100 days. They both lived in a restricted social universe. But in the total number of interpersonal interactions the blue collar workers and housewives differed enormously. Many of the blue collar workers worked in large groups. Their round of life was very repetitive; they saw the same people day in and day out, but at work they saw many of them. Housewives, on the other hand, not only saw few different people, but they saw few people in the course of a day; they had small type–token ratios. They lived in isolation.

In total gregariousness (*i.e.*, number of contact events) there was not much difference among the three working groups. Blue collar workers, white collar workers, and professionals all fell within the same range, and if there is a real difference in the means, our small samples do not justify any conclusions about that. But in the pattern of activity there was a great difference. While blue collar workers were trapped in the round of a highly repetitive life, professionals at the other extreme were constantly seeing new people. They tended to see an average acquaintance only three or four times in the hundred days. One result of this was that the professionals were the persons whose contacts broke out of the confines of social class to some extent. They, like the others (see Table 2) tended to mix to a degree with people like themselves but, to a slightly greater degree than the other classes, they had a chance to meet people in other strata of society.

The tendency of society to cluster itself as like seeks like can also be seen in Tables on contacts by age, sex, and religion (see Tables 3, 4 and 5). These data reflect a society that is very structured indeed. How can we use the data to estimate the acquaintanceship volume of the different respondents? We found that over 100 days the number of different persons they saw ranged between 72 for one housewife and 1043 for one lawyer-politician. Franklin

Table 2. *Number of acquaintances by occupation*

Acquaintances' occupation	Subject's occupation				
	Blue collar (%)	Housewife (%)	White collar (%)	Professional (%)	Entire group (%)
Professional	11	24	20	45	24
Managerial	9	7	19	14	14
Clerical	13	7	13	7	11
Sales worker	5	6	19	4	11
Craftsman, foreman	15	5	6	5	7
Operative	25	1	3	5	8
Service worker	9	2	2	1	3
Laborer	4	1	1	–	1
Housewife	4	35	10	12	13
Student	2	3	1	5	3
Farmer	–	–	–	–	–
Dont' know	4	10	8	3	6
	100*	100*	100*	100*	100*

*Figures may not add up to 100% because of rounding.

Table 3. *Subject's age compared with his acquaintance's age*

Acquaintance's age	Subject's age			
	20 - 30 (%)	31 - 40 (%)	41 - 50 (%)	Over 50 (%)
Under 20	7	2	2	1
20 - 30	[21]	19	11	15
31 - 40	30	[39]	33	20
41 - 50	21	22	[27]	32
Over 50	21	19	27	[33]
	100*	100*	100*	100*

*Figures may not add up to 100% because of rounding.

Roosevelt's presidential appointment book, analyzed by Howard Rosenthal (1960), showed 1404 different persons seeing him. But that leaves us with the question as to what portion of the total acquaintance volume of each of these persons was exhausted.

One of the purposes of the data collection was to enable us to make an estimate of acquaintance volume in a way that has already been described above. With each successive day one would expect fewer people to be added, giving an ogive of persons met to date such as that in Fig. 4. In principle

Table 4. *Sex of subject and sex of acquaintance*

Subject	Acquaintances		
	Male (%)	Female (%)	Total (%)
Blue collar			
Male	83	17	100
White collar			
Male	65	35	100
Female	53	47	100
Professional			
Male	71	29	100
Housewife			
Female	45	55	100

Table 5. *Religion of subject and religion of acquaintance*

Subject's religion	Acquaintance's religion				
	Protestant (%)	Catholic (%)	Christian (didn't know denomination) (%)	Jewish (%)	Religion known (%)
Protestant	46	25	25	4	100*
Catholic	15	57	23	5	100*
Jewish	9	16	27	47	100*

*Figures may not add up because of rounding and omission of other religions.

Figure 4. *Acquaintanceship ogives.*

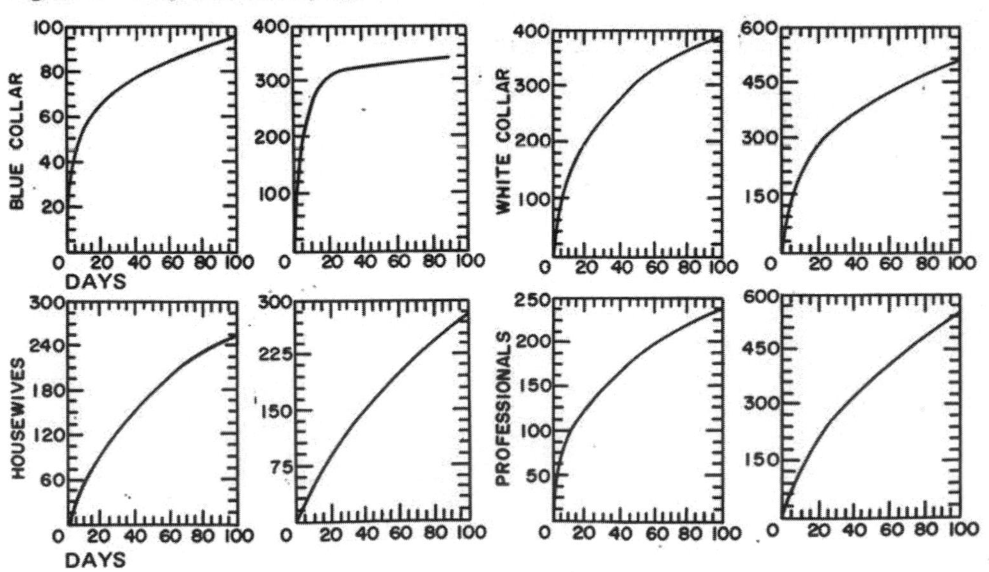

one might hope to extrapolate that curve to a point beyond which net additions would be trivial.

Fitting the 100-day curve for each subject to the equation (acquaintanceship volume) = At^x gave acquaintanceship volumes over 20 years ranging from 122 individuals for a blue collar porter in his fifties to 22 500 persons for Franklin Roosevelt.

However, that estimation procedure does not work with any degree of precision. The explanation is that the estimate of the asymptote is sensitive

Table 6. *Frequency distribution of contacts with acquaintances*

Frequency of contact over 100 days	Blue collar group				
	Case A (%)	Case B (%)	Case C (%)	Case D (%)	Case E (%)
1	4.8	23.9	29.0	9.3	23.5
2	2.4	11.4	11.6	5.0	10.7
3	–	4.1	6.5	3.9	8.4
4	–	4.1	4.3	3.4	4.7
5	1.2	3.1	3.6	3.4	4.9
6 - 10*	2.4	0.4	1.7	3.4	2.2
11 - 20*	0.8	0.5	1.2	2.1	1.3
21 - 30*	1.0	0.6	1.0	1.3	1.0
31 - 40*	1.8	0.6	0.6	0.9	0.7
41 - 50*	1.7	0.3	0.5	0.5	0.4
51 - 60*	1.7	1.4	0.1	0.4	0.2
61 - 70*	0.6	1.1	–	0.7	0.1
71 - 80*	0.1	0.1	0.07	–	0.02
81 - 90*	–	–	–	–	–
91 - 100*	0.2	0.2	0.07	0.05	0.02
	100%	100%	100%	100%	100%

Frequency of contact over 100 days	White collar group					
	Case G (%)	Case H (%)	Case I (%)	Case J (%)	Case K (%)	Case L (%)
1	43.4	44.3	27.2	30.8	47.7	37.7
2	11.5	16.9	20.0	12.4	13.1	12.9
3	7.9	7.5	10.7	9.0	6.5	7.5
4	4.3	3.7	6.1	6.9	7.1	4.5
5	3.2	3.4	6.1	4.0	3.2	3.0
6 - 10*	1.9	1.8	2.3	2.8	1.9	2.3
11 - 20*	0.7	0.8	0.7	1.1	0.6	0.9
21 - 30*	0.4	0.3	0.4	0.4	0.2	0.3
31 - 40*	0.3	–	0.2	0.2	0.2	0,3
41 - 50*	0.5	0.09	0.1	0.2	0.1	0.3
51 - 60*	0.1	0.1	0.2	0.2	0.1	0.4
61 - 70*	–	0.2	–	0.1	0.06	0.1
71 - 80*	–	–	–	–	–	–
81 - 90*	–	–	–	–	–	–
90 - 100*	0.04	0.03	0.03	0.02	0.02	0.02
	100%	100%	100%	100%	100%	100%

(continued on facing page)

Table 6. *(continued)*

Frequency of contact over 100 days	Professionals				Housewives		
	Case M (%)	Case O (%)	Case P (%)	Case Q (%)	Case V (%)	Case W (%)	Case X (%)
1	39.5	53.0	43.3	49.6	56.0	54.6	47.9
2	7.7	12.3	17.5	18.5	18.8	18.9	16.5
3	4.3	7.5	12.2	10.9	7.8	7.8	8.8
4	3.9	4.2	5.9	4.7	1.5	3.2	6.8
5	3.0	3.6	5.2	3.8	3.9	2.5	4.4
6 - 10*	1.2	2.3	1.8	1.3	1.1	1.3	1.6
11 - 20*	1.6	0.4	0.5	0.3	0.3	0.4	0.3
21 - 30*	0.4	0.09	0.07	0.09	0.04	0.04	0.2
31 - 40*	0.4	0.07	0.02	0.06	0.08	0.04	0.03
41 - 50*	0.3	0.05	0.05	0.01	0.04	–	0.1
51 - 60*	0.7	0.07	0.02	0.01	0.08	0.1	–
61 - 70*	0.1	–	–	–	–	–	–
71 - 80*	–	–	–	–	0.04	0.07	–
81 - 90*	–	–	–	–	–	–	0.03
91 - 100*	0.1	0.02	0.02	0.01	0.08	0.04	0.03
	100%	100%	100%	100%	100%	100%	100%

*The percentages in each entry are average percentages for a single day, not for the 5- or 10-day period.

to the tail of the distribution (Granovetter 1976). Such a large proportion of the respondent's acquaintances are seen only once or twice in 100 days that any estimate which we make from such data is very crude. Table 6 shows the figures. Except for blue collar workers, half or more of the acquaintances were seen only once or twice in the period.

One may think that the way around this problem would be to rely more heavily on the shape of the curve in its more rugged region where contact events are more frequent. The problem with that is that the nature of the contacts in the two parts of the curve are really quite dissimilar. To explain that perhaps we should look more closely at a single case; we shall use that of one of the author's own contact lists.

In 100 days he had contact with 685 persons he knew. On any one day the number of contacts ranged from a low of two other persons to a high of 89, the latter in the Christmas season. The mean number of acquaintances with whom he dealt on a day was 22.5. The median number was 19. There were several discreet typical patterns of days, resulting in a multimodal distribution. There was one type of day, including most weekend days, when he would typically meet 7 - 9 people, another type of day with typically around 17 contacts, and a third type of day of highly gregarious activity which involved dealing with about 30 people.

Only about half of the 685 persons were seen more than once in the 100 days. The mean frequency was 3.1 times per person. The distribution, however, is highly skewed (Table 7).

Table 7. *Contact frequency distribution for one person*

Number of days on which contact was had during the 100 days	Number of persons with that frequency of contact	Days	Persons	Days	Persons
1	335	11	4	24	1
2	125	12	4	26	2
3	74	13	1	30	1
4	32	14	2	33	2
5	26	15	4	34	1
6	12	16	2	36	1
7	16	18	1	45	1
8	5	19	1	51	1
9	8	20	4	92	1
10	4	23	2		

These figures, however, are somewhat misleading. It seems that we are actually dealing with two distributions: one which includes those persons living in the author's home and working in his office whom he saw during his regular daily routine, and the other including all his other acquaintances in the seeing of whom all kinds of chance factors operatored. All individuals seen 19 or more times are in the former group; so are all but two individuals seen 13 or more times. Removing 51 such family members and co-workers gives us the data that are really relevant to estimating the large universe of occasional contacts, but in that sample more than half the persons listed were seen only once and 91% five times or less. No easily interpretable distribution (such as Poisson which would imply that there is no structure among these contacts) fits that distribution, and with such small frequencies the shape of the distribution is unstable between respondents. It is possible that the projection of the 100-day data for this author to a year's time could come out at anywhere between 1100 and 1700 persons contacted. That is not a very satisfactory estimate, but it is far better than the estimates we had before.

This estimate is way below our telephone book estimates, which it will be recalled ranged from 3100 to 4250 acquaintances. The discrepancy is more revealing than disturbing. It suggests some hypotheses about the structure of the universe of acquaintances. It suggests that there is a pool of persons with whom one is currently in potential contact, and a larger pool in one's memory, which for the senior author is about 2 - 3 times as large. The active pool consists of acquaintances living in the areas which one frequents, working at the activity related to one's occupation, belonging to the groups to which one belongs. Random factors determine in part which persons out of this pool one happens to meet, or even meet several times during any set period. But in one's memory there are in addition a considerable number of other persons whose names and faces are still effectively stored, but who are not currently moving in the same strata of contacts as oneself. These are recorded by the telephone book measure; they will not appear in the record of meetings except for the rarest kind of purely chance encounter. Needless to say, these two pools are not clearly segregated, but merge into each other. Yet, our data would suggest that they are more segregated than we would otherwise have suspected. The probabilities of encounter with the two types of persons are of quite different orders of magnitude.

We have now established plausible values for some of the parameters of the contact net of one of the authors. He typically deals with about 20 people in a day. These are drawn from a set of some 1500 persons whom he actively knows at the present time. At the same time he remembers many other persons and could still recognize and name perhaps 3500 persons whom he has met at some point in the past. (Incidentally, he has never regarded himself as good at this.)[6]

The remaining parameter which we would wish to estimate is the degree of structuredness in this acquaintanceship universe. The indicator that we proposed to use was the proportion of the acquaintances of the list-keeper who knew each other; *i.e.*, the proportion of triangles in the network graph. When the 100-day data collection was finished, we took the lists of some of the respondents and turned them into a questionnaire. To a sample of the people who appeared on the respondent's list of contacts, we sent a sample of the names on the list and asked, regarding each, "Do you know that person?". This provided a measure of the degree of ingrowth of the contact net. It can be expressed as the percentage of possible triangles that are completed (Wasserman 1977). The values for five subjects from whom we got the data ranged from 8 to 36%, and we would speculate that a typical value lies toward the low end of this range.

We have indicated above that the degree of structure affects how much longer than chance the minimum chain between a pair of randomly chosen persons is apt to be. We can go no further in specifying the effect of structure on the chains in this qualitative verbal discussion. Any more precise conclusion depends on the treatment of this subject in a much more formal mathematical way. We turn, therefore, to a restatement of our presentation in a mathematical model.

3. Mathematical models of social contact

To describe with precision the structure of human acquaintance networks, and the mechanisms by which social and political contacts can be established within them, it is necessary to idealize the empirical situation with a model. Models have been used effectively in a number of related fields. Rapoport

[6] The $n = At^x$ fitted curve for this author's ogive reached that level in just 5 years, but without taking account of forgetting.

and others have modelled the flow of messages in a network (Rapoport and Horvath 1961; Foster *et al.* 1963; Foster and Horvath 1971; Rapoport 1963; Kleinrock 1964). Related models use Markov chains, queuing theory and random walks (White 1970b, 1973). Most such models, however, depend critically upon an assumption that the next step in the flow goes to other units in the model with a probability that is a function of the present position of the wanderer. The problem that we are addressing does not lend itself to that kind of model; the probability of contact between any two persons is a function of a long-established continuing relationship that inheres in them. The model required for our purposes must be one which retains a characterization of the relationship of each pair of individuals.

Nonetheless, it is useful to begin our analysis with the simplest models in order to develop the needed framework within which to formulate the essential problems. Two extreme situations are relatively easy to analyze. The first is one in which the number of individuals is sufficiently small so that combinatorial methods are still feasible. The second is one in which there are so many individuals that we can treat it as an infinite ensemble, applying methods similar to those used in statistical mechanics. The hard problems deal with conditions between these two extremes.

Graph-theoretic models

Let P denote a group of N people. We shall represent the individuals by integers $1,...,i,...,N$. We draw a directed line or arrow from individual i to individual j to indicate that i knows j. This can be presented as a directed graph, shown in Fig. 5 for $N = 5$, and also represented by an incidence matrix in Fig. 6, where a one is entered in the cell of row i and row j if i knows j and a zero otherwise. If we assume the knowing relation to be symmetric, then every arrow from i to j is side by side with an arrow from j to i – and the incidence matrix is symmetric as well – and we may as well use undirected edges. Let M be the total number of edges or mutual knowing-bonds.

Figure 5. *A directed graph.*

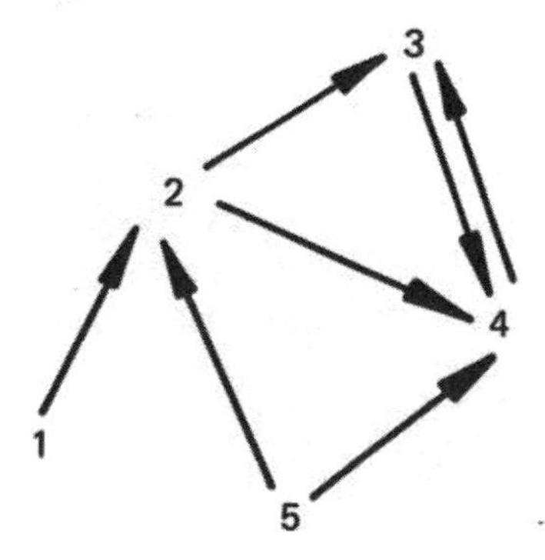

Figure 6. *An incidence matrix.*

	1	2	3	4	5
1	–	1	0	0	0
2	0	–	1	1	0
3	0	0	–	1	0
4	0	0	1	–	0
5	0	1	0	1	0

The incidence matrix has N rows and N columns, but only $(N^2 - N)/2$ of its elements can be chosen freely for a symmetric irreflexive (or reflexive) knowing relation. Thus, there can be at most $(N^2 - N)/2$ pairs or edges. Generally, $0 \leqslant M \leqslant (N^2 - N)/2$. If M takes the largest value possible, then every individual knows every other; if $M = 0$, then no individual knows any other. There is just one structure corresponding to each of these extreme cases. If $M = 1$, there are $(N^2 - N)/2$ possible structures, depending on which pair of people is the one. If $M = 2$, there are $\binom{N^2-N/2}{2}$ possible structures, and there are altogether $2^{(N^2-N)/2}$ possible structures corresponding to $M = 0,1,2,\ldots,(N^2 - N)/2$. The number of possible structures is largest when $M = (N^2 - N)/4$.

Let $\mathbf{U}$ denote the symmetric incidence matrix, and let u_{ij} be (0 or 1) its element in row i, column j. Let $u_{ij}^{(k)}$ denote the corresponding element in the symmetric matrix $\mathbf{U}^k$. This represents the number of different paths of exactly k links between i and j (Luce 1950; Doreian 1974; Peay 1976; Alba 1973). A path is an adjacent series of links that does not cross itself. Two paths are called distinct if not all the links are identical. Thus, there are exactly two 2-step paths from 5 to 3 in Fig. 5, one *via* 4 and one *via* 2; multiplying $\mathbf{U}$ by itself (with 0 in the diagonals) gives

$$\begin{bmatrix} 0 & 0 & 1 & 1 & 0 \\ 0 & 0 & 1 & 1 & 0 \\ 0 & 0 & 1 & 0 & 0 \\ 0 & 0 & 0 & 1 & 0 \\ 0 & 0 & 2 & 1 & 0 \end{bmatrix}$$

and the element in row 5, column 3 is clearly 2, since matrix multiplication calls for the sum of the products of the elements in row 5, (01010), and the elements in column 3, (01010), which is $0 \cdot 0 + 1 \cdot 1 + 0 \cdot 0 + 1 \cdot 1 + 0 \cdot 0 = 2$.

It follows that $u_{ii}^{(3)}$ is the number of triangles that start and end with individual i. Each individual could be the start–end point of as many as $\binom{N-1}{2}$ different triangles, or as few as 0. If $u_{ii}^{(3)} = 0$ for all i, then there cannot be any tightly knit cliques; if $u_{ii}^{(3)} \geqslant 1$ for all i, then there is a considerable degree of connectedness and structure.

Let n denote the number of others each individual knows. This is the number of 1's in each row and each column of the incidence matrix or the number of edges incident on each node of the graph. Let α_k be the sum of all the elements in $\mathbf{U}^k$. It follows that $\alpha_1 = 2M$, and α_2 is twice the number of length-2 paths, which could serve as an index of clustering.

If each of a person's n acquaintances knew one another, $\mathbf{U}$ would consist of N $(n + 1) \times (n + 1)$ matrices consisting of all 1's (except for the diagonal) strung out along the diagonal, assuming that $n + 1$ divides N. Here no individual in one cluster knows anyone in a different cluster.

Such combinatorial, graph-theoretic approaches are intuitively appealing and have considerable descriptive power. There is also a number of theorems for counting the number of different configurations, such as Polya's theorem, as well as computer-based techniques for eliminating structures, such as Lederberg's creation of a language, DENDRAL, for representing the topology of molecules. Graph-theoretic theorems, however, have to ignore reality to introduce assumptions leading to mathematically interesting applications or else follow the scientifically unnatural approach of starting with strong but far-fetched assumptions and relaxing them as little as possible to accommodate reality. The limitations of combinatorial methods become clearest when their computational complexity is studied. Multiplying matrices is of polynomial complexity, requiring of the order of N^3 multiplications; for sparse incidence matrices this can be reduced. But tracing out various configurations or finding a specified path can be much more complex, so that it cannot even be done by computer. Moreover, there is no realistic way that data can be obtained to fill in the elements of **U** for a nation,[7] and different ways of representing acquaintanceship among millions of people must be found. Even storing who knows whom among millions is a non-trivial problem, and more efficient ways of processing such data than are provided by conventional ways of representing sets such as P by ordering its elements $1,\ldots,N$ must be used. The problems of processing data about social networks and drawing inferences from them have received considerable attention, but still face serious obstacles (Wasserman 1977; Holland and Leinhardt 1970; Breiger *et al.* 1975; Granovetter 1974; Newcomb 1961).

Statistical models with independence and no structure

We now take advantage of the large size of N, typically 10^8 or greater, corresponding to the population of a country such as the U.S. We select any two individuals A and B at random from such a large population P. We would like to estimate the distribution of the shortest contact chain necessary for A and B to get in touch.

Let $k = 0$ mean that A and B know one another, that a direct link exists. We have a chain of one link with $k = 0$ intermediaries. But $k = 1$ means that A and B do not know one another, yet have a common acquaintance. It is a chain with two links and one intermediary. $k = 2$ means that A and B do not even have a mutual acquaintance but A knows someone who knows B. It is a chain with three links and two intermediaries.

Let p_k be the probability of a chain with exactly k intermediaries, $k = 1,2,\ldots$. We approximate p_0 by n/N, the ratio of each person's total number of acquaintances to the total population size. Thus, if A knows 1000 people

[7]The use of bibliometric data – for example, who co-authored with whom, who cited whom, which can be obtained in computerized form from the Institute for Scientific Information in Philadelphia for much of the world's scientific literature – may be a practical source. Mathematicians have for some time used the term "Erdös number", which is the distance between any author and Paul Erdös in terms of the number of intermediary co-authors; *e.g.*, A may have co-authored with B who co-authored with C who co-authored with Erdös, making the Erdös distance 2 from A. The use of co-citation and similar data also appears promising (Griffith *et al.* 1973).

out of 100 000 000 Americans (other than A) then the probability of his knowing a randomly chosen person B among the 100 000 000 is $10^3/10^8 = 10^{-5}$.

Let q_0 be the probability that B does not know A. This is $q_0 = 1 - p_0$. It is the probability that one of A's acquaintances is not B. If we now make the strong assumption that the corresponding probability of a second of A's acquaintances is also not B, nor is it affected by knowledge of the probability of the first of A's friends not being B, then the probability that none of the n of A's acquaintances is B is q_0^n. This corresponds to a random or unstructured acquaintance net.

The probability p_1 that A and B are not in direct contact but have at least one common acquaintance is $q_0(1 - q_0^n)$. This assumes that B not being in direct contact with A is also independent of B not being in direct contact with each of the n people whom A knows.

Similarly, we estimate: $p_2 = q_0 q_0^n(1 - q_0^{n^2})$.

This uses another simplifying assumption: each of A's n acquaintances has n *new* acquaintances that will not include any of A's n acquaintances nor any acquaintance of his acquaintances. Thus, there are altogether n^2 different people who are the friends of A's friends. Thus if A knows 1000 people, their friends number a million people not assumed to be counted so far.

If we extend these assumptions for the general case, we have

$$p_k = q_0 q_0^n q_0^{n^2} \ldots q_0^{n^{k-1}} (1 - q_0^{n^k})$$
$$= (1 - p_0)^{(n^k - 1)/(n-1)} [1 - (1 - p_0)^{n^k}] \qquad k = 1, 2, 3, \ldots \qquad (1)$$

Table 8. *Distribution of contact in an unstructured net*

	$n = 500$	$n = 1000$	$n = 2000$
p_0	0.00000500	0.00001000	0.00002000
p_1	0.00249687	0.00995012	0.03921016
p_2	0.71171102	0.98999494	0.96076984
$\sum_{k=3}^{n-1} p_k$	0.28578711	0.00004495	0.00000000
Mean	2.28328023	1.99007483	1.96074984
Variance	0.20805629	0.00993655	0.03774959

Table 8 shows some typical numbers for $N = 10^8$. The numbers were computed using equation 1 on the University of Michigan 470/V6. Note that the average number of intermediaries is 2 (when $n = 1000$), and the average chain is three lengths, with very little variation around that mean. Nor is that average sensitive to n, a person's acquaintance volume. This is not implausible for, according to the above assumptions, if a person knows 1000 people (in one remove), then in two removes he reaches 1000 × 1000, and in three removes 10^9, which exceeds a population of 10^8, according to a simple and intuitive analysis. This result is, however, very sensitive to our independence assumption. The probability of a randomly chosen person C knowing A, given that he knows a friend of A, is almost certainly greater than the unconditional probability that C knows A. (The latter should also exceed the conditional probability of C knowing A, given that C does not know any friend of A.) We turn next to models that do not depend on this independence assumption.

The number of common acquaintances

The independence assumptions of the last section imply that the probability of A having exactly k acquaintances in common with randomly chosen B is $\binom{n}{k}p_0^k q_0^{n-k}$. Here, p_0^k is the probability that k out of the n acquaintances of A each knows B and that each of his remaining $n - k$ acquaintances do not know B; there are $\binom{n}{k}$ ways of selecting these k from the n people whom A knows. The mean of this binomial distribution is np_0 and the variance np_0q_0.

If $n = 10^3$ and $N = 10^8$ then $p_0 = 10^{-5}$, $q_0 = 1 - 10^{-5}$ and the average number of common acquaintances is approximately 0.01 with a variance of 0.01. This is far too small to be realistic, and it points out the weakness of the independence assumption.

One way to replace it is to define p'_0, the conditional probability that a randomly chosen friend of A knows randomly chosen person B′, given that B′ also knows A. This should exceed p_0 or n/N. A plausible estimate for the probability that two of A's friends know each other is $1/(n - 1)$, because there are $n - 1$ people from whom a friend of A could be chosen with whom to form an acquaintance bond. The probability that k of A's friends each knows another friend could now be estimated to be $(p'_0)^k$ or $(n - 1)^{-k}$, if we assume independence of acquaintance among A's friends. Similarly, $(1 - p'_0)^{n-k}$ is an estimate of the probability that $n - k$ of A's friends do not know another of A's friends. As before, the mean number of common acquaintances is np'_0, which is $n/(n - 1)$, or close to 1, with a variance of $np'_0q'_0$, which is close to 0. This, too, is too small for realism, however.

Consider next an approach that relates recursively the average number of acquaintances common to k individuals chosen at random. Call this m_k and assume that

$$m_{k+1} = am_k, \quad m_1 = n, \quad k = 2, 3, \ldots \tag{2a}$$

This means that the average number of acquaintances common to four people is smaller than the average number common to three by a fraction, a, which is the same proportion as the number of friends shared by three is to the number shared by two. This constant a is between 0 and 1 and would have to be statistically estimated. It is assumed to be the same for all $\binom{n}{k}$ groups of k people.

p_0, the probability of A knowing a randomly chosen person B, is n/N or m_1/N, as before. If $m_2 = am_1$, then $n/N = (m_2/a)/N$ and $a = m_2/n$. Thus, if we could estimate the number of acquaintances shared by two people, we could estimate a. Thus, we can set the number of common acquaintances, m_2, to any value we please, and use it to revise the calculation of p_k from what it was in the last section.

p_1, the probability that A does not know randomly chosen B but knows someone who knows B, is $(1 - p_0) \times$ Prob {A and B have at least one common acquaintance}. The latter is the number of ways of choosing a person out of the n people A knows so that he is one of the m_2 common acquaintances, or m_2/n. Thus,

$$p_1 = (1 - p_0)m_2/n$$

and

$$p_2 = (1 - p_0)(1 - p_1)p_2'$$

To calculate p_2', the probability that B knows someone who is a friend of one of A's n acquaintances, we need n', the number of different persons known to the n acquaintances of A. Then we could estimate p_2' by

$$p_2' = \binom{n'}{1}\frac{m_1}{N} - \binom{n'}{2}\frac{m_2}{N} + \binom{n'}{3}\frac{m_3}{N} - \binom{n'}{4}\frac{m_4}{N} + \ldots \pm \binom{n'}{n'}\frac{m_{n'}}{N} \qquad (2b)$$

Here $\binom{n'}{k}m_k$ is the number of ways that B could be one of the m_k acquaintances common to some k of the n' friends of A's friends. It follows from eqn. (2a) that

$$m_2 = am_1$$

$$m_3 = am_2 = a(am_1) = a^2m_1$$

and generally that

$$m_k = a^{k-1}n \qquad (2c)$$

Substituting into eqn. (2b), we can show that

$$p_2' = \frac{n}{aN}[1 - (1 - a)^{n'}]$$

To estimate n', we note that of all A's n friends, m_2 are also known to one other person, m_3 to two others, *etc.* Thus,

$$n' = \binom{n}{1}m_1 - \binom{n}{2}m_2 + \binom{n}{3}m_3 - \binom{n}{4}m_4 + \ldots \pm \binom{n}{n}m_n$$

$$= n\left[\binom{n}{1} - \binom{n}{2}a + \binom{n}{3}a^2 - \binom{n}{4}a^3 + \ldots \pm \binom{n}{n}a^{n-1}\right]$$

$$= \frac{n}{a}\left[\binom{n}{1}a - \binom{n}{2}a^2 + \binom{n}{3}a^3 - \ldots \pm \binom{n}{n}a^n\right]$$

$$= \frac{n}{a}[1 - (1 - a)^n] \text{ by tne binomial theorem} \tag{2d}$$

Hence,

$$p_2 = (1 - p_0)(1 - p_1)(n/aN)[1 - (1 - a)^{n'}]$$

and

$$p_3 = (1 - p_0)(1 - p_1)(1 - p_2)(n/aN)[1 - (1 - a)^{n''}]$$

where

$$n'' = (n/a)[1 - (1 - a)^{n'}]$$

We can set up a recursive equation for p_k in general. We can also require it to hold for $k = 1$, in which case we should expect that

$$m_2/n = (n/aN)[1 - (1 - a)^n] = a \tag{2e}$$

If $n = 10^3$ and $N = 10^8$, then a should be such that $(10^{-5}/a)[1 - (1 - a)^{1000}] = a$. This is a transcendental equation to be solved for a, and the value of $a = 0.003$ is an approximate solution because $10^{-5}[1 - (1 - 0.003)^{1000}]$ is approximately $(0.003)^2$ or 9×10^{-6}, which is reasonably close. A value for $a = 0.003$ or $m_2 = 3$ is no longer so unreasonable for the number of acquaintances common to two people chosen at random. The assumption expressed in eqn. (2a) now implies that m_3, the number of acquaintances common to three people, is $(0.003) \times 3$ or 0.009, which is effectively zero. This is too small to be realistic. Using these values, we obtain,

$p_0 = 0.00001$, as before
$p_1 = 0.003$ compared with 0.009949
$p_2 = 0.00332$ compared with 0.99001
$p_3 = 0.00330$
$n' = 381\,033$, $n'' = 333\,333$

The distribution of k is now considerably flattened, with chains of short length no less improbable than chains of greater length. This is due to a value of a greater than 10^{-5}, as specified by a chosen value of m_2 and eqn. (2e).

The above analysis, though more realistic, is still limited by an independence assumption and the low value of $m_3, m_4, \ldots$. Yet it may be fruitful to explore it further by exploiting the sensitivity of these results to m_2, or replacing eqn. (2a) by one in which a is not constant. We now proceed, however, to replace this approach by defining the following conditional probabilities.

Let K_A be A's circle of acquaintances, with $\overline{K}_A$ its complement. Let $A_1, \ldots, A_n$ denote the individuals in it. Consider:

$\text{Prob}(B \in \overline{K}_{A_1})$, $\text{Prob}(B \in \overline{K}_{A_2} | B \in \overline{K}_{A_1})$, $\text{Prob}(B \in \overline{K}_{A_3} | B \in \overline{K}_{A_2}, B \in \overline{K}_{A_1})$, *etc.* The product of these probabilities is $\text{Prob}(B \in \overline{K}_{A_1} \cap \overline{K}_{A_2} \cap \overline{K}_{A_3} \cap \ldots)$, the probability that a randomly chosen B is not known to each of A's acquaintances.

A simple and perhaps plausible assumption other than independence is that of a Markov chain:

$$\text{Prob}(B \in \bar{K}_{A_k} | B \in \bar{K}_{A_{k-1}}, ..., B \in \bar{K}_{A_1}) = \text{Prob}(\bar{K}_{A_k} | \bar{K}_{A_{k-1}}) = b$$

where b is a constant to be statistically estimated.

Thus,

$$\text{Prob}\,(\bar{K}_{A_n}, \bar{K}_{A_{n-1}}, ..., \bar{K}_{A_1}) = \text{Prob}(\bar{K}_{A_1})b^{n-1} = (1 - n/N)b^{n-1}$$

For $k = 2$,

$$\text{Prob}(\bar{K}_{A_2}, \bar{K}_{A_1}) = (1 - n/N)b = 1 - 2n/N + m_2/N$$

Hence

$$b = \frac{1 - 2n/N + m_2/N}{1 - n/N}$$

This gives more freedom to choose m_2. If $m_2 = 10$, $n = 10^3$, $N = 10^8$, then $b = 0.9999900999$.

Now

$$p_0 = n/N = 0.00001 \quad \text{as before}$$

and

$$p_1 = (1 - p_0)[1 - (1 - n/N)b^{n-1}] = 0.001$$

$$p_2 = (1 - p_0)(1 - p_1)p_2'$$

where

$$p_2' = \text{Prob(B knows at least one of the } n' \text{ friends of A's friends)}$$
$$= 1 - (1 - n/N)b^{n'-1}$$

$$n' = \binom{n}{1}m_1 - \binom{n}{2}m_2 + \binom{n}{3}m_3 - \binom{n}{4}m_4 + ... \pm \binom{n}{n}m_n \quad \text{as before}$$

To estimate m_k we need Prob(K_1, ..., K_k), the probability of B being known to k randomly chosen people, and we shall assume this to be Prob$(K_1) \cdot c^{k-1}$, where $c = \text{Prob}(K_k | K_{k-1})$. If $k = 2$, then

$$\text{Prob}(K_1, K_2) = m_2/N = \text{Prob}(K_1) \cdot c = (n/N)c$$

so that $c = m_2/n$. Hence,

$$m_k = N \cdot (n/N)\,(m_2/n)^{k-1} = n(m_2/n)^{k-1} \qquad k = 1, 2, ...$$

Therefore,

$$n' = \sum_{k=1}^{n} (-1)^{k-1} \binom{n}{k} \cdot n(m_2/n)^{k-1}$$

$$= \frac{n}{m_2/n} \sum_{k=1}^{n} (-1)^{k-1} \binom{n}{k} (m_2/n)^k$$

$$= (n^2/m_2)[1 - (1 - m_2/n)^n]$$

If $m_2 = 10$, then $c - 10/1000 = 0.01$ and $n' = 10^5(1 - e^{-10}) = 99996$. Thus

$$p_2' = 1 - (1 - 0.00001)(0.9999900999)^{99996}$$
$$\simeq 1 - (0.99999)(0.3716)$$
$$\simeq 0.6278$$

and

$$p_2 = (0.99999)(0.999)(0.6278) \simeq 0.627$$

To compute p_3 we shall need p'', the number of different people who are the friends of the acquaintances of the n people whom A knows.

$$n'' = \sum_{k=1}^{n'} (-1)^{k-1}\binom{n'}{k} n(m_2/n)^{k-1} = (n^2/m_2)[1 - (1 - m_2/n)^{n'}]$$

$$\simeq (10^6/10)[1 - (1 - 10^{-2})^{10^5}] \simeq 10^5(1 - e^{-1000}) \simeq 10^5 \simeq n'$$

$$p_3' = 1 - (1 - n/N)b^{n''-1} = 1 - (1 - 10^{-5})(0.9999900999)^{10^5} \simeq 0.6278$$

$$p_3 = (1 - p_0)(1 - p_1)(1 - p_2)p_3' \simeq (0.999)(0.373)(0.6278) = 0.234$$

This calculation leads to more plausible results, but it still does not have an underlying rationale to warrant attempts to fit data.

Contact probabilities in the presence of social strata

In a model of acquaintanceship structure it is desirable to be able to characterize persons as belonging to subsets in the population which can be interpreted as social strata. We show how the distribution for the length of minimal contact chains can be computed when strata are introduced. We begin by partitioning the entire population into r strata, with the ith stratum containing m_i members. Let h_{ij} denote the mean number of acquaintances which a person who is in stratum i has in stratum j. The mean number of acquaintances of a person in stratum i is then $n_i = \Sigma_{j=1}^{r} h_{ij}$. The conditional probability p_{ij} that a person picked at random in stratum j is known to someone in stratum i, given j, is $h_{ij}/m_j = p_{ij}$. The $r \times r$ matrix (p_{ij}) is symmetric and doubly stochastic because we have assumed that the "knowing" relation is symmetric.

We now select two people, A and C, with A in stratum i and C in stratum j. To obtain the probability that there is no 2-link contact chain from A to C, with the intermediary being in a specified stratum k, let K_i be the set of A's h_{ik} friends in stratum k. Combinatorially, Prob$\{K_i \cap K_j = \phi\}$ is the number of ways of selecting h_{ik} and h_{jk} out of m_k elements such that $K_i \cap K_j = \phi$, divided by the total number of ways of selecting h_{ik}, h_{jk} out of m_k elements, assuming independent trials without replacement. Thus,

$$\mathrm{Prob}\{K_i \cap K_j = \phi\} = \frac{m_k!/[h_{ik}!h_{jk}!(m_k - h_{ik} - h_{jk})!]}{\binom{m_k}{h_{ik}}\binom{m_k}{h_{jk}}}$$

$$= \frac{(m_k - h_{ik})!(m_k - h_{jk})!}{m_k!(m_k - h_{ik} - h_{jk})!} \tag{3}$$

The probability that there is no chain from A in stratum i to C in stratum j *via* some mutual acquaintance in any stratum is

$$\prod_{k=1}^{r} \frac{(m_k - h_{ik})!(m_k - h_{jk})!}{m_k!(m_k - h_{ik} - h_{jk})!} \equiv q_{ij}'$$

While data about all the elements of (h_{ij}) are not likely to be readily obtainable, the variables m_i, n_i and h_{ii} for $i = 1, ..., r$ may be estimable. We now make a methodological simplification and assume these variables equal for all i,-with $m_i = m = N/r$, $n_i = n$, $h_{ii} = h$ and

$$h_{ij} = \frac{n - h}{r - 1} = h' \text{ for all } i \neq j \tag{4}$$

To compute q_1', the probability that there is no chain of length 1 – or that there is *no* mutual acquaintance – between two individuals A and C, it is necessary to consider two cases:

(1) that in which A and C are in the same stratum;

(2) that in which A and C are in different strata.

In the first case, $q_1' = uv^{r-1} \equiv q_1'(1)$ (the number in parentheses refers to case 1), where u is the probability that B, the intermediary between A and C, fails to be in the same stratum as A and C, and v is the probability that he fails to be in a different stratum. Using eqn. (3), it is readily seen that

$$u = \frac{(m - h)!^2}{m!(m - 2h)!} \tag{5}$$

$$v = \frac{(m - h')!^2}{m!(m - 2h')!} \tag{6}$$

By similar reasoning,

$$q_1'(2) = w^2 v^{r-2}$$

where w is the probability that the stratum of B is the same as that of A but not of C; this is equal to the probability that the stratum of B is the same as that of C but not of A. This is, by eqn. (3),

$$w = \frac{(m - h)!(m - h')!}{m!(m - h - h')!} \tag{7}$$

With the help of Stirling's formula and series expansions we can derive a useful approximation for w. It is

$$w \simeq (1 + hh'/m^2)e^{-hh'/m} \simeq e^{-hh'/m} \tag{8}$$

As before, let p_1 denote the probability that A and C do not know each other, but that they have at least one common acquaintance. Then

$$p_1(i) \simeq (1 - p_0)[1 - q_1'(i)] \qquad i = 1, 2$$

To estimate p_1, we could take a weighted average,

$$p_1 = (1/r)p_1(1) + (1 - 1/r)p_1(2)$$

The above relation is written as an approximation, because $q_1'(i)$ is not a conditional probability given that A and C do not know each other, but the error is negligible. The number in the parentheses, 1 or 2, refers to whether or not A and C are in the same stratum, respectively. Thus,

$$p_1(1) \simeq (1 - n/N)(1 - uv^{r-1})$$

Because u can also be approximated by $\exp(-h^2/m)$ and v by $\exp(-h'^2/m)$, we can approximate $p_1(1)$ by

$$1 - \exp[-(h^2/m) - (h'^2/m)(r - 1)]$$

Substituting $m = N/r$, this becomes

$$p_1(1) \simeq 1 - \exp\{-(r/N)[h^2 + h'^2(r - 1)]\} \tag{9}$$

If A has more friends in a given stratum not his own than he has in his own stratum, then $h' > h$. If almost all of A's friends are in his own stratum, then $h' \ll h$, and $h \simeq n$. If r is large enough, $p_1(1)$ can be very close to 1. For instance, if $N = 10^8$, $h = 100$, $n = 1000$ and $r = 10$, we have that $h' = 900/9 = 100$, and $p_1(1) \simeq 0.00995$, as in the case of independence.

Next,

$$p_1(2) \simeq (1 - n/N)(1 - w^2v^{r-2})$$

$$\simeq 1 - \exp\{-(2/N)[2hh' + (r - 2)h'^2]\} \tag{10}$$

For the same numerical values as above,

$$p_1(2) \simeq 1 - e^{-10^{-2}} \simeq 0.00995 \text{ also}$$

We now wish to compute $p_2{}^*$, the joint probability that A and C do not know each other, *and* that they have no common friends, *and* that A has some friends, at least one of whom knows some friend of A. As before, we shall compute the conditional probability that A has some friends, at least one of whom knows some friend of C, given that A and C neither know each other, nor have a common acquaintance. We shall denote this conditional probability by $p_2'{}^*$, so that $p_2{}^* = (1 - p_0)(1 - p_1'{}^*)p_2'{}^*$. To say that A has some friends, at least one of whom knows some friend of C, is to say that there is at least one person, B, who knows A *and* who has at least one friend,

D, in common with C. By the assumed symmetry of the knowing relation, this is the same as saying: there exists $B \in K_C$, where K_C is the set of all people who can be linked to C by a minimal chain of length 1 (one intermediary). Select B at random and consider the choice fixed. $\text{Prob}(B \in \bar{K}_C) = 1 - p_1^*$, averaged over all strata. Assuming independence, the probability that any n B's, and in particular the n friends of A, all fail to be connected to C by a minimal chain of length 1 is $(1 - p_1^*)^n$. Hence, neglecting a small correction due to the condition in the definition of $p_1'^*$, we can estimate:

$$p_2'^* = 1 - (1 - p_1^*)^n \simeq 1 - \exp(-p_1^* n)$$

for p_1^* very small.

To obtain a more precise estimate of $p_2'^*$ we proceed as follows. Let s(A) denote the stratum of A. Consider first the case $i = 1$, where s(A) = s(C). Now suppose that s(B) = s(A). Then the probability that no chain of length 1 links B and C is uv^{r-1} as before. If s(B) $\neq$ s(A), however, B can be in any one of $r - 1$ strata, and for each stratum the probability that no chain of length 1 links B and C is $w^2 v^{r-2}$. Hence the probability that no chain of length 2 links A and C with s(A) = s(C) is

$$\begin{aligned} q_2'(1) &= u^h v^{h(r-1)} (w^2 v^{r-2})^{(r-1)h'} \\ &= u^h v^{(r-1)h + (r-1)(r-2)h'} w^{2(r-1)h'} \end{aligned} \tag{11}$$

Consider next the case $i = 2$, where s(A) $\neq$ s(C). If s(B) = s(A), the probability that no chain of length 1 links B and C is $(w^2 v^{r-2})^h$. If s(B) $\neq$ s(A), this probability is the product of:

(a) the probability of no 1-chain linking B and C when s(B) = s(C) – this is $(uv^{r-1})^{h'}$; and

(b) the same probability when s(B) $\neq$ s(C), *i.e.* $(w^2 v^{r-2})^{(r-2)h'}$. Hence, the probability that no chain of length 2 links A and C when s(A) $\neq$ s(C) is

$$\begin{aligned} q_2'(2) &= (w^2 v^{r-2})^h (uv^{r-1})^{h'} (w^2 v^{r-2})^{(r-2)h'} \\ &= u^{h'} v^{h(r-2) + h'(r-1) + (r-2)^2 h'} w^{2h + 2(r-2)h'} \\ &= u^{h'} v^{h(r-2) + h'(r^2 - 3r + 3)} w^{2[h + (r-2)h']} \end{aligned} \tag{12}$$

As before, we may estimate the conditional probability that A and C are linked by at least one 2-chain given that A does not know C or any friend of C by

$$\begin{aligned} 1 - p_2'^* = q_2'^* = (1/r) u^h v^{(r-1)h + (r-1)(r-2)h'} w^{2(r-1)h'} + \\ + (1 - 1/r) u^{h'} v^{h(r-2) - h'(r^2 - 3r + 3)} w^{2[h - (r-2)h']} \end{aligned}$$

Note that effects due to the two conditions have been neglected and that independence has been assumed throughout.

Observe also that we could have written

$$q_2'(1) = [q_1'(1)]^h [q_1'(2)]^{h'(r-1)}$$

$$q_2'(2) = [q_1'(1)]^{h'} [q_1(2)]^h [q_2(2)]^{h'(r-2)}$$

$$q_2'^* = (1/r)q_2'(1) + (1 - 1/r)q_2'(2)$$

The above relation suggests a recursive scheme of generalizing the calculation. That is:

$$p_k = (1 - p_0)(1 - p_1'^*)(1 - p_2'^*) \ldots (1 - p_{k-1}'^*)(1 - q_k'^*)$$

$$q_k'^* = (1/r)q_k'(1) + (1 - 1/r)q_k'(2)$$

$$q_k'(1) = [q_{k-1}'(1)]^h [q_{k-1}'(2)]^{h'(r-1)}$$

$$q_k'(2) = [q_{k-1}'(1)]^{h'} [q_{k-1}'(2)]^{h+h'(r-2)} \qquad k = 2, 3, 4, \ldots$$

Using the cruder method suggested in the first paragraph of the above section,

$$p_k'^* = 1 - (1 - p_{k-1}'^*)^n \qquad k = 2, 3, 4, \ldots$$

There is another iterative method that could be used to compute p_k^*. If k is odd (*e.g.*, $k = 3$), compute $q_k'(1)$ and $q_k(2)$ using formulas (9) and (10) but substituting $p_{k-1}'(1)m$ for h and $p_{k-1}'(2)m$ for h'. Similarly, if k is even, use formulas (11) and (12) with the same substitutions for h and h'.

In the Appendix we develop further approximations to facilitate the calculation of p_0^*, p_1^*, and p_2^*, which we find to be 0.00001, 0.00759, and 0.9924, respectively, with the parameters used previously.

Note the departure from the model without strata is not very great. That is a significant inference. Structuring of the population may have a substantial effect on p_1. (It has no effect, of course, on p_0.) However, in a connected graph (which we believe any society must be) the nuclei get bridged by the longer chains quite effectively, and so the mean length of chains between randomly chosen pairs is only modestly affected by the structuring. We would therefore conjecture that, despite the effects of structure, the modal number of intermediaries in the minimum chain between pairs of Americans chosen at random is 2. We noted above that in an unstructured population with $n \simeq 1000$ it is practically certain that any two individuals can contact one another by means of at least two intermediaries. In a structured population it is less likely, but still seems probable. And perhaps for the whole world's population probably only one more bridging individual should be needed.

Monte-Carlo simulation models

To achieve greater understanding of the structural aspects of acquaintance nets, we approached an explanation of the dynamics of how acquaintance bonds are formed with the help of a stochastic model that was simulated by computer. We regarded each individual to be located as a point in a social space, which we regarded as a square region in the two-dimensional Euclidean plane, to start with. As before, we let N be the number of individuals. Each individual can change his position in time t to time $t + 1$ by $(\Delta x, \Delta y)$ where

$$\Delta x = \begin{cases} s & \text{with probability } p \\ -s & \text{with probability } q \quad \text{where } p + q + r = 1 \\ 0 & \text{with probability } r \end{cases}$$

and with Δy defined similarly, and statistically independent of Δx. Each individual is confined to remain in a $D \times D$ square, so that if his location at t is $z(t) = [x(t), y(t)]$, then in the next simulation cycle it is

$$[x(t) + \Delta x \bmod D, y(t) + \Delta y \bmod D] = [x(t+1), y(t+1)]$$

We now define e_{AB} to be 1 if the line connecting $[x_A(t), y_A(t)]$ and $[x_A(t+1), y_A(t+1)]$ intersects the line from $[x_B(t), y_B(t)]$ to $[x_B(t+1), y_B(t+1)]$, and $e_{AB} = 0$ if these paths do not intersect. The event E_{AB} corresponding to $e_{AB}(t) = 1$ at time t is interpreted as a contact between A and B on day t. $(1/t)\Sigma_{\tau=1}^{t} e_{AB}(\tau)$ denotes the frequency with which A and B have met during the first t days.

Next, let $K_A(t)$ be the set of all people whom A has met by day t, or $\{\text{all B}: e_{AB}(\tau) = 1 \text{ for } \tau \leq t\}$. We now extend $K_A(t)$ to include A and define the center of that group or cohort on day t as follows:

$$c_A(t) = [\bar{x}_A(t), \bar{y}_A(t)]$$

with

$$\bar{x}_A(t) = \frac{x_A(t) + \sum_{B \in K_A(t)} x_B(t) \sum_{\tau=1}^{t} e_{AB}(\tau)}{1 + \sum_B \sum_\tau e_{AB}(\tau)}$$

and $\bar{y}_A(t)$ is similarly defined. The x-coordinate of the center is the average of the x-coordinates of A and all the people he has met, weighted by how frequently they were contacted.

The probabilities p and q also vary with time and with each individual, as follows with $z_A(t) = (x_A(t), y_A(t))$.

If $c_A(t) > z_A(t)$, then
$$p_A(t+1) = p_A(t) + e$$
$$q_A(t+1) = q_A(t) - e/2$$
$$r_A(t+1) = r_A(t) - e/2$$

If $c_A(t) < z_A(t)$, then
$$p_A(t+1) = p_A(t) - e/2$$
$$q_A(t+1) = q_A(t) + e$$
$$r_A(t+1) = r_A(t) - e/2$$

If $c_A(t) = z_A(t)$, then the probabilities do not change. Initially, $[p(0), q(0),$

$r(0)] = (1/3, 1/3, 1/3)$ and no probability must ever fall outside $[\delta, 1-\delta]$ to ensure that the system remains stochastic; when these values are reached, the probabilities stay there until the z's and c's change.

After considerable experimentation with several values of the different parameters, we chose:

Number of individuals	$N = 225$
Size of one side of square grid	$D = 15$
Social responsiveness or elasticity	$e = 0.2$
Lower bound on probability of position change	$\delta = 0.01$
Unit increment in position change	$s = 1$

Well before the 10th iteration, clustering begins and by the 20th iteration it clusters into a single group. For realism, we would expect several clusters to emerge (corresponding to social strata) that exhibit both local and global structure, which are not too rigidly determined by the Euclidean structure of the social space. We have not explored the model sufficiently to determine if it has these properties, if small changes in the model could provide it with these properties, or if this approach should be abandoned. Computation cost increases as N^2 and the number of iterations, and took a few minutes per iteration on the MIT 370-186 system in 1973. This cost could be reduced by sampling, resulting in a fractional decrease that is the sample size divided by N. After enough iterations have produced what appears to be a realistic but scaled-down acquaintance net in such an idealized social space, a second program (also written by Diek Kruyt) to compute the distribution of chain lengths is then applied. Its cost varies as N^3.

Our present decision – held since 1975 – is to explore the use of a computer program that constructs an acquaintance net according to a simulation that uses the data we obtained from the 100-day diaries kept by our 27 respondents (see § 2). The basic inputs to this program are:

The total number of individuals	$N = 1000$
The number of people seen by person A on any f days in 100	$Y(f) =$ data
The number of different people that A did not see in 100 days	$Y_A(0)$
The number of people, each of whom has exactly k acquaintances in common with A	$M_A(k) =$ data

Outputs include the distribution of chain length. The program starts by selecting A and linking to him all the $Y(100)$ people he sees daily (chosen at random from the $N - 1$ in the program). This might, for example, be the nucleus of his circle of acquaintances consisting of $Y(100) = 3$ people. Call them B, C, and D, and we have

A — B, C, D

so far.

Next proceed with the first of A's friends just chosen, say B. Link to him all $Y(100)$ others chosen randomly from $N-1$, but including A. This might generate the following list of B's friends: A, C, F. Repeat this for all people labeled so far, *e.g.*, C, D, F, *etc.*, until there are no more new "target" people. Then repeat this procedure for $Y(99)$ in place of $Y(100)$, but eliminating certain randomly chosen links if they do not satisfy the following constraint.

Our data suggested that there are fewer people who have one acquaintance in common with A than there are who have two acquaintances in common with A, *etc.*, but that only a few people have very many acquaintances in common with A. Thus, there is a value, M, for which $M(k)$ is greatest, where $M = M(K)$. For example, if $M(1) = 2, M(2) = 3, M(4) = 5, M(5) = 4$, *etc.*, then $M = 5, K = 4$. We must ensure that M people among those chosen so far each have K acquaintances in common with A, also with the people he sees daily. We then repeat these steps with $Y(98)$ in place of $Y(99)$ and replace the constraint that M friends have K acquaintances in common with A, *etc.*, by one requiring that $M(K-1)$ people have $K-1$ acquaintances in common with A, B, *etc.* This is continued until $Y(0)$ and $M(1)$ replace $Y(1)$ and $M(2)$, respectively.

Effective and efficient algorithms for making these selections subject to the given constraints have yet to be developed. The computational complexity of this algorithm must also be determined, and hopefully is a polynomial in N. Hopefully also, such a program can be run for N large enough so that distribution of chain length does not change significantly as N is increased. Fruitful next steps seem to us to be the further development and analysis of the models sketched in this section. When these are found to have properties we consider realistic for large social contact nets and are the result of plausible explanatory inferences, then some difficult problems of statistical estimation must be solved. Hopefully, then we will have reached some understanding of contact nets that we have been seeking.

Appendix

Some approximations using Stirling's formula have already been derived and analyzed.

There is another very useful approximation based on a slightly different model in the general case.

Let q'_{ij} be defined as in eqn. (2), but rewrite it as

$$\frac{(m_k - h_{ik})!(m_k - h_{jk})!}{m_k!(m_k - h_{ik} - h_{jk})!} =$$

$$\frac{(m_k - h_{ik})!(m_k - h_{jk})(m_k - h_{jk} - 1)\ldots(m_k - h_{jk} - h_{ik} + 1)(m_k - h_{jk} - h_{ik})!}{m_k(m_k - 1)\ldots(m_k - h_{ik} + 1)(m_k - h_k)!(m_k - h_{jk} - h_{ik})!}$$

$$= \left(1 - \frac{h_{jk}}{m_k}\right)\left(1 - \frac{h_{jk}}{m_k - 1}\right)\left(1 - \frac{h_{jk}}{m_k - 2}\right)\cdots\left(1 - \frac{h_{jk}}{m_k - h_{ik} + 1}\right)$$

(h_{ik} terms)

It is easily seen that this represents the probability of failing to draw a sample of h_{ik} red balls from an urn having m_k balls of which h_{jk} are red, but sampling without replacement. If we sample with replacement, the above formula becomes $q_{jk}{}^{h_{ik}}$, where $q_{jk} = (1 - h_{jk}/m_k)$. This represents the probability that none of A's h_{ik} friends in stratum k is known to C (s(A) = i, s(C) = j), where it is possible to count the same friend more than once. The fractional error committed by this assumption is

$$\epsilon = \left[\prod_{k=1}^{r} q_{jk}{}^{h_{ik}} - \prod_{k=1}^{r}\prod_{l=0}^{h_{ik}-1}\left(1 - \frac{h_{jk}}{m_k - l}\right)\right]\Big/\prod_{k=1}^{r} q_{jk}{}^{h_{ik}}$$

This will be estimated later. Now,

$$\log q'_{ij} \simeq \sum_{k=1}^{r} h_{ik} \log\left(1 - \frac{h_{jk}}{m_k}\right)$$

If $h_{jk} \ll m_k$ for all k, we can further approximate this by

$$-\sum_{k=1}^{r} h_{ik}\frac{h_{jk}}{m_k} = -\sum_{k=1}^{r} h_{ik}\frac{h_{kj}}{m_j} = -\frac{1}{m_j}\sum_{k=1}^{r} h_{ik}h_{kj}$$

with a fractional error of about $h_{jk}/2m_k$, which is less than $(h + h')^2/2m$, as in the previous approximation. Furthermore, this approximation permits matrix multiplication and greater generality than only two values of h_{ij}. If we denote the matrix (h_{ij}) by **H** and $(\log q_{ij})$ by **L**, then $\mathbf{L} = \mathbf{H}\bar{\mathbf{H}}$, $\bar{\mathbf{H}}$ being the transpose of **H**.

To estimate the error, we take

$$\epsilon = 1 = \prod_{k=1}^{r}\prod_{l=0}^{h_{ik}-1}\left[\frac{1 - h_{jk}/(m_k - l)}{1 - h_{jk}/m_k}\right]$$

The term in brackets is approximated by the series

$$\left(1 + \frac{h_{jk}}{m_k} + \frac{h_{jk}^2}{m_k^2} + \ldots\right) - \frac{h_{jk}}{m_k - l}\left(1 + \frac{h_{jk}}{m_k} + \ldots\right)$$

$$= 1 + h_{jk}\left(\frac{1}{m_k} - \frac{1}{m_k - l}\right) + \frac{h_{jk}^2}{m_k}\left(\frac{1}{m_k} - \frac{1}{m_k - l}\right) + \ldots$$

$$= 1 - \frac{l}{m_k(m_k - l)}\left(h_{jk} + \frac{h_{jk}^2}{m_k} + \ldots\right)$$

$$\simeq 1 - \frac{h_{jk}}{m_k^2}\, l$$

$$\epsilon \simeq 1 - \prod_{k=1}^{r} \exp\left(\frac{-h_{jk}}{m_k^2} \sum_{l=0}^{h_{ik}-1} l\right)$$

$$= 1 - \prod_{k=1}^{r} \exp\left[\frac{-h_{jk}}{m_k^2}\,\frac{(h_{ik} - 1)h_{ik}}{2}\right]$$

$$\simeq 1 - \prod_{k=1}^{r} \exp\left(-\frac{h_{jk}h_{ik}^2}{2m_k^2}\right)$$

$$= 1 - \exp\left(-\frac{1}{2m_j^2}\sum_{k=1}^{r} h_{ik}^2 h_{kj}\right)$$

According to this estimate, the approximation is good only when

$$\sum_{k=1}^{r} h_{ik}^2 h_{kj} < m_j^2$$

To compare this with the exponential approximation, let $h_{ik} = h$ if $i = k$, h' if $i \neq k$, so that

$$\sum_k h_{ik}^2 h_{kj} = h^2 h' + hh'^2 (r - 2)h'^3 \qquad i \neq j$$

$$= h^3 + (r - 1)h'^3 \qquad i = j$$

Hence, it would be required that $(h + h')^3 r < m^2$ or $(h + h')^{3/2}\sqrt{r} < m$, compared with $(h + h')^2 < m$.

For the above simplified situation, the replacement model gives

$$q'_{ii} \simeq \exp\left[\frac{-h^2 + (r - 1)h'^2}{m}\right]$$

$$q'_{ij} \simeq \exp\left[\frac{-2hh' + (r - 2)h'^2}{m}\right] \qquad i \neq j$$

As an example where the departure from the results obtained when stratification was disregarded becomes more pronounced than in the illustrations chosen so far, let $N = 10^8$, $n_i = n = 10^3$ for all i, $m_j = m = 10^4$ for all j, $r = 10^4$, $h_{ii} = h = 500$, $h_{ij} = h' = 500/(10^4 - 1) = 5 \times 10^{-2}$ for all $i \neq j$.

(1) $p_0{}^* = n/N = 10^{-5}$

(2) $p_1{}^* = (1 - p_0)p_1{}'^* = (1 - p_0)(1 - q_1{}'^*)$

$$q_1{}'^* = \frac{1}{r} q_1'(1) + \left(1 - \frac{1}{r}\right) q_1'(2)$$

$$q_1'(1) = \exp\left(-\frac{25 \times 10^4}{10^4} + \frac{25 \times 10^{-4}}{10^4} \times 10^4\right) \simeq e^{-25} \simeq 0$$

$$q_1'(2) \simeq \exp\left(-\frac{2 \times 500 \times 5 \times 10^{-2}}{10^4} + \frac{10^4 \times 25 \times 10^{-4}}{10^4}\right) \simeq 0.9925$$

$$q_1{}'^* \simeq 0.99241$$

$$p_1{}^* \simeq 0.00759$$

(3) Recall that $u \simeq \exp(-h^2/m)$, $v \simeq \exp(-h'^2/m)$, $w \simeq \exp(-hh'/m)$, so that

$$q_2'(1) = \exp\left(-\left[\frac{h^3}{m} + \frac{h'^2}{m}(r-1)h + (r-1)(r-2)h' + 2\frac{hh'^2}{m}(r-1)\right]\right)$$

$$= \exp\left(-\left[\frac{h^3}{m} + 3\frac{hh'^2}{m}(r-1) + \frac{h'^3}{m}(r-1)(r-2)\right]\right)$$

$$\simeq \exp\left(-\frac{1}{m}(h^3 + 3rhh'^2 + r^2h'^3)\right)$$

$$q_2'(2) = \exp\left(-\frac{1}{m}\{h^2h' + [h(r-2) + h'(r^2 - 3r + 3)]h'^2 + {} + 2[h + (r-2)h']hh'\}\right)$$

$$\simeq \exp\left(-\frac{1}{m}[h^2h' + hrh'^2 + h'^3r^2 + 2h^2h' + 2(r-2)hh'^2]\right)$$

$$\simeq \exp\left(-\frac{1}{m}(3h^2h' + 3rhh'^2 + r^2h'^3)\right)$$

Then,

$$q_2'(1) = \exp[-10^{-4}(125 \times 10^6 + 3 \times 125 \times 10^{-2} \times 10^4 + 10^8 \times 125 \times 10^{-6})]$$

$$= \exp[-(12500 + 5)] \simeq 0$$

$$q_2'(2) = \exp[-10^{-4}(3 \times 125 \times 10^2 + 3 \times 10^4 \times 125 \times 10^{-2} + + 10^8 \times 125 \times 10^{-6})]$$

$$= \exp(-8.75) \simeq 0.00016$$

Hence,

$$p_2^* \simeq (1 - 10^{-5})(1 - 0.00759)(1 - 0.00016)$$

$$\simeq 0.9924$$

References

Alba, R.
1973 "A graph-theoretic definition of a sociometric clique". *Journal of Mathematical Sociology 3*:113 - 126.

Alba, R. and C. Kadushin
1976 "The intersection of social circles: a new measure of social proximity in networks". *Sociological Methods and Research 5*:77 - 102.

Boissevain, J.
1974 *Friends of Friends: Networks, Manipulators, and Coalitions.* New York: St. Martin's Press.

Boorman, S. and H. White
1976 "Social structures from multiple networks". *American Journal of Sociology 81*:1384 - 1446.

Breiger, R., S. Boorman and P. Arabie
1975 "An algorithm for clustering relational data with applications to social network analysis and comparison with multidimensional scaling". *Journal of Mathematical Psychology 12*: 328 - 383.

de Grazia, A.
1952 *Elements of Political Science.* New York: Free Press.

Deutsch, K.
1956 "Shifts in the balance of communication flows". *Public Opinion Quarterly 20*:143 - 160.
1966 *Nationalism and Social Communication.* Cambridge, Mass.: MIT Press.

Doreian, P.
1974 "On the connectivity of social networks". *Journal of Mathematical Sociology 3*:245 - 258.

Erickson, B. and P. Kringas
1975 "The small world of politics, or, seeking élites from the bottom up". *Canadian Review of Sociology and Anthropology 12*:585 - 593.

Festinger, L., S. Shachter and K. Back
1950 *Social Pressures in Informal Groups.* New York: Harper.

Foster, C. and W. Horvath
1971 "A study of a large sociogram III: reciprocal choice probabilities as a measure of social distance". *Behavioral Science 16*:429 - 435.

Foster, C., A. Rapoport and C. Orwant
1963 A study of large sociogram II: elimination of free parameters". *Behavioral Science 8*:56 - 65.

Granovetter, M.
1974 *Getting a Job: A Study of Contacts and Careers.* Cambridge, Mass.: Harvard University Press.
1976 "Network sampling: some first steps". *American Journal of Sociology 81*:1287 - 1303.
Griffith, B., V. Maier and A. Miller
1973 *Describing Communications Networks Through the Use of Matrix-Based Measures.* Unpublished. Drexel University, Graduate School of Library Science, Philadelphia, Pa.
Gurevich, M.
1961 *The Social Structure of Acquaintanceship Networks.* Cambridge, Mass.: MIT Press.
Gurevich, M. and A. Weingrod
1976 "Who knows whom – contact networks in Israeli National élite". *Megamot 22*:357 - 378.
n.d. *Human Organization.* To be published.
Hallinan, M. and D. Felmlee
1975 "An analysis of intransitivity in sociometric data". *Sociometry 38*:195 - 212.
Hammer, M.
n.d. *Social Access and Clustering of Personal Connections.* Unpublished.
Holland, P. and S. Leinhardt
1970 "A method for detecting structure in sociometric data". *American Journal of Sociology 70*:492 - 513.
Horowitz, A.
1977 "Social networks and pathways to psychiatric treatment". *Social Forces 56*:81 - 105.
Hunter, J. and R. L. Shotland
1974 "Treating data collected by the small world method as a Markov process". *Social Forces 52*:321 - 332.
Jacobson, D.
1970 "Network analysis in East Africa; the social organization of urban transients". *Canadian Review of Sociology and Anthropology 7*:281 - 286.
Jennings, H.
1937 "Structure of leadership – development and sphere of influence". *Sociometry 1*:131.
Katz, E. and P. Lazarsfeld
1955 *Personal Influence.* Glencoe, Ill.: Free Press.
Killworth, P. and B. Russell
1976 "Information accuracy in social network data". *Human Organization 35*:269 - 286.
Kleinrock, L.
1964 *Communication Nets: Stochastic Message Flow and Delay.* New York: McGraw-Hill.
Korte, C. and S. Milgram
1970 "Acquaintanceship networks between racialgroups: application of the small world method". *Journal of Personality and Social Psychology 15*:101 - 108.
Kurtzman, D. H.
1935 *Methods of Controlling Votes in Philadelphia.* Philadelphia: University of Pennsylvania.
Lorrain, F.
1976 *Social Networks and Classification.* Manuscript.
Lorrain, F. and H. White
1971 "Structural equivalence of individuals in social networks". *Journal of Mathematical Sociology 1*:49 - 80.
Luce, R.
1950 "Connectivity and generalized cliques in sociometric group structure". *Psychometrika 15*: 169 - 190.
Lundberg, C.
1975 "Patterns of acquaintanceship in society and complex organization: a comparative study of the small world problem". *Pacific Sociological Review 18*:206 - 222.
McKinlay, J.
1973 "Social networks, lay consultation and help-seeking behavior". *Social Forces 51*:275 - 292.
McLaughlin, E.
1975 "The power network in Phoenix. An application of the smallest space analysis". *The Insurgent Sociologist 5*:185 - 195.
Milgram, S.
1967 "The small world problem". *Psychology Today 22*:61 - 67.
Miller, G.
1956 "The magical number seven plus or minus two". *Psychological Review 63*:81 - 97.

Mitchell, J. C. (Ed.)
1969 *Social Networks in Urban Situations – Analysis of Personal Relationships in Central African Towns.* Manchester: University Press.
Newcomb, T.
1961 *The Acquaintance Process.* New York: Holt, Rinehart, and Winston.
Nutini, H. and D. White
1977 "Community variations and network structure in social functions of Compradrazgo in rural Tlaxcala, Mexico". *Ethnology 16*:353 - 384.
Peay, E.
1976 "A note concerning the connectivity of social networks". *Journal of Mathematical Sociology 4*:319 - 321.
Rapoport, A.
1963 "Mathematical models of social interaction". *Handbook of Mathematical Psychology.* New York: Wiley, pp. 493 - 579.
Rapoport, A. and W. Horvath
1961 "A study of a large sociogram". *Behavioral Science 6*:279 - 291.
Rosenthal, H.
1960 *Acquaintances and Contacts of Franklin Roosevelt.* Unpublished B.S. thesis: MIT.
Saunders, J. and N. Reppucci
1977 "Learning networks among administrators of human service institutions". *American Journal of Community Psychology 5*:269 - 276.
Schulman, N.
1976 "Role differentiation in urban networks". *Sociological Focus 9*:149 - 158.
Travers, J. and S. Milgram
1969 "An experimental study of the small world problem". *Sociometry 32*:425 - 443.
Warner, W. L.
1963 *Yankee City.* New Haven: Yale University Press.
Wasserman, S.
1977 "Random directed graph distributions and the triad census in social networks". *Journal of Mathematical Sociology 5*:61 - 86.
White, H.
1970a "Search parameters for the small world problem". *Social Forces 49*:259 - 264.
1970b *Chains of Opportunity.* Cambridge, Mass.: Harvard University Press.
1973 "Everyday life in stochastic networks". *Sociological Inquiry 43*:43 - 49.
Wolfe, A.
1970 "On structural comparisons of networks". *Canadian Review of Sociology and Anthropology 7*:226 - 244.

An Experimental Study of the Small World Problem*

JEFFREY TRAVERS

Harvard University

AND

STANLEY MILGRAM

The City University of New York

Arbitrarily selected individuals (N=296) in Nebraska and Boston are asked to generate acquaintance chains to a target person in Massachusetts, employing "the small world method" (Milgram, 1967). Sixty-four chains reach the target person. Within this group the mean number of intermediaries between starters and targets is 5.2. Boston starting chains reach the target person with fewer intermediaries than those starting in Nebraska; subpopulations in the Nebraska group do not differ among themselves. The funneling of chains through sociometric "stars" is noted, with 48 per cent of the chains passing through three persons before reaching the target. Applications of the method to studies of large scale social structure are discussed.

The simplest way of formulating the small world problem is "what is the probability that any two people, selected arbitrarily from a large population, such as that of the United States, will know each other?" A more interesting formulation, however, takes account of the fact that, while persons a and z may not know each other directly, they may share one or more mutual acquaintances; that is, there may exist a set of individuals, B, (consisting of individuals $b_1, b_2 \ldots b_n$) who know both a and z and thus link them to one another. More generally, a and z may be connected not by any single common acquaintance, but by a series of such intermediaries, a-b-c- . . . -y-z; i.e., a knows b (and no one else in the chain); b knows a and in addition knows c, c in turn knows d, etc.

To elaborate the problem somewhat further, let us represent the popula-

* The study was carried out while both authors were at Harvard University, and was financed by grants from the Milton Fund and from the Harvard Laboratory of Social Relations. Mr. Joseph Gerver provided invaluable assistance in summarizing and criticizing the mathematical work discussed in this paper.

tion of the United States by a partially connected set of points. Let each point represent a person, and let a line connecting two points signify that the two individuals know each other. (Knowing is here assumed to be symmetric: if *a* knows *b* then *b* knows *a*. Substantively, "knowing" is used to denote a mutual relationship; other senses of the verb, e.g. knowing about a famous person, are excluded.) The structure takes the form of a cluster of roughly 200 million points with a complex web of connections among them. The acquaintance chains described above appear as pathways along connected line segments. Unless some portion of the population is totally isolated from the rest, such that no one in that subgroup knows anyone outside it, there must be at least one chain connecting any two people in the population. In general there will be many such pathways, of various lengths, between any two individuals.

In view of such a structure, one way of refining our statement of the small world problem is the following: given two individuals selected randomly from the population, what is the probability that the minimum number of intermediaries required to link them is 0, 1, 2, . . . *k*? (Alternatively, one might ask not about the minimum chains between pairs of people, but mean chain lengths, median chain lengths, etc.)

Perhaps the most direct way of attacking the small world problem is to trace a number of real acquaintance chains in a large population. This is the technique of the study reported in this paper. The phrase "small world" suggests that social networks are in some sense tightly woven, full of unexpected strands linking individuals seemingly far removed from one another in physical or social space. The principal question of the present investigation was whether such interconnectedness could be demonstrated experimentally.

The only example of mathematical treatment dealing directly with the small world problem is the model provided by Ithiel Pool and Manfred Kochen (unpublished manuscript). Pool and Kochen assume a population of N individuals, each of whom knows, on the average, n others in the population. They attempt to calculate P_k, the probability that two persons chosen randomly from the group can be linked by a chain of k intermediaries. Their basic model takes the form of a "tree" or geometric progression. Using an estimate of average acquaintance volume provided by Gurevitch (1961), they deduce that two intermediaries will be required to link typical pairs of individuals in a population of 200 million. Their model does not take account of social structure. Instead of allowing acquaintance nets to define the boundaries of functioning social groups, Pool and Kochen must, for the purposes of their model, conceive of society as being partitioned into a number of hypothetical groups, each with identical populations. They are then able to devise a way to predict chain lengths within and between such hypothesized groups.

In an empirical study related to the small world problem Rapoport and Horvath (1961) examined sociometric nets in a junior high school of 861 students. The authors asked students to name in order their eight best friends within the school. They then traced the acquaintance chains created by the students' choices. Rapoport was interested in connectivity, i.e. the fraction of the total population that would be contacted by tracing friendship choices from an arbitrary starting population of nine individuals. Rapoport and his associates (Rapoport and Horvath, 1961; Foster et al., 1963; Rapoport, 1953; 1963) have developed a mathematical model to describe this tracing procedure. The model takes as a point of departure random nets constructed in the following manner: a small number of points is chosen from a larger population and a fixed number of "axones" is extended from each of these points to a set of target points chosen at random from the population. The same fixed number of axones is then extended from each of the target points to a set of second generation target points, and the process is repeated indefinitely. A target point is said to be of the tth remove if it is of the tth generation and no lower generation. Rapoport then suggests a formula for calculating the fraction, P_t, of the population points which are targets of the tth remove. He is also able to extend the formula to nonrandom nets, such as those created in the Rapoport and Horvath empirical study, by introducing a number of "biases" into the random net model. Rapoport shows that two parameters, obtainable from the data, are sufficient to produce a close fit between the predictions of the model and the empirical outcome of the trace procedure.[1]

Rapoport's model was designed to describe a trace procedure quite different from the one employed in the present study; however, it has some relation to the small world problem. If we set the number of axones traced from a given individual equal to the total number of acquaintances of an average person, the Rapoport model predicts the total fraction of the population potentially traceable at each remove from the start, serving precisely the aims of the model of Pool and Kochen. (It should, however, be noted that Rapoport's model deals with asymmetric nets, and it would be difficult to modify the model to deal with general symmetric nets, which characterize the small world phenomenon.)

Despite the goodness of fit between Rapoport's model and the data from

[1] There is additional empirical evidence (Fararo and Sunshine, 1964) and theoretical support (Abelson, 1967) for the assumption that two parameters are sufficient to describe the Rapoport tracing procedure, i.e. that more complex biases have minimal effects on connectivity in friendship nets.

two large sociograms, there are unsolved problems in the model, as Rapoport himself and others (Fararo and Sunshine, 1964) have pointed out. The Pool-Kochen model involves assumptions difficult for an empirically oriented social scientist to accept, such as the assumption that society may be partitioned into a set of groups alike in size and in internal and external connectedness. In the absence of empirical data, it is difficult to know which simplifying assumptions are likely to be fruitful. On the other hand, with regard to the empirical study of Rapoport and Horvath, the fact that the total population employed was small, well-defined, and homogeneous leaves open many questions about the nature of acquaintance nets in the larger society.[2] An empirical study of American society as a whole may well uncover phenomena of interest both in their own right and as constraints on the nature of any correct mathematical model of the structure of large-scale acquaintanceship nets.

PROCEDURE

This paper follows the procedure for tracing acquaintance chains devised and first tested by Milgram (1967). The present paper introduces an experimental variation in this procedure, by varying "starting populations"; it also constitutes a first technical report on the small world method.

The procedure may be summarized as follows: an arbitrary "target person" and a group of "starting persons" were selected, and an attempt was made to generate an acquaintance chain from each starter to the target. Each starter was provided with a document and asked to begin moving it by mail toward the target The document described the study, named the target, and asked the recipient to become a participant by sending the document on. It was stipulated that the document could be sent only to a first-name acquaintance of the sender. The sender was urged to choose the recipient in such a way as to advance the progress of the document toward the target; several items of information about the target were provided to guide each new sender in his choice of recipient. Thus, each document made its way along an acquaintance chain of indefinite length, a chain which would end only when it reached the target or when someone along the way declined to participate. Certain basic information, such as age, sex and occupation, was collected for each participant.

[2] In addition to the Pool-Kochen and Rapoport work, there are numerous other studies of social network phenomena tangentially related to the small-world problem. Two well-known examples are Bailey's *The Mathematical Theory of Epidemics* and Coleman, Katz and Menzel's *Medical Innovation*. Bailey's work deals with diffusion from a structured source, rather than with convergence on a target from a set of scattered sources, as in the present study. The Coleman, Katz and Menzel study deals with an important substantive correlate of acquaintance nets, namely information diffusion.

We were interested in discovering some of the internal structural features of chains and in making comparisons across chains as well. Among the questions we hoped to answer were the following: How many of the starters —if any—would be able to establish contact with the target through a chain of acquaintances? How many intermediaries would be required to link the ends of the chains? What form would the distribution of chain lengths take? What degree of homogeneity in age, sex, occupation, and other characteristics of participants would be observed within chains? How would complete chains differ from incomplete on these and other dimensions?

An additional comparison was set up by using three distinct starting subpopulations. The target person was a Boston stockbroker; two of the starting populations were geographically removed from him, selected from the state of Nebraska. A third population was selected from the Boston area. One of the Nebraska groups consisted of bluechip stockholders, while the second Nebraska group and the Boston group were "randomly" selected and had no special access to the investment business. By comparisons across these groups we hoped to assess the relative effects of geographical distance and of contact with the target's occupational group. Moreover we hoped to establish a strategy for future experimental extensions of the procedure, in which the sociological characteristics of the starting and target populations would be systematically varied in order to expose features of social structure.

The primary research questions, then, involved a test of the feasibility and fruitfulness of the method as well as an attempt to discover some elementary features of real social nets. Several experimental extensions of the procedure are already underway. A more detailed description of the current method is given in the following sections.

PARTICIPANTS. *Starting Population.* The starting population for the study was comprised of 296 volunteers. Of these, 196 were residents of the state of Nebraska, solicited by mail. Within this group, 100 were systematically chosen owners of blue-chip stocks; these will be designated "Nebraska stockholders" throughout this paper. The rest were chosen from the population at large; these will be termed the "Nebraska random" group. In addition to the two Nebraska groups, 100 volunteers were solicited through an advertisement in a Boston newspaper (the "Boston random" group). Each member of the starting population became the first link in a chain of acquaintances directed at the target person.

Intermediaries. The remaining participants in the study, who numbered 453 in all, were in effect solicited by other participants; they were acquaintances selected by previous participants as people likely to extend the chain toward the target. Participation was voluntary. Participants were not paid, nor was money or other reward offered as incentive for completion of chains.

THE DOCUMENT. The 296 initial volunteers were sent a document which was the principal tool of the investigation.[8] The document contained:

a. a description of the study, a request that the recipient become a participant, and a set of rules for participation;
b. the name of the target person and selected information concerning him;
c. a roster, to which each participant was asked to affix his name;
d. a stack of fifteen business reply cards asking information about each participant.

Rules for Participation. The document contained the following specific instructions to participants:

a. Add your name to the roster so that the next person who receives this folder will know whom it came from.
b. Detach one postcard from the bottom of this folder. Fill it out and return it to Harvard University. No stamp is needed. The postcard is very important. It allows us to keep track of the progress of the folder as it moves toward the target person.
c. If you know the target person on a personal basis, mail this folder directly to him (her). Do this only if you have previously met the target person and know each other on a first name basis.
d. If you do not know the target person on a personal basis, do not try to contact him directly. Instead, mail this folder to a personal acquaintance who is more likely than you to know the target person. You may send the booklet on to a friend, relative, or acquaintance, but it must be someone you know personally.

Target Person. The target person was a stockholder who lives in Sharon, Massachusetts, a suburb of Boston, and who works in Boston proper. In addition to his name, address, occupation and place of employment, participants were told his college and year of graduation, his military service dates, and his wife's maiden name and hometown. One question under investigation was the type of information which people would use in reaching the target.

Roster. The primary function of the roster was to prevent "looping," i.e., to prevent people from sending the document to someone who had already received it and sent it on. An additional function of the roster was to motivate people to continue the chains. It was hoped that a list of prior participants, including a personal acquaintance who had sent the document to

[8] A photographic reproduction of this experimental document appears in Milgram, 1969: 110–11.

the recipient, would create willingness on the part of those who received the document to send it on.

Tracer Cards. Each participant was asked to return to us a business reply card giving certain information about himself and about the person to whom he sent the document. The name, address, age sex and occupation of the sender and sender's spouse were requested, as were the name, address, sex and age of the recipient. In addition, the nature of the relationship between sender and recipient—whether they were friends, relatives, business associates, etc.—was asked. Finally, participants were asked why they had selected the particular recipient of the folder.

The business reply cards enabled us to keep running track of the progress of each chain. Moreover, they assured us of getting information even from chains which were not completed, allowing us to make comparisons between complete and incomplete chains.

RESULTS

COMPLETIONS. 217 of the 296 starting persons actually sent the document on to friends. Any one of the documents could reach the target person only if the following conditions were met: 1) recipients were sufficiently motivated to send the document on to the next link in the chain; 2) participants were able to adopt some strategy for moving the documents closer to the target (this condition further required that the given information allow them to select the next recipient in a manner that increased the probability of contacting the target); 3) relatively short paths were in fact required to link starters and target (otherwise few chains would remain active long enough to reach completion). Given these contingencies, there was serious doubt in the mind of the investigators whether any of the documents, particularly those starting in an area remote from the target person, could move through interlocking acquaintance networks and converge on him. The actual outcome was that 64 of the folders, or 29 per cent of those sent out by starting persons, eventually reached the target.

DISTRIBUTION OF CHAIN LENGTHS. *Complete Chains.* Figure 1 shows the frequency distribution of lengths of the completed chains. "Chain length" is here defined as the number of intermediaries required to link starters and target. The mean of the distribution is 5.2 links.

It was unclear on first inspection whether the apparent drop in frequency at the median length of five links was a statistical accident, or whether the distribution was actually bimodal. Further investigation revealed that the summary relation graphed in Figure 1 concealed two underlying distributions: when the completed chains were divided into those which approached the target through his hometown and those which approached him via

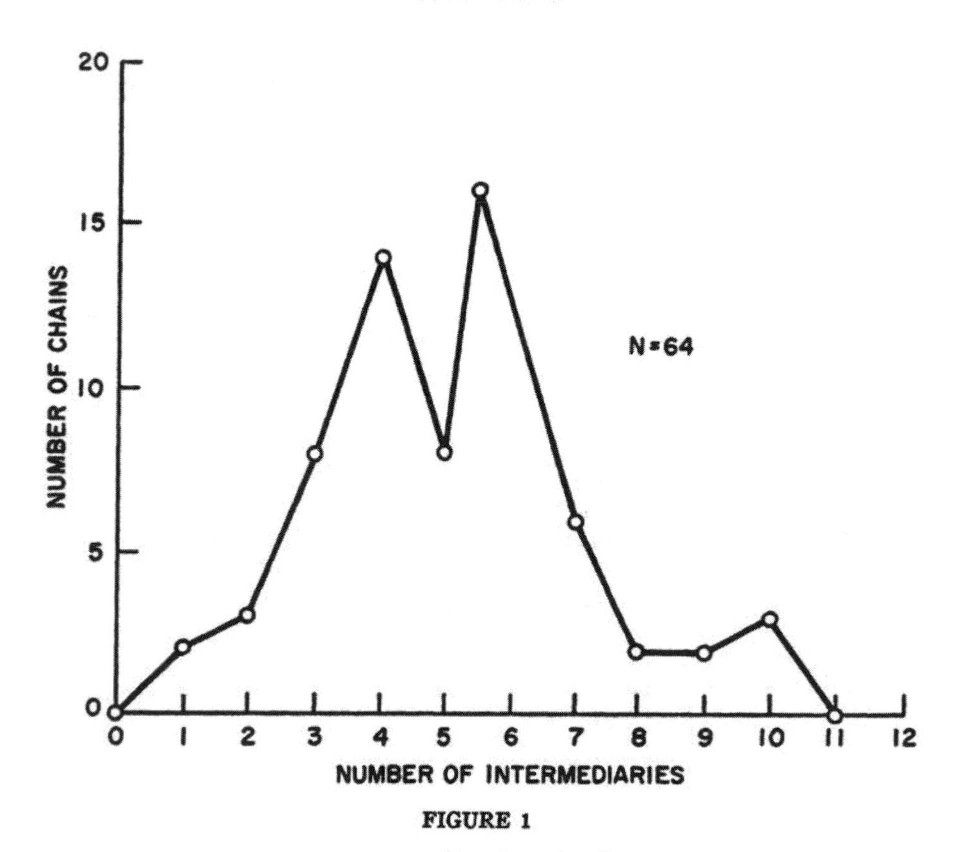

FIGURE 1

Lengths of Completed Chains

Boston business contracts, two distinguishable distributions emerged. The mean of the Sharon distribution is 6.1 links, and that of the Boston distribution is 4.6. The difference is significant at a level better than .0005, as assessed by the distribution-free Mann-Whitney *U* test. (Note that more powerful statistical tests of the significance of differences between means cannot be applied to these data, since those tests assume normality of underlying distributions. The shape of the true or theoretical distribution of lengths of acquaintance chains is precisely what we do not know.)

Qualitatively, what seems to occur is this. Chains which converge on the target principally by using geographic information reach his hometown or the surrounding areas readily, but once there often circulate before entering the target's circle of acquaintances. There is no available information to narrow the field of potential contacts which an individual might have within the town. Such additional information as a list of local organizations of which the target is a member might have provided a natural funnel, facilitating the progress of the document from town to target person. By contrast, those chains which approach the target through occupational channels can take advantage of just such a funnel, zeroing in on him first through the brokerage business, then through his firm.

Incomplete Chains. Chains terminate either through completion or dropout: each dropout results in an incomplete chain. Figure 2 shows the number of chains which dropped out at each "remove" from the starting population. The "0th remove" represents the starting population itself: the "first remove" designates the set of people who received the document directly from members of the starting population. The "second remove" received the document from the starters via one intermediary, the third through two intermediaries, etc. The length of an incomplete chain may be defined as the number of removes from the start at which dropout occurs, or, equivalently, as the number of transmissions of the folder which precede dropout. By this definition, Figure 2 represents a frequency distribution of the lengths of incomplete chains. The mean of the distribution is 2.6 links.

The proportion of chains which drop out at each remove declines as

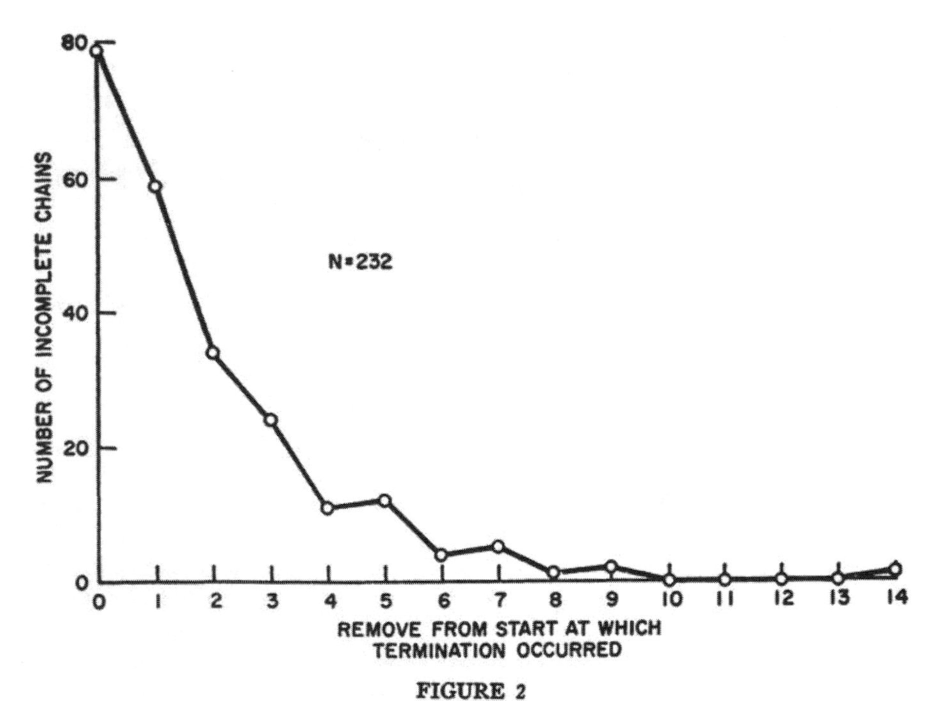

FIGURE 2

Lengths of Incomplete Chains

chains grow in length, if that proportion is based on all chains active at each remove (those destined for completion as well as incompletion). About 27 per cent of the 296 folders sent to the starting population are not sent on. Similarly, 27 per cent of the 217 chains actually initiated by the starters die at the first remove. The percentage of dropouts then appears to fall. It also begins to fluctuate, as the total number of chains in circulation grows small, and an increasing proportion of completions further complicates the picture.

It was argued earlier that, in theory, any two people can be linked by at least one acquaintance chain of finite length, barring the existence of totally isolated cliques within the population under study. Yet, incomplete chains are found in our empirical tracing procedure because a certain proportion of those who receive the document do not send it on. It is likely that this occurs for one of two major reasons: 1) individuals are not motivated to participate in the study; 2) they do not know to whom to send the document in order to advance it toward the target.

For purposes of gauging the significance of our numerical results, it would be useful to know whether the dropouts are random or systematic, i.e., whether or not they are related to a chain's prognosis for rapid completion. It seems possible, for example, that dropouts are precisely those people who are least likely to be able to advance the document toward the target. If so, the distribution of actual lengths of completed chains would understate the true social distance between starters and target by an unknown amount. (Even if dropouts are random, the observed distribution understates the true distribution, but by a potentially calculable amount.) We can offer some evidence, however, that this effect is not powerful.

First, it should be clear that, though people may drop out because they see little possibility that any of their acquaintances can advance the folder toward the target, their subjective estimates are irrelevant to the question just raised. Such subjective estimates may account for individual decisions not to participate; they do not tell us whether chains that die in fact would have been longer than others had they gone to completion. People have poor intuitions concerning the lengths of acquaintance chains. Moreover, people can rarely see beyond their own acquaintances; it is hard to guess the circles in which friends of friends—not to mention people even more remotely connected to oneself—may move.

More direct evidence that dropouts may be treated as "random" can be gleaned from the tracer cards. It will be recalled that each participant was asked for information not only about himself but also about the person to whom he sent the document. Thus some data were available even for dropouts, namely age, sex, the nature of their relationship to the people

TABLE 1

Activity of Chains at Each Remove

All Chains					Incomplete Chains Only			
Remove	Chains Reaching this Remove	Completions at this Remove	Dropouts at this Remove	Per cent Dropouts	Remove	Chains Reaching this Remove	Dropouts at this Remove	Per cent Dropouts
0	296	0	79	27	0	232	79	34
1	217	0	59	27	1	153	59	39
2	158	2	34	22	2	94	34	36
3	122	3	24	20	3	60	24	40
4	95	8	11	12	4	36	11	31
5	76	14	12	16	5	25	12	48
6	50	8	4	8	6	13	4	31
7	38	16	5	13	7	9	5	55
8	17	6	1	6	8	4	1	25
9	10	2	2	20	9	3	2	67
10	6	2	0	0	10	1	0	0
11	4	3	0	0	11	1	0	0
12	1	0	0	0	12	1	0	0
13	1	0	0	0	13	1	0	0
14	1	0	1	100	14	1	1	100

preceding them in the chain, and the reason the dropout had been selected to receive the document. These four variables were tabulated for dropouts versus non-dropouts. None of the resulting contingency tables achieved the .05 level of statistical significance by chi-square test; we are therefore led to accept the null hypothesis of no difference between the two groups, at least on this limited set of variables. Of course, a definitive answer to the question of whether dropouts are really random must wait until the determinants of chain length are understood, or until a way is found to force all chains to completion.[4]

Subpopulation Comparisons. A possible paradigm for future research using the tracing procedure described here involves systematic variation of the relationship between the starting and target populations. One such study, using Negro and White starting and target groups, has already been completed by Korte and Milgram (in press). In the present study, which involved only a single target person, three starting populations were used (Nebraska random, Nebraska stockholders, and Boston random.) The relevant experimental questions were whether the proportion of completed chains or mean chain lengths would vary as a function of starting population.

Chain Length. Letters from the Nebraska subpopulations had to cover a geographic distance of about 1300 miles in order to reach the target, whereas letters originating in the Boston group almost all started within 25 miles of his home and/or place of work. Since social proximity depends in part on geographic proximity, one might readily predict that complete chains originating in the Boston area would be shorter than those originating in Nebraska. This presumption was confirmed by the data. As Table 2 shows, chains originating with the Boston random group showed a mean length of 4.4 intermediaries between starters and target, as opposed to a mean length of 5.7 intermediaries for the Nebraska random group. ($p \leq .001$ by

[4] Professor Harrison White of Harvard University has developed a technique for adjusting raw chain length data to take account of the dropout problem. His method assumes that dropouts are "random," in the following sense. An intermediary who knows the target sends him the folder, completing the chain, with probability 1. Otherwise, an intermediary throws away the folder with fixed probability $1-a$, or sends it on with probability a. If sent on, there is a probability Q_i (which depends on number of removes from the origin) that the next intermediary knows the target. The data is consistent with a value for a of approximately 0.75, independent of remove from the origin, and hence with a "random" dropout rate of 25 per cent. The limited data further suggest that Q_i grows in a "staircase" pattern from zero (at zero removes from the starting population) to approximately one-third at six removes, remaining constant thereafter. Based on these values, the hypothetical curve of completions with no dropouts resembles the observed curve shifted upward; the median length of completed chains rises from 5 to 7, but no substantial alteration is required in conclusions drawn from the raw data.

TABLE 2

Lengths of Completed Chains

	Frequency Distribution												
	Number of Intermediaries												
Population	0	1	2	3	4	5	6	7	8	9	10	11	Total
Nebraska Random	0	0	0	1	4	3	6	2	0	1	1	0	18
Nebraska Stock	0	0	0	3	6	4	6	2	1	1	1	0	24
Boston Random	0	2	3	4	4	1	4	2	1	0	1	0	22
All	0	2	3	8	14	8	16	6	2	2	3	0	64

Means	
Starting Population	Mean Chain Length
Nebraska Random	5.7
Nebraska Stockholders	5.4
All Nebraska	5.5
Boston Random	4.4
All	5.2

a one-tailed Mann-Whitney U test.) Chain length thus proved sensitive to one demographic variable—place of residence of starters and target.

The Nebraska stockholder group was presumed to have easy access to contacts in the brokerage business. Because the target person was a stockbroker, chains originating in this group were expected to reach the target more efficiently than chains from the Nebraska random group. The chain-length means for the two groups, 5.7 intermediaries for the random sample and 5.4 for the stockholders, differed in the expected direction, but the difference was not statistically significant by the Mann-Whitney test. The stockholders used the brokerage business as a communication channel more often than did the random group; 60.7 per cent of all the participants in chains originating with the stockholder group reported occupations connected with finance, while 31.8 per cent of participants in chains originating in the Nebraska random group were so classified.

Proportion of Completions. As indicated in Table 3, the proportions of chains completed for the Nebraska random, Nebraska stockholder, and Boston subpopulations were 24 per cent, 31 per cent and 35 per cent, respectively. Although the differences are not statistically significant, there is a weak tendency for higher completion rates to occur in groups where mean length of completed chains is shorter. This result deserves brief discussion.

Let us assume that the dropout rate is constant at each remove from the start. If, for example, the dropout rate were 25 per cent then any chain would have a 75 per cent probability of reaching one link, $(.75)^2$ of reaching two links, etc. Thus, the longer a chain needed to be in order to reach completion, the less likely that the chain would survive long enough to run its full course. In this case, however, chain-length differences among the three groups were not sufficiently large to produce significant differences in completion rate. Moreover, if the dropout rate declines as chains grow long, such a decrease would off-set the effect just discussed and weaken the observed inverse relation between chain length and proportion of completions.

TABLE 3

Proportion of Completions for Three Starting Populations

	Starting Population							
	Nebraska Random		Nebraska Stock.		Boston		Total	
Complete	18	(24%)	24	(31%)	22	(35%)	64	(29%)
Incomplete	58	(76%)	54	(69%)	41	(65%)	153	(71%)
	76	(100%)	78	(100%)	63	(100%)	217	(100%)

χ^2=2.17, df.=2, p>.3, N.S.

COMMON CHANNELS. As chains converge on the target, common channels appear—that is, some intermediaries appear in more than one chain. Figure 3 shows the pattern of convergence. The 64 letters which reached the target were sent by a total of 26 people. Sixteen, fully 25 per cent, reached the target through a single neighbor. Another 10 made contact through a single business associate, and 5 through a second business associate. These three "penultimate links" together accounted for 48 per cent of the total completions. Among the three, an interesting division of labor appears. Mr. G,

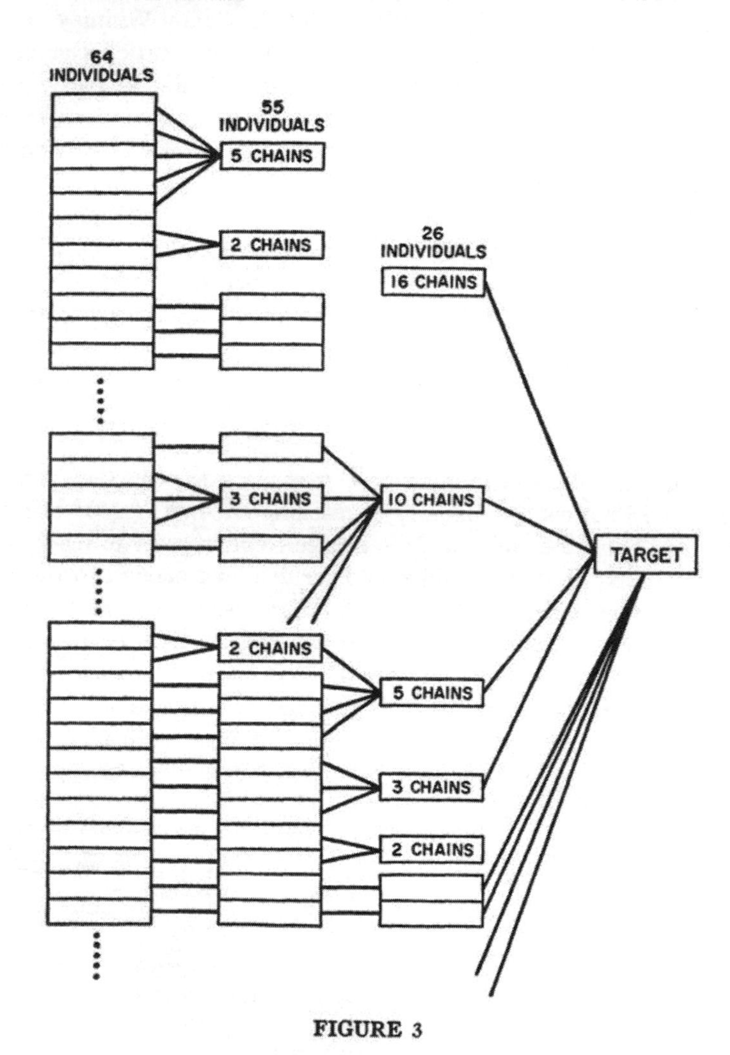

FIGURE 3

Common Paths Appear as Chains Converge on the Target

who accounted for 16 completions, is a clothing merchant in the target's hometown of Sharon; Mr. G funnelled toward the target those chains which were advancing on the basis of the target's place of residence. Twenty-four chains reached the target from his hometown; Mr. G accounted for ⅔ of those completions. All the letters which reached Mr. G came from residents of Sharon. By contrast, Mr. D and Mr. P, who accounted for 10 and 5 completions, respectively, were contacted by people scattered around the Boston area, and in several cases, by people living in other cities entirely. On the other hand, whereas Mr. G received the folder from Sharon residents in a wide variety of occupations, D and P received it almost always from stockbrokers. A scattering of names appear two or three times on the list of penultimate links; seventeen names appear once each.

Convergence appeared even before the penultimate link. Going one step further back, to people two removes from the target, we find that the 64 chains passed through 55 individuals. One man, Mr. B, appeared 5 times, and on all occasions sent the document to Mr. G. Other individuals appeared two or three times each.

Additional Characteristics of Chains. Eighty-six per cent of the participants sent the folder to persons they described as friends and acquaintances; 14 per cent sent it to relatives. The same percentages had been observed in an earlier pilot study.

Data on patterns of age, sex and occupation support the plausible hypothesis that participants select recipients from a pool of individuals similar to themselves. The data on age support the hypothesis unequivocally; the data on sex and occupation are complicated by the characteristics of the target and the special requirement of establishing contact with him.

Age was bracketed into ten-year categories and the ages of those who sent the document tabled against the ages of those to whom they sent it. On inspection the table showed a strong tendency to cluster around the diagonal, and a chi-square test showed the association to be significant at better than the .001 level.

Similarly, the sex of each sender was tabled against the sex of the corresponding recipient. Men were ten times more likely to send the document to other men than to women, while women were equally likely to send the folder to males as to females ($p < .001$). These results were affected by the fact that the target was male. In an earlier pilot study using a female target, both men and women were three times as likely to send the document to members of the same sex as to members of the opposite sex. Thus there appear to be three tendencies governing the sex of the recipient: (1) there is a tendency to send the document to someone of one's own sex, but (2) women are more likely to cross sex lines than men, and (3) there is a tendency to send the document to someone of the same sex as the target person.

The occupations reported by participants were rated on two components—one of social status and one of "industry" affiliation, that is, the subsector of the economy with which the individual would be likely to deal. The coding system was *ad hoc,* designed to fit the occupational titles supplied by participants. Tabling the status and "industry" ratings for all senders of the document against those of respective recipients, we observed a strong tendency for people to select recipients similar to themselves on both measures ($p<.001$ for both tables). However, the strength of the relationship for industry seemed to be largely due to a tendency for the folder to stay within the finance field once it arrived there, obviously because the target was affiliated with that field. Moreover, the participants in the study were a heavily middle-class sample, and the target was himself a member of that class. Thus there was no need for the document to leave middle-class circles in progressing from starters to target.

When separate contingency tables were constructed for complete and incomplete chains, the above results were obtained for both tables. Similarly, when separate tables were constructed for chains originating in the 3 starting populations, the findings held up in all 3 tables. Thus, controlling for completion of chains or for starting population did not affect the finding of demographic homogeneity within chains.

CONCLUSIONS

The contribution of the study lies in the use of acquaintance chains to extend an individual's contacts to a geographically and socially remote target, and in the sheer size of the population from which members of the chains were drawn. The study demonstrated the feasibility of the "small world" technique, and took a step toward demonstrating, defining and measuring inter-connectedness in a large society.

The theoretical machinery needed to deal with social networks is still in its infancy. The empirical technique of this research has two major contributions to make to the development of that theory. First, it sets an upper bound on the minimum number of intermediaries required to link widely separated Americans. Since subjects cannot always foresee the most efficient path to a target, our trace procedure must inevitably produce chains longer than those generated by an accurate theoretical model which takes full account of all paths emanating from an individual. The mean number of intermediaries observed in this study was somewhat greater than five;

additional research (by Korte and Milgram) indicates that this value is quite stable, even when racial crossover is introduced. Both the magnitude and stability of the parameter need to be accounted for. Second, the study has uncovered several phenomena which future models should explain. In particular, the convergence of communication chains through common individuals is an important feature of small world nets, and it should be accounted for theoretically.

There are many additional lines of empirical research that may be examined with the small world method. As suggested earlier, one general paradigm for research is to vary the characteristics of the starting person and the target. Further, one might systematically vary the information provided about the target in order to determine, on the psychological side, what strategies people employ in reaching a distant target, and on the sociological side, what specific variables are critical for establishing contact between people of given characteristics.

REFERENCES

Abelson, R. P.
1967 "Mathematical models in social psychology." Pp. 1–54 in L. Berkowitz (ed.) Advances in Experimental Social Psychology, Vol. III. New York: Academic Press.

Bailey, N. T. J.
1957 The Mathematical Theory of Epidemics. New York: Hafner.

Coleman, J. S., E. Katz and H. Menzel
1966 Medical Innovation: A Diffusion Study. Indianapolis: Bobbs-Merrill.

Fararo, T. J. and M. H. Sunshine
1964 A Study of a Biased Friendship Net. Syracuse: Youth Development Center, Syracuse University.

Foster, C. C., A. Rapoport and C. J. Orwant
1963 "A study of a large sociogram II. Elimination of free parameters." Behavioral Science 8(January):56–65.

Gurevitch, M.
1961 The Social Structure of Acquaintanceship Networks. Unpublished doctoral dissertation, Cambridge: M.I.T.

Korte, C. and S. Milgram
Acquaintance Links Between White and Negro Populations: Application of the Small World Method. Journal of Personality and Social Psychology (in press).

Milgram, S.
1967 "The small world problem." Psychology Today 1(May):61–67.
1969 "Interdisciplinary thinking and the small world problem." Pp. 103–120 in Muzafer Sherif and Carolyn W. Sherif (eds.) Interdisciplinary Relationships in the Social Sciences. Chicago: Aldine Publishing Company.

Pool, I. and M. Kochen
A Non-Mathematical Introduction to a Mathematical Model. Undated mimeo. Cambridge: M.I.T.

Rapoport, A.
1953 "Spread of information through a population with socio-structural bias." Bulletin of Mathematical Biophysics 15(December):523–543.
1963 "Mathematical models of social interaction." Pp. 493–579 in R. D. Luce, R. R. Bush and E. Galanter (eds.) Handbook of Mathematical Psychology, Vol. II. New York: John Wiley and Sons.

Rapoport, A. and W. J. Horvath
1961 "A study of a large sociogram." Behavioral Science 6(October):279–291.

Networks of Scientific Papers

The pattern of bibliographic references indicates the nature of the scientific research front.

Derek J. de Solla Price

This article is an attempt to describe in the broadest outline the nature of the total world network of scientific papers. We shall try to picture the network which is obtained by linking each published paper to the other papers directly associated with it. To do this, let us consider that special relationship which is given by the citation of one paper by another in its footnotes or bibliography. I should make it clear, however, that this broad picture tells us something about the papers themselves as well as something about the practice of citation. It seems likely that many of the conclusions we shall reach about the network of papers would still be essentially true even if citation became much more or much less frequent, and even if we considered links obtained by subject indexing rather than by citation. It happens, however, that we now have available machine-handled citation studies, of large and representative portions of literature, which are much more tractable for such analysis than any topical indexing known to me. It is from such studies, by Garfield (*1, 2*), Kessler (*3*), Tukey (*4*), Osgood (*5*), and others, that I have taken the source data of this study.

Incidence of References

First, let me say something of the incidence of references in papers in serial publications. On the average, there are about 15 references per paper and, of these, about 12 are to other serial publications rather than to books, theses, reports, and unpublished work. The average, of course, gives us only part of the picture. The distribution (see Fig. 1) is such that about 10 percent of the papers contain no references at all; this notwithstanding, 50 percent of the references come from the 85 percent of the papers that are of the "normal" research type and contain 25 or fewer references apiece. The distribution here is fairly flat; indeed about 5 percent of the papers fall in each of the categories of 3, 4, 5, 6, 7, 8, 9, and 10 references each. At the other end of the scale, there are review-type papers with many references each. About 25 percent of all references come from the 5 percent (of all papers) that contain 45 or more references each and average 75 to a paper, while 12 percent of the references come from the "fattest" category—the 1 percent (of all papers) that have 84 or more references each and average about 170 to a paper. It is interesting to note that the number of papers with n references falls off in this "fattest" category as $1/n^2$, up to many hundreds per paper.

These references, of course, cover the entire previous body of literature. We can calculate roughly that, since the body of world literature has been growing exponentially for a few centuries (*6*), and probably will continue at its present rate of growth of about 7 percent per annum, there will be about 7 new papers each year for every 100 previously published papers in a given

The author is Avalon Professor of the History of Science, Yale University, New Haven, Connecticut. This article is based on a paper presented 17 March 1964 at the National Bureau of Standards, Washington, D.C., in a Symposium on Statistical Methods for Mechanized Documentation. Part of this research was supported by grant GN-299 from the National Science Foundation.

field. An average of about 15 references in each of these 7 new papers will therefore supply about 105 references back to the previous 100 papers, which will therefore be cited an average of a little more than once each during the year. Over the long run, and over the entire world literature, we should find that, on the average, *every scientific paper ever published is cited about once a year.*

Incidence of Citations

Now, although the total number of citations must exactly balance the total number of references, the distributions are very different. It seems that, in any given year, about 35 percent of all the existing papers are not cited at all, and another 49 percent are cited only once ($n = 1$) (see Fig. 2). This leaves about 16 percent of the papers to be cited an average of about 3.2 times each. About 9 percent are cited twice; 3 percent, three times; 2 percent, four times; 1 percent, five times; and a remaining 1 percent, six times or more. For large n, the number of papers cited appears to decrease as $n^{2.5}$ or $n^{3.0}$. This is rather more rapid than the decrease found for numbers of references in papers, and indeed the number of papers receiving many citations is smaller than the number carrying large bibliographies. Thus, only 1 percent of the cited papers are cited as many as six or more times each in a year (the average for this top 1 percent is 12 citations), and the maximum likely number of citations to a paper in a year is smaller by about an order of magnitude than the maximum likely number of references in the citing papers. There is, however, some parallelism in the findings that some 5 percent of all papers appear to be review papers, with many (25 or more) references, and some 4 percent of all papers appear to be "classics," cited four or more times in a year.

What has been said of references is true from year to year; the findings for individual cited papers, however, appear to vary from year to year. A paper not cited in one year may well be cited in the next, and one cited often in one year may or may not be heavily cited subsequently. Heavy citation appears to occur in rather capricious bursts, but in spite of that I suspect a strong statistical regularity. I would conjecture that results to date could be explained by the hypotheses that

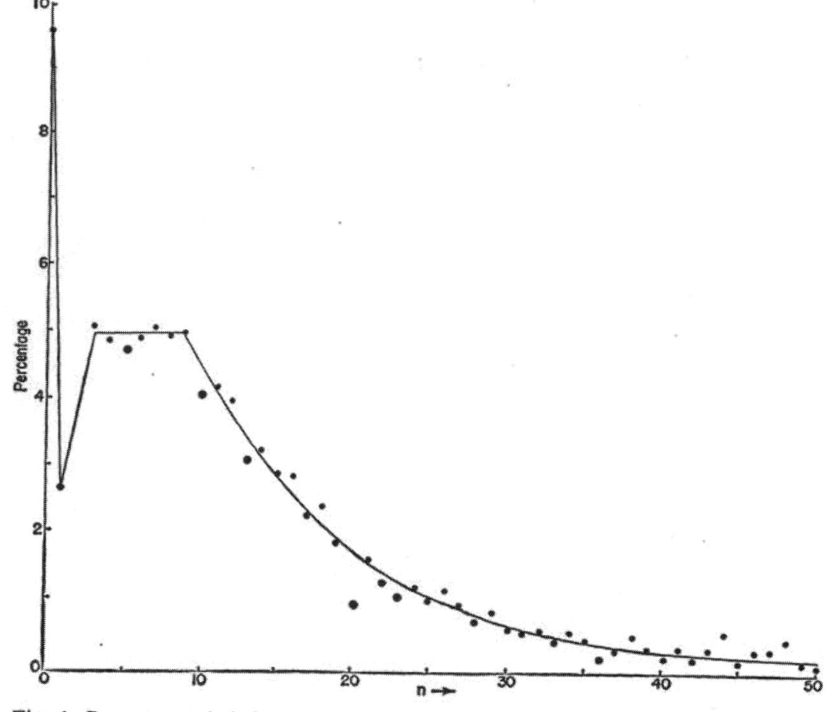

Fig. 1. Percentages (relative to total number of papers published in 1961) of papers published in 1961 which contain various numbers (n) of bibliographic references. The data, which represent a large sample, are from Garfield's 1961 *Index* (2).

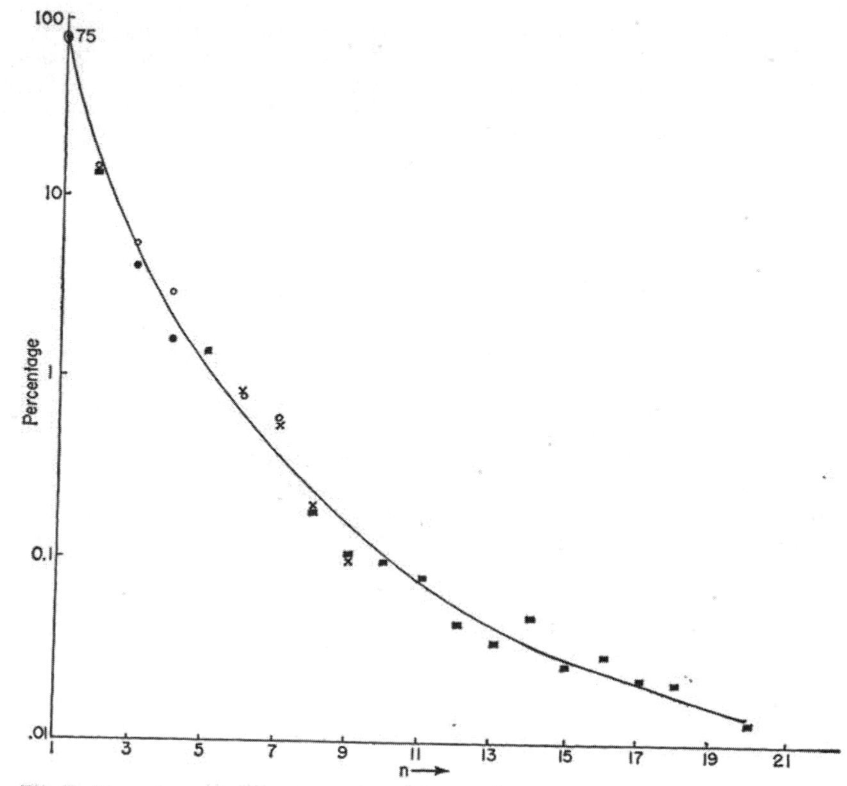

Fig. 2. Percentages (relative to total number of cited papers) of papers cited various numbers (n) of times, for a single year (1961). The data are from Garfield's 1961 *Index* (2), and the points represent four different samples conflated to show the consistency of the data. Because of the rapid decline in frequency of citation with increase in n, the percentages are plotted on a logarithmic scale.

every year about 10 percent of all papers "die," not to be cited again, and that for the "live" papers the chance of being cited at least once in any year is about 60 percent. This would mean that the major work of a paper would be finished after 10 years. The process thus reaches a steady state, in which about 10 percent of all published papers have never been cited, about 10 percent have been cited once, about 9 percent twice, and so on, the percentages slowly decreasing, so that half of all papers will be cited eventually five times or more, and a quarter of all papers, ten times or more. More work is urgently needed on the problem of determining whether there is a probability that the more a paper is cited the more likely it is to be cited thereafter. It seems to me that further work in this area might well lead to the discovery that classic papers could be rapidly identified, and that perhaps even the "superclassics" would prove so distinctive that they could be picked automatically by means of citation-index-production procedures and published as a single *U.S.* (or *World*) *Journal of Really Important Papers.*

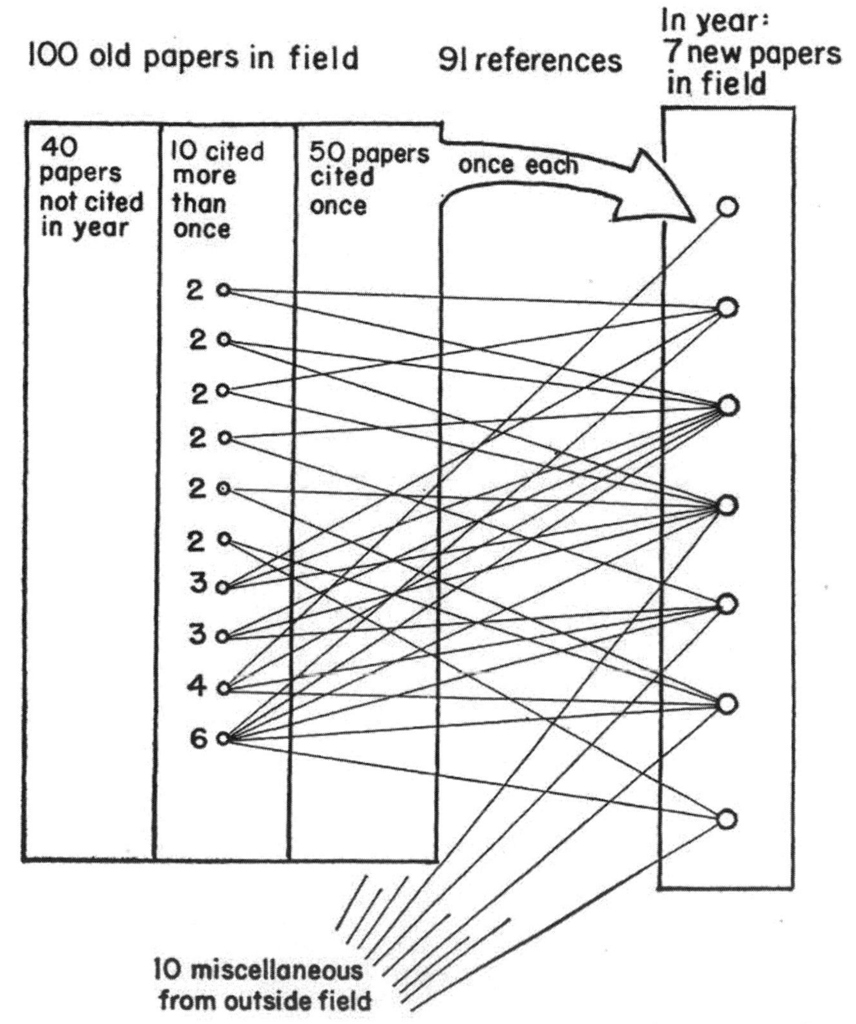

Fig. 3. Idealized representation of the balance of papers and citations for a given "almost closed" field in a single year. It is assumed that the field consists of 100 papers whose numbers have been growing exponentially at the normal rate. If we assume that each of the seven new papers contains about 13 references to journal papers and that about 11 percent of these 91 cited papers (or ten papers) are outside the field, we find that 50 of the old papers are connected by one citation each to the new papers (these links are not shown) and that 40 of the old papers are not cited at all during the year. The seven new papers, then, are linked to ten of the old ones by the complex network shown here.

Unfortunately, we know little about any relationship between the number of times a paper is cited and the number of bibliographic references it contains. Since rough preliminary tests indicate that, for much-cited papers, there is a fairly standard pattern of distribution of numbers of bibliographic references, I conjecture that the correlation, if one exists, is very small. Certainly, there is no strong tendency for review papers to be cited unusually often. If my conjecture is valid, it is worth noting that, since 10 percent of all papers contain no bibliographic references and another, presumably almost independent, 10 percent of all papers are never cited, it follows that there is a lower bound of 1 percent of all papers on the number of papers that are totally disconnected in a pure citation network and could be found only by topical indexing or similar methods; this is a very small class, and probably a most unimportant one.

The balance of references and citations in a single year indicates one very important attribute of the network (see Fig. 3). Although most papers produced in the year contain a near-average number of bibliographic references, half of these are references to about half of all the papers that have been published in previous years. The other half of the references tie these new papers to a quite small group of earlier ones, and generate a rather tight pattern of multiple relationships. Thus each group of new papers is "knitted" to a small, select part of the existing scientific literature but connected rather weakly and randomly to a much greater part. Since only a small part of the earlier literature is knitted together by the new year's crop of papers, we may look upon this small part as a sort of growing tip or epidermal layer, an active research front. I believe it is the existence of a research front, in this sense, that distinguishes the sciences from the rest of scholarship, and, because of it, I propose that one of the major tasks of statistical analysis is to determine the mechanism that enables science to cumulate so much faster than nonscience that it produces a literature crisis.

An analysis of the distribution of publication dates of all papers cited in a single year (Fig. 4) sheds further light on the existence of such a research front. Taking [from Garfield (*2*)] data for 1961, the most numerous count available, I find that papers published in 1961 cite earlier papers at a rate that falls off by a factor of 2 for every 13.5-year interval measured backward from 1961; this rate of decrease must be approximately equal to the exponential growth of numbers of papers published in that interval. Thus, the chance of being cited by a 1961 paper was almost the same for all papers published more than about 15 years before 1961, the rate of citation presumably being the previously computed average rate of one citation per paper per year. It should be noted that, as time goes on, there are more and more papers available to cite each one previously published. Therefore, the chance that any one paper will be cited by any other, later paper decreases exponentially by about a factor of 2 every 13.5 years.

For papers less than 15 years old, the rate of citation is considerably greater than this standard value of one citation per paper per year. The rate increases steadily, from less than twice this value for papers 15 years old to 4 times for those 5 years old; it reaches a maximum of about 6 times the standard value for papers 2½ years old, and of course declines again for papers so recent that they have not had time to be noticed.

Incidentally, this curve enables one to see and dissect out the effect of the wartime declines in production of papers. It provides an excellent indication, in agreement with manpower indexes and other literature indexes, that production of papers began to drop from expected levels at the beginning of World Wars I and II, declining to a trough of about half the normal production in 1918 and mid-1944, respectively, and then recovering in a manner strikingly symmetrical with the decline, attaining the normal rate again by 1926 and 1950, respectively. Because of this decline, we must not take dates in the intervals 1914–25 and 1939–50 for comparison with normal years in determining growth indexes.

The "Immediacy Factor"

The "immediacy factor"—the "bunching," or more frequent citation, of recent papers relative to earlier ones—is, of course, responsible for the well-known phenomenon of papers being considered obsolescent after a decade. A numerical measure of this factor can be derived and is particularly useful. Calculation shows that about 70 percent of all cited papers would account for the normal growth curve, which shows a doubling every 13.5 years, and that about 30 percent would account for the hump of the immediacy curve. Hence, we may say that the 70 percent represents a random distribution of citations of all the scientific papers that have ever been published, regardless of date, and that the 30 percent are highly selective references to recent literature; the distribution of citations of the recent papers is defined by the shape of the curve, half of the 30 percent being papers between 1 and 6 years old.

I am surprised at the extent of this immediacy phenomenon and want to indicate its significance. If all papers followed a standard pattern with respect to the proportions of early and recent papers they cite, then it would follow that 30 percent of all references in all papers would be to the recent research front. If, instead, the papers

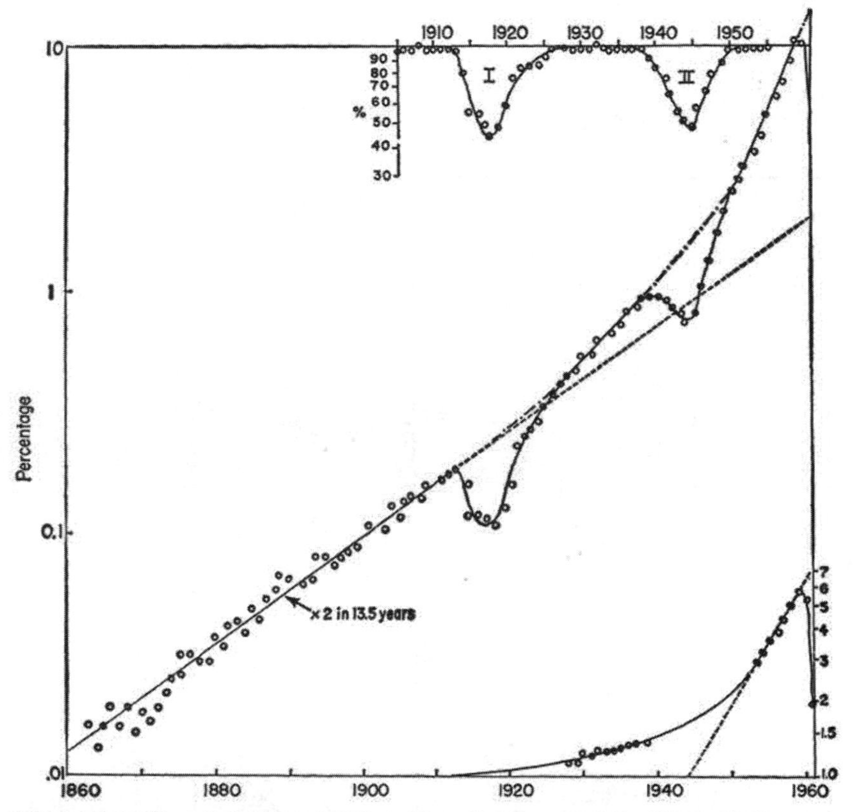

Fig. 4. Percentages (relative to total number of papers cited in 1961) of all papers cited in 1961 and published in each of the years 1862 through 1961 [data are from Garfield's 1961 *Index* (2)]. The curve for the data (solid line) shows dips during world wars I and II. These dips are analyzed separately at the top of the figure and show remarkably similar reductions to about 50 percent of normal citation in the two cases. For papers published before World War I, the curve is a straight line on this logarithmic plot, corresponding to a doubling of numbers of citations for every 13.5-year interval. If we assume that this represents the rate of growth of the entire literature over the century covered, it follows that the more recent papers have been cited disproportionately often relative to their number. The deviation of the curve from a straight line is shown at the bottom of the figure and gives some measure of the "immediacy effect." If, for old papers, we assume a unit rate of citation, then we find that the recent papers are cited at first about six times as much, this factor of 6 declining to 3 in about 7 years, and to 2 after about 10 years. Since it is probable that some of the rise of the original curve above the straight line may be due to an increase in the pace of growth of the literature since World War I, it may be that the curve of the actual "immediacy effect" would be somewhat smaller and sharper than the curve shown here. It is probable, however, that the straight dashed line on the main plot gives approximately the slope of the initial falloff, which must therefore be a halving in the number of citations for every 6 years one goes backward from the date of the citing paper.

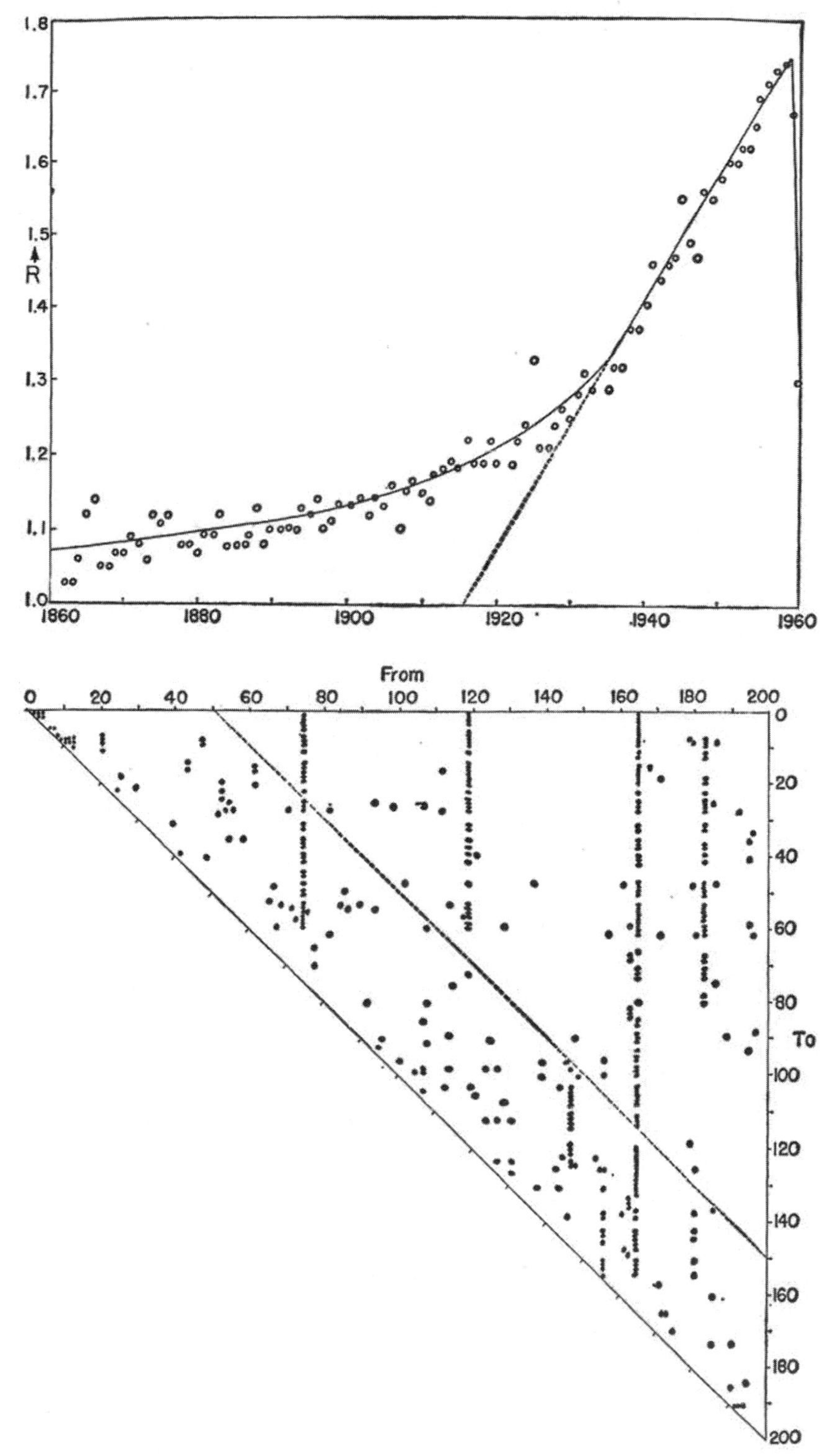

Fig. 5 (top left). Ratios of numbers of 1961 citations to numbers of individual cited papers published in each of the years 1860 through 1960 [data are from Garfield's 1961 *Index* (2)]. This ratio gives a measure of the multiplicity of citation and shows that there is a sharp falloff in this multiplicity with time. One would expect the measure of multiplicity to be also a measure of the proportion of available papers actually cited. Thus, recent papers cited must constitute a much larger fraction of the total available population than old papers cited.

Fig. 6. Matrix showing the bibliographical references to each other in 200 papers that constitute the entire field from beginning to end of a peculiarly isolated subject group. The subject investigated was the spurious phenomenon of N-rays, about 1904. The papers are arranged chronologically, and each column of dots represents the references given in the paper of the indicated number rank in the series, these references being necessarily to previous papers in the series. The strong vertical lines therefore correspond to review papers. The dashed line indicates the boundary of a "research front" extending backward in the series about 50 papers behind the citing paper. With the exception of this research front and the review papers, little background noise is indicated in the figure. The tight linkage indicated by the high density of dots for the first dozen papers is typical of the beginning of a new field.

cited by, say, half of all papers were evenly distributed through the literature with respect to publication date, then it must follow that 60 percent of the papers cited by the other half would be recent papers. I suggest, as a rough guess, that the truth lies somewhere between—that we have here an indication that about half the bibliographic references in papers represent tight links with rather recent papers, the other half representing a uniform and less tight linkage to all that has been published before.

That this is so is demonstrated by the time distribution: much-cited papers are much more recent than less-cited ones. Thus, only 7 percent of the papers listed in Garfield's 1961 *Index* (2) as having been cited four or more times in 1961 were published before 1953, as compared with 21 percent of all papers cited in 1961. This tendency for the most-cited papers to be also the most recent may also be seen in Fig. 5 (based on Garfield's data), where the number of citations per paper is shown as a function of the age of the cited paper.

It has come to my attention that R. E. Burton and R. W. Kebler (7) have already conjectured, though on somewhat tenuous evidence, that the periodical literature may be composed of two distinct types of literature with very different half-lives, the classic and the ephemeral parts. This conjecture is now confirmed by the present evidence. It is obviously desirable to explore further the other tentative finding of Burton and Kebler that the half-lives, and therefore the relative proportions of classic and ephemeral literature, vary considerably from field to field: mathematics, geology, and botany being strongly classic; chemical, mechanical, and metallurgical engineering and physics strongly ephemeral; and chemistry and physiology a much more even mixture.

Historical Examples

A striking confirmation of the proposed existence of this research front has been obtained from a series of historical examples, for which we have been able to set up a matrix (Fig. 6). The dots represent references within a set of chronologically arranged papers which constitute the entire literature in a particular field (the field happens to be very tight and closed over the interval under discussion). In such a matrix there is high probability of citation in a strip near the diagonal and extending over the 30 or 40 papers immediately preceding each paper in turn. Over the rest of the triangular matrix there is much less chance of citation; this remaining part provides, therefore, a sort of background noise. Thus, in the special circumstance of being able to isolate a "tight" subject field, we find that half the references are to a research front of recent papers and that the other half are to papers scattered uniformly through the literature. It also appears that after every 30 or 40 papers there is need of a review paper to replace those earlier papers that have been lost from sight behind the research front. Curiously enough, it appears that classical papers, distinguished by full rows rather than columns, are all cited with about the same frequency, making a rather symmetrical pattern that may have some theoretical significance.

Two Bibliographic Needs

From these two different types of connections it appears that the citation network shows the existence of two different literature practices and of two different needs on the part of the scientist. (i) The research front builds on recent work, and the network becomes very tight. To cope with this, the scientist (particularly, I presume, in physics and molecular biology) needs an alerting service that will keep him posted, probably by citation indexing, on the work of his peers and colleagues. (ii) The random scattering of Fig. 6 corresponds to a drawing upon the totality of previous work. In a sense, this is the portion of the network that treats each published item as if it were truly part of the eternal record of human knowledge. In subject fields that have been dominated by this second attitude, the traditional procedure has been to systematize the added knowledge from time to time in book form, topic by topic, or to make use of a system of classification optimistically considered more or less eternal, as in taxonomy and chemistry. If such classification holds over reasonably long periods, one may have an objective means of reducing the world total of knowledge to fairly small parcels in which the items are found to be in one-to-one correspondence with some natural order.

It seems clear that in any classification into research-front subjects and taxonomic subjects there will remain a large body of literature which is not completely the one or the other. The present discussion suggests that most papers, through citations, are knit together rather tightly. The total research front of science has never, however, been a single row of knitting. It is, instead, divided by dropped stitches into quite small segments and strips. From a study of the citations of journals by journals I come to the conclusion that most of these strips correspond to the work of, at most, a few hundred men at any one time. Such strips represent objectively defined subjects whose description may vary materially from year to year but which remain otherwise an intellectual whole. If one would work out the nature of such strips, it might lead to a method for delineating the topography of current scientific literature. With such a topography established, one could perhaps indicate the overlap and relative importance of journals and, indeed, of countries, authors, or individual papers by the place they occupied within the map, and by their degree of strategic centralness within a given strip.

Journal citations provide the most readily available data for a test of such methods. From a preliminary and very rough analysis of these data I am tempted to conclude that a very large fraction of the alleged 35,000 journals now current must be reckoned as merely a distant background noise, and as very far from central or strategic in any of the knitted strips from which the cloth of science is woven.

References and Notes

1. E. Garfield and I. H. Sher, "New factors in the evaluation of scientific literature through citation indexing," *Am. Doc.* **14**, 191 (1963); ———, *Genetics Citation Index* (Institute for Scientific Information, Philadelphia, 1963). For many of the results discussed in this article I have used statistical information drawn from E. Garfield and I. H. Sher, *Science Citation Index* (Institute for Scientific Information, Philadelphia, 1963), pp. ix, xvii–xviii.
2. I wish to thank Dr. Eugene Garfield for making available to me several machine printouts of original data used in the preparation of the 1961 *Index* but not published in their entirety in the preamble to the index.
3. I am grateful to Dr. M. M. Kessler, Massachusetts Institute of Technology, for data for seven research reports of the following titles and dates: "An Experimental Study of Bibliographic Coupling between Technical Papers" (November 1961); "Bibliographic Coupling Between Scientific Papers" (July 1962); "Analysis of Bibliographic Sources in the *Physical Review* (vol. 77, 1950, to vol. 112, 1958) (July 1962); "Analysis of Bibliographic Sources in a Group of Physics-Related Journals" (August 1962); "Bibliographic Coupling Extended in Time: Ten Case Histories" (August 1962); "Concerning the Probability that a Given Paper will be Cited" (November 1962); "Comparison of the Results of Bibliographic Coupling and Analytic Subject Indexing" (January 1963).
4. J. W. Tukey, "Keeping research in contact with the literature: Citation indices and beyond," *J. Chem. Doc.* **2**, 34 (1962).
5. C. E. Osgood and L. V. Xhignesse, *Characteristics of Bibliographical Coverage in Psychological Journals Published in 1950 and 1960* (Institute of Communications Research, Univ. of Illinois, Urbana, 1963).
6. D. J. de Solla Price, *Little Science, Big Science* (Columbia Univ. Press, New York, 1963).
7. R. E. Burton and R. W. Kebler, "The 'half-life' of some scientific and technical literatures," *Am. Doc.* **11**, 18 (1960).

RODRIGO DE CASTRO AND JERROLD W. GROSSMAN

Famous Trails to Paul Erdős

The notion of Erdős number has floated around the mathematical research community for more than thirty years, as a way to quantify the common knowledge that mathematical and scientific research has become a very collaborative process in the twentieth century, not an activity engaged in solely by isolated individuals. In this paper we explore some (fairly short) collaboration paths that one can follow from Paul Erdős to researchers inside and outside of mathematics.

An Outstanding Component of the Collaboration Graph

The *collaboration graph* C has as vertices all researchers (dead or alive) from all academic disciplines, with an edge joining vertices u and v if u and v have jointly published a research paper or book (with possibly more co-authors). As is the case for any simple (undirected) graph, in C we have a notion of *distance* between two vertices u and v: $d(u,v)$ is the number of edges in the shortest path between u and v, if such a path exists, ∞ otherwise (it is understood that $d(u,u) = 0$).

In this paper we are concerned with the collaboration subgraph centered at Paul Erdős (1913–1996). For a researcher v, the number d(Paul Erdős, v) is called the *Erdős number* of v. That is, Paul Erdős himself has Erdős number 0, and his co-authors have Erdős number 1. People not having Erdős number 0 or 1 who have published with someone with Erdős number 1 have Erdős number 2, and so on. Those who are not linked in this way to Paul Erdős have Erdős number ∞. The collection of all individuals with a finite Erdős number constitutes the *Erdős component* of C.

The Erdős component of C is outstanding for its amazing size and for the manner in which it clusters around Erdős. Almost 500 people have Erdős number 1, and over 5000 have Erdős number 2. In the history of scholarly publishing in mathematics, no one has ever matched Paul Erdős's number of collaborators or number of papers (about 1500, almost 70% of which were joint works). With his recent death the man who inspired so much mathematical thinking has—to use his terminology—left, but his legend lives on (see for example two recent biographies [20], [28]). And part of this legend, inside and outside mathematical circles, is the notion of Erdős numbers.

The first explicit mention in the literature of a person's Erdős number appears to be [11], where the reader is assured that Paul Erdős himself was, for a long time, unaware of this entertainment. But the first systematic attempt to study the Erdős component of C was carried out by the second author in [16] and [18] and continues on the Erdős Number Project World Wide Web site [13]. This Web site contains a list of all people with Erdős number 1 (currently 485) and their other co-authors with Erdős number 2 (currently 5337). The files (available also via anonymous ftp, see [14]) are updated annually.

It has been surmised that most scientists *must* have a finite Erdős number, but the evidence offered in support has not been really abundant. In [5] the first author contributed new information, and the present paper pursues the matter much further. By skimming through several bibliographic sources we have found that many important people in academic areas—other than mathematics proper—as diverse as physics, chemistry, crystallography, economics, finance, biology, medicine, biophysics, genetics, meterology, astronomy, geology, aeronautical engineering, electrical engineering, computer science, linguistics, psychology, and philosophy do indeed have finite Erdős numbers. We report on some of these intriguing connections here; others can be found in an expanded version of this paper, available on-line [13]. Of course, it cannot be immediately inferred that all people in the mentioned disciplines, or related ones, must have finite Erdős numbers. But the names first resulting from this kind of browsing are among the most prominent and productive (including more than 60 Nobel Prize winners), and most have had many collaborators over the years. Thus one is led to believe that the majority of researchers in those fields, except for those working in total isolation, probably have finite Erdős numbers.

When referring to all academic or scientific fields, the last statement should be regarded as bold—though credible—guess. If we restrict ourselves to authors publishing *mathematical research*, then the conjecture

$$(\varepsilon)\ \begin{cases}\text{most active mathematical researchers of}\\ \text{the twentieth century have a finite (and}\\ \text{rather small) Erdős number}\end{cases}$$

seems so plausible that it has been accepted folklore. Looking at the list of those with Erdős number ≤ 2, one sees 5500 people belonging to numerous and varied areas of research in the mathematical sciences; therefore (ε) *should* be true.

We intend to provide some more conclusive or "hard" evidence in support of (ε). To do so we first select a rather high-class sample of the mathematical research community: the winners of the most prestigious awards, namely, the Fields Medal, the Nevanlinna Prize, the Wolf Prize in Mathematics, and the Steele Prize for Lifetime Achievement. By criss-crossing multiple bibliographic references, we have determined that all recipients of these prizes have indeed an Erdős number ≤ 5. Complete tables of upper bounds on these Erdős numbers are presented below[1]. The respective collaboration paths linking all awardees to Paul Erdős are displayed in full detail for the interested reader in the Web site [13]. The individuals belonging to these exclusive lists are especially original, prolific, and influential; most of them have had many disciples, collaborators, and doctoral students. Their impact and influence is not limited to one institution or even to one country or particular epoch (Paul Erdős himself was given the Wolf Prize in 1983–84). Furthermore, these distinctions are conferred with no exclusion of research area (except the Nevanlinna Prize, which is in computer science). The fact that all these big names are in the Erdős component of C is strong evidence for (ε).

Next, we go two steps further: we trace the subject matter of the papers by some researchers known to have a small Erdős number, and we branch out into other academic disciplines. This will give us a more concrete idea of how far the Erdős connection really extends within the mathematical sciences and beyond. Lastly, we pose some open questions. Obviously this work is incomplete (for instance, we have not traced any recent Nobel laureates in physics), and it should not be hard to establish further links with important mathematicians and scientists.

For brevity, we say that a person is Erdős-n if his or her Erdős number is $\leq n$. Thus Erdős's co-authors are Erdős-1 and their co-authors who are not Erdős-0 or Erdős-1 are Erdős-2. The list [14], containing all Erdős-2 individuals and their respective Erdős-1 co-authors, is referred to as the *Erdős-2 list.*

Interesting Connections

Without intending to be 100% exhaustive, we have examined several bibliographic databases and historical accounts (*e.g.*, [3], [4], [19], [22], [24], [25], [26], [31], [33], and Internet sites too numerous to list) and discovered that some very conspicuous thinkers and researchers from manifold academic branches are in the Erdős component of C.

The following examples evince the amazing diversity of the scientific collaboration network directly linked to the name of Paul Erdős, providing—in passing—an ample glimpse into the practice of academic collaboration, an aspect of scientific research that has become essential in the twentieth century and has not been systematically addressed in the literature. Due to space limitations we cannot list at the end of the present paper the 120+ bibliographic references corresponding to the cited collaborative works. We shall use double brackets [[ ]] for those references, which the reader can find in full detail on the Erdős Number Project World Wide Web site [7].

- Albert Einstein has Erdős number 2 due to the two joint papers with his Princeton assistant (in the years 1944–48) Ernst G. Straus, with whom Erdős wrote 20 papers (the first in 1953)[2]. Einstein wrote jointly with about 25 collaborators (see [27]), among them Nobel laureates in physics Wolfgang Pauli and Otto Stern.

  At age 20 Pauli had surprised the physics establishment with his 200-page encyclopedia article on the theory of relativity, a piece of which Einstein wrote a laudatory review. Not surprisingly, their joint paper [[55]] of 1943 (their only joint paper, written during Pauli's stay in Princeton) deals with technical aspects of the general theory of relativity. Pauli received the 1945 Nobel Prize

[1]Upper bounds for the Erdős numbers for all Fields Medalists up to 1994 had already been presented by the authors in [5], [13], and [17], but many bounds have been lowered for the present paper.

[2]A complete bibliography of Erdős's works through about 1996 has been prepared by the second author [15], with annual updates posted on the Erdős Number Project Web site [13].

for the so-called *Pauli exclusion principle*. With Stern, Einstein also wrote only one joint article [[57]], when they were both in Prague. Stern was awarded the 1943 Nobel Prize for his discovery of the magnetic moment of the proton.

Einstein also published with Russian Boris Podolsky and Austrian Paul Ehrenfest [[52]], one of his closest friends. The well-known Einstein-Podolsky-Rosen paradox, conceived as a thought experiment against the quantum-mechanical conception, originated in their 1935 joint paper [[56]]. Co-authors of Podolsky include at least two Nobel laureates: the great British theoretical physicist Paul Dirac [[48]] and American chemist Linus Pauling [[111]]; hence, they are both at most Erdős-4. Pauling received the 1954 chemistry prize for his research on chemical bonding. As a result of his campaign for an international control of nuclear weapons, Pauling was awarded the 1962 Nobel Peace Prize.

There are, moreover, two very curious non-technical joint publications by Einstein. The first is a report about an international bureau of meteorology, written in 1927 with Marie Curie and Hendrik A. Lorentz and published in the journal *Science* [[39]]. The second is a booklet entitled *Why War?*, which he wrote in 1933 with Sigmund Freud [[53]] (see also [27]). It appeared in German, French, and English, and was published by the International Institute of Intellectual Cooperation of the League of Nations.

- Hendrik A. Kramers, a Dutch physicist, was one of Pauli's collaborators [[88]] and also wrote with Danish Nobel laureate Niels Bohr [[22]], one of the pillars of twentieth-century scientific thought. Therefore, Bohr is at most Erdős-5. In 1923 Bohr published jointly with Dirk Coster [[21]], another Dutch physicist, who in the same year co-authored a research paper with George C. De Hevesy[3], a Hungarian chemist who went on to receive the Nobel Prize in chemistry in 1943 for his use of isotopes as tracers. A distinguished collaborator of Bohr was John A. Wheeler. In 1939 they wrote the seminal work *The mechanism of nuclear fission* [[23]], which made Wheeler the first American involved in the theoretical development of nuclear weapons; in that memoir uranium-235 was singled out for use in a possible atomic bomb.

  Another of Kramers's co-authors is Leonard S. Ornstein [[108]], in turn linked with fellow Dutchman Frits Zernike [[109]], winner of the 1953 Nobel Prize in physics (for his invention of the phase-contrast microscope).
- J. Robert Oppenheimer is among Ehrenfest's co-authors [[50]], which collaboration makes him at most Erdős-4. Oppenheimer is remembered as director of the Los Alamos laboratory during development of the atomic bomb (1943–45) and as director of the Institute for Advanced Study at Princeton (1947–66). Robert Serber, Oppenheimer's former student and close collaborator [[106]], is linked to at least two Nobel laureates, American nuclear physicists Ernest O. Lawrence and Edwin M. McMillan [[28]]. Serber, Lawrence, and McMillan were indispensable members of the Los Alamos scientific team. Lawrence was the winner of the 1939 Nobel Prize in physics for his invention of the cyclotron; chemical element 103, lawrencium, is named after him. McMillan shared the 1951 chemistry Nobel Prize for his discovery of element 93, neptunium, the first element heavier than uranium. The above links show that both Lawrence and McMillan have an Erdős number of at most 6.
- Max Born, a Nobel laureate in physics (1954), is at most Erdős-3 through his collaboration with Norbert Wiener, the creator of cybernetics, whose Erdős number is 2. Their only joint paper [[27]] was written during Born's visit to MIT in 1925. Among Born's co-authors we find fellow Germans Werner Heisenberg, Pascual Jordan [[24]]—the three are founders of modern quantum mechanics—and Max von Laue [[26]] (the last collaboration might be considered a bit of a stretch, a jointly written technical obituary for Max Abraham). For his preeminent role in the foundation of quantum mechanics Heisenberg was the sole winner of the 1932 Nobel Prize in physics. Laue had been awarded the 1914 Nobel Prize for his research on the diffraction of X-rays in crystals.

  Furthermore, Heisenberg published with the director of his doctoral dissertation, German Arnold Sommerfeld [[81]] (also Pauli's thesis advisor in Munich), who is remembered for his successful modifications of Bohr's atomic model. One of Sommerfeld's many co-authors is Peter J. Debye (also spelled Debije) [[45]], a Dutch scientist and Nobel laureate in chemistry (1936). Debye's Erdős number is at least one lower than implied by this collaboration, however, since he wrote a joint paper with Pauling [[46]].

  We should mention another very famous co-author of Born, Theodore von Kármán [[25]], the Hungarian-born American research aeronautical engineer.
- John von Neumann and Erdős never wrote jointly although they were both Hungarian by birth and just 10 years apart in age. Actually, von Neumann did not write with any of Erdős's almost 500 co-authors; his Erdős number stands at 3 through his varied collaborations with individuals in the Erdős-2 list (*e.g.*, Salomon Bochner, Paul Halmos, Herman H. Goldstine). In turn, von Neumann had very illustrious co-authors, notably David Hilbert, Oswald Veblen, Garrett Birkhoff, Pascual Jordan and Nobel laureate physicists Eugene Wigner and Subrahmanyan Chandrasekhar (see [30]).

  With Hilbert, von Neumann wrote about the mathematical foundations of quantum mechanics [[82]] shortly after Heisenberg had proposed his quantum scheme, known as matrix mechanics. Hilbert, von Neumann, and Heisenberg were at Göttingen at that time. Wigner was also a Hungarian and a friend of von Neumann since childhood; most of their joint papers deal with quantum mechanics as well. Through Wigner we find a path to one of

[3] In that paper [[37]] they reported the discovery of a new chemical element, hafnium.

The Erdős Graph and the Beast

Paul M. B. Vitanyi

> Let him that hath understanding count the number of the beast: for it is the number of a man; and his number is six hundred three score and six.
>
> St. John, Book of Revelations

The Number of the Beast, 666, has been interpreted as standing for Pope Leo X (by Michael Stifel (1486–1567): he considered the name LEO DECIMUS X and added the values of the constituent letters L, D, C, I, V, X in their meaning as Roman numerals, discarding the letters which do not denote Roman numerals and also the letter M because its meaning "Mystery" disqualifies it). Apart from this, 666 has been variously interpreted to signify the Pope of Rome in general (by Napier); Martin Luther (by Napier's contemporary the Jesuit Father Bongus S.J.); Kaiser Wilhelm (during World War I); Adolf Hitler (during World War II); and earlier, the Roman Emperor Nero (because 666 was claimed to spell Nero when expressed in the symbols of Aramaic, the language in which the Book of Revelations was written [3]). The present article continues the hoary tradition of finding contrived interpretations for the Number of the Beast.

The Erdős graph—the Erdős component of the collaboration graph—has been the subject of some papers [4, 2, 1] and certainly of many well-lubricated conversations. In [2] it is conjectured that as time goes by, the graph will contain arbitrarily large cliques. Exploring consequences of this conjecture [5], I was led to an unexpected occurrence of the number 666. Let me explain this matter in more detail.

Recall that the graph has mathematicians (nowadays perhaps mathematicians and computer scientists) connected by an edge if they have co-authored a paper. A *clique* is, as usual, a subgraph in which each pair of vertices is connected by an edge. Now I want to consider cliques which (beside being arbitrarily large) have arbitrary male–female ratios. Noting that every clique with more than 2 members must have at least one edge joining vertices of like sex, we may investigate whether that kind of coauthorship is preponderant.

For any p, then, consider a clique K_p consisting of p_1 male mathematicians and p_2 female. For the partition (p_1, p_2) of p, let $\alpha(p_1, p_2)$ denote the *ratio* between the number of edges joining vertices of equal sex and the number of edges joining vertices of opposite sex in K_p:

$$\alpha(p_1, p_2) = \frac{\binom{p_1}{2} + \binom{p_2}{2}}{p_1 p_2} \qquad (*)$$

the biggest names in quantum physics, Austrian Erwin Schrödinger, who shared the 1933 Nobel Prize with Dirac for their introduction of wave equations in quantum mechanics. This is the route: Wigner with R. F. O'Connell [[83]] with John Trevor Lewis [[65]] with James McConnell [[94]] with Schrödinger [[100]]. Thus, Schrödinger becomes at most Erdős-8.

Another of von Neumann's collaborators was the Austrian economist Oskar Morgenstern, with whom he wrote in 1944 the very influential work *Theory of Games and Economic Behaviour* [[126]]. This book stimulated a worldwide development of the mathematical aspects of game theory and its applications (see [29]). At least three Nobel Prizes in economics have been awarded to game-theorists; all of them are in the Erdős component of C (as we shall demonstrate below), even though they are not directly linked with either von Neumann or Morgenstern.

Morgenstern, in turn, wrote jointly with John G. Kemeny [[86]], the creator (along with Thomas E. Kurtz) in the mid-1960s of BASIC, a very popular general-purpose programming language.

- George Uhlenbeck, the noted Dutch-American physicist, has Erdős number 2. He is best known for having postulated, along with Samuel Goudsmit, the concept of electron spin, which led to major changes in atomic theory and quantum mechanics. Their famous joint paper [[125]] was published in 1925 when they were graduate students in physics at the University of Leiden in the Netherlands (both of them were pupils of Ehrenfest).

Among Uhlenbeck's co-authors we encounter at least two Nobel physicists: American Willis E. Lamb (1955 prize) [[105]], whose experimental work spurred refinements in the quantum theories of electromagnetic phenomena, and Italian-born Enrico Fermi (1938 prize) [[61]], one of the chief architects of the nuclear age. Fermi had legions of co-authors and collaborators in Europe and the United States; one of them was his former student in Rome Emilio Segrè [[59]]. Segrè and his colleague at the University of California, Berkeley, American Owen Chamberlain, discovered the antiproton in 1955 and for that feat were awarded the Nobel Prize in physics in 1959 (they also published jointly [[31]]). The above links show that the Erdős numbers of Fermi, Segrè and Chamberlain are at most 3, 4, and 5, respectively. (Actually, if we are willing to use technical reports in establishing collaboration links, then we can lower Fermi's Erdős number to 2, for he published a Los Alamos technical report with Stanislaw Ulam.) Another co-author of Segrè is the American nuclear chemist Glenn T. Seaborg [[120]], who received one half of the chemistry Nobel Prize in 1951 for

Let $M(p)$ be the *minimum* of $\alpha(p_1,p_2)$ over all partitions of p. The following equation is easily proven.

$$M(p) = \frac{p^2 - 2p + \text{remainder } (p/2)}{p^2 - \text{remainder } (p/2)}. \quad (**)$$

It follows that, for all $p > 1$, we have $M(p) < 1$. Moreover, $M(p)$ is monotone increasing and tends to 1.

Thus (**) tells us that for each p, some $\alpha(p_1,p_2)$ is <1 (and for large p, close to 1); yet for many arguments, $\alpha(p_1,p_2) >> 1$. We may ask for what clique size and sex ratio it comes closest to 1. It is well known that the male/female ratio amongst mathematicians is far greater than 1; in the general human population this ratio varies from one ethnic group to another and over time, but is usually between 100/100 and 112/100.

It is easily found from (*) that the condition for α to be exactly 1 is $p_1 - p_2 = \sqrt{p_1 + p_2}$. This can be realized for moderate-sized p and for realistic male/female ratios, but the solutions for p_1 and p_2 will usually not be integers. One might talk one's way out of this by allowing one of the mathematicians to be of indeterminate sex, but that would seem evasive, so let us stick to the hypothesis that everyone's chromosomes are unambiguous. Now (*) gives

$$p_1^2 - 2\alpha(p_1,p_2)p_1p_2 + p_2^2 - p_1 - p_2 = 0.$$

Set the sex ratio at $p_1/p_2 = 173/160$. This is a realistic value, $\approx 1.08^+$, and 173 is a prime, so to end with integer values we need only look at values of p_2 which are multiples of 160. We have

$$\alpha(p_1,p_2) \approx 1.00305 - 0.96243/p_2$$

monotonic in p_2. Now $p_2 = 160$ yields $\alpha(p_1,p_2) \approx 0.997$ and $p_2 = 320$ yields $\alpha(p_1,p_2) \approx 1.000045$. The nearest we can get to $\alpha(p_1,p_2) = 1$ is by taking $p_2 = 320$, but then $p_1 = 346$ and we find the ominous clique K_{666}.

REFERENCES

1. De Castro, R., & Grossman, J., Famous trails to Paul Erdős, *Mathematical Intelligencer* 21 (1999), no. 3, 51–63.
2. Erdős, P., On the fundamental problem of mathematics, *American Mathematical Monthly* 79 (1972), 149–150.
3. Eves, H., *An Introduction to the History of Mathematics*. Holt, Rinehart & Winston, New York, 1967.
4. Goffman, C., And what is your Erdős number? *American Mathematical Monthly* 76 (1969), 791.
5. Vitanyi, P.M.B., How well can a graph be *n*-colored? *Discrete Mathematics* 34 (1981), 69–80.

CWI
Kruislaan 413
1098SJ Amsterdam
The Netherlands
e-mail: paulv@cwi.nl

his research on transuranium elements (the other co-winner was McMillan, whom we mentioned above).

Edward Teller, the Hungarian-born American nuclear physicist who led the development of the world's first thermonuclear weapon, is another co-author of Fermi [[60]]. One of Teller's doctoral students in Chicago was the Chinese-born physicist Chen Ning Yang, who later became assistant to Fermi, publishing joint research work with him [[62]]. Yang and fellow Chinese Tsung-dao Lee received the 1957 Nobel Prize in physics for their work [[91]], [[92]] in discovering violations of the principle of parity conservation, a major discovery in particle physics theory. The just cited collaborations make Yang at most Erdős-4 and Lee at most Erdős-5.

- Freeman J. Dyson, the British-American physicist known by the general public for his writings on extraterrestrial civilizations and his advocacy of space exploration, is a conspicuous member of the Erdős-2 list. He is linked, by way of Richard H. Dalitz [[40]], with the German-American physicist Hans A. Bethe [[42]], one of the main figures in twentieth-century atomic physics. Bethe was head of the Theoretical Physics Division of the Manhattan Project and was honored with the 1967 Nobel Prize for his explanation of the energy production in the Sun and other stars. The prominent Austrian astrophysicist Edwin E. Salpeter is included in the large group of Bethe's co-authors and collaborators [[118]].

Bethe is one of the protagonists of a joint publication [[2]] which compels attention for the unique combination of names in its byline: Alpher, Bethe, Gamow. The third author is George Gamow, the Ukranian-born nuclear physicist and cosmologist who also made contributions to modern genetic theory; the first author is Ralph Alpher, one of his students. The paper itself (*The origin of chemical elements*) is actually very important; in it the authors advanced the idea that the chemical elements were synthesized by thermonuclear reactions which took place in a primeval explosion. It was Gamow who coined the expression "big bang."

From Dalitz we find a path to another Nobel physicist, American Robert Hofstadter, a co-recipient of the 1961 prize for his investigations of protons and neutrons. This path makes him at most Erdős-5: Dalitz with D. G. Ravenhall [[41]] with R. Hofstadter [[77]].

- Sheldon Lee Glashow, an American theoretical physicist and Nobel laureate (1979), has Erdős number 2 for his collaboration [[71]] with the Erdős-1 combinatorialist Daniel Kleitman, his brother-in-law. Glashow shares with Einstein the distinction of being, up until now, the only Nobel-winning physicists with Erdős number ≤ 2.

Table 1. Upper bounds on Erdős numbers of some Nobel Prize winners.

Nobel Prize in physics	Year	Erdős number
Max von Laue	1914	4
Albert Einstein	1921	2
Niels Bohr	1922	5
Louis de Broglie	1929	5
Werner Heisenberg	1932	4
Paul A. Dirac	1933	4
Erwin Schrödinger	1933	8
Enrico Fermi	1938	3
Ernest O. Lawrence	1939	6
Otto Stern	1943	3
Isidor I. Rabi	1944	4
Wolfgang Pauli	1945	3
Frits Zernike	1953	6
Max Born	1954	3
Willis E. Lamb	1955	3
John Bardeen	1956	5
Walter H. Brattain	1956	6
William B. Shockley	1956	6
Chen Ning Yang	1957	4
Tsung-dao Lee	1957	5

Nobel Prize in physics	Year	Erdős number
Emilio Segrè	1959	4
Owen Chamberlain	1959	5
Robert Hofstadter	1961	5
Eugene Wigner	1963	4
Richard P. Feynman	1965	4
Julian S. Schwinger	1965	4
Hans A. Bethe	1967	4
Luis W. Alvarez	1968	6
Murray Gell-Mann	1969	3
John Bardeen	1972	5
Leon N. Cooper	1972	6
John R. Schrieffer	1972	5
Aage Bohr	1975	5
Ben Mottelson	1975	5
Leo J. Rainwater	1975	7
Steven Weinberg	1979	4
Sheldon Lee Glashow	1979	2
Abdus Salam	1979	3
S. Chandrasekhar	1983	4
Norman F. Ramsey	1989	3

Nobel Prize in economics	Year	Erdős number
Paul A. Samuelson	1970	6
Kenneth J. Arrow	1972	3
Tjalling C. Koopmans	1975	4
Gerard Debreu	1983	3
Franco Modigliani	1985	5
Robert M. Solow	1987	6
Harry M. Markowitz	1990	2
Merton H. Miller	1990	4
John C. Harsanyi	1994	8
John F. Nash	1994	4
Reinhard Selten	1994	7
Robert C. Merton	1997	7

Nobel Prize in chemistry	Year	Erdős number
Peter J. Debye	1936	5
George De Hevesy	1943	7
Otto Diels	1950	7
Kurt Alder	1950	6
Edwin M. McMillan	1951	6
Glenn T. Seaborg	1951	5
Linus Pauling*	1954	4
Walter Gilbert	1980	4
Jerome Karle	1985	4
Herbert A. Hauptman	1985	3

*Also received the 1962 Nobel Peace Prize

Nobel Prize in physiology/medicine	Year	Erdős number
Francis H. C. Crick	1962	7
James D. Watson	1962	8

A co-author of Glashow is another Nobel-winning physicist (1969): fellow American Murray Gell-Mann [[70]], who introduced the concept and the word *quark* for a basic subatomic particle. Gell-Mann also collaborated with the American physicist and Nobel Prize winner Richard Feynman [[63]]. Feynman participated in the Manhattan Project (he was only 25 when he was recruited) and wrote jointly with Bethe [[9]]. The two of them devised the formula for predicting the energy yield of a nuclear explosive.

Feynman, who had been a pupil of Wheeler at Princeton and published with him [[127]], became a salient figure of postwar physics, receiving the Nobel Prize in 1965 for his quantum electrodynamics theory. Sharing the prize with Feynman was Julian S. Schwinger, who independently formulated a theory of quantum electrodynamics, unaware that Feynman in the United States and Sin-Itiro Tomonaga in Japan were working on the same problem. The equivalent theories reconcile quantum mechanics with the special theory of relativity. We can link Schwinger to Erdős via this path: Schwinger with Norman F. Ramsey [[112]] with W. H. Furry [[68]], the last named being Erdős-2. Thus, Schwinger is at most Erdős-4. Curiously enough, Feynman and Schwinger were born in the same year, 1918, in the same city, New York, received the Nobel Prize the same year for the

Table 2. Upper bounds on Erdős numbers of some distinguished scholars.

	Main research field	Erdős number
Walter Alvarez	Geology	7
Rudolf Carnap	Philosophy	4
Jule G. Charney	Meteorology	4
Noam Chomsky	Linguistics	4
Freeman J. Dyson	Quantum physics	2
George Gamow	Nuclear physics and cosmology	5
Stephen Hawking	Relativity and cosmology	7
Pascual Jordan	Quantum physics	4
Theodore von Kármán	Aeronautical engineering	4
John Maynard Smith	Biology	4
Oskar Morgenstern	Economics	4
J. Robert Oppenheimer	Nuclear physics	4
Roger Penrose	Relativity and cosmology	8
Jean Piaget	Psychology	3
Karl Popper	Philosophy	5
Edwin E. Salpeter	Astrophysics	5
Claude E. Shannon	Electrical engineering	3
Arnold Sommerfeld	Atomic physics	5
Edward Teller	Nuclear physics	4
George Uhlenbeck	Atomic physics	2
John A. Wheeler	Nuclear physics	5

same achievement, and—as far as we know—they also have the same Erdős number, namely 4.

The two co-authors of Schwinger in the above-cited paper [[112]], Americans Isidor I. Rabi and Ramsey, are themselves Nobel laureates. Rabi was given the 1944 Nobel Prize in physics for his 1937 invention of the magnetic resonance method. Ramsey received one half of the 1989 physics award for his development of a technique called the separated oscillatory fields method, which provides the basis for the cesium atomic clock. Hence, Ramsey is at most Erdős-3 and Rabi at most Erdős-4.

- David Pines, an American physicist who publishes on condensed matter theory and theoretical astrophysics, is a key figure in the collaboration graph. As a co-author of Gell-Mann [[58]], he is at most Erdős-4. Pines has co-authored research papers with six Nobel Prize winners and is at distance 2 from eight more (see Figure 1). None of the scientists we mention in the present article (including Erdős himself) clusters so closely around so many Nobel winners.

Two of the co-authors of Pines are John Bardeen [[14]] and John Robert Schrieffer [[110]], who, along with fellow American Leon N. Cooper, received the 1972 Nobel Prize in physics for their joint theory [[12]], [[13]], known as BCS theory for their surname initials, which was the first successful microscopic theory of superconductivity. When he made his principal contribution to the BCS theory, Schrieffer was a 26-year-old graduate student at the University of Illinois, were Bardeen was a professor in the physics and electrical engineering departments.

It should be recalled that Bardeen had been a co-winner of another Nobel Prize in physics, that of 1956, which he shared with Walter H. Brattain and William B. Shockley for their research on semiconductors and their joint invention of the transistor. Therefore, both Brattain and Shockley become at most Erdős-6 because of their joint papers with Bardeen [[11]], [[15]].

Two additional co-authors of Pines are Nobel winners Danish Aage Bohr (the son of Niels Bohr) and Danish-American Ben Mottelson [[19]]. They shared the 1975 physics award for work in the early 1950s in determining the asymmetrical shapes of certain atomic nuclei. Their experiments had been inspired by the theories of the American physicist Leo James Rainwater, who was also a co-recipient of the 1975 Nobel Prize. Rainwater's Erdős number is finite too, at most 7, via Tsung-dao Lee, who, as we saw, is at most Erdős-5: Lee with C. S. Wu [[93]] with Rainwater [[113]].

A co-author of Aage Bohr is Kurt Alder [[1]], a German chemist, former student and assistant of Otto Diels, along with whom he received the 1950 Nobel Prize in chemistry for their joint method of preparing cyclic organic compounds.

Apart from father Niels and son Aage, there is another member of the Bohr family in the Erdős component of C, namely, Niels's younger brother Harald, known especially for the theory of almost periodic functions. Harald's Erdős number is 3 due to his joint paper [[20]] with Borge Jessen, whose Erdős number is 2. (This shows that the brothers Niels and Harald are at distance ≤ 8 in the graph C, and the distance between the two Nobel laureates, Niels and his son Aage, is ≤ 10; most likely these bounds can be improved.)

- Abdus Salam and Steven Weinberg shared with the aforementioned Glashow the 1979 Nobel Prize in physics for their theoretical work linking the electromagnetic interaction and the so-called weak interaction. Salam became the first Pakistani to win a Nobel Prize (in any category);

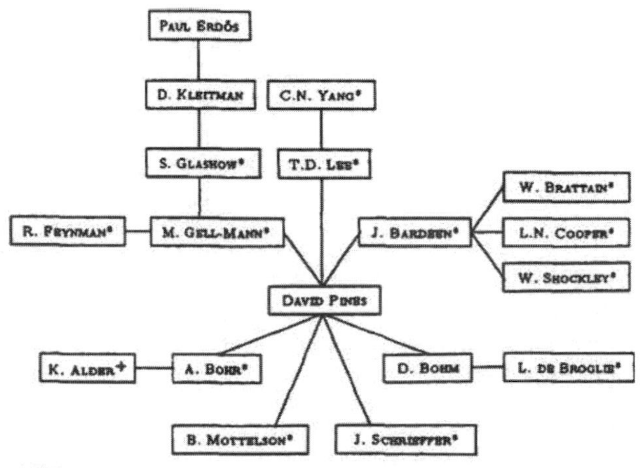

Figure 1. Clustering of Nobel laureates within distance 2 of David Pines.

he is at most Erdős-3 due to his joint paper [[117]] with J. C. Ward who is Erdős-2. Weinberg is at most Erdős-4 for his many collaborations with Salam (*e.g.*, [[73]]).

Among Salam's many co-authors we are able to find another Nobel winner, American molecular biologist Walter Gilbert[4] [[116]], who shared (with Paul Berg and Frederick Sanger) the 1980 chemistry award for their chemical and biological analyses of DNA structure.

- Edward Witten, the outstanding American theoretical physicist and Fields Medalist in 1990, is at most Erdős-5, as will be shown later. Among his co-authors we find the American physicist Luis W. Alvarez [[4]], who received the Nobel Prize in 1968 for his work on subatomic particles.

Another co-author of Witten is Gary Horowitz [[30]], who in turn has collaborated with Stephen Hawking [[69]], the English theoretical physicist. Hawking has also published with fellow British mathematician and physicist Sir Roger Penrose [[80]].

- Jean-Pierre Vigier, a distinguished French physicist, can be linked to some important scientists; his Erdős number is at most 4 as the following collaborations show: Vigier with Constantin Piron [[64]] with Stanley P. Gudder [[76]], Gudder being an Erdős-2 researcher.

Among Vigier's co-authors stands out fellow French physicist Prince Louis de Broglie [[29]], who in the mid-1920s developed (in his doctoral dissertation) a revolutionary theory of electron waves, enthusiastically defended by Einstein. Experimental evidence of his theory came a few years afterwards, and de Broglie was awarded the Nobel Prize in 1929. Also a co-author of Vigier in the work we have just cited is the American physicist and philosopher David Bohm, the last doctoral student of Oppenheimer at Berkeley and the originator of the causal interpretation of quantum theory. Bohm is also acknowledged for his joint research with David Pines (see Figure 1); their joint papers [[18]], published under the title *A collective description of electron interactions*, underlie all current research in plasma-state physics.

Sir Karl R. Popper, the eminent Austrian-born British philosopher, is another co-author of Vigier [[95]].

- Claude E. Shannon, an American electrical engineer, became famous for his elegant and general mathematical model of "communication," known today as *information theory*.

Shannon's Erdős number is at most 3 because of his collaboration [[122]] with Elwyn R. Berlekamp, whose Erdős number is 2.

- Francis H. C. Crick, a British biophysicist, and James D. Watson, an American geneticist and biophysicist, determined the molecular structure of deoxyribonucleic acid (DNA)—as a double-helix polymer—for which accomplishment they were awarded the 1962 Nobel Prize for Physiology/Medicine.

In 1957 Crick published a short paper [[38]] on information theory, thereby entering the Erdős component of *C*. Indeed, Crick's Erdős number is at most 7 due to the following chain of joint works: Crick with J. S. Griffith [[38]] with I. W. Roxburgh [[114]] with P. G. Saffman [[115]] with H. B. Keller [[36]] with K. O. Friedrichs [[66]]. The last named is in the Erdős-2 list.

Hence, James Watson would be Erdős-8, and from this connection we can conclude that many other active researchers in genetics, biophysics, biochemistry, and related fields have finite Erdős numbers as well.

- Herbert A. Hauptman, a mathematician, and Jerome Karle, a chemist and crystallographer (both American), were awarded the Nobel Prize in chemistry in 1985 for their development of mathematical methods for deducing the molecular structure of biological molecules from the patterns formed when X-rays are diffracted by their crystals.

Hauptman's Erdős number is at most 3 for his joint publication [[75]] with Fred Gross, who appears in the Erdős-2 list. This makes Karle an Erdős-4 researcher by way of his numerous joint articles with Hauptman (*e.g.*, [[79]]).

- John Maynard Smith, a British biologist, initiated a whole new area of research by his unusual applications of game theory to animal behavior and evolution, with works like *The theory of games and the evolution of animal conflict* [[98]] and *Evolution and the Theory of Games* [[97]]. It turns out that Maynard Smith has a small Erdős number, at most 4, through Josef Hofbauer [[99]] and Hal L. Smith [[84]], who is Erdős-2.
- Harry M. Markowitz, the American finance expert, is in the Erdős-2 list (because of his collaboration with Alan J. Hoffman [[85]]), and is the only Nobel economist with such a low Erdős number. He shared the 1990 prize with Merton H. Miller and William F. Sharpe for their study of financial markets and investment decision-making.

Miller is at most Erdős-4, through Abraham Charnes [[32]] and Fred Glover [[33]], the latter being Erdős-2. Miller is also linked [[102]], [[103]] with another Nobel laureate, the Italian economist Franco Modigliani, who received the 1985 award for his mathematical analysis of household savings and the dynamics of financial markets.

- Herbert Scarf, an American economist with Erdős number 2, has published articles on economic analysis with many other renowned economists, such as Kenneth J. Arrow [[6]] and Gérard Debreu [[44]], both winners of the Nobel Prize in economics (1972 and 1983, respectively). A co-author of the latter is the Dutch economist Tjalling C. Koopmans [[43]], also a Nobel laureate (1975). Additionally, Scarf has published with Lloyd S. Shapley [[124]], one of the major contributors to the development of game theory and a co-author of the American mathematician John F. Nash [[104]], a co-recipient of the 1994 Nobel Prize in economics. Nash shared his prize with the

[4]Gilbert's Ph.D. degree is actually in mathematics (from Cambridge University).

Hungarian-born economist John C. Harsanyi and the German mathematician Reinhard Selten, for their beneficial use of game theory in economics (more precisely, for "their pioneering analysis of equilibria in the theory of non-cooperative games"). As can be expected, Harsanyi and Selten also have a finite Erdős number. This is the path: Koopmans with Beckman [[87]] with Marschak [[16]] with Selten [[96]] with Harsanyi [[78]]. Therefore, Selten is at most Erdős-7 and Harsanyi at most Erdős-8.

From Arrow we can find a path leading to two more Nobel laureates in economics, the American economists Paul A. Samuelson and Robert M. Solow. This is the path: Arrow with Edward W. Barankin [[5]] with Robert Dorfman [[10]] with Samuelson and Solow [[49]].

An additional and more recent Nobel connection should be cited: Through his collaboration with Samuelson [[101]], the American economist Robert C. Merton, one of the recipients of the 1997 Nobel Prize in economics, is at most Erdős-7. Merton expanded the work of Myron S. Scholes and Fisher Black, who had advanced in 1973 a pioneering formula for the valuation of stock options. Scholes shared the prize with Merton, not so Black, due to his untimely death in 1995[5].

- Noam Chomsky, the American linguist and political activist, is one of the most influential figures of twentieth-century linguistics. The following path shows that Chomsky is at most Erdős-4: Chomsky with M. P. Schutzenberger [[35]] with S. Eilenberg [[51]], the latter being Erdős-2.
- Rudolf Carnap, the German-born philosopher and member of the Vienna Circle, is at most Erdős-4, as shown by this path: Carnap with Yehoshua Bar-Hillel [[7]] with M. Perles [[8]], the latter being in the Erdős-2 list.

A student of Carnap in Prague was Willard V. Quine, an American logician and philosopher, known for undertaking a systematic constructivist analysis of philosophy. He is at most Erdős-3 due to his collaboration with J. C. C. McKinsey [[89]], an Erdős-2 individual and a renowned philosopher himself.

Erdős Numbers of the Fields Medalists

The *Fields Medal* was established by John Charles Fields (1863–1932), a Canadian mathematician. It has always been granted to mathematicians not older than 40, although the age limit was neither demanded nor suggested by Fields himself (see [21]). A minimum of two and a maximum of four medals are awarded on the occasion of the quadrennial International Congress of Mathematicians.

The first two medals were conferred at the Oslo (Norway) Congress in 1936 to Finnish mathematician Lars Ahlfors and New Yorker Jesse Douglas, but due to the Second World War no medals were awarded during the next 14 years. The academic distinction resumed in 1950; to date there have been 42 awardees from 14 different countries.

Table 3. Upper bounds on Erdős numbers of the Fields Medalists.

Fields Medal	Year	Country of origin	Erdős number
Lars Ahlfors	1936	Finland	5
Jesse Douglas	1936	USA	4
Laurent Schwartz	1950	France	5
Atle Selberg	1950	Norway	2
Kunihiko Kodaira	1954	Japan	2
Jean-Pierre Serre	1954	France	3
Klaus Roth	1958	Germany	2
Rene Thom	1958	France	4
Lars Hormander	1962	Sweden	3
John Milnor	1962	USA	3
Michael Atiyah	1966	Great Britain	4
Paul Cohen	1966	USA	5
Alexander Grothendieck	1966	Germany	5
Stephen Smale	1966	USA	5
Alan Baker	1970	Great Britain	2
Heisuke Hironaka	1970	Japan	4
Serge Novikov	1970	Russia	3
John G. Thompson	1970	USA	3
Enrico Bombieri	1974	Italy	2
David Mumford	1974	Great Britain	2
Pierre Deligne	1978	Belgium	3
Charles Fefferman	1978	USA	2
Gregori Margulis	1978	Russia	5
Daniel Quillen	1978	USA	3
Alain Connes	1982	France	5
William Thurston	1982	USA	4
Shing-Tung Yau	1982	China	2
Simon Donaldson	1986	Great Britain	5
Gerd Faltings	1986	Germany	4
Michael Freedman	1986	USA	4
Vladimir Drinfeld	1990	Russia	5
Vaughan Jones	1990	New Zealand	4
Shigemufi Mori	1990	Japan	3
Edward Witten	1990	USA	3
Pierre-Louis Lions	1994	France	4
Jean Christophe Yoccoz	1994	France	5
Jean Bourgain	1994	Belgium	2
Efim Zelmanov	1994	Russia	4
Richard Borcherds	1998	S. Africa/Great Britain	2
William T. Gowers	1998	Great Britain	4
Maxim L. Kontsevich	1998	Russia	4
Curtis McMullen	1998	USA	3

Table 3 shows that although Erdős never wrote jointly with any of the 42 Medalists (a fact perhaps worthy of further contemplation), 10 of them have Erdős number 2 and for none is the number greater than 5. The collaboration paths from which these numbers have been obtained are presented in the Web site [6]. It is possible that some paths can be lowered still more, but with these data the average Erdős number of the Fields Medalists is 3.52.

[5]For the mathematics behind the 1997 Nobel Prize in economics the reader is referred to [8] and [9].

Table 4. Upper bounds on Erdős numbers of the Nevanlinna Prize winners.

Nevanlinna Prize	Year	Country of origin	Erdős number
Robert Tarjan	1982	USA	2
Leslie Valiant	1986	Hungary/Great Britain	3
Alexander Razborov	1990	Russia	2
Avi Wigderson	1994	Israel	2
Peter Shor	1998	USA	2

Erdős Numbers of the Steele, Nevanlinna, and Wolf Prize Winners

The Fields Medal carries the prestige of a Nobel Prize, but there are many other important international awards for mathematicians. Perhaps the three most renowned, which are acquiring more and more prominence over the years, are the Rolf Nevanlinna Prize, the Wolf Prize in Mathematics, and the Leroy P. Steele Prizes. These prizes were established within a span of 12 years, beginning in 1970.

- Since 1982 the *Rolf Nevanlinna Prize* has been presented, along with the Fields Medal, at the International Congress of Mathematicians every four years [21]. The funds for the award are granted by the University of Helsinki. This distinction is given only to young mathematicians who deal with the mathematical aspects of information science, and only one prize is bestowed per congress.
- The *Wolf Prize* is awarded by the Wolf Foundation, based in Israel [32]. Each year (since 1978) it gives prizes of $100,000 for outstanding achievements in agriculture, chemistry, medicine, and the arts, as well as mathematics and physics. The Wolf Prize in Mathematics was conferred (in 1984) on Paul Erdős himself, and in addition to his contributions to many fields, the citation extols him "for personally stimulating mathematicians the world over."
- The *Leroy P. Steele Prizes* are awarded by the American Mathematical Society. From 1970 to 1976 one or more prizes were awarded each year for outstanding published mathematical research; in 1977 the Council of the AMS modified the terms under which the prizes are awarded (see [1]). Since then, up to three prizes have been awarded each year in the following categories: (1) Lifetime Achievement (for the cumulative influence of the total mathematical work of the recipient), (2) Mathematical Exposition (for a book or substantial survey or expository-research paper), and (3) Seminal Contribution to Research (for a paper, whether recent

Table 5. Upper bounds on Erdős numbers of the winners of the Wolf Prize in Mathematics.

Wolf Prize in Mathematics	Year	Country of origin	Erdős number
Izrail M. Gelfand	1978	Russia	4
Carl L. Siegel	1978	Germany	3
Jean Leray	1979	France	3
André Weil (SP)	1979	France	4
Henri Cartan	1980	France	3
Andrei N. Kolmogorov	1980	Russia	5
Lars Ahlfors (FM)	1981	Finland	5
Oscar Zariski (SP)	1981	Poland	3
Hassler Whitney (SP)	1982	USA	2
Mark G. Krein	1982	Ukraine	4
Shiing Shen Chern (SP)	1983–84	China	2
Paul Erdős	1983–84	Hungary	0
Kunihiko Kodaira (FM)	1984–85	Japan	2
Hans Lewy	1984–85	Germany	3
Samuel Eilenberg (SP)	1986	Poland	2
Atle Selberg (FM)	1986	Norway	2
Kiyoshi Ito	1987	Japan	3
Peter D. Lax (SP)	1987	Hungary/USA	3
Friedrich E. Hirzebruch	1988	Germany	3
Lars Hörmander (FM)	1988	Sweden	3
Alberto Calderón	1989	Argentina	3
John Milnor (FM)	1989	USA	3
Ennio De Giorgi	1990	Italy	3
Ilya Piatetski-Shapiro	1990	Russia	5
Lennart A. Carleson	1992	Sweden	4
John G. Thompson (FM)	1992	USA	3
Mikhael Gromov	1993	Russia	3
Jacques Tits	1993	Belgium	4
Jurgen K. Moser	1994–95	Germany	3
Robert Langlands	1995–96	Canada	2
Andrew Wiles	1995–96	Great Britain	3
Joseph B. Keller	1997	USA	3
Yakov G. Sinai	1997	Russia	4

(FM): Fields Medalist
(SP): Steele Prize

Table 6. Upper bounds on Erdos numbers of the Steele Prize (Lifetime Achievement Award) winners.

Steele Prize (Lifetime Achievement)	Year	Country of origin	Erdős number
Salomon Bochner	1979	Poland	2
Antoni Zygmund	1979	Poland	2
André Weil	1980	France	4
Gerhard P. Hochschild	1980	Germany	4
Oscar Zariski	1981	Poland	3
Fritz John	1982	Germany	4
Shiing Shen Chern	1983	China	2
Joseph L. Doob	1984	USA	2
Hassler Whitney	1985	USA	2
Saunders Mac Lane	1986	USA	3
Samuel Eilenberg	1987	Poland	2
Deane Montgomery	1988	USA	3
Irving Kaplansky	1989	Canada	1
Raoul Bott	1990	Hungary	3
Armand Borel	1991	Switzerland	4
Peter D. Lax	1992	Hungary/USA	3
Eugene B. Dynkin	1993	Russia	3
Louis Nirenberg	1994	Canada	3
John T. Tate	1995	USA	3
Goro Shimura	1996	Japan	2
Ralph S. Phillips	1997	USA	2
Nathan Jacobson	1998	USA	3

or not, that has proved to be of fundamental or lasting importance in its field.

We have compiled tables of Erdős numbers for all recipients of the Nevanlinna Prize (Table 4), the Wolf Prize in Mathematics (Table 5), and the Steele Prize for Lifetime Achievement (Table 6), and have found that all these numbers are ≤ 5. Again, one may wonder why Kaplansky is the only recipient of any of these prizes who collaborated with Paul Erdős. (As before, the collaboration paths from which these numbers have been obtained are presented in the Web site [6].)[6]

How Far Does the Erdős Connection Extend?

In this section we first look at the various branches of mathematics to see how they are represented in the Erdős component of *C*. Next, we consider many other academic disciplines in an effort to determine the scope of the Erdős connection beyond mathematics.

Both *Mathematical Reviews* (MR) [25] and *Zentralblatt für Mathematik* (Zbl) [33] assign a number to each published work representing its primary subject area. For example, combinatorics is 05 and number theory is 11 (to mention the areas in which about 80% of Paul Erdős's works appear). A total of 6 broad categories are currently in use [2]. It turns out—not surprisingly—that all 61 subject classifications are represented in the Erdős component of *C*. In fact, we can say much more: Erdős himself published in at least 27 of these categories, his co-authors published in at least 32 more, and there are people with Erdős number 2 who have published in the remaining two (*K*-theory and geophysics).

We showed above that some outstanding scientists from myriad fields have finite Erdős numbers. We now extend the reach even further. Sophisticated mathematical models and tools have become standard in many fields outside the natural sciences, computer science, and engineering.

It is not hard to find researchers with fairly small Erdős numbers publishing in social sciences. For example, Scott A. Boorman, who has Erdős number at most 7, has papers

[6]Note added in proof: The obvious mutability of this article's results, and the consequent usefulness of the regularly updated Web sites, have been illustrated most happily by the recent awarding of the 1999 Wolf Prize to László Lovász, with Erdős number 1 (the same Lovász who is a Correspondent of *The Mathematical Intelligencer*).

in both the *Journal of Mathematical Psychology* and the *Journal of Mathematical Sociology*. Certainly hundreds, if not thousands, of statisticians have small Erdős numbers, and they often become co-authors on papers growing out of their consulting work. As another example, Peter C. Fishburn, whose Erdős number is 1 and who works in a variety of mathematical disciplines, has published in *Management Science* and *Theory and Decision*.

Frank Harary, a co-author of Paul Erdős who himself has over 270 co-authors, reports[6] that he has published with anthropologists, architects, biologists, chemists, economists, engineers, geographers, journalists (including the grand-nephew of the writer James Joyce), philosophers, physicians, physicists (including George Uhlenbeck), political scientists, psychologists, scientific writers (including Martin Gardner), and sociologists, among others.

Clearly much work remains to be done in exploring collaborations in other disciplines.

Final Remarks and Open Questions

It could be thought *a priori* that in order for a mathematician to make his or her entrance into the Erdős component of C it is necessary to have many co-authors. But one of the conclusions we draw from the compilation of data for this article is that what really matters is not *how many* people you publish with but *whom* you publish with.

A more dramatic example than any presented thus far is the great Austrian logician Kurt Gödel. In regard to the number of joint papers, Gödel is at the other end of the spectrum from Erdős: He wrote only one (see [10]), and that is a one-page note (in German) with Karl Menger and Abraham Wald [[72]] concerning Menger's approach to differential and projective geometry. It turns out that Wald's Erdős number is 2. Hence, despite his paucity of joint papers, Gödel still makes his way into the Erdős component of C with a rather small Erdős number.

We close with some open questions which, even in the era of supercomputing and worldwide information networks, are extremely difficult to answer. The first two were already put forward in [18], but no hint to their possible solution has yet surfaced.

- In the collaboration graph C, what is the second largest component (measured by the number of its vertices)? If we restrict ourselves to looking only at mathematicians, then the second largest component is probably not very large, but it is conceivable that there are large components in other disciplines.
- What are the radius and the diameter of the Erdős component of C (in graph-theoretical terms)? Again, the question would be interesting both as applied to all researchers and when restricted to mathematicians.
- The *Nobel-Erdős* number is, at a given moment, the number of Nobel Prize laureates having a finite Erdős number. This number changes as new prizes are awarded and more people enter into the Erdős component of C. We have established that the *Nobel-Erdős* number is ≥ 63 but its exact value (as of, say, the end of 1998) is unknown. Surely our bound is not nearly the best possible.
- The *Erdős span* measures how far back in time the connection with Paul Erdős extends. More precisely, we can define the Erdős span as the smallest number representing the year of birth of a person with a finite Erdős number. All we can say for now is that this number is no greater than 1849, which is the year of birth of Georg Frobenius (1849–1917), the German algebraist who made major contributions to group theory. He developed the theory of finite groups of linear substitutions mostly in collaboration with Issai Schur (1875–1941) [[67]]. It turns out that Schur's Erdős number is 2 because of his 1925 joint paper [[119]] with Gabor Szegö, a co-author of Erdős.

We do not know whether the Erdős span can be traced further back into the early 1800s. What we can be sure of is that the Erdős connection will extend forever into the future.

REFERENCES

[1] American Mathematical Society, *The Leroy P. Steele Prizes*, Internet page: http://www.ams.org/secretary/prizes.html#steele.

[2] American Mathematical Society, *1991 Mathematics Subject Classification*, Internet page: http://www.ams.org/msc/.

[3] Nicolas Bourbaki, *Eléments d'histoire des mathématiques*, Hermann Editeurs, 1969.

[4] *Britannica CD-97*, Encyclopaedia Britannica, Inc., 1997.

[5] Rodrigo De Castro, *Sobre el número de Erdős*, Lect. Mat. **17** (1996), 163–179.

[6] Rodrigo De Castro & Jerrold W. Grossman, *Collaboration paths for this paper*, Internet page: http://www.oakland.edu/~grossman/collabpaths.html.

[7] Rodrigo De Castro & Jerrold W. Grossman, *Primary references for this paper*, Internet page: http://www.oakland.edu/~grossman/erdosrefs.html.

[8] Keith Devlin, *A Nobel formula*, Internet page: http://www.maa.org/devlin/devlin_11_97.html.

[9] Guillermo Ferreyra, *The Mathematics Behind the 1997 Nobel Prize in Economics*, Internet page: http://www.ams.org/new-in-math/black-scholes-ito.html.

[10] Kurt Gödel, *Collected Works*, edited by S. Feferman, Oxford University Press, 1986.

[11] Casper Goffman, *And what is your Erdős number?*, Amer. Math. Monthly **76** (1969), 791.

[12] Ronald L. Graham & Jaroslav Nešetřil, editors, *The Mathematics of Paul Erdős*, vols. I–II, Algorithms and Combinatorics **13–14**, Springer-Verlag, 1997.

[13] Jerrold W. Grossman, *The Erdős Number Project World Wide Web Site*, http://www.oakland.edu/~grossman/erdoshp.html.

[14] Jerrold W. Grossman, *List of people with Erdős Number at most 2*, available in [13] and via anonymous ftp to **vela.acs.oakland.edu** in directory pub/math/erdos, Oakland University, Rochester, MI, 1998 (updated annually).

[15] Jerrold W. Grossman, preparer, *List of publications of Paul Erdős*, in [12], pp. 477–573.

[16] Jerrold W. Grossman, *Paul Erdős: the master of collaboration*, in [12], pp. 467–475.

[17] Jerrold W. Grossman, *Review of* [5], *Mathematical Reviews*, 98h:01041.

[18] Jerrold W. Grossman & Patrick D. F. Ion, *On a portion of the well-known collaboration graph*, Proc. 26th Southeastern Inter. Conf. on Combinatorics, Graph Theory and Computing (Boca Raton, FL, 1995), Congr. Numer. **108** (1995), 129–131.

[19] Steve J. Heims, *John von Neumann and Norbert Wiener*, MIT Press, 1980.

[20] Paul Hoffman, *The Man Who Loved Only Numbers*, Hyperion, 1998.

[21] International Mathematical Union, *Fields Medals and Rolf Nevanlinna Prize*, Internet page: http://elib.zib.de/IMU/medals.

[22] *Jahrbuch über die Fortschritte der Mathematik*, 1868–1942, Berlin.

[23] John H. Kagel & Alvin E. Roth, editors, *The Handbook of Experimental Economics*, Princeton University Press, 1995.

[24] *MacTutor History of Mathematics*, Internet page: http://www-groups.dcs.st-and.ac.uk/~history/.

[25] *Mathematical Reviews*, American Mathematical Society, 1940–.

[26] Gert H. Müller, Wolfgang Lenski, et al., editors, *Ω-Bibliography of Mathematical Logic*, vols. I–VI, Springer-Verlag, 1987.

[27] Abraham Pais, *'Subtle is the Lord . . .': The Science and Life of Albert Einstein*, Oxford University Press, 1982.

[28] Bruce Schechter, *My Brain is Open: The Mathematical Journeys of Paul Erdős*, Simon & Schuster, 1998.

[29] L. C. Thomas, *Games, Theory and Applications*, Ellis Horwood Ltd., 1984.

[30] John von Neumann, *Collected Works*, Pergamon Press, 1961–1963.

[31] Edmund Whittaker, *A History of the Theories of Aether and Electricity*, Dover Publications Inc., 1989.

[32] The Wolf Foundation, *The Wolf Foundation*, Internet page: http://www.aquanet.co.il/wolf.

[33] *Zentralblatt für Mathematik und ihre Grenzgebiete*, Springer-Verlag, 1931–.

Chapter Three
Empirical Studies

The transformation of the computer into an affordable household item and the subsequent rise of the Internet were probably the two most important factors contributing to the recent explosion of interest in network research. The Internet itself is a network of premier importance in modern civilization, and many of thc typcs of information it conveys can also be represented in network form. The new ease of availability of data provides unprecedented opportunities for studying the topology of large networks: online film databases have allowed us to map out networks of movie actors; dictionaries chart the networks of words and language; databases of scientific papers allow us to study citation and coauthorship networks of scientists; and, most prominently, the World Wide Web, a network of electronic "pages" of text and images, has in just a few years become the definitive example of an information network.

Empirical studies of these and other networks have provided convincing evidence that real networks are very different in their structure from simple mathematical models of networks such as the random graph (Section 4.1), and have catalyzed a huge surge of interest in quantifying and understanding this structure. In this chapter we look at some of the empirical studies that have most deeply influenced our understanding of network topology. A number of early works in this area appear in the preceding chapter, such as the papers by **Travers and Milgram (1969)** on acquaintance networks and by **Price (1965)** on citation networks. The papers in this chapter are much more recent. We start with four important papers that focus on degree distributions in the World Wide Web, the Internet, and social networks, followed by six more focusing on other networks of particular interest: biochemical networks, collaboration networks, networks of sexual contacts, and cooccurrence networks of words.

Another paper that might well have appeared in this chapter is the paper by **Watts and Strogatz (1998)**, which was probably the first paper to compare networks from radically different fields and show that they have some of the same properties. However, since this paper also introduced for the first time the small-world model, we have included it instead in Chapter 4 on network models. Similarly, the paper by **Barabasi and Albert (1999)** was the first to point out the ubiquity of power-law degree distributions in networks of many different kinds, but since it also introduced the preferential attachment model of a growing network we have placed it in Chapter 4 as well. The reader with a particular interest in empirical studies of networks might wish to study both of these papers as companions to the papers in the present chapter.

Albert, Jeong, and Barabási (1999)

In parallel with the studies of networks by **Watts and Strogatz (1998)**, **Albert *et al.* (1999)** were also investigating the structure of large networks. Their motive at the time was to test a real-world network against the famous random graph model studied by **Erdős and Rényi** (1959, **1960**, 1961) among others. They struck on the idea of studying the World Wide Web, the network whose vertices are Web pages and whose edges are the hyperlinks that allow us to navigate with a click from one Web page to another. The Web is probably the largest network ever created by mankind. In 1998, Lawrence and Giles estimated that it had over 800 million documents (Lawrence and Giles 1998, 1999), and at the time of writing some search engines have cataloged over 8 billion, with the number still growing.[1] Using a Web "crawler," a program that recursively follows Web links to find all pages within a subset of the Web graph, Albert *et al.* constructed maps of the Web domains `nd.edu`, `mit.edu`, and `whitehouse.gov`. For the domain `nd.edu`, for example, they found 325 729 Web pages (vertices) and close to 1.5 million links (edges).

Albert *et al.* were interested in discovering the "diameter" of the Web—the typical distance between Web pages, or how many clicks it would take a web "surfer" to get from one place to another on the Web.[2] They were interested in whether the Web shows the same small-world effect that is observed in social networks (Chapter 2). Using their data for the domain `nd.edu`, they showed that in fact it does. They found a short vertex-vertex distance within this domain of 11.2, and then, by scaling this result up to the (then) size of the entire Web, they predicted a typical path length of about 19 clicks between any two pages on the Web. One must interpret this result carefully since, as shown by **Broder *et al.* (2000)** and discussed later in this chapter, most pairs of pages on the Web are not actually reachable from one another. However, the figure for the path length found by Albert *et al.* has turned out to be a remarkably good one considering the small data sample from which it was derived, and the basic result that the Web, like social networks, displays the small-world effect is now well established.

Interestingly, however, although path lengths were the main focus of the paper by Albert *et al.*, it is another result from the paper that has attracted most attention. This paper was one of the first to show that the World Wide Web has a degree distribution that approximately follows a power law—the Web is what is now called a *scale-free network*. (The same result was also discovered independently by Kleinberg *et al.* (1999).)

The degree of a vertex in a network is the number of edges connected to it. For instance, in a social network in which the edges represent friendships, the

---

[1] Since much of the Web consists now of dynamically created pages, it makes less sense to talk about the number of pages on the Web—many pages only come into existence when you ask for them. One should perhaps not think of the Web itself as the network, but instead consider only the networks formed by the snapshots created when a search engine crawls the Web. For more information see, for example, `http://www.searchenginewatch.com`.

[2] Strictly the diameter is the *maximum* distance between nodes in the network, but Albert *et al.* used the term loosely to refer to the mean distance.

degree of an individual would be the number of friends he or she has. A quantity of much interest in the study of networks is the frequency distribution of degrees (Rapoport and Horvath 1961; **Price 1965**). If we denote degree by k, then the degree distribution can be represented by the quantity p_k, which is the fraction of vertices in the network that have degree k.

The World Wide Web, like the citation networks discussed in the previous chapter, is a directed network in that each link in the network points only in one direction from one Web page to another. This means that each page on the Web has two different degrees, the *in-degree*, which is the number of other pages that link to it, and the *out-degree*, which is the number that it links to. Correspondingly there are also two degree distributions for in- and out-degree. If the Web were a random (but directed) graph of the Erdős–Rényi type (Section 4.1), then these distributions would both take the simple Poissonian form $p_k \sim z^k/k!$, where z is the mean degree of the network. Albert *et al.*, however, found nothing like a Poisson form in the Web. Instead they found that the tails of both in- and out-degree distributions follow power laws:

$$p_{k_{\text{in}}} \sim k_{\text{in}}^{-\gamma_{\text{in}}}, \tag{3.1}$$

$$p_{k_{\text{out}}} \sim k_{\text{out}}^{-\gamma_{\text{out}}}, \tag{3.2}$$

where $\gamma_{\text{in}} \simeq 2.1$ and $\gamma_{\text{out}} \simeq 2.45$.

While the difference between Poisson and power-law distributions may not at first appear particularly important, it actually has far-reaching implications. The large-k tail of the power law decays much more slowly than the tail of the Poissonian, so that there are a small but significant number of vertices in the network with very high degree. These high-degree vertices are often called *hubs*. As described in a number of the papers in this book, hubs can play an important role in shaping the properties of a network.

The findings of Albert *et al.* have been confirmed and expanded by a number of other studies. Of particular interest is the study by **Broder *et al.* (2000)**, described below, which verified the power-law behavior and the small-world nature of the Web for a very large crawl. The small-world behavior was also independently discovered by Adamic (1999), although she studied a slightly different system. While Albert *et al.* looked at the Web at the level of individual Web pages, Adamic looked at the level of domains. A large Web domain such as `cnn.com` can host hundreds of thousands of Web pages, a nontrivial aggregation of the document level data (Huberman and Adamic 1999). Adamic studied a sample of over 50 million documents grouped into 259 794 domains. Two domains were connected if there was a link from any document in one domain to any document in the other. Adamic found that the average distance between any two nodes in her domain map was 3.1. The reason for this very small figure is that the mean degree of a vertex in the domain network is very high, since each vertex represents many Web pages, and mean vertex–vertex distance is expected to decrease (logarithmically) with increasing degree, at least in most models of networks (Bollobás 2001;

Newman *et al.* 2001; Chung and Lu 2002a; Bollobás and Riordan 2002; Fronczak *et al.* 2002a; Cohen and Havlin 2003). Although the studies by Albert *et al.* and by Adamic thus reflect rather different representations of the Web graph, they concur on the fundamental point that the Web appears to be a small world.

Broder et al. (2000)

Since the pioneering work of **Albert *et al.* (1999)**, Kleinberg *et al.* (1999), and Adamic (1999), the graph structure of the Web has been studied by many authors. One of the most influential studies has been that of Andrei Broder and collaborators at Altavista and IBM **(Broder *et al.* 2000)**. Broder *et al.* studied the largest Web graph available for research to date, consisting of over 200 million documents. They studied the in- and out-degree distributions, as Albert *et al.* and Kleinberg *et al.* did also, and again found power-law forms for both distributions. However, the important new results in this paper concern not the degree distributions but the component structure of the network.

The Web, as mentioned earlier, is a directed graph. Each link points in one direction from one page to another. Just because it is possible to get from page A to page B by following a single link does not mean that there is another link that will take you back again. More generally if you can surf from A to B by a succession of links, there is no guarantee that you will be able to surf back again, and even if you can you are unlikely to take the same path going back as you did getting there. As pointed out by Broder *et al.*, this feature of the network affects its navigability in a fundamental way. Their most striking finding is that most pairs of pages on the Web are not in fact reachable from one another by surfing links. If you choose an A and B at random from the pages of the Web, there is only about a 1 in 4 chance that you can surf from A to B by any path.

Broder *et al.* found that their Web sample was divided into four general regions. First, there is the core, or *giant strongly connected component*, whose distinguishing feature is that there is a path in each direction between any two pages belonging to it. That is, starting from any document in the core, one can click one's way, through some set of hyperlinks, to any other core document. The second and third regions of the graph are called the giant in- and out-components. The *giant in-component* is that set of vertices from which one can reach the core, but which cannot be reached from the core. The *giant out-component* is that set which can be reached from the core, but from which the core cannot be reached. The fourth region of the network contains everything not in the other three regions, meaning *islands* of vertices disconnected from the bulk of the graph and *tendrils*, which are vertices connected to the giant in- or out-components but not belonging to them by virtue of not also being connected to the core. (One can also define the *giant weakly connected component* of the graph to be the giant component one finds if one ignores the directed nature of the edges. The tendrils are that part of the giant weakly connected component that do not belong to any of the other giant components.)

Broder *et al.* drew their now-famous "bow tie" diagram to represent the

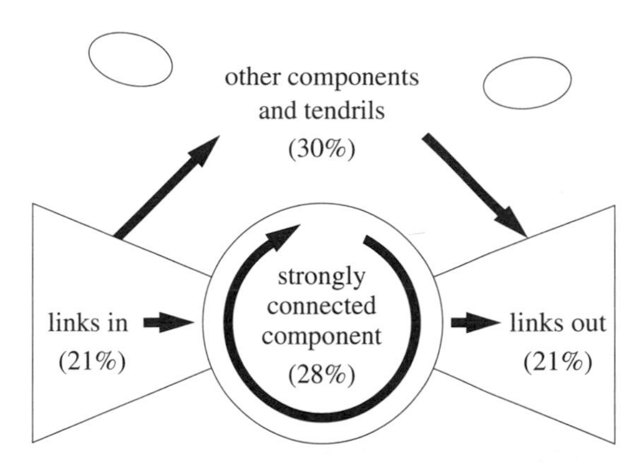

FIGURE 3.1 The "bow tie" diagram representing the structure of the World Wide Web. The center of the bow tie represents the giant strongly connected component, and the left and right loops of the bow the giant in- and out-components. The remainder of the graph consists of the disconnected components and tendrils, as described in the text. Each of the four regions fills about a quarter of the network, the exact percentages being shown in the figure.

relation between the different regions of the graph. We show a version of this diagram in Figure 3.1. Broder *et al.* found that each of the four regions in their Web sample occupied about a quarter of the whole graph. In order to be able to reach (most) other pages, a page must be in the giant strongly connected component or the giant in-component, which between them cover about half the network. Similarly, to be reachable from (most) others, a page must be in the giant strongly connected component or the giant out-component, which again cover about half the network. Thus, as stated above, paths exist between only about $\frac{1}{2} \times \frac{1}{2} = \frac{1}{4}$ of all vertex pairs on the Web. Broder *et al.* also measured the mean length of the paths connecting pairs for which a path exists and found it to be about 16, fairly close to the value of 19 predicted by **Albert *et al.* (1999)** by extrapolation from their smaller data set (see above). In fact, the extrapolation method used by Albert *et al.* predicts a mean distance of 17.9 for a graph the size of the sample used by Broder *et al.*, which is in even better agreement.

Many other networks, both artificial and naturally occurring, are also directed, including metabolic networks, neural networks, food webs, and email networks.[3] Many of the general ideas highlighted by Broder *et al.* for the World Wide Web can be applied to these networks as well, and a number of more recent studies have done exactly this (**Newman *et al.* 2001**; Dorogovtsev *et al.* 2001a; Sánchez *et al.* 2002; Schwartz *et al.* 2002).

[3]Citation networks, which were discussed in the previous chapter, are also directed, but are in addition acyclic, which makes their structure quite unlike that of the Web.

Faloutsos, Faloutsos, and Faloutsos (1999)

In popular parlance the terms "Internet" and "World Wide Web" are often used interchangeably, but the two networks are in fact quite distinct. The World Wide Web is a virtual network of documents connected by hyperlinks, while the Internet is a physical network of computers and routers connected by cables. At around the same time that Albert *et al.* were studying the network structure of the Web, three brothers in California were doing the same for the physical Internet **(Faloutsos *et al.* 1999)**.

The fundamental vertices of the Internet are computers and the edges are the physical connections between them. Viewed at this level, however, the network is enormous, difficult to study, and constantly changing. Practical studies of the graph structure of the Internet have therefore tended to examine the network at one of two possible levels of coarse-graining.[4] At the router level the vertices are routers, the powerful computers that form the switching centers of the network, and the edges are the cables that run between them. At the autonomous system level the vertices are groups of computers—autonomous systems—into which the Internet is divided for administrative purposes. The computers belonging to a single company or university will often form an autonomous system, for instance, and many though not all domains are single autonomous systems. An edge between two autonomous systems indicates that at least one computer in one of them has a direct data connection to at least one computer in the other.[5]

The Faloutsos brothers were among the first to study the graph structure of the Internet, and their principal finding was that the degree distribution p_k of the network approximately follows a power law, both at the router level and at the level of autonomous systems **(Faloutsos *et al.* 1999)**. The Internet, like the World Wide Web, appears to be a scale-free network, with a few hub nodes that are linked to many other nodes. A number of the hubs are prominently visible if we inspect a plot of the Internet graph such as that shown in Figure 3.2.

The finding that the Internet is a scale-free network was influential in the computer science community and motivated a considerable amount of fundamental and applied research. (For a recent review see Pastor-Satorras and Vespignani 2004.) In particular, new data protocols need to be tested on realistic network topologies, so in the absence of detailed Internet maps computer scientists have designed *topology generators*, algorithms that generate networks whose structure is expected to resemble the Internet's real topology. Before 1999 topology generators were mostly based on variants of the Erdős–Rényi random graph model (Section 4.1), and thus generated random networks with Poisson or similar degree distributions. But recent studies have demonstrated that protocols tested and optimized on such networks often perform poorly on networks with topology more like that found by Falout-

[4]In addition to router-level and autonomous system studies, the Internet has also occasionally been studied at the domain level or at the level of class C subnets.

[5]Technically, an autonomous system is defined as a group of computers that advertises a common routing policy using the Border Gateway Protocol (BGP), and the connections between autonomous systems are direct peering relations under BGP.

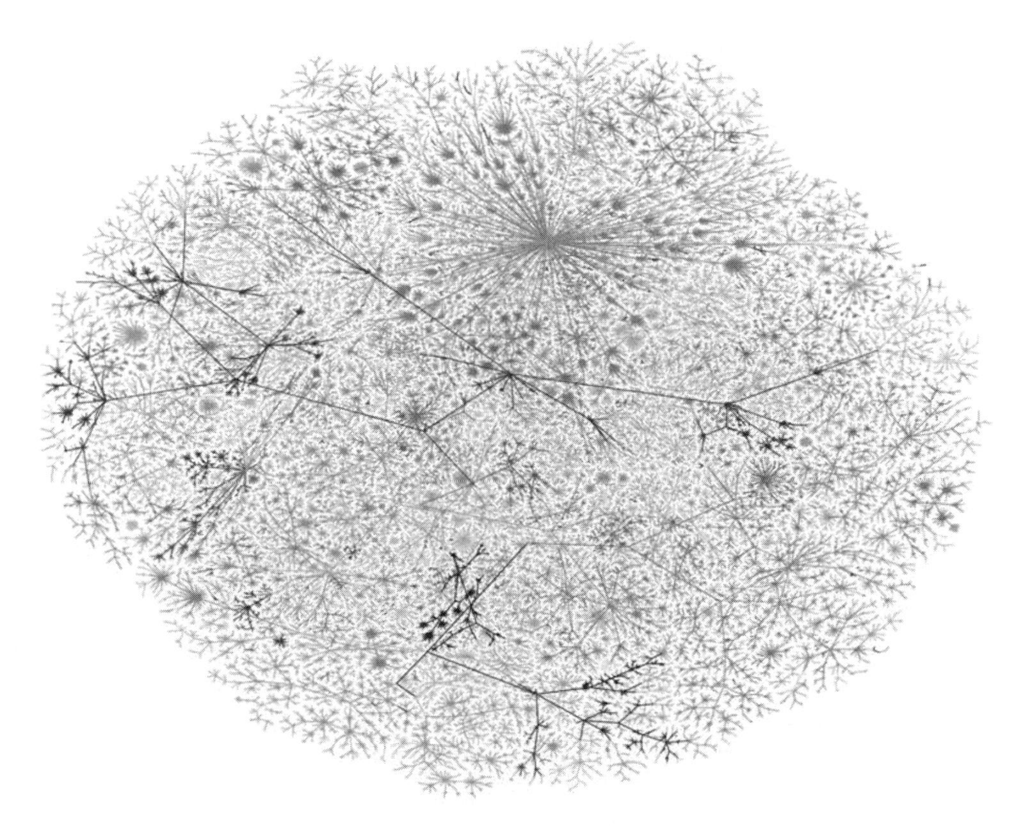

FIGURE 3.2 A map of the Internet at the autonomous system level. Many hubs, vertices with an unusually large number of connections, are clearly visible in the figure. Figure created by Bill Cheswick and Hal Burch, and reproduced by permission of Lumeta Corp.

sos *et al.* for the true Internet. Their findings have led to the construction of a new breed of network topology generators that mimic the important features of the net's structure more faithfully (Medina *et al.* 2000; Yook *et al.* 2001; Li *et al.* 2004a).

A number of other more recent studies have extended and refined the findings of Faloutsos *et al.* Govindan and Tangmunarunkit (2000) have suggested some new and more accurate algorithms for probing the Internet and used them to confirm that the degree distribution of the router-level map is indeed approximately power-law in form. Chen *et al.* (2002) have performed a very thorough study of the autonomous system map, including careful control of sampling errors, and find that the degree distribution of this network does not exactly follow a power law, although they confirm the basic finding of Faloutsos *et al.* that the distribution has a heavy tail of hub nodes.

Faloutsos *et al.* also investigated the spectral properties of the Internet graph, finding that the eigenvalue spectrum of the network has a power law tail (see also **Farkas *et al.* (2001)** and Goh *et al.* (2001a)). They also found that the total number of pairs of nodes that are a distance h or less from one another through the network varies as a power of h. This result, however, has not been confirmed by later work, and is thought to be an artifact of the limited data set that they had at their disposal.

Amaral, Scala, Barthélémy, and Stanley (2000)

The appearance of power laws in the degree distributions of citation networks, the Internet, and the World Wide Web is an important result that has far-reaching implications. However, there are also many networks whose degree distribution does not follow a power law. Consider, for example, the network of atoms connected by bonds in an amorphous material. There are clear physical constraints to the number of bonds an atom can have, and hence one does not expect to see hubs with very high degree in such a network. For other networks, such as social or communication networks, there are no obvious physical constraints like this, but nonetheless the power law is only one of many functional forms that the degree distributions in such networks take. **Amaral *et al.* (2000)** highlighted this observation in an influential paper in the *Proceedings of the National Academy* in which they calculated degree distributions for a variety of different kinds of networks.

Amaral *et al.* give degree distributions for five different networks: an airline network in which the vertices are airports and the edges are passengers flying from one airport to another;[6] a social network of friendships between Mormons in Provo, Utah; a friendship network of 417 junior high school students; and the networks previously studied by **Watts and Strogatz (1998)** of the power grid and of movie actor collaborations (see Section 4.2). Amaral *et al.* found that none of these networks have power-law degree distributions,[7] although none of them follow the Poisson distribution expected for a random graph either. Instead, all of them are right-skewed but with non-power-law distributions: the power grid and air traffic networks have exponential distributions, the high school and Mormon networks have Gaussian distributions, and the movie actor network has an exponentially truncated power-law distribution.[8]

Amaral *et al.* suggest that, as in the example of atoms in an amorphous material, the explanation for the exponential or Gaussian tails seen in their networks is one of constraints, although the constraints now are social or technological rather than physical. They argue that the degrees of vertices are limited if there are costs associated with maintaining links or if nodes show aging or time constraints. For example, if nodes fail to acquire links after a certain age, then their degree has an upper limit. This is a common situation in professional networks such as collaboration networks, where an individual's professional contacts diminish after retirement. Similarly, if there is a cost for adding a new link to an existing node,

[6]This is distinct from the network one sees on the maps in in-flight magazines, in which the edges are routes between airports.

[7]A more recent paper, by Guimera and Amaral, however, claims conversely that the airline network *does* have a power-law degree distribution (Guimerà and Amaral 2003).

[8]The result for the movie actor network appears to be at odds with that of **Barabási and Albert (1999)**, who found a power-law distribution for this network. However, Barabási and Albert studied a different version of the network from which they removed TV series. TV actors can have very high degree, appearing in many episodes of a drama or serial, so certainly it is plausible that their removal could affect the tail of the degree distribution, although it is surprising that their removal appears to *create* a power-law tail; one might have expected the reverse.

then limits on available resources can translate such costs into limits on the degree. Using numerical simulations, Amaral *et al.* demonstrate that effects like these can impose an exponential cutoff on the degree distribution of a network, so that, for instance, a network that might normally have a power-law distribution will actually show an exponentially truncated power-law distribution instead.[9]

Since the publication of the paper by Amaral *et al.*, exponential cutoffs have been observed in a variety of other networks, such as protein interaction networks (Jeong *et al.* 2001) and collaboration networks **(Newman 2001a)**. Cutoffs can have important practical effects, on the spread of infections over contact networks, for example **(Pastor-Satorras and Vespignani 2001)**, and on a network's robustness to vertex deletion (**Albert *et al.* 2000; Cohen *et al.* 2000**, 2001; **Callaway *et al.* 2000**).

Jeong, Tombor, Albert, Oltvai, and Barabási (2000)

We now look at a number of other empirical studies of networks of particular scientific interest, starting with three papers that examine biochemical networks.

The many biochemical reactions that power the cells of living organisms form a network in which the vertices are chemicals, either simple ones like water or carbon dioxide, or more complex biological molecules like the energy-storing molecule adenosine triphosphate (ATP). The directed edges between vertices represent reactions that turn one chemical into another. **Jeong *et al.* (2000)** performed a comparative analysis of the metabolic networks of 43 different organisms from all domains of life, using data drawn from the publicly available WIT database (Overbeek *et al.* 2000). They found that, in common with many of the other networks described in this chapter, metabolic networks appear to have scale-free degree distributions and short vertex-vertex path lengths.

Since metabolic networks are directed, they, like the World Wide Web discussed earlier, have both an in- and an out-degree for each vertex. The distributions of both were found roughly to follow power laws for all the organisms studied, with degree exponents γ_{in} and γ_{out} in the range from 2.0 to 2.4. Thus, while the vast majority of metabolites take part in only one or two reactions, a few participate in hundreds, serving as the hubs of the metabolism.

Jeong *et al.* also found that the average path length in their networks was small: most pairs of molecules can be linked by a path of only three reactions. This finding has biological significance. If one were to find that the shortest chemical path between two molecules was 100, then any change in the concentration of the first molecule would have to go through 100 intermediate reactions before reaching the second molecule. Such a long path would certainly put a strong damper on any correlations between the concentrations of the two metabolites, with perturbations decaying quickly as they moved along the path. By contrast, the short path lengths

---

[9]The presence of aging effects in scale-free networks can also affect the exponent of the degree distribution. Dorogovtsev and Mendes (2000a) studied a model network in which vertices gradually lose their ability to attract connections over time. Their calculations indicate that in this case there is no exponential cutoff, but the degree exponent depends continuously on the decay rate of the aging effect.

found by Jeong *et al.* indicate that changes in concentration of a molecule should reach others quickly.

While unexpectedly short, the precise value of the mean path length was not the most intriguing aspect of the results for path lengths. Jeong *et al.* also looked at how path lengths vary between organisms. Since the networks of the different organisms studied vary considerably in size, one might expect the mean path length to vary also, as it does in other networks. Surprisingly, however, measurements indicated that there was very little variation in the path length between organisms. The tiny network of a parasitic bacterium has the same mean path length as the highly developed network of a large multicellular organism. The measurements indicated that most cells have the same hubs as well. For the vast majority of organisms, the ten most connected molecules are the same. ATP is almost always the biggest hub, followed closely by adenosine diphosphate (ADP) and water. The role of ATP, ADP, and water as prominent hubs is certainly not surprising. Their roles in the storage and release of the energy that powers most functions of the cell makes them crucial to almost every biochemical process.

Wagner and Fell (2001)

Many of the issues concerning metabolic networks addressed by **Jeong *et al.* (2000)** were investigated independently by Wagner and Fell (2001). Their work focused on only a single organism, the gut bacterium *E. coli*, but looked at similar issues of mean path lengths between metabolites and the presence of hubs in the network.

Wagner and Fell studied two different versions of the metabolic network of *E. coli*: the full network and a version in which they removed the highest degree nodes in the network, such as ATP, reasoning that these are connected to essentially all other nodes and hence that calculating path lengths would be meaningless with them present. They then tested the networks for the two properties highlighted by **Watts and Strogatz (1998)** in their comparative analysis of networks which we discuss in Section 4.2, namely short paths lengths and a high clustering coefficient. They confirm that their metabolic network has both of these properties, although the removal of the high-degree vertices increases path lengths substantially from 2.9 (which agrees with the findings of Jeong *et al.*) to 3.9. They also confirm the power-law degree distribution of the network, even with ATP and other high-degree nodes removed, and find an exponent of about 2.3 for the power law, which is in good agreement with that found by Jeong *et al.*

Wagner and Fell also studied a reaction network of metabolites, in which each reaction is denoted by a node, two nodes being connected if the corresponding reactions share at least one metabolite. In contrast with the metabolic network, the reaction network is not scale-free, a finding which corresponds to the well-known fact that most reactions involve only a few metabolites.

Wagner and Fell offer some speculations about the possible evolutionary origins of the properties they observe. For example, it has been pointed out that in some models of scale-free networks the vertices with high degree are also the oldest (Adamic and Huberman 2000; Barabási *et al.* 2000b; Krapivsky and Redner 2001).

Wagner and Fell's investigation supports this picture for metabolic networks; they find that the best connected molecules tend to be those that arose early in evolutionary history, representing the building blocks of the most ancient metabolic pathways. They speculate that the small-world properties of the network may be a consequence of natural selection favoring networks in which the transition time between metabolic states is minimized. This argument is perhaps also supported by the finding of Jeong *et al.* that mean path lengths did not vary much from one organism to another, suggesting that there may be some advantage to deliberately keeping metabolic path lengths short even as networks become large.

Metabolic networks are by no means the only networks characterizing the cellular environment. Another important subcellular network describes physical protein-protein interactions. In this network the nodes are proteins and two proteins are connected if there is experimental evidence that they bind together. Recent experiments have shown that it is possible to reconstruct this network in some detail (Uetz *et al.* 2000; Ito *et al.* 2000, 2001; Giot *et al.* 2003; Li *et al.* 2004b), and statistical analyses of the results have been carried out by several groups (Jeong *et al.* 2001; Wagner 2001; Maslov and Sneppen 2002). Other recent studies have also looked at genetic regulatory networks (Agrawal 2002; Shaw 2003; Provero 2002; Farkas *et al.* 2003) and protein domain networks (Wuchty 2001, 2002), focusing again on degree distributions and path lengths, and confirming their scale-free and small-world nature.

Milo, Shen-Orr, Itzkovitz, Kashtan, Chklovskii, and Alon (2002)

The preceding two papers on biochemical networks focused on some relatively conventional network properties such as path lengths and degree distributions. A very different approach has been taken in the work of **Milo *et al.* (2002)**, who developed a numerical method to search networks for repeated subgraphs.

One of the most striking features of many real-world networks, including social, biological, and technological networks, is the appearance of regular patterns of ties that cannot be attributed to mere random chance. A simple case is the triad or triangle, a fully connected subgraph of three nodes. The importance of the triad in social networks was first pointed out by the great German sociologist Georg Simmel (Simmel 1950), and there has been renewed interest in triads with studies of clustering coefficients in networks (see Section 4.2). The clustering coefficient as usually defined, however, applies only to undirected networks and only accounts for patterns of ties that involve three nodes. As various authors have pointed out, many other small subgraphs may be of importance too (Gleiss *et al.* 2001; Fronczak *et al.* 2002b; Newman 2003a; Bianconi and Capocci 2003; Caldarelli *et al.* 2004). In their paper, **Milo *et al.* (2002)** have substantially extended the exploration of local structure in real-world networks with their studies of what they call *network motifs*.

Milo *et al.* commence by identifying all possible arrangements of directed ties in connected subgraphs of three or four nodes, where each topologically distinct arrangement (invariant under relabeling of the nodes) constitutes a single motif. There are 13 three-node motifs and 199 four-node motifs, each occurrence of which the authors enumerate in a variety of networks, including but not limited to biological ones: two genetic regulatory networks; seven food webs; the neural network of the worm *C. elegans*; eight electronic circuits (five forward logic chips and three digital fractional multipliers); and the Web domain `nd.edu`. They then compare the frequency of occurrence of each motif with the frequency in a simulated random network that they constrain to have the same single-node characteristics as the real networks. That is, each node in the simulated network is constrained to have the same number of incoming and outgoing links as in the corresponding real network. Furthermore, when looking for four-node motifs, Milo *et al.* generate test networks that have the same density of three-node motifs as the real examples, but are in all other respects random, thus controlling for the presence of four-node motifs that are generated by randomly arranged combinations of three-node motifs.

Milo *et al.* present a number of findings that appear to delineate some distinct classes of networks. For example, networks that process information, like the gene networks, the neural network, and the forward logic chips, share basically the same motifs, almost all of which are quite distinct from the food web motifs, perhaps because the food webs have evolved to pump energy (from lower to higher trophic levels) rather than to process information. And the motifs of both food webs and information processing networks are distinct from those characterizing the World Wide Web, which neither processes nor pumps anything, but instead represents communities of knowledge. Milo *et al.* do not consider any social networks, the inclusion of which might have resulted in yet more interesting comparisons, but their methods are quite general (applying both to directed and undirected networks) and could certainly be applied to social networks by others.

Newman (2001a)

The study of social networks is made difficult by the arduous business of collecting social data, by the inaccuracy and subjectivity of those data, and by the limited size of the data sets it is possible to assemble with reasonable expenditure of money and effort. There are, however, a few examples of large social networks (or proxies for social networks) that, for one reason or another, are both copiously and accurately documented. Networks of computer-mediated interactions are one example: email networks (Ebel *et al.* 2002; Newman *et al.* 2002a; Tyler *et al.* 2003) and online communities (Smith 2002; Holme *et al.* 2004; Csányi and Szendrői 2004) fall into this category. The widely studied network of movie actors is another example **(Watts and Strogatz 1998; Barabási and Albert 1999; Amaral *et al.* 2000)**. And one of the largest and best documented networks of all is the network of collaborations of scientists.

It has long been realized that published papers provide a potential window on the collaboration patterns of scientists (Kretschmer 1994; Persson and Beckmann 1995; Melin and Persson 1996; Ding *et al.* 1999; Bordens and Gómez 2000). In particular, one can define and study a network in which the nodes are authors and two authors are connected if they have coauthored a paper. This is the basis, for example, of the mathematician's concept of the "Erdős number," which is one's distance from the mathematician Paul Erdős through the collaboration network **(de Castro and Grossman 1999)**.

Newman (2001a) constructed relatively complete collaboration networks for a variety of different scientific disciplines using bibliographic databases covering publications in physics, high-energy physics, biomedical research, and computer science over a five-year window from 1995 to 1999. He measured a variety of standard (and some not-so-standard) structural quantities for these networks, including path lengths, clustering coefficients, and degree distributions. He found that path lengths were short, with typically fewer than six steps between pairs of scientists, a result that could have implications for the spread of scientific knowledge. Furthermore, he showed that the variation between the different networks in the mean path length is well described by a logarithmic function of the number of vertices in the network, as predicted by random-graph theories but rarely verified empirically. The networks also showed high clustering, a property typical of social networks and very different from random graphs **(Watts and Strogatz 1998)**. Newman found that most of the degree distributions in his networks were well fit by power laws with exponential cutoffs. Following **Amaral *et al.* (2000)** he speculated that the cutoff was due to the "cost" of producing a scientific paper, which prevents authors from writing papers with an arbitrary number of coauthors in a finite time. An interesting exception to the truncated power-law form was the network of collaborations in high-energy physics. Much of experimental high-energy physics is carried out by very large collaborations of scientists who produce papers with tens or hundreds of coauthors; in this field it is possible, with relatively little effort, to collaborate with a very large number of coauthors, and the resulting degree distribution of the coauthorship network appears to follow a pure power law quite closely.

Newman published a number of follow-on papers about collaboration networks (Newman 2001b,c,d) that explored the same data further, treating such issues as component sizes, frequency of collaborations, various measures of centrality, and evolution of the networks over time.

A number of other authors have looked at coauthorship networks as well. Grossman and collaborators (Grossman and Ion 1995; **de Castro and Grossman 1999**; Grossman 2002) started a project in the 1990s to look at the statistics of Erdős numbers, and in recent years, using data drawn from the archives of the journal *Mathematical Reviews*, they have expanded this effort to a study of the entire collaboration network of mathematicians. A recent paper by Newman (2004a) compares Grossman's results to his own.

Collaboration networks were also investigated by Barabási *et al.* (2002), who assembled networks for mathematics and neuroscience from publications appearing over the eight-year period from 1991 to 1998. Barabási *et al.* were interested primarily in the time evolution of the networks, and in particular used them to test the "preferential attachment" hypothesis that individuals accrue new collaborators at a rate proportional to the number they already have. They also presented a simple theoretical model of the growth of a collaboration network that predicts a degree distribution showing a crossover between two power-law regimes: for small k it is predicted that $p_k \sim k^{-2/3}$, while for large k we should see $p_k \sim k^{-3}$. The position k_c at which the crossover takes place increases linearly with time, indicating that in the asymptotic limit $t \to \infty$ only the $k^{-2/3}$ behavior should be visible. For short time windows, however, the k^{-3} behavior may be present. In some cases this theory agrees quite well with the observed degree distributions (Newman 2001b).

Liljeros, Edling, Amaral, Stanley, and Åberg (2001)

An important motivation for the study of social networks has been to understand the epidemiology of diseases that spread over those networks (see Section 5.1). Sexually transmitted diseases (STDs), for example, spread on the network in which the nodes are people and the edges represent sexual contacts between them. In traditional mathematical epidemiology such networks are usually approximated by random mixing of individuals, or at least random mixing within subpopulations, meaning that all individuals have equal likelihood of contact with all others in the same subpopulation and crucially that all individuals have roughly the same number of contacts. As the previous papers in this chapter show, however, this approximation is rarely valid for real-world networks.

Various studies have been performed over the years of networks of sexual contacts (Laumann *et al.* 1994; Potterat *et al.* 2002). Such networks are, in some respects, easier to reconstruct than other social networks because the seriousness of STD infections has encouraged governments to permit and fund extensive studies of a kind that have not been possible in other areas. On the other hand, there are problems with these studies as well. In addition to the obvious reluctance of people to talk about sexual practices (Morris 1993), most of the studies have been carried out using "contact tracing" methods in which individuals are interviewed and asked to name their contacts, who are then interviewed in turn. This method is known to introduce substantial biases into the population sample, and it presents problems, for instance, if we wish to discover the degree distribution of the contact network.

The alternative approach is to conduct so-called sociometric studies, that is, studies in which all members of a community or a random sample are polled to give more statistically reliable data. The recent study by Bearman *et al.* (2004) of high school students falls in this category and reveals some interesting features of the network that contact-tracing might miss.

The last paper reproduced in this section, by **Liljeros *et al.* (2001)**, analyzes data from a sociometric study of sexual partnerships carried out in Sweden. This study differs from the others cited here in that it did not probe the complete network structure, but it did probe the degree distribution of the network. To do this, it is sufficient to ask a representative subset of the population how many partners they had over a given period of time. In the study 4781 individuals of ages 18 to 74 were polled, and, with a response rate of 59%, the experimenters obtained degree information for 2810 nodes in the sexual contact network. Participants were asked both how many sexual partners they had had in the previous year, and how many they had had in their lifetime. The results indicated that the majority of respondents had between one and ten sexual partners during their lifetime, but a few had several hundred. Moreover, a simple cumulative plot appears to show that the distribution follows a power law—sexual contacts form a scale-free network—with exponent 3.5 ± 0.2 for women and 3.3 ± 0.2 for men (i.e., the exponents were the same within measurement error for men and women).

This finding could be relevant to the spread of sexually transmitted diseases. In traditional mathematical epidemiology, diseases display an epidemic threshold, as a function of their probability of transmission, between a regime in which only small local outbreaks of the disease occur and a regime in which the disease shows epidemic behavior, reaching a significant fraction of the entire population in question. In an important paper a few months earlier, however, **Pastor-Satorras and Vespignani (2001)** had demonstrated that networks with power-law degree distributions may not have an epidemic threshold at all: in such networks it is possible for diseases always to spread, regardless of the probability of transmission. Could this be the case with sexually transmitted diseases? The results of Liljeros *et al.* suggest that it could, and this was the primary concern driving their work. Some caveats are in order, however. The results of Pastor-Satorras and Vespignani imply that the epidemic threshold is only absent for power laws with exponent less than 3. Since the measured exponent is greater than 3 in the data of Liljeros *et al.*, the simple interpretation is that a threshold still exists. Also, as other authors have since pointed out, there are other circumstances that may make a threshold appear even for exponents less than 3, such as correlations between the degrees of vertices (Boguñá *et al.* 2003b). Statisticians have also questioned whether the power law apparent in the data of Liljeros *et al.* is a true one (Jones and Handcock 2003a,b). Although the debate on this question is far from settled at the time of writing, recent studies focusing on contact patterns in Britain and Zimbabwe have found power-law degree distributions in sexual interaction networks (Schneeberger *et al.* 2004).

Nonetheless, the general findings of Liljeros *et al.* represent an important contribution to our understanding of sexual contact networks, particularly the finding that the hubs in the network are not outliers but form part of a possibly power-law, and certainly heavy-tailed, degree distribution. The hope is that these findings will help in the discovery of new strategies for avoiding or containing current and future epidemics.

Internet

Diameter of the World-Wide Web

Despite its increasing role in communication, the World-Wide Web remains uncontrolled: any individual or institution can create a website with any number of documents and links. This unregulated growth leads to a huge and complex web, which becomes a large directed graph whose vertices are documents and whose edges are links (URLs) that point from one document to another. The topology of this graph determines the web's connectivity and consequently how effectively we can locate information on it. But its enormous size (estimated to be at least 8×10^8 documents[1]) and the continual changing of documents and links make it impossible to catalogue all the vertices and edges.

The extent of the challenge in obtaining a complete topological map of the web is illustrated by the limitations of the commercial search engines: Northern Light, the search engine with the largest coverage, is estimated to index only 38% of the web[1]. Although much work has been done to map and characterize the Internet's infrastructure[2], little is known about what really matters in the search for information — the topology of the web. Here we take a step towards filling this gap: we have used local connectivity measurements to construct a topological model of the World-Wide Web, which has enabled us to explore and characterize its large-scale properties.

To determine the local connectivity of the web, we constructed a robot that adds to its database all URLs found on a document and recursively follows these to retrieve the related documents and URLs. We used the data collected to determine the probabilities $P_{out}(k)$ and $P_{in}(k)$ that a document has k outgoing and incoming links, respectively. We find that both $P_{out}(k)$ and $P_{in}(k)$ follow a power law over several orders of magnitude, remarkably different not only from the Poisson distribution predicted by the classical theory of random graphs[3,4], but also from the bounded distribution found in models of random networks[5].

The power-law tail indicates that the probability of finding documents with a large number of links is significant, as the network connectivity is dominated by highly connected web pages. Similarly, for incoming links, the probability of finding very popular addresses, to which a large number of other documents point, is non-negligible, an indication of the flocking nature of the web. Furthermore, while the owner of each web page has complete freedom in choosing the number of links on a document and the addresses to which they point, the overall system obeys scaling laws characteristic only of highly interactive self-organized systems and critical phenomena[6].

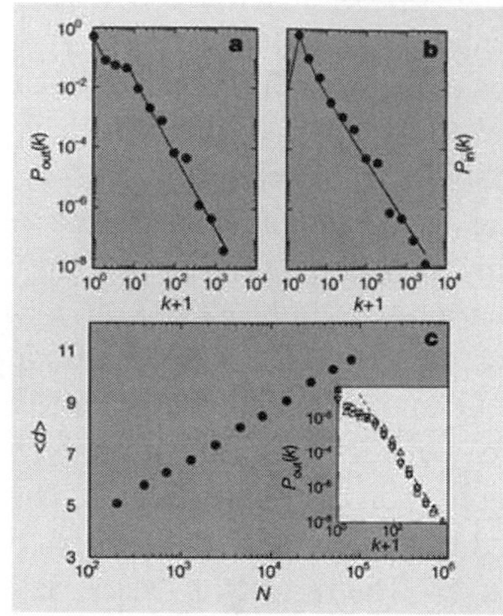

Figure 1 Distribution of links on the World-Wide Web. **a**, Outgoing links (URLs found on an HTML document); **b**, incoming links (URLs pointing to a certain HTML document). Data were obtained from the complete map of the nd.edu domain, which contains 325,729 documents and 1,469,680 links. Dotted lines represent analytical fits used as input distributions in constructing the topological model of the web; the tail of the distributions follows $P(k)\approx k^{-\gamma}$, with $\gamma_{out}=2.45$ and $\gamma_{in}=2.1$. **c**, Average of the shortest path between two documents as a function of system size, as predicted by the model. To check the validity of our predictions, we determined d for documents in the domain nd.edu. The measured $\langle d_{nd.edu}\rangle=11.2$ agrees well with the prediction $\langle d_{3\times10^5}\rangle=11.6$ obtained from our model. To show that the power-law tail of $P(k)$ is a universal feature of the web, the inset shows $P_{out}(k)$ obtained by starting from whitehouse.gov (squares), yahoo.com (triangles) and snu.ac.kr (inverted triangles). The slope of the dashed line is $\gamma_{out}=2.45$, as obtained from nd.edu in **a**.

To investigate the connectivity and the large-scale topological properties of the web, we constructed a directed random graph consisting of N vertices, assigning to each vertex k outgoing (or incoming) links, such that k is drawn from the power-law distribution of Fig. 1a,b. To achieve this, we randomly selected a vertex i and increased its outgoing (or incoming) connectivity to k_i+1 if the total number of vertices with k_i+1 outgoing (or incoming) links is less than $NP_{out}(k_i+1)$ (or $NP_{in}(k_i+1)$).

A particularly important quantity in a search process is the shortest path between two documents, d, defined as the smallest number of URL links that must be followed to navigate from one document to the other. We find that the average of d over all pairs of vertices is $\langle d\rangle=0.35+2.06\log(N)$ (Fig. 1c), indicating that the web forms a small-world network[5,7], which characterizes social or biological systems. For $N=8\times10^8$, $\langle d_{web}\rangle=18.59$; that is, two randomly chosen documents on the web are on average 19 clicks away from each other.

For a given N, d follows a gaussian distribution so $\langle d\rangle$ can be interpreted as the diameter of the web, a measure of the shortest distance between any two points in the system. Despite its huge size, our results indicate that the web is a highly connected graph with an average diameter of only 19 links. The logarithmic dependence of $\langle d\rangle$ on N is important to the future potential of the web: we find that the expected 1,000% increase in the size of the web over the next few years will change $\langle d\rangle$ very little, from 19 to only 21.

The relatively small value of $\langle d\rangle$ indicates that an intelligent agent, who can interpret the links and follow only the relevant one, can find the desired information quickly by navigating the web. But this is not the case for a robot that locates the information based on matching strings. We find that such a robot, aiming to identify a document at distance $\langle d\rangle$, needs to search $M(\langle d\rangle)\approx0.53\times N^{0.92}$ documents, which, with $N=8\times10^8$, leads to $M=8\times10^7$, or 10% of the whole web. This indicates that robots cannot benefit from the highly connected nature of the web, their only successful strategy being to index as much of the web as possible.

The scale-free nature of the link distributions indicates that collective phenomena play a previously unsuspected role in the development of the web[8], forcing us to look beyond the traditional random graph models[3-5,7]. A better understanding of the web's topology, aided by modelling efforts, is crucial in developing search algorithms or designing strategies for making information widely accessible on the World-Wide Web. Fortunately, the surprisingly small diameter of the web means that all that information is just a few clicks away.

Réka Albert, Hawoong Jeong, Albert-László Barabási
Department of Physics, University of Notre Dame, Notre Dame, Indiana 46556, USA
e-mail:alb@nd.edu

1. Lawrence, S. & Giles, C. L. *Nature* **400**, 107–109 (1999).
2. Claffy, K., Monk, T. E. & McRobb, D. Internet tomography. *Nature* [online] <http://helix.nature.com/webmatters/tomog/tomog.html> (1999).
3. Erdős, P. & Rényi, A. *Publ. Math. Inst. Hung. Acad. Sci.* **5**, 17–61 (1960).
4. Bollobás, B. *Random Graphs* (Academic, London, 1985).
5. Watts, D. J. & Strogatz, S. H. *Nature* **393**, 440–442 (1998).
6. Bunde, A. & Havlin, S. *Fractals in Science* (Springer, Berlin, 1994).
7. Barthélémy, M. & Amaral, L. A. N. *Phys. Rev. Lett.* **82**, 3180–3183 (1999).
8. Barabási, A.-L., Albert, R. & Jeong, H. <http://www.nd.edu/~networks>.

Computer Networks 33 (2000) 309–320

COMPUTER NETWORKS

www.elsevier.com/locate/comnet

Graph structure in the Web

Andrei Broder [a], Ravi Kumar [b,*], Farzin Maghoul [a], Prabhakar Raghavan [b], Sridhar Rajagopalan [b], Raymie Stata [c], Andrew Tomkins [b], Janet Wiener [c]

[a] *AltaVista Company, San Mateo, CA, USA*
[b] *IBM Almaden Research Center, San Jose, CA, USA*
[c] *Compaq Systems Research Center, Palo Alto, CA, USA*

Abstract

The study of the Web as a graph is not only fascinating in its own right, but also yields valuable insight into Web algorithms for crawling, searching and community discovery, and the sociological phenomena which characterize its evolution. We report on experiments on local and global properties of the Web graph using two AltaVista crawls each with over 200 million pages and 1.5 billion links. Our study indicates that the macroscopic structure of the Web is considerably more intricate than suggested by earlier experiments on a smaller scale.

Keywords: Graph structure; Diameter; Web measurement

1. Introduction

Consider the directed graph whose nodes correspond to static pages on the Web, and whose arcs correspond to links between these pages. We study various properties of this graph including its diameter, degree distributions, connected components, and macroscopic structure. There are several reasons for developing an understanding of this graph.

(1) Designing crawl strategies on the Web [15].
(2) Understanding of the sociology of content creation on the Web.
(3) Analyzing the behavior of Web algorithms that make use of link information [9–11,20,26]. To take just one example, what can be said of the distribution and evolution of PageRank [9] values on graphs like the Web?
(4) Predicting the evolution of Web structures such as bipartite cores [21] and Webrings, and developing better algorithms for discovering and organizing them.
(5) Predicting the emergence of important new phenomena in the Web graph.

We detail a number of experiments on a Web crawl of approximately 200 million pages and 1.5 billion links; the scale of this experiment is thus five times larger than the previous biggest study [21] of structural properties of the Web graph, which used a pruned data set from 1997 containing about 40 million pages. Recent work ([21] on the 1997 crawl, and [5] on the approximately 325 thousand node nd.edu subset of the Web) has suggested that the distribution of degrees (especially in-degrees, i.e., the number of links to a page) follows a *power law*.

The power law for in-degree: the probability

* Corresponding author. E-mail: ravi@almaden.ibm.com

1389-1286/00/$ – see front matter
PII: S1389-1286(00)00083-9

that a node has in-degree i is proportional to $1/i^x$, for some $x > 1$.

We verify the power law phenomenon in current (considerably larger) Web crawls, confirming it as a basic Web property.

In other recent work, [4] report the intriguing finding that most pairs of pages on the Web are separated by a handful of links, almost always under 20, and suggest that this number will grow logarithmically with the size of the Web. This is viewed by some as a 'small world' phenomenon. Our experimental evidence reveals a rather more detailed and subtle picture: most ordered pairs of pages cannot be bridged at all and there are significant numbers of pairs that can be bridged, but only using paths going through hundreds of intermediate pages. Thus, the Web is not the ball of highly connected spaghetti we believed it to be; rather, the connectivity is strongly limited by a high-level global structure.

1.1. Our main results

We performed three sets of experiments on Web crawls from May 1999 and October 1999. Unless otherwise stated, all results described below are for the May 1999 crawl, but all conclusions have been validated for the October 1999 crawl as well. First, we generated the in- and out-degree distributions, confirming previous reports on power laws; for instance, the fraction of Web pages with i in-links is proportional to $1/i^{2.1}$. The constant 2.1 is in remarkable agreement with earlier studies at varying scales [5,21]. In our second set of experiments we studied the directed and undirected connected components of the Web. We show that power laws also arise in the distribution of *sizes* of these connected components. Finally, in our third set of experiments, we performed a number of breadth-first searches from randomly chosen start nodes. We detail these experiments in Section 2.

Our analysis reveals an interesting picture (Fig. 9) of the Web's macroscopic structure. Most (over 90%) of the approximately 203 million nodes in our May 1999 crawl form a single connected component if links are treated as *undirected* edges. This connected Web breaks naturally into four pieces. The first piece is a central core, all of whose pages can reach one another along directed links; this 'giant strongly connected component' (*SCC*) is at the heart of the Web. The second and third pieces are called *IN* and *OUT*. *IN* consists of pages that can reach the *SCC*, but cannot be reached from it; possibly new sites that people have not yet discovered and linked to. *OUT* consists of pages that are accessible from the *SCC*, but do not link back to it, such as corporate Websites that contain only internal links. Finally, the *TENDRILS* contain pages that cannot reach the *SCC*, and cannot be reached from the *SCC*. Perhaps the most surprising fact is that the size of the *SCC* is relatively small; it comprises about 56 million pages. Each of the other three sets contain about 44 million pages, thus, all four sets have roughly the same size.

We show that the diameter of the central core (*SCC*) is at least 28, and that the diameter of the graph as a whole is over 500 (see Section 1.3 for definitions of diameter). We show that for randomly chosen source and destination pages, the probability that any path exists from the source to the destination is only 24%. We also show that, if a directed path exists, its average length will be about 16. Likewise, if an undirected path exists (i.e., links can be followed forwards or backwards), its average length will be about 6. These analyses appear in the Section 3. These results are remarkably consistent across two different, large AltaVista crawls. This suggests that our results are relatively insensitive to the particular crawl we use, provided it is large enough. We will say more about crawl effects in Section 3.4.

In a sense the Web is much like a complicated organism, in which the local structure at a microscopic scale looks very regular like a biological cell, but the global structure exhibits interesting morphological structure (body and limbs) that are not obviously evident in the local structure. Therefore, while it might be tempting to draw conclusions about the structure of the Web graph from a local picture of it, such conclusions may be misleading.

1.2. Related prior work

Broadly speaking, related prior work can be classified into two groups: (1) observations of the power law distributions on the Web; and (2) work on applying graph theoretic methods to the Web.

1.2.1. Zipf–Pareto–Yule and power laws

Distributions with an inverse polynomial tail have been observed in a number of contexts. The earliest observations are due to Pareto [27] in the context of economic models. Subsequently, these statistical behaviors have been observed in the context of literary vocabulary [32], sociological models [33], and even oligonucleotide sequences [24] among others. Our focus is on the closely related power law distributions, defined on the positive integers, with the probability of the value i being proportional to $1/i^k$ for a small positive number k. Perhaps the first rigorous effort to define and analyze a model for power law distributions is due to Simon [30].

More recently, power law distributions have been observed in various aspects of the Web. Two lines of work are of particular interest to us. First, power laws have been found to characterize user behavior on the Web in two related but dual forms:

(1) access statistics for Web pages, which can be easily obtained from server logs (but for caching effects); see [1,2,17,19];

(2) numbers of times users at a single site access particular pages, as verified by instrumenting and inspecting logs from Web caches, proxies, and clients (see [6] and references therein, as well as [23]).

Second, and more relevant to our immediate context, is the distribution of degrees on the Web graph. In this context, recent work (see [5,21]) suggests that both the in- and the out-degrees of vertices on the Web graph have power laws. The difference in scope in these two experiments is noteworthy. The first [21] examines a Web crawl from 1997 due to Alexa, Inc., with a total of over 40 million nodes. The second [5], examines Web pages from the University of Notre Dame domain, *.nd.edu, as well as a portion of the Web reachable from three other URLs. In this paper, we verify these power laws on more recent (and considerably larger) Web crawls. This collection of findings reveals an almost fractal-like quality for the power law in-degree and out-degree distributions, in that it appears both as a macroscopic phenomenon on the entire Web, as a microscopic phenomenon at the level of a single university Website, and at intermediate levels between these two.

There is no evidence that users' browsing behavior, access statistics and the linkage statistics on the Web graph are related in any fundamental way, although it is very tempting to conjecture that this is indeed the case. It is usually the case, though not always so, that pages with high in-degree will also have high PageRank [9]. Indeed, one way of viewing PageRank is that it puts a number on how easy (or difficult) it is to find particular pages by a browsing-like activity. Consequently, it is plausible that the in-degree distributions induce a similar distribution on browsing activity and consequently, on access statistics.

Faloutsos et al. [16] observe Zipf–Pareto distributions (power law distributions on the *ranks* of values) on the Internet network topology using a graph of the network obtained from the routing tables of a backbone BGP router.

1.2.2. Graph theoretic methods

Much recent work has addressed the Web as a graph and applied algorithmic methods from graph theory in addressing a slew of search, retrieval, and mining problems on the Web. The efficacy of these methods was already evident even in early local expansion techniques [10]. Since then, the increasing sophistication of the techniques used, the incorporation of graph theoretical methods with both classical and new methods which examine context and content, and richer browsing paradigms have enhanced and validated the study and use of such methods. Following Butafogo and Schneiderman [10], the view that connected and strongly connected components represent meaningful entities has become accepted. Pirolli et al. [28] augment graph theoretic analysis to include document content, as well as usage statistics, resulting in a rich understanding of domain structure and a taxonomy of roles played by Web pages.

Graph theoretic methods have been used for search [8,9,12,13,20], browsing and information foraging [10,11,14,28,29], and Web mining [21,22,25, 26]. We expect that a better structural characterization of the Web will have much to say in each of these contexts.

In this section we formalize our view of the Web as a graph; in this view we ignore the text and other content in pages, focusing instead on the links between pages. Adopting the terminology of graph theory [18], we refer to pages as *nodes*, and to links as *arcs*. In this framework, the Web becomes a large graph containing several hundred million nodes, and a few billion arcs. We will refer to this graph as the *Web graph*, and our goal in this paper is to understand some of its properties. Before presenting our model for Web-like graphs, we begin with a brief primer on graph theory, and a discussion of graph models in general.

1.3. A brief primer on graphs and terminology

The reader familiar with basic notions from graph theory may skip this primer.

A *directed graph* consists of a set of *nodes*, denoted V and a set of *arcs*, denoted E. Each arc is an ordered pair of nodes (u, v) representing a directed connection from u to v. The *out-degree* of a node u is the number of distinct arcs $(u, v_1)\ldots(u, v_k)$ (i.e., the number of links from u), and the *in-degree* is the number of distinct arcs $(v_1, u)\ldots(v_k, u)$ (i.e., the number of links to u). A path from node u to node v is a sequence of arcs $(u, u_1), (u_1, u_2), \ldots(u_k, v)$. One can follow such a sequence of arcs to 'walk' through the graph from u to v. Note that a path from u to v does not imply a path from v to u. The *distance* from u to v is one more than the smallest k for which such a path exists. If no path exists, the distance from u to v is defined to be infinity. If (u, v) is an arc, then the distance from u to v is 1.

Given a directed graph, a *strongly connected component* (strong component for brevity) of this graph is a set of nodes such that for any pair of nodes u and v in the set there is a path from u to v. In general, a directed graph may have one or many strong components. The strong components of a graph consist of disjoint sets of nodes. One focus of our studies will be in understanding the distribution of the sizes of strong components on the Web graph.

An *undirected graph* consists of a set of nodes and a set of *edges*, each of which is an unordered pair $\{u, v\}$ of nodes. In our context, we say there is an edge between u and v if there is a link between u and v, without regard to whether the link points from u to v or the other way around. The *degree* of a node u is the number of edges incident to u. A path is defined as for directed graphs, except that now the existence of a path from u to v implies a path from v to u. A *component* of an undirected graph is a set of nodes such that for any pair of nodes u and v in the set there is a path from u to v. We refer to the components of the undirected graph obtained from a directed graph by ignoring the directions of its arcs as the *weak components* of the directed graph. Thus two nodes on the Web may be in the same weak component even though there is no *directed* path between them (consider, for instance, a node u that points to two other nodes v and w; then v and w are in the same weak component even though there may be no sequence of links leading from v to w or vice versa). The interplay of strong and weak components on the (directed) Web graph turns out to reveal some unexpected properties of the Web's connectivity.

A *breadth-first search* (BFS) on a directed graph begins at a node u of the graph, and proceeds to build up the set of nodes reachable from u in a series of layers. Layer 1 consists of all nodes that are pointed to by an arc from u. Layer k consists of all nodes to which there is an arc from some vertex in layer $k - 1$, but are not in any earlier layer. Notice that by definition, layers are disjoint. The distance of any node from u can be read out of the breadth-first search. The shortest path from u to v is the index of the layer v belongs in, i.e., if there is such a layer. On the other hand, note that a node that cannot be reached from u does not belong in any layer, and thus we define the distance to be infinity. A BFS on an undirected graph is defined analogously.

Finally, we must take a moment to describe the exact notions of diameter we study, since several have been discussed informally in the context of the Web. Traditionally, the *diameter* of a graph, directed or undirected, is the maximum over all ordered pairs (u, v) of the shortest path from u to v. Some researchers have proposed studying the *average distance* of a graph, defined to be the length of the shortest path from u to v, averaged over all ordered pairs (u, v); this is referred to as diameter in [4]. The difficulty with this notion is that even a single pair (u, v) with no path from u to v results in an infinite average distance. In fact, as we show from our experiments below, the Web is rife with such pairs (thus it is not merely a matter of discarding a few outliers before taking this average). This motivates the following revised definition: let P be the set of all ordered pairs (u, v) such that there is a path from u to v. The *average connected distance* is the expected length of the shortest path, where the expectation is over uniform choices from P.

2. Experiments and results

2.1. Infrastructure

All experiments were run using the Connectivity Server 2 (CS2) software built at Compaq Systems Research Center using data provided by AltaVista. CS2 provides fast access to linkage information on the Web. A build of CS2 takes a Web crawl as input and creates a representation of the entire Web graph induced by the pages in the crawl, in the form of a database that consists of all URLs that were crawled together with all in-links and out-links among those URLs. In addition, the graph is extended with those URLs referenced at least five times by the crawled pages. (Experimentally, we have determined that the vast majority of URLs encountered fewer than five times but not crawled turn out to be invalid URLs.)

CS2 improves on the original connectivity server (CS1) described in [7] in two important ways. First, it significantly increases the compression of the URLs and the links to data structures. In CS1, each compressed URL is, on average, 16 bytes. In CS2, each URL is stored in 10 bytes. In CS1, each link requires 8 bytes to store as both an in-link and out-link; in CS2, an average of only 3.4 bytes are used. Second, CS2 provides additional functionality in the form of a host database. For example, in CS2, it is easy to get all the in-links for a given node, or just the in-links from remote hosts.

Like CS1, CS2 is designed to give high-performance access to all this data on a high-end machine with enough RAM to store the database in memory. On a 465 MHz Compaq AlphaServer 4100 with 12 GB of RAM, it takes 70–80 μs to convert an URL into an internal id or vice versa, and then only 0.15 μs/link to retrieve each in-link or out-link. On a uniprocessor machine, a BFS that reaches 100 million nodes takes about 4 minutes; on a 2-processor machine we were able complete a BFS every 2 minutes.

In the experiments reported in this paper, CS2 was built from a crawl performed at AltaVista in May, 1999. The CS2 database contains 203 million URLs and 1466 million links (all of which fit in 9.5 GB of storage). Some of our experiments were repeated on a more recent crawl from October, 1999 containing 271 million URLs and 2130 million links.

In general, the AltaVista crawl is based on a large set of starting points accumulated over time from various sources, including voluntary submissions. The crawl proceeds in roughly a BFS manner, but is subject to various rules designed to avoid overloading Web servers, avoid robot traps (artificial infinite paths), avoid and/or detect spam (page flooding), deal with connection time outs, etc. Each build of the AltaVista index is based on the crawl data after further filtering and processing designed to remove duplicates and near duplicates, eliminate spam pages, etc. Then the index evolves continuously as various processes delete dead links, add new pages, update pages, etc. The secondary filtering and the later deletions and additions are not reflected in the connectivity server. But overall, CS2's database can be viewed as a superset of all pages stored in the index at one point in time. Note that due to the multiple starting points, it is possible for the resulting graph to have many connected components.

2.2. Experimental data

The following basic algorithms were implemented using CS2: (1) a BFS algorithm that performs a breadth-first traversal; (2) a WCC algorithm that finds the weak components; and (3) an SCC algorithm that finds the strongly connected components. Recall that both WCC and SCC are simple generalizations of the BFS algorithm. Using these three basic algorithms, we ran several interesting experiments on the Web graph.

2.2.1. Degree distributions

The first experiment we ran was to verify earlier observations that the in- and out-degree distributions on the Web are distributed according to power laws. We ran the experiment on both the May and October crawls of the Web. The results, shown in Figs. 1 and 3, show remarkable agreement with each other, and with similar experiments from data that is over two years old [21]. Indeed, in the case of in-degree, the exponent of the power law is consistently around 2.1, a number reported in [5,21]. The anomalous bump at 120 on the x-axis is due to a large clique formed by a single spammer. In all our log–log plots, straight lines are linear regressions for the best power law fit.

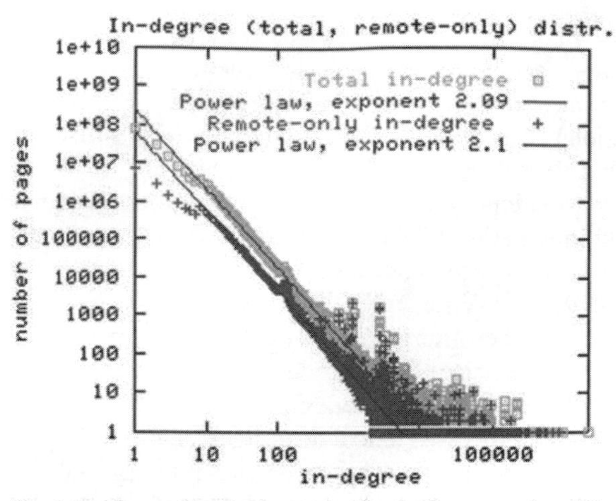

Fig. 1. In-degree distributions subscribe to the power law. The law also holds if only off-site (or 'remote-only') edges are considered.

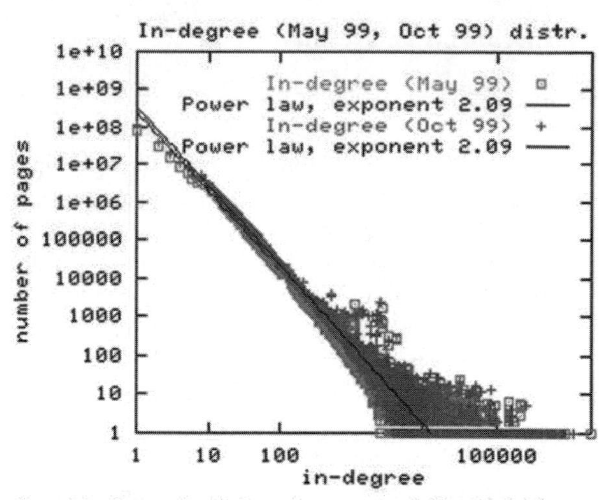

Fig. 3. In-degree distributions show a remarkable similarity over two crawls, run in May and October 1999. Each crawl counts well over 1 billion distinct edges of the Web graph.

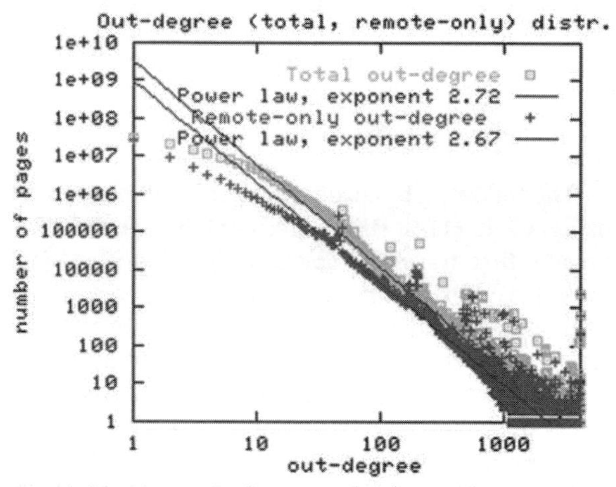

Fig. 2. Out-degree distributions subscribe to the power law. The law also holds if only off-site (or 'remote-only') edges are considered.

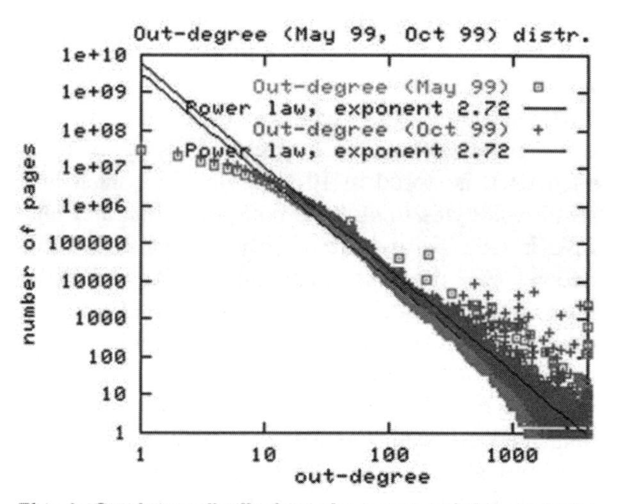

Fig. 4. Out-degree distributions show a remarkable similarity over two crawls, run in May and October 1999. Each crawl counts well over 1 billion distinct edges of the Web graph.

Out-degree distributions also exhibit a power law, although the exponent is 2.72, as can be seen in Figs. 2 and 4. It is interesting to note that the initial segment of the out-degree distribution deviates significantly from the power law, suggesting that pages with low out-degree follow a different (possibly Poisson or a combination of Poisson and power law, as suggested by the concavity of the deviation) distribution. Further research is needed to understand this combination better.

2.2.2. *Undirected connected components*

In the next set of experiments we treat the Web graph as an undirected graph and find the sizes of the undirected components. We find a giant component of 186 million nodes in which fully 91% of the nodes in our crawl are reachable from one another by following either forward or backward links. This is done by running the WCC algorithm which simply finds all connected components in the undirected Web graph. Thus, if one could browse along both

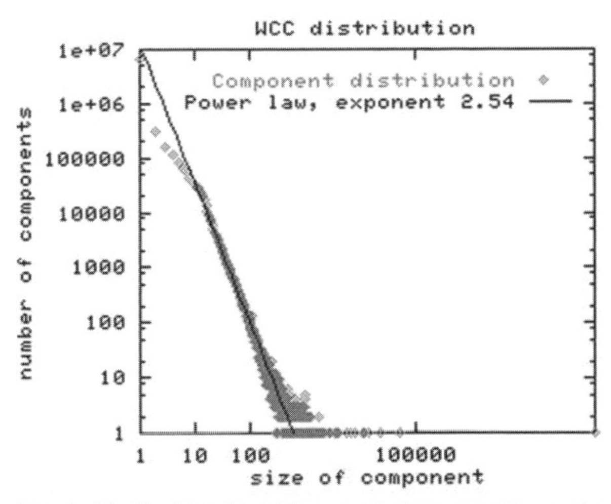

Fig. 5. Distribution of weakly connected components on the Web. The sizes of these components also follow a power law.

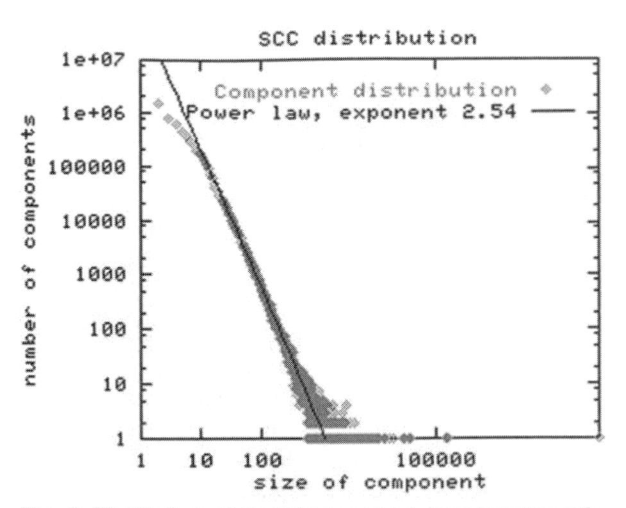

Fig. 6. Distribution of strongly connected components on the Web. The sizes of these components also follow a power law.

forward and backward directed links, the Web is a very well connected graph. Surprisingly, even the distribution of the sizes of WCCs exhibits a power law with exponent roughly 2.5 (Fig. 5).

Does this widespread connectivity result from a few nodes of large in-degree acting as 'junctions'? Surprisingly, this turns out not to be the case. Indeed, even if all links to pages with in-degree 5 or higher are removed (certainly including links to every well-known page on the Web), the graph still contains a giant weak component of size 59 million (see Table 1). This provides us with two interesting and useful insights. First, the connectivity of the Web graph as an undirected graph is extremely resilient and does not depend on the existence of nodes of high in-degree. Second, such nodes, which are very useful and tend to include nodes with high PageRank or nodes that are considered good hubs and authorities, are embedded in a graph that is well connected without them. This last fact may help understand why algorithms such as HITS [20] converge quickly.

Table 1
Size of the largest surviving weak component when links to pages with in-degree at least *k* are removed from the graph.

k	1000	100	10	5	4	3
Size (millions)	177	167	105	59	41	15

2.2.3. *Strongly connected components*

Motivated in part by the intriguing prediction of [4] that the average distance (referred to in their paper as diameter) of the Web is 19 (and thus it should be possible to get from any page to any other in a small number of clicks), we turned to the strongly connected components of the Web as a directed graph. By running the strongly connected component algorithm, we find that there is a single large SCC consisting of about 56 million pages, all other components are significantly smaller in size. This amounts to barely 28% of all the pages in our crawl. One may now ask: where have all the other pages gone? The answer to this question reveals some fascinating detailed structure in the Web graph; to expose this and to further study the issues of the diameter and average distance, we conducted a further series of experiments. Note that the distribution of the sizes of SCCs also obeys a power law (Fig. 6).

2.2.4. *Random-start BFS*

We ran the BFS algorithm twice from each of 570 randomly chosen starting nodes: once in the *forward* direction, following arcs of the Web graph as a browser would, and once *backward* following links in the reverse direction. Each of these BFS traversals (whether forward or backward) exhibited a sharp bimodal behavior: it would either 'die out' after reaching a small set of nodes (90% of the time this set has fewer than 90 nodes; in extreme cases it has a few hundred thousand), or it would 'explode' to cover about 100 million nodes (but never the entire 186 million). Further, for a fraction of the starting nodes, both the forward and the backward BFS runs would 'explode', each covering about 100 million nodes (though not the same 100 million in the two runs). As we show below, these are the starting points that lie in the SCC.

The cumulative distributions of the nodes covered in these BFS runs are summarized in Fig. 7. They reveal that the true structure of the Web graph must be somewhat subtler than a 'small world' phenomenon in which a browser can pass from any Web page to any other with a few clicks. We explicate this structure in Section 3.

2.2.5. *Zipf distributions vs power law distributions*

The *Zipf distribution* is an inverse polynomial function of *ranks* rather than magnitudes; for example, if only in-degrees 1, 4, and 5 occurred then a power law would be inversely polynomial in those values, whereas a Zipf distribution would be inversely polynomial in the ranks of those values: i.e., inversely polynomial in 1, 2, and 3. The in-degree distribution in our data shows a striking fit with a Zipf (more so than the power law) distribution; Fig. 8 shows the in-degrees of pages from the May 1999 crawl plotted against both ranks and magnitudes (corresponding to the Zipf and power law cases). The plot against ranks is virtually a straight line in the log–log plot, without the flare-out noticeable in the plot against magnitudes.

3. Interpretation and further work

Let us now put together the results of the connected component experiments with the results of the random-start BFS experiments. Given that the set SCC contains only 56 million of the 186 million nodes in our giant weak component, we use the BFS runs to estimate the positions of the remaining nodes. The

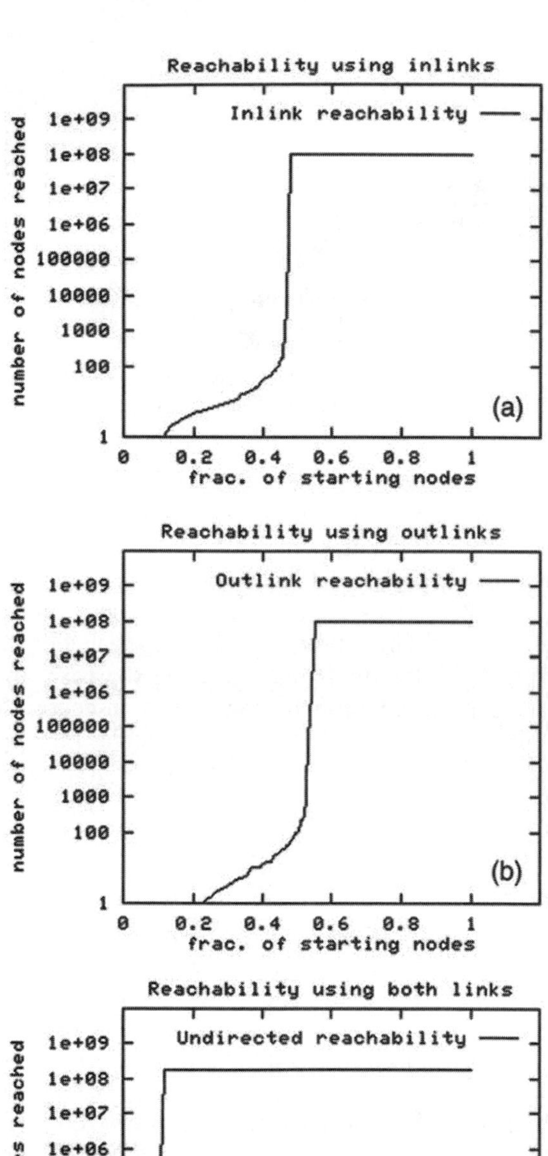

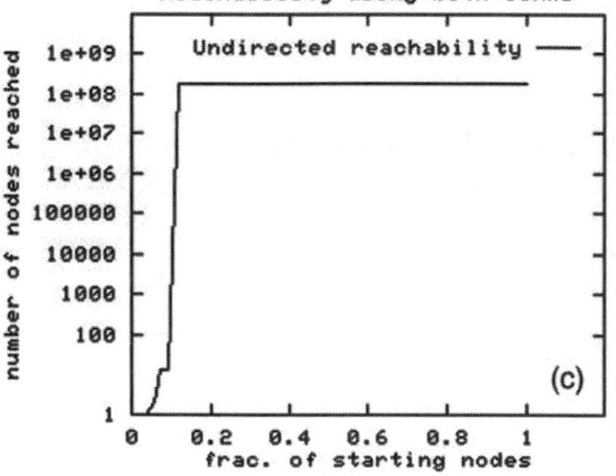

Fig. 7. Cumulative distribution on the number of nodes reached when BFS is started from a random node: (a) follows in-links, (b) follows out-links, and (c) follows both in- and out-links. Notice that there are two distinct regions of growth, one at the beginning and an 'explosion' in 50% of the start nodes in the case of in- and out-links, and for 90% of the nodes in the undirected case. These experiments form the basis of our structural analysis.

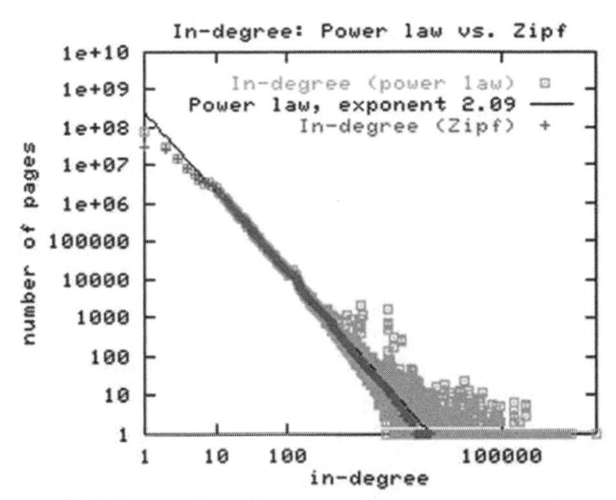

Fig. 8. In-degree distributions plotted as a power law and as a Zipf distribution.

starting points for which the forward BFS 'explodes' are either in SCC, or in a set we call IN, that has the following property: there is a directed path from each node of IN to (all the nodes of) SCC. Symmetrically, there is a set we call OUT containing all starting points for which the backward BFS 'explodes'; there is a directed path from any node in the SCC to every node in OUT. Thus a forward BFS from any node in either the SCC or IN will explode, as will a backward BFS from any node in either the SCC or OUT. By analyzing forward and backward BFS from 570 random starting points, we can compute the number of nodes that are in SCC, IN, OUT or none of these. Fig. 9 shows the situation as we can now infer it.

We now give a more detailed description of the structure in Fig. 9. The sizes of the various components are as follows:

Region	SCC	IN	OUT
Size	56,463,993	43,343,168	43,166,185
Region	TENDRILS	DISC.	Total
Size	43,797,944	16,777,756	203,549,046

These sizes were determined as follows. We know the total number of nodes in our crawl, so by subtracting the size of the giant weak component we determine the size of DISCONNECTED. Then our strong-component algorithm gives us the size of SCC. We turn to our breadth-first search data. As noted, searching from a particular start node following a particular type of edges (in-edges or out-edges) would either terminate quickly, or grow the search to about 100 million nodes. We say that a node *explodes* if it falls into the latter group. Thus, if a node explodes following in-links, and also explodes following out-links, it must be a member of a strong component of size at least $100 + 100 - 186 = 14$ million. Since the second largest strong component is of size 150 thousand, we infer that SCC is the unique strong component that contains all nodes exploding following in- as well as out-links. In fact, this observation contains two corroborating pieces of evidence for the structure in the table above: first, it turns out that the fraction of our randomly chosen BFS start nodes that explode under in- and out-links is the same as the fraction of nodes in the SCC as returned by our SCC algorithm. Second, every BFS start node in the SCC reaches exactly the same number of nodes under in-link expansion; this number is 99,807,161. Likewise, under out-link expansion every node of SCC reaches exactly 99,630,178 nodes.

Thus, we know that SCC + IN = 99,807,161, and similarly SCC + OUT = 99,630,178. Having already found the size of SCC, we can solve for IN and OUT. Finally, since we know the size of the giant weak component, we can subtract SCC, IN, and OUT to get TENDRILS. We now discuss each region in turn.

3.1. TENDRILS and DISCONNECTED

We had 172 samples from TENDRILS and DISCONNECTED; our BFS measurements cannot be used to differentiate between these two regions. By following out-links from a start point in this region, we encounter an average of 20 nodes before the exploration stops. Likewise, by following in-links we encounter an average of 52 nodes.

3.2. IN and OUT

Our sample contains 128 nodes from IN and 134 from OUT. We ask: when following out-links from nodes in OUT, or in-links from nodes in IN, how many nodes do we encounter before the BFS terminates? That is, how large a neighborhood do points in these

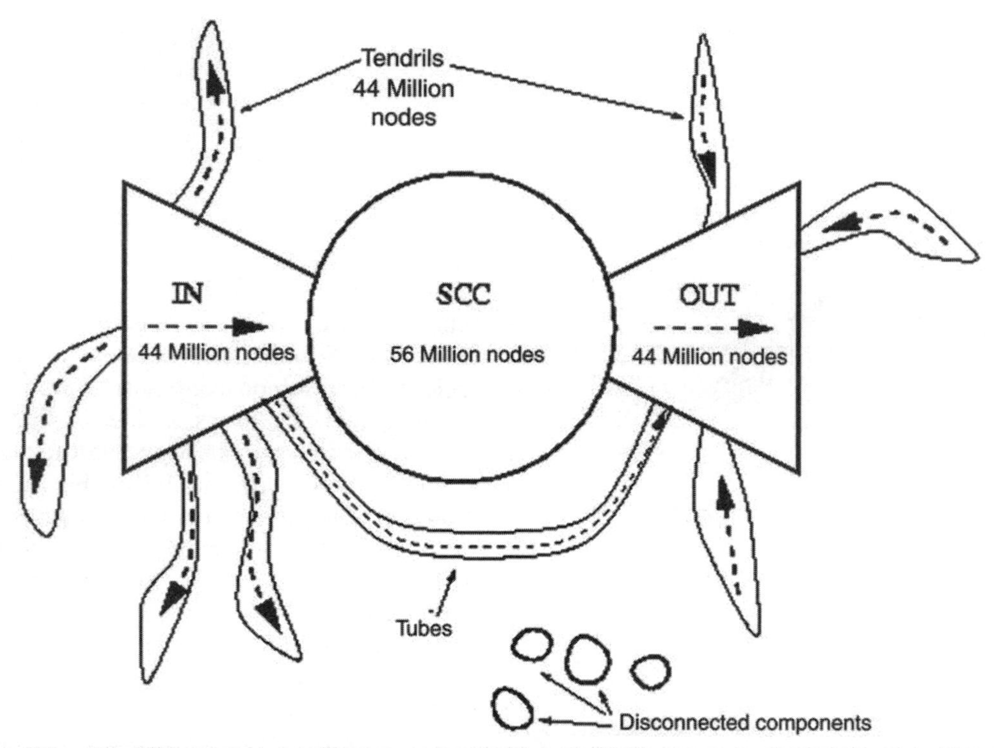

Fig. 9. Connectivity of the Web: one can pass from any node of IN through SCC to any node of OUT. Hanging off IN and OUT are TENDRILS containing nodes that are reachable from portions of IN, or that can reach portions of OUT, without passage through SCC. It is possible for a TENDRIL hanging off from IN to be hooked into a TENDRIL leading into OUT, forming a TUBE: i.e., a passage from a portion of IN to a portion of OUT without touching SCC.

regions have, if we explore in the direction 'away' from the center? The results are shown below in the row labeled 'exploring outward – all nodes'.

Similarly, we know that if we explore in-links from a node in OUT, or out-links from a node in IN, we will encounter about 100 million other nodes in the BFS. Nonetheless, it is reasonable to ask: how many other nodes will we encounter? That is, starting from OUT (or IN), and following in-links (or out-links), how many nodes of TENDRILS and OUT (or IN) will we encounter? The results are shown below in the row labeled 'exploring inwards – unexpected nodes'. Note that the numbers in the table represent averages over our sample nodes.

Starting point	OUT	IN
Exploring outwards – all nodes	3093	171
Exploring inwards – unexpected nodes	3367	173

As the table shows, OUT tends to encounter larger neighborhoods. For example, the second largest strong component in the graph has size approximately 150 thousand, and two nodes of OUT encounter neighborhoods a few nodes larger than this, suggesting that this component lies within OUT. In fact, considering that (for instance) almost every corporate Website not appearing in SCC will appear in OUT, it is no surprise that the neighborhood sizes are larger.

3.3. SCC

Our sample contains 136 nodes from the SCC. To determine other properties of SCC, we require a useful property of IN and OUT: each contains a few long paths such that, once the BFS proceeds beyond a certain depth, only a few paths are being explored, and the last path is much longer than any of the others. We can therefore explore the radius at which the BFS completes, confident that the last long path will be the same no matter which node of SCC we start from. The following table shows the depth at which the BFS terminates in each direction (following in-links or out-links) for nodes in the SCC.

Measure	Minimum depth	Average depth	Maximum depth
In-links	475	482	503
Out-links	430	434	444

As the table shows, from some nodes in the SCC it is possible to complete the search at distance 475, while from other nodes distance 503 is required. This allows us to conclude that the directed diameter of SCC is at least 28.

3.4. Other observations

As noted above, the (min, average, max) depths at which the BFS from SCC terminates following in-links are (475, 482, 503). For IN, we can perform the same analysis, and the values are: (476, 482, 495). These values, especially the average, are so similar that nodes of IN appear to be quite close to SCC.

Likewise, for SCC the (min, average, max) depths for termination under out-links are (430, 434, 444). For OUT, the values are (430, 434, 444).

Now, consider the probability that an ordered pair (u, v) has a path from u to v. By noting that the average in-size of nodes in IN is very small (171) and likewise the average out-size of nodes in OUT is very small (3093), the pair has a path with non-negligible probability if and only if u is in SCC + IN, and v is in SCC + OUT. The probability of this event for node pairs drawn uniformly from our crawl is only 24%; for node pairs drawn from the weak component it is only 28%. This leads to the somewhat surprising conclusion that, given a random start and finish page on the Web, one can get from the start page to the finish page by traversing links barely a quarter of the time.

The structure that is now unfolding tells us that it is relatively insensitive to the particular large crawl we use. For instance, if AltaVista's crawler fails to include some links whose inclusion would add one of the tendrils to the SCC, we know that the resulting change in the sizes of SCC and TENDRIL will be small (since any individual tendril is small). Likewise, our experiments in which we found that large components survived the deletion of nodes of large in-degree show that the connectivity of the Web is resilient to the removal of significant portions.

3.5. Diameter and average connected distance

As we discussed above, the directed diameter of the SCC is at least 28. Likewise, the maximum finite shortest path length is at least 503, but is probably substantially more than this: unless a short tube connects the most distant page of IN to the most distant page of OUT without passing through the SCC, the maximum finite shortest path length is likely to be close to $475 + 430 = 905$.

We can estimate the average connected distance using our 570 BFS start points, under both in-links and out-links. The values are shown below; the column headed 'Undirected' corresponds to the average undirected distance.

Edge type	In-links (directed)	Out-links (directed)	Undirected
Average connected distance	16.12	16.18	6.83

These results are in interesting contrast to those of [4], who predict an average distance of 19 for the Web based on their crawl of the nd.edu site; it is unclear whether their calculations consider directed or undirected distances. Our results on the other hand show that over 75% of time there is no directed path from a random start node to a random finish node; when there *is* a path, the figure is roughly 16. However, if links can be traversed in either direction, the distance between random pairs of nodes can be much smaller, around 7, on average.

4. Further work

Further work can be divided into three broad classes.

(1) More experiments aimed at exposing further details of the structures of *SCC*, *IN*, *OUT*, and the *TENDRILS*. Would this basic structure, and the relative fractions of the components, remain stable over time?

(2) Mathematical models for evolving graphs, motivated in part by the structure of the Web; in addition, one may consider the applicability of such models to other large directed graphs such as the phone-call graph, purchase/transaction graphs, etc. [3].

(3) What notions of connectivity (besides weak and strong) might be appropriate for the Web graph? For instance, what is the structure of the undirected graph induced by the co-citation relation or by bibliographic coupling [31].

Acknowledgements

We thank Keith Randall for his insights into our SCC algorithm and implementation.

References

[1] L. Adamic and B. Huberman, The nature of markets on the World Wide Web, Xerox PARC Technical Report, 1999.

[2] L. Adamic and B. Huberman, Scaling behavior on the World Wide Web, Technical comment on [5].

[3] W. Aiello, F. Chung and L. Lu, A random graph model for massive graphs, ACM Symposium on the Theory and Computing, 2000.

[4] R. Albert, H. Jeong and A.-L. Barabasi, Diameter of the World Wide Web, Nature 401 (1999) 130–131.

[5] A. Barabasi and R. Albert, Emergence of scaling in random networks, Science 286 (509) (1999).

[6] P. Barford, A. Bestavros, A. Bradley and M.E. Crovella, Changes in Web client access patterns: characteristics and caching implications, World Wide Web, Special Issue on Characterization and Performance Evaluation, 2 (1999) 15–28.

[7] K. Bharat, A. Broder, M. Henzinger, P. Kumar and S. Venkatasubramanian, The connectivity server: fast access to linkage information on the web, in: Proc. 7th WWW, 1998.

[8] K. Bharat and M. Henzinger, Improved algorithms for topic distillation in hyperlinked environments, in: Proc. 21st SIGIR, 1998.

[9] S. Brin and L. Page, The anatomy of a large scale hypertextual web search engine, in: Proc. 7th WWW, 1998.

[10] R.A. Butafogo and B. Schneiderman, Identifying aggregates in hypertext structures, in: Proc. 3rd ACM Conference on Hypertext, 1991.

[11] J. Carriere and R. Kazman, WebQuery: searching and visualizing the Web through connectivity, in: Proc. 6th WWW, 1997.

[12] S. Chakrabarti, B. Dom, D. Gibson, J. Kleinberg, P. Raghavan and S. Rajagopalan, Automatic resource compilation by analyzing hyperlink structure and associated text, in: Proc. 7th WWW, 1998.

[13] S. Chakrabarti, B. Dom, D. Gibson, S. Ravi Kumar, P. Raghavan, S. Rajagopalan and A. Tomkins, Experiments in topic distillation, in: Proc. ACM SIGIR Workshop on Hypertext Information Retrieval on the Web, 1998.

[14] S. Chakrabarti, D. Gibson and K. McCurley, Surfing the Web backwards, in: Proc. 8th WWW, 1999.

[15] J. Cho, H. Garcia-Molina, Synchronizing a database to improve freshness, To appear in 2000 ACM International Conference on Management of Data (SIGMOD), May 2000.

[16] M. Faloutsos, P. Faloutsos and C. Faloutsos, On power law relationships of the internet topology, ACM SIGCOMM, 1999.

[17] S. Glassman, A caching relay for the world wide web, in: Proc. 1st WWW, 1994.

[18] F. Harary, Graph Theory, Addison-Wesley, Reading, MA, 1975.

[19] B. Huberman, P. Pirolli, J. Pitkow and R. Lukose, Strong regularities in World Wide Web surfing, Science 280 (1998) 95–97.

[20] J. Kleinberg, Authoritative sources in a hyperlinked environment, in: Proc. 9th ACM–SIAM SODA, 1998.

[21] R. Kumar, P. Raghavan, S. Rajagopalan and A. Tomkins, Trawling the Web for cyber communities, in: Proc. 8th WWW, April 1999.

[22] R. Kumar, P. Raghavan, S. Rajagopalan and A. Tomkins, Extracting large scale knowledge bases from the Web, in: Proc. VLDB, July 1999.

[23] R.M. Lukose and B. Huberman, Surfing as a real option, in: Proc. 1st International Conference on Information and Computation Economies, 1998.

[24] C. Martindale and A.K. Konopka, Oligonucleotide frequencies in DNA follow a Yule distribution, Computer and Chemistry 20 (1) (1996) 35–38.

[25] A. Mendelzon, G. Mihaila and T. Milo, Querying the World Wide Web, Journal of Digital Libraries 1 (1) (1997) 68–88.

[26] A. Mendelzon and P. Wood, Finding regular simple paths in graph databases, SIAM J. Comp. 24 (6) (1995) 1235–1258.

[27] V. Pareto, Cours d'économie politique, Rouge, Lausanne et Paris, 1897.

[28] P. Pirolli, J. Pitkow and R. Rao, Silk from a sow's ear: extracting usable structures from the Web, in: Proc. ACM SIGCHI, 1996.

[29] J. Pitkow and P. Pirolli, Life, death, and lawfulness on the electronic frontier, in: Proc. ACM SIGCHI, 1997.

[30] H.A. Simon, On a class of stew distribution functions, Biometrika 42 (1955) 425–440.

[31] H.D. White and K.W. McCain, Bibliometrics, in: Annual Review of Information Science and Technology, Vol. 24, Elsevier, Amsterdam, 1989, pp. 119–186.

[32] G.U. Yule, Statistical Study of Literary Vocabulary, Cambridge University Press, New York, 1944.

[33] G.K. Zipf, Human Behavior and the Principle of Least Effort, Addison-Wesley, Reading, MA, 1949.

On Power-Law Relationships of the Internet Topology

Michalis Faloutsos
U.C. Riverside
Dept. of Comp. Science
michalis@cs.ucr.edu

Petros Faloutsos
U. of Toronto
Dept. of Comp. Science
pfal@cs.toronto.edu

Christos Faloutsos *
Carnegie Mellon Univ.
Dept. of Comp. Science
christos@cs.cmu.edu

Abstract

Despite the apparent randomness of the Internet, we discover some surprisingly simple power-laws of the Internet topology. These power-laws hold for three snapshots of the Internet, between November 1997 and December 1998, despite a 45% growth of its size during that period. We show that our power-laws fit the real data very well resulting in correlation coefficients of 96% or higher.

Our observations provide a novel perspective of the structure of the Internet. The power-laws describe concisely skewed distributions of graph properties such as the node outdegree. In addition, these power-laws can be used to estimate important parameters such as the average neighborhood size, and facilitate the design and the performance analysis of protocols. Furthermore, we can use them to generate and select realistic topologies for simulation purposes.

1 Introduction

"What does the Internet look like?" "Are there any topological properties that don't change in time?" "How will it look like a year from now?" "How can I generate Internet-like graphs for my simulations?" These are some of the questions motivating this work.

In this paper, we study the topology of the Internet and we identify several power-laws. Furthermore, we discuss multiple benefits from understanding the topology of the Internet. First, we can design more efficient protocols that take advantage of its topological properties. Second, we can create more accurate artificial models for simulation purposes. And third, we can derive estimates for topological parameters (e.g. the average number of neighbors within h hops) that are useful for the analysis of protocols and for speculations of the Internet topology in the future.

Modeling the Internet topology[1] is an important open problem despite the attention it has attracted recently. Paxson and Floyd consider this problem as a major reason "Why We Don't Know How To Simulate The Internet" [16]. Several graph-generator models have been proposed [23] [5] [27], but the problem of creating realistic topologies is not yet solved; the selection of several parameter values are left to the intuition and the experience of each researcher.

As our primary contribution, we identify three power-laws for the topology of the Internet over the duration of a year in 1998. Power-laws are expressions of the form $y \propto x^a$, where a is a constant, x and y are the measures of interest, and $\propto$ stands for "proportional to". Some of those exponents do not change significantly over time, while some exponents change by approximately 10%. However, the important observation is the existence of power-laws, i.e., the fact that there is *some* exponent for each graph instance. During 1998, these power-laws hold in three Internet instances with good linear fits in log-log plots; the correlation coefficient of the fit is at least 96% and usually higher than 98%. In addition, we introduce a graph metric to quantify the density of a graph and propose a rough power-law approximation of that metric. Furthermore, we show how to use our power-laws and our approximation to estimate useful parameters of the Internet, such as the average number of neighbors within h hops. Finally, we focus on the generation of realistic graphs. Our power-laws can help verify the realism of synthetic topologies. In addition, we measure several crucial parameters for the most recent graph generator [27].

Our work in perspective. Our work is based on three Internet instances over a one-year period. During this time, the size of the network increased substantially (45%). Despite this, the sample space is rather limited, and making any generalizations would be premature until additional studies are conducted. However, the authors believe that these power-laws characterize the dynamic equilibrium of the Internet growth in the same way power-laws appear to describe various natural networks such as the the human respiratory system [12], and automobile networks [6]. At a more practical level, the regularities characterize the topology concisely during 1998. If this time period turns out to be a transition phase for the Internet, our observations will obviously be valid only for 1998. In absence of revolutionary

*This research was partially funded by the National Science Foundation under Grants No. IRI-9625428 and DMS-9873442. Also, by the National Science Foundation, ARPA and NASA under NSF Cooperative Agreement No. IRI-9411299, and by DARPA/ITO through Order F463, issued by ESC/ENS under contract N66001-97-C-851. Additional funding was provided by donations from NEC and Intel. Views and conclusions contained in this document are those of the authors and should not be interpreted as representing official policies, either expressed or implied, of the Defense Advanced Research Projects Agency or of the United States Government.

[1] In this paper, we use the expression "the topology of the Internet", although the topology changes and it would be more accurate to talk about "Internet topologies". We hope that this does not mislead or confuse the reader.

SIGCOMM '99 8/99 Cambridge, MA, USA
© 1999 ACM 1-58113-135-6/99/0008...$5.00

changes, it is reasonable to expect that our power-laws will continue to hold in the future.

The rest of this paper is structured as follows. In Section 2, we present some definitions and previous work on measurements and models for the Internet. In Section 3, we present our Internet instances and provide useful measurements. In Section 4, we present our three observed power-laws and our power-law approximation. In Section 5, we explain the intuition behind our power-laws, discuss their use, and show how we can use them to predict the growth of the Internet. In Section 6, we conclude our work and discuss future directions.

2 Background and Previous Work

The Internet can be decomposed into connected subnetworks that are under separate administrative authorities, as shown in Figure 1. These subnetworks are called *domains* or *autonomous systems*[2]. This way, the topology of the Internet can be studied at two different granularities. At the **router level**, we represent each router by a node [14]. At the **inter-domain level**, each domain is represented by a single node [10] and each edge is an inter-domain interconnection. The study of the topology at both levels is equally important. The Internet community develops and employs different protocols inside a domain and between domains. An intra-domain protocol is limited within a domain, while an inter-domain protocol runs between domains treating each domain as one entity.

Symbol	Definition
G	An undirected graph.
N	Number of nodes in a graph.
E	Number of edges in a graph.
δ	The diameter of the graph.
d_v	Outdegree of node v.
$\bar{d}$	The average outdegree of the nodes of a graph: $\bar{d} = 2\,E/N$

Table 1: Definitions and symbols.

Metrics. The metrics that have been used so far to describe graphs are mainly the node outdegree, and the distances between nodes. Given a graph, the outdegree of a node is defined as the number of edges incident to the node (see Table 1). The distance between two nodes is the number of edges of the shortest path between the two nodes. Most studies report minimum, maximum, and average values and plot the outdegree and distance distribution. We denote the number of nodes of a graph by N, the number of edges by E, and the diameter of the graph by δ.

Real network studies. Govindan and Reddy [10] study the growth of the inter-domain topology of the Internet between 1994 and 1995. The graph is sparse with 75% of the nodes having outdegrees less or equal to two. They distinguish four groups of nodes according to their outdegree. The authors observe an increase in the connectivity over time. Pansiot and Grad [14] study the topology of the Internet in 1995 at the router level. The distances they report are approximately two times larger compared to those of Govindan and Reddy. This leads to the interesting observation that, on average, one hop at the inter-domain level corresponded to two hops at the router level in 1995.

Generating Internet Models. Regarding the creation of realistic graphs, Waxman introduced what seems to be one of the most popular network models [23]. These graphs are created probabilistically considering the distance between nodes in a Euclidean sense. This model was successful in representing small early networks such as the ARPANET. As the size and the complexity of the network increased more detailed models were needed [5] [27]. In the most recent work, Zegura et al. [27] introduce a comprehensive model that includes several previous models [3]. They call their model transit-stub, which combines simple topologies (e.g. Waxman graphs and trees) in a hierarchical structure. There are several parameters that control the structure of the graph. For example, parameters define the total number and the size of the stubs. An advantage of this model lies in its ability to describe a number of topologies. At the same time, a researcher needs experimental estimates to set values to the parameters of the model.

Power-laws in communication networks. Power-laws have been used to describe the traffic in communications networks, but not their topology. Actually, both self-similarity, and heavy tails appear in network traffic and are both related to power-laws. A variable X follows a heavy tail distribution if $P[X > x] = k^a x^{-a}\ L(x)$, where $k \in \Re^+$ and $L(x)$ is a slowly varying function: $lim_{t \to \infty}[L(tx)/L(x)] = 1$ [20] [24]. A Pareto distribution is a special case of a heavy tail distribution with $P[X > x] = k^a\ x^{-a}$. It is easy to see that power-laws, Pareto and heavy-tailed distributions are intimately related. In a pioneering work, Leland et al. [11] show the self-similar nature of Local Area Network (LAN) traffic. Second, Paxson and Floyd [15] provide evidence of self similarity in Wide Area Network (WAN) traffic. In modeling the traffic, Willinger et al. [25] provide structural models that describe LAN traffic as a collective effect of simple heavy-tailed ON-OFF sources. Finally, Willinger et al. [24] bring all of the above together by describing LAN and WAN traffic through structural models and showing the relation of the self-similarity at the macroscopic level of WANs with the heavy-tailed behavior at the microscopic level of individual sources. In addition, Crovella and Bestavros use power-laws to describe traffic patterns in the World Wide Web [3]. At an intuitive level, the previous works seem to attribute the heavy-tailed behavior of the traffic to the heavy-tailed distribution of the size of the transmitted data files, and to the heavy-tailed characteristics of the human-computer interaction. Recently, Chuang and Sirbu [2] use a power-law to estimate the size of multicast distribution trees. Note that in a follow-up work, Philips et al. [17] verify the reasonable accuracy of the Chuang-Sirbu scaling law for practical purposes, but they also propose an estimate that does not follow a power-law.

3 Internet Instances

In this section, we present the Internet instances we acquired and we study their evolution in time. We examine the inter-domain topology of the Internet from the end of 1997 until the end of 1998. We use three real graphs that correspond to six-month intervals approximately. The data

[2]The definition of an autonomous system can vary in the literature, but it usually coincides with that of the domain [10].

[3]The graph generator software is publicly available [27].

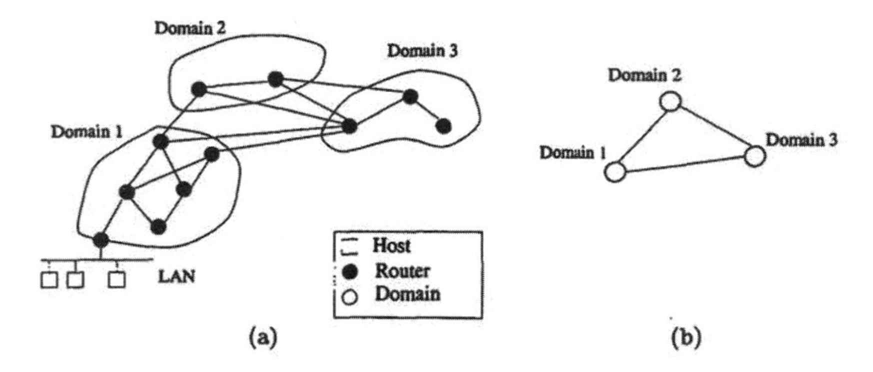

Figure 1: The structure of Internet at a) the router level and b) the inter-domain level. The hosts connect to routers in LANs.

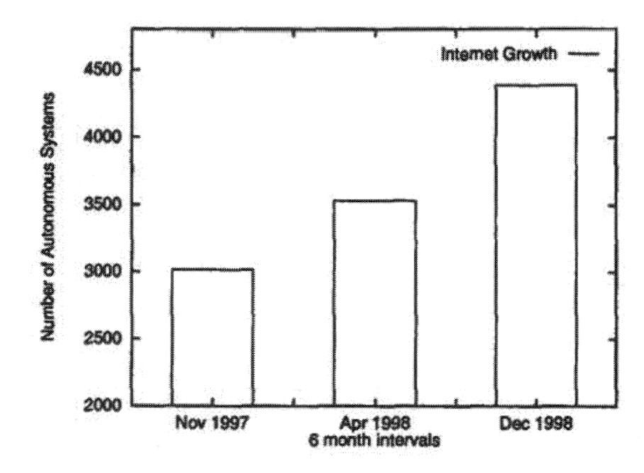

Figure 2: The growth of the Internet: the number of domains versus time between the end of 1997 until the end of 1998.

is provided by the National Laboratory for Applied Network Research [9]. The information is collected by a route server from BGP[4] routing tables of multiple geographically distributed routers with BGP connections to the server. We list the three datasets that we use in our paper, and we present more information in Appendix A.

- Int-11-97: the inter-domain topology of the Internet in November of 1997 with 3015 nodes, 5156 edges, and 3.42 avg. outdegree.
- Int-04-98: the inter-domain topology of the Internet in April of 1998 with 3530 nodes, 6432 edges, and 3.65 avg. outdegree.
- Int-12-98: the inter-domain topology of the Internet in December of 1998 with 4389 nodes, 8256 edges, and 3.76 avg. outdegree.

Note that the growth of the Internet in the time period we study is 45% (see Figure 2). The change is significant, and it ensures that the three graphs reflect different instances of an evolving network.

Although we focus on the Internet topology at the inter-domain level, we also examine an instance at the router level. The graph represents the topology of the routers of the Internet in 1995, and was tediously collected by Pansiot and Grad [14].

- Rout-95: the routers of the Internet in 1995 with 3888 nodes, 5012 edges, and an average outdegree of 2.57.

Clearly, the above graph is considerably different from the first three graphs. First of all, the graphs model the topology at different levels. Second, the Rout-95 graph comes from a different time period, in which Internet was in a fairly early phase.

To facilitate the graph generation procedures, we analyze the Internet in a way that suits the graph generator models [27]. Namely, we decompose each graph in two components: the *tree component* that contains all nodes that belong exclusively to trees and the *core* component that contains the rest of the nodes including the roots of the trees. We report several interesting measurements in Appendix A. For example, we find that 40-50% of the nodes belong to trees. Also, 80% the trees have a depth of one, while the maximum tree depth is three.

4 Power-Laws of the Internet

In this section, we observe three of power-laws of the Internet topology. Namely, we propose and measure graph properties, which demonstrate a regularity that is unlikely to be a coincidence. The exponents of the power-laws can be used to characterize graphs. In addition, we introduce a graph metric that is tailored to the needs of the complexity analysis of protocols. The metric reflects the density or the connectivity of nodes, and we offer a rough approximation of its value through a power-law. Finally, using our observations and metrics, we identify a number of interesting relationships between important graph parameters.

In our work, we want to find metrics or properties that quantify topological properties and describe concisely skewed data distributions. Previous metrics, such as the average outdegree, fail to do so. First, metrics that are based on minimum, maximum and average values are not good descriptors of skewed distributions; they miss a lot of information and probably the "interesting" part that we would want to capture. Second, the plots of the previous metrics are difficult to quantify, and this makes difficult the comparison of graphs. Ideally, we want to describe a plot or a distribution with one number.

[4]BGP stands for the Border Gateway Protocol [19], and it is the inter-domain routing protocol.

Symbol	Definition
f_d	The frequency of an outdegree, d, is the number of nodes with outdegree d.
r_v	The rank of a node, v, is its index in the order of decreasing outdegree.
$P(h)$	The *number of pairs* of nodes is the total number of pairs of nodes within less or equal to h hops, including self-pairs, and counting all other pairs twice.
NN(h)	The average number of nodes in a neighborhood of h hops.
λ	The eigen value of a square matrix A: $\exists x \in \mathcal{R}^N$ *and* $Ax = \lambda x$.
i	The *order* of λ_i in $\lambda_1 \geq \lambda_2 \ldots \geq \lambda_N$

Table 2: Novel definitions and their symbols.

To express our power-laws, we introduce several graph metrics that we show in Table 2. We define frequency, f_d, of some outdegree, d, to be the number of nodes that have this outdegree. If we sort the nodes in decreasing outdegree sequence, we define rank, r_v, to be the index of the node in the sequence, while ties in sorting are broken arbitrarily. We define the number of pairs of nodes $P(h)$ to be the total number of pairs of nodes within less or equal to h hops, including self-pairs, and counting all other pairs twice. The use of this metric will become apparent later. We also define NN(h) to be the average number of nodes in a neighborhood of h hops. Finally, we recall the definition of the eigenvalues of a graph, which are the eigenvalues of its adjacency matrix.

In this section, we use linear regression to fit a line in a set of two-dimensional points [18]. The technique is based on the least-square errors method. The validity of the approximation is indicated by the correlation coefficient which is a number between -1.0 and 1.0. For the rest of this paper, we use the absolute value of the correlation coefficient, ACC. An ACC value of 1.0 indicates perfect linear correlation, i.e., the data points are exactly on a line.

4.1 The rank exponent $\mathcal{R}$

In this section, we study the outdegrees of the nodes. We sort the nodes in decreasing order of outdegree, d_v, and plot the (r_v, d_v) pairs in log-log scale. The results are shown in Figures 3 and 4. The measured data is represented by diamonds, while the solid line represents the least-squares approximation.

A striking observation is that the plots are approximated well by the linear regression. The correlation coefficient is higher than 0.974 for the inter-domain graphs and 0.948 for the Rout-95 graph. This leads us to the following power-law and definition.

Power-Law 1 (rank exponent) *The outdegree, d_v, of a node v, is proportional to the rank of the node, r_v, to the power of a constant, $\mathcal{R}$:*

$$d_v \propto r_v^{\mathcal{R}}$$

Definition 1 *Let us sort the nodes of a graph in decreasing order of outdegree. We define the rank exponent, $\mathcal{R}$, to be the slope of the plot of the outdegrees of the nodes versus the rank of the nodes in log-log scale.*

For the three inter-domain instances, the rank exponent, $\mathcal{R}$, is -0.81, -0.82 and -0.74 in chronological order as we see in Appendix B. The rank exponent of the Rout-95 graph, -0.48, is different compared to that of the first three graphs. This is something that we expected, given the differences in the nature of the graphs. On the other hand, this difference suggests that the rank exponent can distinguish graphs of different nature, although they both follow Power-Law 1. This property can make the rank exponent a powerful metric for characterizing families of graphs, see Section 5.

Intuitively, Power-Law 1 most likely reflects a principle of the way domains connect; the linear property observed in our four graph instances is unlikely to be a coincidence. The power-law seems to capture the equilibrium of the trade-off between the gain and the cost of adding an edge from a financial and functional point of view, as we discuss in Section 5.

Extended Discussion - Applications. We can estimate the proportionality constant for Power-Law 1, if we require that the minimum outdegree of the graph is one ($d_N = 1$). This way, we can refine the power-law as follows.

Lemma 1 *The outdegree, d_v, of a node v, is a function of the rank of the node, r_v and the rank exponent, $\mathcal{R}$, as follows*

$$d_v = \frac{1}{N^{\mathcal{R}}} \; r_v^{\mathcal{R}}$$

Proof. The proof can be found in Appendix C.

Finally, using lemma 1, we relate the number of edges with the number of nodes and the rank exponent.

Lemma 2 *The number of edges, E, of a graph can be estimated as a function of the number of nodes, N, and the rank exponent, $\mathcal{R}$, as follows:*

$$E = \frac{1}{2\,(\mathcal{R}+1)} \; (1 - \frac{1}{N^{\mathcal{R}+1}}) \; N$$

Proof. The proof can be found in Appendix C.

Note that Lemma 2 can give us the number of edges as a function of the number of nodes for a given rank exponent. We tried the lemma in our datasets and the estimated number of edges differed by 9% to 20% from the actual number of edges. More specifically for the Int-12-98, the lemma underestimates the number of edges by 10%. We can get a closer estimate (3.6%) by using a simple linear interpolation in the number of edges given the number of nodes. Note that the two prediction mechanisms are different: our lemma does not need previous network instances, but it needs to know the rank exponent. However, given previous network instances, we seem to be better off using the linear interpolation according to the above analysis. We examined the sensitivity of our lemma with respect to the value of rank exponent. A 5% increase (decrease) in the absolute value of the rank exponent increases (decreases) the number of edges by 10% for the number of nodes in Int-12-98.

4.2 The outdegree exponent $\mathcal{O}$

In this section, we study the distribution of the outdegree of the graphs, and we manage to describe it concisely by a single number. Recall that the frequency, f_d, of an outdegree, d, is the number of nodes with outdegree d. We plot the frequency f_d versus the outdegree d in log-log scale in

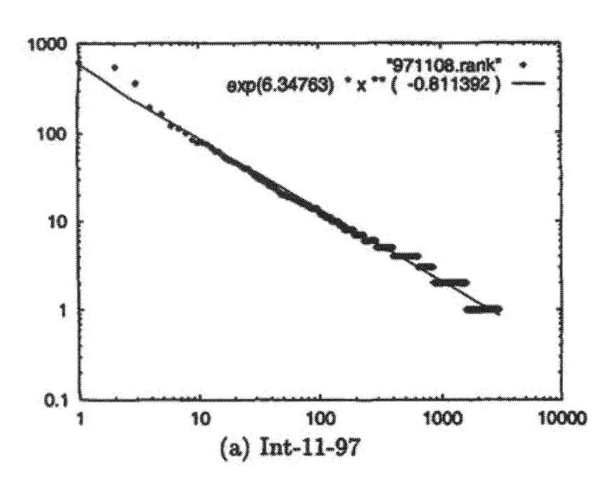

(a) Int-11-97

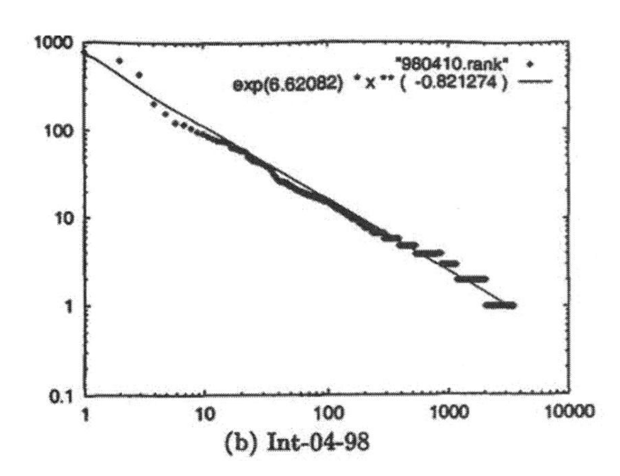

(b) Int-04-98

Figure 3: The rank plots. Log-log plot of the outdegree d_v versus the rank r_v in the sequence of decreasing outdegree.

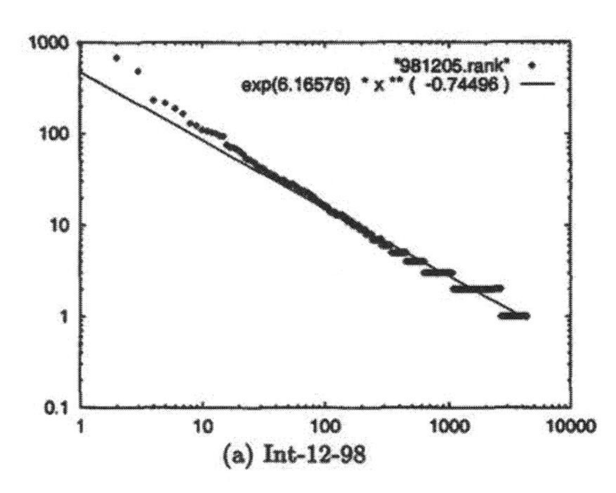

(a) Int-12-98

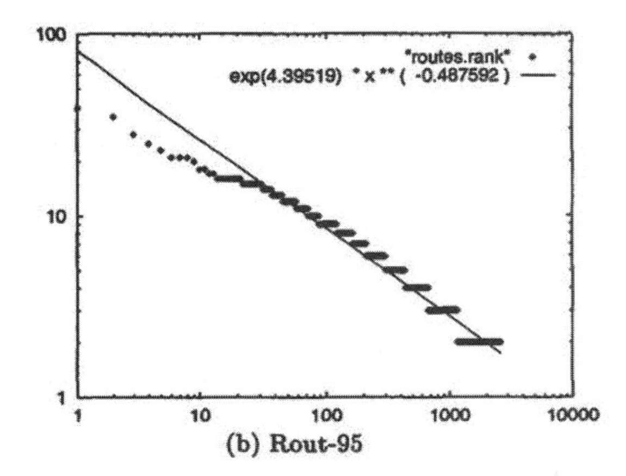

(b) Rout-95

Figure 4: The rank plots. Log-log plot of the outdegree d_v versus the rank r_v in the sequence of decreasing outdegree.

figures 5 and 6. In these plots, we exclude a small percentage of nodes of higher outdegree that have frequency of one. Specifically, we plot the outdegrees starting from one until we reach an outdegree that has frequency of one. As we saw earlier, the higher outdegrees are described and captured by the rank exponent. In any case, we plot more than 98% of the total number of nodes. The solid lines are the result of the linear regression.

The major observation is that the plots are approximately linear (see Table 8). The correlation coefficients are between 0.968-0.99 for the inter-domain graphs and 0.966 for the Rout-95. This leads us to the following power-law and definition.

> **Power-Law 2 (outdegree exponent)**
> *The frequency, f_d, of an outdegree, d, is proportional to the outdegree to the power of a constant, $\mathcal{O}$:*
>
> $$f_d \propto d^{\mathcal{O}}$$

Definition 2 *We define the outdegree exponent, $\mathcal{O}$, to be the slope of the plot of the frequency of the outdegrees versus the outdegrees in log-log scale.*

The second striking observation is that the value of the outdegree exponent is practically constant in our graphs of the inter-domain topology. The exponents are -2.15, -2.16 and -2.2, as shown in Appendix B. It is interesting to note that even the Rout-95 graph obeys the same power-law (Figure 6.b) with an outdegree exponent of -2.48. These facts suggest that Power-Law 2 describes a fundamental property of the network.

The intuition behind this power-law is that the distribution of the outdegree of Internet nodes is not arbitrary. The qualitative observation is that lower degrees are more frequent. Our power-law manages to quantify this observation by a single number, the outdegree exponent. This way, we can test the realism of a graph with a simple numerical comparison. If a graph does not follow Power-Law 2, or if its outdegree exponent is considerably different from the real exponents, it probably does not represent a realistic topology.

4.3 The hop-plot exponent $\mathcal{H}$

In this section, we quantify the connectivity and distances between the Internet nodes in a novel way. We choose to

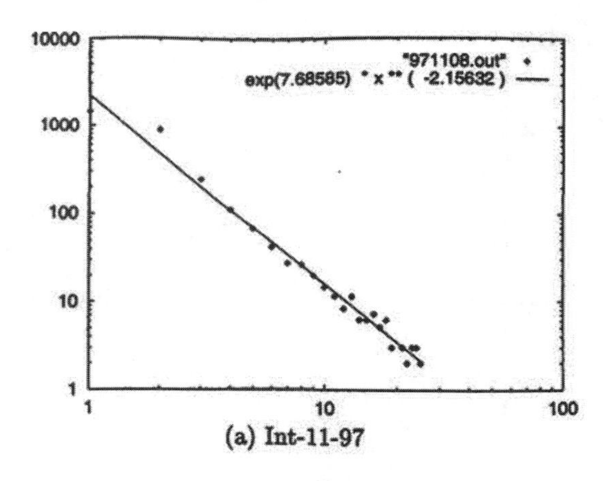

(a) Int-11-97

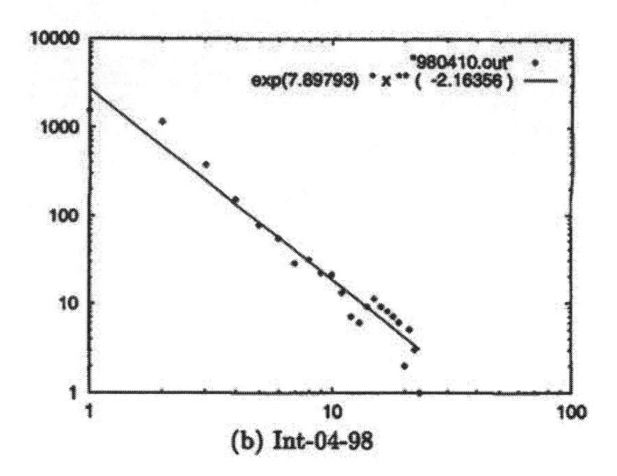

(b) Int-04-98

Figure 5: The outdegree plots: Log-log plot of frequency f_d versus the outdegree d.

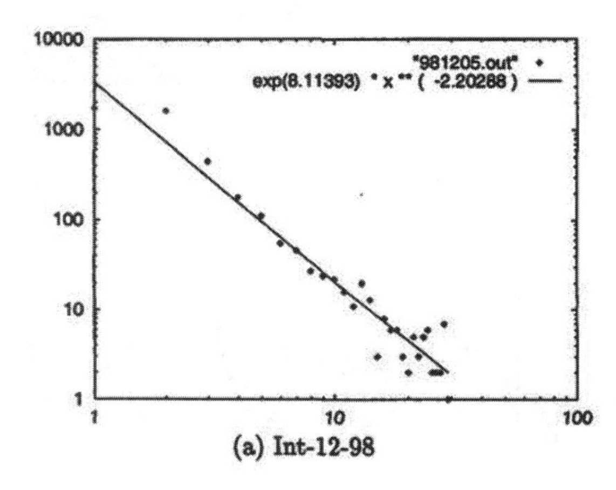

(a) Int-12-98

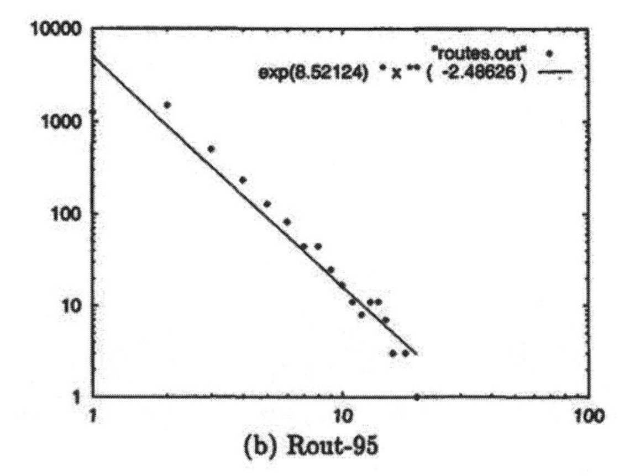

(b) Rout-95

Figure 6: The outdegree plots: Log-log plot of frequency f_d versus the outdegree d.

study the size of the neighborhood within some distance, instead of the distance itself. Namely, we use the total number of pairs of nodes $P(h)$ within h hops, which we define as the total number of pairs of nodes within less or equal to h hops, including self-pairs, and counting all other pairs twice.

Let us see the intuition behind the number of pairs of nodes $P(h)$. For $h = 0$, we only have the self-pairs: $P(0) = N$. For the diameter of the graph δ, $h = \delta$, we have the self-pairs plus all the other possible pairs: $P(\delta) = N^2$, which is the maximum possible number of pairs. For a hypothetical ring topology, we have $P(h) \propto h^1$, and, for a 2-dimensional grid, we have $P(h) \propto h^2$, for $h \ll \delta$. We examine whether the number of pairs $P(h)$ for the Internet follows a similar power-law.

In figures 7 and 8, we plot the number of pairs $P(h)$ as a function of the number of hops h in log-log scale. The data is represented by diamonds, and the dotted horizontal line represents the maximum number of pairs, which is N^2. We want to describe the plot by a line in least-squares fit, for $h \ll \delta$, shown as a solid line in the plots. We approximate the first 4 hops in the inter-domain graphs, and the first 12 hops in the Rout-95. The correlation coefficients are is 0.98 for inter-domain graphs and 0.96, for the Rout-95, as we see in Appendix B. Unfortunately, four points is a rather small number to verify or disprove a linearity hypothesis experimentally. However, even this rough approximation has several useful applications as we show later in this section.

Approximation 1 (hop-plot exponent) *The total number of pairs of nodes, $P(h)$, within h hops, is proportional to the number of hops to the power of a constant, $\mathcal{H}$:*

$$P(h) \propto h^{\mathcal{H}}, \quad h \ll \delta$$

Definition 3 *Let us plot the number of pairs of nodes, $P(h)$, within h hops versus the number of hops in log-log scale. For $h \ll \delta$, we define the slope of this plot to be the* hop-plot exponent, $\mathcal{H}$.

Observe that the three inter-domain datasets have practically equal hop-plot exponents; 4.6, 4.7, and 4.86 in chronological order, as we see in Appendix B. This shows that the hop-plot exponent describes an aspect of the connectivity of the graph in a single number. The Rout-95 plot, in fig. 8.b,

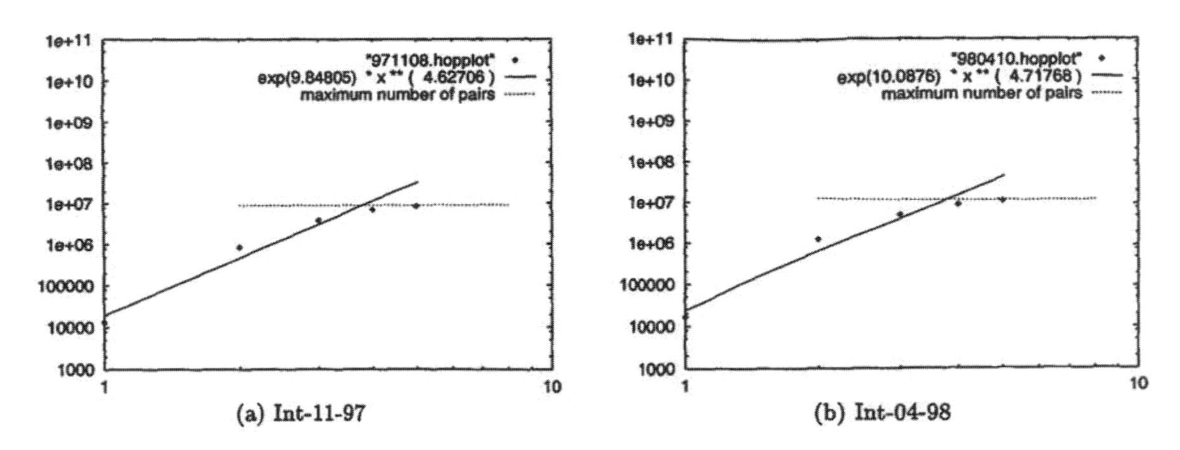

(a) Int-11-97 (b) Int-04-98

Figure 7: The hop-plots: Log-log plots of the number of pairs of nodes $P(h)$ within h hops versus the number of hops h.

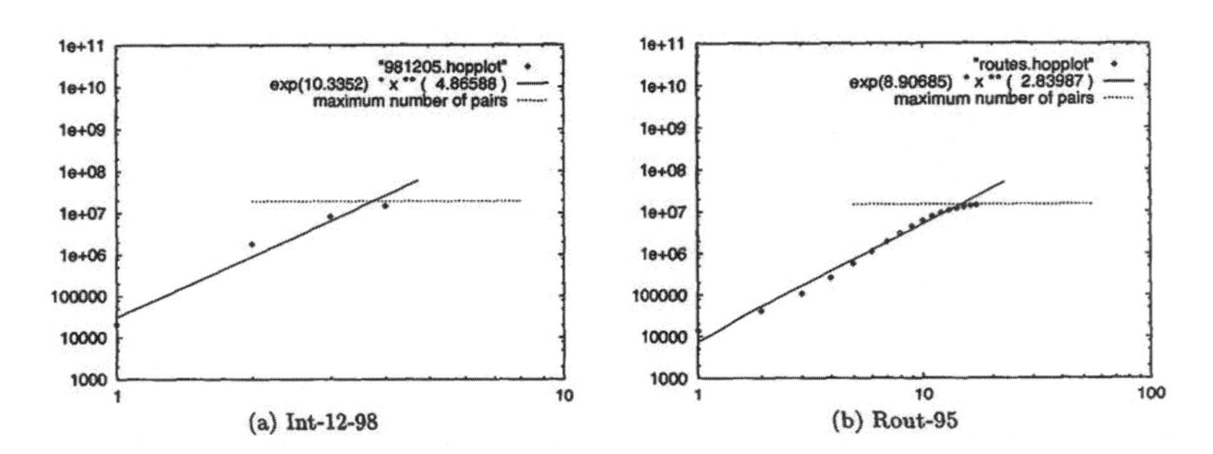

(a) Int-12-98 (b) Rout-95

Figure 8: The hop-plots: Log-log plots of the number of pairs of nodes $P(h)$ within h hops versus the number of hops h.

has more points, and thus, we can argue for its linearity with more confidence. The hop-plot exponent of Rout-95 is 2.8, which is much different compared to those of the inter-domain graphs. This is expected, since the Rout-95 is a sparser graph. Recall that for a ring topology, we have $\mathcal{H} = 1$, and, for a 2-dimensional grid, we have $\mathcal{H} = 2$. The above observations suggest that the hop-plot exponent can distinguish families of graphs efficiently, and thus, it is a good metric for characterizing the topology.

Extended Discussion - Applications. We can refine Approximation 1 by calculating its proportionality constant. Let us recall the definition of the number of pairs, $P(h)$. For $h = 1$, we consider each edge twice and we have the self-pairs, therefore: $P(1) = N + 2\,E$. We demand that Approximation 1 satisfies the previous equation as an initial condition.

Lemma 3 *The number of pairs within h hops is*

$$P(h) \;=\; \begin{cases} c\,h^{\mathcal{H}}, & h \ll \delta \\ N^2, & h \geq \delta \end{cases}$$

where $c = N + 2\,E$ *to satisfy initial conditions.*

In networks, we often need to reach a target without knowing its exact position [7] [1]. In these cases, selecting the extent of our broadcast or search is an issue. On the one hand, a small broadcast will not reach our target. On the other hand, an extended broadcast creates too many messages and takes a long time to complete. Ideally, we want to know how many hops are required to reach a "sufficiently large" part of the network. In our hop-plots, a promising solution is the intersection of the two asymptote lines: the horizontal one at level N^2 and the asymptote with slope $\mathcal{H}$. We calculate the intersection point using Lemma 3, and we define:

Definition 4 (effective diameter) *Given a graph with N nodes, E edges, and* $\mathcal{H}$ *hop-plot exponent, we define the effective diameter,* δ_{ef}*, as:*

$$\delta_{ef} = \left(\frac{N^2}{N + 2\,E}\right)^{1/\mathcal{H}}$$

Intuitively, the effective diameter can be understood as follows: any two nodes are within δ_{ef} hops from each other with high probability. We verified the above statement experimentally. The effective diameters of our inter-domain

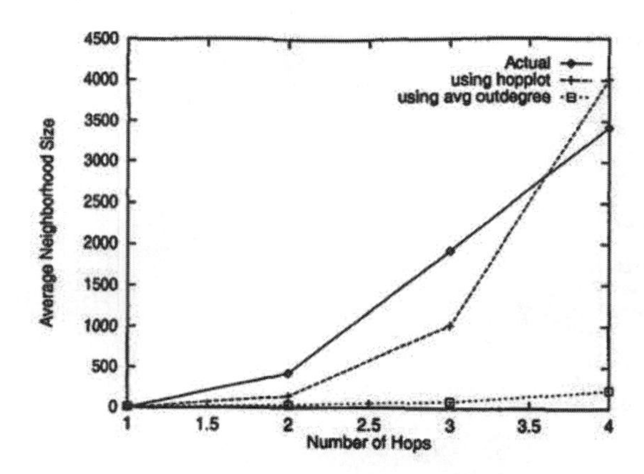

Figure 9: Average neighborhood size versus number of hops the actual, and estimated size a) using hop-plot exponent, b) using the average outdegree for Int-12-98.

Hops	hop-plot	avg. outdegree
1	0.02	1.82
2	-0.66	-0.93
3	-0.47	-0.95
4	0.17	-0.93

Table 3: The relative error of the two estimates for the average neighborhood size with respect to the real value. Negative error means under-estimate.

graphs was slightly over four. Rounding the effective diameter to four, approximately 80% of the pairs of nodes are within this distance. The ceiling of the effective diameter is five, which covers more than 95% of the pairs of nodes.

An advantage of the effective diameter is that it can be calculated easily, when we know N, and $\mathcal{H}$. Recall that we can calculate the number of edges from Lemma 2. Given that the hop-plot exponent is practically constant, we can estimate the effective diameter of future Internet instances as we do in Section 5.

Furthermore, we can estimate the average size of the neighborhood, NN(h), within h hops using the number of pairs $P(h)$. Recall that $P(h) - N$ is the number of pairs without the self-pairs.

$$\mathrm{NN}(h) = \frac{P(h)}{N} - 1 \qquad (1)$$

Using Equation 1 and Lemma 3, we can estimate the average neighborhood size.

Lemma 4 *The average size of the neighborhood, NN(h), within h hops as a function of the hop-plot exponent, $\mathcal{H}$, for $h \ll \delta$, is*

$$NN(h) = \frac{c}{N}\, h^{\mathcal{H}} - 1$$

where $c = N + 2\,E$ to satisfy initial conditions.

The average neighborhood is a commonly used parameter in the performance of network protocols. Our estimate is an improvement over the commonly used estimate that uses the average outdegree [26] [7] which we call **average-outdegree estimate**:

$$\mathrm{NN}'(h) = \bar{d}\,(\bar{d} - 1)^{h-1}$$

In figure 9, we plot the actual and both estimates of the average neighborhood size versus the number of hops for the Int-12-98 graph. In Table 3, we show the normalized error of each estimate: we calculate the quantity: $(p-r)/r$ where p the prediction and r the real value. The results for the other inter-domain graphs are similar. The superiority of the hop-plot exponent estimate is apparent compared to the average-outdegree estimate. The discrepancy of the average-outdegree estimate can be explained if we consider that the estimate does not comply with the real data; it implicitly assumes that the outdegree distribution is uniform. In more detail, it assumes that each node in the periphery of the neighborhood adds $\bar{d} - 1$ new nodes at the next hop. Our data shows that the outdegree distribution is highly skewed, which explains why the use of the hop-plot estimate gives a better approximation.

The most interesting difference between the two estimates is qualitative. The previous estimate considers the neighborhood size exponential in the number of hops. Our estimate considers the neighborhood as an $\mathcal{H}$-dimensional sphere with radius equal to the number of hops, which is a novel way to look at the topology of a network[5]. Our data suggests that the hop-plot exponent-based estimate gives a closer approximation compared to the average-outdegree-based metric.

4.4 The eigen exponent $\mathcal{E}$

In this section, we identify properties of the eigenvalues of our Internet graphs. There is a rich literature that proves that the eigenvalues of a graph are closely related to many basic topological properties such as the diameter, the number of edges, the number of spanning trees, the number of connected components, and the number of walks of a certain length between vertices, as we can see in [8] and [4]. All of the above suggest that the eigenvalues intimately relate to topological properties of graphs.

We plot the eigenvalue λ_i versus i in log-log scale for the first 20 eigenvalues. Recall that i is the order of λ_i in the decreasing sequence of eigenvalues. The results are shown in Figure 10 and Figure 11. The eigenvalues are shown as diamonds in the figures, and the solid lines are approximations using a least-squares fit.

Observe that in all graphs, the plots are practically linear with a correlation coefficient of 0.99, as we see in Appendix B. It is rather unlikely that such a canonical form of the eigenvalues is purely coincidental, and we therefore conjecture that it constitutes an empirical power-law of the Internet topology.

Power-Law 3 (eigen exponent) *The eigenvalues, λ_i, of a graph are proportional to the order, i, to the power of a constant, $\mathcal{E}$:*

$$\lambda_i \propto i^{\mathcal{E}}$$

Definition 5 *We define the eigen exponent, $\mathcal{E}$, to be the slope of the plot of the sorted eigenvalues versus their order in log-log scale.*

[5]Note that our results focus on relatively small neighborhoods compared to the diameter $h \ll \delta$. Other experimental studies focus on neighborhoods of larger radius [17].

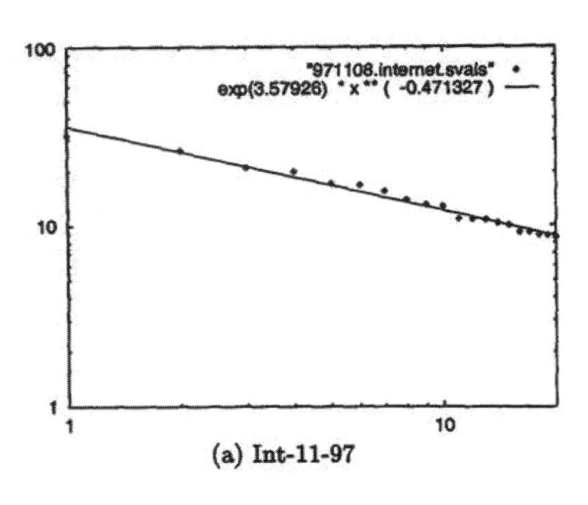

(a) Int-11-97

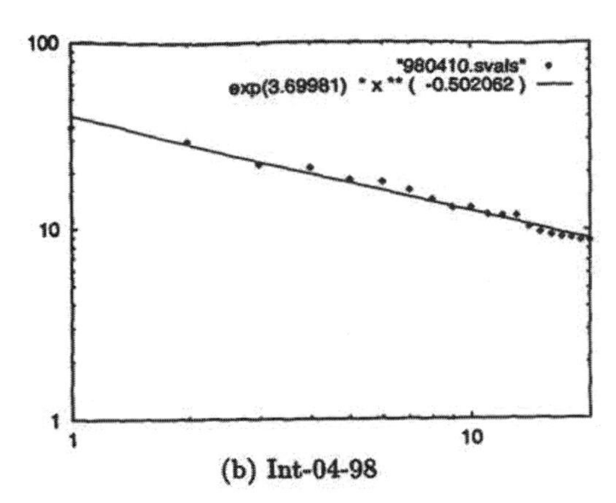

(b) Int-04-98

Figure 10: The eigenvalue plots: Log-log plot of eigenvalues in decreasing order.

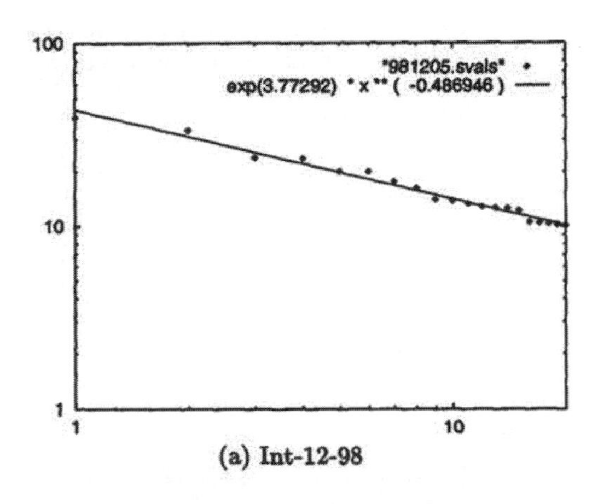

(a) Int-12-98

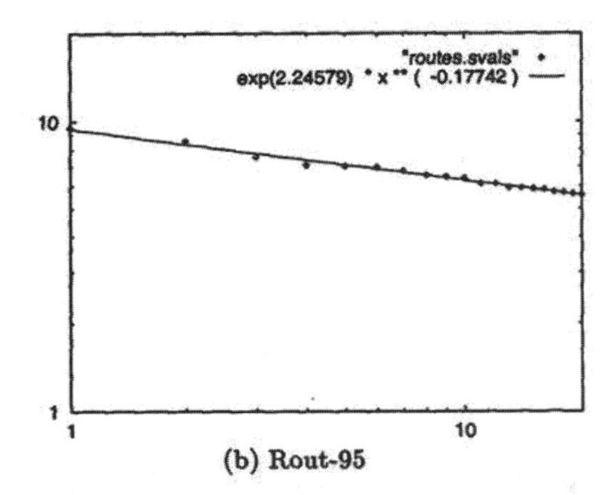

(b) Rout-95

Figure 11: The eigenvalue plots: Log-log plot of eigenvalues in decreasing order.

A surprising observation is that the eigen exponents of the three inter-domain graphs are practically equal: -0.47, -0.50 and -0.48 in chronological order. This means that the eigen exponent captures a property of the Internet that characterizes all three instances despite the increase in size. On the other hand, the eigen exponent of the routers graph is significantly different -0.177, from the previous slopes. This shows that the eigen exponent can distinguish differences between families of graphs.

5 Discussion

In this section, we discuss the practical uses of our power-laws and our approximation. We also present the intuition behind the existence of such power-laws in a chaotic environment such as the Internet. In addition, we discuss the scope of the predictions that are based on our work.

Describing Graphs: Exponents versus Averages. We propose a new way to describe topological properties using power-laws. Our observations show that most of the distributions of interest are *skewed*, typically following a power-law. Average values falsely imply a uniform distribution, and they can be misleading. For example, 85% of the nodes in Int-12-98 have outdegree less than the average outdegree! We propose to use the exponents of power-laws, which manage to capture the trend of a property in a single number.

Protocol Performance. Our work can facilitate the design, and the performance analysis of protocols. As we saw, our power-laws help us estimate useful graph metrics. We provide formulas for the effective diameter, the average neighborhood size, and the number of edges, in Definition 4, Lemma 4 and Lemma 2 respectively. Our $O(\bar{d}\ h^{\mathcal{H}})$ estimate for the average neighborhood size is a fundamental improvement over the commonly used $O(\bar{d}^{h})$. This way, we can fine-tune and analyze the performance and the complexity of several protocols[6].

Predictions and Extrapolations. Our power-laws offer guidelines for answering "what-if" questions. First, we can scrutinize the plausibility of a hypothesis, if they contradict our power-laws. Second, we can predict useful parameters of the Internet under different hypotheses and assumptions. Actually, given just a hypothesis for the number of nodes, we can estimate the number of edges from Lemma 2, and

[6]Some protocols that employ broadcasting or flooding techniques are the link-state protocols OSPF and MOSPF [13], and the multicast protocols DVMRP [22], QoSMIC [7], YAM [1].

Year	1999	2000	2001	2002
Nodes	4389	5763	7137	8511
Edges	8256	12639	15301	18384
Effective diameter	4.26	4.39	4.61	4.78

Table 4: Internet prediction assuming linear node increase. We predict the number of edges and effective diameter of the Internet at the inter-domain level at the beginning of each year.

Year	1999	2000	2001	2002
Nodes	4389	6364	9227	13380
Edges	8256	13576	19996	29421
Effective diameter	4.26	4.51	4.86	5.25

Table 5: Internet prediction assuming 45% increase in the number of nodes every year. We predict the number of edges and effective diameter of the Internet at the inter-domain level at the beginning of each year.

the effective diameter using Definition 4. Note that our tools do not predict the number of nodes of the Internet, but for the sake of the example we will examine two possible growth patterns. We can assume that the number of nodes increases a) linearly, or b) by 1.45 each year. The results our shown in Table 4 for the linear growth and Table 5 for the 1.45 growth. Given the number of nodes, we calculate the number of edges using Lemma 2 with rank exponent of -0.81, which is the median of the three observed rank exponents. We calculate the effective diameter using Definition 4 with a hop-plot exponent of 4.71, the median of the observed values.

Predicting the evolution of a dynamic system such as the Internet is not trivial. There are many social, economical, and technological factors that can alter significantly the topology of the network. Furthermore, systems often evolve in bursts following social and technological breakthroughs. In this paper, we claim that our power-laws characterize the Internet topology during the year 1998. However, given the large number of natural distributions that follow power-laws, the Internet topology will likely be described by power-laws even in the future. In the absence of any other information, a practitioner would reasonably conjecture that our power-laws might continue to hold, at least for the near future. We elaborate further on our intuition regarding power-laws and natural systems in section 5.1.

Graph Generation and Selection. Our power-laws can be used to characterize graph topologies. This way, the power-laws can be used as a composite "qualifying exam" for the realism of a graph. Recall that some power-laws showed significantly different exponents in the inter-domain and the router-level graphs. We conducted some preliminary experiments with some artificial topologies and some real graphs of different nature (e.g. web-site topology). Some graphs did not comply to the power-laws at all, while some others showed large differences in the values of the exponents. The observations for these graphs and the Internet graphs in this paper suggest that our power-laws could be used to characterize and distinguish graphs.

In addition, we provide measurements that are targeted towards the current graph models [27], as we saw in Section 3 and Appendix A. In an overview, we list the following guidelines for creating inter-domain topologies. First, a large but decreasing percentage of the nodes(50%, 45%, and 40%) belong to trees. Second, more than 80% of the trees have depth one, and the maximum depth is three. Third, the outdegree distribution is skewed following our power-laws 1 and 2 within a range of 1 to 1000 approximately. As a final step, the realism of the resulting graph can be tested using our power-laws.

5.1 Finding Order in Chaos

Why would such an unruly entity like the Internet follow any statistical regularities? Note that the high correlation coefficients exclude the role of coincidence. Intrigued by the previous question, we attempt an intuitive explanation. The topological structure of the Internet is the collective result of many small forces in antagonistic and cooperative relationships. These forces find an equilibrium in a state, and it is this state that our power-laws capture. Let us think of how change happens. New nodes are not just "glued" on the existing graph; they trigger a chain of restructuring changes. If many new nodes connect to an existing node, it will probably have to increase its connectivity to accommodate the new demand in traffic. In other words, the change propagates to the rest of the network like a fading wave. Therefore, at any time the topology is characterized by the same fundamental properties. As an analogy, we can think of a heap of sand that we create by dropping sand from one point. At any given moment, the heap is a cone, though its size changes and the grains are just dropped unorderly.

The above intuitive understanding of the network topology is reinforced by the fact that this kind of dynamic equilibrium, and power-laws characterize many natural systems. First, power-laws govern the nature of various networks. The traffic of the Internet and the World Wide Web is characterized by power-laws, as we already saw in section 2. Furthermore, power-laws describe the topology of multiple real networks of biological and geographical nature such as the human respiratory system [12] with a scaling factor of 2.9, and automobile networks [6] with an exponent of 1.6. Second, power-laws are obeyed in diverse settings, like income distribution (the "Pareto law"), and the frequency distribution of words in natural text (the "Zipf distribution" [28]).

6 Conclusions

Our main contribution is a novel way to study the Internet topology, namely through power-laws. These power-laws capture concisely the highly skewed distributions of the graph properties and quantify them by single numbers, the power-law exponents. Our contributions can be summarized in the following points:

- We discover three power-laws that characterize the inter-domain Internet topology during the year of 1998.
- Our power-laws hold for three Internet instances with high correlation coefficients.
- We propose the number of pairs, $P(h)$, within h hops, as a metric of the density of the graph and approximate it with the use of the hop-plot exponent, $\mathcal{H}$.
- We derive formulas that link the exponents of our power-laws with vital graph metrics such as the num-

ber of nodes, the number of edges, and the average neighborhood size.

- We propose power-law exponents, instead of averages, as an efficient way to describe the highly-skewed graph metrics which we examined.

Apart from their theoretical interest, we showed a number of practical applications of our power-laws. First, our power-laws can assess the realism of synthetic graphs, and enhance the validity of our simulations. Second, they can help analyze the average-case behavior of network protocols. For example, we can estimate the message complexity of protocols using our estimate for the neighborhood size. Third, the power-laws can help answer "what-if" scenarios like *"what will be the diameter of the Internet, when the number of nodes doubles?" "what will be the number of edges then?"*

In addition, we decompose and measure the Internet in a way that relates to the state-of-the-art graph generation models. This decomposition provides measurements that facilitate the selection of parameters for the graph generators.

For the future, we believe that our suggestion to look for power-laws will open the floodgates to discovering many additional power-laws of the Internet topology. Our optimism is based on two facts: (a) power-laws are intimately related to fractals, chaos and self-similarity [21] and (b) there is overwhelming evidence that self-similarity appears in a large number of settings, ranging from traffic patterns in networks [24], to biological and economical systems [12].

ACKNOWLEDGMENTS. We would like to thank Mark Craven, Daniel Zappala, and Adrian Perrig for their help in earlier phases of this work. The authors are grateful to Pansiot and Grad for providing the Rout-95 routers data. We would also like to thank Vern Paxson, and Ellen Zegura for the thorough review and valuable feedback. Finally, we would like to thank our mother Sofia Faloutsou-Kalamara and dedicate this work to her.

References

[1] K. Carlberg and J. Crowcroft. Building shared trees using a one-to-many joining mechanism. *ACM Computer Communication Review*, pages 5–11, January 1997.

[2] J. Chuang and M. Sirbu. Pricing multicast communications: A cost based approach. *In Proc. of the INET'98*, 1998.

[3] M. Crovella and A. Bestavros. Self-similarity in World Wide Web traffic, evidence and possible causes. *SIGMETRICS*, pages 160–169, 1996.

[4] D. M. Cvetkovič, M. Boob, and H. Sachs. *Spectra of Graphs*. Academic press, 1979.

[5] M. Doar. A better model for generating test networks. *Proc. Global Internet, IEEE*, Nov. 1996.

[6] Christos Faloutsos and Ibrahim Kamel. Beyond uniformity and independence: Analysis of R-trees using the concept of fractal dimension. In *Proc. ACM SIGACT-SIGMOD-SIGART PODS*, pages 4–13, Minneapolis, MN, May 24-26 1994. Also available as CS-TR-3198, UMIACS-TR-93-130.

[7] M. Faloutsos, A. Banerjea, and R. Pankaj. QoSMIC: a QoS Multicast Internet protoCol. *ACM SIGCOMM. Computer Communication Review.*, Sep 2-4, Vancouver BC 1998.

[8] M. Faloutsos, P. Faloutsos, and C. Faloutsos. Power-laws of the Internet topology. Technical Report UCR-CS-99-01, University of California Riverside, Computer Science, 1999.

[9] National Laboratory for Applied Network Research. Routing data. Supported by NSF, http://moat.nlanr.net/Routing/rawdata/, 1998.

[10] R. Govindan and A. Reddy. An analysis of internet inter-domain topology and route stability. *Proc. IEEE INFOCOM*, Kobe, Japan, April 7-11 1997.

[11] W.E. Leland, M.S. Taqqu, W. Willinger, and D.V. Wilson. On the self-similar nature of ethernet traffic. *IEEE Transactions on Networking*, 2(1):1–15, February 1994. (earlier version in SIGCOMM '93, pp 183-193).

[12] B. Mandelbrot. *Fractal Geometry of Nature*. W.H. Freeman, New York, 1977.

[13] J. Moy. Multicast routing extensions for OSPF. *ACM Connunications*, 37(8):61–66, 1994.

[14] J.-J. Pansiot and D Grad. On routes and multicast trees in the Internet. *ACM Computer Communication Review*, 28(1):41–50, January 1998.

[15] V. Paxson and S. Floyd. Wide-area traffic: The failure of poisson modeling. *IEEE/ACM Transactions on Networking*, 3(3):226–244, June 1995. (earlier version in SIGCOMM'94, pp. 257-268).

[16] V. Paxson and S. Floyd. Why we don't know how to simulate the internet. *Proceedings of the 1997 Winter Simulation Conference*, December 1997.

[17] G. Philips, S. Shenker, and H. Tangmunarunkit. Scaling of multicast trees: Comments on the chuang-sirbu scaling law. *ACM SIGCOMM. Computer Communication Review.*, Sep 1999.

[18] William H. Press, Saul A. Teukolsky, William T. Vetterling, and Brian P. Flannery. *Numerical Recipes in C*. Cambridge University Press, 2nd edition, 1992.

[19] Y. Rekhter and T. Li (Eds). A Border Gateway Protocol 4 (BGP-4). Internet-Draft:draft-ietf-idr-bgp4-08.txt available from ftp://ftp.ietf.org/internet-drafts/, 1998.

[20] S. R. Resnick. Heavy tail modeling and teletraffic data. *Annals of Statistics*, 25(5):1805–1869, 1997.

[21] Manfred Schroeder. *Fractals, Chaos, Power Laws: Minutes from an Infinite Paradise*. W.H. Freeman and Company, New York, 1991.

[22] D. Waitzman, C. Partridge, and S. Deering. Distance vector multicast routing protocol. IETF RFC 1075, 1998.

[23] B. M. Waxman. Routing of multipoint connections. *IEEE Journal of Selected Areas in Communications*, pages 1617–1622, 1988.

[24] W. Willinger, V. Paxson, and M.S. Taqqu. Self-similarity and heavy tails: Structural modeling of network traffic. *In A Practical Guide to Heavy Tails: Statistical Techniques and Applications*, 1998. Adler, R., Feldman, R., and Taqqu, M.S., editors, Birkhauser.

[25] Walter Willinger, Murad Taqqu, Robert Sherman, and Daniel V. Wilson. Self-similarity through high variability: statistical analysis of ethernet LAN traffic at the source level. *ACM SIGCOMM'95. Computer Communication Review*, 25:100–113, 1995.

[26] D. Zappala, D. Estrin, and S. Shenker. Alternate path routing and pinning for interdomain multicast routing. Technical Report USC CS TR 97-655, U. of South California, 1997.

[27] E. W. Zegura, K. L. Calvert, and M. J. Donahoo. A quantitative comparison of graph-based models for internetworks. *IEEE/ACM Transactions on Networking*, 5(6):770–783, December 1997. http://www.cc.gatech.edu/projects/gtitm/.

[28] G.K. Zipf. *Human Behavior and Principle of Least Effort: An Introduction to Human Ecology*. Addison Wesley, Cambridge, Massachusetts, 1949.

	Int-11-97	Int-04-98	Int-12-98
nodes	3015	3530	4389
edges	5156	6432	8256
avg. outdegree	3.42	3.65	3.76
max. outdegree	590	745	979
diameter	9	11	10
avg. distance	3.76	3.77	3.75

Table 6: The evolution of the Internet at the inter-domain level.

	Int-11-97	Int-04-98	Int-12-98
#nodes in trees (%)	50.05	45.05	40.76
#trees over #nodes (%)	10.12	10.26	9.4
max depth	3	3	3
avg. tree size	4.9	4.4	4.3
core outdegree	4.7	4.9	4.9

Table 7: The evolution of the Internet considering the core and the trees.

A Decomposing the Internet

We analyze the Internet in a way that suits the graph generator models [27]. The measurements we present can facilitate the selection of parameters for these generators.

We study the graphs through their decomposition into two components: the *tree* component that contains all nodes that belong exclusively to trees and the *core* component that contains the rest of the nodes including the roots of the trees. We measure several parameters from this decomposition that are shown in Table 7. These results leads to the following observations.

- Approximately half of the nodes are in trees 40-50%
- The number of nodes in trees decreased with time by 10% means that the Internet becomes more connected all around.
- The maximum tree depth is 3, however more than 80% of the trees have depth one.
- More than 95% of the tree-nodes have a degree of one. This leads to the following interesting observation: *if we remove the nodes with outdegree one from the original graph, we practically get the core component.*

These observations can help users select appropriate values for the parameters used in various graph generation techniques [27].

B The Exponents of Our Power-Laws

We present the exponents of our power-laws in Table 8.

Exponent	Int-11-97	Int-04-98	Int-12-98	Rout-95
rank	-0.81	-0.82	-0.74	-0.48
ACC	0.981	0.979	0.974	0.948
outdegree	-2.15	-2.16	-2.20	-2.48
ACC	0.991	0.979	0.968	0.966
hop-plot	4.62	4.71	4.86	2.83
ACC	0.983	0.981	0.980	0.991
eigen	-0.471	-0.502	-0.487	-0.17
ACC	0.990	0.989	0.991	0.994

Table 8: An overview of all the exponents for all our graphs. Note that ACC is the absolute value of the correlation coefficient.

C The Proofs

Here we prove the Lemmas we present in our paper.

Lemma 1. *The outdegree, d_v, of a node v, is a function of the rank of the node, r_v and the rank exponent, $\mathcal{R}$, as follows*

$$d_v = \frac{1}{N^{\mathcal{R}}}\, r_v^{\mathcal{R}}$$

Proof. We can estimate the proportionality constant, C, for Power-Law 1, if we require that the outdegree of the N-th node is one, $d_N = 1$.

$$\begin{aligned} d_N &= C\,N^{\mathcal{R}} \Rightarrow \\ C &= 1/N^{\mathcal{R}} \end{aligned} \tag{2}$$

We combine Power-Law 2 with Equation 2, and conclude the proof. ■

Lemma 2. *The number of edges, E, of a graph can be estimated as a function of the number of nodes, N, and the rank exponent, $\mathcal{R}$, as follows:*

$$E = \frac{1}{2\,(\mathcal{R}+1)}\,(1 - \frac{1}{N^{\mathcal{R}+1}})\,N$$

Proof: The sum of all the outdegrees for all the ranks is equal to two times the number of edges, since we count each edge twice.

$$\begin{aligned} 2\,E &= \sum_{r_v=1}^{N} d_v \\ 2\,E &= \sum_{r_v=1}^{N} (r_v/N)^{\mathcal{R}} = (1/N)^{\mathcal{R}} \sum_{r_v=1}^{N} r_v^{\mathcal{R}} \\ E &\approx \frac{1}{2\,N^{\mathcal{R}}} \int_1^N r_v^{\mathcal{R}}\, dr_v \end{aligned} \tag{3}$$

In the last step, above we approximate the summation with an integral. Calculating the integral concludes the proof. ■

Classes of small-world networks

L. A. N. Amaral*, A. Scala, M. Barthélémy†, and H. E. Stanley

Center for Polymer Studies and Department of Physics, Boston University, Boston, MA 02215

Communicated by Herman Z. Cummins, City College of the City University of New York, New York, NY, July 13, 2000 (received for review April 20, 2000)

We study the statistical properties of a variety of diverse real-world networks. We present evidence of the occurrence of three classes of small-world networks: (*a*) scale-free networks, characterized by a vertex connectivity distribution that decays as a power law; (*b*) broad-scale networks, characterized by a connectivity distribution that has a power law regime followed by a sharp cutoff; and (*c*) single-scale networks, characterized by a connectivity distribution with a fast decaying tail. Moreover, we note for the classes of broad-scale and single-scale networks that there are constraints limiting the addition of new links. Our results suggest that the nature of such constraints may be the controlling factor for the emergence of different classes of networks.

Disordered networks, such as small-world networks are the focus of recent interest because of their potential as models for the interaction networks of complex systems (1–7). Specifically, neither random networks nor regular lattices seem to be an adequate framework within which to study "real-world" complex systems (8) such as chemical-reaction networks (9), neuronal networks (2), food webs (10–12), social networks (13, 14), scientific-collaboration networks (15), and computer networks (4, 16–19).

Small-world networks (2), which emerge as the result of randomly replacing a fraction P of the links of a d dimensional lattice with new random links, interpolate between the two limiting cases of a regular lattice ($P = 0$) and a random graph ($P = 1$). A small-world network is characterized by the following properties: (*i*) the local neighborhood is preserved (as for regular lattices; ref. 2); and (*ii*) the diameter of the network, quantified by average shortest distance between two vertices (20), increases logarithmically with the number of vertices n (as for random graphs; ref. 21). The latter property gives the name small-world to these networks, because it is possible to connect any two vertices in the network through just a few links, and the local connectivity would suggest the network to be of finite dimensionality.

The structure of small-world networks and of real networks has been probed through the calculation of their diameter as a function of network size (2). In particular, networks such as (*a*) the electric power grid for Southern California, (*b*) the network of movie-actor collaborations, and (*c*) the neuronal network of the worm *Caenorhabditis elegans* seem to be small-world networks (2). Further, it was proposed (5) that these three networks (*a*–*c*) as well as the world-wide web (4) and the network of citations of scientific papers (22, 23) are scale-free—that is, they have a distribution of connectivities that decays with a power law tail.

Scale-free networks emerge in the context of a growing network in which new vertices connect preferentially to the more highly connected vertices in the network (5). Scale-free networks are also small-world networks, because (*i*) they have clustering coefficients much larger than random networks (2) and (*ii*) their diameter increases logarithmically with the number of vertices n (5).

Herein, we address the question of the conditions under which disordered networks are scale-free through the analysis of several networks in social, economic, technological, biological, and physical systems. We identify a number of systems for which there is a single scale for the connectivity of the vertices. For all these networks, there are constraints limiting the addition of new links. Our results suggest that such constraints may be the controlling factor for the emergence of scale-free networks.

Empirical Results

First, we consider two examples of technological and economic networks: (*i*) the electric power grid of Southern California (2), the vertices being generators, transformers, and substations and the links being high-voltage transmission lines; and (*ii*) the network of world airports (24), the vertices being the airports and the links being nonstop connections. For the case of the airport network, we have access to data on number of passengers in transit and of cargo leaving or arriving at the airport, instead of data on the number of distinct connections. Working under some reasonable assumptions,‡ one can expect that the number of distinct connections from a major airports is proportional to the number of passengers in transit through that airport, making the two examples, *i* and *ii*, comparable. Fig. 1 shows the connectivity distribution for these two examples. It is visually apparent that neither case has a power law regime and that both have exponentially decaying tails, implying that there is a single scale for the connectivity k.

Second, we consider three examples of "social" networks: (*iii*) the movie-actor network (2), the links in this network indicating that the two actors were cast at least once in the same movie; (*iv*) the acquaintance network of Mormons (25), the vertices being 43 Utah Mormons and the number of links the number of other Mormons they know; and (*v*) the friendship network of 417 Madison Junior High School students (26). These three examples describe apparently distinct types of social networks with very different sample sizes. In fact it can be argued that the network of movie-actor collaborations is not really a social network but is instead an economic network. However, because it was considered in other publications (1, 2, 5) as a social network, we classify it similarly here. We feel that the acquaintance and friendship networks may be better proxies of real social networks and, as such, expect similar results from the analysis of both networks. Fig. 2 shows the connectivity distribution for these social networks. The scale-free (power law) behavior of the movie-actor network (5) is truncated by an exponential tail. In contrast, the network of acquaintances of the Utah Mormons and the friendship network of the high school students display no power law regime, but instead we find results

*To whom reprint requests should be addressed. E-mail: amaral@buphy.bu.edu.

†Present address: CEA, Service de Physique de la Matière Condensée, 91680 Bruyeres-le-Chatel, France.

‡To be able to compare the two types of distributions, one must make two assumptions. The first assumption is that there is a typical number of passengers per flight. This assumption is reasonable, because the number of seats in airplanes does not follow a power law distribution. The second assumption is that there is a typical number of flights per day between two cities. This assumption is also reasonable, because at most there will be about 20 flights per day and per airline between any two cities; thus, the distribution of number of flights per day between two cities is bounded.

The publication costs of this article were defrayed in part by page charge payment. This article must therefore be hereby marked "*advertisement*" in accordance with 18 U.S.C. §1734 solely to indicate this fact.

Article published online before print: *Proc. Natl. Acad. Sci. USA*, 10.1073/pnas.200327197. Article and publication date are at www.pnas.org/cgi/doi/10.1073/pnas.200327197

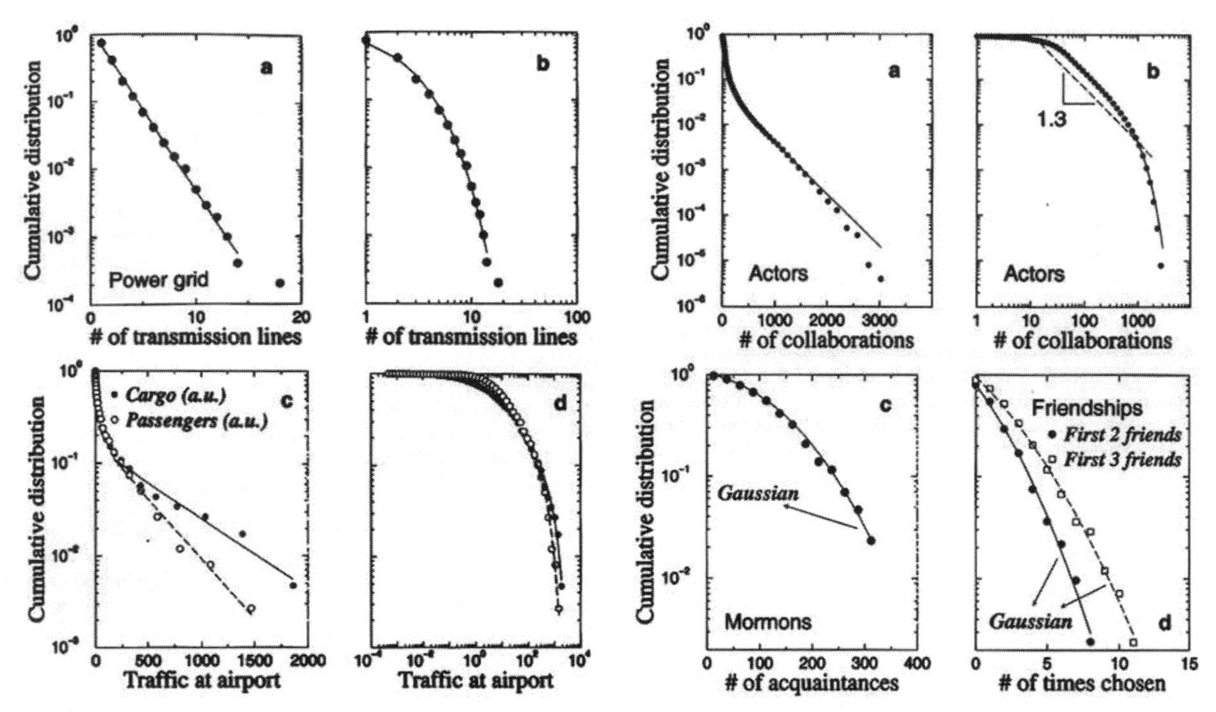

Fig. 1. Technological and economic networks. (*a*) Linear-log plot of the cumulative distribution of connectivities for the electric power grid of Southern California (2). For this type of plot, the distribution falls on a straight line, indicating an exponential decay of the distribution of connectivities. The full line, which is an exponential fit to the data, displays good agreement with the data. (*b*) Log-log plot of the cumulative distribution of connectivities for the electric power grid of Southern California. If the distribution would have a power law tail then it would fall on a straight line in a log-log plot. Clearly, the data reject the hypothesis of power law distribution for the connectivity. (*c*) Linear-log plot of the cumulative distribution of traffic at the world's largest airports for two measures of traffic, cargo, and number of passengers. The network of world airports is a small-world network; one can connect any two airports in the network by only one to five links. To study the distribution of connectivities of this network, we assume that, for a given airport, cargo and number of passengers are proportional to the number of connections of that airport with other airports. The data are consistent with a decay of the distribution of connectivities for the network of world airports that decays exponentially or faster. The full line is an exponential fit to the cargo data for values of traffic between 500 and 1,500. For values of traffic larger than 1,500, the distribution seems to decay even faster than an exponential. The long-dashed line is an exponential to the passenger data for values of traffic between 500 and 1,500. a.u., arbitrary units. (*d*) Log-log plot of the cumulative distribution of traffic at the world's largest airports. This plot confirms that the tails of the distributions decay faster than a power law would.

Fig. 2. Social networks. (*a*) Linear-log plot of the cumulative distribution of connectivities for the network of movie actors (2). The full line is a guide for the eye of what an exponential decay would be. The data seem to fall faster in the tail than they would for an exponential decay, suggesting a Gaussian decay. Both exponential and Gaussian decays indicate that the connectivity distribution is not scale-free. (*b*) Log-log plot of the cumulative distribution of connectivities for the network of movie actors. This plot suggests that, for values of number of collaborations between 30 and 300, the data are consistent with a power law decay. The apparent exponent of this cumulative distribution, $\alpha - 1 \sim 1.3$, is consistent with the value $\alpha = 2.3 \pm 0.1$ reported for the probability density function (5). For larger numbers of collaborations, the power law decay is truncated. (*c*) Linear-log plot of the cumulative distribution of connectivities for the network of acquaintances of 43 Utah Mormons (25). The full line is the fit to the cumulative distribution of a Gaussian. The tail of the distribution seems to fall off as a Gaussian, suggesting that there is a single scale for the number of acquaintances in social networks. (*d*) Linear-log plot of the cumulative distribution of connectivities for the friendship network of 417 high school students (26). The number of links is the number of times a student is chosen by another student as one of his/her two (or three) best friends. The lines are Gaussian fits to the empirical distributions.

consistent with a Gaussian distribution of connectivities, indicating the existence of a single scale for k.[§]

Third, we consider two examples of networks from the natural sciences: (*vi*) the neuronal network of the worm *C. elegans* (2, 27, 28), the vertices being the individual neurons and the links being connections between neurons; and (*vii*) the conformation space of a lattice polymer chain (29), the vertices being the possible conformations of the polymer chain and the links being the possibility of connecting two conformations through local movements of the chain (29). The conformation space of a linear polymer chain seems to be well described (29) by the small-world networks of ref. 2. Fig. 3 *a* and *b* shows for *C. elegans* the cumulative distribution of k for both incoming and outgoing neuronal links. The tails of both distributions are well approximated by exponential decays, consistent with a single scale for the connectivities. For the network of conformations of a polymer chain, the connectivity follows a binomial distribution, which converges to the Gaussian (29); thus, we also find a single scale for the connectivity of the vertices (Fig. 3*c*).

Discussion

Thus far, we presented empirical evidence for the occurrence of three structural classes of small-world networks: (*a*) scale-free networks, characterized by a connectivity distribution with a tail that decays as a power law (4, 22, 23); (*b*) broad-scale or truncated scale-free networks, characterized by a connectivity distribution that has a power law regime followed by a sharp

[§]Note that even though the sample sizes of these two networks is rather small, the agreement with the Gaussian distribution is very good, suggesting that our results are reliable. Moreover, a power law distribution would curve the opposite way in the semilog plot.

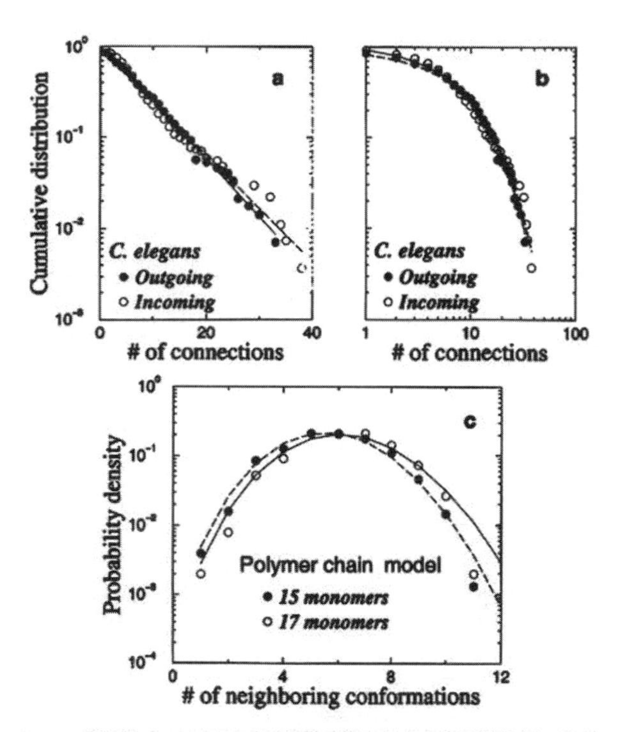

Fig. 3. Biological and physical networks. (*a*) Linear-log plot of the cumulative distribution of outgoing (i.e., connections by axons to other cells) and incoming (i.e., connections by axons from other cells) connections for the neuronal network of the worm *C. elegans* (27, 28). The full and long-dashed lines are exponential fits to the distributions of outgoing and incoming connections, respectively. The tails of the distributions seem consistent with an exponential decay. (*b*) Log-log plot of the cumulative distribution of outgoing and incoming connections for the neuronal network of the worm *C. elegans*. If the distribution would have a power law tail, then it would fall on a straight line in a log-log plot. The data seem to reject the hypothesis of a power law distribution for the connectivity. (*c*) Linear-log plot of the probability density function of connectivities for the network of conformations of a lattice polymer chain (29). A simple argument suggests that the connectivities follow a binomial distribution. The full and dashed lines are fits of a binomial probability density function to the data for polymer chains of different lengths. For the values of the parameters obtained in the fit, the binomial closely resembles the Gaussian, indicating that there is a single scale for the connectivities of the conformation space of polymers.

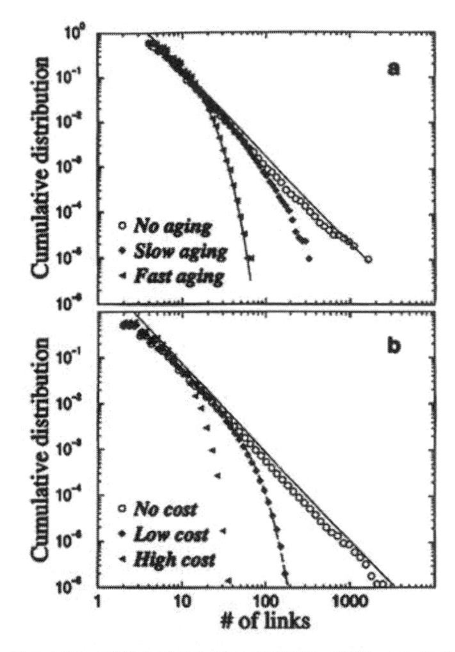

Fig. 4. Truncation of scale-free connectivity by adding constraints to the model of ref. 5. (*a*) Effect of aging of vertices on the connectivity distribution. We see that aging leads to a cutoff of the power law regime in the connectivity distribution. For sufficient aging of the vertices, the power law regime disappears altogether. (*b*) Effect of cost of adding links on the connectivity distribution. Our results indicate that the cost of adding links also leads to a cutoff of the power law regime in the connectivity distribution and that, for a sufficiently large cost, the power law regime disappears altogether.

cutoff, like an exponential or Gaussian decay of the tail (see example *iii*); and (*c*) single-scale networks, characterized by a connectivity distribution with a fast decaying tail, such as exponential or Gaussian (see examples *i*, *ii*, and *iv–vii*).

A natural question is "what are the reasons for such a rich range of possible structures for small-world networks?" To answer this question, let us recall that preferential attachment in growing networks gives rise to a power law distribution of connectivities (5). However, preferential attachment can be hindered by two classes of factors.

Aging of the vertices. This effect can be pictured for the network of actors; in time, every actor will stop acting. For the network, this fact implies that even a very highly connected vertex will, eventually, stop receiving new links. The vertex is still part of the network and contributes to network statistics, but it no longer receives links. The aging of the vertices thus limits the preferential attachment preventing a scale-free distribution of connectivities.

Cost of adding links to the vertices or the limited capacity of a vertex. This effect is exemplified by the network of world airports: for reasons of efficiency, commercial airlines prefer to have a small number of hubs where all routes connect. In fact, this situation is, to a first approximation, indeed what happens for individual airlines; however, when we consider all airlines together, it becomes physically impossible for an airport to become a hub to all airlines. Because of space and time constraints, each airport will limit the number of landings/departures per hour and the number of passengers in transit. Hence, physical costs of adding links and limited capacity of a vertex (30, 31) will limit the number of possible links attaching to a given vertex.

Modeling. To test numerically the effect of aging and cost constraints on the local structure of networks with preferential attachment, we simulate the scale-free model of ref. 5 but introduce aging and cost constraints of varying strength. In the original scale-free model, a network grows over time by the addition of new vertices and links. A vertex newly added to the network randomly selects m other vertices to establish new links, with a selection probability that increases with the number of links of the selected vertex. This mechanism generates faster growth of the most connected vertices—in a process identical to the city growth model of Simon and Bonini (32)—and it is well-known that the mechanism leads to a steady state with a power law distribution of connectivities (33).

We generalize this model by classifying vertices into one of two groups: active or inactive. Inactive vertices cannot receive new links. All new vertices are created active but in time may become inactive. We consider two types of constraints that are responsible for the transition from active to inactive. In the first, which we call "aging," vertices may become inactive each time step with a constant probability P_i. This fact implies that the time a vertex may remain active decays exponentially. In the second, which we call "cost," a vertex becomes inactive when it reaches a maximum number of links k_{max}. Fig. 4 shows our results for both types of constraint. It is clear that both lead to cutoffs on the power law decay of the tail of connectivity distribution and that, for strong enough constraints, no power law region is visible.

Analogy with Critical Phenomena. We note that the possible distributions of connectivity of the small-world networks have an analogy in the theory of critical phenomena (34). At the gas-liquid critical point, the distribution of sizes of the droplets of the gas (or of the liquid) is scale-free, as there is no free-energy cost in their formation (34). As for the case of a scale-free network, the size s of a droplet is power law distributed: $P(s) \approx s^{-\alpha}$. As we move away from the critical point, the appearance of a non-negligible surface tension introduces a free-energy cost for droplets that limits their sizes such that their distribution becomes broad-scale: $P(s) \approx s^{-\alpha}f(s/\xi)$, where ξ is the typical size for which surface tension starts to be significant, and the function $f(s/\xi)$ introduces a sharp cutoff for droplet sizes $s > \xi$. Far from the critical point, the scale ξ becomes so small that no power law regime is observed and the droplets become single-scale distributed: $P(s) \approx f(s/\xi)$. Often, the distribution of sizes in this regime is exponential or Gaussian.

We thank J. S. Andrade, Jr., R. Cuerno, N. Dokholyan, P. Gopikrishnan, C. Hartley, E. LaNave, K. B. Lauritsen, F. Liljeros, H. Orland, F. Starr, and S. Zapperi for stimulating discussions and helpful suggestions. The Center for Polymer Studies is funded by the National Science Foundation and National Institutes of Health (NCRR P41 RP13622).

1. Watts, D. J. & Strogatz, D. H. (1998) *Nature (London)* **393,** 440–442.
2. Watts, D. J. (1999) *Small Worlds: The Dynamics of Networks Between Order and Randomness* (Princeton Univ. Press, Princeton, NJ).
3. Barthélémy, M. & Amaral, L. A. N. (1999) *Phys. Rev. Lett.* **82,** 3180–3183.
4. Albert, R., Jeong, H. & Barabási, A.-L. (1999) *Nature (London)* **401,** 130.
5. Barabási, A.-L. & Albert, R. (1999) *Science* **286,** 509–512.
6. Lago-Fernandez, L. F., Huerta, R., Corbacho, F. & Siguenza, J. A. (2000) *Phys. Rev. Lett.* **84,** 2758–2761.
7. Newman, M. E. J. (2000) *J. Stat. Phys.*, in press.
8. Kochen, M., ed. (1989) *The Small World* (Ablex, Norwood, NJ).
9. Alon, U., Surette, M. G., Barkai, N. & Leibler, S. (1999) *Nature (London)* **397,** 168–171.
10. Pimm, S. L., Lawton, J. H. & Cohen, J. E. (1991) *Nature (London)* **350,** 669–674.
11. Paine, R. T. (1992) *Nature (London)* **355,** 73–75.
12. McCann, K., Hastings, A. & Huxel, G. R. (1998) *Nature (London)* **395,** 794–798.
13. Wasserman, S. & Faust, K. (1994) *Social Network Analysis* (Cambridge Univ. Press, Cambridge, U.K.).
14. Axtell, R. (1999) in *Behavioral Dimensions of Retirement Economics*, ed. Aaron, H. J. (Brookings Instit., Washington, DC), pp. 161–183.
15. Van Raan, A. F. J. (1990) *Nature (London)* **347,** 626.
16. Adamic, L. A. (1999) *The Small World Web Res. Adv. Tech. Digit. Libr. Proc.* **1696,** 443–452.
17. Huberman, B. A., Pirolli, P. L. T., Pitkow, J. E. & Lukose, R. J. (1999) *Science* **280,** 95–97.
18. Huberman, B. A. & Adamic, L. A. (1999) *Nature (London)* **401,** 131.
19. Adamic, L. A., Huberman, B. A., Barabási, A.-L., Albert, R., Jeong, H. & Bianconi, G. (2000) *Science* **287,** 2115.
20. van Leeuwen, J., ed. (1999) *Handbook of Theoretical Computer Science. Volume A: Algorithms and Complexity* (Elsevier, Amsterdam).
21. Bollobás, B. (1985) *Random Graphs* (Academic, London).
22. Seglen, P. O. (1992) *J. Am. Soc. Inf. Sci.* **43,** 628–638.
23. Redner, S. (1998) *Eur. Phys. J.* **4,** 131–134.
24. Airport Council International (1999) ACI Annual Worldwide Airports Traffic Reports (Airport Counc. Int., Geneva).
25. Bernard, H. R., Kilworth, P. D., Evans, M. J., McCarty, C. & Selley, G. A. (1988) *Ethnology* **27,** 155–179.
26. Fararo, T. J. & Sunshine, M. H. (1964) *A Study of a Biased Friendship Net* (Syracuse Univ. Press, Syracuse, NY).
27. White, J. G., Southgate, E., Thomson, J. N. & Brenner, S. (1986) *Philos. Trans. R. Soc. London B* **314,** 1–340.
28. Koch, C. & Laurent, G. (1999) *Science* **284,** 96–98.
29. Scala, A., Amaral, L. A. N. & Barthélémy, M. (1999) *Small-World Networks and the Conformation Space of a Lattice Polymer Chain* (cond-mat/0004380).
30. Bonney, M. E. (1956) in *Sociometry and the Science of Man*, ed. Moreno, J. L. (Beacon House, New York), pp. 275–286.
31. Moreno, J. L. (1956) *Sociometry and the Science of Man* (Beacon House, New York).
32. Simon, H. A. & Bonini, C. P. (1958) *Am. Econ. Rev.* **48,** 607–617.
33. Ijiri, Y. & Simon, H. A. (1977) *Skew Distributions and the Sizes of Business Firms* (North-Holland, Amsterdam).
34. Stanley, H. E. (1971) *Introduction to Phase Transitions and Critical Phenomena* (Oxford Univ. Press, Oxford).

The large-scale organization of metabolic networks

H. Jeong*, B. Tombor†, R. Albert*, Z. N. Oltvai† & A.-L. Barabási*

* *Department of Physics, University of Notre Dame, Notre Dame, Indiana 46556, USA*

† *Department of Pathology, Northwestern University Medical School, Chicago, Illinois 60611, USA*

In a cell or microorganism, the processes that generate mass, energy, information transfer and cell-fate specification are seamlessly integrated through a complex network of cellular constituents and reactions[1]. However, despite the key role of these networks in sustaining cellular functions, their large-scale structure is essentially unknown. Here we present a systematic comparative mathematical analysis of the metabolic networks of 43 organisms representing all three domains of life. We show that, despite significant variation in their individual constituents and pathways, these metabolic networks have the same topological scaling properties and show striking similarities to the inherent organization of complex non-biological systems[2]. This may indicate that metabolic organization is not only identical for all living organisms, but also complies with the design principles of robust and error-tolerant scale-free networks[2–5], and may represent a common blueprint for the large-scale organization of interactions among all cellular constituents.

An important goal in biology is to uncover the fundamental design principles that provide the common underlying structure and function in all cells and microorganisms[6–13]. For example, it is increasingly appreciated that the robustness of various cellular processes is rooted in the dynamic interactions among its many constituents[14–16], such as proteins, DNA, RNA and small molecules. Scientific developments have improved our ability to identify the design principles that integrate these interactions into a complex system. Large-scale sequencing projects have not only provided complete sequence information for a number of genomes, but also allowed the development of integrated pathway–genome databases[17–19] that provide organism-specific connectivity maps of metabolic and, to a lesser extent, other cellular networks. However, owing to the large number and diversity of the constituents and reactions that form such networks, these maps are extremely complex, offering only limited insight into the organizational principles of these systems. Our ability to address in quantitative terms the structure of these cellular networks has benefited from advances in understanding the generic properties of complex networks[2].

Until recently, complex networks have been modelled using the classical random network theory introduced by Erdös and Rényi[20,21]. The Erdös–Rényi model assumes that each pair of nodes (that is, constituents) in the network is connected randomly with probability p, leading to a statistically homogeneous network in which, despite the fundamental randomness of the model, most nodes have the same number of links, $\langle k \rangle$ (Fig. 1a). In particular, the connectivity follows a Poisson distribution that peaks strongly at $\langle k \rangle$ (Fig. 1b), implying that the probability of finding a highly connected node decays exponentially ($P(k) \approx e^{-k}$ for $k \gg \langle k \rangle$). On the other hand, empirical studies on the structure of the World-Wide Web[22], Internet[23] and social networks[2] have reported serious deviations from this random structure, showing that these systems are described by scale-free networks[2] (Fig. 1c), for which $P(k)$ follows a power-law, $P(k) \approx k^{-\gamma}$ (Fig. 1d). Unlike exponential networks, scale-free networks are extremely heterogeneous, their topology being dominated by a few highly connected nodes (hubs) which link the rest of the less connected nodes to the system (Fig. 1c). As the distinction between scale-free and exponential networks emerges as a result of simple dynamical principles[24,25], understanding the large-scale structure of cellular networks can not only provide valuable and perhaps universal structural information, but could also lead to a better understanding of the dynamical processes that generated these networks. In this respect the emergence of power-law distribution is intimately linked to the growth of the network in which new nodes are preferentially attached to already established nodes[2], a property that is also thought to characterize the evolution of biological systems[1].

To begin to address the large-scale structural organization of cellular networks, we have examined the topological properties of the core metabolic network of 43 different organisms based on data deposited in the WIT database[19]. This integrated pathway–genome database predicts the existence of a given metabolic pathway on the basis of the annotated genome of an organism combined with firmly established data from the biochemical literature. As 18 of the 43 genomes deposited in the database are not yet fully sequenced, and a substantial portion of the identified open reading frames are

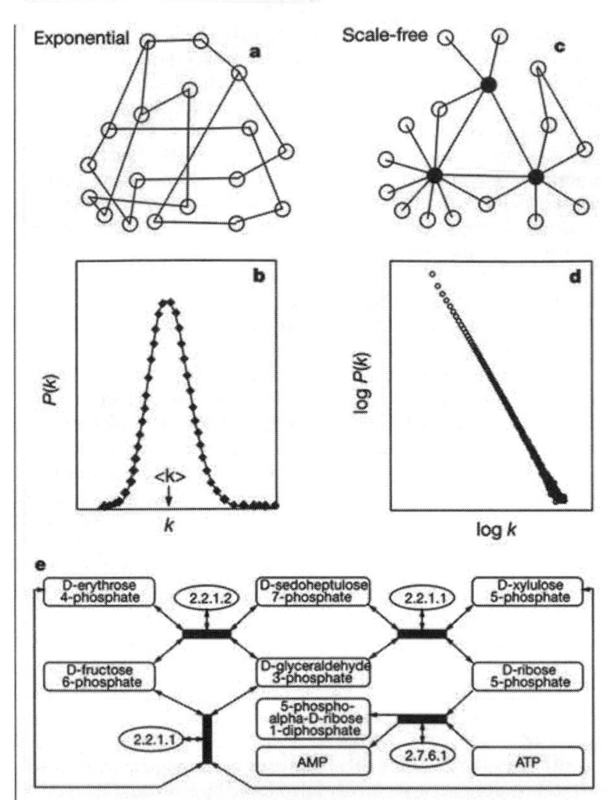

Figure 1 Attributes of generic network structures. **a**, Representative structure of the network generated by the Erdős–Rényi network model[20,21]. **b**, The network connectivity can be characterized by the probability, $P(k)$, that a node has k links. For a random network $P(k)$ peaks strongly at $k = \langle k \rangle$ and decays exponentially for large k (that is, $P(k) \approx e^{-k}$ for $k \gg \langle k \rangle$ and $k \ll \langle k \rangle$). **c**, In the scale-free network most nodes have only a few links, but a few nodes, called hubs (red), have a very large number of links. **d**, $P(k)$ for a scale-free network has no well-defined peak, and for large k it decays as a power-law, $P(k) \approx k^{-\gamma}$, appearing as a straight line with slope $-\gamma$ on a log–log plot. **e**, A portion of the WIT database for *E. coli*. Each substrate can be represented as a node of the graph, linked through temporary educt–educt complexes (black boxes) from which the products emerge as new nodes (substrates). The enzymes, which provide the catalytic scaffolds for the reactions, are shown by their EC numbers.

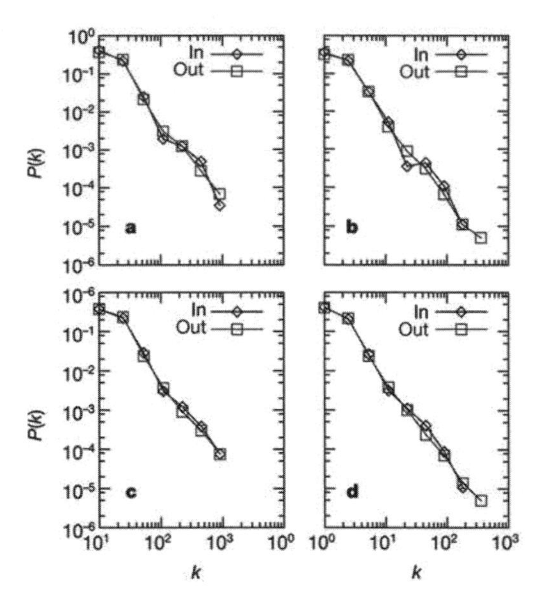

Figure 2 Connectivity distributions $P(k)$ for substrates. **a**, *Archaeoglobus fulgidus* (archae); **b**, *E. coli* (bacterium); **c**, *Caenorhabditis elegans* (eukaryote), shown on a log–log plot, counting separately the incoming (In) and outgoing links (Out) for each substrate. k_{in} (k_{out}) corresponds to the number of reactions in which a substrate participates as a product (educt). The characteristics of the three organisms shown in **a–c** and the exponents γ_{in} and γ_{out} for all organisms are given in Table 1 of the Supplementary Information. **d**, The connectivity distribution averaged over all 43 organisms.

functionally unassigned, the list of enzymes, and consequently the list of substrates and reactions (see Table 1 in Supplementary Information), will certainly be expanded in the future. Nevertheless, this publicly available database represents our best approximation for the metabolic pathways in 43 organisms and provides sufficient data for their unambiguous statistical analysis (see Methods and Supplementary Information).

As we show in Fig. 1e, we first established a graph theoretic representation of the biochemical reactions taking place in a given metabolic network. In this representation, a metabolic network is built up of nodes, the substrates, that are connected to one another through links, which are the actual metabolic reactions. The physical entity of the link is the temporary educt–educt complex itself, in which enzymes provide the catalytic scaffolds for the reactions yielding products, which in turn can become educts for subsequent reactions. This representation allows us systematically to investigate and quantify the topologic properties of various metabolic networks using the tools of graph theory and statistical mechanics[21].

Our first goal was to identify the structure of the metabolic networks: that is, to establish whether their topology is best described by the inherently random and uniform exponential model[21] (Fig. 1a, b), or the highly heterogeneous scale-free model[2] (Fig. 1c, d). As illustrated in Fig. 2, our results convincingly indicate that the probability that a given substrate participates in k reactions follows a power-law distribution; in other words, metabolic networks belong to the class of scale-free networks. As under physiological conditions a large number of biochemical reactions (links) in a metabolic network are preferentially catalysed in one direction (the links are directed), for each node we distinguish between incoming and outgoing links (Fig. 1e). For instance, in *Escherichia coli* the probability that a substrate participates as an educt in k metabolic reactions follows $P(k) \approx k^{-\gamma_{in}}$, with $\gamma_{in} = 2.2$, and the probability that a given substrate is produced by k different metabolic reactions follows a similar distribution, with $\gamma_{out} = 2.2$ (Fig. 2b). We find that scale-free networks describe the metabolic networks in all organisms in all three domains of life (Fig. 2a–c; see Supplementary Information, also available at www.nd.edu/~networks/cell), indicating the generic nature of this structural organization (Fig. 2d).

A general feature of many complex networks is their small-world character[26], meaning that any two nodes in the system can be connected by relatively short paths along existing links. In metabolic networks these paths correspond to the biochemical pathway connecting two substrates (Fig. 3a). The degree of interconnectivity of a metabolic network can be characterized by the network diameter, defined as the shortest biochemical pathway averaged over all pairs of substrates. For all non-biological networks examined, the average connectivity of a node is fixed, which implies that the diameter of a network increases logarithmically with the addition of new nodes[2,26,27]. For metabolic networks this implies that a more complex bacterium with more enzymes and substrates, such as *E. coli*, would have a larger diameter than a simple bacterium, such as *Mycoplasma genitalium*. We find, however, that the diameter of the metabolic network is the same for all 43 organisms,

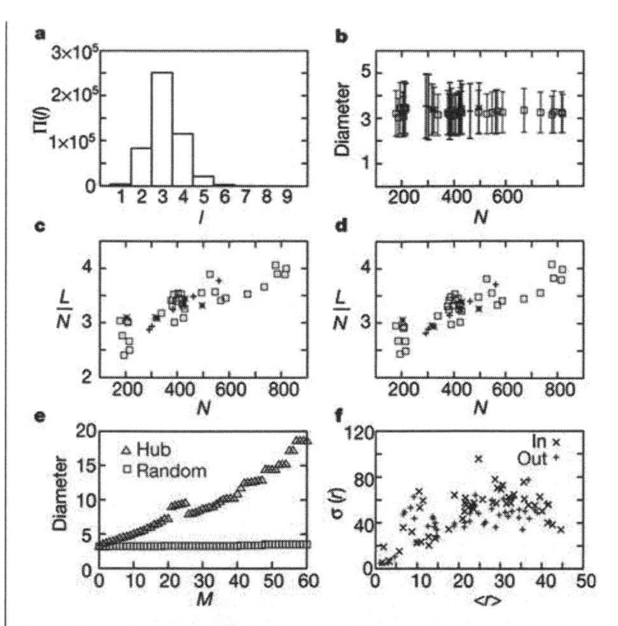

Figure 3 Properties of metabolic networks. **a**, The histogram of the biochemical pathway lengths, l, in *E. coli*. **b**, The average path length (diameter) for each of the 43 organisms. Error bars represent standard deviation $\sigma \approx \langle l^2 \rangle - \langle l \rangle^2$ as determined from $\Pi(l)$ (shown in **a** for *E. coli*). **c**, **d**, Average number of incoming links (**c**) or outgoing links (**d**) per node for each organism. **e**, The effect of substrate removal on the metabolic network diameter of *E. coli*. In the top curve (red) the most connected substrates are removed first. In the bottom curve (green) nodes are removed randomly. $M = 60$ corresponds to ~8% of the total number of substrates in found in *E. coli*. **f**, Standard deviation of the substrate ranking (σ_r) as a function of the average ranking $\langle r \rangle_o$ for substrates present in all 43 organisms investigated. The horizontal axis in **b–d** denotes the number of nodes in each organism. **b–d**, Archaea (magenta), bacteria (green) and eukaryotes (blue) are shown.

irrespective of the number of substrates found in the given species (Fig. 3b). This is unexpected, and is possible only if with increasing organism complexity individual substrates are increasingly connected to maintain a relatively constant metabolic network diameter. We find that the average number of reactions in which a certain substrate participates increases with the number of substrates found within a given organism (Fig. 3c, d).

An important consequence of the power-law connectivity distribution is that a few hubs dominate the overall connectivity of the network (Fig. 1c), and upon the sequential removal of the most connected nodes the diameter of the network rises sharply, the network eventually disintegrating into isolated clusters that are no longer functional. But scale-free networks also demonstrate unexpected robustness against random errors[5]. To investigate whether metabolic networks display a similar error tolerance we performed computer simulations on the metabolic network of *E. coli*. Upon removal of the most connected substrates the diameter increases rapidly, illustrating the special role of these metabolites in maintaining a constant metabolic network diameter (Fig. 3e). However, when a randomly chosen M substrates are removed—mimicking the consequence of random mutations of catalysing enzymes—the average distance between the remaining nodes is not affected, indicating a striking insensitivity to random errors. Indeed, *in silico* and *in vivo* mutagenesis studies indicate remarkable fault tolerance upon removal of a substantial number of metabolic enzymes from the *E. coli* metabolic network[28]. Data similar to those shown in Fig. 3e have been obtained for all organisms investigated, without detectable correlations with their evolutionary position.

As the large-scale architecture of the metabolic network rests on the most highly connected substrates, we need to investigate whether the same substrates act as hubs in all organisms, or whether there are organism-specific differences in the identity of the most connected substrates. When we rank all the substrates in a given organism on the basis of the number of links they have (Table 1; see Supplementary Information), we find that the ranking of the most connected substrates is practically identical for all 43 organisms. Also, only around 4% of all substrates that are found in all 43 organisms are present in all species. These substrates represent the most highly connected substrates found in any individual organism, indicating the generic utilization of the same substrates by each species. In contrast, species-specific differences among organisms emerge for less connected substrates. To quantify this observation, we examined the standard deviation (σ_r) of the rank for substrates that are present in all 43 organisms. As shown in Fig. 3f, σ_r increases with the average rank order $\langle r \rangle$, implying that the most connected substrates have a relatively fixed position in the rank order, but the ranking of less connected substrates is increasingly species-specific. Thus, the large-scale structure of the metabolic network is identical for all 43 species, being dominated by the same highly connected substrates, while less connected substrates preferentially serve as the educts or products of species-specific enzymatic activities.

The contemporary topology of a metabolic network reflects a long evolutionary process moulded in general for a robust response towards internal defects and environmental fluctuations and in particular to the ecological niche occupied by a specific organism. As a result, we would expect that these networks are far from random, and our data show that the large-scale structural organization of metabolic networks is indeed very similar to that of robust and error-tolerant networks[2,5]. The uniform network topology observed in all 43 organisms indicates that, irrespective of their individual building blocks or species-specific reaction pathways, the large-scale structure of metabolic networks may be identical in all living organisms, in which the same highly connected substrates may provide the connections between modules responsible for distinct metabolic functions[1].

A unique feature of metabolic networks, as opposed to non-biological scale-free networks, is the apparent conservation of the network diameter in all living organisms. Within the special characteristics of living systems this attribute may represent an additional survival and growth advantage, as a larger diameter would attenuate the organism's ability to respond efficiently to external changes or internal errors. For example, if the concentration of a substrate were to suddenly diminish owing to a mutation in its main catalysing enzyme, offsetting the changes would involve the activation of longer alternative biochemical pathways, and consequently the synthesis of more new enzymes, than within a metabolic network with a smaller diameter.

How generic are these principles for other cellular networks (for example, apoptosis or cell cycle)? Although the current mathematical tools do not allow unambiguous statistical analysis of the topology of other networks owing to their relatively small size, our preliminary analysis indicates that connectivity distribution of non-metabolic pathways may also follow a power-law distribution, indicating that cellular networks as a whole are scale-free networks. Therefore, the evolutionary selection of a robust and error-tolerant architecture may characterize all cellular networks, for which scale-free topology with a conserved network diameter appears to provide an optimal structural organization. □

Methods

Database preparation

For our analyses of core cellular metabolisms we used the 'Intermediate metabolism and bioenergetics' portions of the WIT database[19] (http://igweb.integratedgenomics.com/IGwit/), which predicts the existence of a metabolic pathway in an organism on the basis of its annotated genome (on the presence of the presumed open reading frame of an enzyme that catalyses a given metabolic reaction), in combination with firmly established data from the biochemical literature. As of December 1999, this database provides descriptions for 6 archaea, 32 bacteria and 5 eukaryotes. The downloaded data were manually rechecked, removing synonyms and substrates without defined chemical identity.

Construction of metabolic network matrices

Biochemical reactions described within a WIT database are composed of substrates and enzymes connected by directed links. For each reaction, educts and products were considered as nodes connected to the temporary educt–educt complexes and associated enzymes. Bidirectional reactions were considered separately. For a given organism with N substrates, E enzymes and R intermediate complexes the full stoichiometric interactions were compiled into an $(N+E+R) \times (N+E+R)$ matrix, generated separately for each of the 43 organisms.

Connectivity distribution $P(k)$

Substrates generated by a biochemical reaction are products, and are characterized by incoming links pointing to them. For each substrate we have determined k_{in}, and prepared a histogram for each organism, showing how many substrates have exactly $k_{in} = 0,1,\ldots$. Dividing each point of the histogram with the total number of substrates in the organism provided $P(k_{in})$, or the probability that a substrate has k_{in} incoming links. Substrates that participate as educts in a reaction have outgoing links. We have performed the analysis described above for k_{in}, determining the number of outgoing links (k_{out}) for each substrate. To reduce noise logarithmic binning was applied.

Biochemical pathway lengths $[\Pi(l)]$

For all pairs of substrates, the shortest biochemical pathway, $\Pi(l)$ (that is, the smallest number of reactions by which one can reach substrate B from substrate A) was determined using a burning algorithm. From $\Pi(l)$ we determined the diameter, $D = \Sigma_l l \cdot \Pi(l) / \Sigma_l \Pi(l)$, which represents the average path length between any two substrates.

Substrate ranking $\langle r \rangle_o$, $\sigma(r)$

Substrates present in all 43 organisms (a total of 51 substrates) were ranked on the basis of the number of links each had in each organisms, having considered incoming and outgoing links separately ($r = 1$ was assigned for the substrate with the largest number of connections, $r = 2$ for the second most connected one, and so on). This gave a well defined r value in each organism for each substrate. The average rank $\langle r \rangle_o$ for each substrate was determined by averaging r over the 43 organisms. We also determined the standard deviation, $\sigma(r) = \langle r^2 \rangle_o - \langle r \rangle_o^2$ for all 51 substrates present in all organisms.

Analysis of the effect of database errors

Of the 43 organisms whose metabolic network we have analysed, the genomes of 25 have been completely sequenced (5 archaea, 18 bacteria and 2 eukaryotes), whereas the remaining 18 are only partially sequenced. Therefore two main sources of possible errors in the database could affect our analysis: the erroneous annotation of enzymes and, consequently, biochemical reactions (the likely source of error for the organisms with completely sequenced genomes); and reactions and pathways missing from the database (for organisms with incompletely sequenced genomes, both sources of error are possible). We investigated the effect of database errors on the validity of our findings. The data, presented in Supplementary Information, indicate that our results are robust to these errors.

Received 3 April; accepted 18 July 2000.

1. Hartwell, L. H., Hopfield, J. J., Leibler, S. & Murray, A. W. From molecular to modular cell biology. *Nature* **402**, C47–52 (1999).
2. Barabási, A.-L. & Albert, R. Emergence of scaling in random networks. *Science* **286**, 509–512 (1999).
3. West, G. B., Brown, J. H. & Enquist, B. J. The fourth dimension of life: fractal geometry and allometric scaling of organisms. *Science* **284**, 1677–1679 (1999).
4. Banavar, J. R., Maritan, A. & Rinaldo, A. Size and form in efficient transportation networks. *Nature* **399**, 130–132 (1999).
5. Albert, R., Jeong, H. & Barabási, A.-L. Error and attack tolerance of complex networks. *Nature* **406**, 378–382 (2000).
6. Ingber, D. E. Cellular tensegrity: defining new rules of biological design that govern the cytoskeleton. *J. Cell Sci.* **104**, 613–627 (1993).
7. Bray, D. Protein molecules as computational elements in living cells. *Nature* **376**, 307–312 (1995).
8. McAdams, H. H. & Arkin, A. It's a noisy business! Genetic regulation at the nanomolar scale. *Trends Genet.* **15**, 65–69 (1999).
9. Gardner, T. S., Cantor, C. R. & Collins, J. J. Construction of a genetic toggle switch in *Escherichia coli*. *Nature* **403**, 339–342 (2000).
10. Elowitz, M. B. & Leibler, S. A synthetic oscillatory network of transcriptional regulators. *Nature* **403**, 335–338 (2000).
11. Hasty, J., Pradines, J., Dolnik, M. & Collins, J. J. Noise-based switches and amplifiers for gene expression. *Proc. Natl Acad. Sci. USA* **97**, 2075–2080 (2000).
12. Becskei, A. & Serrano, L. Engineering stability in gene networks by autoregulation. *Nature* **405**, 590–593 (2000).
13. Kirschner, M., Gerhart, J. & Mitchison, T. Molecular 'vitalism'. *Cell* **100**, 79–88 (2000).
14. Barkai, N. & Leibler, S. Robustness in simple biochemical networks. *Nature* **387**, 913–917 (1997).
15. Yi, T. M., Huang, Y., Simon, M. I. & Doyle, J. Robust perfect adaptation in bacterial chemotaxis through integral feedback control. *Proc. Natl Acad. Sci. USA* **97**, 4649–4653 (2000).
16. Bhalla, U. S. & Iyengar, R. Emergent properties of networks of biological signaling pathways. *Science* **283**, 381–387 (1999).
17. Karp, P. D., Krummenacker, M., Paley, S. & Wagg, J. Integrated pathway–genome databases and their role in drug discovery. *Trends Biotechnol.* **17**, 275–281 (1999).
18. Kanehisa, M. & Goto, S. KEGG: Kyoto encyclopedia of genes and genomes. *Nucleic Acids Res.* **28**, 27–30 (2000).
19. Overbeek, R. *et al.* WIT: integrated system for high-throughput genome sequence analysis and metabolic reconstruction. *Nucleic Acids Res.* **28**, 123–125 (2000).
20. Erdös, P. & Rényi, A. On the evolution of random graphs. *Publ. Math. Inst. Hung. Acad. Sci.* **5**, 17–61 (1960).
21. Bollobás, B. *Random Graphs* (Academic, London, 1985).
22. Albert, R., Jeong, H. & Barabási, A.-L. Diameter of the World-Wide Web. *Nature* **400**, 130–131 (1999).
23. Faloutsos, M., Faloutsos, P. & Faloutsos, C. On power-law relationships of the internet topology. *Comp. Comm. Rev.* **29**, 251 (1999).
24. Amaral, L. A. N., Scala, A., Barthelemy, M. & Stanley, H. E. Classes of behavior of small-world networks. (cited 31 January 2000) ⟨http://xxx.lanl.gov/abs/cond-mat/0001458⟩ (2000).
25. Dorogovtsev, S. N. & Mendes, J. F. F. Evolution of reference networks with aging (cited 28 January 2000) ⟨http://xxx.lanl.gov/abs/cond-mat/0001419⟩ (2000).
26. Watts, D. J. & Strogatz, S. H. Collective dynamics of 'small-world' networks. *Nature* **393**, 440–442 (1998).
27. Barthelemy, M. & Amaral, L. A. N. Small-world networks: Evidence for a crossover picture. *Phys. Rev. Lett.* **82**, 3180–3183 (1999).
28. Edwards, J. S. & Palsson, B. O. The *Escherichia coli* MG1655 *in silico* metabolic genotype: its definition, characteristics, and capabilities. *Proc. Natl Acad. Sci. USA* **97**, 5528–5533 (2000).

Supplementary information is available on *Nature*'s World-Wide Web site (http://www.nature.com) or as paper copy from the London editorial office of *Nature*.

Acknowledgements

We thank all members of the WIT project for making this invaluable database publicly available. We also thank C. Waltenbaugh and H. S. Seifert for comments on the manuscript. Research at the University of Notre Dame was supported by the National Science Foundation, and at Northwestern University by grants from the National Cancer Institute.

Correspondence and requests for materials should be addressed to A.-L.B. (e-mail: alb@nd.edu) or Z.N.O. (e-mail: zno008@northwestern.edu).

METABOLIC ENGINEERING

The small world of metabolism

David A. Fell and Andreas Wagner

Genome sequencing is now advancing at a frenetic pace, which has the consequence that many organisms now being sequenced have not had their biochemistry extensively studied. Thus, the metabolic phenotype of these organisms has to be determined using annotated genome sequence data. Ideally, this determination should be automated, but that would require clear criteria and algorithms for identifying and classifying metabolism. Also, traditional textbook representations of metabolic pathways may neither capture the full number of potential network functions nor the network's resilience to disruption[1–3]. Whereas algorithmic approaches to these latter problems have been proposed, many aspects of metabolic network function remain to be clearly delineated. For example, in stoichiometric network analysis[4], it is convenient to define a subset of central metabolic intermediates that represent the products of catabolism that are used to initiate anabolism. However, even for *Escherichia coli*, there is no agreement on the identity of this subset[5,6].

In seeking to establish a firm basis for identifying a set of central metabolites defining the core of metabolism[7,8], we have taken advantage of analysis tools used by mathematicians to understand the structure of sociological networks. These include networks of personal and professional relations, such as collaboration networks of film actors or scientists. The aspect of such analyses germane to our endeavor is best illustrated with the example of the prolific Hungarian graph theorist Paul Erdös. He is the center of a graph of mathematical collaboration. Coauthors of a paper with Erdös are one step from Erdös himself, and have Erdös number 1. Coauthors of mathematicians with Erdös number 1 have Erdös number 2, and so on. Most mathematicians active this century can be connected to Erdös in a small number of steps. In this sense, he is the undisputed center of the mathematical world[9].

A similar principle underlies the Kevin Bacon game, which has the aim of connecting an arbitrarily chosen movie actor with the actor Kevin Bacon by the shortest sequence of actor-pairs who have appeared together in a film. The average Bacon number of a randomly chosen actor, representing the mean minimum number of actors connecting the actor to Kevin Bacon is only 2.87 (ref. 10). (However, Kevin Bacon is not even the center of this small world of film actor collaborations, defined as having on average the shortest distance to all the other stars. This center is Christopher Lee, with a mean path length of 2.60.)

We recently have analyzed the structure of the *E. coli* core metabolism with the following goal in mind: to identify metabolites central to metabolism in this sense, without relying on subjective criteria. To this end, we assembled a list of 317 stoichiometric equations involving 275 substrates that represent the central routes of energy metabolism and small-molecule building block synthesis in *E. coli*[11–15] under aerobic growth, with glucose as sole carbon source and O_2 as electron acceptor. From these reaction equations, we generated a connection matrix where two metabolites were regarded as connected if they appeared in the same reaction, whether as substrate or product. We did not include the common coenzymes, such as ATP, ADP, or NAD, because they are evidently ubiquitous. On this basis, the center of the *E. coli* metabolic map is glutamate, with a mean path length of 2.46, followed by pyruvate with a value of 2.59.

The analogy between metabolism and the collaboration networks of mathematicians and film stars does not end there. The question of how networks that are both large and sparse can nevertheless be traversed in very few steps (cf. "six degrees of separation"[16]) has been analyzed by Watts and Strogatz[17]. Uniform, latticelike networks tend to have long path lengths because pairs that are connected tend to be connected to the same other members of the set, that is, they are clustered. Randomly generated networks can have short path lengths, but also show little clustering because connected pairs show little similarity in their other connections. However, a number of the sparse, natural networks studied by Watts and Strogatz showed short path lengths but high clustering. They named such networks "small-world" networks. The friendship networks studied in sociology as well as mathematical and acting collaboration networks are of this type. We found that the *E. coli* metabolic network falls into the same category[7,8]. This implies that, in modeling the properties of metabolism, neither very regular structures nor completely random networks would be very faithful representations of metabolism in general.

What is the biological relevance of the "small-worldness" of the *E. coli* metabolism? One way to generate a small-world network is to take a regular network and randomly reassign some of the connections. Another way is by accretion, where new members are added by preferentially making connections to existing members that already have large numbers of connections. Barabási and Albert[18] have shown that the latter led to a small-world network where the number of connections of the members fell off in a power–law relationship (i.e., a small number of members have a large number of connections, and this falls off smoothly so that the larger number of members has few connections). Randomly reassigned regular networks in contrast have a notably peaked distribution of connections. Film collaborations, hyperlinks in the worldwide web, and the US power grid all show the power–law connectivity[18].

The *E. coli* network does so as well[7,8]. Here, glutamate followed by pyruvate were the most connected metabolites (again omitting common coenzymes). After that, the lists of metabolites ranked by their number of connections and by their minimum mean path length to other metabolites were not exactly congruent, although both lists featured tricarboxylic acid cycle intermediates, and associated amino acids, in highly ranked positions. Recently, Jeong et al.[19] have reached a similar conclusion for the metabolic networks of a number of microorganisms by a related though slightly different analysis.

If, early in the evolution of life, metabolic networks grew by adding new metabolites, then the most highly connected metabolites should also be the phylogenetically oldest. Glycolysis and the tricarboxylic acid cycle are perhaps the most ancient metabolic pathways, and various of their intermediates (e.g., 2-oxoglutarate, succinate, pyruvate, and 3-phosphoglycerate) occur near the top of our lists, along with the amino acids thought to be used earliest (glutamine, glutamate, aspartate, and serine). This potential link with evolutionary history is consistent with Morowitz's[20] claim that intermediary metabolism recapitulates the evolution of biochemistry.

David A. Fell is professor of biochemistry at the School of Biological and Molecular Sciences, Oxford Brookes University, Headington, Oxford, OX3 0BP, UK (daf@bms.brookes.ac.uk), and Andreas Wagner is assistant professor of biology at the Department of Biology, University of New Mexico, Albuquerque, NM 87131-1091.

Of course, metabolism might have evolved small-world characteristics to optimize metabolic function in some way. Watts and Strogatz[17] have studied how fast perturbations spread through small-world networks. They concluded that the time required for spreading of a perturbation in a small-world network is close to the theoretically possible minimum. The importance of minimizing the transition time between metabolic states is recognized[21,22], and small-worldness may be a factor in allowing a metabolism to react rapidly to perturbations, although this requires further investigation because metabolic dynamics are more complicated than the simple kinetics used by Watts and Strogatz.

In conclusion, a purely structural analysis of a metabolic network may be able to teach us about the network's evolutionary history and design principles. Some of our propositions are speculative at this stage, but our analysis has revealed aspects of the network structure that were not previously apparent. The wealth of metabolic information about to become available from the genomes of ecologically diverse microbes will undoubtedly help to test these propositions.

1. Schilling, C.H., Schuster, S., Palsson, B.O. & Heinrich, R. Metabolic pathway analysis: basic concepts and scientific applications in the post-genomic era. *Biotechnol. Prog.* **15**, 296–303 (1999).
2. Edwards, J.S. & Palsson, B.O. The *Escherichia coli* MG1655 *in silico* metabolic genotype: Its definition, characteristics, and capabilities. *Proc. Natl. Acad. Sci. USA* **97**, 5528–5533 (2000).
3. Schuster, S., Fell, D.A. & Dandekar, T. A general definition of metabolic pathways useful for systematic organization and analysis of complex metabolic networks. *Nat. Biotechnol.* **18**, 326–332 (2000).
4. Varma, A. & Palsson, B.O. Metabolic capabilities of Escherichia coli. 1. Synthesis of biosynthetic precursors and cofactors. *J. Theoret. Biol.* **165**, 477–502 (1993).
5. Ingraham, J.L., Maaloe, O.E. & Neidhardt, F.C. *Growth of the bacterial cell.* (Sinauer Associates Inc, Sunderland, MA; 1983).
6. Holmes, W.H. The central metabolic pathways of Escherichia coli: relationship between flux and control at a branch point, efficiency of conversion to biomass, and excretion of acetate. *Curr. Top. Cellul. Regul.* **28**, 69–105 (1986).
7. Fell, D.A. & Wagner, A. Structural properties of metabolic networks: Implications for evolution and modelling of metabolism. In *Animating the cellular map.* (eds Hofmeyr, J.-H.S., Rohwer, J.M. & Snoep, J.L) 79–85 (Stellenbosch University Press, Stellenbosch, South Africa; 2000).
8. Wagner, A. & Fell, D.A. The small world inside large metabolic networks. Working paper 00-07-041 The Santa Fe Institute (2000). http://www.santafe.edu/sfi/publications/00wplist.html.
9. The Erdös number project (2000). http://www.acs.oakland.edu/ grossman/erdoshp.html.
10. The Oracle of Bacon at Virginia (2000). http://www.cs.virginia.edu/oracle/.
11. Neidhardt, F.C (ed.). *Escherichia coli* and *Salmonella*: *Molecular and cellular biology.* (ASM Press, Washington DC; 1996).
12. Karp, P., Riley, M., Paley, S., Pellegrini-Toole, A. & Krummenacker, M. Eco Cyc: Encyclopedia of *Escherichia coli* genes and metabolism. *Nucleic Acids Res.* **27**, 55–58 (1999).
13. Pramanik, J. & Keasling, J.D. Stoichiometric model of Escherichia coli metabolism: incorporation of growth-rate dependent biomass composition and mechanistic energy requirements. *Biotechnol. Bioeng.* **56**, 398–421 (1997).
14. Selkov, E. *et al.* The metabolic pathway collection from EMP—the enzymes and metabolic pathways database. *Nucleic Acids Res.* **24**, 26–28 (1996).
15. Bairoch, A. The ENZYME data bank in 1999. *Nucleic Acids Res.* **27**, 310–311 (1999).
16. Migram, S. The small—world problem. *Psychology Today* **2**, 60–67 (1967).
17. Watts, D.J. & Strogatz, S.H. Collective dynamics of 'small-world' networks. *Nature* **393**, 440–442 (1998).
18. Barabási, A.L. & Albert, R. Emergence of scaling in random networks. *Science* **286**, 509–512 (1999).
19. Jeong, H., Tombor, B., Albert, R., Oltvai, Z.N. & Barabási, A.-L. The large-scale organization of metabolic networks. *Nature* **407**, 651–654 (2000).
20. Morowitz, H.J. *Beginnings of cellular life: metabolism recapitulates biogenesis.* (Yale University Press, New Haven, CT; 1992).
21. Easterby, J.S. The effect of feedback on pathway transient response. *Biochem. J.* **233**, 871–875 (1986).
22. Cascante, M., Meléndez-Hevia, E., Kholodenko, B.N., Sicilia, J. & Kacser, H. Control analysis of transit-time for free and enzyme-bound metabolites—physiological and evolutionary significance of metabolic response-times. *Biochem. J.* **308**, 895–899 (1995).

Network Motifs: Simple Building Blocks of Complex Networks

R. Milo,[1] S. Shen-Orr,[1] S. Itzkovitz,[1] N. Kashtan,[1] D. Chklovskii,[2] U. Alon[1*]

Complex networks are studied across many fields of science. To uncover their structural design principles, we defined "network motifs," patterns of interconnections occurring in complex networks at numbers that are significantly higher than those in randomized networks. We found such motifs in networks from biochemistry, neurobiology, ecology, and engineering. The motifs shared by ecological food webs were distinct from the motifs shared by the genetic networks of *Escherichia coli* and *Saccharomyces cerevisiae* or from those found in the World Wide Web. Similar motifs were found in networks that perform information processing, even though they describe elements as different as biomolecules within a cell and synaptic connections between neurons in *Caenorhabditis elegans*. Motifs may thus define universal classes of networks. This approach may uncover the basic building blocks of most networks.

Many of the complex networks that occur in nature have been shown to share global statistical features (*1–10*). These include the "small world" property (*1–9*) of short paths between any two nodes and highly clustered connections. In addition, in many natural networks, there are a few nodes with many more connections than the average node has. In these types of networks, termed "scale-free networks" (*4, 6*), the fraction of nodes having k edges, $p(k)$, decays as a power law $p(k) \sim k^{-\gamma}$ (where γ is often between 2 and 3). To go beyond these global features would require an understanding of the basic structural elements particular to each class of networks (*9*). To do this, we developed an algorithm for detecting network motifs: recurring, significant patterns of interconnections. A detailed application to a gene regulation network has been presented (*11*). Related methods were used to test hypotheses on social networks (*12, 13*). Here we generalize this approach to virtually any type of connectivity graph and find the striking appearance of

[1]Departments of Physics of Complex Systems and Molecular Cell Biology, Weizmann Institute of Science, Rehovot, Israel 76100. [2]Cold Spring Harbor Laboratory, Cold Spring Harbor, NY 11724, USA.

*To whom correspondence should be addressed. E-mail: urialon@weizmann.ac.il

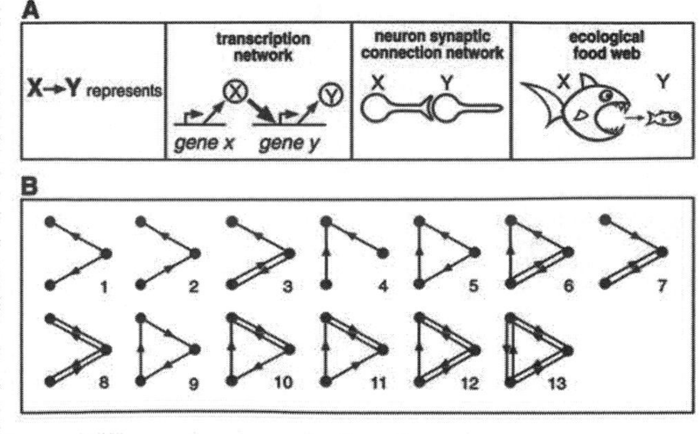

Fig. 1. (**A**) Examples of interactions represented by directed edges between nodes in some of the networks used for the present study. These networks go from the scale of biomolecules (transcription factor protein X binds regulatory DNA regions of a gene to regulate the production rate of protein Y), through cells (neuron X is synaptically connected to neuron Y), to organisms (X feeds on Y). (**B**) All 13 types of three-node connected subgraphs.

motifs in networks representing a broad range of natural phenomena.

We started with networks where the interactions between nodes are represented by directed edges (Fig. 1A). Each network was scanned for all possible n-node subgraphs (in the present study, $n = 3$ and 4), and the number of occurrences of each subgraph was recorded. Each network contains numerous types of n-node subgraphs (Fig. 1B). To focus on those that are likely to be important, we compared the real network to suitably randomized networks (*12–16*) and only selected patterns appearing in the real network at numbers significantly higher than those in the randomized networks (Fig. 2). For a stringent comparison, we used randomized networks that have the same single-node characteristics as does the real network: Each node in the randomized networks has the same number of incoming and outgoing edges as the corresponding node has in the real network. The comparison to this randomized ensemble accounts for patterns that appear only because of the single-node characteristics of the network (e.g., the presence of nodes with a large number of edges). Furthermore, the randomized networks used to calculate the significance of n-node subgraphs were generated to preserve the same number of appearances of all $(n-1)$-node subgraphs as in the real network (*17, 18*). This ensures that a high significance was not assigned to a pattern only because it has a highly significant subpattern. The "network motifs" are those patterns for which the probability P of appearing in a randomized network an equal or greater number of times than in the real network is lower than a cutoff value (here $P = 0.01$). Patterns that are functionally important but not statistically significant could exist, which would be missed by our approach.

We applied the algorithm to several networks from biochemistry (transcriptional gene regulation), ecology (food webs), neurobiology (neuron connectivity), and engineering (electronic circuits, World Wide Web). The network motifs found are shown in Table 1. Transcription networks are biochemical networks responsible for regulating the expression of genes in cells (*11, 19*). These are directed graphs, in which the nodes represent genes (Fig. 1A). Edges are directed from a gene that encodes for a transcription factor protein to a gene transcriptionally regulated by that transcription factor. We analyzed the two best characterized transcriptional regulation networks, corresponding to organisms from different kingdoms: a eukaryote (the yeast *Saccharomyces cerevisiae*) (*20*) and a bacterium (*Escherichia coli*) (*11, 19*). The two transcription networks show the same motifs: a three-node motif termed "feedforward loop" (*11*) and a four-node motif termed "bi-fan." These motifs appear numerous times in each network (Table 1), in nonhomologous gene systems that perform diverse biological functions. The number of times they appear is more than 10 standard deviations greater than their mean number of appearances in randomized networks. Only these subgraphs, of the 13 possible different three-node subgraphs (Fig. 1B) and 199 different four-node subgraphs, are significant and are therefore considered network motifs. Many other three- and four-node subgraphs recur throughout the networks, but at numbers that are less than the mean plus 2 standard deviations of their appearance in randomized networks.

We next applied the algorithm to ecosystem food webs (*21, 22*), in which nodes represent groups of species. Edges are directed from a node representing a predator to the node representing its prey. We analyzed data collected by different groups at seven distinct ecosystems (*22*), including both aquatic and terrestrial habitats. Each of the food webs displayed one or two three-node network motifs and one to five four-node network motifs. One can define the "consensus motifs" as the motifs shared by networks of a given type. Five of the seven food webs shared one three-node motif, and all seven shared one four-node motif (Table 1). In contrast to the three-node motif (termed "three chain"), the three-node feedforward loop was underrepresented in the food webs. This suggests that direct interactions between species at a separation of two layers [as in the case of omnivores (*23*)] are selected against. The bi-parallel motif indicates that two species that are prey of the same predator both tend to share the same prey. Both network motifs may thus represent general tendencies of food webs (*21, 22*).

We next studied the neuronal connectivity network of the nematode *Caenorhabditis elegans* (*24*). Nodes represent neurons (or neuron

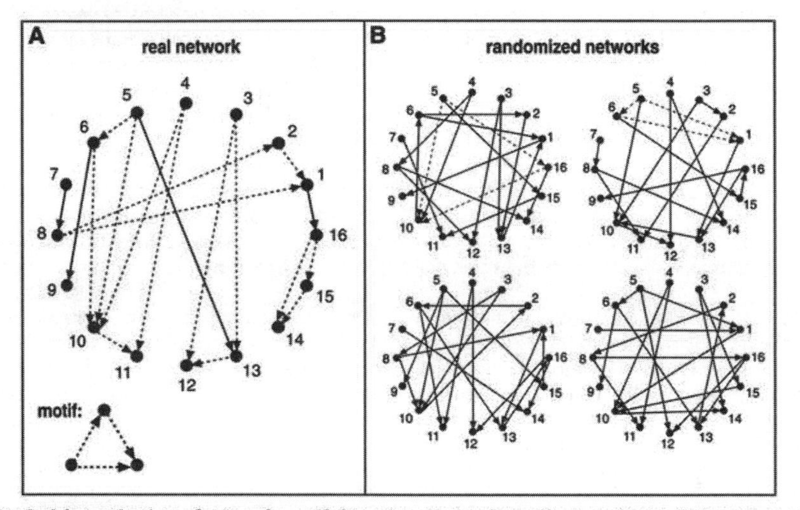

Fig. 2. Schematic view of network motif detection. Network motifs are patterns that recur much more frequently (**A**) in the real network than (**B**) in an ensemble of randomized networks. Each node in the randomized networks has the same number of incoming and outgoing edges as does the corresponding node in the real network. Red dashed lines indicate edges that participate in the feedforward loop motif, which occurs five times in the real network.

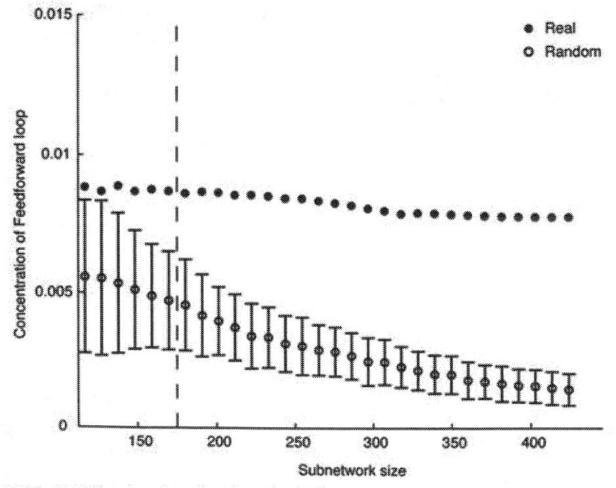

Fig. 3. Concentration C of the feedforward loop motif in real and randomized subnetworks of the *E. coli* transcription network (*11*). C is the number of appearances of the motif divided by the total number of appearances of all connected three-node subgraphs (Fig. 1B). Subnetworks of size S were generated by choosing a node at random and adding to it nodes connected by an incoming or outgoing edge, until S nodes were obtained, and then including all of the edges between these S nodes present in the full network. Each of the subnetworks was randomized (*17, 18*) (shown are mean and SD of 400 subnetworks of each size).

classes), and edges represent synaptic connections between the neurons. We found the feedforward loop motif in agreement with anatomical observations of triangular connectivity structures (*24*). The four-node motifs include the bi-fan and the bi-parallel (Table 1). Two of these motifs (feedforward loop and bi-fan) were also found in the transcriptional gene regulation networks. This similarity in motifs may point to a fundamental similarity in the design constraints of the two types of networks. Both networks function to carry information from sensory components (sensory neurons/transcription factors regulated by biochemical signals) to effectors (motor neurons/structural genes). The feedforward loop motif common to both types of networks may play a functional role in information processing. One possible function of this circuit is to activate output only if the input signal is persistent and to allow a rapid deactivation when the input goes off (*11*). Indeed, many of the input nodes in the neural feedforward loops are sensory neurons, which may require this type of information processing to reject transient input fluctuations that are inherent in a variable or noisy environment.

Table 1. Network motifs found in biological and technological networks. The numbers of nodes and edges for each network are shown. For each motif, the numbers of appearances in the real network (N_{real}) and in the randomized networks ($N_{rand} \pm$ SD, all values rounded) (*17, 18*) are shown. The P value of all motifs is $P < 0.01$, as determined by comparison to 1000 randomized networks (100 in the case of the World Wide Web). As a qualitative measure of statistical significance, the Z score $= (N_{real} - N_{rand})/\text{SD}$ is shown. NS, not significant. Shown are motifs that occur at least $U = 4$ times with completely different sets of nodes. The networks are as follows (*18*): transcription interactions between regulatory proteins and genes in the bacterium *E. coli* (*11*) and the yeast *S. cerevisiae* (*20*); synaptic connections between neurons in *C. elegans*, including neurons connected by at least five synapses (*24*); trophic interactions in ecological food webs (*22*), representing pelagic and benthic species (Little Rock Lake), birds, fishes, invertebrates (Ythan Estuary), primarily larger fishes (Chesapeake Bay), lizards (St. Martin Island), primarily invertebrates (Skipwith Pond), pelagic lake species (Bridge Brook Lake), and diverse desert taxa (Coachella Valley); electronic sequential logic circuits parsed from the ISCAS89 benchmark set (*7, 25*), where nodes represent logic gates and flip-flops (presented are all five partial scans of forward-logic chips and three digital fractional multipliers in the benchmark set); and World Wide Web hyperlinks between Web pages in a single domain (*4*) (only three-node motifs are shown). e, multiplied by the power of 10 (e.g., 1.46e6 $= 1.46 \times 10^6$).

Network	Nodes	Edges	N_{real}	$N_{rand} \pm$ SD	Z score	N_{real}	$N_{rand} \pm$ SD	Z score	N_{real}	$N_{rand} \pm$ SD	Z score
Gene regulation (transcription)					**Feed-forward loop**			**Bi-fan**			
E. coli	424	519	40	7 ± 3	10	203	47 ± 12	13			
*S. cerevisiae**	685	1,052	70	11 ± 4	14	1812	300 ± 40	41			
Neurons					**Feed-forward loop**			**Bi-fan**			**Bi-parallel**
C. elegans†	252	509	125	90 ± 10	3.7	127	55 ± 13	5.3	227	35 ± 10	20
Food webs					**Three chain**			**Bi-parallel**			
Little Rock	92	984	3219	3120 ± 50	2.1	7295	2220 ± 210	25			
Ythan	83	391	1182	1020 ± 20	7.2	1357	230 ± 50	23			
St. Martin	42	205	469	450 ± 10	NS	382	130 ± 20	12			
Chesapeake	31	67	80	82 ± 4	NS	26	5 ± 2	8			
Coachella	29	243	279	235 ± 12	3.6	181	80 ± 20	5			
Skipwith	25	189	184	150 ± 7	5.5	397	80 ± 25	13			
B. Brook	25	104	181	130 ± 7	7.4	267	30 ± 7	32			
Electronic circuits (forward logic chips)					**Feed-forward loop**			**Bi-fan**			**Bi-parallel**
s15850	10,383	14,240	424	2 ± 2	285	1040	1 ± 1	1200	480	2 ± 1	335
s38584	20,717	34,204	413	10 ± 3	120	1739	6 ± 2	800	711	9 ± 2	320
s38417	23,843	33,661	612	3 ± 2	400	2404	1 ± 1	2550	531	2 ± 2	340
s9234	5,844	8,197	211	2 ± 1	140	754	1 ± 1	1050	209	1 ± 1	200
s13207	8,651	11,831	403	2 ± 1	225	4445	1 ± 1	4950	264	2 ± 1	200
Electronic circuits (digital fractional multipliers)					**Three-node feedback loop**			**Bi-fan**			**Four-node feedback loop**
s208	122	189	10	1 ± 1	9	4	1 ± 1	3.8	5	1 ± 1	5
s420	252	399	20	1 ± 1	18	10	1 ± 1	10	11	1 ± 1	11
s838‡	512	819	40	1 ± 1	38	22	1 ± 1	20	23	1 ± 1	25
World Wide Web					**Feedback with two mutual dyads**			**Fully connected triad**			**Uplinked mutual dyad**
nd.edu§	325,729	1.46e6	1.1e5	2e3 ± 1e2	800	6.8e6	5e4±4e2	15,000	1.2e6	1e4 ± 2e2	5000

*Has additional four-node motif: (X→Z, W; Y→Z, W; Z→W), $N_{real} = 150$, $N_{rand} = 85 \pm 15$, $Z = 4$. †Has additional four-node motif: (X→Y, Z; Y→Z; Z→W), $N_{real} = 204$, $N_{rand} = 80 \pm 20$, $Z = 6$. The three-node pattern (X→Y, Z; Y→Z; Z→Y) also occurs significantly more than at random. It is not a motif by the present definition because it does not appear with completely distinct sets of nodes more than $U = 4$ times. ‡Has additional four-node motif: (X→Y; Y→Z, W; Z→X; W→X), $N_{real} = 914$, $N_{rand} = 500 \pm 70$, $Z = 6$. §Has two additional three-node motifs: (X→Y, Z; Y→Z; Z→Y), $N_{real} = 3e5$, $N_{rand} = 1.4e3 \pm 6e1$, $Z = 6000$, and (X→Y, Z; Y→Z), $N_{real} = 5e5$, $N_{rand} = 9e4 \pm 1.5e3$, $Z = 250$.

We also studied several technological networks. We analyzed the ISCAS89 benchmark set of sequential logic electronic circuits (*7, 25*). The nodes in these circuits represent logic gates and flip-flops. These nodes are linked by directed edges. We found that the motifs separate the circuits into classes that correspond to the circuit's functional description. In Table 1, we present two classes, consisting of five forward-logic chips and three digital fractional multipliers. The digital fractional multipliers share three motifs, including three- and four-node feedback loops. The forward logic chips share the feedforward loop, bi-fan, and bi-parallel motifs, which are similar to the motifs found in the genetic and neuronal information-processing networks. We found a different set of motifs in a network of directed hyperlinks between World Wide Web pages within a single domain (*4*). The World Wide Web motifs may reflect a design aimed at short paths between related pages. Application of our approach to nondirected networks shows distinct sets of motifs in networks of protein interactions and Internet router connections (*18*).

None of the network motifs shared by the food webs matched the motifs found in the gene regulation networks or the World Wide Web. Only one of the food web consensus motifs also appeared in the neuronal network. Different motif sets were found in electronic circuits with different functions. This suggests that motifs can define broad classes of networks, each with specific types of elementary structures. The motifs reflect the underlying processes that generated each type of network; for example, food webs evolve to allow a flow of energy from the bottom to the top of food chains, whereas gene regulation and neuron networks evolve to process information. Information processing seems to give rise to significantly different structures than does energy flow.

We further characterized the statistical significance of the motifs as a function of network size, by considering pieces of various sizes (subnetworks) of the full network. The concentration of motifs in the subnetworks is about the same as that in the full network (Fig. 3). In contrast, the concentration of the corresponding subgraphs in the randomized versions of the subnetworks decreases sharply with size. In analogy with statistical physics, the number of appearances of each motif in the real networks appears to be an extensive variable (i.e., one that grows linearly with the system size). These variables are nonextensive in the randomized networks. The existence of such variables may be a unifying property of evolved or designed systems. The decrease of the concentration C with randomized network size S (Fig. 3) qualitatively agrees with exact results (*2*, *26*) on Erdos-Renyi random graphs (random graphs that preserve only the number of nodes and edges of the real network) in which $C \sim 1/S$. In general, the larger the network is, the more significant the motifs tend to become. This trend can also be seen in Table 1 by comparing networks of different sizes. The network motif detection algorithm appears to be effective even for rather small networks (on the order of 100 edges). This is because three- or four-node subgraphs occur in large numbers even in small networks. Furthermore, our approach is not sensitive to data errors; for example, the sets of significant network motifs do not change in any of the networks upon addition, removal, or rearrangement of 20% of the edges at random.

In information-processing networks, the motifs may have specific functions as elementary computational circuits (*11*). More generally, they may be interpreted as structures that arise because of the special constraints under which the network has evolved (*27*). It is of value to detect and understand network motifs in order to gain insight into their dynamical behavior and to define classes of networks and network homologies. Our approach can be readily generalized to any type of network, including those with multiple "colors" of edges or nodes. It would be fascinating to see what types of motifs occur in other networks and to understand the processes that yield given motifs during network evolution.

References and Notes

1. S. H. Strogatz, *Nature* **410**, 268 (2001).
2. B. Bollobas, *Random Graphs* (Academic, London, 1985).
3. D. Watts, S. Strogatz, *Nature* **393**, 440 (1998).
4. A.-L. Barabási, R. Albert, *Science* **286**, 509 (1999).
5. M. Newman, *Proc. Natl. Acad. Sci. U.S.A.* **98**, 404 (2001).
6. H. Jeong, B. Tombor, R. Albert, Z. N. Oltvai, A. L. Barabasi, *Nature* **407**, 651 (2000).
7. R. F. Cancho, C. Janssen, R. V. Sole, *Phys. Rev. E* **64**, 046119 (2001).
8. R. F. Cancho, R. V. Sole, *Proc. R. Soc. London Ser. B* **268**, 2261 (2001).
9. L. Amaral, A. Scala, M. Barthelemy, H. Stanley, *Proc. Natl. Acad. Sci. U.S.A.* **97**, 11149 (2000).
10. B. Huberman, L. Adamic, *Nature* **401**, 131 (1999).
11. S. Shen-Orr, R. Milo, S. Mangan, U. Alon, *Nature Genet.* **31**, 64 (2002).
12. P. Holland, S. Leinhardt, in *Sociological Methodology*, D. Heise, Ed. (Jossey-Bass, San Francisco, 1975), pp. 1–45.
13. S. Wasserman, K. Faust, *Social Network Analysis* (Cambridge Univ. Press, New York, 1994).
14. N. Guelzim, S. Bottani, P. Bourgine, F. Kepes, *Nature Genet.* **31**, 60 (2002).
15. M. Newman, S. Strogatz, D. Watts, *Phys. Rev. E* **64**, 6118 (2001).
16. S. Maslov, K. Sneppen, *Science* **296**, 910 (2002).
17. The randomized networks used for detecting three-node motifs preserve the numbers of incoming, outgoing, and double edges with both incoming and outgoing arrows for each node. The randomized networks used for detecting four-node motifs preserve the above characteristics as well as the numbers of all 13 three-node subgraphs as in the real network. Algorithms for constructing these randomized network ensembles are described (*18*). Additional information is available at www.weizmann.ac.il/mcb/UriAlon.
18. Methods are available as supporting material on *Science* Online.
19. D. Thieffry, A. M. Huerta, E. Perez-Rueda, J. Collado-Vides, *Bioessays* **20**, 433 (1998).
20. M. C. Costanzo *et al.*, *Nucleic Acids Res.* **29**, 75 (2001).
21. J. Cohen, F. Briand, C. Newman, *Community Food Webs: Data and Theory* (Springer, Berlin, 1990).
22. R. Williams, N. Martinez, *Nature* **404**, 180 (2000).
23. S. Pimm, J. Lawton, J. Cohen, *Nature* **350**, 669 (1991).
24. J. White, E. Southgate, J. Thomson, S. Brenner, *Philos. Trans. R. Soc. London Ser. B* **314**, 1 (1986).
25. F. Brglez, D. Bryan, K. Kozminski, *Proc. IEEE Int. Symp. Circuits Syst.*, 1929 (1989).
26. In Erdos-Renyi randomized networks with a fixed connectivity (*2*), the concentration of a subgraph with n nodes and k edges scales with network size as $C \sim S^{n-k-1}$ (thus, $C \sim 1/S$ for the feedforward loop of Fig. 3 where $n = k = 3$). The sole exception in Table 1 in which C should not vanish at large S is the three-chain pattern in food webs where $n = 3$ and $k = 2$.
27. D. Callaway, J. Hopcroft, J. Kleinberg, M. Newman, S. Strogatz, *Phys. Rev. E* **64**, 041902 (2001).

28. We thank S. Maslov and K. Sneppen for valuable discussions. We thank J. Collado-Vides, N. Martinez, R. Govindan, R. Durbin, L. Amaral, R. Cancho, S. Maslov, and K. Sneppen for kindly providing data, as well as D. Alon, E. Domany, M. Elowitz, I. Kanter, O. Hobart, M. Naor, D. Mukamel, A. Murray, S. Quake, R. Raz, M. Reigl, M. Surette, K. Sneppen, P. Sternberg, E. Winfree, and all members of our lab for comments. We thank Caltech and the Aspen Center for Physics for their hospitality during part of this work. We acknowledge support from the Israel Science Foundation, the Human Frontier Science Program, and the Minerva Foundation.

Supporting Online Material

www.sciencemag.org/cgi/content/full/298/5594/824/DC1
Methods
Table S1

1 May 2002; accepted 10 September 2002

The structure of scientific collaboration networks

M. E. J. Newman*

Santa Fe Institute, 1399 Hyde Park Road, Santa Fe, NM 87501

Communicated by Murray Gell-Mann, Santa Fe Institute, Santa Fe, NM, November 13, 2000 (received for review July 12, 2000)

The structure of scientific collaboration networks is investigated. Two scientists are considered connected if they have authored a paper together and explicit networks of such connections are constructed by using data drawn from a number of databases, including MEDLINE (biomedical research), the Los Alamos e-Print Archive (physics), and NCSTRL (computer science). I show that these collaboration networks form "small worlds," in which randomly chosen pairs of scientists are typically separated by only a short path of intermediate acquaintances. I further give results for mean and distribution of numbers of collaborators of authors, demonstrate the presence of clustering in the networks, and highlight a number of apparent differences in the patterns of collaboration between the fields studied.

A social network is a collection of people, each of whom is acquainted with some subset of the others. Such a network can be represented as a set of points (or vertices) denoting people, joined in pairs by lines (or edges) denoting acquaintance. One could, in principle, construct the social network for a company or firm, for a school or university, or for any other community up to and including the entire world.

Social networks have been the subject of both empirical and theoretical study in the social sciences for at least 50 years (1–3), partly because of inherent interest in the patterns of human interaction, but also because their structure has important implications for the spread of information and disease. It is clear, for example, that variation in just the average number of acquaintances that individuals have (also called the average degree of the network) might substantially influence the propagation of a rumor, a fashion, a joke, or this year's flu.

A famous early empirical study of the structure of social networks, conducted by Stanley Milgram (4), asked test subjects, chosen at random from a Nebraska telephone directory, to get a letter to a target subject in Boston, a stockbroker friend of Milgram's. The instructions were that the letters were to be sent to their addressee (the stockbroker) by passing them from person to person, but that they could be passed only to someone whom the passer knew on a first-name basis. Because it was not likely that the initial recipients of the letters were on a first-name basis with a Boston stockbroker, their best strategy was to pass their letter to someone whom they felt was nearer to the stockbroker in some sense, either social or geographical: perhaps someone they knew in the financial industry, or a friend in Massachusetts.

A moderate number of Milgram's letters did eventually reach their destination, and Milgram discovered that the average number of steps taken to get them there was only about six, a result that has since passed into folklore and was immortalized by John Guare in the title of his 1990 play, *Six Degrees of Separation* (5). Although there were certainly biases present in Milgram's experiment—letters that took a longer path were perhaps more likely to get lost or forgotten, for instance (6)—his result is usually taken as evidence of the "small-world hypothesis," that most pairs of people in a population can be connected by only a short chain of intermediate acquaintances, even when the size of the population is very large.

Milgram's work, although cleverly conducted and in many ways revealing, does not, however, tell us much about the detailed structure of social networks, data that are crucial to the understanding of information or disease propagation. Many other studies have addressed this problem (discussions can be found in refs. 1–3). Foster *et al.* (7), Fararo and Sunshine (8), and Moody and White (9), for instance, all conducted studies of friendship networks among middle- or high-school students, Bernard *et al.* (10) did the same for communities of Utah Mormans, Native Americans, and Micronesian islanders, and there are many other examples to be found in the literature. Surveys or interviews were used to determine friendships.

Although these studies directly probe the structure of the relevant social network, they suffer from two substantial shortcomings that limit their usefulness. First, the studies are labor intensive, and the size of the network that can be mapped is therefore limited—typically to a few tens or hundreds of people. Second, these studies are highly sensitive to subjective bias on the part of interviewees; what is considered to be an "acquaintance" can differ considerably from one person to another. To avoid these issues, a number of researchers have studied networks for which there exist more numerous data and more precise definitions of connectedness. Examples of such networks are the electric power grid (3, 11), the Internet (12, 13), and the pattern of air traffic between airports (14). These networks, however, suffer from a different problem: although they may loosely be said to be social networks in the sense that their structure in some way reflects features of the society that built them, they do not directly measure actual contact between people. Many researchers, of course, are interested in these networks for their own sake, but to the extent that we want to know about human acquaintance patterns, power grids and computer networks are a poor proxy for the real thing.

Perhaps the nearest that studies of this kind have come to looking at a true acquaintance network is in studies of the network of movie actors (11, 14). In this network, which has been thoroughly documented and contains nearly half a million people, two actors are considered connected if they have been credited with appearance in the same film. However, although this is genuinely a network of people, it is far from clear that the appearance of two actors in the same movie implies that they are acquainted in any but the most cursory fashion, or that their acquaintance extends off screen. To draw conclusions about patterns of everyday human interaction from the movies would, it seems certain, be a mistake.

In this paper, I present a study of a genuine network of human acquaintances that is large—containing over a million people—and for which a precise definition of acquaintance is possible. That network is the network of scientific collaboration, as documented in the papers scientists write.

Scientific Collaboration Networks

I study networks of scientists in which two scientists are considered connected if they have coauthored a paper. This seems a reasonable definition of scientific acquaintance: most people who have written a paper together will know one another quite

*E-mail: mark@santafe.edu.

Article published online before print: *Proc. Natl. Acad. Sci. USA*, 10.1073/pnas.021544898. Article and publication date are at www.pnas.org/cgi/doi/10.1073/pnas.021544898

Table 1. Summary of results of the analysis of seven scientific collaboration networks

	MEDLINE	Los Alamos e-Print Archive: Complete	astro-ph	cond-mat	hep-th	SPIRES	NCSTRL
Total papers	2,163,923	98,502	22,029	22,016	19,085	66,652	13,169
Total authors	1,520,251	52,909	16,706	16,726	8,361	56,627	11,994
First initial only	1,090,584	45,685	14,303	15,451	7,676	47,445	10,998
Mean papers per author	6.4 (6)	5.1 (2)	4.8 (2)	3.65 (7)	4.8 (1)	11.6 (5)	2.55 (5)
Mean authors per paper	3.754 (2)	2.530 (7)	3.35 (2)	2.66 (1)	1.99 (1)	8.96 (18)	2.22 (1)
Collaborators per author	18.1 (1.3)	9.7 (2)	15.1 (3)	5.86 (9)	3.87 (5)	173 (6)	3.59 (5)
Cutoff z_c	5,800 (1,800)	52.9 (4.7)	49.0 (4.3)	15.7 (2.4)	9.4 (1.3)	1,200 (300)	10.7 (1.6)
Exponent τ	2.5 (1)	1.3 (1)	0.91 (10)	1.1 (2)	1.1 (2)	1.03 (7)	1.3 (2)
Size of giant component	1,395,693	44,337	14,845	13,861	5,835	49,002	6,396
First initial only	1,019,418	39,709	12,874	13,324	5,593	43,089	6,706
As a percentage	92.6 (4)%	85.4 (8)%	89.4 (3)	84.6 (8)%	71.4 (8)%	88.7 (1.1)%	57.2 (1.9)%
Second largest component	49	18	19	16	24	69	42
Mean distance	4.6 (2)	5.9 (2)	4.66 (7)	6.4 (1)	6.91 (6)	4.0 (1)	9.7 (4)
Maximum distance	24	20	14	18	19	19	31
Clustering coefficient C	0.066 (7)	0.43 (1)	0.414 (6)	0.348 (6)	0.327 (2)	0.726 (8)	0.496 (6)

Numbers in parentheses are standard errors on the least significant figures.

well. It is a moderately stringent definition, since there are many scientists who know one another to some degree but have never collaborated on the writing of a paper. Stringency, however, is not inherently a bad thing. A stringent condition of acquaintance is perfectly acceptable, provided, as in this case, that it can be applied consistently.

I have constructed collaboration graphs for scientists in a variety of fields. The data come from four databases: MEDLINE (which covers published papers on biomedical research), the Los Alamos e-Print Archive (preprints primarily in theoretical physics), SPIRES (published papers and preprints in high-energy physics), and NCSTRL (preprints in computer science). In each case, I have examined papers that appeared in a 5-year window, from 1995 to 1999 inclusive. The sizes of the databases range from 2 million papers for MEDLINE to 13,000 for NCSTRL.

That some of the databases used contain unrefereed preprints should not be regarded negatively. Although unrefereed preprints may be of lower average scientific quality than papers in peer-reviewed journals, as an indicator of social connection, they are every bit as good as their refereed counterparts.

The idea of studying collaboration patterns by using data drawn from the publication record is not new. There is a substantial body of literature in information science dealing with coauthorship patterns (15–19) and cocitation patterns (20–22) (i.e., connections between authors established via the citation of their works in the same literature). However, to our knowledge, no detailed reconstruction of an actual collaboration network has previously been attempted. Indeed, the nearest thing to such a reconstruction comes not from information science at all, but from the mathematics community, within which the concept of the Erdös number has a long history. Paul Erdös was an influential but itinerant Hungarian mathematician, who apparently spent a large portion of his later life living out of a suitcase and writing papers with those of his colleagues willing to give him room and board (23). He published at least 1,401 papers during his life, more than any other mathematician in history, except perhaps Leonhard Euler. The Erdös number measures a mathematician's proximity, in bibliographical terms, to the great man. Those who have published a paper with Erdös have an Erdös number of 1. Those have published with a coauthor of Erdös have an Erdös number of 2, and so on. An exhaustive list exists of all mathematicians with Erdös numbers of 1 and 2 (24).

In addition to distance between authors, there are many other interesting quantities to be measured on collaboration networks, including the number of collaborators of scientists, the numbers of papers they write, and the degree of "clustering," which is the probability that two of a scientist's collaborators have themselves collaborated. All of these quantities and several others are considered in this paper.

Results

Table 1 gives a summary of some of the results of the analysis of databases described in the previous section. In addition to results for the four complete databases, results are also given for three subject-specific subsets of the Los Alamos Archive, covering astrophysics (denoted astro-ph), condensed matter physics (cond-mat), and theoretical high-energy physics (hep-th). In this section, I highlight some of these results and discuss their implications.

Number of Authors. Estimating the true number of distinct authors in a database is complicated by two problems. First, two authors may have the same name. Second, authors may identify themselves in different ways on different papers, e.g., by using first initial only, by using all initials, or by using full name. To estimate the size of the error introduced by these effects, all analyses reported here have been carried out twice. The first time, all initials of each author are used. This will rarely confuse two different authors for the same person (although this will still happen occasionally) but sometimes misidentifies the same person as two different people, thereby overestimating the total number of authors. The second analysis is carried out using only the first initial of each author, which will ensure that different publications by the same author are almost always identified as such, but will with some regularity confuse distinct authors for the same person. Thus these two analyses give upper and lower bounds on the number of authors and also give an indication of the expected precision

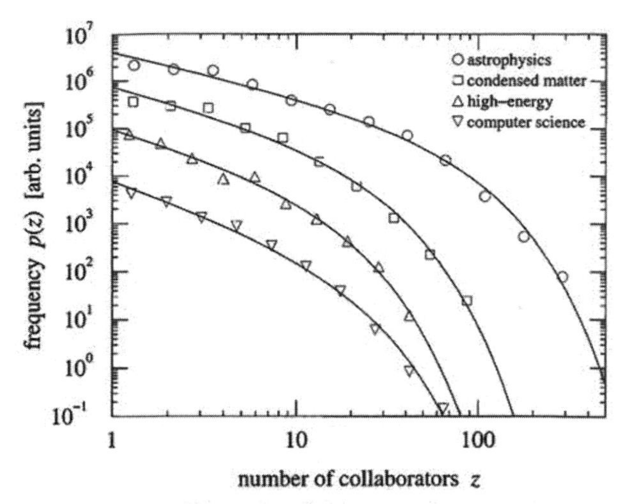

Fig. 1. Histograms of the number of collaborators of scientists in four of the databases studied here. The solid lines are least-squares fits to Eq. 1.

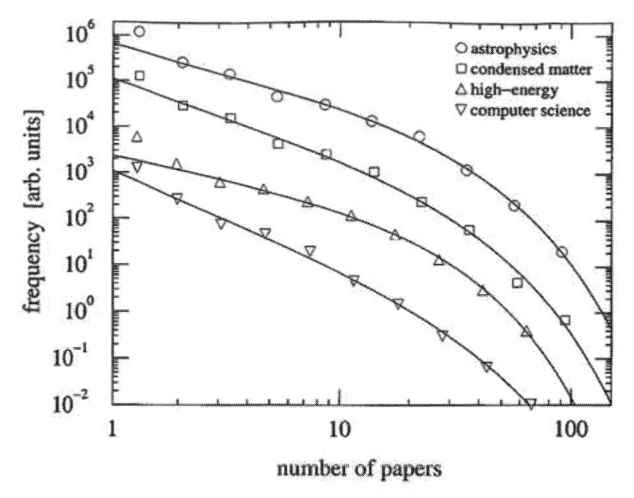

Fig. 2. Histograms of the number of papers written by scientists in four of the databases. As with Fig. 1, the solid lines are least-squares fits to Eq. 2.

of many of our other measurements. In Table 1, both estimates of the number of authors for each database are quoted. For most other quantities, only an error estimate based on the separation of the upper and lower bounds is quoted.

Mean Papers per Author and Authors per Paper. Authors typically wrote about four papers in the 5-year period covered by this study. The average paper had about three authors. Notable exceptions are in theoretical high-energy physics and computer science, in which smaller collaborations are the norm (an average of two people), and the SPIRES high-energy physics database, with an average of nine authors per paper. The reason for this last impressive figure is that the SPIRES database contains data on experimental as well as theoretical work. High-energy experimental collaborations can run to hundreds or thousands of people, the largest author list in the SPIRES database giving the names of a remarkable 1,681 authors on a single paper.

Number of Collaborators. The striking difference in collaboration patterns in high-energy physics is highlighted further by the results for the average number of collaborators of an author. This is the average total number of people with whom a scientist collaborated during the period of study—the average degree, in the graph theorist's language. For purely theoretical databases, such as the hep-th subset of the Los Alamos Archive (covering high-energy physics theory) and NCSTRL (computer science), this number is low, on the order of four. For partly or wholly experimental databases [condensed matter physics and astrophysics at Los Alamos and MEDLINE (biomedicine)], the degree is significantly higher, as high as 18 for MEDLINE. But high-energy experiment easily takes the prize, with an average of 173 collaborators per author.

There is more to the story of numbers of collaborators, however. In Fig. 1, histograms of the numbers of collaborators of scientists in four of the smaller databases are shown. There has been a significant amount of recent discussion of this distribution for a variety of networks in the literature. A number of authors (12, 13) have pointed out that if one makes a similar plot for the number of connections (or "links") z to or from sites on the World Wide Web, the resulting distribution closely follows a power law: $P(z) \approx z^{-\tau}$, where τ is a constant exponent with (in that case) a value of about 2.5. Barabási and Albert have suggested (25) that a similar power-law result may apply to all or at least most other networks of interest, including social networks. Others have presented a variety of evidence to the contrary (14). My data do not follow a power-law form perfectly. If they did, the curves in Fig. 2 would be straight lines on the logarithmic scales used. However, these data are well fitted by a power-law form with an exponential cutoff:

$$P(z) \sim z^{-\tau} e^{-z/z_c}, \qquad [1]$$

where τ and z_c are constants. Fits to this form are shown as the solid lines in Fig. 2. In each case, the fit has an R^2 of better than 0.99 and P values for both power-law and exponential terms of less than 10^{-3} (except for the "all-initials" version of the MEDLINE network, for which the exponential term has $P = 0.17$, indicating that this distribution is moderately well fit by a pure power-law form).

This form is commonly seen in physical systems and suggests an underlying degree distribution that follows a power law, but with some imposed constraint that places a limit on the maximum value of z. One possible explanation of this cutoff in the present case is that it arises as a result of the finite (5-year) window of data used. If this were the case, we would expect the cutoff to increase with increasing window size. But even in the (impractical) limit of infinite window size, a cutoff would still be imposed by the finite working lifetime of a professional scientist (about 40 years).

The values of τ and z_c are given in the table for each database. The value of the cutoff size, z_c, varies considerably. For the mostly theoretical condensed matter, high-energy theory, and computer science databases, it takes small values on the order of 10, indicating that theorists rarely had more than this many collaborators during the 5-year period. In other cases, such as SPIRES and MEDLINE, it takes much larger values. In the case of SPIRES, this is probably again because of the presence of very large experimental collaborations in the data. MEDLINE is more interesting. There are few very large collaborations in the MEDLINE database, and yet there are a small number of individuals with very large numbers of collaborators. How does this arise? One possibility is that it is the result of the practice in the biomedical research community of laboratory directors signing their name to all (or most) papers emerging from their laboratories. One can well imagine that, with some individuals directing very large laboratories, this could generate authors with a very high apparent number of collaborators. [It is possible that a similar mechanism is at work in the SPIRES data also.] This hypothesis could be checked by verifying whether the individuals with the largest numbers of collaborators are indeed lab directors or principal investigators and might make an interesting topic for further study.

The exponent τ of the power-law distribution is also interesting. We note that in all of the "hard sciences," this exponent takes values close to 1. In the MEDLINE (biomedicine) database, however, its value is 2.5, similar to that noted for the World Wide Web. The value $\tau = 2$ forms a dividing line between two fundamentally different behaviors of the network. For $\tau < 2$, the average properties of the network are dominated by the few individuals who have a large number of collaborators, whereas networks with $\tau > 2$ are dominated by the "little people"—those with few collaborators. Thus, one finds that in biomedical research, highly connected individuals do not determine the average characteristics of their field, despite their names appearing on a lot of papers. In physics and computer science, on the other hand, it appears that such individuals do determine these characteristics.

In Fig. 2, histograms are shown of the number of papers that authors have written in the same four databases. As the figure shows, the distribution of papers follows a similar form to the distribution of collaborators. The solid lines are again fits to Eq. 1 and again match the data well in all cases. This form may be regarded as a generalization of the well-known Lotka law of scientific productivity, which states that the distribution of numbers of papers written should follow a power law (16, 26). The clear exponential cutoff seen in the distribution is again presumably a result of the finite time window used in this study. It would be interesting to test this hypothesis by varying the window size, although a thorough test may have to wait until more years of data are available; most of the databases studied here have not been in existence long enough to give good statistics on this point. The SPIRES database, which has been in existence for more than a quarter of a century, is an exception and might make an interesting case study (27).

The Giant Component. In all social networks, there is the possibility of a percolation transition (28). In networks with very small numbers of connections between individuals, all individuals belong only to small islands of collaboration or communication. As the total number of connections increases, however, there comes a point at which a giant component forms—a large group of individuals who are all connected to one another by paths of intermediate acquaintances. It appears that all of the databases considered here are connected in this sense. Measuring the size of groups of connected authors in each database, we find (see Table 1) that in most of the databases, the largest such group occupies around 80 or 90% of all authors: almost everyone in the community is connected to almost everyone else by some path (probably many paths) of intermediate coauthors. In high-energy theory and computer science, the fraction is smaller but still more than half the total size of the network. (These two databases may, it appears, give a less complete picture of their respective fields than the others, because of the existence of competing databases with overlapping coverage. The small size of the giant component may in part be attributable to this.)

I have also calculated the size of the second-largest group of connected authors for each database. In each case, this group is far smaller than the largest. This is a characteristic signature of networks that are well inside the percolating regime. In other words, it appears that scientific collaboration networks are not on the borderline of connectedness—they are very highly connected and in no immediate danger of fragmentation. This is a good thing. Science would probably not work at all if scientific communities were not densely interconnected.

Average Degrees of Separation. I have calculated exhaustively the minimum distance, in terms of numbers of links in the network, between all pairs of scientists in our databases for whom a connection exists. I find that the typical distance between a pair of scientists is about six; there are six degrees of separation in science, just as there are in the larger world of human acquaintance. Even in very large communities, such as the biomedical research community documented by MEDLINE, it takes an average of only about six steps to reach a randomly chosen scientist from any other, of the more than one million who have published. We conjecture that this has a profound effect on the way the scientific community operates. Despite the importance of written communication in science as a document and archive of work carried out, and of scientific conferences as a broadcast medium for summary results, it is probably safe to say that the majority of scientific communication still takes place by private conversation. The existence of a large giant component, as discussed in the previous section, allows news of important discoveries and scientific information to reach most members of the network via such private conversations, and clearly information can circulate far faster in a world where the typical separation of two scientists is six than it can in one where it is a thousand or a million.

The variation of average vertex–vertex distances from one database to another also shows interesting behavior. The simplest model of a social network is the random graph—a network in which people are connected to one another uniformly at random (29). For a given number N of scientists with a given mean number z of collaborators, the average vertex–vertex distance on a random graph varies as the logarithm of N according to $\log N/\log z$. Social networks are measurably different from random graphs (3), but the random graph nonetheless provides a useful benchmark against which to compare them. Watts and Strogatz (11) defined a social network as being "small" if typical distances were comparable to those on a random graph. This implies that such networks should also have typical distances that grow roughly logarithmically in N, and indeed some authors (e.g., ref. 14) have used this logarithmic growth as the defining criterion for a "small world." In Fig. 3, the average distance between all pairs of scientists for each of the networks studied here is shown, including separate calculations for eight subject divisions of the Los Alamos Archive. In total, there are 12 points, which have been plotted against $\log N/\log z$ using the appropriate values of N and z from Table 1. As the figure shows, there is a strong correlation ($R^2 = 0.83$) between the measured distances and the expected $\log N$ behavior, indicating that distances do indeed vary logarithmically with the number of scientists in a community. As far as I am aware, this is the first empirical demonstration of logarithmic variation with network size for any real social network.

Also quoted in Table 1 are figures for the maximum separation of pairs of scientists in each database, which tells us the greatest distance we will ever have to go to connect two people. This quantity is often referred to as the diameter of the network. For all of the networks examined here, it is on the order of 20; there is a chain of at most about 20 acquaintances connecting any two scientists. (This result, of course, excludes pairs of scientists who are not connected at all, as will often be the case for the 10 or 20% who fall outside the giant component.)

Clustering. Real social networks have another important property that is absent from many network models. Real networks are clustered, meaning they possess local communities in which a

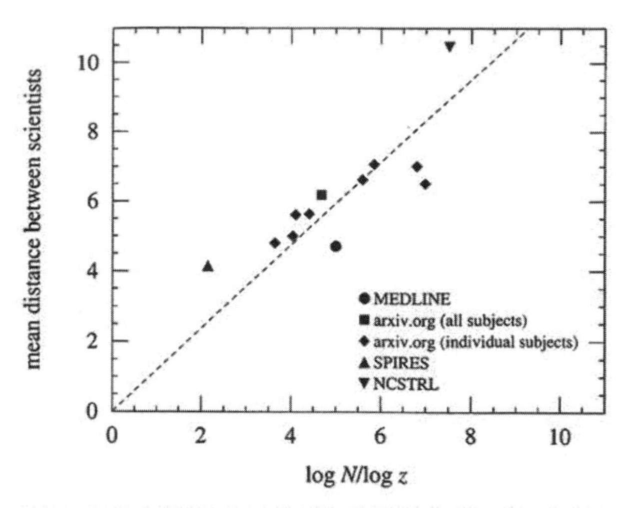

Fig. 3. Average distance between pairs of scientists in the various communities, plotted against the average distance on a random graph of the same size and average coordination number. The dotted line is the best fit to the data that also passes through the origin.

higher than average number of people know one another. A laboratory or university department might form such a community in science, as might the set of researchers who work in a particular subfield. One way of probing for the existence of such clustering in network data is to measure the fraction of "transitive triples" in a network (1), also called the clustering coefficient C (11), which for a collaboration graph is the average fraction of pairs of a person's collaborators who have also collaborated with one another. Mathematically,

$$C = \frac{3\times \text{number of triangles on the graph}}{\text{number of connected triples of vertices}}. \qquad [2]$$

Here a "triangle" is a trio of authors, each of whom is connected to both of the others, and a "connected triple" is a single author connected to two others. $C = 1$ for a fully connected graph and for a random graph, tends to zero as $1/N$ as the graph becomes large.

In Table 1, values of C are given for each of the networks studied here, and we can see that there is a very strong clustering effect in the scientific community: two scientists typically have a 30% or greater probability of collaborating if both have collaborated with a third scientist. A number of explanations of this result are possible. To some extent, it is certainly the result of the appearance of papers with three or more authors: such papers clearly contain trios of scientists who have all collaborated with one another. However, the values measured here cannot be entirely accounted for in this way (30) and indicate also that scientists either introduce their collaborators to one another, thereby engendering new collaborations, or perhaps that institutions bring sets of collaborators together to form a variety of new collaborations. Processes such as these have been discussed extensively in the social networks literature, in the context of structural balance within networks (1).

The MEDLINE database is interesting in that it possesses a much lower value of the clustering coefficient than the "hard science" databases. This appears to indicate that it is significantly less common in biological research for scientists to broker new collaborations between their acquaintances than it is in physics or computer science. This could again be a result of the "top-down" organization of laboratories under laboratory directors, which tends to produce "tree-like" collaboration networks, with many branches but few short loops. Such tree-like networks are known to possess low clustering coefficients.

Conclusions

The collaboration networks of scientists in biology and medicine, various subdisciplines of physics, and computer science have been analyzed, by using author attributions from papers or preprints appearing in those areas over a 5-year period from 1995 to 1999. We find a number of interesting properties of these networks. In all cases, scientific communities seem to constitute a "small world," in which the average distance between scientists via a line of intermediate collaborators varies logarithmically with the size of the relevant community. Typically, we find that only about five or six steps are necessary to get from one randomly chosen scientist in a community to another. It is conjectured that this smallness is a crucial feature of a functional scientific community.

We also find that the networks are highly clustered, meaning that two scientists are much more likely to have collaborated if they have a third common collaborator than are two scientists chosen at random from the community. This may indicate that the process of scientists introducing their collaborators to one another is an important one in the development of scientific communities.

We have studied the distributions of both the number of collaborators of scientists and the numbers of papers they write. In both cases, we find these distributions are well fit by power-law forms with an exponential cutoff. This cutoff may be caused by the finite time window used in the study.

We find a number of significant statistical differences between different scientific communities. Some of these are obvious: experimental high-energy physics, for example, which is famous for the staggering size of its collaborations, has a vastly higher average number of collaborators per author than any other field examined. Other differences are less obvious, however. Biomedical research, for example, shows a much lower degree of clustering than any of the other fields examined. In other words, it is less common in biomedicine for two scientists to start a collaboration if they have another collaborator in common. Biomedicine is also the only field in which the exponent of the distribution of numbers of collaborators is greater than 2, implying that the average properties of the collaboration network are dominated by the many people with few collaborators, rather than, as in other fields, by the few people with many.

The work reported in this paper represents, inevitably, only a first look at the collaboration networks described. Many theoretical measures have been discussed elsewhere, in addition to the distances and clustering studied here, which reflect socially important structure in such networks. I hope that academic collaboration networks will prove a reliable and copious source of data for testing theories about such measures, as well as being interesting in their own right, especially to ourselves, the scientists whom they describe.

I am indebted to Paul Ginsparg and Geoffrey West (Los Alamos e-Print Archive), Carl Lagoze (NCSTRL), Oleg Khovayko, David Lipman and Grigoriy Starchenko (MEDLINE), and Heath O'Connell (SPIRES), for making available the publication data used for this study. I also thank Dave Alderson, Paul Ginsparg, Laura Landweber, Ronald Rousseau, Steve Strogatz, and Duncan Watts for illuminating conversations. This work was funded in part by a grant from Intel Corporation to the Santa Fe Institute Network Dynamics Program. The NCSTRL digital library was made available through the Defense Advanced Research Planning Agency (DARPA)/Corporation for National Research Initiatives test suites program funded under DARPA Grant N66001–98-1–8908. The Los Alamos e-Print archive is funded by the National Science Foundation under Grant PHY-9413208.

1. Wasserman, S. & Faust, K. (1994) *Social Network Analysis* (Cambridge Univ. Press, Cambridge).
2. Scott, J. (2000) *Social Network Analysis* (Sage Publications, London).
3. Watts, D. J. (1999) *Small Worlds* (Princeton Univ. Press, Princeton, NJ).
4. Milgram, S. (1967) *Psychol. Today* **2,** 60–67.
5. Guare, J. (1990) *Six Degrees of Separation* (Vintage, New York).
6. White, H. C. (1970) *Social Forces* **49,** 259–264.
7. Foster, C. C., Rapoport, A. & Orwant, C. J. (1963) *Behav. Sci.* **8,** 56–65.
8. Fararo, T. J. & Sunshine, M. (1964) *A Study of a Biased Friendship Network* (Syracuse Univ. Press, Syracuse, NY).
9. Moody, J. & White, D. R. (2000) *Social Cohesion and Embeddedness: A Hierarchical Conception of Social Groups* (Santa Fe Institute working paper 00-07-49).
10. Bernard, H. R., Kilworth, P. D., Evans, M. J., McCarty, C. & Selley, G. A. (1988) *Ethnology* **2,** 155–179.
11. Watts, D. J. & Strogatz, S. H. (1998) *Nature (London)* **393,** 440–442.
12. Albert, R., Jeong, H. & Barabási, A.-L. (1999) *Nature (London)* **401,** 130–131.
13. Broder, A., Kumar, R., Maghoul, F., Raghavan, P., Rajagopalan, S., Stata, R., Tomkins, A. & Wiener, J. (2000) in *Computer Networks* **33,** 309–320.
14. Amaral, L. A. N., Scala, A., Barthélémy. M. & Stanley, H. (2000) *Proc. Natl. Acad. Sci. USA* **97,** 11149–11152. (First Published September 26, 2000; 10.1073/pnas.200327197)
15. de Solla Price, D. (1965) *Science* **149,** 510–515.
16. Egghe, L. & Rousseau, R. (1990) *Introduction to Informetrics* (Elsevier, Amsterdam).
17. Melin, G. & Persson, O. (1996) *Scientometrics* **36,** 363–377.
18. Kretschmer, H. (1998) *Z. Sozialpsychol.* **29,** 307–324.
19. Ding, Y., Foo, S. & Chowdhury, G. (1999) *Int. Inform. Lib. Rev.* **30,** 367–376.
20. Crane, D. (1972) *Invisible Colleges* (Univ. of Chicago Press, Chicago).
21. van Raan, A. F. J. (1990) *Science* **347,** 626.
22. Persson, O. & Beckmann, M. (1995) *Scientometrics* **33,** 351–366.
23. Hoffman, P. (1998) *The Man Who Loved Only Numbers* (Hyperion, New York).
24. Grossman, J. W. & Ion, P. D. F. (1995) *Congressus Numerantium* **108,** 129–131.
25. Barabási, A. L. & Albert, R. (1999) *Science* **286,** 509–512.
26. Lotka, A. J. (1926) *J. Wash. Acad. Sci.* **16,** 317–323.
27. O'Connell, H. B. (2000) *Physicists Thriving with Paperless Publishing* (physics/0007040).
28. Stauffer, D. & Aharony, A. (1991) *Introduction to Percolation Theory* (Taylor and Francis, London), 2nd Ed.
29. Bollobás, B. (1985) *Random Graphs* (Academic, New York).
30. Newman, M. E. J., Strogatz, S. H. & Watts, D. J. (2000) *Random Graphs with Arbitrary Degree Distribution and Their Applications*, preprint, cond-mat/0007235.

The web of human sexual contacts

Promiscuous individuals are the vulnerable nodes to target in safe-sex campaigns.

Unlike clearly defined 'real-world' networks[1], social networks tend to be subjective to some extent[2,3] because the perception of what constitutes a social link may differ between individuals. One unambiguous type of connection, however, is sexual contact, and here we analyse the sexual behaviour of a random sample of individuals[4] to reveal the mathematical features of a sexual-contact network. We find that the cumulative distribution of the number of different sexual partners in one year decays as a scale-free power law that has a similar exponent for males and females. The scale-free nature of the web of human sexual contacts indicates that strategic safe-sex campaigns are likely to be the most efficient way to prevent the spread of sexually transmitted diseases.

Many real-world networks[1] typify the 'small-world' phenomenon[5], so called because of the surprisingly small average path lengths between nodes[6,7] in the presence of a large degree of clustering[3,6] (Fig. 1). Small-world networks are classed as single-scale, broad-scale or scale-free, depending on their connectivity distribution, $P(k)$, where k is the number of links connected to a node[8]. Scale-free networks, which are characterized by a power-law decay of the cumulative distribution $P(k) \approx k^{-\alpha}$, may be formed as a result of preferential attachment of new links between highly connected nodes[9,10].

We analysed the data gathered in a 1996 Swedish survey of sexual behaviour[4]. The survey involved a random sample of 4,781 Swedes (aged 18–74 years) and used structured personal interviews and questionnaires. The response rate was 59%, which corresponds to 2,810 respondents. Two independent analyses of non-response error revealed that elderly people, particularly women, are under-represented in the sample; apart from this skew, the sample is representative in all demographic dimensions.

Connections in the network of sexual contacts appear and disappear as sexual relations are initiated and terminated. To investigate the connectivity of this dynamic network, in which links may be short-lived, we first analysed the number, k, of sex partners over a relatively short time period — the 12 months before the survey. Figure 2a shows the cumulative distribution, $P(k)$, for female and male respondents. The data closely follow a straight line in a double-logarithmic plot, which is consistent with a power-law dependence. Males report a larger number of sexual partners than females[11], but both show the same scaling properties.

These results contrast with the exponential or gaussian distributions — for which there is a well-defined scale — found for friendship networks[8]. Plausible explanations for the structure of the sexual-contact network described here include increased skill in acquiring new partners as the number of previous partners grows, varying degrees of attractiveness, and the motivation to have many new partners to sustain self-image. Our results are consistent with the preferential-attachment mechanism of scale-free networks: evidently, in sexual-contact networks, as in other scale-free networks, 'the rich get richer'[9,10].

We next analysed the total number of partners, k_{tot}, in the respondent's life up to the time of the survey. This value is not relevant to the instantaneous structure of the network, but may help to elucidate the mechanisms responsible for the distribution of number of partners. Figure 2b shows the cumulative distribution, $P(k_{tot})$: for $k_{tot} > 20$, the data follow a straight line in a double-logarithmic plot, which is consistent with a power-law dependence in the tails of the distribution.

Figure 1 It's a small world: social networks have small average path lengths between connections and show a large degree of clustering. Painting by Idahlia Stanley.

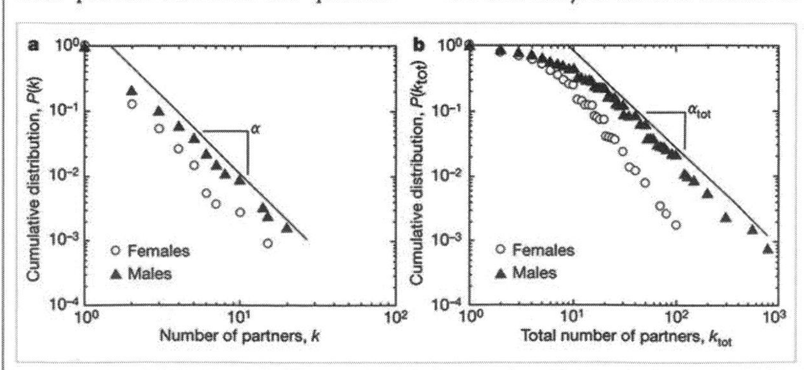

Figure 2 Scale-free distribution of the number of sexual partners for females and males. **a**, Distribution of number of partners, k, in the previous 12 months. Note the larger average number of partners for male respondents: this difference may be due to 'measurement bias' — social expectations may lead males to inflate their reported number of sexual partners. Note that the distributions are both linear, indicating scale-free power-law behaviour. Moreover, the two curves are roughly parallel, indicating similar scaling exponents. For females, $\alpha = 2.54 \pm 0.2$ in the range $k > 4$, and for males, $\alpha = 2.31 \pm 0.2$ in the range $k > 5$. **b**, Distribution of the total number of partners k_{tot} over respondents' entire lifetimes. For females, $\alpha_{tot} = 2.1 \pm 0.3$ in the range $k_{tot} > 20$, and for males, $\alpha_{tot} = 1.6 \pm 0.3$ in the range $20 < k_{tot} < 400$. Estimates for females and males agree within statistical uncertainty.

Our most important finding is the scale-free nature of the connectivity of an objectively defined, non-professional social network. This result indicates that the concept of the 'core group' considered in epidemiological studies[12] must be arbitrary, because there is no well-defined threshold or boundary that separates the core group from other individuals (as there would be for a bimodal distribution).

Our results may have epidemiological implications, as epidemics arise and propagate much faster in scale-free networks than in single-scale networks[6,13]. Also, the measures adopted to contain or stop the propagation of diseases in a network need to be radically different for scale-free networks. Single-scale networks are not susceptible to attack at even the most connected nodes, whereas scale-free networks are resilient to random failure but are highly susceptible to destruction of the best-connected nodes[14]. The possibility that the web of sexual contacts has a scale-free structure indicates that strategic targeting of safe-sex education campaigns to those individuals with a large number of partners may significantly reduce the propagation of sexually transmitted diseases.

Fredrik Liljeros*, Christofer R. Edling*, Luís A. Nunes Amaral†, H. Eugene Stanley†, Yvonne Åberg*

**Department of Sociology, Stockholm University, S-106 91 Stockholm, Sweden*
e-mail: liljeros@sociology.su.se
†Center for Polymer Studies and Department of Physics, Boston University, Boston, Massachusetts 02215, USA

1. Strogatz, S. H. *Nature* **410**, 268–276 (2001).
2. Kochen, M. (ed.) *The Small World* (Ablex, Norwood, New Jersey, 1989).
3. Wasserman, S. & Faust, K. *Social Network Analysis* (Cambridge Univ. Press, Cambridge, 1994).
4. Lewin, B. (ed.) *Sex i Sverige. Om Sexuallivet i Sverige 1996* [Sex in Sweden. On the Sexual Life in Sweden 1996] (Natl Inst. Pub. Health, Stockholm, 1998).
5. Milgram, S. *Psychol. Today* **2**, 60–67 (1967).
6. Watts, D. J. & Strogatz, S. H. *Nature* **393**, 440–442 (1998).
7. Barthelemy, M. & Amaral, L. A. N. *Phys. Rev. Lett.* **82**, 3180–3183 (1999).
8. Amaral, L. A. N., Scala, A., Barthélémy, M. & Stanley, H. E. *Proc. Natl Acad. Sci. USA* **97**, 11149–11152 (2000).
9. Simon, H. A. *Biometrika* **42**, 425–440 (1955).
10. Barabási, A.-L. & Albert, R. *Science* **286**, 509–512 (1999).
11. Laumann, E. O., Gagnon, J. H., Michael, R. T. & Michaels, S. *The Social Organization of Sexuality* (Univ. Chicago Press, Chicago, 1994).
12. Hethcote, H. W. & Yorke, J. A. *Gonorrhea Transmission Dynamics and Control* (Springer, Berlin, 1984).
13. Pastor-Satorras, R. & Vespignani, A. *Phys. Rev. Lett.* **86**, 3200–3203 (2001).
14. Albert, R., Jeong, H. & Barabási, A.-L. *Nature* **406**, 378–382 (2000).

Chapter Four
Models of networks

In this chapter we turn our attention to theoretical models of network structure. We divide these models into three basic classes. In Section 4.1 we look at one of the oldest models of networks, the random graph of **Solomonoff and Rapoport (1951)** and **Erdős and Rényi (1960)**, which has been extended in recent work to include graphs with the highly skewed degree distributions seen in many of the papers of Chapter 3, as well as graphs with directed edges or bipartite structure. In Section 4.2, we look at the so-called "small-world model," a model of the structure of social networks that was introduced by **Watts and Strogatz (1998)**. In Section 4.3 we look at models of network growth, which, unlike the models of the preceding sections, explicitly incorporate an element of time evolution in the network structure. In particular, we look at the class of models known as "scale-free models," which were introduced by **Barabási and Albert (1999)**. These models offer a plausible explanation, in terms of the growth dynamics of the network, for the skewed degree distributions seen in the World Wide Web and many other networks.

4.1 RANDOM GRAPH MODELS

One of the simplest and oldest of network models is the *random graph* (Bollobás 2001; Janson *et al.* 1999), which was introduced by **Solomonoff and Rapoport (1951)** in their paper reproduced in Chapter 2, and studied extensively by Paul Erdős and Alfréd Rényi in a series of classic papers published in the 1950s and 1960s (**Erdős and Rényi** 1959, **1960**, 1961). (This is the same Paul Erdős after whom the Erdős number of Chapter 3 is named—Erdős was both an influential early participant in the study of networks and an influential vertex in one of the networks studied.) In fact, Erdős and Rényi looked at two different models in their work, both of which are referred to as random graphs. The two models are normally distinguished from one another by their symbolic names: they are called $G_{n,m}$ and $G_{n,p}$:

- $G_{n,m}$ is the set of all graphs consisting of n vertices and m edges. In order to generate a graph sampled uniformly at random from the set $G_{m,n}$, we simply throw down m edges between vertex pairs chosen at random from n initially unconnected vertices (see Figure 4.1).

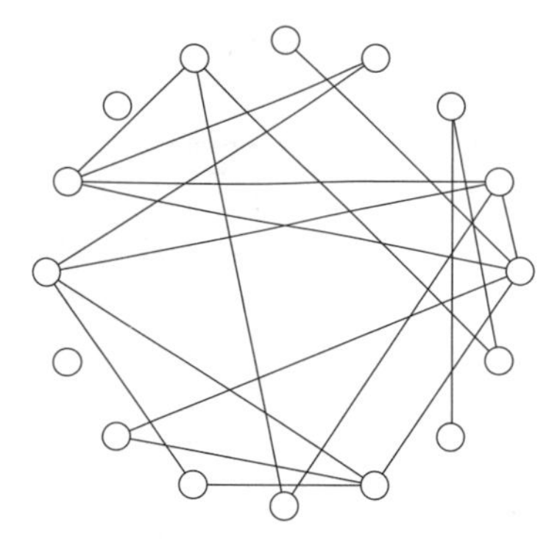

FIGURE 4.1 An example of a random graph with 16 vertices and 19 edges. The edges are placed between pairs of vertices chosen uniformly at random. The resulting degree distribution is binomial, or Poissonian in the limit of large graph size.

- $G_{n,p}$ is the set of all graphs consisting of n vertices, where each pair is connected together with independent probability p. In order to generate a graph sampled uniformly at random from the set $G_{n,p}$, we take n initially unconnected vertices and go through each pair of them, joining the pair with an edge with probability p, or not with probability $1 - p$.

Notice that it is in theory possible for graphs in $G_{n,m}$ to have more than one edge between any pair of vertices. However, we are usually interested in the case where the graphs in question are large, i.e., $n \to \infty$, in which case the probability of such a multiple edge appearing goes as m/n^2, which vanishes in the limit of large size as long as m increases more slowly than n^2. (Graphs for which this condition holds are said to be *sparse*.)

A physicist might think of the two random graph models as being "microcanonical" and "canonical" versions of the same model. In $G_{n,m}$ the number of edges is fixed, in a way similar to the fixing of the energy in the standard microcanonical ensemble of statistical mechanics. In $G_{n,p}$ the number of edges can fluctuate, but its average is fixed, similar to the fixing of the average energy in the canonical ensemble. A physicist might also guess that the average properties of the two ensembles of graphs would be similar in the thermodynamic limit, i.e., in the limit of large n, and indeed he or she would be correct. For our purposes (though not for a mathematician's more exacting ones) the two models are essentially equivalent in the limit of large n. Since $G_{n,p}$ is somewhat simpler to work with than $G_{n,m}$, we will concentrate our attention here primarily on $G_{n,p}$.

The most striking property of the ensemble of graphs $G_{n,p}$ is that it shows a phase transition at which a so-called *giant component* forms when the average degree z of a vertex is $z = np = 1$. The giant component is a subset of vertices each of which is reachable from the others along some path, and which contains a number of vertices which increases with the size of the graph as some positive power of n. (In fact, its size increases linearly with n except when we are precisely at the phase transition.) The phase transition is an ordinary continuous phase transition of the type physicists are familiar with from other systems such as the Ising model; the order parameter for the transition is the fraction S of the graph occupied by the giant component, which varies continuously from zero for $z \leq 1$ to nonzero for

$z > 1$. The transition is in the same universality class as the percolation transition in infinite dimension (or on a Bethe lattice) if the spanning cluster in a percolation system is equated with the giant component (Stauffer and Aharony 1992).

To clarify the basic behavior of the random graph $G_{n,p}$, consider the following simple argument. Imagine that we focus on a subset of vertices somewhere on our graph that are connected together, and we consider all the edges that lead away from that subset to other neighboring vertices. Now imagine that we follow one of these edges to the vertex at its other end and add that vertex to our subset. When we do this, the number of unexplored edges leading away from our subset goes down by one because we have explored one of them. However, it also goes up because the new vertex we find may have other edges leading out of it. Provided our subset of connected vertices is small compared with the size of the graph as a whole, the average number of new edges leading from our new vertex will be approximately $z = np$, and thus the average net change in the number of unexplored edges leading from our subset is $z - 1$.[1] If this average change is positive, $z - 1 > 0$, then our subset of connected vertices will (on average) continue to grow as we explore along more and more edges, leading to a giant component that ultimately fills a large fraction of the graph. Conversely, if it is less than zero, then the number of unexplored edges will dwindle to zero and the subset will stop growing at some finite size. The phase transition between the two regimes takes place when $z - 1 = 0$, or when $z = 1$. While this argument is not a rigorous one, it can be made rigorous, and the result that the transition occurs at $z = 1$ turns out to be exact.

Note that our argument assumes that all the vertices we find by following unexplored edges are unique: we don't encounter any vertex during the process that we have encountered before. If we did, we would overcount vertices and misestimate the rate at which our subset of vertices grows or shrinks. The assumption of uniqueness among vertices is equivalent to saying that the paths we follow through the graph from our initial subset contain no closed loops. This "no loops" criterion is the crucial feature that makes the properties of the random graph calculable. It is true in the limit of large graph size *except* for the giant component, which does contain loops. However, the result that the transition takes place at $z = 1$ still holds, since we can derive this result from considering only the behavior below the transition where there is no giant component.

We notice also that there is only ever one giant component in a random graph. We can see this by the following argument. Suppose there were two giant components containing N_1 and N_2 vertices each. Now consider any one edge in the graph. The probability that that edge's two ends do not join together the two giant components is $1 - 2N_1N_2/n^2$. Since there are $\frac{1}{2}n(n-1)p \simeq \frac{1}{2}n^2p$ edges in the graph in all, the probability that in the limit of large graph size there really are two giant components, in other words that the none of the edges join the two

[1] One might imagine that the average number of new edges leading from the vertex would be $z - 1$ rather than z, since one of the edges attached to the vertex is the one we arrived along and we shouldn't count that one as a new edge. This, however, would be a mistake. All the edges in a $G_{n,p}$ random graph are present with independent probability p, so that the number of new edges outgoing from the vertex we just arrived at is unaffected by the presence of the edge we arrived along—the presence or absence of different edges is entirely uncorrelated.

together, is

$$P_2 = \lim_{n\to\infty}\left[1 - \frac{2N_1N_2}{n^2}\right]^{\frac{1}{2}n^2p} = \mathrm{e}^{-N_1N_2p}, \tag{4.1}$$

and this probability clearly goes to zero as N_1 and N_2 become large for fixed p. In the more common case where p is not fixed, but $z = np$ is, then $P_2 \to 0$ as long as N_1N_2 grows faster than n. In fact, as we mentioned above, the size of the giant component usually grows as n, and hence this condition is easily satisfied and there cannot be two giant components. The only exception to the linear growth of the giant component is when we are precisely at the phase transition $z = 1$, where it can be shown that the giant component grows as $n^{2/3}$. Even in this case, however, $N_1N_2 \sim n^{4/3}$, which is still fast enough to make $P_2 \to 0$. Similar arguments show that there can never be three or more giant components either. We can only ever have one. Practical evidence of this is seen in many of the networks discussed in Chapter 3, where there is clearly only one large component.

Some other results about random graphs are reminiscent of behavior seen in real-world networks also (Bollobás 2001):

- Below the phase transition there is no giant component, only a large number of small components disconnected from one another. The average size of the component to which a randomly chosen vertex belongs is $1/(1-z)$, which is just a constant as the graph becomes large if we keep z constant.

- Above the phase transition the giant component fills a fraction S of the graph given by the larger solution to the transcendental equation $S = 1 - \mathrm{e}^{-zS}$. This equation has no closed-form solutions other than $S = 0$, but it is clear that S increases monotonically once we pass the threshold at $z = 1$. Above the transition, there are also many smaller components that fill the portion of the graph not occupied by the giant component. These have average size $1/(1-z+zS)$, which is again a constant independent of n if we keep z fixed.

The basic properties of the random graph are a good guide to the way real-world networks behave. The existence (or not) of a giant component, the small non-giant components, the phase transition, and so forth all seem to be typical of many networks. However, there is one glaring problem with the random graph as a model for practical networks: the degree distribution.

As emphasized in many of the papers reproduced in Chapters 2 and 3, the degree distribution is an important property of networks **(Price 1965; Albert *et al.* 1999; Faloutsos *et al.* 1999; Broder *et al.* 2000; Newman 2001a; Liljeros *et al.* 2001)**. Many networks are found to have a power-law degree distribution (scale-free networks), while others have truncated power laws, exponentials, or strongly peaked distributions **(Amaral *et al.* 2000)**. The properties and behavior of a network are affected in many ways by its degree distribution **(Albert *et al.* 2000; Cohen *et al.* 2000; Callaway *et al.* 2000)**.

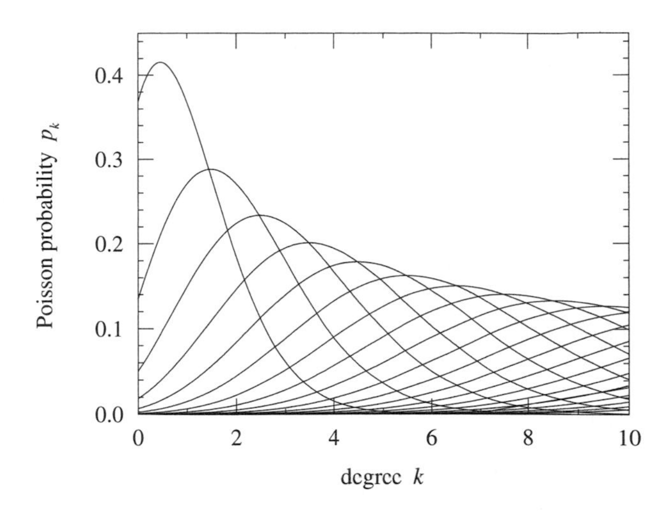

FIGURE 4.2 The Poisson probability distribution $p_k = z^k e^{-z}/k!$ as a function of k for $z = 1, 2, 3, \ldots$ For each curve the distribution has a clear peak close to $k = z$, followed by a rapidly decaying tail.

The random graph $G_{n,p}$ has a binomial degree distribution. The probability p_k that a randomly chosen vertex is connected to exactly k others is

$$p_k = \binom{n}{k} p^k (1-p)^{n-k}. \tag{4.2}$$

In the limit where n becomes large, this becomes

$$p_k = \lim_{n\to\infty} \frac{n^k}{k!}\left(\frac{p}{1-p}\right)^k (1-p)^n = \frac{z^k e^{-z}}{k!}, \tag{4.3}$$

which is the *Poisson distribution*. In Figure 4.2 we show examples of the Poisson distribution as a function of k for a variety of values of the mean degree z. As the figure shows, they are all sharply peaked, and they have a tail that decays as $1/k!$, which is considerably quicker than any exponential. This distribution is very far from the degree distributions we see in most of the networks of Chapter 3. Certainly it is entirely unlike a power-law distribution with its "fat tail" of vertices with very high degree. The Poisson distribution of degrees is a fundamental problem with the simple random graph when it comes to modeling real-world networks. To rectify this problem, a number of authors have turned to other random graph models with different degree distributions.

It is possible to define random graphs with any degree distribution we please. This development has been discussed by a number of different authors (Bender and Canfield 1978; Łuczak 1992; **Molloy and Reed 1995**). In fact, these authors considered graphs with a given degree *sequence* rather than degree distribution, because the former are better defined mathematically. A degree sequence is a specified set of degrees $k_1, k_2, k_3, \ldots$ which the vertices $1, 2, 3, \ldots$ must take. A degree distribution, on the other hand, merely stipulates on average what fraction of vertices should have particular degrees. The distinction between the two is not of great importance to us, however, since one can simply take the degree distribution one is interested in and draw a degree sequence from it and then use that sequence to create a network.

We consider then the set of all graphs of n vertices in which the vertices have the specified degrees $\{k_i\}$, with $i = 1, \ldots, n$. By calculating properties averaged uniformly over this ensemble of graphs, we can get an idea of how the degree sequence (or distribution) affects the properties of a graph. But before we do this, we need to consider how one can generate graphs drawn uniformly from this ensemble. **Molloy and Reed (1995)** suggest the following algorithm:

i) Create a list in which the label i of each vertex appears exactly k_i times.

ii) Pair up elements from this list uniformly at random until none remain.

iii) Add an edge to the graph joining the two vertices in each pair.

Provided that the sum $\sum_i k_i$ of the degrees of all vertices is even, this procedure can always be completed. (If it is not even, no graph exists with the given set of degrees, and we must throw them away and start again.) However, it is not clear that the procedure gives precisely the graph that we want. Note that there is usually more than one way to generate a given graph using this procedure. If, for example, vertex 1 has degree 3 and vertex 2 has degree 2, then there are $3 \times 2 = 6$ different ways in which they might be connected together. In general there are $k_i k_j$ ways of connecting vertices i and j, and this number is higher for high-degree vertices, implying that such vertices are more likely to be connected to one another.

In fact, however, this turns out not to be a problem. It takes only a moment to see that the number of different ways of deriving a particular pattern of connections from the pairing procedure of Molloy and Reed is $\prod_i k_i!$, which is a constant for a given degree sequence. Thus it is true that there are many ways of generating each pattern of connections, but the number of such ways is the same for all patterns, and hence the procedure does indeed uniformly sample the ensemble of graphs with the given degree sequence. An exception to this rule is the case of graphs in which there is more than one edge between any pair of vertices. In that case, permuting those edges does not produce a new graph, and the number of ways of generating such graphs is smaller. Normally this does not matter, since we are interested only in the properties of networks in the limit of large system size, in which limit the probability of having two edges between the same two vertices tends to zero. However, if we are interested in networks of finite size there are corrections due to this effect, and they may not be negligible, even for quite large networks (Park and Newman 2003; Maslov *et al.* 2004).

In this section we present three theoretical papers on random graphs with arbitrary degree sequences. The first **(Molloy and Reed 1995)** gives a rigorous treatment of the phase transition in these graphs. The second **(Aiello *et al.* 2000)** applies the results of Molloy and Reed to graphs with power-law degree distributions similar to those seen in many of the networks of Chapter 3. The third **(Newman**

***et al.* 2001)** gives a less rigorous physics-style derivation of the properties of graphs with general degree distributions, including a number of new results, particularly for directed and bipartite graphs.

Molloy and Reed (1995)

This paper is an example of a mathematically rigorous treatment of random graphs with arbitrary degree sequences. In it, Molloy and Reed show that these graphs possess a phase transition at which a giant component appears, just as in the standard Poisson random graph. In the notation used here, their primary result is that with probability 1 the giant component exists in the thermodynamic limit if and only if

$$\frac{1}{n}\sum_{i=1}^{n} k_i(k_i - 2) > 0. \tag{4.4}$$

The point at which the left-hand side crosses zero marks the phase transition for the graph. Molloy and Reed's paper is quite technical, although for the benefit of the reader not interested in the fine print they provide a less technical outline of their methods and results in Section 1. They also give a simple and nonrigorous heuristic argument to explain Eq. (4.4), which can be paraphrased as follows.

Imagine again, as we did for the standard $G_{n,p}$ random graph model, that we focus on a connected subset of vertices somewhere on our graph. As before, we consider the neighboring vertices that can be reached by following a single edge from one of the vertices in our subset. When we explore one of these vertices by following a single edge, the number of unexplored neighboring vertices of our subset decreases by one. But we also gain some new unexplored vertices—those that are neighbors of the vertex we just found. If the vertex we just found has degree k, then there are $k - 1$ such new vertices. (One of its edges is the one we arrived along, which does not lead to a new vertex.[2]) Thus, the net increase in the number of unexplored neighboring vertices is $k - 2$. However, there is another important bias that we must take into account: the probability that upon following an edge we arrive at a vertex with degree k is proportional to k, since there are k times as many ways of arriving at such a vertex than there are of arriving at a vertex of degree 1. Thus, the average net increase in the number of unexplored neighbors when we add a new vertex to our subset of vertices is $\sum_i k_i(k_i - 2)/\sum_i k_i$. And this quantity is proportional to the left-hand side of Eq. (4.4). If this number is greater than zero, then the number of unexplored vertices increases in size at every step, and our subset will grow to form the nucleus of a giant component, just as in the normal random graph. If it is less than zero, then the number of unexplored vertices will dwindle until it reaches zero and our subset grows no more, and thus a giant component will never form.

[2]This might appear to be at odds with our previous contention for the Poisson random graph that we can ignore the presence of the edge we arrived along. That, however, was a special result that applies only for the Poisson case, with its independently chosen edges. In graphs with a given degree sequence, the edges are no longer independent, and we must count them with more care.

It is interesting to consider why vertices of degree 0 and vertices of degree 2 do not contributed to the sum in Eq. (4.4). No matter how many vertices of degree 0 or 2 we add to our graph, it makes no difference to the position of the phase transition. Why is this? For vertices of degree 0 the answer is obvious: such vertices are not connected to anything, so adding them to the graph makes no difference to whether there is a giant component. Vertices of degree 2 are a little more subtle. A vertex of degree 2 has two edges emerging from it, or alternatively it can be thought of as sitting in the middle of an edge between two other vertices. Adding vertices of degree 2 to a graph is therefore equivalent to adding extra vertices to the middle of edges that already exist. However, such vertices clearly have no effect on whether the graph contains a giant component, so Eq. (4.4) correctly ignores them.

Molloy and Reed wrote a follow-up paper that is not reproduced in this book (Molloy and Reed 1998). We encourage the reader who finds their 1995 paper useful to look at this second paper also. In it, the authors extend their rigorous treatment to the calculation of the size of the giant component in a graph with an arbitrary degree sequence.

Aiello, Chung, and Lu (2000)

The work of Molloy and Reed is about graphs as mathematical entities. Their results, however, have been applied to real-world networks by William Aiello and collaborators (**Aiello *et al.* 2000**). Motivated by the power-law degree distributions seen in the Web (**Albert *et al.* 1999**; Kleinberg *et al.* 1999; **Broder *et al.* 2000**), the Internet (**Faloutsos *et al.* 1999**; Chen *et al.* 2002), and call graphs of telephone calls (Abello and Buchsbaum, unpublished), they asked what happens if one applies the general results of Molloy and Reed to the specific case of a power-law degree sequence. They consider graphs in which the number n_k of vertices with degree k obeys

$$\log n_k = \alpha - \beta \log k, \tag{4.5}$$

with α and β being constants. No vertices are allowed to have degree zero in this model.

In our notation, Eq. (4.5) is equivalent to a power-law degree distribution of the form

$$p_k = \frac{\mathrm{e}^{\alpha}}{n} k^{-\beta}. \tag{4.6}$$

Since this distribution must be normalized, $\sum_k p_k = 1$, we can sum both sides to show that the constant α is related to the graph size n according to $n = \mathrm{e}^{\alpha}\zeta(\beta)$, where $\zeta(\beta)$ is the Riemann ζ-function. Because the distribution is a power law, with a long tail, some of the moments of the distribution (4.6), and possibly all of them, will diverge. To prevent this and make the calculations more realistic, Aiello *et al.* truncated the distribution at the point where $n_k = 1$, which is the point $k = \mathrm{e}^{\alpha/\beta}$. No vertices are allowed to have degree greater than this.

The principal findings reported in the paper are that the behavior of power-law graphs of this type has three general regimes as a function of the exponent β. If $\beta < 1$, there is a giant component in the graph, and that giant component contains a fraction of the vertices which tends to 1 as the graph becomes large. For $1 \leq \beta < \beta_c$ there is also a giant component, whose size scales linearly with the size of the system but which no longer fills the whole graph. The critical value β_c is given implicitly by the equation $\zeta(\beta_c - 2) = 2\zeta(\beta_c - 1)$, and is found numerically to be about $\beta_c = 3.47875\ldots$. For $\beta \geq \beta_c$ there is no giant component, only a large number of smaller components. Aiello *et al.* also show that the interval $1 \leq \beta < \beta_c$ divides further into the portion below $\beta = 2$, where the components other than the giant component have an average size that is a constant as $n \to \infty$, and the portion above $\beta = 2$, where the average size of non-giant components increases as $\log n$ as the graph becomes large.

Aiello *et al.* suggest a possible application of their results to call graphs of telephone calls. These are networks in which the vertices represent telephone numbers and the (directed) edges represent calls from one number to another during a specified period of time. Aiello *et al.* use data compiled by Abello and Buchsbaum (unpublished) for telephone calls occurring during a single day. (Presumably the calls in question were restricted to a certain geographical region also, although this is not discussed in the paper.) They deliberately exclude from the graph all telephone numbers that did not originate or receive a call during the day in question, so that there are no vertices with degree zero, in keeping with the model. They also ignore the directed nature of the edges. The resulting graph has about 70 million vertices in it.

Constructing a histogram of vertex degrees in this call graph, Aiello *et al.* find that the degree distribution is roughly power-law in form with an exponent $\beta \simeq 2.1$. This puts the graph in the regime in which it is expected to possess a giant component of size order n, plus a large number of smaller components of typical size order $\log n$. And roughly speaking this appears to be true. The largest component appears to fill nearly all of the network, while the next largest has size around 200—much smaller but not of order unity, and plausibly therefore of order $\log n$. Thus, the results appear to suggest that the real-world call graph has properties statistically similar to those of a random graph with the same degree distribution. This is somewhat surprising; one might imagine that the calling patterns of real human beings are anything but random. Viewed on a very large scale however, the work of Aiello *et al.* suggests that this is not in fact the case.

Newman, Strogatz, and Watts (2001)

The third paper in this section on random graphs comes from a physics rather than a mathematics journal, and has a distinctly physics flavor. In this paper, the authors show that it is possible to derive many of the properties of random graphs with arbitrary degree distributions using the mathematics of probability generating functions (Wilf 1994). Given a degree distribution p_k, they define the probability generating function

$$G_0(x) = \sum_k p_k x^k, \tag{4.7}$$

which encapsulates the entire distribution in a single function. They then show that, by considering suitable combinations of generating functions, many properties of the graph, including component sizes and similar quantities, can be written in closed form in terms of various fixed points or derivatives of G_0. While these derivations are not as rigorous of those of **Molloy and Reed (1995**, 1998), they have the advantage of being simple and relatively easy to understand. Molloy and Reed's results about the position of the phase transition and the size of the giant component, for instance, are rederived in just a few lines. The condition for the position of the transition is expressed simply as $G_0''(1) = G_0'(1)$ in the generating function language. The results of **Aiello *et al.* (2000)** can also be rederived in this way (Newman *et al.* 2002c).

Because of the relative simplicity of the method, the authors were able to extend their calculations to derive a variety of additional results that were not previously known. For example, they derived exact expressions for the average size of a component below the phase transition, the average size of components other than the giant component above the transition, and the average number of vertices a given distance away from a randomly chosen vertex. They also extended their calculations to properties of directed graphs and bipartite graphs. The directed graphs might be useful as models of, for example, the Web, the telephone call graphs studied by Aiello *et al.* (see above), or food webs of predator-prey interactions. Bipartite graphs can be used as models of collaboration networks, such as the networks of scientists and movie stars discussed in Chapter 3.

Like Aiello *et al.*, Newman *et al.* gave some examples of applications of their results to real-world networks. For instance, they constructed bipartite random graphs with the same degree distributions as known collaboration networks of physicists, biologists, and movie stars. In each of these cases, the models predicted the statistical properties of the corresponding real-world networks to reasonable accuracy, but were typically off by a factor of two or so when examined in detail. The authors suggested that this discrepancy was due to social influences on the formation of the networks which are not taken into account in a simple random graph model. However, for one network they looked at—the network of the boards of directors of the Fortune 1000 companies (the 1000 U.S. companies with the largest revenues)—they found startlingly good agreement between theory and empirical results, at least for some measures, suggesting perhaps that in this case the assumption of randomness is a reasonable one. They did not, however, propose an explanation for why this should be so.

Newman *et al.* also applied their directed random graph model to modeling the World Wide Web. However, they were hampered in this case by lack of appropriate data. Their model requires a knowledge of the joint degree distribution p_{jk} of pages on the Web—the probability that a page simultaneously has in-degree j and out-degree k. This distribution does not appear to have been measured; studies of the Web so far have only measured the in- and out-degree distributions separately. This makes the results for the Web model somewhat inconclusive. In more recent work, Camacho *et al.* (2002) have applied the random directed graph model to modeling food webs, with some apparent success. Generating function methods have also been applied to a number of other questions, particularly network robustness (**Callaway *et al.* 2000**; Cohen *et al.* 2001), and epidemiology (Newman 2002b). Some of these applications are discussed later in this book.

A Critical Point for Random Graphs with a Given Degree Sequence

Michael Molloy
Department of Mathematics, Carnegie Mellon University, Pittsburgh, PA 15213

Bruce Reed
Equipe Combinatoire, CNRS, Université Pierre et Marie Curie, Paris, France

ABSTRACT

Given a sequence of nonnegative real numbers $\lambda_0, \lambda_1, \ldots$ which sum to 1, we consider random graphs having approximately $\lambda_i n$ vertices of degree i. Essentially, we show that if $\Sigma\, i(i-2)\lambda_i > 0$, then such graphs almost surely have a giant component, while if $\Sigma\, i(i-2)\lambda_i < 0$, then almost surely all components in such graphs are small. We can apply these results to $G_{n,p}$, $G_{n,M}$, and other well-known models of random graphs. There are also applications related to the chromatic number of sparse random graphs. © 1995 John Wiley & Sons, Inc.

1. INTRODUCTION AND OVERVIEW

In this paper we consider two parameters of certain random graphs: the number of vertices and the number of cycles in the largest component. Of course, the behavior of these parameters depends on the probability distribution from which the graphs are picked. In one standard model we pick a random graph $G_{n,M}$ with n vertices and M edges where each graph is equally likely. We are interested in what happens when we choose M as a function of n and let n go to infinity. The point $M = \frac{1}{2}n$ is referred to as the *critical point* or the *double-jump threshold* because of classical results due to Erdös and Rényi [8] concerning the dramatic changes which occur to these parameters at this point. If $M = cn + o(n)$ for $c < \frac{1}{2}$, then almost surely (i.e., with probability tending to 1 as n tends to infinity) $G_{n,M}$

Random Structures and Algorithms, Vol. 6, Nos. 2 and 3 (1995) has no component of size greater than $O(\log n)$, and no component has more than one cycle. If $M = \frac{1}{2}n + o(n)$, then almost surely (a.s.) the largest component of $G_{n,M}$ has size $\Theta(n^{2/3})$. If $M = cn$ for $c > \frac{1}{2}$, then there are constants $\epsilon, \delta > 0$ dependent on c such that a.s. $G_{n,M}$ has a component on at least ϵn vertices with at least δn cycles, and no other component has more than $O(\log n)$ vertices or more than one cycle. This component is referred to as the *giant component* of $G_{n,M}$. For more specifics on these two parameters at and around $M = \frac{1}{2}n$, see [3], [11], or [14].

In this paper, we are interested in random graphs with a fixed degree sequence where each graph with that degree sequence is chosen with equal probability. Of course, we have to say what we mean by a degree sequence. If the number of vertices in our graph, n, is fixed, then a degree sequence is simply a sequence of n numbers. However, we are concerned here with what happens asymptotically as n tends to infinity, so we have to look at a "sequence of sequences." Thus, we generalize the definition of degree sequence:

Definition. *An* asymptotic degree sequence *is a sequence of integer-valued functions* $\mathscr{D} = d_0(n), d_1(n), \ldots$ *such that*

1. $d_i(n) = 0$ *for* $i \geq n$;
2. $\sum_{i \geq 0} d_i(n) = n$.

Given an asymptotic degree sequence $\mathscr{D}$, we set $\mathscr{D}_n$ to be the degree sequence $\{c_1, c_2, \ldots, c_n\}$, where $c_j \geq c_{j+1}$ and $|\{j : c_j = i\}| = d_i(n)$ for each $i \geq 0$. Define $\Omega_{\mathscr{D}_n}$ to be the set of all graphs with vertex set $[n]$ with degree sequence $\mathscr{D}_n$. A random graph on n vertices with degree sequence $\mathscr{D}$ is a uniformly random member of $\Omega_{\mathscr{D}_n}$.

Definition. *An asymptotic degree sequence* $\mathscr{D}$ *is* feasible *if* $\Omega_{\mathscr{D}_n} \neq \emptyset$ *for all* $n \geq 1$.

In this paper, we will only discuss feasible degree sequences.

Because we wish to discuss asymptotic properties of random graphs with degree sequence $\mathscr{D}$, we want the sequences $\mathscr{D}_n$ to be in some sense similar. We do this by insisting that for any fixed i, the proportion of vertices of degree i is roughly the same in each sequence.

Definition. *An asymptotic degree sequence* $\mathscr{D}$ *is* smooth *if there exist constants* λ_i *such that* $\lim_{n\to\infty} d_i(n)/n = \lambda_i$.

Throughout this paper, all asymptotics will be taken as n tends to ∞, and we only claim things to be true for sufficiently large n.

In the past, the most commonly studied random graphs of this type have been random regular graphs. Perhaps the most important recent result is by Robinson and Wormald [19, 20], who proved that if G is a random k-regular graph for any constant $k \geq 3$, then G is a.s. Hamiltonian.

Another motivation for studying random graphs on a fixed degree sequence comes from the analysis of the chromatic number of sparse random graphs. This is because a minimally $(r+1)$-chromatic graph must have minimum degree at least

r. In an attempt to determine how many edges were necessary to force a random graph to a.s. be not 3-colorable, Chvátal [7] studied the expected number of subgraphs of minimum degree 3 in random graphs with a linear number of edges. He showed that for $c < c^* = 1.442\ldots$, the expected number of such subgraphs in $G_{n,M=cn}$ is exponentially small, while for $c > c^*$ the expected number of such subgraphs in $G_{n,M=cn}$ is exponentially large. In the work that motivated the results of this paper, the authors used a special case of the main theorem of this paper to show that the probability that a random graph on n vertices with minimum degree three and at most $1.793n$ edges is minimally 4-chromatic is exponentially small [18]. We used this to show that, for c a little bit bigger than c^*, the expected number of minimally 4-chromatic subgraphs of $G_{n,M=cn}$ is exponentially small. This suggests that determining the minimum value of c for which a random graph with cn edges is a.s. 4-chromatic may require more than a study of the subgraphs with minimum degree 3.

Recently Łuczak [14] showed (among other things) that if G is a random graph on a fixed degree sequence*, with no vertices of degree less than 2, and at least $\Theta(n)$ vertices of degree greater than 2, then G a.s. has a unique giant component. Our main theorem also generalizes this result.

We set $Q(\mathscr{D}) = \Sigma_{i \geq 1} i(i-2)\lambda_i$. Essentially, if $Q(\mathscr{D}) > 0$, then a random graph with degree sequence $\mathscr{D}$ a.s. has a giant component, while if $Q(\mathscr{D}) < 0$, then all the components of such a random graph are a.s. quite small. Note how closely this parallels the phenomenon in the more standard model $G_{n,M}$.

Note further that these results allow us to determine a similar threshold for any model of random graphs as long as: (i) We can determine the degree sequence of graphs in the model with reasonable accuracy, and (ii) once the degree sequence is determined, every graph on that degree sequence is equally likely. $G_{n,p}$ is such a model, and thus (as we see later), our results can be used to verify the previously known threshold for $G_{n,p}$.

Before defining the parameter precisely, we give an intuitive explanation of why it determines whether or not a giant component exists. Suppose that $\mathscr{D}_n$ has $(\lambda_i + o(1))n$ vertices of degree i for each $i \geq 0$. Pick a random vertex in our graph and expose the component in which it lies using a branching process. In other words, expose its neighbors, and then the neighbors of its neighbors, repeating until the entire component is exposed. Now when a vertex of degree i is exposed, then the number of "unknown" neighbors increases by $i-2$. The probability that a certain vertex is selected as a neighbor is proportional to its degree. Therefore, the expected increase in the number of unknown neighbors is (roughly) $\Sigma_{i \geq 1} i(i-2)\lambda_i$. This is, of course, $Q(\mathscr{D})$.

Thus, if $Q(\mathscr{D})$ is negative, then the component will a.s. be exposed very quickly. However, if it is positive then the number of unknown neighbors, and thus the size of the component, might grow quite large. This gives the main thrust of our arguments. We will now begin to state all of this more formally.

There are a few caveats, so in order for our results to hold, we must insist that the asymptotic degree sequences we consider are well behaved. In particular, when the maximum degree in our degree sequence grows with n, we can run into some problems if things do not converge uniformly. For example, if $d_1(n) = n -$

* He did not use the asymptotic degree sequence introduced here, but the results translate.

$\lceil n^{.9} \rceil$, $d_i(n) = \lceil n^{.9} \rceil$ if $i = \lceil \sqrt{n} \rceil$, and $d_i(n) = 0$ otherwise, then $\lambda_1 = 1$, and $\lambda_i = 0$ for $i > 1$, and we get $Q(\mathcal{D}) = -1$. However, this is deceiving as there are enough vertices of degree $\sqrt{n}$ to ensure that a giant component containing $n - o(n)$ vertices a.s. exists.

Definition. *An asymptotic degree sequence $\mathcal{D}$ is* well-behaved *if:*

1. *$\mathcal{D}$ is feasible and smooth.*
2. *$i(i-2)d_i(n)/n$ tends uniformly to $i(i-2)\lambda_i$; i.e., for all $\epsilon > 0$ there exists N such that for all $n > N$ and for all $i \geq 0$.*

$$\left| \frac{i(i-2)d_i(n)}{n} - i(i-2)\lambda_i \right| < \epsilon .$$

3. $$L(\mathcal{D}) = \lim_{n\to\infty} \sum_{i\geq 1} i(i-2)d_i(n)/n$$

exists, and the sum approaches the limit uniformly; i.e.:

(a) If $L(\mathcal{D})$ is finite then for all $\epsilon > 0$ there exists i^*, N such that for all $n > N$:

$$\left| \sum_{i=1}^{i^*} i(i-2)d_i(n)/n - L(\mathcal{D}) \right| < \epsilon .$$

(b) If $L(\mathcal{D})$ is finite, then, for all $T > 0$, there exists i^*, N such that for all $n > N$

$$\sum_{i=1}^{i^*} i(i-2)d_i(n)/n > T .$$

We note that it is an easy exercise to show that if $\mathcal{D}$ is well behaved, then

$$L(\mathcal{D}) = Q(\mathcal{D}) .$$

It is not surprising that the threshold occurs when there are a linear number of edges in our degree sequence. We define such a degree sequence as sparse:

Definition. *An asymptotic degree sequence $\mathcal{D}$ is* sparse *if $\Sigma_{i\geq 0}\, id_i(n)/n = K + o(1)$ for some constant K.*

Note that for a well-behaved asymptotic degree sequence $\mathcal{D}$, if $Q(\mathcal{D})$ is finite, then $\mathcal{D}$ *is sparse.*

The main result in this paper is the following:

Theorem 1. *Let $\mathcal{D} = d_0(n), d_1(n), \ldots$ be a well-behaved sparse asymptotic degree sequence for which there exists $\epsilon > 0$ such that for all n and $i > n^{1/4-\epsilon}$, $d_i(n) = 0$. Let G be a graph with n vertices, $d_i(n)$ of which have degree i, chosen uniformly at random from among all such graphs. Then:*

a. *If $Q(\mathcal{D}) > 0$ then there exist constants $\zeta_1, \zeta_2 > 0$ dependent on $\mathcal{D}$ such that G a.s. has a component with at least $\zeta_1 n$ vertices and $\zeta_2 n$ cycles. Furthermore, if*

$Q(\mathscr{D})$ is finite, then G a.s. has exactly one component of size greater than $\gamma \log n$ for some constant γ dependent on $\mathscr{D}$.

b. *If $Q(\mathscr{D})<0$ and for some function $0 \leq \omega(n) \leq n^{1/8-\epsilon}$, $d_i(n)=0$ for all $i \geq \omega(n)$, then, for some constant R dependent on $Q(\mathscr{D})$, G a.s. has no component with at least $Rw(n)^2 \log n$ vertices, and a.s. has fewer than 2 $Rw(n)^2 \log n$ cycles. Also, a.s. no component of G has more than one cycle.*

Consistent with the model $G_{n,M}$, we call the component referred to in Theorem 1a a *giant component*.

Note that if $Q(\mathscr{D})<0$, then $Q(\mathscr{D})$ is finite. Note also that Theorem 1 fails to cover the case where $Q(\mathscr{D})=0$. This is analogous to the case $M=\frac{1}{2}n+o(n)$ in the model $G_{n,M}$, and would be interesting to analyze.

One immediate application of Theorem 1 is that if G is a random graph on a fixed well-behaved degree sequence with $cn+o(n)$ edges for any $c>1$ then G a.s. has a giant component, as there is no solution to

$$\sum_{i \geq 1} i\lambda_i > 2\,, \qquad \sum_{i \geq 1} i(i-2)\lambda_i < 0\,, \qquad \sum_{i \geq 1} \lambda_i = 1\,, \qquad 0 \leq \lambda_i \leq 1\,.$$

A major difficulty in the study of random graphs on fixed degree sequences is that it is difficult to generate such graphs directly. Instead it has become standard to study random configurations on a fixed degree sequence, and use some lemmas which allow us to translate results from one model to the other. The configuration model was introduced by Bender and Canfield [2] and refined by Bollobás [3] and also Wormald [21].

In order to generate a random configuration with n vertices and a fixed degree sequence, we do the following:

1. Form a set L containing $\deg(v)$ distinct copies of each vertex v.
2. Choose a random matching of the elements of L.

Each configuration represents an underlying multigraph whose edges are defined by the pairs in the matching. We say that a configuration has a graphical property P if its underlying multigraph does.

Using the main result in [17], it follows that the underlying multigraph of a random configuration on a degree sequence meeting the conditions of Theorem 1 is simple with probability tending to $e^{-\lambda(\mathscr{D})}$, for some $\lambda(\mathscr{D})<O(n^{1/2-\epsilon})$. The condition $d_i(n)=0$ for all $i>n^{1/4-\epsilon}$ is needed to apply this result. If $Q(\mathscr{D})$ is finite, then $\lambda(\mathscr{D})$ tends to a constant.

Also, any simple graph G can be represented by $\Pi_{v \in V(G)} \deg(v)!$ configurations, which is clearly equal for all graphs on the same degree sequence and the same number of vertices.

This gives us the following very useful lemmas:

Lemma 1. *If a random configuration with a given degree sequence $\mathscr{D}$ meeting the conditions of Theorem 1 [with $Q(\mathscr{D})$ possibly unbounded] has a property P with probability at least $1-z^n$ for some constant $z<1$, then a random graph with the same degree sequence a.s. has P.*

Lemma 2. *If a random configuration with a given degree sequence $\mathscr{D}$ meeting the conditions of Theorem 1 a.s. has a property P, and if $Q(\mathscr{D}) < \infty$, then a random graph with the same degree sequence a.s. has P.*

Using these lemmas, it will be enough to prove Theorem 1 for a random configuration.

The configuration model is very similar to the pseudograph model developed independently by Bollobás and Frieze [5], Flajolet, Knuth, and Pittel [10], and Chvátal [7]. Both models are very useful when working with random graphs on a given degree sequence.

Having defined the precise objects that we are interested in, and the model in which we are studying them, we can now give a more formal overview of the proof. The remainder of this section is devoted to this overview. In the following two sections we give all the details of the proof. In Section 4, we see some applications of Theorem 1: the aforementioned work concerning the chromatic number of sparse random graphs, and a new proof of a classical double-jump theorem, showing that this work generalizes that result. A reader who is not interested in the details of the proof might want to just finish this section and then skip ahead to the last one.

In order to examine the components of our random configuration, we will be more specific regarding the order in which we expose the pairs of the random matching. Given $\mathscr{D}$, we will expose a random configuration F on n vertices, $d_i(n)$ of which have degree i as follows:

At each step, a vertex all of whose copies are in exposed pairs is *entirely exposed*. A vertex some but not all of whose copies are in exposed pairs is *partially exposed*. All other vertices are *unexposed*. The copies of partially exposed vertices which are not in exposed pairs are *open*.

1. Form a set of L consisting of i distinct copies of each of the $d_i(n)$ vertices which have degree i.
2. Repeat until L is empty:
   a. Expose a pair of F by first choosing any member of L, and then choosing its partner at random. Remove them from L.
   b. Repeat until there are no partially exposed vertices:

      Choose an open copy of a partially exposed vertex, and pair it with another randomly chosen member of L. Remove them both from L.

All random choices are made uniformly.

Essentially we are exposing the random configuration one component at a time. When any component is completely exposed, we move on to a new one; i.e., we repeat step 2a.

It is clear that every possible matching among the vertex-copies occurs with the same probability under this procedure, and hence this is a valid way to choose a random configuration.

Note that we have complete freedom as to which vertex we pick in Step 2a. In a few places in this paper, it will be important that we take advantage of this freedom, but in most cases we will pick it randomly in the same manner in which we pick all the other vertex-copies, i.e., unless we state otherwise, we will always just pick a uniformly random member of L.

Now, let X_i represent the number of open vertex-copies after the ith pair is exposed. If the neighbor of v chosen in step 2b is of degree d, then X_i goes up by $d-2$. Each time a component is completely exposed and we repeat step 2a, if the pair exposed in step 2a involves vertices of degree d_1 and d_2, then X_i is set to a value of d_1+d_2-2.

Note that if the number of vertex-copies in L which are copies of vertices of degree d is r_d, then the probability that we pick a copy of a vertex of degree d in step 2b is $r_d/\Sigma_{i\geq 1} r_i$. Therefore, initially the expected change in X_i is approximately

$$\frac{\sum_{i\geq 1} i(i-2)d_i(n)}{\sum_{j\geq 1} jd_j(n)} = \frac{Q(\mathscr{D})}{K}.$$

Therefore, at least initially, if this value is positive then X_i follows a Markov process very close in distribution to the well-studied "drunkard's walk," with an expected change of $Q(\mathscr{D})/K$. Since $X_{i+1} \geq X_i - 1$ always, a standard result of random walk theory (see, for example, [9]) implies that if $Q(\mathscr{D})>0$, then after $\Theta(n)$ steps, X_i is a.s. of order $\Theta(n)$.

It follows that our random configuration a.s. has at least one component on $\Theta(n)$ vertices. We will see that such a component a.s. has at least $\Theta(n)$ cycles in it, and this will give us the first part of Theorem 1. We will also see that if $Q(\mathscr{D})$ is bounded, then this giant component is a.s. unique.

On the other hand, if $Q(\mathscr{D})<0$, then X_i a.s. returns to zero fairly quickly, and this will give us the other part of Theorem 1, as the sizes of the components of F are bounded above by the distances between values of i such that $X_i=0$.

Of course, the random walk followed by X_i is not really as simple as this. There are three major complications:

1. A pure random walk can drop below 0. Whenever X_i reaches 0, it resets itself to a positive number.
2. We neglected to consider that the second vertex-copy chosen in Step 2b might be an open vertex-copy in which case X_i decreases by 2. We will call such a pair of vertex-copies a *backedge*.
3. As more and more vertices are exposed, the ratio of the members of L which are copies of vertices of degree d shifts, and the expected increase of X_i changes.

These complications are handled as follows:

1. This will increase the probability of X_i growing large, and so this only poses a potential problem in proving part (b). In this case, we will show that the probability of a component growing too big is of order $o(n^{-1})$, and hence even if we "try again" n times, this will a.s. never happen.
2. We will see that this a.s. doesn't happen often enough to pose a serious problem, unless the partially exposed component is already of size $\Theta(n)$.

3. In proving part a, we look at our component at a time when the expected increase in X_i is still at least $\frac{1}{2}$ its original value. We will see that the component being exposed at this point is a.s. a giant component. In proving part b, it is enough to consider the configuration after $o(n)$ steps. At this point, the expected increase hasn't changed significantly.

This is a rough outline of the proof. We will fill in the details in the next two sections.

2. GRAPHS WITH NO LARGE COMPONENTS

In this section we will prove that the analogue of Theorem 1b holds for random configurations. Lemma 1 will then imply that it holds for random graphs. We will first prove that if F is a random configuration meeting the conditions given in Theorem 1b, then F a.s. does not have any large components.

Given $Q(\mathscr{D})<0$, set $\nu=-Q(\mathscr{D})/K$ and set $R=150/\nu^2$.

Lemma 3. *Let F be a random configuration with n vertices and degree sequence $\mathscr{D}_n$ meeting the conditions of Theorem 1. If $Q(\mathscr{D})<0$ and if, for some function $0\le\omega(n)\le n^{1/8-\epsilon}$, F has no vertices of degree greater than $\omega(n)$, then F a.s. has no components with more than $\alpha=\lceil R\omega(n)^2\log n\rceil$ vertices.*

The following theorem of Azuma will play an important role:

Azuma's Inequality [1]. *Let $0=X_0,\ldots,X_n$ be a martingale with*

$$|X_{i+1}-X_i|\le 1$$

for all $0\le i<n$. Let $\lambda>0$ be arbitrary. Then

$$\Pr[|X_n|>\lambda\sqrt{n}]<e^{-\lambda^2/2}.$$

This yields the following very useful standard corollary.

Corollary. *Let $\Sigma=\Sigma_1,\Sigma_2,\ldots,\Sigma_n$ be a sequence of random events. Let $f(\Sigma)=f(\Sigma_1,\Sigma_2,\ldots,\Sigma_n)$ be a random variable defined by these Σ_i. If for each i*

$$\max|E(f(\Sigma)\mid\Sigma_1,\Sigma_2,\ldots,\Sigma_{i+1})-E(f(\Sigma)\mid\Sigma_1,\Sigma_2,\ldots,\Sigma_i)|\le c_i,$$

where $E(f)$ denotes the expected value of f, then the probability that $|f-E(f)|>t$ is at most

$$2\exp\left(\frac{-t^2}{2\sum c_i^2}\right).$$

For more details on this corollary and an excellent discussion of martingale arguments see either [16] or [5].

In order to prove Lemma 3, we will analyze the Markov process described in Section 1. Recall that X_i is the number of open vertex-copies after i pairs of our configuration have been exposed. Similarly, we let Y_i be the number of backedges formed, and C_i be the number of components that have been at least partially exposed during the first i steps. We also define W_i to be the sum of $\deg(v)-2$ over all vertices v completely or partially exposed during the first i steps. We note that $W_i = X_i + 2Y_i - 2C_i$.

Now W_i "stalls" whenever a backedge is formed, and only changes whenever a new vertex is completely or partially exposed. For this reason, it is easier to analyze W_i when it is indexed not by the number of pairs exposed, but by the number of new vertices exposed. Thus we introduce another variable which does exactly this. We let Z_j be the sum of $\deg(v)-2$ over the first j new vertices (partially or completely) exposed.

The reason that we are introducing Z_j is that it has the same initial expected increase as X_i, but behaves much more nicely. In particular, it is not affected by the first and second complications discussed at the end of Section 1. Specifically, if, after the first j vertices have been completely or partially exposed, there are exactly $r_i(j)$ unexposed vertices of degree i, then $Z_{j+1} = Z_j + (i-2)$ with probability $ir_i(j)/\Sigma\, ir_i(j)$.

Now in order to discuss X_i and Z_j at the same time, we will introduce the random variable I_j which is the number of pairs exposed by the time that the jth vertex is partially exposed; i.e., $W_I = Z_j$.

Recall that $\alpha = \lceil R\omega(n)^2 \log n \rceil$.

Lemma 4. *Suppose that F is as described in Lemma 3. Given any vertex v in F, probability that v lies on a component of size at least α is less than n^{-2}.*

Proof. Here we will insist that v is the first vertex chosen in Step 2a. Therefore, the probability that v lies on a component that large is at most the probability that $X_i > 0$ for all $1 \le i \le \alpha$. Thus, we will consider the probability of the latter.

Note that for any i, if $C_i = 1$, then $W_i = X_i + 2Y_i - 2 \le X_i - 2$. In fact, we can also get $Z_i \ge X_i - 2$. This is because at each iteration we either have a backedge or expose a new vertex. Thus in iteration i, we have exposed $i - Y_i$ new vertices, therefore, $W_i = Z_{i-Y_i}$. Now Z_i decreases by at most one at each step; therefore, $Z_i \ge Z_{i-Y_i} - Y_i \ge W_i - Y_i \ge X_i + Y_i - 2 \ge X_i - 2$.

Now, if $X_i > 0$ for all $1 \le i \le \alpha$, then $C_\alpha = 1$. Therefore, the probability that $X_i > 0$ for all $1 \le i \le \alpha$ is at most the probability that $Z_\alpha > -2$. We will concentrate on this probability, as Z_i behaves much more predictably than X_i.

Initially the expected increase in Z_j is $\Sigma_{i\ge1}\, i(i-2)d_i(n)/\Sigma_{i\ge1}\, id_i(n) = -\nu + o(1)$. We claim that, for $j \le \alpha$, the expected increase in Z_j is less than $-\nu/2$.

This is true because the expected increase of Z_j would be highest if the first j vertex-copies chosen were all copies of vertices of degree 1. If this were the case then the expected increase in Z_j would be

$$\frac{-(d_1(n)-j) + \sum_{i\ge2} i(i-2)d_i(n)}{(d_1(n)-j) + \sum_{i\ge2} id_i(n)} + o(1) = -\nu + o(1)$$

$$\le -\frac{\nu}{2}$$

for sufficiently large n, as $j = o(n)$ and $id_i(n) \to \lambda_i$ uniformly.

Therefore, the expected value of Z_α is less than $-\frac{\nu}{2}\alpha + \deg(v) < -\frac{\nu}{3}\alpha$. We will use the corollary of Azuma's Inequality to show that Z_α is a.s. very close to its expected value.

Σ_i will indicate the choice of the ith new vertex exposed, $i = 1, \ldots, \alpha$, and $f(\Sigma) = Z_\alpha$. We need to bound

$$|E(f(\Sigma) \mid \Sigma_1, \Sigma_2, \ldots, \Sigma_{i+1}) - E(f(\Sigma) \mid \Sigma_1, \Sigma_2, \ldots, \Sigma_i)| .$$

Suppose that we are choosing the $(i+1)$st vertex to be partially exposed. Let Ω be the set of unexposed vertices at this point. The size of Ω is $n - i$. For each $x \in \Omega$, define $E_{i+1}(x)$ to be $E(Z_\alpha \mid \Sigma_1, \Sigma_2, \ldots, \Sigma_{i+1})$, where Σ_{i+1} is the event that x is the $(i+1)$st new vertex exposed. Consider any two vertices $u, v \in \Omega$. We will bound $|E_{i+1}(u) - E_{i+1}(v)|$. Consider the order that the vertices in $\Omega - \{u, v\}$ are exposed. Note that the distribution of this order is unaffected by the positions of u, v.

Let S be the set of the first $\alpha - 2$ vertices under this order, and let w be the next vertex. Now, $Z_\alpha = Z_{j-1} + (\Sigma_{x \in S} \deg(x) - 2) + \deg(y_1) - 2 + \deg(y_2) - 2$, where y_1 is the jth vertex exposed (either u or v) and y_2 is either u, v, or w. Therefore, the most that choosing between u, v can affect the conditional expected value of Z_α is twice the maximum degree, i.e., $|E_{i+1}(u) - E_{i+1}(v)| \le 2\omega(n)$.

Since

$$E(f(\Sigma) \mid \Sigma_1, \Sigma_2, \ldots, \Sigma_i) = \sum_{x \in \Omega} \Pr\{x \text{ is chosen}\} \times E_{i+1}(x) ,$$

we have that

$$|E(f(\Sigma) \mid \Sigma_1, \Sigma_2, \ldots, \Sigma_{i+1}) - E(f(\Sigma) \mid \Sigma_1, \Sigma_2, \ldots, \Sigma_i)| \le 2\omega(n) .$$

Therefore, by the corollary of Azuma's Inequality, the probability that $Z_\alpha > 0$ is at most

$$2 \exp\left(- \frac{(\frac{\nu}{3} R\omega(n)^2 \log n)^2}{2 \Sigma (2\omega(n))^2} \right) = 2n^{-\frac{\nu^2}{72R}}$$

$$< n^{-2} . \qquad \square$$

And now Lemma 3 follows quite easily:

Proof of Lemma 3. By Lemma 4, the expected number of vertices which lie on components of size at least α is $o(1)$. Therefore a.s. none exist. $\square$

We also get the following corollary:

Corollary 1. *Under the same conditions as Lemma* 3, *a.s.* $X_i < 2\alpha$ *throughout the exposure of our configuration.*

Proof. Because X_i drops by at most 2 at each step, if it ever got that high, it would not be able to reach 0 within $R\omega(n)^2 \log n$ steps. $\square$

We will now show that F a.s. does not have many cycles. First, we will see that it a.s. has no multicyclic components.

Lemma 5. *Let F be a random configuration meeting the same conditions as in Lemma 3. F a.s. has no component with at least 2 cycles.*

Proof. Choose any vertex v. Let E_v be the event that v lies on a component of size at most α with more than one cycle, and that throughout the exposure of this component, $X_i < 2\alpha$.

We will insist that v is the first vertex examined under Step 2a. If the size of the first component is at most α, then the second backedge must be chosen within at most $\alpha + 2$ steps. Therefore, the probability that E holds is less than

$$\binom{\alpha+2}{2}\left(\frac{2\alpha}{M-2\alpha-3}\right)^2 = o(n^{-1})$$

as $\omega(n) < n^{1/8-\epsilon}$.

Therefore, the expected number of vertices for which E_v holds is $o(1)$ and so the probability that E_v holds for any v is $o(1)$. Therefore, by Lemma 3 and Corollary 1, a.s. no components of F have more than one cycle. □

We can now show that F a.s. does not have many cycles, by showing that it a.s. does not have many cyclic components.

Lemma 6. *Let F be a random configuration meeting the same conditions as in Lemma 3. F a.s. has less than $2\alpha \log n$ cycles.*

Proof. We will show that a.s. throughout the exposure of F, at most $2\alpha \log n$ backedges are formed. The rest will then follow, since by Lemma 5, a.s. no component contains more than one cycle, and so a.s. the number of cycles in F is exactly the number of backedges.

First we must define a set B_i of unmatched vertex-copies: For each i, if there are more than 2α open vertex-copies at the ith iteration, then let B_i consist of any 2α of them. Otherwise, let B_i consist of the open vertex-copies and enough arbitrarily chosen members of L to bring the size of B_i up to α. Of course, if L is too small to do this, then we will just add all of L to B_i. Let T_i be the event that a member of B_i is chosen in step i.

Clearly the number of backedges formed is at most the number of successful T_i's, plus the number of backedges formed at times when $X_i > 2\alpha$. Now, by Corollary 1, we know that there are a.s. none of the latter type of backedges, so we will concentrate on the number of the former type.

Now the number of vertex-copies to choose from is $\Sigma_{j\geq 1} jd_j(n) - 2i + 1 = M - 2i + 1$. Therefore, the probability of T_i holding is $\frac{2\alpha}{M-2i+1}$, for $M - 2i + 1 \geq 2\alpha$ and 1 otherwise.

Therefore, the expected value of T, the number of successful T_i's is

$$E(T) = 2\alpha = \sum_{i=1}^{(M-2\alpha)/2} \frac{2\alpha}{M-2i+1} = \alpha \log(M)(1 + o(1)) .$$

Now we will use a second moment argument to show that T is a.s. not much bigger than $E(T)$:

$$E(T^2) = \sum_{i \neq j} \frac{\alpha^2}{(M-2i+1)(M-2j+1} + E(T)$$

$$= (E(T)^2 + E(T))(1 + o(1)) .$$

Therefore, by Chebyshev's inequality, the probability that $T > 1.5\alpha \log(M)$ is at most $1/(4E(T))(1 + o(1)) = o(1)$.

Therefore, a.s. the number of backedges formed is less than $1.5\alpha \log(M) < 2\alpha \log n$, proving the result.

And now we can prove Theorem 1b.

Proof of Theorem 1*b*. This clearly follows from Lemmas 2, 3, 5, and 6. □

3. GRAPHS WITH GIANT COMPONENTS

In this section we will prove the analogue of Theorem 1a for random configurations. Lemmas 1 and 2 will then imply that Theorem 1a holds.

First we will show that a giant component exists with high probability:

Lemma 7. *Let F be a random configuration with n vertices and degree sequence $\mathcal{D}_n$ meeting the conditions of Theorem* 1. *If $Q(\mathcal{D}) > 0$, then there exist constants $\zeta_1, \zeta_2 > 0$ dependent on $\mathcal{D}$ such that F a.s. has a component with at least $\zeta_1 n$ vertices and $\zeta_2 n$ cycles. Moreover, the probability of the converse is at most z^n, for some fixed $0 < z < 1$.*

Throughout this section we will assume that the conditions of Lemma 7 hold. As in Section 2, we will prove Lemma 7 by analyzing the Markov process discussed in the previous section. Again, the key will be to concentrate on the random variable Z_j.

Lemma 8. *There exists $0 < \epsilon < 1$, $0 < \Delta < \min(\frac{1}{4}, \frac{K}{4})$ such that for all $0 < \delta < \Delta$ a.s. $Z_{\lceil \delta n \rceil} > \epsilon \delta n$. Moreover, the probability of the converse is at most $(z_1)^n$, for some fixed $0 < z_1 < 1$.*

Proof. For simplicity, we will assume that δn is an integer. Initially, the probability that a vertex-copy of degree i is chosen as a partner is $p_i(n) = id_i(n)/\sum_{j \geq 1} jd_j(n) = i\lambda_i/K + o(1)$.

Unlike in Section 2, we have to consider the behavior of our walk after $\Theta(n)$ steps. Thus we have to worry about the third complication described at the end of Section 2, i.e., the fact that the ratios of unexposed vertices of different degrees are shifting.

It turns out that this problem is much less serious if we can ignore vertices of high degree. So what we will do is show that we can find a value i^*, such that if we change Z_j slightly by saying that every time a vertex of degree $i > i^*$ is chosen, we subtract 1 from Z_j instead of adding $i-2$ to it, then we will still have positive expected increase.

We will then show that we can find a sequence $\phi_1, \ldots, \phi_{i^*}$ summing to one, such that for each $2 \le i \le i^*$, ϕ_i is a little less than the initial probability of a vertex of degree i being chosen. However, if we were to adjust Z_j a little further by selecting a vertex of degree i with probability ϕ_i at each step, then we would still have a positive expected increase.

We will call this "adjusted Z_j" Z_j^*. Clearly, if we find some J such that after J steps, the probability of choosing a vertex of degree i is still at least ϕ_i for $2 \le i \le i^*$, then the probability that $Z_J > R$ for any R is at least as big as the probability that $Z_j^* > R$. We will concentrate on the second probability as Z_j^* is much simpler to analyze.

More formally, what we wish to do is choose a sequence $\phi_1, \ldots, \phi_{i^*}$ such that:

1. $\Sigma\, \phi_i = 1$;
2. $0 < \phi_i < i\lambda_i/K$, for $2 \le i \le i^*$, unless $0 = \phi_i = i\lambda_i/K$;
3. $\Sigma_{i\ge1}\, i(i-2)\phi_i > 0$.

Note that

$$\sum_{i\ge1} (i-2)p_i(n) = \sum_{i\ge1} (i-1)p_i(n) - \sum_{i\ge1} p_i(n)$$
$$= \sum_{i\ge2} (i-1)p_i(n) - 1\,.$$

Set $p_i = i\lambda_i/K$. Since $\mathcal{D}$ is well behaved and $Q(\mathcal{D}) > 0$, there exists i^* such that $\Sigma_{i=2}^{i^*} (i-1)p_i > 1 + \epsilon'$, for some $\epsilon' > 0$ and sufficiently large n.

Therefore, we can choose a sequence $\phi_1, \ldots, \phi_{i^*}$ such that for all $2 \le i \le i^*$, $p_i > \phi_i > 0$ unless $p_i = \phi_i = 0$, $\phi_1 = 1 - \phi_2 - \phi_3 - \cdots - \phi_{i^*}$, and $\Sigma_{i=2}^{i^*} (i-1)\phi_i = 1 + \epsilon'/2$. It follows that $\Sigma_{i\ge1} (i-2)\phi_i = \epsilon'/2$.

Consider the random variable Z_j^* which follows the following random walk:

- $Z_0^* = 0$
- $Z_{j+1}^* = Z_j^* + (i-2)$ with probability ϕ_i, $1 \le i \le i^*$.

For $i = 2, \ldots, i^*$, choose any $\Delta_i > 0$ such that $\frac{i\lambda_i - \Delta i}{K} < \pi_i$, and set $\Delta = \min\{\Delta_2, \ldots, \Delta_{i^*}, \frac{K}{4}\}$. Clearly, after at most Δ iterations, the probability of choosing a copy of a vertex of degree $i \ge 2$ is at least ϕ_i. Therefore, for $0 \le j \le \Delta n$, the random variable Z_j majorizes Z_j^*; i.e., for any R,

$$\Pr[Z_j > R] \ge \Pr[Z_j^* > R.]$$

Now the expected increase in Z_j^* at any step is $\epsilon'/2$. Thus the lemma follows by letting $\epsilon = \epsilon'/4$, as is well known (see, for example [9]) that $Z_{\delta n}^*$ is a.s. concentrated around its expected value which is $2\delta\epsilon n$, and that the probability of deviating from the expected value by more than $\Theta(n)$ is as low as claimed. □

We have just shown that Z_j a.s. grows large. However, we really want to analyze X_j. In order to do this, recall that the random variable I_j is defined to be the number of pairs exposed by the time that the jth vertex is partially exposed; i.e., $W_{I_j} = Z_j$.

Lemma 9. *There exists $0 < \delta' < \Delta$ such that for any $0 < \delta \leq \delta'$ there a.s. exists some $1 \leq I \leq I_{\lceil \delta n \rceil}$ such that $X_I > \gamma n$, where $\gamma = \min(\epsilon\delta/2, \frac{1}{4})$. Moreover, the probability of the converse is at most $(z_2)^n$ for some $0 < z_2 < 1$, dependent on δ.*

Proof. For simplicity, we will assume that δn is an integer. We will count W, the number of backedges formed before either $X_i > \gamma n$ or $I_{\delta n}$ pairs have been exposed. We claim that we can choose δ' such that a.s. $W < \frac{\gamma}{2} n$ for $\delta \leq \delta'$.

At any step i, $1 \leq i \leq I_{\delta n}$, the probability that an open vertex-copy is chosen is $\frac{X_i}{Kn - 2i} + o(1)$, regardless of the choices made previous to that step. Now $I_j \leq j + Y_{I_j} \leq j + \frac{Z_i}{2} \leq (\delta + \epsilon\delta)n$.

Therefore, at each step, the probability that such a backedge is formed is 0 if $X_j > \gamma n$ and at most

$$p = \frac{\frac{1}{2}\delta\epsilon}{K - 2\delta - 2\delta\epsilon}$$

if $X_j \leq \gamma n$.

Thus the number of such copies chosen is majorized by the binomial variable $\mathrm{BIN}(p, I_{\delta n})$.

Therefore the lemma follows so long as $pI_{\delta n} \leq p(\delta + \epsilon\delta)n < \frac{\gamma}{2} n$, which is equivalent to $4\delta + 4\delta\epsilon < K$, yielding δ'.

Now if $X_i \leq \gamma n$ for all $1 \leq i < I_{\delta n}$ then W is equal to $Y_{I_{\delta n}}$.

Therefore $X_{I_{\delta n}} = Z_{\delta n} - 2Y_{I_{\delta n}}$ which with probability at least $1 - (z_1)^n$ is at least $\delta\epsilon/2n$, which yields our result. □

Now that we know that X_i a.s. gets to be as large as $\Theta(n)$, we can show that there is a.s. a giant component:

Lemma 10. *There exists $\zeta_1, \zeta_2 > 0$ such that the component being exposed at step $I = I_{\lceil \delta' n \rceil}$ will a.s. have at least $\zeta_1 n$ vertices and $\zeta_2 n$ cycles. Moreover, the probability of the converse is at most $(z_3)^n$, for some fixed $0 < z_3 < 1$.*

Proof. Note that at this point there are a.s. at least $n - 2\delta' n - \gamma n > n/5$ unexposed vertices. Form a set β consisting of exactly one copy of each of them.

There is a set χ of X_I open vertex-copies whose partners must be exposed before this component is entirely exposed. We will show that a.s. at least $\zeta_1 n$ of these will be matched with members of β, and at least $\zeta_2 n$ of these will be matched with other open vertex-copies from χ. Clearly this will prove the lemma.

Now there are $M - 2I$ vertex-copies available to be matched. Our procedure for exposing F simply generates a random matching among them where each matching is equally probable. The expected number of pairs containing one vertex from each of χ, β is at least $\frac{n}{5}(\frac{X_I}{2M - I})$, and the expected number of pairs of open vertex-copies which form an edge of F is $(\frac{M}{2} - I)(\frac{X_I}{M - 2I})^2$.

The previous lemmas give us a lower bound of $2\zeta_1 n, 2\zeta_2 n$ on these numbers, and it follows from the Chernoff bounds that these numbers are a.s. at least half of their expected values with the probability of the converse as low as claimed. Therefore, the component a.s. has at least $\zeta_1 n$ vertices and at least $\zeta_2 n$ cycles. □

And now Lemma 7 follows quite easily:

Proof of Lemma 7. This is clearly a corollary of Lemma 10. □

We will now see that F a.s. has only one large component.

Lemma 11. *If F is a random configuration as described in Lemma* 7, *then F a.s. has exactly one component on more than $T \log n$ vertices, for some constant T dependent on the degree sequence.*

Proof. We have already shown that F a.s. has at least one giant component of size at least $\zeta_1 n$. We will see here that no other components of F are large.

Consider any ordered pair of vertices (u, v). We say that (u, v) has *property A* if u and v lie on components of size at least $\zeta_1 n$ and $T \log n$, respectively. We will show that for an appropriate choice of T, the probability that (u, v) has property A is $o(n^{-2})$, which is enough to prove the lemma.

Recall that we may choose any vertex-copy we wish to start the exposure with. We will choose u.

By Lemma 9, there a.s. exists some $I \leq I_{\lceil \zeta_1 n \rceil}$, such that $X_I > \gamma n$, where $\gamma = \min(\frac{\epsilon\delta'}{2}, \frac{\epsilon\zeta_1}{2}, \frac{1}{4})$, and so we can assume this is to be the case. Note that if after I steps, we are not still exposing the first component, C_1, or if we have exposed a copy of v, then (u, v) does not have property A, so we will assume the contrary. Define χ to be the set of open vertex-copies after I steps.

Here we will break from the standard method of exposure. We will start exposing v's component, C_2, immediately, and put off the exposure of the rest of C_1 until later. We will see that if C_2 gets too big, then it will a.s. include a member of χ.

We expose C_2 in the following way. We start by picking any copy of v, and exposing its partner. We continue exposing pairs, always choosing a copy of partially-exposed vertex is known to be in C_2 (if one is available), and exposing its partner. We check to see if this partner lies in χ. This would imply that v lies in C_1. Once C_2 is entirely exposed, if it is disjoint from C_1, then we return to exposing the rest of C_1 and continue to expose F in the normal manner. Note that this is a valid way to expose F.

At each step, the probability that a member of χ is chosen is at least γ/K. Also, if v lies on a component of size greater than $T \log n$, then it must take at least $T \log n$ steps to expose this component. Therefore, the probability that v lies on a component of size greater than $T \log n$ which is not C_1 is at most

$$\left(1 - \frac{\gamma}{K}\right)^{T \log n} = o(n^{-2})$$

for a suitable value of T.

Therefore, the expected number of pairs of vertices with property A tends to zero as $n \to \infty$, so a.s. none exist. □

It only remains to be shown that F a.s. has no small components with more than one cycle.

Lemma 12. *If F is a random configuration as described in Lemma 7, then F a.s. has no multicyclic component on at most $T \log n$ vertices, for any constant T.*

Proof. Consider the probability of some vertex v lying on such a component. We will insist that we expose an edge containing v in the first execution of Step 2a. Now, if this component contains at most $T \log n$ vertices, then it is entirely exposed after at most $o(n^{1/4})$ steps as the maximum degree is $n^{1/4-\epsilon}$.

At each point during the exposure we can assume $X_i < n^{1/4}$, as otherwise X_i would not be able to return to zero quickly enough. Therefore, at each step, the probability that a backedge is formed is at most $o(1)n^{1/4}/(M - 2n^{1/4}) = o(n^{-3/4})$. Therefore, the probability that at least 2 cycles are formed is at most

$$o(1) \times \binom{n^{1/4}}{2}(n^{-3/4})^2 = o(n^{-1}) .$$

Therefore the expected number of vertices lying on such components is $o(1)$, and hence a.s. none exist. □

And now we have Theorem 1a:

Proof of Theorem 1a. This clearly follows from Lemmas 1, 2, 7, 11, and 12. □

It is worth noting that by analyzing the number of open vertices of each degree more carefully throughout the exposure of F, it is possible to compute the size of the giant component more precisely. In fact, we can find a $\kappa(\mathscr{D})$ such that the size of the giant component is a.s. $(1 + o(1)\kappa(\mathscr{D})n$. Details will appear in a future paper.

4. APPLICATIONS

Here are a few applications of Theorem 1. The first is a new proof of a classical result concerning the double-jump threshold:

Theorem 2. *For $c > \frac{1}{2}$, $G_{n,M=cn}$ a.s. has a giant component, while for $c < \frac{1}{2}$, $G_{n,M=cn}$ a.s. does not have one.*

Proof. It is well known (see, for example, [7]) that such a graph a.s. has

$$\frac{(2c)^i}{i!} e^{-2c} n + o(n^{.51})$$

vertices of degree i for each $i \leq O(\log n / \log \log n)$, and no vertices of higher degree.

Now expose G by first exposing its degree sequence, and then choosing a random graph on that degree sequence. We will a.s. get a sequence $\mathscr{D}$ which satisfies all the conditions of Theorem 1 and for which

$$Q(\mathcal{D}) = \sum_{i \geq 1} i(i-2) \frac{(2c)^i}{i!} e^{-2c},$$

which is positive for $c > \frac{1}{2}$ and negative for $c < \frac{1}{2}$. □

Note that this only gives an upper bound of $O(\frac{(\log n)^3}{(\log \log n)^2})$ for the size of the largest component of G for $c < \frac{1}{2}$, rather than the proper upper bound of $O(\log n)$.

As mentioned earlier, this work was motivated by the study of minimally 4-chromatic subgraphs of a random graph G. Recall that such a subgraph must have minimum degree at least 3. This is of interest mainly in the study of the chromatic number of sparse random graphs, as if $\chi(G) \geq 4$, then G must have a minimally 4-chromatic subgraph, H.

Chvatál [7] showed that if G is a random graph on n vertices and cn edges, then for $c < c^* = 1.442\ldots$, the expected number of subgraphs of G with minimum degree at least 3 tends to 0 with n, while, for $c > c^*$, the expected number of such subgraphs is exponentially large in n.

The authors wished to find which such subgraphs could actually be minimally 4-chromatic graphs. We looked at the following condition of Gallai [12]:

Definition. *If H is a graph with minimum degree r, then the* low graph *of H ($L(H)$) is the subgraph induced by the vertices of degree r.*

Theorem 3. *If H is a minimally k-chromatic graph with minimum degree $k-1$, then $L(H)$ has no even cycles whose vertices do not induce a clique.*

We used this to prove the following:

Theorem 4. *Let H be a random graph on n vertices and at most $1.793n$ edges with minimal degree 3. H is a.s. not a minimally 4-chromatic graph. Moreover, the probability of failure is at most z^n, for some fixed $0 < z < 1$.*

Outline of Proof. $L(H)$ is a graph whose vertices are all of degree $0, 1, 2, 3$. We showed that the degree sequence of $L(H)$ could be approximately determined by the edge-density of H, and that all graphs on that degree sequence were equally likely to appear as $L(H)$. It then followed from Theorem 1 that if H has edge-density at most 1.793, then $L(H)$ a.s. has $O(n)$ cycles. We then showed that a.s. at least one of these cycles was even and of length at least 6, and the result followed from the fact that $L(H)$ has no cliques on more than four vertices.

We used Theorem 4 to show:

Corollary 2. *There exists $\delta > 0$ such that if G is a random graph on n vertices and $\lceil \delta n \rceil$ edges, then the expected number of minimally 4-chromatic subgraphs of G is exponentially small, while the expected number of subgraphs of G with minimum degree at least 3 is exponentially large in n.*

Outline of Proof. It follows from the results of [7] that for c slightly larger than

c^*, the expected number of such subgraphs of G with edge-density at least 1.793 is exponentially small. □

The details to Theorem 4 and Corollary 2 will appear in a future paper, and can also be found in [18].

It is worth noting that Frieze, Pittel, and the authors [18], used a different type of argument to show that for $c < 1.756$ a random graph with edge-density c a.s. has no subgraph with minimum degree at least 3, and hence is a.s. 3-colorable.

ACKNOWLEDGMENTS

The authors would like to thank Avrim Blum, Alan Frieze, Colin McDiarmid, Chris Small, and Mete Soner for their helpful comments and advice, and two anonymous referees for several suggestions of improvement.

REFERENCES

[1] K. Azuma, Weighted sums of certain dependent random variables, *Tokuku Math. J.*, **19**, 357–367 (1967).

[2] E. A. Bender and E. R. Canfield, The asymptotic number of labelled graphs with given degree sequences, *J. Combinat. Theory* (A), **24**, 296–307 (1978).

[3] B. Bollobás, *Random Graphs*, Academic, New York, 1985.

[4] B. Bollobás, Martingales, Isoperimetric inequalities and random graphs, *Colloq. Math. Soc. Janós Bolyai*, **52**, 113–139 (1987).

[5] B. Bollobás and A. Frieze, On matchings and Hamiltonian cycles in random graphs, *Ann. Discrete Math.*, **28**, 23–46 (1985).

[6] B. Bollobás and A. Thompson, Random graphs of small order, Random Graphs '83, *Ann. Discrete Math.*, **28**, 47–97 (1985).

[7] V. Chvátal, Almost all graphs with 1.44*n* edges are 3-colorable, *Random Struct. Alg.*, **2**, 11–28 (1991).

[8] P. Erdös and A. Rényi, On the evolution of random graphs, *Magayr Tud. Akad. Mat. Kutato Int. Kozl.*, **5**, 17–61 (1960).

[9] W. Feller, *An Introduction to Probability Theory and its Applications*, Vol. 1. Wiley, New York, 1966.

[10] P. Flajolet, D. Knuth, and B. Pittel, The first cycles in an evolving graph, *Discrete Math.* **75**, 167–215 (1989).

[11] P. Flajolet, D. Knuth, T. Łuczak, and B. Pittel, The birth of the giant component, *Random Struct. Alg.*, **4**, (1993).

[12] T. Gallai, Kritische graphen I, *Magayr Tud. Akad. Mat. Kutato Int. Kozl.*, **8** 165–192 (1963).

[13] T. Łuczak, Component behaviour near the critical point of the random graph process, *Random Struct. Alg.*, I, 287–310 (1990).

[14] T. Łuczak, Sparse random graphs with a given degree sequence, *Random Graphs*, **2**, 165–182 (1992).

[15] T. Łuczak and J. Wierman, The chromatic number of random graphs at the double-jump threshold, *Combinatorica*, **9**, 39–50 (1989).

[16] C. McDiarmid, On the method of bounded differences, *Surveys in Combinatorics, Proc. Twelfth British Combinatorial Conference*, 1989, pp. 148–188.

[17] B. D. McKay, Asymptotics for symmetric 0–1 matrices with prescribed row sums, *Ars Combinat.*, **19A**, 15–25 (1985).

[18] M. Molloy, The chromatic number of sparse random graphs, Masters thesis, University of Waterloo, 1992.

[19] R. W. Robinson and N. C. Wormald, Almost all cubic graphs are Hamiltonian, *Random Struct. Alg.* **3**, 117–126 (1992).

[20] R. W. Robinson and N. C. Wormald, Almost all regular graphs are Hamiltonian, manuscript.

[21] N. C. Wormald, Some problems in the enumeration of labelled graphs, Doctoral thesis, Newcastle University, 1978.

Received August 6, 1993
Accepted November 24, 1994

A Random Graph Model for Massive Graphs

William Aiello
AT&T Labs
Florham Park, New Jersey
aiello@research.att.com

Fan Chung
University of California,
San Diego
fan@ucsd.edu

Linyuan Lu
University of California,
San Diego
llu@math.ucsd.edu

ABSTRACT

We propose a random graph model which is a special case of sparse random graphs with given degree sequences. This model involves only a small number of parameters, called logsize and log-log growth rate. These parameters capture some universal characteristics of massive graphs. Furthermore, from these parameters, various properties of the graph can be derived. For example, for certain ranges of the parameters, we will compute the expected distribution of the sizes of the connected components which almost surely occur with high probability. We will illustrate the consistency of our model with the behavior of some massive graphs derived from data in telecommunications. We will also discuss the threshold function, the giant component, and the evolution of random graphs in this model.

1. INTRODUCTION

Is the World Wide Web completely connected? If not, how big is the largest component, the second largest component, etc.? Anyone who has "surfed" the Web for any length of time will undoubtedly come away feeling that if there are disconnected components at all, then they must be small and few in number. Is the Web too large, dynamic and structureless to answer these questions?

Probably yes, if the answers for the sizes of the largest components are required to be exact. Recently, however, some structure of the Web has come to light which may enable us to describe graph properties of the Web qualitatively. Kumar et al. [11; 12] and Kleinberg et al. [10] have measured the degree sequences of the Web and shown that it is well approximated by a power law distribution. That is, the number of nodes, y, of a given degree x is proportional to $x^{-\beta}$ for some constant $\beta > 0$. This was reported independently by Albert, Barabási and Jeong in [3; 5; 6]. The power law distribution of the degree sequence appears to be a very robust property of the Web despite its dynamic nature. In fact, the power law distribution of the degree sequence may be a ubiquitous characteristic, applying to many massive real world graphs. Indeed, Abello et al. [1] have shown that the degree sequence of so called *call graphs* is nicely approximated by a power law distribution. Call graphs are graphs of calls handled by some subset of telephony carriers for a specific time period. In addition, Faloutsos, et al. [9] have shown that the degree sequence of the Internet router graph also follows a power law.

Just as many other real world processes have been effectively modeled by appropriate random models, in this paper we propose a parsimonious random graph model for graphs with a power law degree sequence. We then derive connectivity results which hold with high probability in various regimes of our parameters. And finally, we compare the results from the model with the exact connectivity structure for some call graphs computed by Abello et al. [1].

STOC 2000 Portland Oregon USA

1.1 Power-Law Random Graphs

The study of random graphs dates back to the work of Erdős and Rényi whose seminal papers [7; 8] laid the foundation for the theory of random graphs. There are three standard models for what we will call in this paper *uniform* random graphs [4]. Each has two parameters. One parameters controls the number of nodes in the graph and one controls the density, or number of edges. For example, the random graph model $G(n, m)$ assigns uniform probability to all graphs with n nodes and m edges while in the random graph model $\mathcal{G}(n,p)$ each edge in an n node graph is chosen with probability p.

Our *power law* random graph model also has two parameters. The two parameters only roughly delineate the size and density but they are natural and convenient for describing a power law degree sequence. The power law random graph model $P(\alpha, \beta)$ is described as follows. Let y be the number of nodes with degree x. $P(\alpha,\beta)$ assigns uniform probability to all graphs with $y = e^\alpha/x^\beta$ (where self loops are allowed). Note that α is the intercept and β is the (negative) slope when the degree sequence is plotted on a log-log scale.

We remark that there is also an alternative power law random graph model analogous to the uniform graph model $\mathcal{G}(n,p)$. Instead of having a fixed degree sequence, the random graph has an expected degree sequence distribution. The two models are basically asymptotically equivalent, subject to bounding error estimates of the variances (which will be further described in a subsequent paper).

1.2 Our Results

Just as for the uniform random graph model where graph properties are studied for certain regimes of the density parameter and shown to hold with high probability asymptotically in the size parameter, in this paper we study the connectivity properties of $P(\alpha, \beta)$ as a function of the power β which hold almost surely for sufficiently large graphs. Briefly, we show that when $\beta < 1$, the graph is almost surely connected. For $1 < \beta < 2$ there is a giant component, i.e., a component of size $\Theta(n)$. Moreover, all smaller components are of size $O(1)$. For $2 < \beta < \beta_0 = 3.4785$ there is a giant component and all smaller components are of size $O(\log n)$. For $\beta = 2$ the smaller components are of size $O(\log n / \log\log n)$. For $\beta > \beta_0$ the graph almost surely has no giant component. In addition we derive several results on the sizes of the second largest component. For example, we show that for $\beta > 4$ the numbers of components of given sizes can be approximated by a power law as well.

1.3 Previous Work

Strictly speaking our model is a special case of random graphs with a given degree sequence for which there is a large literature. For example, Wormald [17] studied the connectivity of graphs whose degrees are in an interval $[r, R]$, where $r \geq 3$. Luczak [13] considered the asymptotic behavior of the largest component of a random graph with given degree sequence as a function of the number of vertices of degree 2. His result was further improved by Molloy and Reed [14; 15]. They consider a random graph on n vertices with the following degree distribution. The fraction of vertices of degree $0, 1, 2, \ldots$ is asymptotically $\lambda_0, \lambda_1, \ldots$, respectively, where the λ's sum to 1. It is shown in [14] that if $Q = \sum_i i(i-2)\lambda_i > 0$ (and the maximum degree is not too large), then such random graphs have a giant component with probability tending to 1 as n goes to infinity, while if $Q < 0$ (and the maximum degree is not too large), then all components are small with probability tending to 1 as $n \to \infty$. They also examined the threshold behavior of such graphs. In this paper, we will apply these techniques to deal with the special case that applies to our model.

Several other papers have taken a different approach to modeling power law graphs than the one taken here [2; 5; 6; 10; 12]. The essential idea of these papers is to define a random process for growing a graph by adding nodes and edges. The intent is to show that the defined processes asymptotically yield graphs with a power law degree sequence with very high probability. While this approach is interesting and important it has several difficulties. First, the models are difficult to analyze rigorously since the transition probabilities are themselves dependent on the current state. For example, [5; 6] implicitly assume that the probability that a node has a given degree is a continuous function. The authors of [10; 12] will offer an improved analysis in an upcoming paper [16]. In [2] we derive a power law degree sequence for several graph evolution models for asymptotically large graphs by explicitly solving the recurrence relations given by the random evolution process for the expected degree sequence and showing tight concentration around the mean using Azuma's inequality for martingales. We also derive results for the distribution of connected component sizes, but not for the entire range of powers given in this paper. Second, while the models may generate graphs with power law degree sequences, it remains to be seen if they generate graphs which duplicate other structural properties of the Web, the Internet, and call graphs. For example, the model in [5; 6] cannot generate graphs with a power law other than c/x^3. Moreover, all the graphs can be decomposed into m disjoint trees, where m is a parameter of the model. The (α, β) model in [12] is able to generate graphs for which the power law for the indegree is different than the power law for the outdegree as is the case for the Web. However, to do so, the model requires that there be a constant fraction of nodes that have only indegree and no outdegree and visa versa. While this may be appropriate for call graphs (e.g., customer service numbers) it remains to be seen whether it models the Web. Thus, while the random graph generation approach holds the promise of accurately predicting a wide variety a structural properties of many real world massive graphs much work remains to be done.

In this paper we take a different approach. We do not attempt to answer how a graph comes to have a power law degree sequence. Rather, we take that as a given. In our model, all graphs with a given power law degree sequence are equi-probable. The goal is to derive structural properties which hold with probability asymptotically approaching 1. Such an approach, while potentially less accurate than the detailed modeling approach above, has the advantage of being robust: the structural properties derived in this model will be true for the vast majority of graphs with the given degree sequence. Thus, we believe that this model will be an important complement to random graph generation models.

The power law random graph model will be described in detail in the next section. In Sections 3 and 4, our results on connectivity will be derived. In section 5, we discuss the sizes of the second largest components. In section 6, we compare the results of our model to exact connectivity data for call graphs.

2. A RANDOM GRAPH MODEL

We consider a random graph with the following degree distribution depending on two given values α and β. Suppose there are y vertices of degree $x > 0$ [1] where x and y satisfy

$$\log y = \alpha - \beta \log x.$$

In other words, we have

$$|\{v | deg(v) = x\}| = y = \frac{e^\alpha}{x^\beta}.$$

Basically, α is the logarithm of the number of nodes of degree 1 and β is the log-log rate of decrease of the number of nodes a given degree.

We note that the number of edges should be an integer. To be precise, the above expression for y should be rounded down to $\lfloor \frac{e^\alpha}{x^\beta} \rfloor$. If we use real numbers instead of rounding down to integers, it may cause some error terms in further computation. However, we will see that the error terms can be easily bounded. For simplicity and convenience, we will use real numbers with the understanding the actual numbers are their integer parts. Another constraint is that the sum of the degrees should be even. This can be assured by adding a vertex of degree 1 if the sum is old if needed. Furthermore, for simplicity, we here assume that there is no isolated vertices.

We can deduce the following facts for our graph:

[1]There are several ways to deal with nodes with zero degree. For simplicity, here we simply exclude such isolated nodes from the graph.

(1) The maximum degree of the graph is $e^{\frac{\alpha}{\beta}}$. Note that $0 \le \log y = \alpha - \beta \log x$.

(2) The vertices number n can be computed as follows: By summing $y(x)$ for x from 1 to $e^{\frac{\alpha}{\beta}}$, we have

$$n = \sum_{x=1}^{e^{\frac{\alpha}{\beta}}} \frac{e^\alpha}{x^\beta} \approx \begin{cases} \zeta(\beta)e^\alpha & \text{if } \beta > 1 \\ \alpha e^\alpha & \text{if } \beta = 1 \\ \frac{e^{\frac{\alpha}{\beta}}}{1-\beta} & \text{if } 0 < \beta < 1 \end{cases}$$

where $\zeta(t) = \sum_{n=1}^{\infty} \frac{1}{n^t}$ is the Riemann Zeta function.

(3) The number of edges E can be computed as follows:

$$E = \frac{1}{2}\sum_{x=1}^{e^{\frac{\alpha}{\beta}}} x\frac{e^\alpha}{x^\beta} \approx \begin{cases} \frac{1}{2}\zeta(\beta-1)e^\alpha & \text{if } \beta > 2 \\ \frac{1}{4}\alpha e^\alpha & \text{if } \beta = 2 \\ \frac{1}{2}\frac{e^{\frac{2\alpha}{\beta}}}{2-\beta} & \text{if } 0 < \beta < 2 \end{cases}$$

(4) The differences of the real numbers in (1)-(3) and their integer parts can be estimated as follows: For the number n of vertices, the error term is at most $e^{\frac{\alpha}{\beta}}$. For $\beta \ge 1$, it is $o(n)$, which is a lower order term. For $0 < \beta < 1$, the error term for n is relatively large. In this case, we have

$$n \ge \frac{e^{\frac{\alpha}{\beta}}}{1-\beta} - e^{\frac{\alpha}{\beta}} = \frac{\beta e^{\frac{\alpha}{\beta}}}{1-\beta}$$

Therefore, n has the same magnitude as $\frac{e^{\frac{\alpha}{\beta}}}{1-\beta}$. The number E of edges can be treated in a similarly way. For $\beta \ge 2$, the error term of E is $o(E)$, a lower order term. For $0 < \beta < 2$, E has the same magnitude as in formula of item (3). In this paper, we mainly deal with the case $\beta > 2$. The only place that we deal with the case $0 < \beta < 2$ is in the next section where we refer to $2-\beta$ as a constant. By using real numbers instead of rounding down to their integer parts, we simplify the arguments without affecting the conclusions.

In order to consider the random graph model, we will need to consider large n. We say that some property almost surely (a. s.) happens if the probability that the property holds tends to 1 as the number n of the vertices goes to infinity. Thus we consider α to be large but where β is fixed.

We use the following random graph model for a given degree sequence:

The model:

1. Form a set L containing $deg(v)$ distinct copies of each vertex v.
2. Choose a random matching of the elements of L.
3. For two vertices u and v, the number of edges joining u and v is equal to the number of edges in the matching of L joining copies of u to copies of v.

We remark that the graphs that we are considering are in fact multi-graphs, possibly with loops. This model is a natural extension of the model for k-regular graphs, which is formed by combining k random matching. For references and undefined terminology, the reader is referred to [4; 18].

We note that this random graph model is slightly different from the uniform selection model $P(\alpha, \beta)$ as described in section 1.1. However, by using techniques in Lemma 1 of [15], it can be shown that if a random graph with a given degree sequence a. s. has property P under one of these two models, then it a. s. has property P under the other model, provided some general conditions are satisfied.

3. THE CONNECTED COMPONENTS

Molloy and Reed [14] showed that for a random graph with $(\lambda_i + o(1))n$ vertices of degree i, where λ_i are non-negative values which sum to 1, the giant component emerges a. s. when $Q = \sum_{i \ge 1} i(i-2)\lambda_i > 0$, provided that the maximum degree is less than $n^{1/4-\epsilon}$. They also show that almost surely there is no giant component when $Q = \sum_{i \ge 1} i(i-2)\lambda_i < 0$ and maximum degree less than $n^{1/8-\epsilon}$.

Here we compute Q for our (α, β)-graphs.

$$\begin{aligned} Q &= \sum_{x=1}^{e^{\frac{\alpha}{\beta}}} x(x-2)\lfloor \frac{e^\alpha}{x^\beta} \rfloor \\ &\approx \sum_{x=1}^{e^{\frac{\alpha}{\beta}}} \frac{e^\alpha}{x^{\beta-2}} - 2\sum_{x=1}^{e^{\frac{\alpha}{\beta}}} \frac{e^\alpha}{x^{\beta-1}} \\ &\approx (\zeta(\beta-2) - 2\zeta(\beta-1))e^\alpha \text{ if } \beta > 3 \end{aligned}$$

Hence, we consider the value $\beta_0 = 3.47875\ldots$, which is a solution to

$$\zeta(\beta-2) - 2\zeta(\beta-1) = 0$$

If $\beta > \beta_0$, we have

$$\sum_{x=1}^{e^{\frac{\alpha}{\beta}}} x(x-2)\lfloor \frac{e^\alpha}{x^\beta} \rfloor < 0$$

We first summarize the results here:

1. When $\beta > \beta_0 = 3.47875\ldots$, the random graph a. s. has no giant component. When $\beta < \beta_0 = 3.47875\ldots$, there is a. s. a unique giant component.
2. When $2 < \beta < \beta_0 = 3.47875\ldots$, the second largest components are a. s. of size $\Theta(\log n)$. For any $2 \le x < \Theta(\log n)$, there is almost surely a component of size x.
3. When $\beta = 2$, a. s. the second largest components are of size $\Theta(\frac{\log n}{\log\log n})$. For any $2 \le x < \Theta(\frac{\log n}{\log\log n})$, there is almost surely a component of size x.
4. When $1 < \beta < 2$, the second largest components are a. s. of size $\Theta(1)$. The graph is a. s. not connected.
5. When $0 < \beta < 1$, the graph is a. s. connected.
6. For $\beta = \beta_0 = 3.47875\ldots$, this is a very complicated case. It corresponds to the double jump of random graph $\mathcal{G}(n,p)$ with $p = \frac{1}{n}$. For $\beta = 1$, there is a non-trivial probability for either cases that the graph is connected or disconnected.

Before proceeding to state the main theorems, here are some general discussions:

For $\beta > 8$, Molloy and Reed's result immediately implies that almost surely there is no giant component. When $\beta \le 8$, additional analysis is needed to deal with the degree constraints. We will prove in Theorem 2 that almost surely there is no giant component when $\beta > \beta_0$. Also, almost surely there is a unique giant component when $\beta < \beta_0$ (The proof will be given in the full paper).

For $2 \le \beta \le \beta_0$, we will consider the sizes of the second largest component in section 5. It can be shown that the second largest component almost surely has size $O(\log n)$.

In the other direction, we will show that the second largest component has size at least $\Theta(\log n)$.

For $0 < \beta < 2$, the graph has $\Theta(e^{\frac{2\alpha}{\beta}})$ edges. We expect that the giant exponent is very large. For some constants T and C, a. s. every vertex of degree greater that $T \log n \approx C\alpha$ belongs to the giant component. That is, the number of edges which do not belong to the giant component is quite small. It is at most

$$\frac{1}{2}\sum_{x=1}^{C\alpha} x \lfloor \frac{e^\alpha}{x^\beta} \rfloor \approx \Theta((C\alpha)^{2-\beta}e^\alpha) \approx O(E^{\frac{\beta}{2}} \log^{2-\beta} E)$$

Now we consider the second largest component. For any pair (u, v), the probability that u belongs to the giant component while v belongs to the other component of size greater than $M = O(1)$ is at most

$$(E^{\frac{\beta}{2}-1} \log^{2-\beta} E)^M = o(n^{-2})$$

for some large constant M, which only depends on β. This implies that all components except for the giant component a. s. have size at most M. Therefore, a. s. the second largest component has size $O(1)$.

For $1 < \beta < 2$, fix a vertex v of degree 1. The probability that the other vertex that connects to v is also of degree 1 is about

$$\Theta(\frac{e^\alpha}{e^{\frac{2\alpha}{\beta}}})$$

Therefore the probability that no component has size of 2 is at most

$$(1 - \Theta(\frac{e^\alpha}{e^{\frac{2\alpha}{\beta}}}))^{e^\alpha} \approx e^{-\Theta(e^{2\alpha - \frac{2\alpha}{\beta}})} \approx o(1)$$

In other words, the graph a. s. has at least one component of size 2.

For $0 < \beta < 1$, the random graph is a. s. connected. Here we sketch the ideas. Since the size of the possible second largest component is bounded by a constant M, all vertices of degree $\geq M$ are almost surely in the giant component. We only need to show the probability that there is an edge connecting two small degree vertices is small. There are only

$$\sum_{x=1}^{M} x \lfloor \frac{e^\alpha}{x^\beta} \rfloor \approx Ce^\alpha$$

vertices with degree less than M. For any random pair of vertices (u, v), the probability that there is an edges connecting them is about

$$\frac{1}{E} = \Theta(e^{-\frac{2\alpha}{\beta}})$$

Hence the probability that there is edge connecting two small degree vertices is at most

$$\sum_{u,v} \frac{1}{E} = (Ce^\alpha)^2 \Theta(e^{\frac{2\alpha}{\beta}}) = o(1)$$

Hence, every vertex is a. s. connected to a vertex with degree $\geq M$, which a. s. belongs to the giant exponent. Hence, the random graph is a. s. connected.

The case of $\beta = 2$ is quite interesting. In this case, the graph has $\frac{1}{4}\alpha e^\alpha$ edges. Since a. s. all other components except for the giant one has size at most $O(\log n) = M\alpha \log \alpha$ for some constant M. Hence, a. s. all vertices with degree at least $M\alpha \log \alpha$ are in the giant component. Hence, the giant component is so large that only a small portion of vertices (as bounded below) are not in it.

$$\sum_{x=1}^{M\alpha \log \alpha} x \lfloor \frac{e^\alpha}{x^2} \rfloor \approx (\log \alpha) e^\alpha$$

For any pair of vertices (u, v), the probability that u is in the giant component while v is in other component of size at least $2.1\frac{\alpha}{\log \alpha}$ is at most

$$(\frac{\log \alpha}{\alpha})^{2.1\frac{\alpha}{\log \alpha}} = e^{2.1\frac{\alpha \log \log \alpha}{\log \alpha} - 2.1\alpha} = o(n^{-2})$$

Again, this almost surely is not likely to happen. Hence, we prove that the size of the second largest components is at most $2.1\frac{\alpha}{\log \alpha}$.

Now we find a vertex v of degree $x = 0.9\frac{\alpha}{\log \alpha}$. The probability that all its neighbors are of degree 1 is $(\frac{1}{\alpha})^x$. The probability that no such vertex exists is at most

$$(1 - (\frac{1}{\alpha})^x)^{\frac{e^\alpha}{x^2}} \approx e^{-(\frac{1}{\alpha})^x \frac{e^\alpha}{x^2}} = e^{-\frac{e^{0.1\alpha}}{x^2}} = o(1)$$

Hence, a. s. there is a vertex of degree $0.9\frac{\alpha}{\log \alpha}$, which forms a connected component of size $0.9\frac{\alpha}{\log \alpha} + 1$. Again, when x is smaller, almost surely there is a component of size x.

4. THE SIZES OF CONNECTED COMPONENTS IN CERTAIN RANGES FOR β

For $\beta > \beta_0 = 3.47875\ldots$, almost surely there is no giant component. This range is of special interest since it is quite useful later for describing the distribution of small components. We will prove the following:

THEOREM 1. *For (α, β)-graphs with $\beta > 4$, the distribution of the number of connected components is as follows:*

1. *For each vertex v of degree $d = \Omega(1)$, let τ be the size of connected component containing v. Then*

$$Pr\left(|\tau - \frac{d}{c_1}| > \frac{2\lambda}{c_1}\sqrt{\frac{dc_2}{c_1}}\right) \leq \frac{2}{\lambda^2}$$

where $c_1 = 2 - \frac{\zeta(\beta-2)}{\zeta(\beta-1)}$ and $c_2 = \frac{\zeta(\beta-3)}{\zeta(\beta-1)} - \left(\frac{\zeta(\beta-2)}{\zeta(\beta-1)}\right)^2$ are two constants. In other words, for d a (slowly) increasing function and $\lambda = d^\epsilon$, for some arbitrarily small postive constant ϵ, the vertex v a. s. belongs to a connected component of size $\frac{d}{c_1} + O(d^{\frac{1}{2}+\epsilon})$.

2. *The number of connected components of size x is a. s. at least*

$$(1 + o(1))\frac{e^\alpha}{c_1^{\beta-1} x^\beta}$$

and at most

$$c_3 \frac{e^\alpha \log^{\frac{\beta}{2}-1} n}{x^{\frac{\beta}{2}+1}}$$

where $c_3 = \frac{4^{1+\beta} c_2}{(\beta-2)c_1^{\frac{1}{2}+\beta}}$ is a constant only depending on β.

3. A connected component of the (α,β)-graph a. s. has the size at most

$$e^{\frac{2\alpha}{\beta+2}}\alpha = \Theta(n^{\frac{2}{\beta+2}}\log n)$$

In our proof we use the second moment whose convergence depends on $\beta > 4$. In fact for $\beta \le 4$ the second moment diverges as the size of the graph goes to infinity so that our method no longer applies.

Theorem 1 strengthens the following result (which can be derived from Lemma 3 in [14]) for the range of $\beta > 4$.

THEOREM 2. *For $\beta > \beta_0 = 3.47875\ldots$, a connected component of the (α,β)-graph a. s. has the size at most*

$$Ce^{\frac{2\alpha}{\beta}}\alpha = \Theta(n^{\frac{2}{\beta}}\log n)$$

where $C = \frac{16}{c_1^2}$ is a constant only depending on β.

The proof for Theorem 2 is by using branching process method. We here briefly describe the proof since it is needed for the proof of Theorem 1. We start by "exposing" any vertex v_0 in our graph, then we expose its neighbors, and then the neighbors of its neighbors, repeating until the entire component is exposed. At any stage of the process the entire component will have some nodes which are marked "live," some which are marked "dead," and some which are not marked at all. At stage i, we choose an arbitrary live vertex v to expose. Then we mark v dead and, for each neighbor u of v, we mark u live if u is unmarked so far. Let L_i be the set of marked vertices at stage i and X_i be the random variable that denotes the number of vertices in L_i. We note that all vertices in L_i are marked by either "live" or "dead". Let O_i be the set of live vertices and Y_i be the random variable that is the number of vertices of O_i. At each step we mark exact one dead vertex, so the total number of dead vertices at i-th step is i. We have $X_i = Y_i + i$. Initially we assign $L_0 = O_0 = \{v_0\}$. Then at stage $i \ge 1$, we do the following:

1. If $Y_{i-1} = 0$, then we stop and output X_{i-1}.
2. Otherwise, randomly choose a live vertex u from O_{i-1} and expose its neighbors in N_u. Then mark u dead and mark each vertex live if it is in N_u but not in L_{i-1}. We have $L_i = L_{i-1} \cup N_u$, and $O_i = (O_{i-1} \setminus \{u\}) \cup (N_u \setminus L_{i-1})$.

Suppose that v has degree d. Then $X_1 = d + 1$, and $Y_1 = d$. Eventually Y_i will hit 0 if i is large enough. Let τ denote the stopping time of Y, namely, $Y_\tau = 0$. Then $X_\tau = Y_\tau + \tau = \tau$ measures the size of the connected component. We first compute the expected value of Y_i and then use Azuma's Inequality [14] to prove Theorem 2.

Suppose that the vertex u is exposed at stage i. Then $N_u \cap L_{i-1}$ contains at least one vertex, which was exposed to reach u. However, $N_u \cap L_{i-1}$ may contain more than one vertex. We call them "backedges". We note that "backedges" causes the exploration to stop more quickly, especially when the component is large. However in our case $\beta > \beta_0 = 3.47875\ldots$, the contribution of "backedges" is quite small. We denote $Z_i = \#\{N_u\}$ and $W_i = \#\{N_u \cap L_{i-1}\} - 1$. Z_i measures the degree of the vertex exposed at stage i, while W_i measures the number of "backedges". By definition, we have

$$Y_i - Y_{i-1} = Z_i - 2 - W_i.$$

We have

$$\begin{aligned} E(Z_i) &= \textstyle\sum_{x=1}^{e^{\frac{\alpha}{\beta}}} x\frac{x\frac{e^\alpha}{x^\beta}}{E} = \frac{e^\alpha}{E}\sum_{x=1}^{e^{\frac{\alpha}{\beta}}} x^{2-\beta} \\ &= \frac{\zeta(\beta-2)+O(n^{\frac{3}{\beta}-1})}{\zeta(\beta-1)+O(n^{\frac{2}{\beta}-1})} \\ &= \frac{\zeta(\beta-2)}{\zeta(\beta-1)} + O(n^{\frac{3}{\beta}-1}) \end{aligned}$$

Now we will bound W_i. Suppose that there are m edges exposed at stage $i-1$. Then the probability that a new neighbor is in L_{i-1} is at most $\frac{m}{E}$. We have

$$\begin{aligned} E(W_i) &\le \sum_{x=1}^{\infty} x\left(\frac{m}{E}\right)^x \\ &= \frac{\frac{m}{E}}{(1-\frac{m}{E})^2} \qquad (*) \\ &= \frac{m}{E} + O((\frac{m}{E})^2) \end{aligned}$$

provided $\frac{m}{E} = o(1)$.

When $i \le Ce^{\frac{2\alpha}{\beta}}\alpha$, m is at most $ie^{\frac{\alpha}{\beta}} \le Ce^{\frac{3\alpha}{\beta}}\alpha$. Hence,

$$\frac{m}{E} = O(n^{\frac{3}{\beta}-1}\log n) = o(1)$$

We have

$$\begin{aligned} E(Y_i) &= Y_1 + \sum_{j=2}^{i} E(Y_j - Y_{j-1}) \\ &= d + \sum_{j=2}^{i} E(Z_j - 2 - W_j) \\ &= d + (i-1)\left(\frac{\zeta(\beta-2)}{\zeta(\beta-1)} - 2\right) - iO(n^{\frac{3}{\beta}-1}\log n) \\ &= d - c_1(i-1) + io(1) \end{aligned}$$

Proof of Theorem 2: Since $|Y_j - Y_{j-1}| \le e^{\frac{\alpha}{\beta}}$, by Azuma's martingale inequality, we have

$$Pr(|Y_i - E(Y_i)| > t) \le 2\exp\left(\frac{-t^2}{2ie^{2\alpha/\beta}}\right)$$

By taking $i = \frac{16}{c_1^2}e^{\frac{2\alpha}{\beta}}\log n$, and $t = \frac{c_1}{2}i$. Since

$$E(Y_i)+t = d-c_1(i-1)+io(1)+\frac{c_1}{2}i = -\frac{c_1}{2}i+d+c_1+io(1) < 0$$

We have

$$\begin{aligned} Pr(\tau > \tfrac{16}{c_1^2}e^{\frac{\alpha}{\beta}}\log n) &= Pr(\tau > i) \le Pr(Y_i \ge 0) \\ &\le Pr(Y_i > E(Y_i) + t) \\ &\le 2\exp\left(\frac{-t^2}{2ie^{2\alpha/\beta}}\right) = \frac{2}{n^2} \end{aligned}$$

Hence, the probability that there exists a vertex v such that v lies in a component of size greater than $\frac{16}{c_1^2}e^{\frac{2\alpha}{\beta}}\log n$ is at most

$$n\frac{2}{n^2} = \frac{2}{n} = o(1). \qquad \square$$

vertices

The proof of Theorem 1 uses the methodology above as a starting point while introducing the calculation of the variance of the above random variables.

Proof of Theorem 1

We follow the notation and previous results of Section 4. Under the assumption $\beta > 4$, we consider the following:

$$\begin{aligned}
Var(Z_i) &= \sum_{x=1}^{e^{\frac{\alpha}{\beta}}} x^2 \frac{x \frac{e^\alpha}{x^\beta}}{E} - E(Z_i)^2 \\
&= \frac{e^\alpha}{E} \sum_{x=1}^{e^{\frac{\alpha}{\beta}}} x^{3-\beta} - E(Z_i)^2 \\
&= \frac{\zeta(\beta-3) + O(n^{\frac{4}{\beta}-1})}{\zeta(\beta-1) + O(n^{\frac{2}{\beta}-1})} - \left(\frac{\zeta(\beta-2)}{\zeta(\beta-1)}\right)^2 \\
&\quad + O(n^{\frac{3}{\beta}-1}) \\
&= \frac{\zeta(\beta-3)}{\zeta(\beta-1)} - \left(\frac{\zeta(\beta-2)}{\zeta(\beta-1)}\right)^2 + O(n^{\frac{4}{\beta}-1}) \\
&= c_2 + o(1)
\end{aligned}$$

since $\beta > 4$.

We need to compute the covariants. There are models for random graphs in which the edges are in dependently chosen. Then, Z_i and Z_j are independent. However, in the model based on random matchings, there is a small correlation. For example, $Z_i = x$ slightly effects the probability of $Z_j = y$. Namely, $Z_j = x$ has slightly less chance, while $Z_j = y \neq x$ has slightly more chance. Both differences can be bounded by

$$\frac{1}{E-1} - \frac{1}{E} \leq \frac{2}{E^2}$$

Hence $CoVar(Z_i, Z_j) \leq E(Z_i)E\frac{2}{E^2} = O(\frac{1}{n})$ if $i \neq j$.

Now we will bound W_i. Suppose that there are m edges exposed at stage $i-1$. Then the probability that a new neighbor is in L_{i-1} is at most $\frac{m}{E}$. We have

$$\begin{aligned}
Var(W_i) &\leq \sum_{x=1}^{\infty} x^3 \left(\frac{m}{E}\right)^x - E(W_i)^2 \\
&= \frac{\frac{m}{E}(\frac{m}{E}+1)}{(1-\frac{m}{E})^3} - O((\frac{m}{E})^2) \\
&= \frac{m}{E} + O((\frac{m}{E})^2)
\end{aligned}$$

$$CoVar(W_i, W_j) \leq \sqrt{Var(W_i)Var(W_j)} \leq \frac{m}{E} + O((\frac{m}{E})^2)$$

$$CoVar(Z_i, W_j) \leq \sqrt{Var(Z_i)Var(W_j)} = O(\sqrt{\frac{m}{E}})$$

When $i = O(e^{\frac{\alpha}{\beta}})$, $m \leq ie^{\frac{\alpha}{\beta}} = O(e^{\frac{2\alpha}{\beta}})$, we have

$$\begin{aligned}
E(Y_i) &= d + (i-1)\left(\frac{\zeta(\beta-2)}{\zeta(\beta-1)} - 2\right) + iO(n^{\frac{3}{\beta}-1}) + i\frac{m}{E} \\
&= d - (i-1)c_1 + O(n^{\frac{4}{\beta}-1}) \\
&= d - (i-1)c_1 + o(1)
\end{aligned}$$

$$\begin{aligned}
Var(Y_i) &= Var(d + \sum_{j=2}^{i}(Y_j - Y_{j-1})) \\
&= Var(\sum_{j=2}^{i}(Z_j - W_j)) \\
&= \sum_{j=2}^{i}(Var(Z_j) + Var(W_j)) \\
&\quad + \sum_{2 \leq j \neq k \leq i}(CoVar(Z_j, Z_k) \\
&\quad - CoVar(Z_j, W_k) + CoVar(W_j, W_k)) \\
&= ic_2 + io(1) + i^2(O(\frac{1}{n}) + O(\sqrt{e^{(\frac{2}{\beta}-1)\alpha}}) \\
&\quad + O(e^{(\frac{2}{\beta}-1)\alpha})) \\
&= ic_2 + io(1) + i(O(e^{(\frac{2}{\beta}-\frac{1}{2})\alpha}) + O(e^{(\frac{3}{\beta}-1)\alpha})) \\
&= ic_2 + io(1)
\end{aligned}$$

Chebyshev's inequality gives

$$Pr(|Y_i - E(Y_i)| > \lambda\sigma) < \frac{1}{\lambda^2}$$

where σ is the standard deviation of Y_i, $\sigma = \sqrt{ic_2} + o(\sqrt{i})$
Let
$i_1 = \lfloor \frac{d}{c_1} - \frac{2\lambda}{c_1}\sqrt{\frac{dc_2}{c_1}} \rfloor$ and $i_2 = \lceil \frac{d}{c_1} + \frac{2\lambda}{c_1}\sqrt{\frac{dc_2}{c_1}} \rceil$. We have

$$\begin{aligned}
E(Y_{i_1}) - \lambda\sigma &= d - (i_1 - 1)c_1 + o(1) - (\lambda\sqrt{c_2 i_1} + o(\sqrt{i_1})) \\
&\geq 2\lambda\sqrt{\frac{dc_2}{c_1}} - \lambda\sqrt{c_2\frac{d}{c_1}} - o(\sqrt{d}) \\
&= \lambda\sqrt{\frac{dc_2}{c_1}} - o(\sqrt{d}) \\
&> 0
\end{aligned}$$

Hence,

$$Pr(\tau < i_1) \leq Pr(Y_{i_1} \leq 0) \leq Pr(Y_{i_1} < E(Y_{i_1}) - \lambda\sigma) \leq \frac{1}{\lambda^2}$$

Similarly,

$$\begin{aligned}
E(Y_{i_2}) + \lambda\sigma &= d - (i_2 - 1)c_1 + o(1) + (\lambda\sqrt{c_2 i_2} + o(\sqrt{i_2})) \\
&\geq -2\lambda\sqrt{\frac{dc_2}{c_1}} + \lambda\sqrt{c_2\frac{d}{c_1}} + o(\sqrt{d}) \\
&= -\lambda\sqrt{\frac{dc_2}{c_1}} + o(\sqrt{d}) \\
&< 0
\end{aligned}$$

Hence,

$$Pr(\tau > i_2) \leq Pr(Y_{i_2} > 0) \leq Pr(Y_{i_2} > E(Y_{i_2}) + \lambda\sigma) \leq \frac{1}{\lambda^2}$$

Therefore

$$Pr\left(|\tau - \frac{d}{c_1}| > \frac{2\lambda}{c_1}\sqrt{\frac{dc_2}{c_1}}\right) \leq \frac{2}{\lambda^2}$$

For a fixed v and λ a slowly increasing function to infinity, above inequality implies that almost surely we have $\tau = \frac{d}{c_1} + O(\lambda\sqrt{d})$.

We note that almost all components generated by vertices of degree x is about the size of $\frac{d}{c_1}$. One such component can have at most about $\frac{1}{c_1}$ vertices of degree d. Hence, the number of component of size $\frac{d}{c_1}$ is at least $\frac{c_1 e^{\frac{\alpha}{\beta}}}{d^\beta}$. Let $d = c_1 x$. Then the number of components of size x is at least

$$\frac{e^{\frac{\alpha}{\beta}}}{c_1^{\beta-1} x^\beta}(1+o(1))$$

The above proof actually gives the following result. The size of every component, whose vertices have degree at most d_0, is almost surely $Cd_0^2 \log n$ where $C = \frac{16}{c_1^2}$ is the same constant as in Theorem 2. Let $x = Cd_0^2 \log n$ and consider the number of components of size x. A component of size x almost surely contains at least one vertex of degree greater than d_0.

For each vertex v with degree $d \geq d_0$, by part 1, we have

$$Pr\left(|\tau - \frac{d}{c_1}| > \frac{2\lambda_d}{c_1}\sqrt{\frac{dc_2}{c_1}}\right) \leq \frac{2}{\lambda_d^2}$$

Let $\lambda_d = \frac{c_1 C d_0^2 \log n}{4}\sqrt{\frac{c_1}{c_2 d}}$, we have

$$\begin{aligned} Pr(\tau \geq Cd_0^2 \log n) &\leq Pr\left(\tau > \frac{d}{c_1} + \frac{2\lambda_d}{c_1}\sqrt{\frac{dc_2}{c_1}}\right) \\ &\leq C_3 \frac{d}{d_0^4 \log^2 n} \end{aligned}$$

where $C_3 = \frac{32c_2}{c_1^3 C^2} = \frac{c_1 c_2}{8}$ is constant depending only on β. Since there are only $\frac{e^\alpha}{d^\beta}$ vertices of degree d, the number of components of size at least x is at most

$$\begin{aligned} \sum_{d=d_0}^{e^{\frac{\alpha}{\beta}}} \frac{e^\alpha}{d^\beta} C_3 \frac{d}{d_0^4 \log^2 n} &\leq \frac{C_3 e^\alpha}{d_0^4 \log^2 n} \sum_{d=d_0}^{\infty} \frac{1}{d^{\beta-1}} \\ &\leq \frac{C_3 e^\alpha}{d_0^4 \log^2 n} \frac{2}{\beta-2} \frac{1}{d_0^{\beta-2}} \\ &= \frac{2C_3 e^\alpha}{(\beta-2) d_0^{\beta+2} \log^2 n} \\ &= c_3 \frac{e^\alpha \log^{\frac{\beta}{2}-1} n}{x^{\frac{\beta}{2}+1}} \end{aligned}$$

where $c_3 = \frac{2C_3}{(\beta-2)} C^{1+\frac{\beta}{2}} = \frac{4^{1+\beta} c_2}{(\beta-2)c_1^{1+\beta}}$. For $x = e^{\frac{2\alpha}{\beta+2}}\alpha$, the above inequality implies that the number of components of size at least x is at most $o(1)$. In other words, almost surely no component has size greater than $e^{\frac{2\alpha}{\beta+2}}\alpha$. This completes the proof of Theorem 1.

5. ON THE SIZE OF THE SECOND LARGEST COMPONENT

For the range of $2 < \beta < \beta_0$, we want to show that the second largest components almost surely have size of at most $O(\log n)$. However, we can not apply Azuma's martingale inequality directly as in the proofs of previous sections. For example, the branching process method is no longer feasible when vertices of large degrees are involved. We will modify the branching process method as follows:

For $2 < \beta < \beta_0$, we consider $Q = \frac{1}{E}\sum_{x=1}^{e^{\frac{\alpha}{\beta}}} x(x-2)\lfloor \frac{e^\alpha}{x^\beta} \rfloor$. (Note that Q is a positive constant.) There is a constant integer x_0 satisfying $\frac{1}{E}\sum_{x=1}^{x_0} x(x-2)\lfloor \frac{e^\alpha}{x^\beta} \rfloor > \frac{Q}{2}$. We choose δ satisfying:

$$\frac{\delta}{(1-\delta)^2} = \frac{Q}{4}.$$

If the component has more than δE edges, it must have $\Theta(n)$ vertices. So it is a giant component and we are done. We may assume that the component has no more than δE edges.

We now consider the following modified branching process: We start with Y_0^* live vertices and $Y_0^* \geq C \log n$. At the i-th step, we choose one live vertex u and exposed its neighbors. If the degree of u is less than or equal to x_0, we proceed as in section 4, by marking u dead and all vertices $v \in N(u)$ live (provided v is not marked before). If the degree of u is greater than x_0, we will mark exactly one vertex $v \in N(u)$ live and others dead, provided v is not marked before. In both case u is marked dead. Let O_i^* be the set of live vertices at i-th step (in contrast to the live set O_i). We denote by Y_i^* the new random variable (in contrast to Y_i) that is the number of vertices in O_i^*. Our main idea is to show that Y_i^*, a truncated version of Y_i, is well-concentrated around $E(Y_i^*)$. Although it is difficult to directly derive such result for Y_i because of vertices of large degrees, we will be able to bound the distribution Y_i^*.

To be precise, Y_i^* satisfies the following:

- $Y_0^* \geq \lceil C \log n \rceil$, where $C = \frac{130x_0^2}{Q}$ is a constant only depending on β.
- $-1 \leq Y_i^* - Y_{i-1}^* \leq x_0$.
- Let W_i be the number of "backedges" as defined in section 4. By inequality (*) and the assumption that the number of edges m in the component is at most δn, we have $E(W_i) \leq \frac{\delta}{(1-\delta)^2} = \frac{Q}{4}$. Hence, we have

$$\begin{aligned} E(Y_i^* - Y_{i-1}^*) &\approx \frac{1}{E}\sum_{x=1}^{x_0} x(x-2)\lfloor \frac{e^\alpha}{x^\beta} \rfloor - E(W_i) \\ &\geq \frac{Q}{2} - \frac{Q}{4} = \frac{Q}{4}. \end{aligned}$$

By Azuma's martingale inequality, we have

$$\begin{aligned} Pr(Y_i^* \leq \frac{Qi}{8}) &\leq Pr(Y_i^* - E(Y_i^*) \leq -\frac{Qi}{8}) \\ &< e^{-\frac{(Qi/8)^2}{2ix_0^2}} = o(n^{-1}) \end{aligned}$$

provided $i > C \log n$.

The above inequality implies that with probability at least $1 - o(n^{-1})$, $Y_i^* > \frac{Qi}{8} > 0$ when $i > \lceil C \log n \rceil$. Since Y_i^* decreases at most by 1 at each step, Y_i^* can not be zero if $i \leq \lceil C \log n \rceil$. So $Y_i^* > 0$ for all i. In other words, a. s. the branching process will not stop. However, it is impossible to have $Y_n^* > 0$, that is a contradiction. Thus we conclude that the component must have at least δn edges. So it is a giant component. We note that if a component has more than $\lceil C \log n \rceil$ edges exposed, then almost surely it is a giant component. In particular, any vertex with degree more than

$\lceil C \log n \rceil$ is almost surely in the giant component. Hence, the second component have size of at most $\Theta(\log n)$.

Next, we will show that the second largest has size at least $\Theta(\log n)$. We consider the vertices v of degree $x = c\alpha$, where c is some constant. There is a positive probability that all neighboring vertices of v have degree 1. In this case, we get a connected component of size $x+1 = \Theta(\log n)$. The probability of this is about

$$(\frac{1}{\zeta(\beta-1)})^{c\alpha}$$

Since there are $\frac{e^\alpha}{(c\alpha)^\beta}$ vertices of degree x, the probability that none of them has the above property is about

$$\begin{aligned}(1-\frac{1}{\zeta(\beta-1)^{c\alpha}})^{\frac{e^\alpha}{(c\alpha)^\beta}} &\approx e^{-\frac{1}{\zeta(\beta-1)^{c\alpha}}\frac{e^\alpha}{(c\alpha)^\beta}} \\ &= e^{-\frac{(\frac{e}{\zeta(\beta-1)^c})^\alpha}{(c\alpha)^\beta}} = o(1)\end{aligned}$$

where we have

$$c = \begin{cases} 1 & \text{if } \beta \geq 3 \\ \frac{1}{-2\log(\beta-2)} & \text{if } 3 > \beta > 2 \end{cases}$$

In other words, a. s. there is a component of size $c\alpha + 1 = \Theta(\log n)$. Therefore, the second largest component has size $\Theta(\log n)$. Moreover, the above argument holds if we replace $c\alpha$ by any small number. Hence, small components exhibit a continuous behavior.

We remark that the methods described in this section can be extended to deal with the case of $0 \leq \beta \leq 2$. The detailed treatment will be left to the full paper.

6. COMPARISONS WITH REALISTIC MASSIVE GRAPHS

Our (α, β)-random graph model was originally derived from massive graphs generated by long distance telephone calls. These so-called *call graphs* are taken over different time intervals. For the sake of simplicity, we consider all the calls made in one day. Every completed phone call is an edge in the graph. Every phone number which either originates or receives a call is a node in the graph. When a node originates a call, the edge is directed out of the node and contributes to that node's outdegree. Likewise, when a node receives a call, the edge is directed into the node and contributes to that node's indegree.

In Figure 2, we plot the number of vertices versus the indegree for the call graph of a particular day. Let $y(i)$ be the number of vertices with indegree i. For each i such that $y(i) > 0$, a $\times$ is marked at the coordinate $(i, y(i))$. As similar plot is shown in Figure 1 for the outdegree. Plots of the number of vertices versus the indegree or outdegree for the call graphs of other days are very similar. For the same call graph in Figure 3 we plot the number of connected components for each possible size.

The degree sequence of the call graph does not obey perfectly the (α, β)-graph model. The number of vertices of a given degree does not even monotonically decrease with increasing degree. Moreover, the call graph is directed, i.e., for each edge there is a node that originates the call and a node that receives the call. The indegree and outdegree of a node need not be the same. Clearly the (α, β)-random graph model does not capture all of the random behavior of the real world call graph.

Nonetheless, our model does capture some of the behavior of the call graph. To see this we first estimate α and β of Figure 2. Recall that for an (α, β)-graph, the number of vertices as a function of degree is given by $\log y = \alpha - \beta \log x$. By approximating Figure 2 by a straight line, β can be estimated using the slope of the line to be approximately 2.1. The value of e^α for Figure 2 is approximately 30×10^6. The total number of nodes in the call graph can be estimated by $\zeta(2.1) * e^\alpha = 1.56 * e^\alpha \approx 47 \times 10^6$

For β between 2 and β_0, the (α, β)-graph will have a giant component of size $\Theta(n)$. In addition, a. s. , all other components are of size $O(\log n)$. Moreover, for any $2 \geq x \geq O(\log n)$, a component of size x exists. This is qualitatively true of the distribution of component sizes of the call graph in Figure 3[2]. The one giant component contains nearly all of the nodes. The maximum size of the next largest component is indeed exponentially smaller than the size of the giant component. Also, a component of nearly every size below this maximum exists. Interestingly, the distribution of the number of components of size smaller than the giant component is nearly log-log linear. This suggests that after removing the giant component, one is left with an (α, β)-graph with $\beta > 4$ (Theorem 1 yields a log-log linear relation between number of components and component size for $\beta > 4$.) This intuitively seems true since the greater the degree, the fewer nodes of that degree we expect to remain after deleting the giant component. This will increase the value of β for the resulting graph.

There are numerous questions that remain to be studied. For example, what is the effect of time scaling? How does it correspond with the evolution of β? What are the structural behaviors of the call graphs? What are the correlations between the directed and undirected graphs? It is of interest to understand the phase transition of the giant component in the realistic graph. In the other direction, the number of tiny components of size 1 is leading to many interesting questions as well. Clearly, there is much work to be done in our understanding of massive graphs.

Acknowledgments. We are grateful to J. Feigenbaum, J. Abello, A. Buchsbaum, J. Reeds, and J. Westbrook for their assistance in preparing the figures and for many interesting discussions on call graphs.

7. REFERENCES

[1] J. Abello, A. Buchsbaum, and J. Westbrook, A functional approach to external graph algorithms, *Proc. 6th European Symposium on Algorithms*, pp. 332–343, 1998.

[2] W. Aiello, F. Chung, L. Lu, Random evolution of power law graphs, manuscript.

[3] R. Albert, H. Jeong and A. Barabási, Diameter of the World Wide Web, *Nature*, **401**, September 9, 1999.

[4] N. Alon and J. H. Spencer, *The Probabilistic Method*, Wiley and Sons, New York, 1992.

[5] A. Barabási, and R. Albert, Emergence of scaling in random networks, *Science*, **286**, October 15, 1999.

[2] This data was compiled by J. Abello and A. Buchsbaum of AT&T Labs from raw phone call records using, in part, the external memory algorithm of Abello, Buchsbaum, and Westbrook [1] for computing connected components of massive graphs.

[6] A. Barabási, R. Albert, and H. Jeong Scale-free characteristics of random networks: the topology of the world wide web, *Elsevier Preprint* August 6, 1999.

[7] P. Erdős and A. Rényi, On the evolution of random graphs, *Publ. Math. Inst. Hung. Acad. Sci.* 5 (1960), 17–61.

[8] P. Erdős and A. Rényi, On the strength of connectedness of random graphs, *Acta Math. Acad. Sci. Hungar.* **12** (1961), 261-267.

[9] M. Faloutsos, P. Faloutsos, and C. Faloutsos, On power-law relationships of the internet topology, *Proceedings of the ACM SIGCOM Conference*, Cambridge, MA, 1999.

[10] J. Kleinberg, S. R. Kumar, P. Raphavan, S. Rajagopalan and A. Tomkins, The web as a graph: Measurements, models and methods, *Proceedings of the International Conference on Combinatorics and Computing*, July 26–28, 1999.

[11] S. R. Kumar, P. Raphavan, S. Rajagopalan and A. Tomkins, Trawling the web for emerging cyber communities, *Proceedings of the 8th World Wide Web Conference*, Edinburgh, Scotland, May 15–19, 1999.

[12] S. R. Kumar, P. Raghavan, S. Rajagopalan and A. Tomkins, Extracting large-scale knowledge bases from the web, *Proceedings of the 25th VLDB Conference*, Edinburgh, Scotland, September 7–10, 1999.

[13] Tomasz Luczak, Sparse random graphs with a given degree sequence, *Random Graphs*, vol 2 (Poznań, 1989), 165-182, Wiley, New York, 1992.

[14] Michael Molloy and Bruce Reed, A critical point for random graphs with a given degree sequence. *Random Structures and Algorithms*, Vol. 6, no. 2 and 3 (1995). 161-179.

[15] Michael Molloy and Bruce Reed, The size of the giant component of a random graph with a given degree sequence, *Combin. Probab. Comput.* 7, no. (1998), 295-305.

[16] P. Raghavan, personal communication.

[17] N. C. Wormald, The asymptotic connectivity of labeled regular graphs, *J. Comb. Theory* (B) **31** (1981), 156-167.

[18] N. C. Wormald, Models of random regular graphs, *Surveys in Combinatorics*, 1999 (LMS Lecture Note Series 267, Eds J.D.Lamb and D.A.Preece), 239–298.

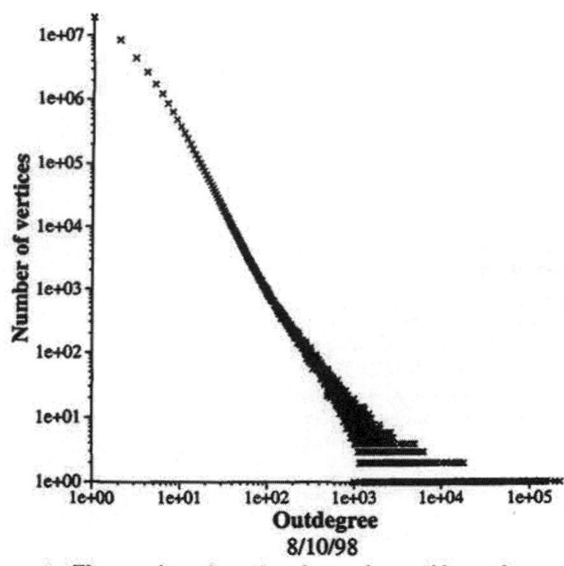

Figure 1: *The number of vertices for each possible outdegree for the call graph of a typical day.*

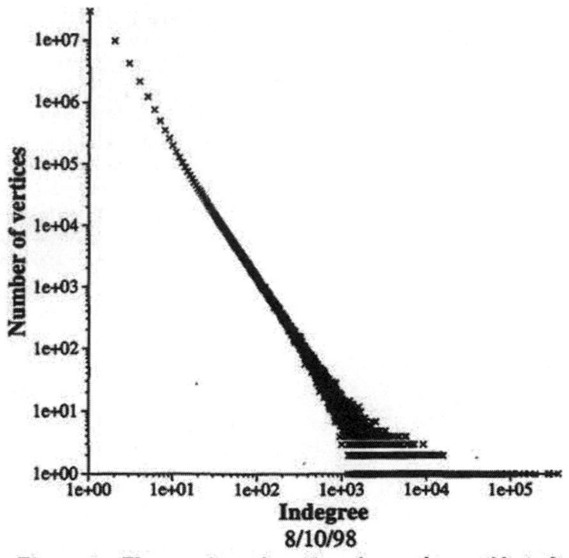

Figure 2: *The number of vertices for each possible indegree for the call graph of a typical day.*

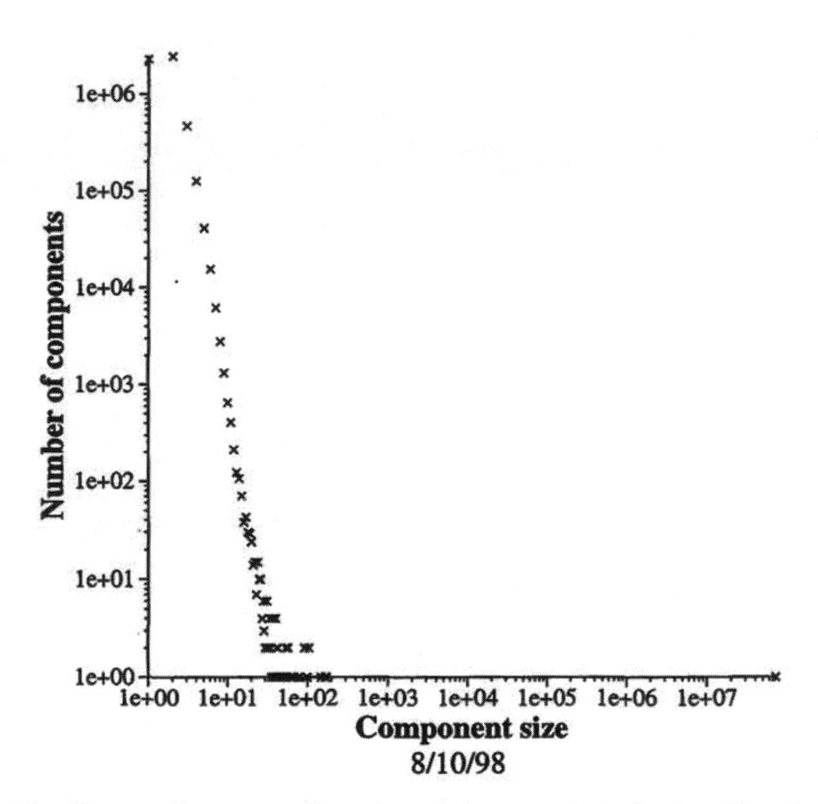

Figure 3: *The number of connected components for each possible component size for the call graph of a typical day.*

PHYSICAL REVIEW E, VOLUME 64, 026118

Random graphs with arbitrary degree distributions and their applications

M. E. J. Newman,[1,2] S. H. Strogatz,[2,3] and D. J. Watts[1,4]
[1]*Santa Fe Institute, 1399 Hyde Park Road, Santa Fe, New Mexico 87501*
[2]*Center for Applied Mathematics, Cornell University, Ithaca, New York 14853-3401*
[3]*Department of Theoretical and Applied Mechanics, Cornell University, Ithaca, New York 14853-1503*
[4]*Department of Sociology, Columbia University, 1180 Amsterdam Avenue, New York, New York 10027*
(Received 19 March 2001; published 24 July 2001)

Recent work on the structure of social networks and the internet has focused attention on graphs with distributions of vertex degree that are significantly different from the Poisson degree distributions that have been widely studied in the past. In this paper we develop in detail the theory of random graphs with arbitrary degree distributions. In addition to simple undirected, unipartite graphs, we examine the properties of directed and bipartite graphs. Among other results, we derive exact expressions for the position of the phase transition at which a giant component first forms, the mean component size, the size of the giant component if there is one, the mean number of vertices a certain distance away from a randomly chosen vertex, and the average vertex-vertex distance within a graph. We apply our theory to some real-world graphs, including the world-wide web and collaboration graphs of scientists and Fortune 1000 company directors. We demonstrate that in some cases random graphs with appropriate distributions of vertex degree predict with surprising accuracy the behavior of the real world, while in others there is a measurable discrepancy between theory and reality, perhaps indicating the presence of additional social structure in the network that is not captured by the random graph.

DOI: 10.1103/PhysRevE.64.026118 PACS number(s): 89.75.Hc, 87.23.Ge, 05.90.+m

I. INTRODUCTION

A random graph [1] is a collection of points, or vertices, with lines, or edges, connecting pairs of them at random [Fig. 1(a)]. The study of random graphs has a long history. Starting with the influential work of Erdős and Rényi in the 1950s and 1960s [2–4], random graph theory has developed into one of the mainstays of modern discrete mathematics, and has produced a prodigious number of results, many of them highly ingenious, describing statistical properties of graphs, such as distributions of component sizes, existence and size of a giant component, and typical vertex-vertex distances.

In almost all of these studies the assumption has been made that the presence or absence of an edge between two vertices is independent of the presence or absence of any other edge, so that each edge may be considered to be present with independent probability p. If there are N vertices in a graph, and each is connected to an average of z edges, then it is trivial to show that $p=z/(N-1)$, which for large N is usually approximated by z/N. The number of edges connected to any particular vertex is called the degree k of that vertex, and has a probability distribution p_k given by

$$p_k=\binom{N}{k}p^k(1-p)^{N-k}\simeq\frac{z^k e^{-z}}{k!}, \tag{1}$$

where the second equality becomes exact in the limit of large N. This distribution we recognize as the Poisson distribution: the ordinary random graph has a Poisson distribution of vertex degrees, a point which turns out to be crucial, as we now explain.

Random graphs are not merely a mathematical toy; they have been employed extensively as models of real-world networks of various types, particularly in epidemiology. The passage of a disease through a community depends strongly on the pattern of contacts between those infected with the disease and those susceptible to it. This pattern can be depicted as a network, with individuals represented by vertices and contacts capable of transmitting the disease by edges. A large class of epidemiological models known as susceptible/infectious/recovered models [5–7] makes frequent use of the so-called fully mixed approximation, which is the assumption that contacts are random and uncorrelated, i.e., they form a random graph.

Random graphs however turn out to have severe shortcomings as models of such real-world phenomena. Although it is difficult to determine experimentally the structure of the network of contacts by which a disease is spread [8], studies have been performed of other social networks such as networks of friendships within a variety of communities [9–11], networks of telephone calls [12,13], airline timetables [14], and the power grid [15], as well as networks in physical or

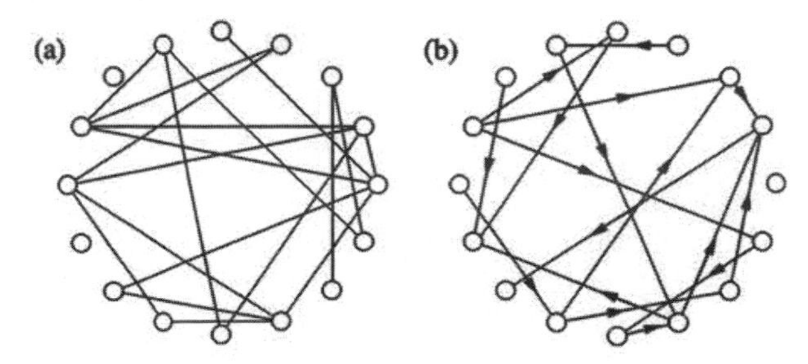

FIG. 1. (a) A schematic representation of a random graph, the circles representing vertices and the lines representing edges. (b) A directed random graph, i.e., one in which each edge runs in only one direction.

biological systems, including neural networks [15], the structure and conformation space of polymers [16,17], metabolic pathways [18,19], and food webs [20,21]. It is found [13,14] that the distribution of vertex degrees in many of these networks is measurably different from a Poisson distribution—often wildly different—and this strongly suggests, as has been emphasized elsewhere [22], that there are features of such networks that we would miss if we were to approximate them by an ordinary (Poisson) random graph.

Another very widely studied network is the internet, whose structure has attracted an exceptional amount of scrutiny, academic and otherwise, following its meteoric rise to public visibility starting in 1993. Pages on the world-wide web may be thought of as the vertices of a graph and the hyperlinks between them as edges. Empirical studies [23–26] have shown that this graph has a distribution of vertex degree which is heavily right skewed and possesses a fat (power law) tail with an exponent between -2 and -3. (The underlying physical structure of the internet also has a degree distribution of this type [27].) This distribution is very far from Poisson, and therefore we would expect that a simple random graph would give a very poor approximation of the structural properties of the web. However, the web differs from a random graph in another way also: it is directed. Links on the web lead from one page to another in only one direction [see Fig. 1(b)]. As discussed by Broder *et al.* [26], this has a significant practical effect on the typical accessibility of one page from another, and this effect also will not be captured by a simple (undirected) random graph model.

A further class of networks that has attracted scrutiny is the class of collaboration networks. Examples of such networks include the boards of directors of companies [28–31], co-ownership networks of companies [32], and collaborations of scientists [33–37] and movie actors [15]. As well as having strongly non-Poisson degree distributions [14,36], these networks have a bipartite structure; there are two distinct kinds of vertices on the graph with links running only between vertices of unlike kinds [38]—see Fig. 2. In the case of movie actors, for example, the two types of vertices are movies and actors, and the network can be represented as a graph with edges running between each movie and the actors that appear in it. Researchers have also considered the projection of this graph onto the unipartite space of actors only, also called a one-mode network [38]. In such a projection two actors are considered connected if they have appeared in a movie together. The construction of the one-mode network however involves discarding some of the information contained in the original bipartite network, and for this reason it is more desirable to model collaboration networks using the full bipartite structure.

Given the high current level of interest in the structure of many of the graphs described here [39], and given their substantial differences from the ordinary random graphs that have been studied in the past, it would clearly be useful if we could generalize the mathematics of random graphs to non-Poisson degree distributions, and to directed and bipartite graphs. In this paper we do just that, demonstrating in detail how the statistical properties of each of these graph types can be calculated exactly in the limit of large graph size. We also give examples of the application of our theory to the modeling of a number of real-world networks, including the world-wide web and collaboration graphs.

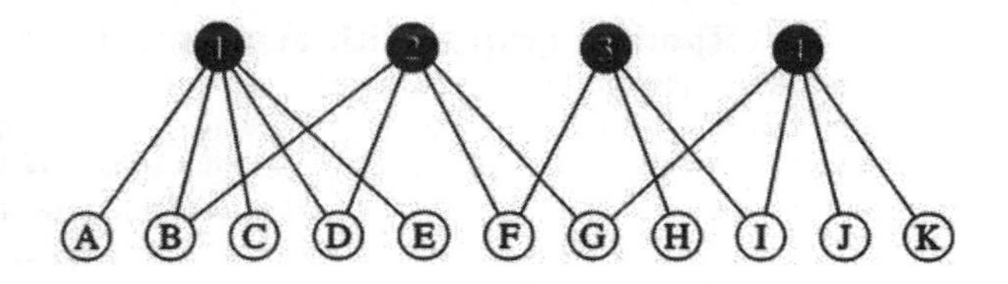

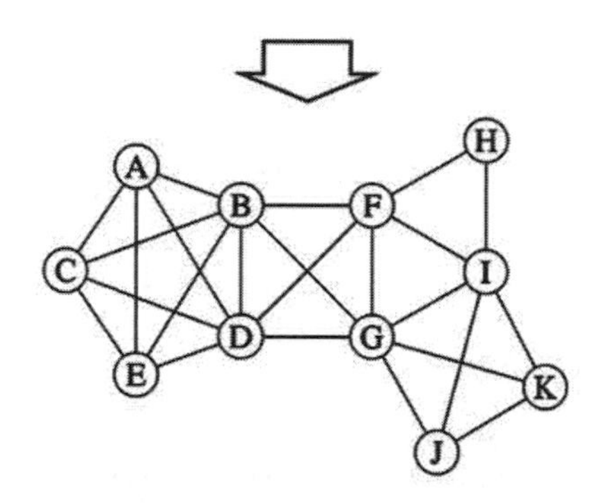

FIG. 2. A schematic representation (top) of a bipartite graph, such as the graph of movies and the actors who have appeared in them. In this small graph we have four movies, labeled 1 to 4, and 11 actors, labeled *A* to *K*, with edges joining each movie to the actors in its cast. In the lower part of the picture we show the one-mode projection of the graph for the 11 actors.

II. RANDOM GRAPHS WITH ARBITRARY DEGREE DISTRIBUTIONS

In this section we develop a formalism for calculating a variety of quantities, both local and global, on large unipartite undirected graphs with arbitrary probability distribution of the degrees of their vertices. In all respects other than their degree distribution, these graphs are assumed to be entirely random. This means that the degrees of all vertices are independent identically distributed random integers drawn from a specified distribution. For a given choice of these degrees, also called the "degree sequence," a graph is chosen uniformly at random from the set of all graphs with that degree sequence. All properties calculated in this paper are averaged over the ensemble of graphs generated in this way. In the limit of large graph size an equivalent procedure is to study only one particular degree sequence, averaging uniformly over all graphs with that sequence, where the sequence is chosen to approximate as closely as possible the desired probability distribution. The latter procedure can be thought of as a "microcanonical ensemble" for random graphs, where the former is a "canonical ensemble."

Some results are already known for random graphs with arbitrary degree distributions: in two beautiful recent papers [40,41], Molloy and Reed have derived formulas for the position of the phase transition at which a giant component first appears, and the size of the giant component. (These results are calculated within the microcanonical ensemble, but apply equally to the canonical one in the large system size limit.) The formalism we present in this paper yields an alternative derivation of these results and also provides a framework for obtaining other quantities of interest, some of which we calculate. In Secs. III and IV we extend our formalism to the case of directed graphs (such as the world-wide web) and bipartite graphs (such as collaboration graphs).

A. Generating functions

Our approach is based on generating functions [42], the most fundamental of which, for our purposes, is the generating function $G_0(x)$ for the probability distribution of vertex degrees k. Suppose that we have a unipartite undirected graph—an acquaintance network, for example—of N vertices, with N large. We define

$$G_0(x)=\sum_{k=0}^{\infty} p_k x^k, \tag{2}$$

where p_k is the probability that a randomly chosen vertex on the graph has degree k. The distribution p_k is assumed correctly normalized, so that

$$G_0(1)=1. \tag{3}$$

The same will be true of all generating functions considered here, with a few important exceptions, which we will note at the appropriate point. Because the probability distribution is normalized and positive definite, $G_0(x)$ is also absolutely convergent for all $|x|\leqslant 1$, and hence has no singularities in this region. All the calculations of this paper will be confined to the region $|x|\leqslant 1$.

The function $G_0(x)$, and indeed any probability generating function, has a number of properties that will prove useful in subsequent developments.

Derivatives. The probability p_k is given by the kth derivative of G_0 according to

$$p_k=\frac{1}{k!}\left.\frac{d^k G_0}{dx^k}\right|_{x=0}. \tag{4}$$

Thus the one function $G_0(x)$ encapsulates all the information contained in the discrete probability distribution p_k. We say that the function $G_0(x)$ "generates" the probability distribution p_k.

Moments. The average over the probability distribution generated by a generating function—for instance, the average degree z of a vertex in the case of $G_0(x)$—is given by

$$z=\langle k\rangle=\sum_k k p_k = G_0'(1). \tag{5}$$

Thus if we can calculate a generating function we can also calculate the mean of the probability distribution which it generates. Higher moments of the distribution can be calculated from higher derivatives also. In general, we have

$$\langle k^n\rangle=\sum_k k^n p_k=\left[\left(x\frac{d}{dx}\right)^n G_0(x)\right]_{x=1}. \tag{6}$$

Powers. If the distribution of a property k of an object is generated by a given generating function, then the distribution of the total of k summed over m independent realizations of the object is generated by the mth power of that generating function. For example, if we choose m vertices at random from a large graph, then the distribution of the sum of the degrees of those vertices is generated by $[G_0(x)]^m$. To see why this is so, consider the simple case of just two vertices. The square $[G_0(x)]^2$ of the generating function for a single vertex can be expanded as

$$\begin{aligned}[G_0(x)]^2&=\left[\sum_k p_k x^k\right]^2\\ &=\sum_{jk} p_j p_k x^{j+k}\\ &=p_0p_0x^0+(p_0p_1+p_1p_0)x^1\\ &\quad+(p_0p_2+p_1p_1+p_2p_0)x^2\\ &\quad+(p_0p_3+p_1p_2+p_2p_1+p_3p_0)x^3+\cdots. \end{aligned} \tag{7}$$

It is clear that the coefficient of the power of x^n in this expression is precisely the sum of all products $p_j p_k$ such that $j+k=n$, and hence correctly gives the probability that the sum of the degrees of the two vertices will be n. It is straightforward to convince oneself that this property extends also to all higher powers of the generating function.

All of these properties will be used in the derivations given in this paper.

Another quantity that will be important to us is the distribution of the degree of the vertices that we arrive at by following a randomly chosen edge. Such an edge arrives at a vertex with probability proportional to the degree of that vertex, and the vertex therefore has a probability distribution of degree proportional to kp_k. The correctly normalized distribution is generated by

$$\frac{\sum_k k p_k x^k}{\sum_k k p_k}=x\frac{G_0'(x)}{G_0'(1)}. \tag{8}$$

If we start at a randomly chosen vertex and follow each of the edges at that vertex to reach the k nearest neighbors, then the vertices arrived at each have the distribution of remaining outgoing edges generated by this function, less one power of x, to allow for the edge that we arrived along. Thus the distribution of outgoing edges is generated by the function

$$G_1(x)=\frac{G_0'(x)}{G_0'(1)}=\frac{1}{z}G_0'(x), \tag{9}$$

where z is the average vertex degree, as before. The probability that any of these outgoing edges connects to the original vertex that we started at, or to any of its other immediate neighbors, goes as N^{-1} and hence can be neglected in the limit of large N. Thus, making use of the "powers" property of the generating function described above, the generating function for the probability distribution of the number of *second* neighbors of the original vertex can be written as

$$\sum_k p_k[G_1(x)]^k = G_0(G_1(x)). \tag{10}$$

Similarly, the distribution of third-nearest neighbors is generated by $G_0(G_1(G_1(x)))$, and so on. The average number z_2 of second neighbors is

$$z_2 = \left[\frac{d}{dx}G_0(G_1(x))\right]_{x=1} = G_0'(1)G_1'(1) = G_0''(1), \tag{11}$$

where we have made use of the fact that $G_1(1)=1$. (One might be tempted to conjecture that since the average number of first neighbors is $G_0'(1)$, Eq. (5), and the average number of second neighbors is $G_0''(1)$, Eq. (11), then the average number of mth neighbors should be given by the mth derivative of G_0 evaluated at $x=1$. As we show in Sec. II F, however, this conjecture is wrong.)

B. Examples

To make things more concrete, we immediately introduce some examples of specific graphs to illustrate how these calculations are carried out.

(a) Poisson-distributed graphs. The simplest example of a graph of this type is one for which the distribution of degree is binomial, or Poisson in the large N limit. This distribution yields the standard random graph studied by many mathematicians and discussed in Sec. I. In this graph the probability $p=z/N$ of the existence of an edge between any two vertices is the same for all vertices, and $G_0(x)$ is given by

$$G_0(x) = \sum_{k=0}^{N}\binom{N}{k}p^k(1-p)^{N-k}x^k = (1-p+px)^N = e^{z(x-1)}, \tag{12}$$

where the last equality applies in the limit $N\to\infty$. It is then trivial to show that the average degree of a vertex is indeed $G_0'(1)=z$ and that the probability distribution of degree is given by $p_k = z^k e^{-z}/k!$, which is the ordinary Poisson distribution. Notice also that for this special case we have $G_1(x)=G_0(x)$, so that the distribution of outgoing edges at a vertex is the same, regardless of whether we arrived there by choosing a vertex at random, or by following a randomly chosen edge. This property, which is peculiar to the Poisson-distributed random graph, is the reason why the theory of random graphs of this type is especially simple.

(b) Exponentially distributed graphs. Perhaps the next simplest type of graph is one with an exponential distribution of vertex degrees

$$p_k = (1-e^{-1/\kappa})e^{-k/\kappa}, \tag{13}$$

where κ is a constant. The generating function for this distribution is

$$G_0(x) = (1-e^{-1/\kappa})\sum_{k=0}^{\infty} e^{-k/\kappa}x^k = \frac{1-e^{-1/\kappa}}{1-xe^{-1/\kappa}}, \tag{14}$$

and

$$G_1(x) = \left[\frac{1-e^{-1/\kappa}}{1-xe^{-1/\kappa}}\right]^2. \tag{15}$$

An example of a graph with an exponential degree distribution is given in Sec. V A.

(c) Power-law distributed graphs. The recent interest in the properties of the world-wide web and of social networks leads us to investigate the properties of graphs with a power-law distribution of vertex degrees. Such graphs have been discussed previously by Barabási and co-workers [22,23] and by Aiello *et al.* [13]. In this paper, we will look at graphs with degree distribution given by

$$p_k = Ck^{-\tau}e^{-k/\kappa} \quad \text{for } k\geq 1, \tag{16}$$

where C, τ, and κ are constants. The reason for including the exponential cutoff is twofold: first many real-world graphs appear to show this cutoff [14,36]; second it makes the distribution normalizable for all τ, and not just $\tau\geq 2$.

The constant C is fixed by the requirement of normalization, which gives $C=[\mathrm{Li}_\tau(e^{-1/\kappa})]^{-1}$ and hence

$$p_k = \frac{k^{-\tau}e^{-k/\kappa}}{\mathrm{Li}_\tau(e^{-1/\kappa})} \quad \text{for } k\geq 1, \tag{17}$$

where $\mathrm{Li}_n(x)$ is the nth polylogarithm of x. [For those unfamiliar with this function, its salient features for our purposes are that it is zero at $x=0$ and, real, finite, and monotonically increasing in the range $0\leq x<1$, for all n. It also decreases with increasing n, and has a pole at $x=1$ for $n\leq 1$ only, although it has a valid analytic continuation below $n=1$ which takes the value $\zeta(n)$ at $x=1$.]

Substituting Eq. (17) into Eq. (2), we find that the generating function for graphs with this degree distribution is

$$G_0(x) = \frac{\mathrm{Li}_\tau(xe^{-1/\kappa})}{\mathrm{Li}_\tau(e^{-1/\kappa})}. \tag{18}$$

In the limit $\kappa\to\infty$—the case considered in Refs. [13] and [23]—this simplifies to

$$G_0(x) = \frac{\mathrm{Li}_\tau(x)}{\zeta(\tau)}, \tag{19}$$

where $\zeta(\tau)$ is the Riemann ζ function.

The function $G_1(x)$ is given by

$$G_1(x) = \frac{\mathrm{Li}_{\tau-1}(xe^{-1/\kappa})}{x\,\mathrm{Li}_{\tau-1}(e^{-1/\kappa})}. \tag{20}$$

Thus, for example, the average number of neighbors of a randomly chosen vertex is

$$z=G_0'(1)=\frac{\mathrm{Li}_{\tau-1}(e^{-1/\kappa})}{\mathrm{Li}_{\tau}(e^{-1/\kappa})}, \tag{21}$$

and the average number of second neighbors is

$$z_2=G_0''(1)=\frac{\mathrm{Li}_{\tau-2}(e^{-1/\kappa})-\mathrm{Li}_{\tau-1}(e^{-1/\kappa})}{\mathrm{Li}_{\tau}(e^{-1/\kappa})}. \tag{22}$$

(d) Graphs with arbitrary specified degree distribution. In some cases we wish to model specific real-world graphs that have known degree distributions—known because we can measure them directly. A number of the graphs described in the Introduction fall into this category. For these graphs, we know the exact numbers n_k of vertices having degree k, and hence we can write down the exact generating function for that probability distribution in the form of a finite polynomial

$$G_0(x)=\frac{\sum_k n_k x^k}{\sum_k n_k}, \tag{23}$$

where the sum in the denominator ensures that the generating function is properly normalized. As an example, suppose that in a community of 1000 people, each person knows between zero and five of the others, the exact numbers of people in each category being, from zero to five: {86,150,363,238,109,54}. This distribution will then be generated by the polynomial

$$G_0(x)=\frac{86+150x+363x^2+238x^3+109x^4+54x^5}{1000}. \tag{24}$$

C. Component sizes

We are now in a position to calculate some properties of interest for our graphs. First let us consider the distribution of the sizes of connected components in the graph. Let $H_1(x)$ be the generating function for the distribution of the sizes of components that are reached by choosing a random edge and following it to one of its ends. We explicitly exclude from $H_1(x)$ the giant component, if there is one; the giant component is dealt with separately below. Thus, except when we are precisely at the phase transition where the giant component appears, typical component sizes are finite, and the chances of a component containing a closed loop of edges goes as N^{-1}, which is negligible in the limit of large N. This means that the distribution of components generated by $H_1(x)$ can be represented graphically as in Fig. 3; each component is treelike in structure, consisting of the single site we reach by following our initial edge, plus any number (including zero) of other treelike clusters, with the same size distribution, joined to it by single edges. If we denote by q_k the probability that the initial site has k edges coming out of it other than the edge we came in along, then, making use of the "powers" property of Sec. II A, $H_1(x)$ must satisfy a self-consistency condition of the form

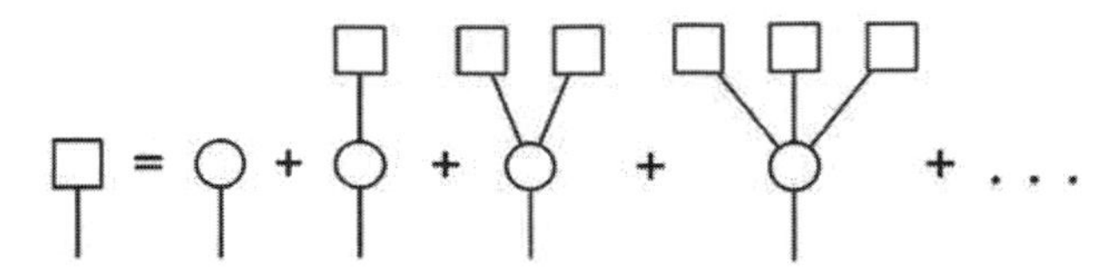

FIG. 3. Schematic representation of the sum rule for the connected component of vertices reached by following a randomly chosen edge. The probability of each such component (left-hand side) can be represented as the sum of the probabilities (right-hand side) of having only a single vertex, having a single vertex connected to one other component, or two other components, and so forth. The entire sum can be expressed in closed form as Eq. (26).

$$H_1(x)=xq_0+xq_1H_1(x)+xq_2[H_1(x)]^2+\cdots. \tag{25}$$

However, q_k is nothing other than the coefficient of x^k in the generating function $G_1(x)$, Eq. (9), and hence Eq. (25) can also be written

$$H_1(x)=xG_1(H_1(x)). \tag{26}$$

If we start at a randomly chosen vertex, then we have one such component at the end of each edge leaving that vertex, and hence the generating function for the size of the whole component is

$$H_0(x)=xG_0(H_1(x)). \tag{27}$$

In principle, therefore, given the functions $G_0(x)$ and $G_1(x)$, we can solve Eq. (26) for $H_1(x)$ and substitute into Eq. (27) to get $H_0(x)$. Then we can find the probability that a randomly chosen vertex belongs to a component of size s by taking the sth derivative of H_0. In practice, unfortunately, this is usually impossible; Equation (26) is a complicated and frequently transcendental equation, which rarely has a known solution. On the other hand, we note that the coefficient of x^s in the Taylor expansion of $H_1(x)$ (and therefore also the sth derivative) are given exactly by only $s+1$ iterations of Eq. (27), starting with $H_1=1$, so that the distribution generated by $H_0(x)$ can be calculated exactly to finite order in finite time. With current symbolic manipulation programs, it is quite possible to evaluate the first one hundred or so derivatives in this way. Failing this, an approximate solution can be found by numerical iteration and the distribution of cluster sizes calculated from Eq. (4) by numerical differentiation. Since direct evaluation of numerical derivatives is prone to machine-precision problems, we recommend evaluating the derivatives by numerical integration of the Cauchy formula, giving the probability distribution P_s of cluster sizes thus:

$$P_s=\frac{1}{s!}\left.\frac{d^sH_0}{dz^s}\right|_{z=0}=\frac{1}{2\pi i}\oint\frac{H_0(z)}{z^{s+1}}dz. \tag{28}$$

The best numerical precision is obtained by using the largest possible contour, subject to the condition that it encloses no poles of the generating function. The largest contour for which this condition is satisfied in general is the unit circle $|z|=1$ (see Sec. II A), and we recommend using this contour for Eq. (28). It is possible to find the first thousand derivatives of a function without difficulty using this method [43].

D. The mean component size, the phase transition, and the giant component

Although it is not usually possible to find a closed-form expression for the complete distribution of cluster sizes on a graph, we can find closed-form expressions for the *average* properties of clusters from Eqs. (26) and (27). For example, the average size of the component to which a randomly chosen vertex belongs, for the case where there is no giant component in the graph, is given in the normal fashion by

$$\langle s\rangle = H_0'(1) = 1 + G_0'(1)H_1'(1). \tag{29}$$

From Eq. (26) we have

$$H_1'(1) = 1 + G_1'(1)H_1'(1), \tag{30}$$

and hence

$$\langle s\rangle = 1 + \frac{G_0'(1)}{1-G_1'(1)} = 1 + \frac{z_1^2}{z_1 - z_2}, \tag{31}$$

where $z_1 = z$ is the average number of neighbors of a vertex and z_2 is the average number of second neighbors. We see that this expression diverges when

$$G_1'(1) = 1. \tag{32}$$

This point marks the phase transition at which a giant component first appears. Substituting Eqs. (2) and (9) into Eq. (32), we can also write the condition for the phase transition as

$$\sum_k k(k-2)p_k = 0. \tag{33}$$

Indeed, since this sum increases monotonically as edges are added to the graph, it follows that the giant component exists if and only if this sum is positive. This result has been derived by different means by Molloy and Reed [40]. An equivalent and intuitively reasonable statement, which can also be derived from Eq. (31), is that the giant component exists if and only if $z_2 > z_1$.

Our generating function formalism still works when there is a giant component in the graph, but, by definition, $H_0(x)$ then generates the probability distribution of the sizes of components *excluding* the giant component. This means that $H_0(1)$ is no longer unity, as it is for the other generating functions considered so far, but instead takes the value $1-S$, where S is the fraction of the graph occupied by the giant component. We can use this to calculate the size of the giant component from Eqs. (26) and (27) thus:

$$S = 1 - G_0(u), \tag{34}$$

where $u \equiv H_1(1)$ is the smallest non-negative real solution of

$$u = G_1(u). \tag{35}$$

This result has been derived in a different but equivalent form by Molloy and Reed [41], using different methods.

The correct general expression for the average component size, excluding the (formally infinite) giant component, if there is one, is

$$\langle s\rangle = \frac{H_0'(1)}{H_0(1)} = \frac{1}{H_0(1)}\left[G_0(H_1(1)) + \frac{G_0'(H_1(1))G_1(H_1(1))}{1-G_1'(H_1(1))}\right]$$

$$= 1 + \frac{zu^2}{[1-S][1-G_1'(u)]}, \tag{36}$$

which is equivalent to Eq. (31) when there is no giant component ($S=0$, $u=1$).

For example, in the ordinary random graph with Poisson degree distribution, we have $G_0(x) = G_1(x) = e^{z(x-1)}$ [Eq. (12)], and hence we find simply that $1-S=u$ is a solution of $u = G_0(u)$, or equivalently that

$$S = 1 - e^{-zS}. \tag{37}$$

The average component size is given by

$$\langle s\rangle = \frac{1}{1 - z + zS}. \tag{38}$$

These are both well-known results [1].

For graphs with purely power-law distributions [Eq. (17) with $\kappa \to \infty$], S is given by Eq. (34) with u the smallest non-negative real solution of

$$u = \frac{\mathrm{Li}_{\tau-1}(u)}{u\zeta(\tau-1)}. \tag{39}$$

For all $\tau \leqslant 2$ this gives $u=0$, and hence $S=1$, implying that a randomly chosen vertex belongs to the giant component with probability tending to 1 as $\kappa \to \infty$. For graphs with $\tau > 2$, the probability of belonging to the giant component is strictly less than 1, even for infinite κ. In other words, the giant component essentially fills the entire graph for $\tau \leqslant 2$, but not for $\tau > 2$. These results have been derived by different means by Aiello *et al.* [13].

E. Asymptotic form of the cluster size distribution

A variety of results are known about the asymptotic properties of the coefficients of generating functions, some of which can usefully be applied to the distribution of cluster sizes P_s generated by $H_0(x)$. Close to the phase transition, we expect the tail of the distribution P_s to behave as

$$P_s \sim s^{-\alpha} e^{-s/s^*}, \tag{40}$$

where the constants α and s^* can be calculated from the properties of $H_0(x)$ as follows.

The cutoff parameter s^* is simply related to the radius of convergence $|x^*|$ of the generating function [42,44], according to

$$s^* = \frac{1}{\ln|x^*|}. \tag{41}$$

The radius of convergence $|x^*|$ is equal to the magnitude of the position x^* of the singularity in $H_0(x)$ nearest to the origin. From Eq. (27) we see that such a singularity may arise either through a singularity in $G_0(x)$ or through one in $H_1(x)$. However, since the first singularity in $G_0(x)$ is known to be outside the unit circle (Sec. II A), and the first singularity in $H_1(x)$ tends to $x=1$ as we go to the phase transition (see below), it follows that, sufficiently close to the phase transition, the singularity in $H_0(x)$ closest to the origin is also a singularity in $H_1(x)$. With this result x^* is easily calculated.

Although we do not in general have a closed-form expression for $H_1(x)$, it is easy to derive one for its functional inverse. Putting $w=H_1(x)$ and $x=H_1^{-1}(w)$ in Eq. (26) and rearranging, we find

$$x = H_1^{-1}(w) = \frac{w}{G_1(w)}. \tag{42}$$

The singularity of interest corresponds to the point w^* at which the derivative of $H_1^{-1}(w)$ is zero, which is a solution of

$$G_1(w^*) - w^* G_1'(w^*) = 0. \tag{43}$$

Then x^* (and hence s^*) is given by Eq. (42). Note that there is no guarantee that Eq. (43) has a finite solution, and that if it does not, then P_s will not in general follow the form of Eq. (40).

When we are precisely at the phase transition of our system, we have $G_1(1)=G_1'(1)=1$, and hence the solution of Eq. (43) gives $w^*=x^*=1$—a result that we used above—and $s^*\to\infty$. We can use the fact that $x^*=1$ at the transition to calculate the value of the exponent α as follows. Expanding $H_1^{-1}(w)$ about $w^*=1$ by putting $w=1+\epsilon$ in Eq. (42), we find that

$$H_1^{-1}(1+\epsilon) = 1 - \frac{1}{2} G_1''(1)\epsilon^2 + O(\epsilon^3), \tag{44}$$

where we have made use of $G_1(1)=G_1'(1)=1$ at the phase transition. So long as $G_1''(1)\neq 0$, which in general it is not, this implies that $H_1(x)$ and hence also $H_0(x)$ are of the form

$$H_0(x) \sim (1-x)^\beta \quad \text{as } x\to 1, \tag{45}$$

with $\beta=\frac{1}{2}$. This exponent is related to the exponent α as follows. Equation (40) implies that $H_0(x)$ can be written in the form

$$H_0(x) = \sum_{s=0}^{a-1} P_s x^s + C \sum_{s=a}^{\infty} s^{-\alpha} e^{-s/s^*} x^s + \epsilon(a), \tag{46}$$

where C is a constant and the last (error) term $\epsilon(a)$ is assumed much smaller than the second term. The first term in this expression is a finite polynomial and therefore has no singularities on the finite plane; the singularity resides in the second term. Using this equation, the exponent β can be written

$$\begin{aligned}\beta &= \lim_{x\to 1}\left[1 + (x-1)\frac{H_0''(x)}{H_0'(x)}\right] \\ &= \lim_{a\to\infty}\lim_{x\to 1}\left[\frac{1}{x} + \frac{x-1}{x}\frac{\sum_{s=a}^{\infty} s^{2-\alpha}x^{s-1}}{\sum_{s=a}^{\infty} s^{1-\alpha}x^{s-1}}\right] \\ &= \lim_{a\to\infty}\lim_{x\to 1}\left[\frac{1}{x} + \frac{1-x}{x\ln x}\frac{\Gamma(3-\alpha,-a\ln x)}{\Gamma(2-\alpha,-a\ln x)}\right],\end{aligned} \tag{47}$$

where we have replaced the sums with integrals as a becomes large, and $\Gamma(\nu,\mu)$ is the incomplete Γ-function. Taking the limits in the order specified and rearranging for α, we then get

$$\alpha = \beta + 1 = \tfrac{3}{2}, \tag{48}$$

regardless of degree distribution, except in the special case where $G_1''(1)$ vanishes [see Eq. (44)]. The result $\alpha=\frac{3}{2}$ was known previously for the ordinary Poisson random graph [1], but not for other degree distributions.

F. Numbers of neighbors and average path length

We turn now to the calculation of the number of neighbors who are m steps away from a randomly chosen vertex. As shown in Sec. II A, the probability distributions for first- and second-nearest neighbors are generated by the functions $G_0(x)$ and $G_0(G_1(x))$. By extension, the distribution of mth neighbors is generated by $G_0(G_1(\ldots G_1(x)\ldots))$, with $m-1$ iterations of the function G_1 acting on itself. If we define $G^{(m)}(x)$ to be this generating function for mth neighbors, then we have

$$G^{(m)}(x) = \begin{cases} G_0(x) & \text{for } m=1, \\ G^{(m-1)}(G_1(x)) & \text{for } m\geq 2. \end{cases} \tag{49}$$

Then the average number z_m of mth-nearest neighbors is

$$z_m = \left.\frac{dG^{(m)}}{dx}\right|_{x=1} = G_1'(1)G^{(m-1)\prime}(1) = G_1'(1)z_{m-1}. \tag{50}$$

Along with the initial condition $z_1=z=G_0'(1)$, this then tells us that

$$z_m = [G_1'(1)]^{m-1} G_0'(1) = \left[\frac{z_2}{z_1}\right]^{m-1} z_1. \quad (51)$$

From this result we can make an estimate of the typical length ℓ of the shortest path between two randomly chosen vertices on the graph. This typical path length is reached approximately when the total number of neighbors of a vertex out to that distance is equal to the number of vertices on the graph, i.e., when

$$1 + \sum_{m=1}^{\ell} z_m = N. \quad (52)$$

Using Eq. (51) this gives us

$$\ell = \frac{\ln[(N-1)(z_2 - z_1) + z_1^2] - \ln z_1^2}{\ln(z_2/z_1)}. \quad (53)$$

In the common case where $N \gg z_1$ and $z_2 \gg z_1$, this reduces to

$$\ell = \frac{\ln(N/z_1)}{\ln(z_2/z_1)} + 1. \quad (54)$$

This result is only approximate for two reasons. First, the conditions used to derive it are only an approximation; the exact answer depends on the detailed structure of the graph. Second, it assumes that all vertices are reachable from a randomly chosen starting vertex. In general however this will not be true. For graphs with no giant component it is certainly not true and Eq. (54) is meaningless. Even when there is a giant component, however, it is usually not the case that it fills the entire graph. A better approximation to ℓ may therefore be given by replacing N in Eq. (54) by NS, where S is the fraction of the graph occupied by the giant component, as in Sec. II D.

Such shortcomings notwithstanding, there are a number of remarkable features of Eq. (54).

(1) It shows that the average vertex-vertex distance for all random graphs, regardless of degree distribution, should scale logarithmically with size N, according to $\ell = A + B \ln N$, where A and B are constants. This result is of course well known for a number of special cases.

(2) It shows that the average distance, which is a global property, can be calculated from a knowledge only of the average numbers of first- and second-nearest neighbors, which are local properties. It would be possible therefore to measure these numbers empirically by purely local measurements on a graph such as an acquaintance network and from them to determine the expected average distance between vertices. For some networks at least, this gives a surprisingly good estimate of the true average distance [37].

(3) It shows that only the average numbers of first- and second-nearest neighbors are important to the calculation of average distances, and thus that two random graphs with completely different distributions of vertex degrees, but the same values of z_1 and z_2, will have the same average distances.

For the case of the purely theoretical example graphs we discussed earlier, we cannot make an empirical measurement of z_1 and z_2, but we can still employ Eq. (54) to calculate ℓ. In the case of the ordinary (Poisson) random graph, for instance, we find from Eq. (12) that $z_1 = z$, $z_2 = z^2$, and so $\ell = \ln N / \ln z$, which is the standard result for graphs of this type [1]. For the graph with degree distributed according to the truncated power law, Eq. (17), z_1 and z_2 are given by Eqs. (21) and (22), and the average vertex-vertex distance is

$$\ell = \frac{\ln N + \ln[\mathrm{Li}_\tau(e^{-1/\kappa})/\mathrm{Li}_{\tau-1}(e^{-1/\kappa})]}{\ln[\mathrm{Li}_{\tau-2}(e^{-1/\kappa})/\mathrm{Li}_{\tau-1}(e^{-1/\kappa}) - 1]} + 1. \quad (55)$$

In the limit $\kappa \to \infty$, this becomes

$$\ell = \frac{\ln N + \ln[\zeta(\tau)/\zeta(\tau-1)]}{\ln[\zeta(\tau-2)/\zeta(\tau-1) - 1]} + 1. \quad (56)$$

Note that this expression does not have a finite positive real value for any $\tau < 3$, indicating that one must specify a finite cutoff κ for the degree distribution to get a well-defined average vertex-vertex distance on such graphs.

G. Simulation results

As a check on the results of this section, we have performed extensive computer simulations of random graphs with various distributions of vertex degree. Such graphs are relatively straightforward to generate. First, we generate a set of N random numbers $\{k_i\}$ to represent the degrees of the N vertices in the graph. These may be thought of as the "stubs" of edges, emerging from their respective vertices. Then we choose pairs of these stubs at random and place edges on the graph joining them up. It is simple to see that this will generate all graphs with the given set of vertex degrees with equal probability. The only small catch is that the sum $\Sigma_i k_i$ of the degrees must be even, since each edge added to the graph must have two ends. This is not difficult to contrive however. If the set $\{k_i\}$ is such that the sum is odd, we simply throw it away and generate a new set.

As a practical matter, integers representing vertex degrees with any desired probability distribution can be generated using the transformation method if applicable, or failing that, a rejection or hybrid method [45]. For example, degrees obeying the power-law-plus-cutoff form of Eq. (17) can be generated using a two-step hybrid transformation/rejection method as follows. First, we generate random integers $k \geq 1$ with distribution proportional to $e^{-k/\kappa}$ using the transformation [46]

$$k = \lceil -\kappa \ln(1-r) \rceil, \quad (57)$$

where r is a random real number uniformly distributed in the range $0 \leq r < 1$. Second, we accept this number with probability $k^{-\tau}$, where by "accept" we mean that if the number is not accepted we discard it and generate another one according to Eq. (57), repeating the process until one is accepted.

In Fig. 4 we show results for the size of the giant component in simulations of undirected unipartite graphs with vertex degrees distributed according to Eq. (17) for a variety of different values of τ and κ. On the same plot we also show

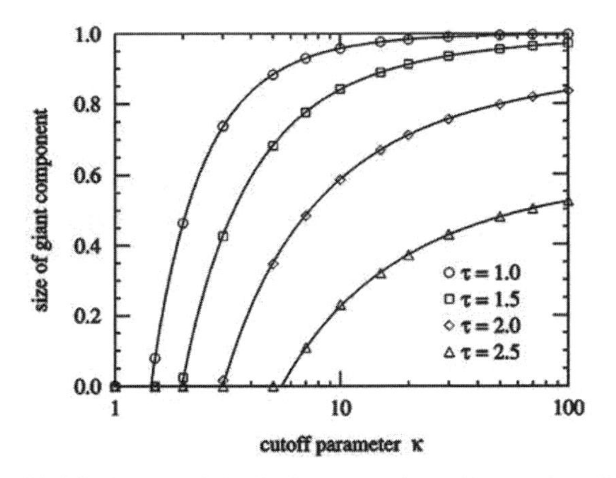

FIG. 4. The size of the giant component in random graphs with vertex degrees distributed according to Eq. (17), as a function of the cutoff parameter κ for five different values of the exponent τ. The points are results from numerical simulations on graphs of $N = 1\,000\,000$ vertices, and the solid lines are the theoretical value for infinite graphs, Eqs. (34) and (35). The error bars on the simulation results are smaller than the data points.

the expected value of the same quantity derived by numerical solution of Eqs. (34) and (35). As the figure shows, the agreement between simulation and theory is excellent.

III. DIRECTED GRAPHS

We turn now to directed graphs with arbitrary degree distributions. An example of a directed graph is the world-wide web, since every hyperlink between two pages on the web goes in only one direction. The web has a degree distribution that follows a power law, as discussed in Sec. I.

Directed graphs introduce a subtlety that is not present in undirected ones, and which becomes important when we apply our generating function formalism. In a directed graph it is not possible to talk about a "component"—i.e., a group of connected vertices—because even if vertex A can be reached by following (directed) edges from vertex B, that does not necessarily mean that vertex B can be reached from vertex A. There are two correct generalizations of the idea of the component to a directed graph: the set of vertices that are reachable from a given vertex, and the set from which a given vertex can be reached. We will refer to these as "out-components" and "in-components," respectively. An in-component can also be thought of as those vertices reachable by following edges backwards (but not forwards) from a specified vertex. It is possible to study directed graphs by allowing both forward and backward traversal of edges (see Ref. [26], for example). In this case, however, the graph effectively becomes undirected and should be treated with the formalism of Sec. II.

With these considerations in mind, we now develop the generating function formalism appropriate to random directed graphs with arbitrary degree distributions.

A. Generating functions

In a directed graph, each vertex has separate in-degree and out-degree for links running into and out of that vertex. Let us define p_{jk} to be the probability that a randomly chosen vertex has in-degree j and out-degree k. It is important to realize that in general this joint distribution of j and k is not equal to the product $p_j p_k$ of the separate distributions of in- and out-degree. In the world-wide web, for example, it seems likely (although this question has not been investigated to our knowledge) that sites with a large number of outgoing links also have a large number of incoming ones, i.e., that j and k are correlated, so that $p_{jk} \neq p_j p_k$. We appeal to those working on studies of the structure of the web to measure the joint distribution of in-degrees and out-degrees of sites; empirical data on this distribution would make theoretical work much easier.

We now define a generating function for the joint probability distribution of in-degrees and out-degrees, which is necessarily a function of two independent variables, x and y, thus:

$$\mathcal{G}(x,y) = \sum_{jk} p_{jk} x^j y^k. \tag{58}$$

Since every edge on a directed graph must leave some vertex and enter another, the net average number of edges entering a vertex is zero, and hence p_{jk} must satisfy the constraint

$$\sum_{jk} (j-k) p_{jk} = 0. \tag{59}$$

This implies that $\mathcal{G}(x,y)$ must satisfy

$$\left.\frac{\partial \mathcal{G}}{\partial x}\right|_{x,y=1} = \left.\frac{\partial \mathcal{G}}{\partial y}\right|_{x,y=1} = z, \tag{60}$$

where z is the average degree (both in and out) of vertices in the graph.

Using the function $\mathcal{G}(x,y)$, we can, as before, define generating functions G_0 and G_1 for the number of out-going edges leaving a randomly chosen vertex, and the number leaving the vertex reached by following a randomly chosen edge. We can also define generating functions F_0 and F_1 for the number arriving at such a vertex. These functions are given by

$$F_0(x) = \mathcal{G}(x,1), \quad F_1(x) = \frac{1}{z} \left.\frac{\partial \mathcal{G}}{\partial y}\right|_{y=1}, \tag{61}$$

$$G_0(y) = \mathcal{G}(1,y), \quad G_1(y) = \frac{1}{z} \left.\frac{\partial \mathcal{G}}{\partial x}\right|_{x=1}. \tag{62}$$

Once we have these functions, many results follow as before. The average numbers of first and second neighbors reachable from a randomly chosen vertex are given by Eq. (60) and

$$z_2 = G_0'(1) G_1'(1) = \left.\frac{\partial^2 \mathcal{G}}{\partial x\, \partial y}\right|_{x,y=1}. \tag{63}$$

These are also the numbers of first and second neighbors from which a random vertex can be reached, since Eqs. (60)

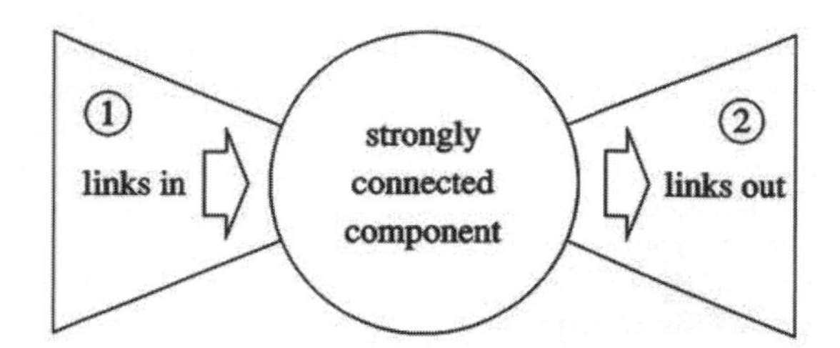

FIG. 5. The "bow-tie" diagram proposed by Broder *et al.* as a representation of the giant component of the world-wide web (although it can be used to visualize any directed graph).

and (63) are manifestly symmetric in x and y. We can also make an estimate of the average path length on the graph from

$$\ell=\frac{\ln(N/z_1)}{\ln(z_2/z_1)}+1, \tag{64}$$

as before. However, this equation should be used with caution. As discussed in Sec. II F, the derivation of this formula assumes that we are in a regime in which the bulk of the graph is reachable from most vertices. On a directed graph however, this may be far from true, as appears to be the case with the world-wide web [26].

The probability distribution of the numbers of vertices reachable from a randomly chosen vertex in a directed graph—i.e., of the sizes of the out-components—is generated by the function $H_0(y)=yG_0(H_1(y))$, where $H_1(y)$ is a solution of $H_1(y)=yG_1(H_1(y))$, just as before. (A similar and obvious pair of equations governs the sizes of the in-components.) The results for the asymptotic behavior of the component size distribution from Sec. II E generalize straightforwardly to directed graphs. The average out-component size for the case where there is no giant component is given by Eq. (31), and thus the point at which a giant component first appears is given once more by $G_1'(1)=1$. Substituting Eq. (58) into this expression gives the explicit condition

$$\sum_{jk}(2jk-j-k)p_{jk}=0 \tag{65}$$

for the first appearance of the giant component. This expression is the equivalent for the directed graph of Eq. (33). It is also possible, and equally valid, to define the position at which the giant component appears by $F_1'(1)=1$, which provides an alternative derivation for Eq. (65).

Just as with the individual in-component and out-components for vertices, the size of the giant component on a directed graph can also be defined in different ways. The giant component can be represented using the "bow-tie" diagram of Broder *et al.* [26], which we depict (in a simplified form) in Fig. 5. The diagram has three parts. The strongly connected portion of the giant component, represented by the central circle, is that portion in which every vertex can be reached from every other. The two sides of the bow tie represent (1) those vertices from which the strongly connected component can be reached but which it is not possible to reach from the strongly connected component and (2) those vertices that can be reached from the strongly connected component but from which it is not possible to reach the strongly connected component. The solution of Eqs. (34) and (35) with $G_0(x)$ and $G_1(x)$ defined according to Eq. (62) gives the number of vertices, as a fraction of N, in the giant strongly connected component plus those vertices from which the giant strongly connected component can be reached. Using $F_0(x)$ and $F_1(x)$ [Eq. (61)] in place of $G_0(x)$ and $G_1(x)$ gives a different solution, which represents the fraction of the graph in the giant strongly connected component plus those vertices that can be reached from it.

B. Simulation results

We have performed simulations of directed graphs as a check on the results above. Generation of random directed graphs with known joint degree distribution p_{jk} is somewhat more complicated than the generation of undirected graphs discussed in Sec. II G. The method we use is as follows. First, it is important to ensure that the averages of the distributions of in-degree and out-degree of the graph are the same, or equivalently that p_{jk} satisfies Eq. (59). If this is not the case, at least to good approximation, then generation of the graph will be impossible. Next, we generate a set of N in/out-degree pairs (j_i,k_i), one for each vertex i, according to the joint distribution p_{jk}, and calculate the sums $\Sigma_i j_i$ and $\Sigma_i k_i$. These sums are required to be equal if there are to be no dangling edges in the graph, but in most cases we find that they are not. To rectify this we use a simple procedure. We choose a vertex i at random, discard the numbers (j_i,k_i) for that vertex and generate new ones from the distribution p_{jk}. We repeat this procedure until the two sums are found to be equal. Finally, we choose random in/out pairs of edges and join them together to make a directed graph. The resulting graph has the desired number of vertices and the desired joint distribution of in- and out-degree.

We have simulated directed graphs in which the distribution p_{jk} is given by a simple product of independent distributions of in-degree and out-degree. (As pointed out in Sec. III A, this is not generally the case for real-world directed graphs, where in-degree and out-degree may be correlated.) In Fig. 6 we show results from simulations of graphs with identically distributed (but independent) in-degree and out-degrees drawn from the exponential distribution, Eq. (13). For this distribution, solution of the critical-point equation $G_1'(1)=1$ shows that the giant component first appears at $\kappa_c=[\ln 2]^{-1}=1.4427$. The three curves in the figure show the distribution of numbers of vertices accessible from each vertex in the graph for $\kappa=0.5$, 0.8, and κ_c. The critical distribution follows a power-law form (see Sec. II C), while the others show an exponential cutoff. We also show the exact distribution derived from the coefficients in the expansion of $H_1(x)$ about zero. Once again, theory and simulation are in good agreement. A fit to the distribution for the case $\kappa=\kappa_c$ gives a value of $\alpha=1.50\pm0.02$, in good agreement with Eq. (48).

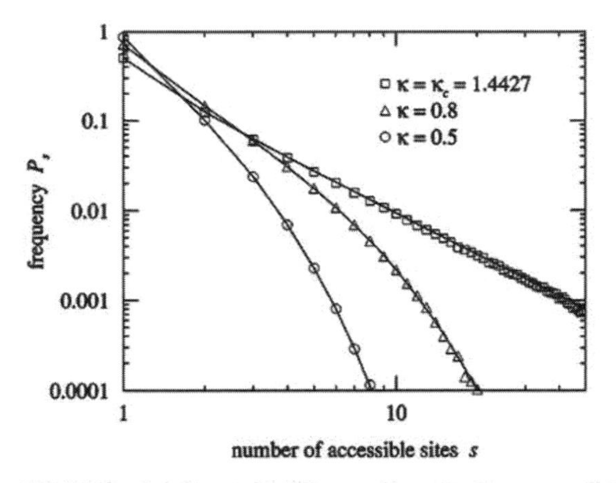

FIG. 6. The distribution P_s of the numbers of vertices accessible from each vertex of a directed graph with identically exponentially distributed in-degree and out-degree. The points are simulation results for systems of $N = 1\,000\,000$ vertices and the solid lines are the analytic solution.

IV. BIPARTITE GRAPHS

The collaboration graphs of scientists, company directors, and movie actors discussed in Sec. I are all examples of bipartite graphs. In this section we study the theory of bipartite graphs with arbitrary degree distributions. To be concrete, we will speak in the language of "actors" and "movies," but clearly all the developments here are applicable to academic collaborations, boards of directors, or any other bipartite graph structure.

A. Generating functions and basic results

Consider then a bipartite graph of M movies and N actors, in which each actor has appeared in an average of μ movies and each movie has a cast of average size ν actors. Note that only three of these parameters are independent, since the fourth is given by the equality

$$\frac{\mu}{M} = \frac{\nu}{N}. \tag{66}$$

Let p_j be the probability distribution of the degree of actors (i.e., of the number of movies in which they have appeared) and q_k be the distribution of degree (i.e., cast size) of movies. We define two generating functions that generate these probability distributions thus:

$$f_0(x) = \sum_j p_j x^j, \quad g_0(x) = \sum_k q_k x^k. \tag{67}$$

(It may be helpful to think of f as standing for "film," in order to keep these two straight.) As before, we necessarily have

$$f_0(1) = g_0(1) = 1, \quad f_0'(1) = \mu, \quad g_0'(1) = \nu. \tag{68}$$

If we now choose a random edge on our bipartite graph and follow it both ways to reach the movie and actor that it connects, then the distribution of the number of other edges leaving those two vertices is generated by the equivalent of Eq. (9):

$$f_1(x) = \frac{1}{\mu} f_0'(x), \quad g_1(x) = \frac{1}{\nu} g_0'(x). \tag{69}$$

Now we can write the generating function for the distribution of the number of co-stars (i.e., actors in shared movies) of a randomly chosen actor as

$$G_0(x) = f_0(g_1(x)). \tag{70}$$

If we choose a random edge, then the distribution of number of co-stars of the actor to which it leads is generated by

$$G_1(x) = f_1(g_1(x)). \tag{71}$$

These two functions play the same role in the one-mode network of actors as the functions of the same name did for the unipartite random graphs of Sec. II. Once we have calculated them, all the results from Sec. II follow exactly as before.

The numbers of first and second neighbors of a randomly chosen actor are

$$z_1 = G_0'(1) = f_0'(1) g_1'(1), \tag{72}$$

$$z_2 = G_0'(1) G_1'(1) = f_0'(1) f_1'(1) [g_1'(1)]^2. \tag{73}$$

Explicit expressions for these quantities can be obtained by substituting from Eqs. (67) and (69). The average vertex-vertex distance on the one-mode graph is given as before by Eq. (54). Thus, it is possible to estimate average distances on such graphs by measuring only the numbers of first and second neighbors.

The distribution of the sizes of the connected components in the one-mode network is generated by Eq. (27), where $H_1(x)$ is a solution of Eq. (26). The asymptotic results of Sec. II E generalize simply to the bipartite case, and the average size of a connected component in the absence of a giant component is

$$\langle s \rangle = 1 + \frac{G_0'(1)}{1 - G_1'(1)}, \tag{74}$$

as before. This diverges when $G_1'(1) = 1$, marking the first appearance of the giant component. Equivalently, the giant component first appears when

$$f_0''(1) g_0''(1) = f_0'(1) g_0'(1). \tag{75}$$

Substituting from Eq. (67), we then derive the explicit condition for the first appearance of the giant component:

$$\sum_{jk} jk(jk - j - k) p_j q_k = 0. \tag{76}$$

The size S of the giant component, as a fraction of the total number N of actors, is given as before by the solution of Eqs. (34) and (35).

Of course, all of these results work equally well if "actors" and "movies" are interchanged. One can calculate the average distance between movies in terms of common actors shared, the size and distribution of connected components of movies, and so forth, using the formulas given above, with only the exchange of f_0 and f_1 for g_0 and g_1. The formula (75) is, not surprisingly, invariant under this interchange, so that the position of the onset of the giant component is the same regardless of whether one is looking at actors or movies.

B. Clustering

Watts and Strogatz [15] have introduced the concept of clustering in social networks, also sometimes called network transitivity. Clustering refers to the increased propensity of pairs of people to be acquainted with one another if they have another acquaintance in common. Watts and Strogatz defined a clustering coefficient that measures the degree of clustering on a graph. For our purposes, the definition of this coefficient is

$$C=\frac{3\times(\text{number of triangles on the graph})}{(\text{number of connected triples of vertices})}=\frac{3N_\Delta}{N_3}. \tag{77}$$

Here "triangles" are trios of vertices each of which is connected to both of the others, and "connected triples" are trios in which at least one is connected to both the others. The factor of 3 in the numerator accounts for the fact that each triangle contributes to three connected triples of vertices, one for each of its three vertices. With this factor of 3, the value of C lies strictly in the range from zero to one. In the directed and undirected unipartite random graphs of Secs. II and III, C is trivially zero in the limit $N\to\infty$. In the one-mode projections of bipartite graphs, however, both the actors and the movies can be expected to have nonzero clustering. We here treat the case for actors. The case for movies is easily derived by swapping f's and g's.

An actor who has $z\equiv z_1$ co-stars in total contributes $\frac{1}{2}z(z-1)$ connected triples to N_3, so that

$$N_3=\tfrac{1}{2}N\sum_z z(z-1)r_z, \tag{78}$$

where r_z is the probability of having z co-stars. As shown above [Eq. (70)], the distribution r_z is generated by $G_0(x)$ and so

$$N_3=\tfrac{1}{2}NG_0''(1). \tag{79}$$

A movie that stars k actors contributes $\frac{1}{6}k(k-1)(k-2)$ triangles to the total triangle count in the one-mode graph. Thus the total number of triangles on the graph is the sum of $\frac{1}{6}k(k-1)(k-2)$ over all movies, which is given by

$$N_\Delta=\tfrac{1}{6}M\sum_k k(k-1)(k-2)q_k=\tfrac{1}{6}Mg_0'''(1). \tag{80}$$

Substituting into Eq. (77), we then get

$$C=\frac{M}{N}\frac{g_0'''(1)}{G_0''(1)}. \tag{81}$$

Making use of Eqs. (66), (67), and (70), this can also be written as

$$\frac{1}{C}-1=\frac{(\mu_2-\mu_1)(\nu_2-\nu_1)^2}{\mu_1\nu_1(2\nu_1-3\nu_2+\nu_3)}, \tag{82}$$

where $\mu_n=\Sigma_k k^n p_k$ is the nth moment of the distribution of numbers of movies in which actors have appeared, and ν_n is the same for cast size (number of actors in a movie).

C. Example

To give an example, consider a random bipartite graph with Poisson-distributed numbers of both movies per actor and actors per movie. In this case, following the derivation of Eq. (12), we find that

$$f_0(x)=e^{\mu(x-1)},\quad g_0(x)=e^{\nu(x-1)}, \tag{83}$$

and $f_1(x)=f_0(x)$ and $g_1(x)=g_0(x)$. Thus

$$G_0(x)=G_1(x)=\exp[\mu(e^{\nu(x-1)}-1]. \tag{84}$$

This implies that $z_1=\mu\nu$ and $z_2=(\mu\nu)^2$, so that

$$\ell=\frac{\ln N}{\ln\mu\nu}=\frac{\ln N}{\ln z}, \tag{85}$$

just as in an ordinary Poisson-distributed random graph. From Eq. (74), the average size $\langle s\rangle$ of a connected component of actors, below the phase transition, is

$$\langle s\rangle=\frac{1}{1-\mu\nu}, \tag{86}$$

which diverges, yielding a giant component, at $\mu\nu=z=1$, also as in the ordinary random graph. From Eqs. (34) and (35), the size S of the giant component as a fraction of N is a solution of

$$S=1-e^{\mu(e^{-\nu S}-1)}. \tag{87}$$

And from Eq. (81), the clustering coefficient for the one-mode network of actors is

$$C=\frac{M\nu^3}{N\nu^2(\mu^2+\mu)}=\frac{1}{\mu+1}, \tag{88}$$

where we have made use of Eq. (66).

Another quantity of interest is the distribution of numbers of co-stars, i.e., of the numbers of people with whom each actor has appeared in a movie. As discussed above, this distribution is generated by the function $G_0(x)$ defined in Eq. (70). For the case of the Poisson degree distribution, we can perform the derivatives, Eq. (4), and setting $x=0$ we find that the probability r_z of having appeared with a total of exactly z co-stars is

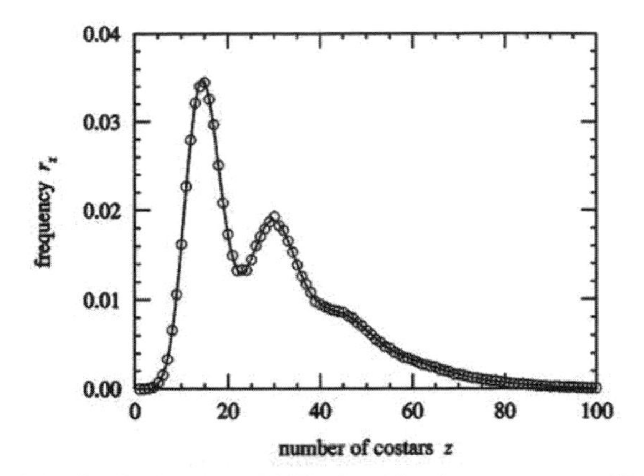

FIG. 7. The frequency distribution of numbers of co-stars of an actor in a bipartite graph with $\mu=1.5$ and $\nu=15$. The points are simulation results for $M=10\,000$ and $N=100\,000$. The line is the exact solution, Eqs. (89) and (90). The error bars on the numerical results are smaller than the points.

$$r_z=\frac{\nu^z}{z!}e^{\mu(e^{-\nu}-1)}\sum_{k=1}^{z}\left\{\begin{matrix}z\\k\end{matrix}\right\}[\mu e^{-\nu}]^k, \qquad (89)$$

where the coefficients $\{\begin{smallmatrix}z\\k\end{smallmatrix}\}$ are the Stirling numbers of the second kind [47]

$$\left\{\begin{matrix}z\\k\end{matrix}\right\}=\sum_{r=1}^{k}\frac{(-1)^{k-r}}{r!(k-r)!}r^z. \qquad (90)$$

D. Simulation results

Random bipartite graphs can be generated using an algorithm similar to the one described in Sec. III B for directed graphs. After making sure that the required degree distributions for both actor and movie vertices have means consistent with the required total numbers of actors and movies according to Eq. (66), we generate vertex degrees for each actor and movie at random and calculate their sum. If these sums are unequal, we discard the degree of one actor and one movie, chosen at random, and replace them with new degrees drawn from the relevant distributions. We repeat this process until the total actor and movie degrees are equal. Then we join vertices up in pairs.

In Fig. 7 we show the results of such a simulation for a bipartite random graph with Poisson degree distribution. (In fact, for the particular case of the Poisson distribution, the graph can be generated simply by joining up actors and movies at random, without regard for individual vertex degrees.) The figure shows the distribution of the number of co-stars of each actor, along with the analytic solution, Eqs. (89) and (90). Once more, numerical and analytic results are in good agreement.

V. APPLICATIONS TO REAL-WORLD NETWORKS

In this section we construct random graph models of two types of real-world networks, namely, collaboration graphs and the world-wide web, using the results of Secs. III and IV to incorporate realistic degree distributions into the models. As we will show, the results are in reasonably good agreement with empirical data, although there are some interesting discrepancies also, perhaps indicating the presence of social phenomena that are not incorporated in the random graph.

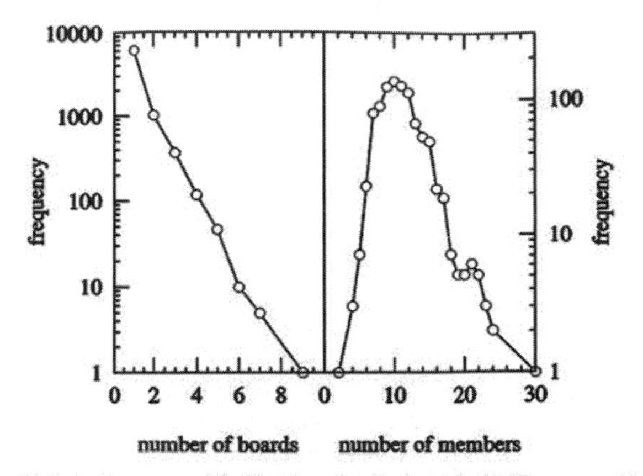

FIG. 8. Frequency distributions for the boards of directors of the Fortune 1000. Left panel: the numbers of boards on which each director sits. Right panel: the numbers of directors on each board.

A. Collaboration networks

In this section we construct random bipartite graph models of the known collaboration networks of company directors [29–31], movie actors [15], and scientists [36]. As we will see, the random graph works well as a model of these networks, giving good order-of-magnitude estimates of all quantities investigated, and in some cases giving results of startling accuracy.

Our first example is the collaboration network of the members of the boards of directors of the Fortune 1000 companies (the 1000 US companies with the highest revenues). The data come from the 1999 Fortune 1000 [29–31] and in fact include only 914 of the 1000, since data on the boards of the remaining 86 were not available. The data form a bipartite graph in which one type of vertex represents the boards of directors, and the other type the members of those boards, with edges connecting boards to their members. In Fig. 8 we show the frequency distribution of the numbers of boards on which each member sits, and the numbers of members of each board. As we see, the former distribution is close to exponential, with the majority of directors sitting on only one board, while the latter is strongly peaked around ten, indicating that most boards have about ten members.

Using these distributions, we can define generating functions $f_0(x)$ and $g_0(x)$ as in Eq. (23), and hence find the generating functions $G_0(x)$ and $G_1(x)$ for the distributions of numbers of co-workers of the directors. We have used these generating functions and Eqs. (72) and (81) to calculate the expected clustering coefficient C and the average number of co-workers z in the one-mode projection of board directors on a random bipartite graph with the same vertex degree distributions as the original dataset. In Table I we show the results of these calculations, along with the same quantities

TABLE I. Summary of results of the analysis of four collaboration networks.

Network	Clustering C Theory	Clustering C Actual	Average degree z Theory	Average degree z Actual
Company directors	0.590	0.588	14.53	14.44
Movie actors	0.084	0.199	125.6	113.4
Physics (arxiv.org)	0.192	0.452	16.74	9.27
Biomedicine (MEDLINE)	0.042	0.088	18.02	16.93

for the real Fortune 1000. As the table shows the two are in remarkable—almost perfect—agreement.

It is not just the average value of z that we can calculate from our generating functions, but the entire distribution: since the generating functions are finite polynomials in this case, we can simply perform the derivatives to get the probability distribution r_z. In Fig. 9, we show the results of this calculation for the Fortune 1000 graph. The points in the figure show the actual distribution of z for the real-world data, while the solid line shows the theoretical results. Again the agreement is excellent. The dashed line in the figure shows the distribution for an ordinary Poisson random graph with the same mean. Clearly this is a significantly inferior fit.

In fact, within the business world, attention has focused not on the collaboration patterns of company directors, but on the "interlocks" between boards, i.e., on the one-mode network in which vertices represent boards of directors and two boards are connected if they have one or more directors in common [28,29]. This is also simple to study with our model. In Fig. 10 we show the distribution of the numbers of interlocks that each board has, along with the theoretical prediction from our model. As we see, the agreement between empirical data and theory is significantly worse in this case than for the distribution of co-directors. In particular, it appears that our theory significantly underestimates the number of boards that are interlocked with very small or very large

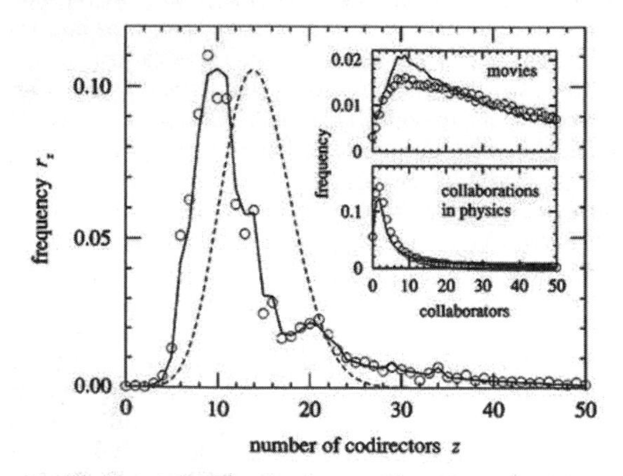

FIG. 9. The probability distribution of numbers of co-directors in the Fortune 1000 graph. The points are the real-world data, the solid line is the bipartite graph model, and the dashed line is the Poisson distribution with the same mean. Insets: the equivalent distributions for the numbers of collaborators of movie actors and physicists.

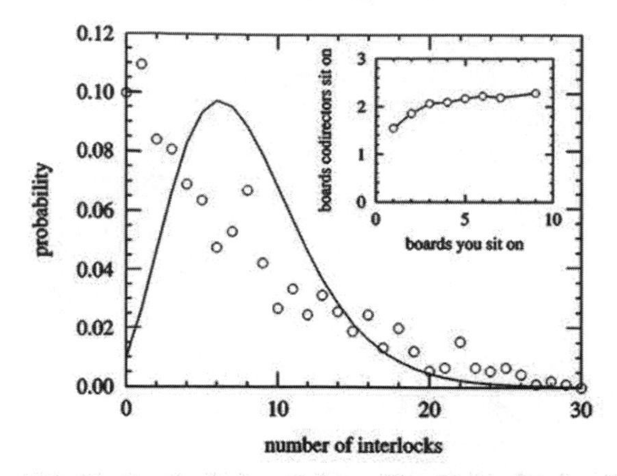

FIG. 10. The distribution of the number of other boards with which each board of directors is "interlocked" in the Fortune 1000 data. An interlock between two boards means that they share one or more common members. The points are the empirical data, the solid line is the theoretical prediction. Inset: the number of boards on which one's codirectors sit, as a function of the number of boards one sits on oneself.

numbers of other boards, while overestimating those with intermediate numbers of interlocks. One possible explanation of this is that "bigshots work with other bigshots." That is, the people who sit on many boards tend to sit on those boards with other people who sit on many boards. And conversely the people who sit on only one board (which is the majority of all directors), tend to do so with others who sit on only one board. This would tend to stretch the distribution of numbers of interlocks, just as seen in figure, producing a disproportionately high number of boards with very many or very few interlocks to others. To test this hypothesis, we have calculated, as a function of the number of boards on which a director sits, the average number of boards on which each of their co-directors sit. The results are shown in the inset of Fig. 10. If these two quantities were uncorrelated, the plot would be flat. Instead, however, it slopes clearly upwards, indicating indeed that on the average the big shots work with other big shots. (This idea is not new. It has been discussed previously by a number of others—see Refs. [48] and [49], for example.)

The example of the boards of directors is a particularly instructive one. What it illustrates is that the cases in which our random graph models agree well with real-world phenomena are not necessarily the most interesting. Certainly it is satisfying, as in Fig. 9, to have the theory agree well with the data. But probably Fig. 10 is more instructive: we have learned something about the structure of the network of the boards of directors by observing the way in which the pattern of board interlocks differs from the predictions of the purely random network. Thus it is perhaps best to regard our random graph as a null model—a baseline from which our expectations about network structure should be measured. It is deviation from the random graph behavior, not agreement with it, that allows us to draw conclusions about real-world networks.

We now look at three other graphs for which our theory also works well, although again there are some noticeable deviations from the random graph predictions, indicating the presence of social or other phenomena at work in the networks.

We consider the graph of movie actors and the movies in which they appear [15,50] and graphs of scientists and the papers they write in physics and biomedical research [36]. In Table I we show results for the clustering coefficients and average coordination numbers of the one-mode projections of these graphs onto the actors or scientists. As the table shows, our theory gives results for these figures that are of the right general order of magnitude, but typically deviate from the empirically measured figures by a factor of 2 or so. In the insets of Fig. 9 we show the distributions of numbers of collaborators in the movie actor and physicist graphs, and again the match between theory and real data is good, but not as good as with the Fortune 1000.

The figures for clustering and mean numbers of collaborators are particularly revealing. The former is uniformly about twice as high in real life as our model predicts for the actor and scientist networks. This shows that there is a significant tendency to clustering in these networks, in addition to the trivial clustering one expects on account of the bipartite structure. This may indicate, for example, that scientists tend to introduce pairs of their collaborators to one another, thereby encouraging clusters of collaboration. The figures for average numbers of collaborators show less deviation from theory than the clustering coefficients, but nonetheless there is a clear tendency for the numbers of collaborators to be smaller in the real-world data than in the models. This probably indicates that scientists and actors collaborate repeatedly with the same people, thereby reducing their total number of collaborators below the number that would naively be expected if we consider only the numbers of papers that they write or movies they appear in. It would certainly be possible to take effects such as these into account in a more sophisticated model of collaboration practices.

B. The world-wide web

In this section we consider the application of our theory of random directed graphs to the modeling of the world-wide web. As we pointed out in Sec. III A, it is not at present possible to make a very accurate random-graph model of the web, because to do so we need to know the joint distribution p_{jk} of in-degree and out-degrees of vertices, which has not to our knowledge been measured. However, we can make a simple model of the web by assuming in-degree and out-degree to be independently distributed according to their known distributions. Equivalently, we assume that the joint probability distribution factors according to $p_{jk}=p_j q_k$.

Broder *et al.* [26] give results showing that the in-degree and out-degree distributions of the web are approximately power law in form with exponents $\tau_{\text{in}}=2.1$ and $\tau_{\text{out}}=2.7$, although there is some deviation from the perfect power law for small degree. In Fig. 11, we show histograms of their data with bins chosen to be of uniform width on the logarithmic scales used. (This avoids certain systematic errors

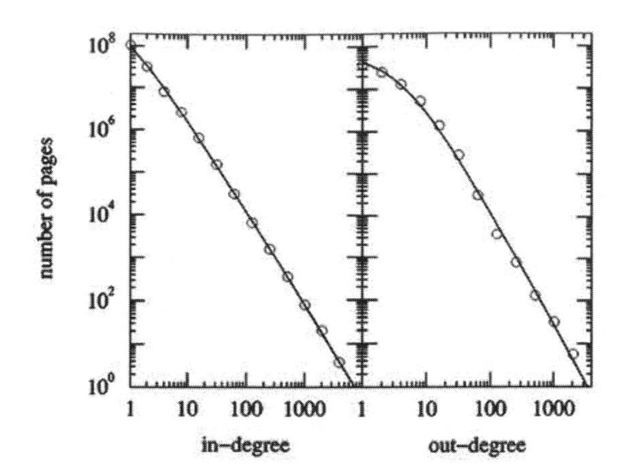

FIG. 11. The probability distribution of in-degree (left panel) and out-degree (right panel) on the world-wide web, rebinned from the data of Broder *et al.* [26]. The solid lines are best fits of the form (91).

known to afflict linearly histogrammed data plotted on log scales.) We find both distributions to be well fitted by the form

$$p_k=C(k+k_0)^{-\tau}, \tag{91}$$

where the constant C is fixed by the requirement of normalization, taking the value $1/\zeta(\tau,k_0)$, were $\zeta(x,y)$ is the generalized ζ function [47]. The constants k_0 and τ are found by least-squares fits, giving values of 0.58 and 3.94 for k_0, and 2.17 and 2.69 for τ, for the in-degree and out-degree distributions, respectively, in reasonable agreement with the fits performed by Broder *et al.* With these choices, the data and Eq. (91) match closely (see Fig. 11).

Neither the raw data nor our fits to them satisfy the constraint (59), that the total number of links leaving pages should equal the total number arriving at them. This is because the data set is not a complete picture of the web. Only about 2×10^8 of the web's 10^9 or so pages were included in the study. Within this subset, our estimate of the distribution of out-degree is presumably quite accurate, but many of the outgoing links will not connect to other pages within the subset studied. At the same time, no incoming links that originate outside the subset of pages studied are included, because the data are derived from "crawls" in which web pages are found by following links from one to another. In such a crawl one only finds links by finding the pages that they originate from. Thus our data for the incoming links is quite incomplete, and we would expect the total number of incoming links in the dataset to fall short of the number of outgoing ones. This indeed is what we see. The totals for incoming and outgoing links are approximately 2.3×10^8 and 1.1×10^9.

The incompleteness of the data for incoming links limits the information we can at present extract from a random graph model of the web. There are however some calculations that only depend on the out-degree distribution.

Given Eq. (91), the generating functions for the out-degree distribution take the form

$$G_0(x)=G_1(x)=\frac{\Phi(x,\tau,k_0)}{\zeta(\tau,k_0)}, \tag{92}$$

where $\Phi(x,y,z)$ is the Lerch Φ function [47]. The corresponding generating functions F_0 and F_1 we cannot calculate accurately because of the incompleteness of the data. The equality $G_0=G_1$ (and also $F_0=F_1$) is a general property of all directed graphs for which $p_{jk}=p_j q_k$ as above. It arises because in such graphs in-degree and out-degree are uncorrelated, and therefore the distribution of the out-degree of a vertex does not depend on whether you arrived at it by choosing a vertex at random, or by following a randomly chosen edge.

One property of the web that we can estimate from the generating functions for out-degree alone is the fraction S_{in} of the graph taken up by the giant strongly connected component plus those sites from which the giant strongly connected component can be reached. This is given by

$$S_{\text{in}}=1-G_0(1-S_{\text{in}}). \tag{93}$$

In other words, $1-S_{\text{in}}$ is a fixed point of $G_0(x)$. Using the measured values of k_0 and τ, we find by numerical iteration that $S_{\text{in}}=0.527$, or about 53%. The direct measurements of the web made by Broder *et al.* show that in fact about 49% of the web falls in S_{in}, in reasonable agreement with our calculation. Possibly this implies that the structure of the web is close to that of a directed random graph with a power-law degree distribution, though it is possible also that it is merely coincidence. Other comparisons between random graph models and the web will have to wait until we have more accurate data on the joint distribution p_{jk} of in-degree and out-degree.

VI. CONCLUSIONS

In this paper we have studied in detail the theory of random graphs with arbitrary distributions of vertex degree, including directed and bipartite graphs. We have shown how, using the mathematics of generating functions, one can calculate exactly many of the statistical properties of such graphs in the limit of large numbers of vertices. Among other things, we have given explicit formulas for the position of the phase transition at which a giant component forms, the size of the giant component, the average and distribution of the sizes of the other components, the average numbers of vertices a certain distance from a given vertex, the clustering coefficient, and the typical vertex-vertex distance on a graph. We have given examples of the application of our theory to the modeling of collaboration graphs, which are inherently bipartite, and the world-wide web, which is directed. We have shown that the random graph theory gives good order-of-magnitude estimates of the properties of known collaboration graphs of business people, scientists, and movie actors, although there are measurable differences between theory and data that point to the presence of interesting sociological effects in these networks. For the web we are limited in what calculations we can perform because of the lack of appropriate data to determine the generating functions. However, the calculations we can perform agree well with empirical results, offering some hope that the theory will prove useful once more complete data become available.

ACKNOWLEDGMENTS

The authors would like to thank Lada Adamic, Andrei Broder, Jon Kleinberg, Cris Moore, and Herb Wilf for useful comments and suggestions, and Jerry Davis, Paul Ginsparg, Oleg Khovayko, David Lipman, Grigoriy Starchenko, and Janet Wiener for supplying data used in this study. This work was funded in part by the National Science Foundation, the Army Research Office, the Electric Power Research Institute, and Intel Corporation.

[1] B. Bollobás, *Random Graphs* (Academic Press, New York, 1985).
[2] P. Erdös and A. Rényi, Publ. Math. **6**, 290 (1959).
[3] P. Erdös and A. Rényi, Publ. Math. Inst. Hung. Acad. Sci. **5**, 17 (1960).
[4] P. Erdös and A. Rényi, Acta Math. Acad. Sci. Hung. **12**, 261 (1961).
[5] L. Sattenspiel and C. P. Simon, Math. Biosci. **90**, 367 (1988).
[6] R. M. Anderson and R. M. May, J. Math. Biol. **33**, 661 (1995).
[7] M. Kretschmar and M. Morris, Math. Biosci. **133**, 165 (1996).
[8] D. D. Heckathorn, Soc. Prob. **44**, 174 (1997).
[9] C. C. Foster, A. Rapoport, and C. J. Orwant, Behav. Sci. **8**, 56 (1963).
[10] T. J. Fararo and M. Sunshine, *A Study of a Biased Friendship Network* (Syracuse University Press, Syracuse, NY, 1964).
[11] H. R. Bernard, P. D. Kilworth, M. J. Evans, C. McCarty, and G. A. Selley, Ethnology **2**, 155 (1988).
[12] J. Abello, A. Buchsbaum, and J. Westbrook, in Proceedings of the 6th European Symposium on Algorithms, 2000 (unpublished).
[13] W. Aiello, F. Chung, and L. Lu, in Proceedings of the 32nd Annual ACM Symposium on Theory of Computing, 2000 (unpublished).
[14] L. A. N. Amaral, A. Scala, M. Barthélémy, and H. E. Stanley, Proc. Natl. Acad. Sci. U.S.A. **97**, 11 149 (2000).
[15] D. J. Watts and S. H. Srogatz, Nature (London) **393**, 440 (1998).
[16] S. Jespersen, I. M. Sokolov, and A. Blumen, J. Chem. Phys. **113**, 7652 (2000).
[17] A. Scala, L. A. N. Amaral, and M. Barthélémy, e-print cond-mat/0004380.
[18] D. Fell and A. Wagner, Nat. Biotechnol. **18**, 1121 (2000).
[19] H. Jeong, B. Tombor, R. Albert, Z. N. Oltvai, and A.-L. Barabási, Nature (London) **407**, 651 (2000).
[20] R. J. Williams and N. D. Martinez, Nature (London) **404**, 180 (2000).

[21] J. M. Montoya and R. V. Solé, e-print cond-mat/0011195.

[22] A.-L. Barabási and R. Albert, Science **286**, 509 (1999).

[23] R. Albert, H. Jeong, and A.-L. Barabási, Nature (London) **401**, 130 (1999).

[24] B. A. Huberman and L. A. Adamic, Nature (London) **401**, 131 (1999).

[25] J. M. Kleinberg, S. R. Kumar, P. Raghavan, S. Rajagopalan, and A. Tomkins, in *The Web as a Graph: Measurements, Models, and Methods*, edited by T. Asano, H. Imai, D. T. Lee, S.-I. Nakano, and T. Tokuyama, Lecture Notes in Computer Science Vol. 1627 (Springer-Verlag, Berlin, 1999).

[26] A. Broder, R. Kumar, F. Maghoul, P. Raghavan, S. Rajagopalan, R. Stata, A. Tomkins, and J. Wiener, Comput. Netw. **33**, 309 (2000).

[27] M. Faloutsos, P. Faloutsos, and C. Faloutsos, Comput. Commun. Rev. **29**, 251 (1999).

[28] P. Mariolis, Soc. Sci. Q. **56**, 425 (1975).

[29] G. F. Davis, Corp. Gov. **4**, 154 (1996).

[30] G. F. Davis and H. R. Greve, Am. J. Sociol. **103**, 1 (1997).

[31] G. F. Davis, M. Yoo, and W. E. Baker (unpublished).

[32] B. Kogut and G. Walker (unpublished).

[33] J. W. Grossman and P. D. F. Ion, Congr. Numer. **108**, 129 (1995).

[34] R. De Castro and J. W. Grossman, Math. Intell. **21**, 51 (1999).

[35] V. Batagelj and A. Mrvar, Soc. Networks **22**, 173 (2000).

[36] M. E. J. Newman, Proc. Natl. Acad. Sci. U.S.A. **98**, 409 (2001).

[37] M. E. J. Newman, Phys. Rev. E **64**, 016131 (2001); **64**, 016132 (2001).

[38] S. Wasserman and K. Faust, *Social Network Analysis* (Cambridge University Press, Cambridge, 1994).

[39] S. H. Strogatz, Nature (London) **410**, 268 (2001).

[40] M. Molloy and B. Reed, Random Struct. Algorithms **6**, 161 (1995).

[41] M. Molloy and B. Reed, Combinatorics, Probab. Comput. **7**, 295 (1998).

[42] H. S. Wilf, *Generatingfunctionology*, 2nd ed. (Academic Press, London, 1994).

[43] C. Moore and M. E. J. Newman, Phys. Rev. E **62**, 7059 (2000).

[44] G. H. Hardy and J. E. Littlewood, Proc. London Math. Soc. **13**, 174 (1914).

[45] M. E. J. Newman and G. T. Barkema, *Monte Carlo Methods in Statistical Physics* (Oxford University Press, Oxford, 1999).

[46] Note that one must use $\ln(1-r)$ in this expression, and not $\ln r$, even though one might expect the two to give the same result. The reason is that r can be zero where $1-r$ cannot. With the standard 32-bit random number generators used on most computers, r will be zero about once in every 4×10^9 calls, and when it is, calculating $\ln r$ will give an error but $\ln(1-r)$ will not. For simulations on large graphs of a few million vertices or more this will happen with some frequency, and calculation of $\ln r$ should therefore be avoided.

[47] M. Abramowitz and I. Stegun, *Handbook of Mathematical Functions* (Dover, New York, 1965).

[48] B. Mintz and M. Schwartz, *The Power Structure of American Business* (University of Chicago Press, Chicago, 1985).

[49] G. F. Davis and M. S. Mizruchi, Adm. Sci. Q. **44**, 215 (1999).

[50] The figures given in our table differ from those given by Watts and Strogatz in Ref. [15] because we use a more recent version of the actor database. Our version dates from May 1, 2000 and contains about 450 000 actors, whereas the 1998 version contained only about 225 000.

4.2 THE SMALL-WORLD MODEL

In this section we look at the *small-world model*, which was introduced by **Watts and Strogatz (1998)** as a simple model of social networks. This model has proved amenable to treatment using a variety of techniques from statistical physics, including Monte Carlo simulation **(Barthélémy and Amaral 1999a; Newman and Watts 1999b)**, scaling and renormalization group methods (**Barthélémy and Amaral 1999a**; Newman and Watts 1999a; **Newman and Watts 1999b**; Kulkarni *et al.* 2000), mean-field methods (Newman *et al.* 2000), and exact treatments (Dorogovtsev and Mendes 2000b), and for this reason has attracted a good deal of attention in the physics community and elsewhere. Although the model has some drawbacks as a model of a real social network, it provides good intuition about the small-world effect as well as demonstrating convincingly the utility of statistical physics techniques in the study of networks.

The small-world model (sometimes also called the Watts–Strogatz model) is motivated by the observation that many real-world networks show the following two properties:

i) The small-world effect, meaning that most pairs of vertices are connected by a short path through the network (see Chapter 2). In recent years, the term "small-world effect" has come to mean specifically that the mean (or sometimes the maximum) vertex-vertex distance in the network (not counting vertex pairs that are not connected at all) increases logarithmically (or more slowly) with the total number of vertices in the network. There is some empirical evidence that this type of logarithmic scaling does occur in real-world networks (**Newman 2001a**; Albert and Barabási 2002), and there are excellent mathematical reasons for believing that it should be so. Watts and Strogatz implicitly assumed this scaling by defining short path length to mean path lengths comparable to those found in a random graph of the same size and average degree. Mean path length in a random graph is known to scale logarithmically with graph size (Bollobás 2001; Chung and Lu 2002a; Bollobás and Riordan 2002; Fronczak *et al.* 2002a; Cohen and Havlin 2003).

ii) High "clustering" or "transitivity," meaning that there is a heightened probability that two vertices will be connected directly to one another if they have another neighboring vertex in common. In the language of social networks, two people are much more likely to be acquainted with one another if they have another common acquaintance.

Given a network, the small-world effect is simple to measure: one just finds the distances between all pairs of vertices in the network and computes their average. Vertex-vertex distances can be found for example by breadth-first search (see Section 5.1).

Measuring clustering is a little more complicated. **Watts and Strogatz (1998)** proposed a measure of clustering, which they called the *clustering coefficient*. Consider Figure 4.3, which depicts the immediate neighborhood of a vertex i in some network. This vertex has six neighbors (connected to it by solid lines), and

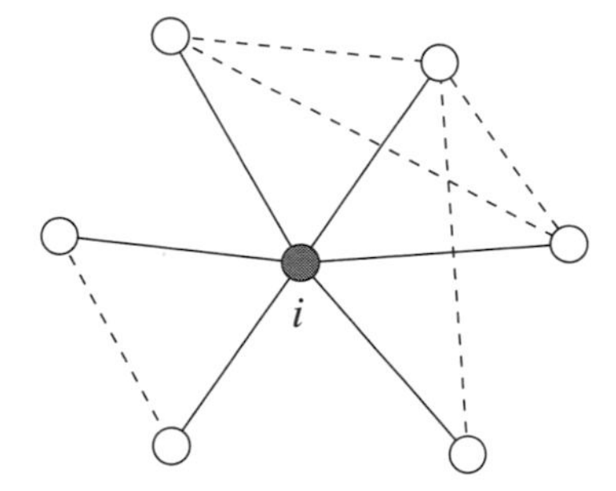

FIGURE 4.3 The central vertex i has six neighbors. Five of the fifteen possible pairs of these neighbors are themselves connected (dotted lines). The resulting clustering coefficient is $\frac{5}{15} = \frac{1}{3}$.

therefore $\frac{1}{2} \times 6 \times 5 = 15$ pairs of neighbors. Of those pairs, five are themselves connected (dotted lines). We define the local clustering coefficient C_i of the vertex i to be the fraction of such connected pairs, i.e., $C_i = \frac{5}{15} = \frac{1}{3}$. Then the clustering coefficient of an entire network is the average of this quantity over all n vertices:

$$C_{ws} = \frac{1}{n}\sum_{i=1}^{n} C_i = \frac{1}{n}\sum_{i=1}^{n} \frac{\text{(number of connected neighbor pairs)}}{\frac{1}{2}k_i(k_i - 1)}, \tag{4.8}$$

where k_i is the degree of vertex i. This quantity, which is the network average of the fraction of pairs of a vertex's neighbors that are connected, is one for a fully connected graph but tends to zero on a random graph as the graph becomes large, since the chance of two vertices being connected is independent of whether they have another neighbor in common and tends to zero as n^{-1} for fixed mean degree z.

The definition (4.8) has some problems however. Conceptually one thinks of it as being the mean probability that two people with a common friend will be friends of each other. Unfortunately, it is not actually equal to this probability. To see this, consider two vertices in a network, one having two neighbors and the other having a hundred. Furthermore, suppose that the two neighbors of the first vertex are themselves connected, but none of the neighbors of the second vertex are connected. Then $C_1 = 1$ for the first vertex and $C_2 = 0$ for the second, so that the average value of C for just these two vertices is 0.5. We can see immediately that this is the wrong value however: the first vertex only has one pair of neighbors, while the second has $\frac{1}{2} \times 99 \times 100 = 4950$, and out of the 4951 total pairs, only one is connected. So the average probability of a pair being connected is not 0.5, but $\frac{1}{4951} = 0.0002\ldots$ The problem is that Eq. (4.8) is heavily biased in favor of vertices with low degree because of the factor $k(k-1)$ in the denominator, and, as we see, this can make a huge difference to the value of C.

The correct way to calculate the average probability of a pair of neighbors being connected is to count up the total number of pairs of vertices on the entire graph that have a common neighbor and the total number of such pairs that are also themselves connected, and divide the one by the other. **Newman *et al.* (2001)** have expressed this as

$$C = \frac{3 \times \text{(number of triangles on a graph)}}{\text{(number of connected triples of vertices)}}, \tag{4.9}$$

where a triangle means three vertices that are each connected to both of the others, and a connected triple means a vertex that is connected to an (unordered) pair of other vertices, which may or may not be connected to each other. (An equivalent definition was also proposed independently by **Barrat and Weigt (2000)**.) The factor of 3 in the numerator accounts for the fact that each triangle contributes three separate connected triples. With this factor the value of C lies strictly in the range from zero to one. The quantity C defined in (4.9) is also sometimes called the *fraction of transitive triples* by social network analysts.

Newman (2003a) has given an alternative definition of the clustering coefficient as

$$C = \frac{6 \times (\text{number of triangles on a graph})}{(\text{number of paths of length 2})}, \tag{4.10}$$

where a path of length 2 is any three distinct vertices A, B, C for which there are edges A $\rightarrow$ B and B $\rightarrow$ C in the network. It is straightforward to show that the definitions (4.9) and (4.10) are equivalent, but the latter gives a different interpretation. The number of paths of length two leading from a vertex is also the number of friends of friends that a vertex has, so the second definition tells us that C is the probability that the friend of your friend is also your friend. The factor of 6 in the numerator again ensures that the value of C lies between zero and one and makes the two definitions numerically equal.

The value of the clustering coefficient on a fully connected graph (i.e., one in which every vertex is connected to every other) is $C = 1$. C also takes large values on some low-dimensional lattices, such as the triangular lattice (though not on others, such as the square lattice with nearest-neighbor bonds, for which $C = 0$). On a random graph, however, the probability that two vertices are connected is simply $p = z/n$ for all vertex pairs, where z is the mean degree of a vertex. Thus $C_{rg} = z/n$ in such a network, and this number is usually quite small for the kinds of values of z and n that occur in real-world networks. **Watts and Strogatz (1998)** defined a network to have high clustering if $C \gg C_{rg}$. In Table 4.1 we show the measured values of C for a number of real-world networks drawn from the literature, compared with the values of C_{rg} for random graphs with the same values of z and n. As the table shows, C is indeed much greater than C_{rg} in all of these cases.[3]

Watts and Strogatz defined a network to be a *small-world network*[4] if it shows both of the properties described above, that is, if the mean vertex-vertex distance ℓ is comparable with that on a random graph, $\ell/\ell_{rg} \sim 1$, and the clustering coefficient

[3]It can be argued that the random graph is not the correct model against which to compare the value of the clustering coefficient. In particular, Newman and Park (2003) have suggested that one should really compare the clustering coefficient to that of a random graph with the same degree distribution, of the type discussed in Section 4.1. On doing this, they find that social networks do, by and large, have higher clustering than such a model, but some other kinds of networks, like the Internet and the World Wide Web, do not.

[4]This expression, however, has been used inconsistently by other authors, probably because of confusion about what it means. It might be natural to assume that a "small-world network" would be one that shows the small-world effect, and indeed some authors use the term in this way. Some other authors use it to mean specifically networks taking the form of the Watts–Strogatz model. One should be careful, therefore, in reading the literature on this subject; the use of the term "small-world network" sometimes but not always implies high clustering, and may or may not refer to a specific network model.

network	n	z	clustering coefficient C measured	clustering coefficient C random graph
Internet[a]	6 374	3.8	0.24	0.00060
World Wide Web[b]	153 127	35.2	0.11	0.00023
power grid[c]	4 941	2.7	0.080	0.00054
biology collaborations[d]	1 520 251	15.5	0.081	0.000010
mathematics collaborations[e]	253 339	3.9	0.15	0.000015
film actor collaborations[f]	449 913	113.4	0.20	0.00025
company directors[f]	7 673	14.4	0.59	0.0019
word cooccurrence[g]	460 902	70.1	0.44	0.00015
neural network[c]	282	14.0	0.28	0.049
metabolic network[h]	315	28.3	0.59	0.090
food web[i]	134	8.7	0.22	0.065

TABLE 4.1 Number of vertices n, mean degree z, and clustering coefficient C for a number of different networks, along with the expected value of the clustering coefficient on a random graph with the same number of vertices and the same mean degree. Numbers are taken from [a]Pastor-Satorras *et al.* (2001), [b]Adamic (1999), [c]**Watts and Strogatz (1998)**, [d]Newman (2001b), [e]Newman (2003a), [f]**Newman *et al.* (2001)**, [g]Ferrer i Cancho and Solé (2001), [h]Fell and Wagner (2000), [i]Montoya and Solé (2002).

is much greater than that for a random graph, $C/C_{rg} \gg 1$. Walsh (1999) used this idea to define the *proximity ratio*

$$\mu = \frac{C/C_{rg}}{\ell/\ell_{rg}}, \tag{4.11}$$

which is of order 1 on a random graph, but much greater than 1 on a network obeying the Watts–Strogatz definition of a small-world network.

The issue addressed by **Watts and Strogatz (1998)** then is the following. High clustering can be found in a simple regular lattice. A one-dimensional lattice with bonds between nearest and next-nearest neighbors out to some maximum range k clearly has significant clustering. Such a lattice is shown in Figure 4.4a, with periodic boundary conditions making the lattice into a ring. If this lattice has linear dimension L, then the number of triangles on it is $\frac{1}{2}Lk(k-1)$ and the number of connected triples is $Lk(2k-1)$. Hence the clustering coefficient is[5]

$$C = \frac{3 \times \frac{1}{2}Lk(k-1)}{Lk(2k-1)} = \frac{3(k-1)}{2(2k-1)}, \tag{4.12}$$

which tends to a maximum value of $\frac{3}{4}$ as k becomes large. But this lattice does not show the small-world effect: the typical vertex-vertex distance is $L/4k$, which grows linearly with system size L rather than logarithmically, and hence can become very large if L is large and k is small.

[5]Strictly this formula is only correct so long as $L > 3k$, but this condition is rarely violated for the typical system sizes that have been investigated.

(a)
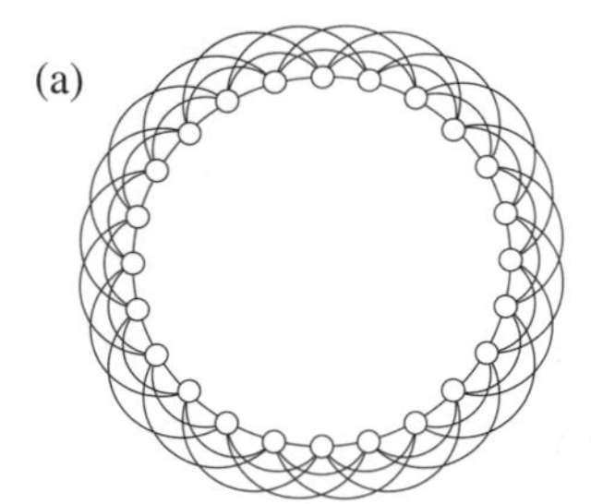
(b)
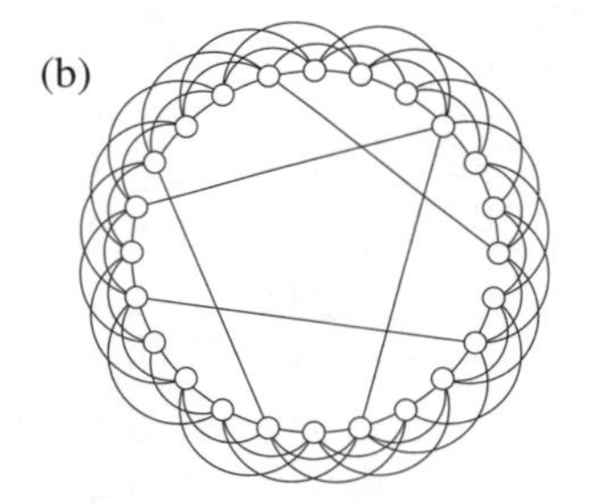
(c)
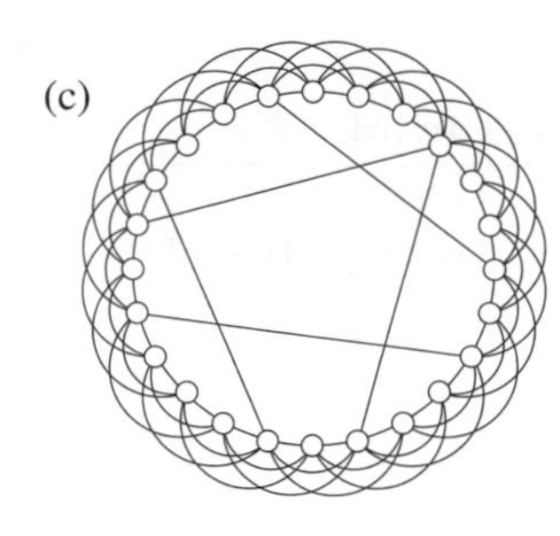

FIGURE 4.4 (a) A one-dimensional lattice with periodic boundary conditions and nearest and next-nearest neighbor bonds out to range k, where $k = 3$ in this case. (b) The original small-world model in which a fraction p of the bonds from (a) are "rewired" by moving one of their ends to a new vertex chosen uniformly at random. (c) The variant of the small-world model described in the text in which "shortcut" bonds are added between randomly chosen vertex pairs in (a), but no bonds are removed.

The random graph of Erdős and Rényi, described in Section 4.1, does show the small-world effect. It has a typical vertex-vertex distance of $\log n/\log z$, or $\log L/\log(2k)$ in the nomenclature of the small-world model. But the random graph, as we saw in Table 4.1, does not show significant clustering. How can we produce a network that shows both of these properties? The answer Watts and Strogatz gave for this question was to combine the regular lattice and the random graph as follows. Starting with the lattice of Figure 4.4a, we go through each bond in turn, and with some probability p we "rewire" that bond, meaning we take one of its ends (say the one that is further clockwise around the ring) and move it to a new location chosen uniformly at random. The result is a network that looks like Figure 4.4b, in which on average Lkp bonds have been rewired, creating "shortcuts" across the ring. This is the small-world model. It interpolates between the regular lattice and the random graph. When $p = 0$, no bonds are rewired; it is simply the regular lattice, with the high clustering coefficient of Eq. (4.12) but high vertex-vertex path lengths compared to the random graph. When $p = 1$, all edges are rewired and we get a random graph (well, almost—see below), which has a low clustering coefficient of $C = n/z = L/(2k)$ but short path lengths. In between, however, there is a considerable range of values of p in which the network has both high clustering and short path lengths. In the upper panel of Figure 4.5, we show the values of C and ℓ for the small-world model as a function of p for $L = 1000$ and $k = 10$, and in the lower panel we show the value of the Walsh proximity ratio μ for the same parameter values. As we can see, μ is small for small p as we would expect, but becomes significantly greater than 1 for $p \gtrsim 0.1$, indicating that the small-world model is a small-world network as defined by Watts and Strogatz for such values of the rewiring probability.

In fact, the description of the small-world model that we have given is somewhat simplified. Watts and Strogatz imposed two additional constraints to the construction of their networks: (1) the rewiring was not completely random, since they forbade any edge from linking a site to itself; (2) they also forbade more than one edge from linking the same pair of sites. In addition, as we have already

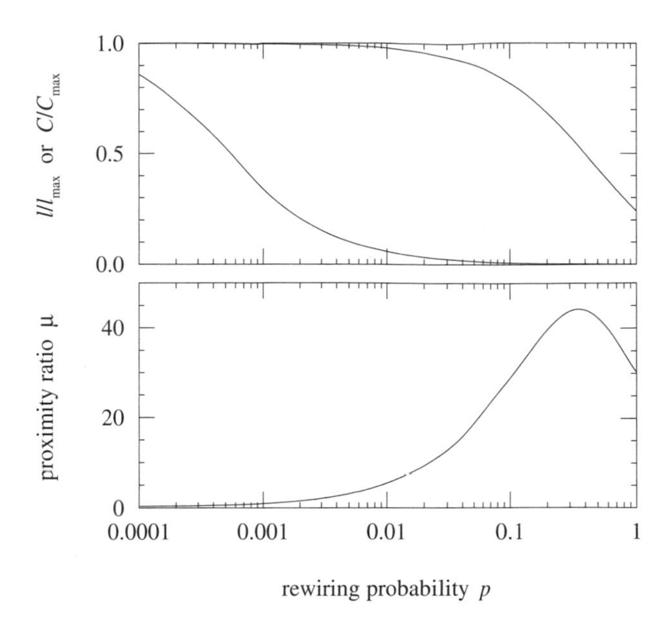

FIGURE 4.5 Top: the clustering coefficient C (upper curve) and the mean path length ℓ (lower curve) between vertices on small-world networks with $L = 1000$ and $k = 10$ as a function of the rewiring probability p. Both C and ℓ are rescaled so that their maximum values (at $p = 0$) are 1. The value of C is taken from Eq. (4.14) and ℓ from numerical simulations. Notice that the horizontal axis is logarithmic. Bottom: the value of the proximity ratio μ (Eq. (4.11)) as a function of p.

mentioned, the rewiring itself only moved *one* end of each rewired bond to a new random position and not both. This added condition prevents the resulting network from being a random graph even in the limit $p = 1$. To see this, notice that the degree of a vertex in the small-world model can never fall below k, which clearly produces a departure from the Poisson distribution of the normal random graph (see Section 4.1).

Although there were reasons for each of these restrictions on the model, they unfortunately give rise to a model whose analytic treatment is far from simple. To obviate this problem, therefore, a simplified version of the model was subsequently proposed by Newman and Watts (1999a) and independently by Monasson (1999). In this version, we again start with the lattice of Figure 4.4a and go through the bonds one by one. But now, with probability p per bond, we *add* another bond to the network between two vertices chosen uniformly at random, but we do not remove the original bond from the underlying lattice. Thus we add on average Lkp new bonds to the starting lattice, for a total of $Lk(1 + p)$ bonds overall. The resulting network is illustrated in Figure 4.4c. (In fact, the results of Figure 4.5 are for this latter version of the model and not for the original version of Watts and Strogatz, although the difference between the two is very slight. The equivalent of Figure 4.5 for the Watts–Strogatz version of the model can be found in **Watts and Strogatz (1998)**.) This model is also equivalent to the "great circle" model of **Ball *et al.* (1997)**, which was proposed earlier in the context of studies of disease spreading in social networks.

Even in this simplified form, however, analytic work on the small-world model has proved difficult. The original studies by **Watts and Strogatz (1998)** were purely numerical, but later authors attempted analytic treatments, some of which are described in the papers reproduced in this section. The clustering coefficient is known exactly. For the original Watts–Strogatz version of the model, **Barrat and Weigt (2000)** showed that

$$C = \frac{3(k-1)}{2(2k-1)}(1-p)^3, \tag{4.13}$$

while for the version of Figure 4.4c, Newman (2002c) showed that

$$C = \frac{3(k-1)}{2(2k-1)+4kp(p+2)}. \tag{4.14}$$

Barthélémy and Amaral (1999a), **Barrat (1999)**, **Newman and Watts (1999b)**, and Kulkarni *et al.* (2000) all derived exact scaling results for the model, while Newman and Watts (1999a) gave a simple real-space renormalization group treatment. Dorogovtsev and Mendes (2000b) solved exactly a model that is not quite the same as either of the versions described here, but is similar in many respects. These developments are described further below. A number of authors have also looked at generalizations of the small-world model built on higher dimensional lattices (Watts 1999; Newman and Watts 1999a; **Newman and Watts 1999b**; de Menezes *et al.* 2000; Ozana 2001; Sander *et al.* 2002; Newman *et al.* 2002b), using a mixture of numerical methods, scaling theory, exact methods, and series expansions.

Before we turn to detailed discussion of the papers in this section, it is worth commenting that the small-world model is not in general expected to be a very good model of real networks, including real social networks. One can give some justification for the model by saying that the underlying lattice represents local geographical constraints—people tend to know others who live close to them—and that the long-range shortcuts in the model represent ties between individuals who are geographically far apart, which occur with lower probability. There is little evidence, however, that these effects lead to network structure substantially similar to the small-world model. One area in which the model may be reasonable is in the study of the spread of plant diseases (Sander *et al.* 2002), but for networks of people it is clearly not ideal.

A number of other models have been proposed that also account for the simultaneous appearance of short path lengths and clustering in networks, which may be more realistic than the small-world model. Perhaps the oldest of these is the random biased net of Rapoport (1957), in which clustering is added to a random graph by "triadic closure"—the deliberate completion of connected triples of vertices to form triangles in the network, thereby increasing the clustering coefficient. Such triadic closure models have been discussed extensively in the social networks literature in the years following Rapoport's work (Wasserman and Faust 1994; Banks and Carley 1996), and recently a number of similar models have been proposed and studied by physicists and others as well (Watts 1999; Jin *et al.* 2001; Davidsen *et al.* 2002). Another attempt to model clustering was made in the 1980s by Holland and Leinhardt (1981), using the class of network models known as "exponential random graphs." Holland and Leinhardt's model has recently been revisited in the physics literature by Burda *et al.* (2004). A number of authors have also proposed simple schemes for growing "scale-free" networks (see Section 4.3) with clustering (see, for example, Klemm and Eguiluz (2002)). Another method for generating clustering in networks may be membership of individuals in groups. If three individuals are all members of a group that is much smaller than the network as a whole, they may have a higher chance of mutual acquaintance (thereby producing clustering) as a result of that membership. This mechanism has been investigated by **Newman *et al.*** (**2001**; Newman 2003c) using so-called bipartite graph models. A more sophisticated and possibly moderately realistic model of a social network based on hierarchical division of communities into groups has been proposed recently by Watts *et al.* (2002) and independently by Kleinberg (2002).

Watts and Strogatz (1998)

The small-world model was first introduced by Duncan Watts and Steven Strogatz in an influential paper published in the journal *Nature* (**Watts and Strogatz 1998**). Their paper is our first in this section.

Although it is short, this paper achieves a lot in its few pages and touches on many of the issues that are central to the research described in this book. In particular, it focuses on four things: the dual properties of short path lengths and high clustering seen in many networks, the empirical documentation of these properties using real network data, the introduction of the small-world model to mimic these properties, and finally the behavior of various dynamical systems on networks with the small-world topology.

After arguing in general that the properties of real-world networks appear to fall somewhere between the regular and the random, Watts and Strogatz introduce the small-world model in the first version described above, that of Figure 4.4b, in which the shortcuts across the lattice are generated by "rewiring" edges rather than by adding new edges, so that the total number of edges in the graph, and hence the mean degree, remains constant. They show by numerical simulation that there is a range of values of the rewiring parameter p where the clustering coefficient is high[6] but the mean path length is low, and present their results in a figure similar to our Figure 4.5. They then present evidence from three real-world networks to back up their claims. First, they study a network of collaborations between film actors, in which two actors are considered connected if they have appeared in a film together. This network was constructed using the resources of the Internet Movie

[6] They use the first definition of clustering coefficient, given in Eq. (4.8).

Database, an online database containing data about actors and the films they have appeared in. At the time of their study, this database contained about a quarter of a million actors (it is much larger now), and the actor network is by far the largest of the networks they looked at. Second, they look at the network formed by the high-voltage transmission lines of the Western States Power Grid of the United States, the network over which electricity is carried around the western part of America. In this network the vertices are generating stations and substations, and the edges are transmission lines. (Only high-voltage transmission lines are included. The lower voltage lines used for electrical delivery are not counted as a part of this network.) Third, they look at the neural network of the worm *C. elegans*. This tiny nematode is one of the best-studied organisms in the whole of biological science, and among other things the structure of its entire neural network—the pattern of synaptic connections between its neurons—has been completely determined in a series of experiments by John White and collaborators in the early 1980s (White *et al.* 1986).

For each of these three networks, Watts and Strogatz demonstrate that the mean vertex-vertex distance is close to that of a random graph with the same size and mean degree, but that the clustering coefficient is much higher—as much as three orders of magnitude higher in the case of the network of film actors.

In the final part of the paper, Watts and Strogatz give a discussion of the behavior of dynamical systems on networks. This is a particularly challenging area of study in which analytic progress has proved elusive. Prudently, Watts and Strogatz picked as a first example one of the most tractable problems in the area, the spread of disease modeled using the SIR epidemic model (see Section 5.1 for a detailed discussion of this model). They measured the position of the epidemic threshold—the critical value of the probability r of infection of a susceptible individual by an infective one at which the disease first becomes an epidemic, spreading through the population rather than dying out. They found a clear decline in the critical r with increasing p, indicating that the small-world topology makes it easier for the disease to spread. They also found that the typical time for the disease to spread through the community decreases with increasing p. Exact analytical solutions of SIR epidemics on the small-world model by others confirm these results (**Ball *et al.* 1997**; Moore and Newman 2000a,b).

Watts and Strogatz also discuss briefly the behavior of three other dynamical systems on networks with the topology of the small-world model:

i) A cellular automaton, attempting to perform the "density classification" task. Starting with a configuration in which each vertex in the network is either "on" or "off," the problem is to construct a local dynamical rule for a cellular automaton that will drive the system to an all-on or all-off state, depending on whether the initial configuration had more on-sites or off-sites.

ii) The iterated Prisoner's Dilemma game, in which players on the vertices of the network can either "cooperate" or "defect" with their network neighbors, with a substantial payoff if they cooperate, but a larger one if they defect and their neighbor cooperates. The idea is to maximize total payoff over repeated rounds of the game.

iii) Coupled oscillators on the vertices, which synchronize or "phase lock" as a result of weak interactions along the edges of the network.

Watts and Strogatz found that in cases (1) and (3) the small-world nature of the network helps the system, improving performance in the density classification problem and making the oscillators synchronize more readily than on a regular lattice. In case (2), curiously, the small-world topology seems to hinder overall performance in the Prisoner's Dilemma; the well-known "tit-for-tat" strategy and similar strategies that favor emergence of the optimal all-cooperate sequence were less likely to achieve this result in the small-world model than on a simple regular lattice. A related result is that of Walsh (1999), who found that coloring problems (NP-hard problems related to satisfiability and ground-state problems for Potts models) were considerably harder to solve on networks with the small-world topology than on either regular lattices or random graphs.

Barthélémy and Amaral (1999a)

The small-world model is well suited to treatment using the tools of statistical physics. It is a lattice model, with a single parameter k defining the lattice and a well-defined ensemble of randomized states parameterized by p. It is not surprising then, that physicists took up the challenge of understanding its behavior. The next paper in this section **(Barthélémy and Amaral 1999a)** was the first on the small-world model to appear in the physics literature. Its primary contribution is in highlighting the scaling properties of the model. In it the authors, Marc Barthélémy and Luis Amaral, studied the small-world model built on a one-dimensional lattice and focused on one fundamental quantity, the typical path length ℓ between vertices. They proposed that this quantity satisfies a scaling law of the form

$$\ell = \xi F(L/\xi), \tag{4.15}$$

where L again is the system size and ξ is a characteristic scale[7] that depends on the parameters p and k. Since ℓ is expected to scale linearly with L for small system sizes and logarithmically for large ones, the scaling function $F(x)$ should have the limiting forms

$$F(x) \sim \begin{cases} x & \text{as } x \to 0, \\ \log x & \text{as } x \to \infty, \end{cases} \tag{4.16}$$

with some crossover between these two regimes occurring in between at a value of L that scales as ξ. They further conjectured that ξ depends on p according to

$$\xi \sim p^{-\tau}, \tag{4.17}$$

for some constant exponent τ.

[7] We use L and ξ for consistency with earlier and later presentation, but Barthélémy and Amaral used the symbols n and n^* for these two quantities.

The scaling law (4.16) is expected to apply only in the limit of large ξ, and hence only for small p, i.e., the regime of low density of shortcuts in the model. Performing computer simulations on systems of size up to $L = 1000$, Barthélémy and Amaral measured the value of ℓ and hence extracted values of ξ for systems of various sizes. From the asymptotic behavior of $\log \xi$ as a function of $\log L$ they then conjectured that $\tau = \frac{2}{3}$.

The most important take-home message of the paper is the scaling relation, Eq. (4.15), which is now known to be correct, as are the limiting forms, Eq. (4.16). Scaling forms for the small-world model are discussed further below.

A substantial portion of the paper is also directed at arguing that the change from "large-world" behavior (the regular lattice behavior seen for small p) to small-world behavior (mean vertex-vertex distances scaling as $\log L$) is a "crossover phenomenon" rather than a phase transition. This is reasonable although uncontroversial—to our knowledge, no one has suggested that this change is a phase transition. There *is* a phase transition in the small-world model, as pointed out later by various authors (Newman and Watts 1999a; **Newman and Watts 1999b**; de Menezes *et al.* 2000), but it occurs at $p = 0$ and not in the region where large-world behavior gives way to small-world, which is around $p = 1/(kL)$.

One issue with this paper is that Barthélémy and Amaral were unable to make an accurate determination of the exponent τ because of the small system sizes they studied, and it turns out that their conjecture of $\tau = \frac{2}{3}$ was not right, the correct value being $\tau = 1$, as the following paper by **Barrat (1999)** shows. In fact, Barthélémy and Amaral suggested in their paper that one might expect $\tau = 1$ from simple physical arguments, but were apparently persuaded by their numerical calculations that this was wrong. They later published an erratum agreeing with Barrat and giving more extensive simulation results confirming the value $\tau = 1$ (Barthélémy and Amaral 1999b).

Barrat (1999)

In this short comment, previously unpublished, Alain Barrat examines the scaling form proposed for the small-world model by **Barthélémy and Amaral (1999a)**, Eq. (4.15), and considers their conjecture that the exponent τ, Eq. (4.17), has the value $\frac{2}{3}$. He rejects this conjecture with the following simple argument. If one assumes $\tau < 1$ and considers the value of ℓ predicted by Eq. (4.15) for a system of size L, at a value of $p \sim L^{-1/\alpha}$, where α is chosen to be greater than τ but less than 1, then it is easy to show that in the limit $L \to \infty$, ℓ goes as $\log L$ (from Eq. (4.16)) and we are therefore in the small-world regime, but that simultaneously the number of shortcuts in the model due to rewiring of bonds goes to zero. This leaves one with the physically unrealistic conclusion that one can have small-world behavior without any rewiring. Hence, we conclude, τ cannot be less than 1. Barrat gives simulation results for systems of size up to $L = 5000$ that appear to show that $\tau = 1$.

Following the appearance of Barrat's comment, Newman and Watts (1999a) gave an exact renormalization group treatment of the small-world model that shows that $\tau = 1$ exactly. They also gave further numerical evidence of this result. The renormalization group treatment also leads to another scaling form,

$$\ell = \frac{L}{k} f(L/\xi), \tag{4.18}$$

for the scaling of ℓ when $\xi \gg 1$. This form is equivalent to the one given by Barthélémy and Amaral, Eq. (4.15), to within a factor of k, by the substitution $F(x) \to xf(x)$. In order that ℓ have the expected behavior for large and small L, the function $f(x)$ should have the limiting forms

$$f(x) \sim \begin{cases} \text{constant} & \text{for } x \ll 1, \\ (\log x)/x & \text{for } x \gg 1. \end{cases} \tag{4.19}$$

Newman and Watts (1999b)

Building on the scaling ideas proposed by **Barthélémy and Amaral (1999a)**, **Newman and Watts (1999b)** published a detailed analysis of the scaling behavior of the small-world model, in the modified version of Figure 4.4c. In this paper, the authors point out that the small-world model has a natural length-scale built into it, which is the typical distance between the ends of shortcuts on the lattice. They also argue that this is the only independent length-scale in the model and hence it must be equivalent to the length-scale denoted ξ above. To within a constant this gives

$$\xi = \frac{1}{kp}. \tag{4.20}$$

(They derived the same result previously using their renormalization group method (Newman and Watts 1999a). This result clearly also implies the result $\tau = 1$ above.) Combining Eqs. (4.18) and (4.20), we then find

$$\ell = \frac{L}{k} f(Lkp). \tag{4.21}$$

They confirmed this form by numerical simulation, making a plot of $k\ell/L$ as a function of the scaling variable $x \equiv Lkp$ and finding a good data collapse over about five orders of magnitude in Lkp. Since the scaling relation is only valid for small p, one must be careful to attempt the scaling collapse only in this regime. Equation (4.21) is only valid when $\xi \gg k$, since k is the basic length-scale on which connections appear on the underlying lattice. With Eq. (4.20), this implies that we must have $p \ll 1/k^2$ to fall in the scaling regime.

Newman and Watts also gave generalizations of Eqs. (4.20) and (4.21) to small-world models built on underlying lattices with dimension greater than 1, which have also been confirmed by numerical simulations (Newman and Watts 1999a; de Menezes *et al.* 2000).

Newman and Watts gave a number of other results in their paper. They gave an approximate calculation of the form of the scaling function $f(x)$ using a combination of series expansions and Padé approximants. Although they used only a low-order Padé approximant (third order), and one moreover that has an incorrect scaling behavior in the limit of large x, their result appears to fit the numerical data quite well. Probably a better Padé approximant fit is possible if one is willing to go to higher order. So far no one has been able to find an exact form for the scaling function, although Newman *et al.* (2000) found a "mean-field" solution that is valid in the limit of low density of shortcuts. This solution was later extended and made more rigorous by Barbour and Reinert (2001), who derived an improved approximation that includes the mean-field result at leading order.

Newman and Watts also described a method for calculating the mean number of vertices a given distance r or less from a randomly chosen vertex on an infinite small-world lattice, the equivalent of the volume of a "sphere" of radius r. They showed that in general dimension d this volume satisfies

$$V(r) = \sum_{r'=0}^{r} a(r')[1 + 2\xi^{-d} V(r - r')], \tag{4.22}$$

where $a(r)$ is the "surface area" of a sphere of radius r on the underlying lattice. They solved this equation for the one-dimensional case by making an integral approximation, giving

$$V_1 = \tfrac{1}{2}\xi(e^{4r/\xi} - 1). \tag{4.23}$$

Moukarzel (1999) later gave a solution for general dimension (also using an integral approximation), which leads to

$$V_2 = \tfrac{1}{2}\xi^2\left[\cosh\left(2\sqrt{\pi}\,\frac{r}{\xi}\right) - 1\right], \tag{4.24}$$

$$V_3 = \tfrac{1}{2}\xi^3\left[\tfrac{1}{3}\exp\left(2\sqrt[3]{\pi}\,\frac{r}{\xi}\right) + \tfrac{2}{3}\exp\left(-\sqrt[3]{\pi}\,\frac{r}{\xi}\right)\cos\left(\sqrt{3}\sqrt[3]{\pi}\,\frac{r}{\xi}\right) - 1\right], \tag{4.25}$$

for two and three dimensions.

In the last part of their paper, Newman and Watts consider percolation on the small-world model, which they regard as a simple model for the spread of information or disease across a social network. (It is known that the SIR model of disease spread (Section 5.1) can be mapped exactly onto a bond percolation model (Mollison 1977; Sander *et al.* 2002). Newman and Watts, however, studied site percolation, which has similar but not identical behavior.) They gave an approximate solution for the position of the percolation threshold for site percolation in the small-world model, and confirmed that it was roughly correct by numerical simulation. In two later papers, Moore and Newman (2000a,b) gave an exact solution for the same problem, and also for the bond percolation problem.

Barrat and Weigt (2000)

Many of the scaling ideas pursued by **Barthélémy and Amaral (1999a)**, **Barrat (1999)**, and **Newman and Watts (1999b)** were explored further by **Barrat and Weigt (2000)**. In this paper the authors looked again at typical vertex-vertex path lengths in the small-world model, in this case the original version of **Watts and Strogatz (1998)**. They showed, as Newman and Watts (1999a) also did, that the mean path length ℓ scales simply with the number of vertices L as L becomes large if the number of shortcuts is fixed, and pointed out that this implies that the *density* of shortcuts needed to achieve small-world behavior in the limit of large L tends to zero. In other words, in the thermodynamic limit, infinitesimal values of p are sufficient to ensure small-world behavior of the form $\ell \sim \log L$. They verified their findings with extensive numerical simulations and scaling analyses.

The paper also includes a variety of other results. For example, the authors computed the degree distribution for the model, finding that it takes roughly the form of a binomial distribution but truncated at a minimum degree of k. (For the other version of the small-world model shown in Figure 4.4c, the degree distribution is different, taking the form of a simple binomial distribution plus a constant.) Barrat and Weigt also proposed an alternative definition of the clustering coefficient equivalent to that of Eq. (4.9), calculated its value, Eq. (4.13), for the model, and performed simulations to check their prediction.

The remainder of the paper by Barrat and Weigt is given over to a solution of the Ising model on networks with the geometry of the small-world model. Although there is no evidence of the existence of real Ising magnets with geometry of this kind, the solution is nonetheless interesting because it emphasizes again the fact that only an infinitesimal value of p is necessary to produce small-world behavior in the limit of large system size. The Ising model on a one-dimensional lattice with finite-range interactions, such as the underlying lattice of the small-world model, has no phase transition at finite temperature; its correlation length only diverges at $T = 0$ (see, for instance, Plischke and Bergersen (1994)). In higher dimensions, however, it has a phase transition to a ferromagnetic state at finite temperature. Clearly, then, when $p = 0$ the Ising model will have no transition on a small-world lattice, but for $p > 0$ we might expect that it would, since, as **Newman and Watts (1999b)** showed, the effective dimension of such a system is infinite in the limit of infinite system size. Furthermore, if the effective dimension is truly infinite, we would expect the Ising model to have a transition in the universality class appropriate to infinite dimension, the mean-field class. And this is in fact what Barrat and Weigt find. Using a combination of the replica trick and a perturbation expansion in powers of p, they show that for any nonzero value of p there is a ferromagnetic transition at finite temperature with critical exponents in the mean-field universality class. The critical temperature has the behavior $T_c \sim -k/\log p$. A number of other studies of the Ising model on small-world-type lattices have been performed by others (Pękalski 2001; Hong *et al.* 2002; Kuperman and Zanette 2002; Herrero 2002; Zhu and Zhu 2002; Viana Lopes *et al.* 2004).

The authors also analyze a slightly modified version of the small-world model in which no bonds connecting vertices farther apart than next-nearest neighbors are allowed on the underlying lattice, but each pair of next-nearest-neighbor vertices is connected by k parallel edges, each of which may be independently rewired with probability p. For this variant model, which has properties similar overall to the original model of **Watts and Strogatz (1998)**, perturbation theory is not needed to solve the Ising model and a complete solution can be found directly by applying the replica trick. Again a phase transition is found at finite temperature for any nonzero value of the rewiring probability p.

Collective dynamics of 'small-world' networks

Duncan J. Watts* & Steven H. Strogatz

Department of Theoretical and Applied Mechanics, Kimball Hall, Cornell University, Ithaca, New York 14853, USA

Networks of coupled dynamical systems have been used to model biological oscillators[1–4], Josephson junction arrays[5,6], excitable media[7], neural networks[8–10], spatial games[11], genetic control networks[12] and many other self-organizing systems. Ordinarily, the connection topology is assumed to be either completely regular or completely random. But many biological, technological and social networks lie somewhere between these two extremes. Here we explore simple models of networks that can be tuned through this middle ground: regular networks 'rewired' to introduce increasing amounts of disorder. We find that these systems can be highly clustered, like regular lattices, yet have small characteristic path lengths, like random graphs. We call them 'small-world' networks, by analogy with the small-world phenomenon[13,14] (popularly known as six degrees of separation[15]). The neural network of the worm *Caenorhabditis elegans*, the power grid of the western United States, and the collaboration graph of film actors are shown to be small-world networks. Models of dynamical systems with small-world coupling display enhanced signal-propagation speed, computational power, and synchronizability. In particular, infectious diseases spread more easily in small-world networks than in regular lattices.

To interpolate between regular and random networks, we consider the following random rewiring procedure (Fig. 1). Starting from a ring lattice with n vertices and k edges per vertex, we rewire each edge at random with probability p. This construction allows us to 'tune' the graph between regularity ($p = 0$) and disorder ($p = 1$), and thereby to probe the intermediate region $0 < p < 1$, about which little is known.

We quantify the structural properties of these graphs by their characteristic path length $L(p)$ and clustering coefficient $C(p)$, as defined in Fig. 2 legend. Here $L(p)$ measures the typical separation between two vertices in the graph (a global property), whereas $C(p)$ measures the cliquishness of a typical neighbourhood (a local property). The networks of interest to us have many vertices with sparse connections, but not so sparse that the graph is in danger of becoming disconnected. Specifically, we require $n \gg k \gg \ln(n) \gg 1$, where $k \gg \ln(n)$ guarantees that a random graph will be connected[16]. In this regime, we find that $L \sim n/2k \gg 1$ and $C \sim 3/4$ as $p \rightarrow 0$, while $L \approx L_{random} \sim \ln(n)/\ln(k)$ and $C \approx C_{random} \sim k/n \ll 1$ as $p \rightarrow 1$. Thus the regular lattice at $p = 0$ is a highly clustered, large world where L grows linearly with n, whereas the random network at $p = 1$ is a poorly clustered, small world where L grows only logarithmically with n. These limiting cases might lead one to suspect that large C is always associated with large L, and small C with small L.

On the contrary, Fig. 2 reveals that there is a broad interval of p over which $L(p)$ is almost as small as L_{random} yet $C(p) \gg C_{random}$. These small-world networks result from the immediate drop in $L(p)$ caused by the introduction of a few long-range edges. Such 'short cuts' connect vertices that would otherwise be much farther apart than L_{random}. For small p, each short cut has a highly nonlinear effect on L, contracting the distance not just between the pair of vertices that it connects, but between their immediate neighbourhoods, neighbourhoods of neighbourhoods and so on. By contrast, an edge

* Present address: Paul F. Lazarsfeld Center for the Social Sciences, Columbia University, 812 SIPA Building, 420 W118 St, New York, New York 10027, USA.

removed from a clustered neighbourhood to make a short cut has, at most, a linear effect on C; hence $C(p)$ remains practically unchanged for small p even though $L(p)$ drops rapidly. The important implication here is that at the local level (as reflected by $C(p)$), the transition to a small world is almost undetectable. To check the robustness of these results, we have tested many different types of initial regular graphs, as well as different algorithms for random rewiring, and all give qualitatively similar results. The only requirement is that the rewired edges must typically connect vertices that would otherwise be much farther apart than L_{random}.

The idealized construction above reveals the key role of short cuts. It suggests that the small-world phenomenon might be common in sparse networks with many vertices, as even a tiny fraction of short cuts would suffice. To test this idea, we have computed L and C for the collaboration graph of actors in feature films (generated from data available at http://us.imdb.com), the electrical power grid of the western United States, and the neural network of the nematode worm *C. elegans*[17]. All three graphs are of scientific interest. The graph of film actors is a surrogate for a social network[18], with the advantage of being much more easily specified. It is also akin to the graph of mathematical collaborations centred, traditionally, on P. Erdös (partial data available at http://www.acs.oakland.edu/~grossman/erdoshp.html). The graph of the power grid is relevant to the efficiency and robustness of power networks[19]. And *C. elegans* is the sole example of a completely mapped neural network.

Table 1 shows that all three graphs are small-world networks. These examples were not hand-picked; they were chosen because of their inherent interest and because complete wiring diagrams were available. Thus the small-world phenomenon is not merely a curiosity of social networks[13,14] nor an artefact of an idealized model—it is probably generic for many large, sparse networks found in nature.

We now investigate the functional significance of small-world connectivity for dynamical systems. Our test case is a deliberately simplified model for the spread of an infectious disease. The population structure is modelled by the family of graphs described in Fig. 1. At time $t = 0$, a single infective individual is introduced into an otherwise healthy population. Infective individuals are removed permanently (by immunity or death) after a period of sickness that lasts one unit of dimensionless time. During this time, each infective individual can infect each of its healthy neighbours with probability r. On subsequent time steps, the disease spreads along the edges of the graph until it either infects the entire population, or it dies out, having infected some fraction of the population in the process.

Table 1 Empirical examples of small-world networks

	L_{actual}	L_{random}	C_{actual}	C_{random}
Film actors	3.65	2.99	0.79	0.00027
Power grid	18.7	12.4	0.080	0.005
C. elegans	2.65	2.25	0.28	0.05

Characteristic path length L and clustering coefficient C for three real networks, compared to random graphs with the same number of vertices (n) and average number of edges per vertex (k). (Actors: $n = 225{,}226$, $k = 61$. Power grid: $n = 4{,}941$, $k = 2.67$. *C. elegans*: $n = 282$, $k = 14$.) The graphs are defined as follows. Two actors are joined by an edge if they have acted in a film together. We restrict attention to the giant connected component[16] of this graph, which includes ~90% of all actors listed in the Internet Movie Database (available at http://us.imdb.com), as of April 1997. For the power grid, vertices represent generators, transformers and substations, and edges represent high-voltage transmission lines between them. For *C. elegans*, an edge joins two neurons if they are connected by either a synapse or a gap junction. We treat all edges as undirected and unweighted, and all vertices as identical, recognizing that these are crude approximations. All three networks show the small-world phenomenon: $L \gtrsim L_{random}$ but $C \gg C_{random}$.

Regular Small-world Random

$p = 0$ Increasing randomness $p = 1$

Figure 1 Random rewiring procedure for interpolating between a regular ring lattice and a random network, without altering the number of vertices or edges in the graph. We start with a ring of n vertices, each connected to its k nearest neighbours by undirected edges. (For clarity, $n = 20$ and $k = 4$ in the schematic examples shown here, but much larger n and k are used in the rest of this Letter.) We choose a vertex and the edge that connects it to its nearest neighbour in a clockwise sense. With probability p, we reconnect this edge to a vertex chosen uniformly at random over the entire ring, with duplicate edges forbidden; otherwise we leave the edge in place. We repeat this process by moving clockwise around the ring, considering each vertex in turn until one lap is completed. Next, we consider the edges that connect vertices to their second-nearest neighbours clockwise. As before, we randomly rewire each of these edges with probability p, and continue this process, circulating around the ring and proceeding outward to more distant neighbours after each lap, until each edge in the original lattice has been considered once. (As there are $nk/2$ edges in the entire graph, the rewiring process stops after $k/2$ laps.) Three realizations of this process are shown, for different values of p. For $p = 0$, the original ring is unchanged; as p increases, the graph becomes increasingly disordered until for $p = 1$, all edges are rewired randomly. One of our main results is that for intermediate values of p, the graph is a small-world network: highly clustered like a regular graph, yet with small characteristic path length, like a random graph. (See Fig. 2.)

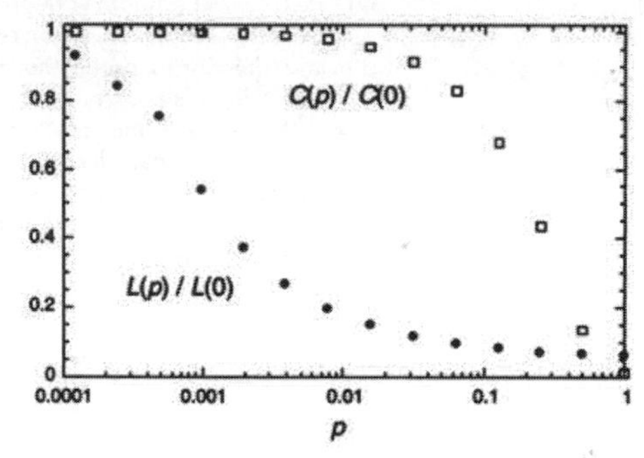

Figure 2 Characteristic path length $L(p)$ and clustering coefficient $C(p)$ for the family of randomly rewired graphs described in Fig. 1. Here L is defined as the number of edges in the shortest path between two vertices, averaged over all pairs of vertices. The clustering coefficient $C(p)$ is defined as follows. Suppose that a vertex v has k_v neighbours; then at most $k_v(k_v - 1)/2$ edges can exist between them (this occurs when every neighbour of v is connected to every other neighbour of v). Let C_v denote the fraction of these allowable edges that actually exist. Define C as the average of C_v over all v. For friendship networks, these statistics have intuitive meanings: L is the average number of friendships in the shortest chain connecting two people; C_v reflects the extent to which friends of v are also friends of each other; and thus C measures the cliquishness of a typical friendship circle. The data shown in the figure are averages over 20 random realizations of the rewiring process described in Fig. 1, and have been normalized by the values $L(0)$, $C(0)$ for a regular lattice. All the graphs have $n = 1{,}000$ vertices and an average degree of $k = 10$ edges per vertex. We note that a logarithmic horizontal scale has been used to resolve the rapid drop in $L(p)$, corresponding to the onset of the small-world phenomenon. During this drop, $C(p)$ remains almost constant at its value for the regular lattice, indicating that the transition to a small world is almost undetectable at the local level.

Two results emerge. First, the critical infectiousness r_{half}, at which the disease infects half the population, decreases rapidly for small p (Fig. 3a). Second, for a disease that is sufficiently infectious to infect the entire population regardless of its structure, the time $T(p)$ required for global infection resembles the $L(p)$ curve (Fig. 3b). Thus, infectious diseases are predicted to spread much more easily and quickly in a small world; the alarming and less obvious point is how few short cuts are needed to make the world small.

Our model differs in some significant ways from other network models of disease spreading[20-24]. All the models indicate that network structure influences the speed and extent of disease transmission, but our model illuminates the dynamics as an explicit function of structure (Fig. 3), rather than for a few particular topologies, such as random graphs, stars and chains[20-23]. In the work closest to ours, Kretschmar and Morris[24] have shown that increases in the number of concurrent partnerships can significantly accelerate the propagation of a sexually-transmitted disease that spreads along the edges of a graph. All their graphs are disconnected because they fix the average number of partners per person at $k = 1$. An increase in the number of concurrent partnerships causes faster spreading by increasing the number of vertices in the graph's largest connected component. In contrast, all our graphs are connected; hence the predicted changes in the spreading dynamics are due to more subtle structural features than changes in connectedness. Moreover, changes in the number of concurrent partners are obvious to an individual, whereas transitions leading to a smaller world are not.

We have also examined the effect of small-world connectivity on three other dynamical systems. In each case, the elements were coupled according to the family of graphs described in Fig. 1. (1) For cellular automata charged with the computational task of density classification[25], we find that a simple 'majority-rule' running on a small-world graph can outperform all known human and genetic algorithm-generated rules running on a ring lattice. (2) For the iterated, multi-player 'Prisoner's dilemma'[11] played on a graph, we find that as the fraction of short cuts increases, cooperation is less likely to emerge in a population of players using a generalized 'tit-for-tat'[26] strategy. The likelihood of cooperative strategies evolving out of an initial cooperative/non-cooperative mix also decreases with increasing p. (3) Small-world networks of coupled phase oscillators synchronize almost as readily as in the mean-field model[2], despite having orders of magnitude fewer edges. This result may be relevant to the observed synchronization of widely separated neurons in the visual cortex[27] if, as seems plausible, the brain has a small-world architecture.

We hope that our work will stimulate further studies of small-world networks. Their distinctive combination of high clustering with short characteristic path length cannot be captured by traditional approximations such as those based on regular lattices or random graphs. Although small-world architecture has not received much attention, we suggest that it will probably turn out to be widespread in biological, social and man-made systems, often with important dynamical consequences. □

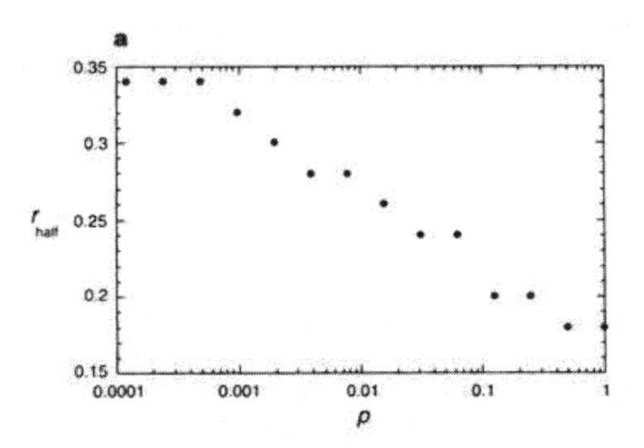

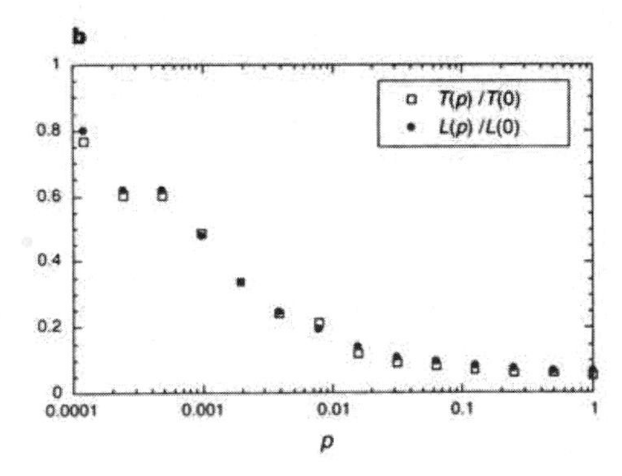

Figure 3 Simulation results for a simple model of disease spreading. The community structure is given by one realization of the family of randomly rewired graphs used in Fig. 1. **a**, Critical infectiousness r_{half}, at which the disease infects half the population, decreases with p. **b**, The time $T(p)$ required for a maximally infectious disease ($r = 1$) to spread throughout the entire population has essentially the same functional form as the characteristic path length $L(p)$. Even if only a few per cent of the edges in the original lattice are randomly rewired, the time to global infection is nearly as short as for a random graph.

Received 27 November 1997; accepted 6 April 1998.

1. Winfree, A. T. *The Geometry of Biological Time* (Springer, New York, 1980).
2. Kuramoto, Y. *Chemical Oscillations, Waves, and Turbulence* (Springer, Berlin, 1984).
3. Strogatz, S. H. & Stewart, I. Coupled oscillators and biological synchronization. *Sci. Am.* **269**(6), 102–109 (1993).
4. Bressloff, P. C., Coombes, S. & De Souza, B. Dynamics of a ring of pulse-coupled oscillators: a group theoretic approach. *Phys. Rev. Lett.* **79**, 2791–2794 (1997).
5. Braiman, Y., Lindner, J. F. & Ditto, W. L. Taming spatiotemporal chaos with disorder. *Nature* **378**, 465–467 (1995).
6. Wiesenfeld, K. New results on frequency-locking dynamics of disordered Josephson arrays. *Physica B* **222**, 315–319 (1996).
7. Gerhardt, M., Schuster, H. & Tyson, J. J. A cellular automaton model of excitable media including curvature and dispersion. *Science* **247**, 1563–1566 (1990).
8. Collins, J. J., Chow, C. C. & Imhoff, T. T. Stochastic resonance without tuning. *Nature* **376**, 236–238 (1995).
9. Hopfield, J. J. & Herz, A. V. M. Rapid local synchronization of action potentials: Toward computation with coupled integrate-and-fire neurons. *Proc. Natl Acad. Sci. USA* **92**, 6655–6662 (1995).
10. Abbott, L. F. & van Vreeswijk, C. Asynchronous states in neural networks of pulse-coupled oscillators. *Phys. Rev. E* **48**(2), 1483–1490 (1993).
11. Nowak, M. A. & May, R. M. Evolutionary games and spatial chaos. *Nature* **359**, 826–829 (1992).
12. Kauffman, S. A. Metabolic stability and epigenesis in randomly constructed genetic nets. *J. Theor. Biol.* **22**, 437–467 (1969).
13. Milgram, S. The small world problem. *Psychol. Today* **2**, 60–67 (1967).
14. Kochen, M. (ed.) *The Small World* (Ablex, Norwood, NJ, 1989).
15. Guare, J. *Six Degrees of Separation: A Play* (Vintage Books, New York, 1990).
16. Bollabás, B. *Random Graphs* (Academic, London, 1985).
17. Achacoso, T. B. & Yamamoto, W. S. *AY's Neuroanatomy of C. elegans for Computation* (CRC Press, Boca Raton, FL, 1992).
18. Wasserman, S. & Faust, K. *Social Network Analysis: Methods and Applications* (Cambridge Univ. Press, 1994).
19. Phadke, A. G. & Thorp, J. S. *Computer Relaying for Power Systems* (Wiley, New York, 1988).
20. Sattenspiel, L. & Simon, C. P. The spread and persistence of infectious diseases in structured populations. *Math. Biosci.* **90**, 341–366 (1988).
21. Longini, I. M. Jr A mathematical model for predicting the geographic spread of new infectious agents. *Math. Biosci.* **90**, 367–383 (1988).
22. Hess, G. Disease in metapopulation models: implications for conservation. *Ecology* **77**, 1617–1632 (1996).
23. Blythe, S. P., Castillo-Chavez, C. & Palmer, J. S. Toward a unified theory of sexual mixing and pair formation. *Math. Biosci.* **107**, 379–405 (1991).
24. Kretschmar, M. & Morris, M. Measures of concurrency in networks and the spread of infectious disease. *Math. Biosci.* **133**, 165–195 (1996).
25. Das, R., Mitchell, M. & Crutchfield, J. P. in *Parallel Problem Solving from Nature* (eds Davido, Y., Schwefel, H.-P. & Manner, R.) 344–353 (Lecture Notes in Computer Science 866, Springer, Berlin, 1994).
26. Axelrod, R. *The Evolution of Cooperation* (Basic Books, New York, 1984).
27. Gray, C. M., Konig, P., Engel, A. K. & Singer, W. Oscillatory responses in cat visual cortex exhibit inter-columnar synchronization which reflects global stimulus properties. *Nature* **338**, 334–337 (1989).

Acknowledgements. We thank B. Tjaden for providing the film actor data, and J. Thorp and K. Bae for the Western States Power Grid data. This work was supported by the US National Science Foundation (Division of Mathematical Sciences).

Correspondence and requests for materials should be addressed to D.J.W. (e-mail: djw24@columbia.edu).

VOLUME 82, NUMBER 15 PHYSICAL REVIEW LETTERS 12 APRIL 1999

Small-World Networks: Evidence for a Crossover Picture

Marc Barthélémy* and Luís A. Nunes Amaral

Center for Polymer Studies and Department of Physics, Boston University, Boston, Massachusetts 02215

(Received 8 December 1998)

Watts and Strogatz [Nature (London) **393**, 440 (1998)] have recently introduced a model for disordered networks and reported that, even for very small values of the disorder p in the links, the network behaves as a "small world." Here, we test the hypothesis that the appearance of small-world behavior is not a phase transition but a crossover phenomenon which depends both on the network size n and on the degree of disorder p. We propose that the average distance ℓ between any two vertices of the network is a scaling function of n/n^*. The crossover size n^* above which the network behaves as a small world is shown to scale as $n^*(p \ll 1) \sim p^{-\tau}$ with $\tau \approx 2/3$. [S0031-9007(99)08892-4]

PACS numbers: 84.35.+i, 05.40.−a, 05.50.+q, 87.18.Sn

Two limiting-case topologies have been extensively considered in the literature. The first is the regular lattice, or regular network, which has been the chosen topology of innumerable physical models such as the Ising model or percolation [1–3]. The second is the random graph, or random network, which has been studied in mathematics and used in both natural and social sciences [4–16].

Erdös and co-workers studied extensively the properties of random networks—see [17] for a review. Most of this work concentrated on the case in which the number of vertices is kept constant but the total number of links between vertices increases [17]: The Erdös-Rényi result [18] states that for many important quantities there is a percolationlike transition at a specific value of the average number of links per vertex. In physics, random networks are used, for example, in studies of dynamical problems [19,20], spin models and thermodynamics [20,21], random walks [22], and quantum chaos [23]. Random networks are also widely used in economics and other social sciences [8,24,25] to model, for example, interacting agents.

In contrast to these two limiting topologies, empirical evidence [26,27] suggests that many biological, technological, or social networks appear to be somewhere in between these extremes. Specifically, many real networks seem to share with regular networks the concept of neighborhood, which means that if vertices i and j are neighbors then they will have many common neighbors—which is obviously not true for a random network. On the other hand, studies on epidemics [14,15,26] show that it can take only a few "steps" on the network to reach a given vertex from any other vertex. This is the foremost property of random networks, which is not fulfilled by regular networks.

To bridge the two limiting cases, and to provide a model for real-world systems [28,29], Watts and Strogatz [26,27] have recently introduced a new type of network which is obtained by randomizing a fraction p of the links of the regular network. As in Ref. [26], we consider as an initial structure ($p = 0$) the one-dimensional regular network where each vertex is connected to its z nearest neighbors. For $0 < p < 1$, we denote these networks disordered, and keep the name random network for the case $p = 1$. Reference [26] reports that for a small value of the parameter p—which interpolates between the regular ($p = 0$) and random ($p = 1$) networks—there is an onset of "small-world" behavior. The small-world behavior is characterized by the fact that the distance between any two vertices is of the order of that for a random network and, at the same time, the concept of neighborhood is preserved, as for regular lattices (Fig. 1). The effect of a change in p is extremely nonlinear as is visually demonstrated by the difference between Figs. 1a and 1d and Figs. 1b and 1e where a very small change in the adjacency matrix leads to a dramatic change in the distance between different pairs of vertices.

Here, we study the origins of the small-world behavior [28,29]. In particular, we investigate if the onset of small-world networks is a phase transition or a crossover phenomena. To answer this question, we consider not only changes in the value of p but also in the system size n.

The motivation for this study is the following. In a regular one-dimensional network with n vertices and z links per vertex, the average distance ℓ between two vertices increases as $n/(2z)$—the distance is defined as the minimum number of steps between the two vertices. The regular network is similar to the streets of Manhattan: Walking along 5th Avenue from Washington Square Park on 4th Street to Central Park on 59th Street, we have to go past 55 blocks. On the other hand, for a random network, each "block" brings us to a point with z new neighbors. Hence, the number of vertices increases with the number of steps k as $n \sim z^k$, which implies that ℓ increase as $\ln n / \ln z$. The random network is then similar to a strange subway system that would directly connect different parts of Manhattan and enable us to go from Washington Square Park to Central Park in just one stop. In view of these facts, it is natural to enquire if the change from large world ($\ell \sim n$) to small world ($\ell \sim \ln n$) in disordered networks occurs through a phase transition for some given value of p [30] or if, for any value of p, there

 0031-9007/99/82(15)/3180(4)$15.00

VOLUME 82, NUMBER 15 PHYSICAL REVIEW LETTERS 12 APRIL 1999

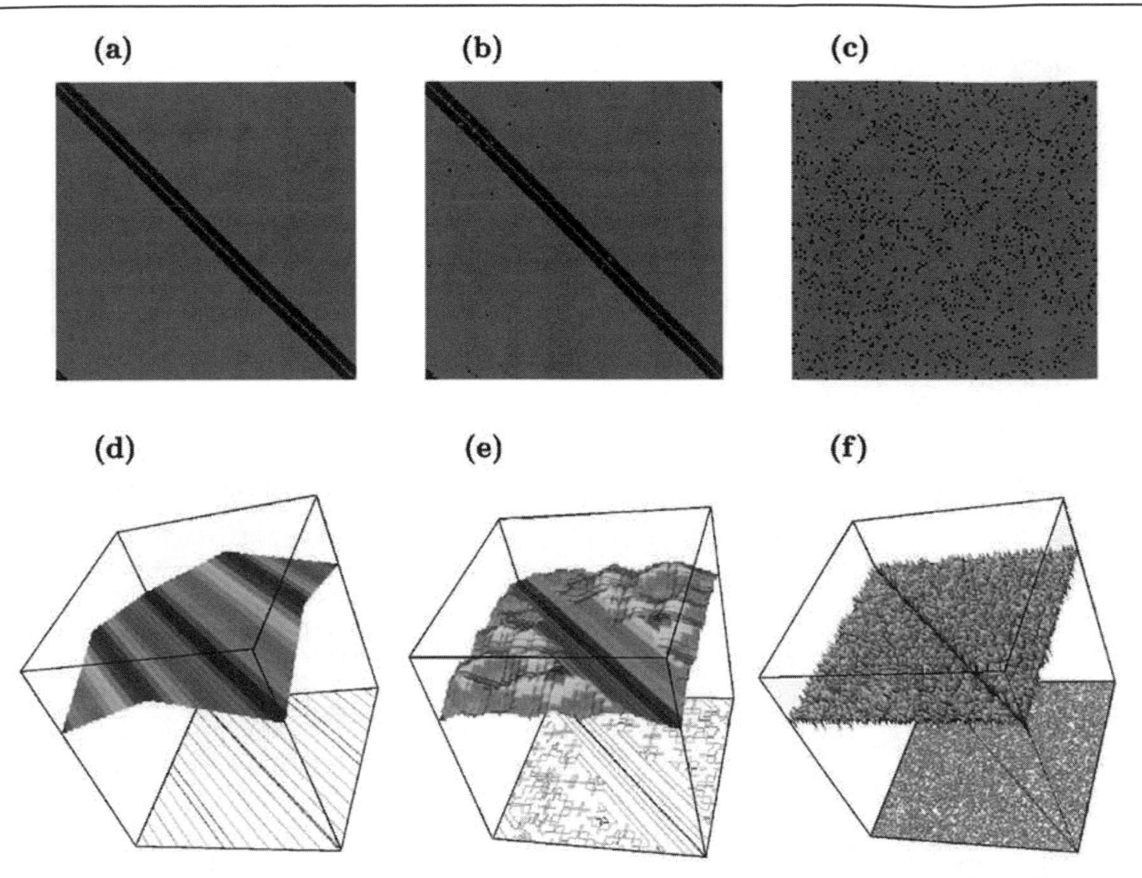

FIG. 1. Effect of disorder on the distance between vertices of the network (go to http://polymer.bu.edu/~amaral/Networks.html for color pictures). We consider here matrices with $z = 10$, $n = 128$, and with periodic boundary conditions, that is, vertex 1 follows vertex n. Adjacency matrices for (a) a regular one-dimensional network where each vertex is connected to its z nearest neighbors, (b) a disordered network with $p = 0.01$, and (c) a random network. Black indicates that a link is present between the two vertices while gray indicates the absence of a link. Note that (a) and (b) are nearly identical. Distance matrices for (d) the regular network, (e) the disordered network with $p = 0.01$, and (f) the random network. We use the relief of the surface and a gray scale to represent the distance between two vertices. Greater height indicates larger distance. The gray scale is the same for the relief and for the contour lines: Distance increases from very dark gray to gray to light gray to dark gray. For the regular network, the contour lines are parallel to the diagonal. On the other hand, for the disordered network the contour lines "circle" around specific links that act as "throughways" of the network. This effect prevents the distance between any two vertices from ever becoming large, that is, of the order of the system size.

is a crossover size $n^*(p)$ below which our network is a large world and above which it is a small world.

In the present Letter, we report that the appearance of the small-world behavior is not a phase transition but a crossover phenomena. We propose the scaling ansatz,

$$\ell(n, p) \sim n^* F\left(\frac{n}{n^*}\right), \tag{1}$$

where $F(u \ll 1) \sim u$, $F(u \gg 1) \sim \ln u$, and n^* is a function of p [31]. Naively, we would expect that, when the average number of rewired links, $pnz/2$, is much less than one, the network should be in the large-world regime. On the other hand, when $pnz/2 \gg 1$, the network should be a small world [32]. Hence, the crossover size should occur for $n^* p = O(1)$, which implies $n^* \sim p^{-\tau}$ with $\tau = 1$. This result relies on the fact that the crossover from large to small worlds is obtained with only a small but finite fraction of rewired links. We find that the scaling ansatz (1) is indeed verified by the average distance ℓ between any two vertices of the network. We also identify the crossover size n^* above which the network behaves as a small world, and find that it scales as $n^* \sim p^{-\tau}$ with $\tau \approx 2/3$, distinct from the trivial expectation $\tau = 1$.

Next, we define the model and present our results. We start from a regular one-dimensional network with n vertices, each connected to z neighbors. We then apply the "rewiring" algorithm of [26] to this network. The algorithm prescribes that every link has a probability p of being broken and replaced by a new random link. We replace the broken link by a new one connecting one of the original vertices to a new randomly selected vertex. Each of the other $n - 2$ vertices—we exclude the other vertex of the broken link—has an *a priori* equal probability of being selected, but we then make sure that there are no duplicate links. Hence, the algorithm preserves the total number of links which is equal to $nz/2$. A quantity that is affected by the rewiring algorithm is the probability distribution of local connectivities. For $p \simeq 0$, this probability is narrowly peaked around z, but it gets broader with increasing p. For $p = 1$, the average and the standard deviation of the local connectivity are of the same order of magnitude and equal to z.

Once the disordered network is created, we calculate the distance between any two vertices of the network and its average value ℓ. To calculate the distance for each pair of vertices, we use the Moore-Dijkstra algorithm [33] whose execution time scales with network size as $n^3 \ln n$. We perform between 100 and 300 averages over realizations of the disorder for each pair of values of n and p.

Here, we present results for three values of connectivity $z = 10$, 20, and 30 and system sizes up to 1000. The scaling ansatz (1) enables us to determine $n^*(p)$ from $\ell(n)$ at fixed p. Indeed, $\ell(n \gg n^*) \sim n^* \ln n$ which implies that n^* is the asymptotic value of $d\ell/d(\ln n)$ [Fig. 2(a)]. Figure 2(b) shows the dependence of n^* on p for different values z. We hypothesize that

$$n^* - \frac{1}{\ln z} \sim p^{-\tau} g(p), \tag{2}$$

where the term in z arises from the fact that $\ell = \ln n / \ln z$

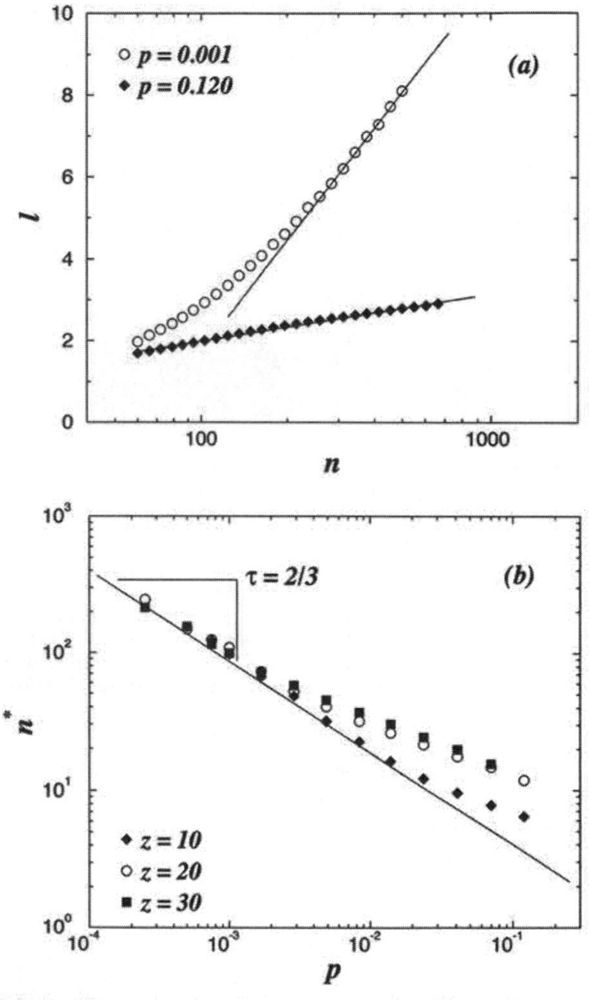

FIG. 2. Determination of the crossover size n^*. (a) Semilog plot of ℓ versus network size for two representative values of p and for $z = 20$. Following Eqs. (1)–(3), we can determine n^*—apart from a multiplicative constant—from the asymptotic slope of ℓ against $\ln n$. (b) Scaling of n^* with p for the three values of z discussed in the text. The curves for $z = 20$ and 30 have been shifted up so as to coincide in the region where they scale as a power law. Following Eq. (3), we make a power-law fit to $n^*(p)$ for $p \ll 1$ and obtain $\tau \approx 2/3$.

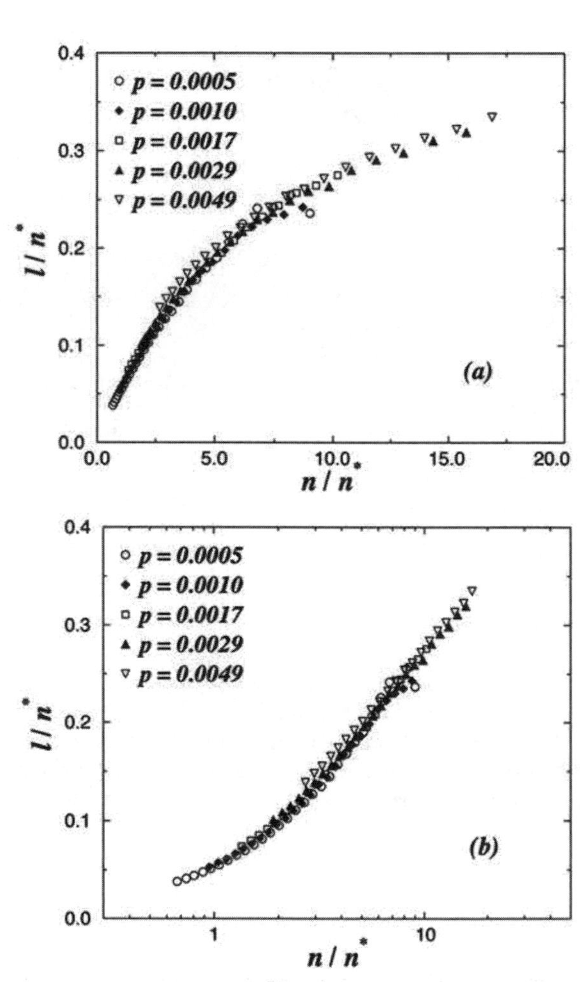

FIG. 3. Data collapse of $\ell(n, p)$ for $z = 10$ and different values of p and n. (a) Plot of the scaled average distance between vertices ℓ/n^* versus scaled system size n/n^*. (b) Same data as in (a) but in a semilog plot. Note the linear behavior of the data for $n < n^*$ and the logarithmic increase of ℓ for large system sizes.

for a random network ($p = 1$), and $g(p \to 1) \to 0$. Moreover, $g(p)$ approaches a constant as $p \to 0$, leading to

$$n^* \sim p^{-\tau}, \tag{3}$$

for small p. Because of the effect of g and the fact that $n < 1000$ in our numerical simulations, we are constrained to estimate τ from the region $2.5 \times 10^{-4} < p < 2 \times 10^{-2}$. For all values of z, we obtain $\tau = 0.67 \pm 0.10$ (Fig. 2).

Using this value of τ and the scaling form (1), we are able to collapse all the values of $\ell(n,p)$ onto a single curve (Fig. 3). This data collapse confirms our scaling ansatz and estimate of τ.

In summary, we have shown that the onset of small-world behavior is a crossover phenomena and not a phase transition from a large world to a small one. The crossover size scales as $p^{-\tau}$ with $\tau \simeq 2/3$. The surprising fact that $\tau < 1$ shows that the rewiring process is highly nonlinear and can have dramatic consequences on the global behavior of the network. This implies that in order to *decrease* the radius of a network it is necessary to rewire only a few links. We also note that the value of the exponent τ will likely depend on the dimensionality of the initial regular network. This point will be addressed in future work.

We believe that the disordered networks introduced in [26] may constitute a promising topology for more realistic studies of many important problems such as flow in electric power or information networks, spread of epidemics, or financial systems. The results reported here support this hypothesis because they suggest that, for *any* given degree of disorder of the network, if the system is larger than the crossover size, the network will be in the small-world regime.

We thank S. V. Buldyrev, L. Cruz, P. Gopikrishnan, P. Ivanov, H. Kallabis, E. La Nave, T. J. P. Penna, A. Scala, and H. E. Stanley for stimulating discussions. L. A. N. A. thanks the FCT/Portugal and M. B. thanks the DGA for financial support.

---

*Permanent address: CEA-BIII, Service de Physique de la Matière Condensée, France.

[1] H. E. Stanley, *Introduction to Phase Transitions and Critical Phenomena* (Oxford University Press, New York, 1971).
[2] *Correlations and Connectivity: Geometry Aspects of Physics, Chemistry and Biology*, edited by H. E. Stanley and N. Ostrowsky (Kluwer Academic Publishers, Dordrecht, 1990).
[3] *Fractals and Disordered Systems*, edited by A. Bunde and S. Havlin (Springer-Verlag, Berlin, 1996), 2nd ed.
[4] A. T. Winfree, *The Geometry of Biological Time* (Springer-Verlag, New York, 1980).
[5] Y. Kuramoto, *Chemical Oscillations, Waves, and Turbulence* (Springer-Verlag, Berlin, 1984).
[6] M. Gerhardt, H. Schuster, and J. J. Tyson, Science **247**, 1563 (1990).
[7] M. A. Nowak and R. M. May, Nature (London) **359**, 826 (1992).
[8] S. Wasserman and K. Faust, *Social Network Analysis: Methods and Applications* (Cambridge University Press, Cambridge, England, 1994).
[9] Y. Braiman, J. F. Lindner, and W. L. Ditto, Nature (London) **378**, 465 (1995).
[10] J. J. Collins, C. C. Chow, and T. T. Imhoff, Nature (London) **376**, 236 (1995).
[11] J. J. Hopfield and A. V. M. Hertz, Proc. Natl. Acad. Sci U.S.A. **92**, 6655 (1995).
[12] J. P. Crutchfield and M. Mitchell, Proc. Natl. Acad. Sci. U.S.A. **92**, 10 742 (1995).
[13] K. Wiesenfeld, Physica (Amsterdam) **222B**, 315 (1996).
[14] G. Hess, Ecology **77**, 1617 (1996).
[15] M. Kretschmar and M. Morris, Math. Biosci. **133**, 165 (1996).
[16] P. C. Bressloff, S. Comber, and B. De Souza, Phys. Rev. Lett. **79**, 2791 (1997).
[17] B. Bollobás, *Random Graphs* (Academic Press, London, 1985).
[18] P. Erdös and A. Rényi, Publ. Math. Debrecen **6**, 290 (1959); Publ. Math. Inst. Hungar. Acad. Sci. **5**, 17 (1960); Bull. Inst. Int. Statis. Tokyo **38**, 343 (1961).
[19] K. Christensen, R. Donangelo, B. Koiller, and K. Sneppen, Phys. Rev. Lett. **81**, 2380 (1998).
[20] A. Barrat and R. Zecchina, Phys. Rev. B **59**, R1299 (1999); cond-mat/9811033.
[21] B. Luque and R. V. Solé, Phys. Rev. E **55**, 257 (1997).
[22] D. Cassi, Phys. Rev. Lett. **76**, 2941 (1996).
[23] T. Kottos and U. Smilansky, Phys. Rev. Lett. **79**, 4794 (1997).
[24] R. Axelrod, *The Evolution of Cooperation* (Basic Books, New York, 1984).
[25] S. Jain and S. Krishna, Phys. Rev. Lett. **81**, 5684 (1998); adap-org/9810005; adap-org/9809003.
[26] D. J. Watts and S. H. Strogatz, Nature (London) **393**, 440 (1998).
[27] J. J. Collins and C. C. Chow, Nature (London) **393**, 409 (1998).
[28] S. Milgram, Psychology Today **2**, 60 (1967).
[29] *The Small World*, edited by M. Kochen (Ablex, Norwood, NJ, 1989).
[30] Reference [26] describes an "onset of the small-world phenomena" which suggests a transition between different "phases." This impression is reinforced by the fact that in [26] there is no mention of a dependence of the results on network size.
[31] Note that there are factors depending on z which we will ignore here.
[32] H. Herzel, Fractals **6**, 301 (1998).
[33] M. Gondran and M. Minoux, *Graphs and Algorithms* (Wiley, New York, 1984). See page 48 for details of the Moore-Dijkstra algorithm.

Comment on "Small-world networks: Evidence for a crossover picture"

In a recent letter, Barthélémy and Nunes Amaral [1] examine the crossover behaviour of networks known as "small-world". They claim that, for an initial network with n vertices and z links per vertex, each link being rewired according to the procedure of [2] with a probability p, the average distance ℓ between two vertices scales as

$$\ell(n,p) \sim n^* F\left(\frac{n}{n^*}\right) \quad (1)$$

where $F(u \ll 1) \sim u$, and $F(u \gg 1) \sim \ln u$, and n^* $n^* \sim p^{-\tau}$ with $\tau = 2/3$ as p goes to zero.

Other quantities can be of interest in small-world networks, and will be discussed in [3]. In this comment however, we concentrate like [1] on ℓ and we show, using analytical arguments and numerical simulations with larger values of n, that: (i) the proposed scaling form $\ell(n,p) \sim n^*F(n/n^*)$ seems to be valid, BUT (ii) the value of τ *cannot* be lower than 1, and therefore the value found in [1] is clearly wrong.

The naive argument developed in [1] uses the mean number of rewired links, $N_r = pnz/2$. According to [1], one could expect that the crossover happens for $N_r = O(1)$, which gives $\tau = 1$ [4]. However they find $\tau = 2/3$. Let us suppose that $\tau < 1$. Then, if we take α such that $\tau < \alpha < 1$, according to eq (1), we obtain that

$$\ell(n, n^{-1/\alpha}) \sim n^{\tau/\alpha} F(n^{1-\tau/\alpha}) \sim n^{\tau/\alpha} \ln(n^{1-\tau/\alpha}) \quad (2)$$

since $\tau/\alpha < 1$ and $n^{1-\tau/\alpha} \gg 1$ for large n. However, the mean number of rewired links in this case is $N_r = n^{1-1/\alpha}z/2$, which goes to zero for large n. The immediate conclusion is that a change in the behaviour of ℓ (from $\ell \sim n$ to $\ell \sim n^{\tau/\alpha} \ln(n)$) could occur by the rewiring of a vanishing *number* of links! This is a physical nonsense, showing that, if $n^* \sim p^{-\tau}$, τ cannot be lower than 1.

We know present our numerical simulations. The value of $n^*(p)$ is obtained by studying, at fixed p (we take $p = 2^k/2^{20}$, $k = 0, \cdots 20$), the crossover between $\ell \sim n$ at small n to $\ell \sim \ln(n)$ at large n [1]. For small values of p, it is difficult to reach large enough values of n to accurately determine n^*, and we think that the underestimation of n^* given by [1] comes from this problem. We here simulate networks with $z = 4$, 6, 10 up to sizes $n = 11000$, and find that n^* behaves like $1/p$ for small p (Inset of fig. (1)). We moreover show the collapse of the curves ℓ/n^* versus n/n^* in figure (1), for $z = 4$ and $z = 10$: note that we obtain the collapse over a much wider rabge than [1].

Besides, we present results for another quantity: at fixed n we evaluate $\ell(n,p)$ and look for the value $p_{1/2}(n)$ of p such that $\ell(n, p_{1/2}(n)) = \ell(n,0)/2$. This value of p corresponds to the rapid drop in the plot of ℓ versus p at fixed N, and can therefore also be considered as a crossover value. If we note u^* the number such that $F(u^*) = u^*/2$, then we obtain, since $\ell(n,0) \sim n$, that $n^*(p_{1/2}(n)) = n/u^*$. If $n^*(p) \sim p^{-\tau}$, this implies $p_{1/2}(n) \sim n^{-1/\tau}$. We show in fig (2) that $p_{1/2}(n) \sim 1/n$ (and clearly not $\sim n^{-3/2}$ like the results of [1] would imply), meaning that $\tau = 1$: a finite number of rewired links already has a strong influence on ℓ. Again the $\tau = 2/3$ result of [1] is clearly ruled out.

It is a pleasure to thank G. Biroli, R. Monasson and M. Weigt for discussions.

A. Barrat

Laboratoire de Physique Théorique [*], Université de Paris-Sud, 91405 Orsay cedex, France

* Unité Mixte de Recherche UMR 8627.

[1] M. Barthélémy and L.A. Nunes Amaral, preprint cond-mat/9903108, Phys. Rev. Lett. to appear (Volume 82, Number 15).

[2] D.J. Watts and S.H. Strogatz, Nature **393**, 440 (1998).

[3] A. Barrat, M. Weigt, in preparation; R. Monasson, in preparation.

[4] Note that $N_r = O(1)$ gives a finite *number* of rewired links, and not a "small but finite fraction" like written in [1].

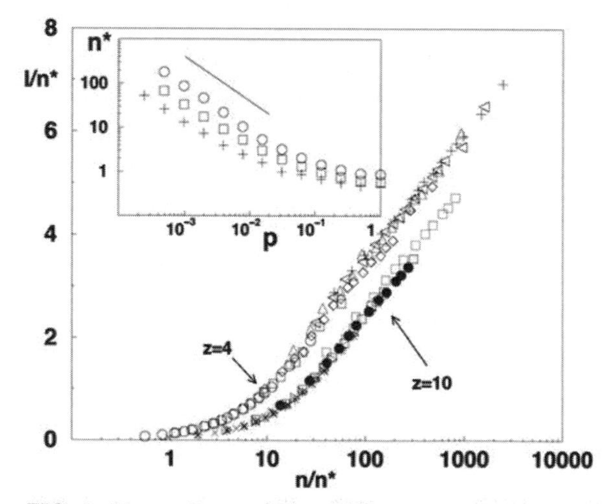

FIG. 1. Data collapse of $\ell(n,p)/n^*$ versus n/n^* for $z = 4$ and $z = 10$, for various values of p and n from 100 to 5000. Inset: n^* versus p for $z = 4$ (circles), 6 (squares), 10 (crosses); the straight line has slope -1.

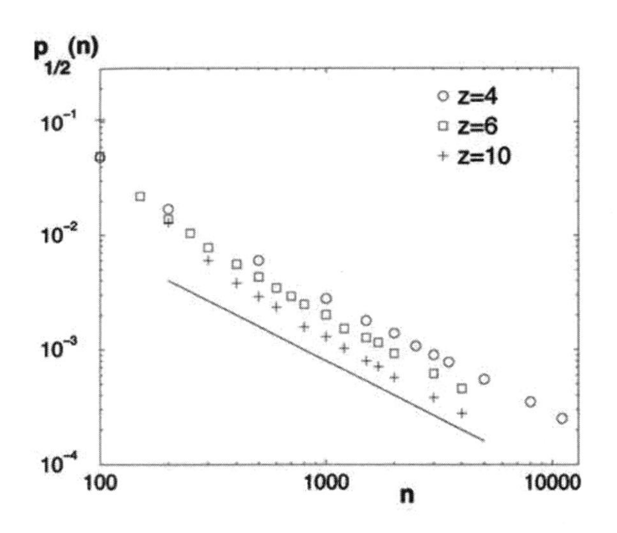

FIG. 2. $p_{1/2}(n)$ such that $\ell(n, p_{1/2}(n)) = \ell(n,0)/2$, for $z = 4$, 6, 10, and values of n ranging from 100 to 11000; the straight line has slope -1.

PHYSICAL REVIEW E VOLUME 60, NUMBER 6 DECEMBER 1999

Scaling and percolation in the small-world network model

M. E. J. Newman and D. J. Watts
Santa Fe Institute, 1399 Hyde Park Road, Santa Fe, New Mexico 87501
(Received 7 May 1999)

In this paper we study the small-world network model of Watts and Strogatz, which mimics some aspects of the structure of networks of social interactions. We argue that there is one nontrivial length-scale in the model, analogous to the correlation length in other systems, which is well-defined in the limit of infinite system size and which diverges continuously as the randomness in the network tends to zero, giving a normal critical point in this limit. This length-scale governs the crossover from large- to small-world behavior in the model, as well as the number of vertices in a neighborhood of given radius on the network. We derive the value of the single critical exponent controlling behavior in the critical region and the finite size scaling form for the average vertex-vertex distance on the network, and, using series expansion and Padé approximants, find an approximate analytic form for the scaling function. We calculate the effective dimension of small-world graphs and show that this dimension varies as a function of the length-scale on which it is measured, in a manner reminiscent of multifractals. We also study the problem of site percolation on small-world networks as a simple model of disease propagation, and derive an approximate expression for the percolation probability at which a giant component of connected vertices first forms (in epidemiological terms, the point at which an epidemic occurs). The typical cluster radius satisfies the expected finite size scaling form with a cluster size exponent close to that for a random graph. All our analytic results are confirmed by extensive numerical simulations of the model. [S1063-651X(99)12412-7]

PACS number(s): 87.23.Ge, 05.40.−a, 05.70.Jk, 64.60.Fr

I. INTRODUCTION

Networks of social interactions between individuals, groups, or organizations have some unusual topological properties which set them apart from most of the networks with which physics deals. They appear to display simultaneously properties typical both of regular lattices and of random graphs. For instance, social networks have well-defined locales in the sense that if individual A knows individual B and individual B knows individual C, then it is likely that A also knows C—much more likely than if we were to pick two individuals at random from the population and ask whether they are acquainted. In this respect social networks are similar to regular lattices, which also have well-defined locales, but very different from random graphs, in which the probability of connection is the same for any pair of vertices on the graph. On the other hand, it is widely believed that one can get from almost any member of a social network to any other via only a small number of intermediate acquaintances, the exact number typically scaling as the logarithm of the total number of individuals comprising the network. Within the population of the world, for example, it has been suggested that there are only about "six degrees of separation" between any human being and any other [1]. This behavior is not seen in regular lattices but is a well-known property of random graphs, where the average shortest path between two randomly chosen vertices scales as $\log N/\log z$, where N is the total number of vertices in the graph and z is the average coordination number [2].

Recently, Watts and Strogatz [3] have proposed a model which attempts to mimic the properties of social networks. This "small-world" model consists of a network of vertices whose topology is that of a regular lattice, with the addition of a low density ϕ of connections between randomly chosen pairs of vertices [4]. Watts and Strogatz showed that graphs of this type can indeed possess well-defined locales in the sense described above while at the same time possessing average vertex-vertex distances which are comparable with those found on true random graphs, even for quite small values of ϕ.

In this paper we study in detail the behavior of the small-world model, concentrating particularly on its scaling properties. The outline of the paper is as follows. In Sec. II we define the model. In Sec. III we study the typical length scales present in the model and argue that the model undergoes a continuous phase transition as the density of random connections tends to zero. We also examine the crossover between large- and small-world behavior in the model, and the structure of "neighborhoods" of adjacent vertices. In Sec. IV we derive a scaling form for the average vertex-vertex distance on a small-world graph and demonstrate numerically that this form is followed over a wide range of the parameters of the model. In Sec. V we calculate the effective dimension of small-world graphs and show that this dimension depends on the length scale on which we examine the graph. In Sec. VI we consider the properties of site percolation on these systems, as a model of the spread of information or disease through social networks. Finally, in Sec. VII we give our conclusions.

II. SMALL-WORLD MODEL

The original small-world model of Watts and Strogatz, in its simplest incarnation, is defined as follows. We take a one-dimensional lattice of L vertices with connections or bonds between nearest neighbors and periodic boundary conditions (the lattice is a ring). Then we go through each of the bonds in turn and independently with some probability ϕ

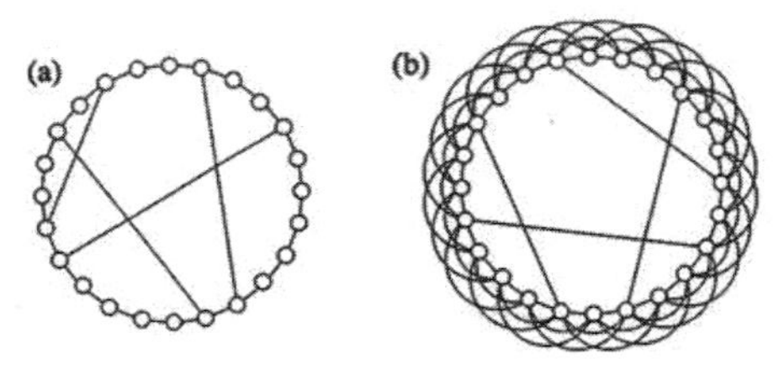

FIG. 1. (a) An example of a small-world graph with $L=24$, $k=1$ and, in this case, four shortcuts. (b) An example with $k=3$.

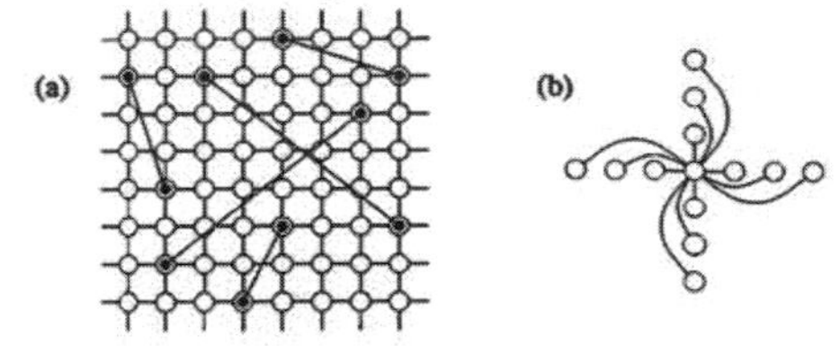

FIG. 2. (a) An example of a $k=1$ small-world graph with an underlying lattice of dimension $d=2$. (b) The pattern of bonds around a vertex on the $d=2$ lattice for $k=3$.

"rewire" it. Rewiring in this context means shifting one end of the bond to a new vertex chosen uniformly at random from the whole lattice, with the exception that no two vertices can have more than one bond running between them, and no vertex can be connected by a bond to itself. In this model the average coordination number z remains constant ($z=2$) during the rewiring process, but the coordination number of any particular vertex may change. The total number of rewired bonds, which we will refer to as "shortcuts," is ϕL on average.

For the purposes of analytic treatment the Watts-Strogatz model has a number of problems. One problem is that the distribution of shortcuts is not completely uniform; not all choices of the positions of the rewired bonds are equally probable. For example, configurations with more than one bond between a particular pair of vertices are explicitly forbidden. This nonuniformity of the distribution makes an average over different realizations of the randomness hard to perform.

A more serious problem is that one of the crucial quantities of interest in the model, the average distance between pairs of vertices on the graph, is poorly defined. The reason is that there is a finite probability of a portion of the lattice becoming detached from the rest in this model. Formally, we can represent this by saying that the distance from such a portion to a vertex elsewhere on the lattice is infinite. However, this means that the average vertex-vertex distance on the lattice is then itself infinite, and hence that the vertex-vertex distance averaged over all realizations is also infinite. For numerical studies such as those of Watts and Strogatz this does not present any substantial difficulties, but for analytic work it results in a number of quantities and expressions being poorly defined.

Both of these problems can be circumvented by a slight modification of the model. In our version of the small-world model we again start with a regular one-dimensional lattice, but now instead of rewiring each bond with probability ϕ, we add shortcuts between pairs of vertices chosen uniformly at random but we do not remove any bonds from the regular lattice. We also explicitly allow there to be more than one bond between any two vertices, or a bond which connects a vertex to itself. In order to preserve compatibility with the results of Watts and Strogatz and others, we add with probability ϕ one shortcut for each bond on the original lattice, so that there are again ϕL shortcuts on average. The average coordination number is $z=2(1+\phi)$. This model is equivalent to the Watts-Strogatz model for small ϕ, whilst being better behaved when ϕ becomes comparable to 1. Figure 1(a) shows one realization of our model for $L=24$.

Real social networks usually have average coordination numbers z significantly higher than 2, and we can arrange for higher z in our model in a number of ways. Watts and Strogatz [3] proposed adding bonds to next-nearest or further neighbors on the underlying one-dimensional lattice up to some fixed range which we will call k [5]. In our variation on the model we can also start with such a lattice and then add shortcuts to it. The mean number of shortcuts is then ϕkL and the average coordination number is $z=2k(1+\phi)$. Figure 1(b) shows a realization of this model for $k=3$.

Another way of increasing the coordination number, suggested first by Watts [6,7], is to use an underlying lattice for the model with dimension greater than one. In this paper we will consider networks based on square and (hyper)cubic lattices in d dimensions. We take a lattice of linear dimension L, with L^d vertices, nearest-neighbor bonds and periodic boundary conditions, and add shortcuts between randomly chosen pairs of vertices. Such a graph has ϕdL^d shortcuts and an average coordination number $z=2d(1+\phi)$. An example is shown in Fig. 2(a) for $d=2$. We can also add bonds between next-nearest or further neighbors to such a lattice. The most straightforward generalization of the one-dimensional case is to add bonds along the principal axes of the lattice up to some fixed range k, as shown in Fig. 2(b) for $k=3$. Graphs of this type have ϕkdL^d shortcuts on average and a mean coordination number of $z=2kd(1+\phi)$.

Our main interest in this paper is with the properties of the small-world model for small values of the shortcut probability ϕ. Watts and Strogatz [3] found that the model displays many of the characteristics of true random graphs even for $\phi \ll 1$, and it seems to be in this regime that the model's properties are most similar to those of real-world social networks.

III. LENGTH-SCALES IN SMALL-WORLD GRAPHS

A fundamental observable property of interest on small-world lattices is the shortest path between two vertices—the number of degrees of separation—measured as the number of bonds traversed to get from one vertex to another, averaged over all pairs of vertices and over all realizations of the randomness in the model. We denote this quantity l. On ordinary regular lattices l scales linearly with the lattice size L. On the underlying lattices used in the models described here for instance, it is equal to $\frac{1}{4}dL/k$. On true random graphs, in which the probability of connection between any two vertices is the same, l is proportional to $\log N/\log z$, where N is the number of vertices on the graph [2]. The small-world model interpolates between these extremes, showing linear scaling $l \sim L$ for small ϕ, or on systems small enough that there are very few shortcuts, and logarithmic scaling $l \sim \log N = d \log L$ when ϕ or L is large enough. In this section and the following one we study the nature of the crossover between these two regimes, which we refer to as "large-world" and "small-world" regimes, respectively. For simplicity we will work mostly with the case $k=1$, although we will quote results for $k>1$ where they are of interest.

When $k=1$ the small-world model has only one independent parameter—the probability ϕ—and hence can have only one nontrivial length scale other than the lattice constant of the underlying lattice. This length scale, which we will denote ξ, can be defined in a number of different ways, all definitions being necessarily proportional to one another. One simple way is to define ξ to be the typical distance between the ends of shortcuts on the lattice. In a one-dimensional system with $k=1$, for example, there are on average ϕL shortcuts and therefore $2\phi L$ ends of shortcuts. Since the lattice has L vertices, the average distance between ends of shortcuts is $L/(2\phi L) = 1/(2\phi)$. In fact, it is more convenient for our purposes to define ξ without the factor of 2 in the denominator, so that $\xi = 1/\phi$, or for general d

$$\xi = \frac{1}{(\phi d)^{1/d}}. \tag{1}$$

For $k>1$ the appropriate generalization is [8]

$$\xi = \frac{1}{(\phi k d)^{1/d}}. \tag{2}$$

As we see, ξ diverges as $\phi \to 0$ according to [9]

$$\xi \sim \phi^{-\tau}, \tag{3}$$

where the exponent τ is

$$\tau = \frac{1}{d}. \tag{4}$$

A number of authors have previously considered a divergence of the kind described by Eq. (3) with ξ defined not as the typical distance between the ends of shortcuts, but as the system size L at which the crossover from large- to small-world scaling occurs [10–13]. We will shortly argue that in fact the length-scale ξ defined here is precisely equal to this crossover length, and hence that these two divergences are the same.

The quantity ξ plays a role similar to that of the correlation length in an interacting system in standard statistical physics. Its divergence leaves the system with no length scale other than the lattice spacing, so that at long distances we expect all spatial distributions to be scale-free. This is precisely the behavior one sees in an interacting system undergoing a continuous phase transition, and it is reasonable to regard the small-world model as having a continuous phase transition at this point. Note that the transition is a one-sided one since ϕ is a probability and cannot take values less than zero. In this respect the transition is similar to that seen in the one-dimensional Ising model, or in percolation on a one-dimensional lattice. The exponent τ plays the part of a critical exponent for the system, similar to the correlation length exponent ν for a thermal phase transition.

De Menezes *et al.* [13] have argued that the length scale ξ can *only* be defined in terms of the crossover point between large- and small-world behavior, that there is no definition of ξ which can be made consistent in the limit of large system size. For this reason they argue that the transition at $\phi=0$ should be regarded as first-order rather than continuous. In fact, however, the arguments of de Menezes *et al.* show only that one particular definition of ξ is inconsistent; they show that ξ cannot be consistently defined in terms of the mean vertex-vertex distance between vertices in finite regions of infinite small-world graphs. This does not prove that no definition of ξ is consistent in the $L \to \infty$ limit and, as we have demonstrated here, consistent definitions do exist. Thus it seems appropriate to consider the transition at $\phi=0$ to be a continuous one.

Barthélémy and Amaral [10] have conjectured on the basis of numerical simulations that $\tau = \frac{2}{3}$ for $d=1$. As we have shown here, τ is in fact equal to $1/d$, and specifically $\tau=1$ in one dimension. We have also demonstrated this result previously using a renormalization group (RG) argument [12], and it has been confirmed by extensive numerical simulations [11–13].

The length scale ξ governs a number of other properties of small-world graphs. First, as mentioned above, it defines the point at which the average vertex-vertex distance l crosses over from linear to logarithmic scaling with system size L. This statement is necessarily true, since ξ is the only nontrivial length scale in the model, but we can demonstrate it explicitly by noting that the linear scaling regime is the one in which the average number of shortcuts on the lattice is small compared with unity and the logarithmic regime is the one in which it is large [6]. The crossover occurs in the region where the average number of shortcuts is about one, or in other words when $\phi k d L^d = 1$. Rearranging for L, the crossover length is

$$L = \frac{1}{(\phi k d)^{1/d}} = \xi. \tag{5}$$

The length scale ξ also governs the average number $V(r)$ of neighbors of a given vertex within a neighborhood of radius r. The number of vertices in such a neighborhood increases as r^d for $r \ll \xi$ while for $r \gg \xi$ the graph behaves as a random graph and the size of the neighborhood must increase exponentially with some power of r/ξ. To derive the specific functional form of $V(r)$ we consider a small-world graph in the limit of infinite L. Let $a(r)$ be the surface area of a "sphere" of radius r on the underlying lattice of the model, i.e., it is the number of points which are exactly r steps away from any vertex. [For $k=1$, $a(r) = 2^d r^{d-1}/\Gamma(d)$ when $r \gg 1$.] The volume $V(r)$ has two contributions: the first comes from sites on the underlying lattice, whose number is given by the sum of $a(r)$ over r; the second comes from the sites which can be reached via shortcuts. These latter contribute a volume $V(r-r')$ for every shortcut encountered at a distance r', of which there are on average $2\xi^{-d}a(r')$. Thus $V(r)$ is in general the solution of the equation

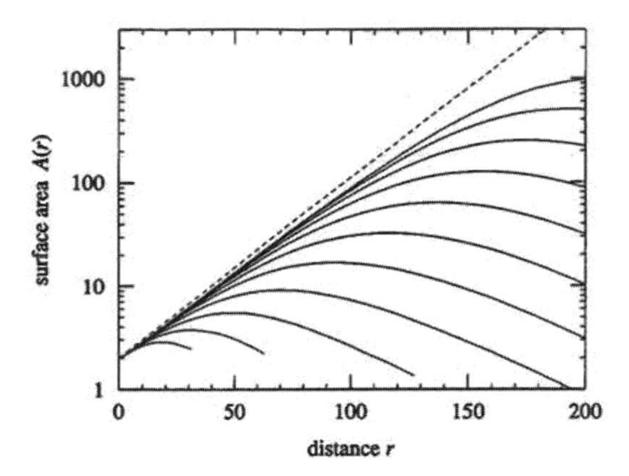

FIG. 3. The mean surface area $A(r)$ of a neighborhood of radius r on a $d=1$ small-world graph with $\phi=0.01$ for $L=128\cdots131\,072$ (solid lines). The measurements are averaged over 1000 realizations of the system each. The dotted line is the theoretical result for $L=\infty$, Eq. (9).

$$V(r)=\sum_{r'=0}^{r} a(r')[1+2\xi^{-d}V(r-r')]. \tag{6}$$

In one dimension with $k=1$, for example, $a(r)=2$ for all r and, approximating the sum with an integral and then differentiating with respect to r, we get

$$\frac{dV}{dr}=2[1+2V(r)/\xi], \tag{7}$$

which has the solution

$$V(r)=\tfrac{1}{2}\xi(e^{4r/\xi}-1). \tag{8}$$

Note that for $r\ll\xi$ this scales as r, independent of ξ, and for $r\gg\xi$ it grows exponentially, as expected. Equation (8) also implies that the surface area of a sphere of radius r on the graph, which is the derivative of $V(r)$, should be

$$A(r)=2e^{4r/\xi}. \tag{9}$$

These results are easily checked numerically and give us a simple independent measurement of ξ which we can use to confirm our earlier arguments. In Fig. 3 we show curves of $A(r)$ from computer simulations of systems with $\phi=0.01$ for values of L equal to powers of two from 128 up to 131 072 (solid lines). The dotted line is Eq. (9) with ξ taken from Eq. (1). The convergence of the simulation results to the predicted exponential form as the system size grows confirms our contention that ξ is well defined in the limit of large L. Figure 4 shows $A(r)$ for $L=100\,000$ for various values of ϕ. Equation (9) implies that the slope of the lines in the limit of small r is $4/\xi$. In the inset we show the values of ξ extracted from fits to the slope as a function of ϕ on logarithmic scales, and a straight-line fit to these points gives us an estimate of $\tau=0.99\pm0.01$ for the exponent governing the transition at $\phi=0$ [Eq. (3)]. This is in good agreement with our theoretical prediction that $\tau=1$.

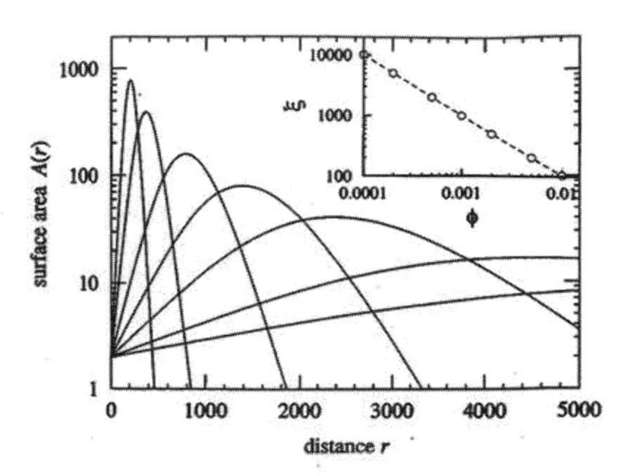

FIG. 4. The mean surface area $A(r)$ of a neighborhood of radius r on a $d=1$ small-world graph with $L=100\,000$ for $\phi=10^{-4}\ldots10^{-2}$. The measurements are averaged over 1000 realizations of the system each. Inset: the value of ξ extracted from the curves in the main figure, as a function of ϕ. The gradient of the line gives the value of the exponent τ, which is found by a least squares fit (the dotted line) to be 0.99 ± 0.01.

IV. SCALING IN SMALL-WORLD GRAPHS

Given the existence of the single nontrivial length scale ξ for the small-world model, we can also say how the mean vertex-vertex distance l should scale with system size and other parameters near the phase transition. In this regime the dimensionless quantity l/L can be a function only of the dimensionless quantity L/ξ, since no other dimensionless combinations of variables exist. Thus we can write

$$l=Lf(L/\xi), \tag{10}$$

where $f(x)$ is an unknown but universal scaling function. A scaling form similar to this was suggested previously by Barthélémy and Amaral [10] on empirical grounds. Substituting from Eq. (1), we then get for the $k=1$ case

$$l=Lf(\phi^{1/d}L). \tag{11}$$

[We have absorbed a factor of $d^{1/d}$ into the definition of $f(x)$ here to make it consistent with the definition we used in Ref. [12].] The usefulness of this equation derives from the fact that the function $f(x)$ contains no dependence on ϕ or L other than the explicit dependence introduced through its argument. Its functional form can, however, change with dimension d and indeed it does. In order to obey the known asymptotic forms of l for large and small systems, the scaling function $f(x)$ must satisfy

$$f(x)\sim\frac{\log x}{x} \quad \text{as } x\to\infty \tag{12}$$

and

$$f(x)\to\frac{1}{4}d \quad \text{as } x\to 0. \tag{13}$$

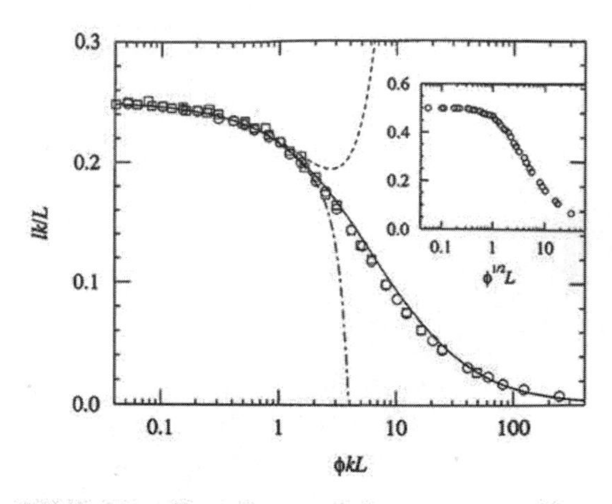

FIG. 5. Data collapse for numerical measurements of the mean vertex-vertex distance on small-world graphs with $d=1$. Circles and squares are results for $k=1$ and $k=5$, respectively, for values of L between 128 and 32 768 and values of ϕ between 1×10^{-6} and 3×10^{-2}. Each point is averaged over 1000 realizations of the randomness. In all cases the errors on the points are smaller than the points themselves. The dashed line is the second-order series approximation with exact coefficients given in Eq. (18), while the dot-dashed line is the fifth-order approximation using numerical results for the last three coefficients. The solid line is the third-order Padé approximant, Eqs. (21) and (23). Inset: data collapse for two-dimensional systems with $k=1$ for values of L from 64 to 1024 and ϕ from 3×10^{-6} up to 1×10^{-3}.

When $k>1$, l tends to $\frac{1}{4}dL/k$ for small values of L and ξ is given by Eq. (2), so the appropriate generalization of the scaling form is

$$l=\frac{L}{k}f((\phi k)^{1/d}L), \tag{14}$$

with $f(x)$ taking the same limiting forms (12) and (13). Previously we derived this scaling form in a more rigorous way using an RG argument [12].

We can again test these results numerically by measuring l on small-world graphs for various values of ϕ, k, and L. Eq. (14) implies that if we plot the results on a graph of lk/L against $(\phi k)^{1/d}L$, they should collapse onto a single curve for any given dimension d. In Fig. 5 we have done this for systems based on underlying lattices with $d=1$ for a range of values of ϕ and L, for $k=1$ and 5. As the figure shows, the collapse is excellent. In the inset we show results for $d=2$ with $k=1$, which also collapse nicely onto a single curve. The lower limits of the scaling functions in each case are in good agreement with our theoretical predictions of $\frac{1}{4}$ for $d=1$ and $\frac{1}{2}$ for $d=2$.

We are not able to solve exactly for the form of the scaling function $f(x)$, but we can express it as a series expansion in powers of ϕ as follows. Since the scaling function is universal and has no implicit dependence on k, it is adequate to calculate it for the case $k=1$; its form is the same for all other values of k. For $k=1$ the probability of having exactly m shortcuts on the graph is

$$P_m=\binom{dL^d}{m}\phi^m(1-\phi)^{dL^d-m}. \tag{15}$$

TABLE I. Average vertex-vertex distances per vertex l_m/L on $d=1$ small-world graphs with exactly m shortcuts and $k=1$. Values up to $m=2$ are the exact results of Strang and Eriksson [14]. Values for $m=3\cdots5$ are our numerical results.

m	l_m/L
0	1/4
1	5/24
2	131/720
3	0.1549±0.0003
4	0.1365±0.0003
5	0.1232±0.0003

Let l_m be the mean vertex-vertex distance on a graph with m shortcuts in the limit of large L, averaged over all such graphs. Then the mean vertex-vertex distance averaged over all graphs regardless of the number of shortcuts is

$$l=\sum_{m=0}^{dL^d}P_m l_m. \tag{16}$$

Note that in order to calculate l up to order ϕ^m we only need to know the behavior of the model when it has m or fewer shortcuts. For the $d=1$ case the values of the l_m have been calculated up to $m=2$ by Strang and Eriksson [14] and are given in Table I. Substituting these into Eq. (16) and collecting terms in ϕ, we then find that

$$\frac{l}{L}=\frac{1}{4}-\frac{1}{24}\phi L+\frac{11}{1440}\phi^2L^2-\frac{11}{1440}\phi^2L+O(\phi^3). \tag{17}$$

The term in ϕ^2L can be dropped when L is large or ϕ small, since it is negligible by comparison with at least one of the terms before it. Thus the scaling function is

$$f(x)=\frac{1}{4}-\frac{1}{24}x+\frac{11}{1440}x^2+O(x^3). \tag{18}$$

This form is shown as the dotted line in Fig. 5 and agrees well with the numerical calculations for small values of the scaling variable x, but deviates badly for large values.

Calculating the exact values of the quantities l_m for higher orders is an arduous task and probably does not justify the effort involved. However, we have calculated the values of the l_m numerically up to $m=5$ by evaluating the average vertex-vertex distance l on graphs which are constrained to have exactly 3, 4, or 5 shortcuts. Performing a Taylor expansion of l/L about $L=\infty$, we get

$$\frac{l}{L}=\frac{l_m}{L}\left[1+\frac{c_m}{L}+O(L^{-2})\right], \tag{19}$$

where c_m is a constant. Thus we can estimate l_m/L from the vertical-axis intercept of a plot of l/L against L^{-1} for large L. The results are shown in Table I. Calculating higher orders still would be straightforward.

Using these values we have evaluated the scaling function $f(x)$ up to fifth order in x; the result is shown as the dot-dashed line in Fig. 5. As we can see the range over which it matches the numerical results is greater than before, but not by much, indicating that the series expansion converges only slowly as extra terms are added. It appears therefore that series expansion would be a poor way of calculating $f(x)$ over the entire range of interest.

A much better result can be obtained by using our series expansion coefficients to define a Padé approximant to $f(x)$ [15,16]. Since we know that $f(x)$ tends to a constant $f(0) = \frac{1}{4}d$ for small x and falls off approximately as $1/x$ for large x, the appropriate Padé approximants to use are odd-order approximants where the approximant of order $2n+1$ (n integer) has the form

$$f(x)=f(0)\frac{A_n(x)}{B_{n+1}(x)}, \tag{20}$$

where $A_n(x)$ and $B_n(x)$ are polynomials in x of degree n with constant term equal to 1. For example, to third order we should use the approximant

$$f(x)=f(0)\frac{1+a_1x}{1+b_1x+b_2x^2}. \tag{21}$$

Expanding about $x=0$ this gives

$$\frac{f(x)}{f(0)}=1+(a_1-b_1)x+(b_1^2-a_1b_1-b_2)x^2 +[(a_1-b_1)(b_1^2-b_2)+b_1b_2]x^3+O(x^4). \tag{22}$$

Equating coefficients order by order in x and solving for the a's and b's, we find that

$$a_1=1.825\pm0.075,$$
$$b_1=1.991\pm0.075, \tag{23}$$
$$b_2=0.301\pm0.012.$$

Substituting these back into Eq. (21) and using the known value of $f(0)$ then gives us our approximation to $f(x)$. This approximation is plotted as the solid line in Fig. 5 and, as the figure shows, is an excellent guide to the value of $f(x)$ over a large range of x. In theory it should be possible to calculate the fifth-order Padé approximant using the numerical results in Table I, although we have not done this here. Substituting $f(x)$ back into the scaling form, Eq. (14), we can also use the Padé approximant to predict the value of the mean vertex-vertex distance for any values of ϕ, k, and L within the scaling regime. We will make use of this result in the next section to calculate the effective dimension of small-world graphs.

V. EFFECTIVE DIMENSION

The calculation of the volumes and surface areas of neighborhoods of vertices on small-world graphs in Sec. III leads us naturally to the consideration of the dimension of these systems. On a regular lattice of dimension D, the volume $V(r)$ of a neighborhood of radius r increases in proportion to r^D, and hence one can calculate D from [17]

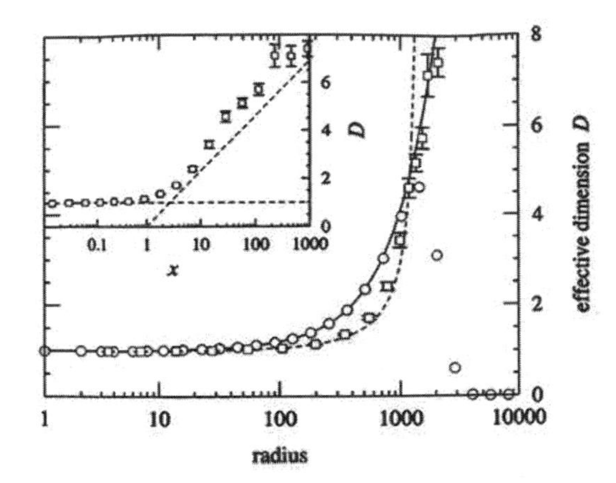

FIG. 6. Effective dimension D of small-world graphs. The circles are results for D from numerical calculations on an $L = 1\,000\,000$ system with $d=1$, $k=1$, and $\phi=10^{-3}$ using Eq. (24). The errors on the points are in all cases smaller than the points themselves. The solid line is Eq. (25). The squares are calculated from Eq. (27) by numerical differentiation of simulation results for the scaling function $f(x)$ of one-dimensional systems. The dotted line is Eq. (27) evaluated using the third-order Padé approximant to the scaling function derived in Sec. IV. Inset: effective dimension from Eq. (27) plotted as a function of the scaling variable x. The dotted lines represent the asymptotic forms for large and small x discussed in the text.

$$D=\frac{d\log V}{d\log r}=\frac{rA(r)}{V(r)}, \tag{24}$$

where $A(r)$ is the surface area of the neighborhood, as previously. We can use the same expression to calculate the effective dimension of our small-world graphs. Thus in the case of an underlying lattice of dimension $d=1$, the effective dimension of the graph is

$$D=\frac{4r}{\xi}\frac{e^{4r/\xi}}{e^{4r/\xi}-1}, \tag{25}$$

where we have made use of Eqs. (8) and (9). For $r\ll\xi$ this tends to one, as we would expect, and for $r\gg\xi$ it tends to $4r/\xi$, increasing linearly with the radius of the neighborhood. Thus the effective dimension of a small-world graph depends on the length scale on which we look at it, in a way reminiscent of the behavior of multifractals [18,19]. This result will become important in Sec. VI when we consider site percolation on small-world graphs.

In Fig. 6 we show the effective dimension of neighborhoods on a large graph measured in numerical simulations (circles), along with the analytic result, Eq. (25) (solid line). As we can see from the figure, the numerical and analytic results are in good agreement for small radii r, but the numerical results fall off sharply for larger r. The reason for this is that Eq. (24) breaks down as $V(r)$ approaches the volume of the entire system; $V(r)$ must tend to L^d in this limit and hence the derivative in Eq. (24) tends to zero. The same effect is also seen if one tries to use Eq. (24) on ordinary regular lattices of finite size. To characterize the dimension of an entire system therefore, we use another measure of D as follows.

On a regular lattice of finite linear size l, the number of vertices N scales as l^D and hence we can calculate the dimension from

$$D=\frac{d\log N}{d\log l}. \tag{26}$$

We can apply the same formula to the calculation of the effective dimension of small-world graphs putting $N=L^d$, although, since we don't have an analytic solution for l, we cannot derive an analytic solution for D in this case. On the other hand, if we are in the scaling regime described in Sec. IV, then Eq. (14) applies, along with the limiting forms, Eqs. (12) and (13). Substituting into Eq. (26), this gives us

$$\frac{1}{D}=\frac{d\log l}{d\log L^d}=\frac{1}{d}\left[1+\frac{d\log f(x)}{d\log x}\right], \tag{27}$$

where $x=(\phi k)^{1/d}L\propto L/\xi$. In other words D is a universal function of the scaling variable x. We know that $f(x)$ tends to a constant for small x (i.e., $\xi\gg L$), so that $D=d$ in this limit, as we would expect. For large x (i.e., $\xi\ll L$), Eq. (12) applies. Substituting into Eq. (27) this gives us $D=d\log x$. In the inset of Fig. 6 we show D from numerical calculations as a function of x in one-dimensional systems of a variety of sizes, along with the expected asymptotic forms, which it follows reasonably closely. In the main figure we also show this second measure of D (squares with error bars) as a function of the system radius l (with which it should scale linearly for large l, since $l\sim\log x$ for large x). As the figure shows, the two measures of effective dimension agree reasonably well. The numerical errors on the first measure, Eq. (24) are much smaller than those on the second, Eq. (26) (which is quite hard to calculate numerically), but the second measure is clearly preferable as a measure of the dimension of the entire system, since the first fails badly when r approaches l. We also show the value of our second measure of dimension calculated using the Padé approximant to $f(x)$ derived in Sec. IV (dotted line in the main figure). This agrees well with the numerical evaluation for radii up to about 1000 and has significantly smaller statistical error, but overestimates D somewhat beyond this point because of inaccuracies in the approximation; the Padé approximant scales as $1/x$ for large values of x rather than $(\log x)/x$, which means that D will scale as x rather than $\log x$ for large x.

VI. PERCOLATION

In the previous sections of this paper we have examined statistical properties of small-world graphs such as typical length scales, vertex-vertex distances, scaling of volumes and areas, and effective dimension of graphs. These are essentially static properties of the networks; to the extent that small-world graphs mimic social networks, these properties tell us about the static structure of those networks. However, social science also deals with dynamic processes going on within social networks, such as the spread of ideas, information, or diseases. This leads us to the consideration of dynamical models defined on small-world graphs. A small amount of research has already been conducted in this area. Watts [6,7], for instance, has considered the properties of a number of simple dynamical systems defined on small-world graphs, such as networks of coupled oscillators and cellular automata. Barrat and Weigt [20] have looked at the properties of the Ising model on small-world graphs and derived a solution for its partition function using the replica trick. Monasson [21] looked at the spectral properties of the Laplacian operator on small-world graphs, which tells us about the time evolution of a diffusive field on the graph. There is also a moderate body of work in the mathematical and social sciences which, although not directly addressing the small-world model, deals with general issues of information propagation in networks, such as the adoption of innovations [22–25], human epidemiology [26–28], and the flow of data on the Internet [29,30].

In this section we discuss the modeling of information or disease propagation specifically on small-world graphs. Suppose for example that the vertices of a small-world graph represent individuals and the bonds between them represent physical contact by which a disease can be spread. The spread of ideas can be similarly modeled; the bonds then represent information connections between individuals which could include letters, telephone calls, or email, as well as physical contacts. The simplest model for the spread of disease is to have the disease spread between neighbors on the graph at a uniform rate, starting from some initial carrier individual. From the results of Sec. IV we already know what this will look like. If, for example, we wish to know how many people in total have contracted a disease, that number is just equal to the number $V(r)$ within some radius r of the initial carrier, where r increases linearly with time. (We assume that no individual can catch the disease twice, which is the case with most common diseases.) Thus, Eq. (8) tells us that, for a $d=1$ small-world graph, the number of individuals who have had a particular disease increases exponentially, with a time-constant governed by the typical length scale ξ of the graph. Since all real-world social networks have a finite number of vertices N, this exponential growth is expected to saturate when $V(r)$ reaches $N=L^d$. This is not a particularly startling result; the usual model for the spread of epidemics is the logistic growth model, which shows initial exponential spread followed by saturation.

For a disease such as influenza, which spreads fast but is self-limiting, the number of people who are ill at any one time should be roughly proportional to the area $A(r)$ of the neighborhood surrounding the initial carrier, with r again increasing linearly in time. This implies that the epidemic should have a single humped form with time, similar to the curves of $A(r)$ plotted in Fig. 4. Note that the vertical axis in this figure is logarithmic; on linear axes the curves are bell shaped rather than quadratic. In the context of the spread of information or ideas, similar behavior might be seen in the development of fads. By a fad we mean an idea which is catchy and therefore spreads fast, but which people tire of quickly. Fashions, jokes, toys, or buzzwords might be expected to show popularity profiles over time similar to the curves in Fig. 4.

However, for most real diseases (or fads) this is not a very good model of how they spread. For real diseases it is commonly the case that only a certain fraction p of the population is susceptible to the disease. This can be mimicked in our model by placing a two-state variable on each vertex which denotes whether the individual at that vertex is susceptible. The disease then spreads only within the local "cluster" of connected susceptible vertices surrounding the initial carrier. One question which we can answer with such a model is how high the density p of susceptible individuals can be before the largest connected cluster covers a significant fraction of the entire network and an epidemic ensues [31].

Mathematically, this is precisely the problem of site percolation on a social network, at least in the case where the susceptible individuals are randomly distributed over the vertices. To the extent that small-world graphs mimic social networks, therefore, it is interesting to look at the percolation problem. The transition corresponds to the point on a regular lattice at which a percolating cluster forms whose size increases with the size L of the lattice for arbitrarily large L [32]. On random graphs there is a similar transition, marked by the formation of a so-called "giant component" of connected vertices [33]. On small-world graphs we can calculate approximately the percolation probability $p=p_c$ at which the transition takes place as follows.

Consider a $d=1$ small-world graph of the kind pictured in Fig. 1. For the moment let us ignore the shortcut bonds and consider the percolation properties just of the underlying regular lattice. If we color in a fraction p of the sites on this underlying lattice, the occupied sites will form a number of connected clusters. In order for two adjacent parts of the lattice not to be connected, we must have a series of at least k consecutive unoccupied sites between them. The probability that we have such a series starting at a particular site, followed by an occupied site is $p(1-p)^k$, and the number n of such series in the whole system is

$$n=Lp(1-p)^k. \tag{28}$$

For this one-dimensional system, the percolation transition occurs when we have just one break in the chain, i.e., when $n=1$. This gives us a kth order equation for p_c which is in general not exactly soluble, but we can find its roots numerically if we wish.

Now consider what happens when we introduce shortcuts into the graph. The number of breaks n, Eq. (28), is also the number of connected clusters of occupied sites on the underlying lattice. Let us for the moment suppose that the size of each cluster can be approximated by the average cluster size. A number ϕkL of shortcuts are now added to the graph between pairs of vertices chosen uniformly at random. A fraction p^2 of these will connect two occupied sites and therefore can connect together two clusters of occupied sites. The problem of when the percolation transition occurs is then precisely that of the formation of a giant component on an ordinary random graph with n vertices. It is known that such a component forms when the mean coordination number of the random graph is one [33], or alternatively when the number of bonds on the graph is half the number of vertices. In other words, the transition probability p_c must satisfy

$$p_c^2\phi kL=\tfrac{1}{2}Lp_c(1-p_c)^k \tag{29}$$

or

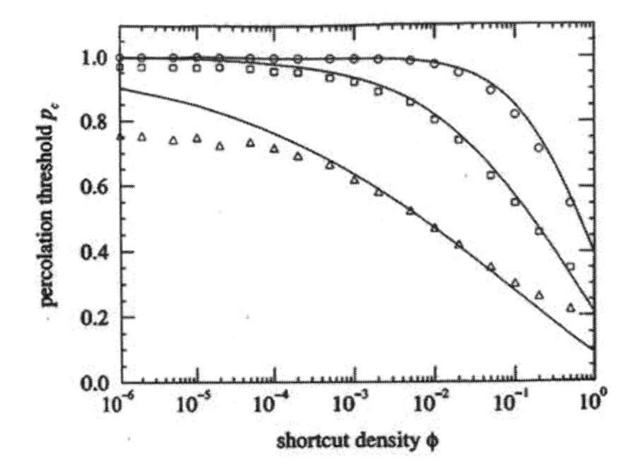

FIG. 7. Numerical results for the percolation threshold on $L=10\,000$ small-world graphs with $k=1$ (circles), 2 (squares), and 5 (triangles) as a function of the shortcut density ϕ. The solid lines are the analytic approximation to the same quantity, Eq. (30).

$$\phi=\frac{(1-p_c)^k}{2kp_c}. \tag{30}$$

We have checked this result against numerical calculations. In order to find the value of p_c numerically, we employ a tree-based invasion algorithm similar to the invaded cluster algorithm used to find the percolation point in Ising systems [34,35]. This algorithm can calculate the entire curve of average cluster size versus p in time which scales as $L\log L$ [36]. We define p_c to be the point at which the average cluster size divided by L rises above a certain threshold. For systems of infinite size the transition is instantaneous and hence the choice of threshold makes no difference to p_c, except that p_c can never take a value lower than the threshold itself, since even in a fully connected graph the average cluster size per vertex can be no greater than the fraction p_c of occupied vertices. Thus it makes sense to choose the threshold as low as possible. In real calculations, however, we cannot use an infinitesimal threshold because of finite size effects. For the systems studied here we have found that a threshold of 0.2 works well.

Figure 7 shows the critical probability p_c for systems of size $L=10\,000$ for a range of values of ϕ for $k=1$, 2, and 5. The points are the numerical results and the solid lines are Eq. (30). As the figure shows the agreement between simulation and theory is good although there are some differences. As ϕ approaches one and the value of p_c drops, the two fail to agree because, as mentioned above, p_c cannot take a value lower than the threshold used in its calculation, which was 0.2 in this case. The results also fail to agree for very low values of ϕ where p_c becomes large. This is because Eq. (28) is not a correct expression for the number of clusters on the underlying lattice when $n<1$. This is clear since when there are no breaks in the sequence of connected vertices around the ring it is not also true that there are no connected clusters. In fact there is still one cluster; the equality between number of breaks and number of clusters breaks down at $n=1$. The value of p at which this happens is given by putting $n=1$ in Eq. (28). Since p is close to one at this point its value is well approximated by

$$p \simeq 1 - L^{-1/k}, \tag{31}$$

and this is the value at which the curves in Fig. 7 should roll off at low ϕ. For $k=5$ for example, for which the roll-off is most pronounced, this expression gives a value of $p \simeq 0.8$, which agrees reasonably well with what we see in the figure.

There is also an overall tendency in Fig. 7 for our analytic expression to overestimate the value of p_c slightly. This we put down to the approximation we made in the derivation of Eq. (30) that all clusters of vertices on the underlying lattice can be assumed to have the size of the average cluster. In actual fact, some clusters will be smaller than the average and some larger. Since the shortcuts will connect to clusters with probability proportional to the cluster size, we can expect percolation to set in within the subset of larger-than-average clusters before it would set in if all clusters had the average size. This makes the true value of p_c slightly lower than that given by Eq. (30). In general however, the equation gives a good guide to the behavior of the system.

We have also examined numerically the behavior of the mean cluster radius ρ for percolation on small-world graphs. The radius of a cluster is defined as the average distance between vertices within the cluster, along the edges of the graph within the cluster. This quantity is small for small values of the percolation probability p and increases with p as the clusters grow larger. When we reach percolation and a giant component forms it reaches a maximum value and then drops as p increases further. The drop happens because the percolating cluster is most filamentary when percolation has only just set in and so paths between vertices are at their longest. With further increases in p the cluster becomes more highly connected and the average shortest path between two vertices decreases.

By analogy with percolation on regular lattices we might expect the average cluster radius for a given value of ϕ to satisfy the scaling form [32]

$$\rho = l^{\gamma/\nu} \tilde{\rho}((p-p_c) l^{1/\nu}), \tag{32}$$

where $\tilde{\rho}(x)$ is a universal scaling function, l is the radius of the entire system, and γ and ν are critical exponents. In fact this scaling form is not precisely obeyed by the current system because the exponents ν and γ depend in general on the dimension of the lattice. As we showed in Sec. V, the dimension D of a small-world graph depends on the length scale on which you look at it. Thus the value of D "felt" by a cluster of radius ρ will vary with ρ, implying that ν and γ will vary both with the percolation probability and with the system size. If we restrict ourselves to a region sufficiently close to the percolation threshold, and to a sufficiently small range of values of l, then Eq. (32) should be approximately correct.

In Fig. 8 we show numerical data for ρ for small-world graphs with $k=1$, $\phi=0.1$, and L equal to a power of two from 512 up to 16 384. As we can see, the data show the expected peaked form, with the peak in the region of $p=0.8$, close to the expected position of the percolation transition. In order to perform a scaling collapse of these data we need first to extract a suitable value of p_c. We can do this by performing a fit to the positions of the peaks in ρ [37]. Since the scaling function $\tilde{\rho}(x)$ is (approximately) universal, the positions of these peaks all occur at the same value of the

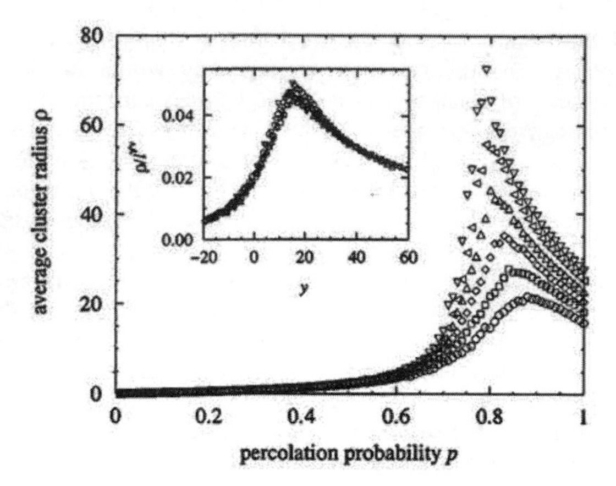

FIG. 8. Average cluster radius ρ as a function of the percolation probability p for site percolation on small-world graphs with $k=1$, $\phi=0.1$, and L equal to a power from 512 up to 16 384 (circles, squares, diamonds, upward-pointing triangles, left-pointing triangles, and downward-pointing triangles, respectively). Each set of points is averaged over 100 realizations of the corresponding graph. Inset: the same data collapsed according to Eq. (32) with $\nu=0.59$, $\gamma=1.3$, and $p_c=0.74$.

scaling variable $y=(p-p_c)l^{1/\nu}$. Calling this value y_0 and the corresponding percolation probability p_0, we can rearrange for p_0 as a function of l to get

$$p_0 = p_c + y_0 l^{-1/\nu}. \tag{33}$$

Thus if we plot the measured positions p_0 as a function of $l^{-1/\nu}$, the vertical-axis intercept should give us the corresponding value of p_c. We have done this for a single value of ν in the inset to Fig. 9, and in the main figure we show the resulting values of p_c as a function of $1/\nu$. If we now perform our scaling collapse, with the restriction that the values of ν and p_c fall on this line, then the best coincidence of the curves for ρ is obtained when $p_c=0.74$ and $\nu=0.59\pm0.05$—see the inset to Fig. 8. The value of γ can be found separately by requiring the heights of the peaks to match up, which gives $\gamma=1.3\pm0.1$. The collapse is noticeably poorer when we include systems of size smaller than

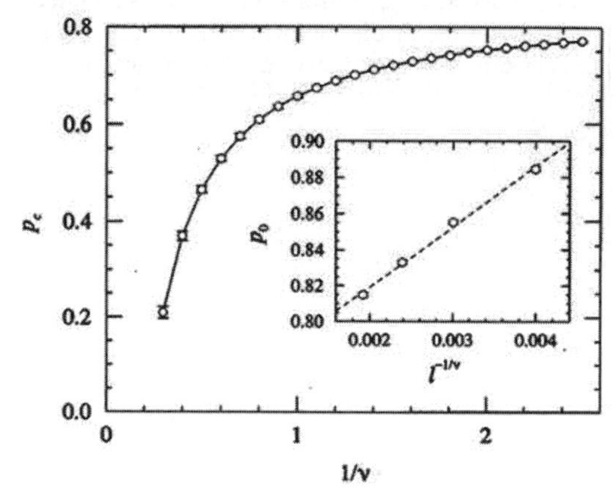

FIG. 9. Best fit values of p_c as a function of $1/\nu$. Inset: the values are calculated from the vertical-axis intercept of a plot of the position p_0 of the peak of ρ against $l^{-1/\nu}$ [see Eq. (33)].

$L=512$, and we attribute this not merely to finite size corrections to the scaling form, but also to variation in the values of the exponents γ and ν with the effective dimension of the percolating cluster.

We note that ν is expected to tend to $\frac{1}{2}$ in the limit of an infinite-dimensional system. The value $\nu=0.59$ found here therefore confirms our contention that small-world graphs have a high effective dimension even for quite moderate values of ϕ, and thus are in some sense close to being random graphs. (On a two-dimensional lattice by contrast $\nu=\frac{4}{3}$.)

VII. CONCLUSIONS

In this paper we have studied the small-world network model of Watts and Strogatz, which mimics the behavior of networks of social interactions. In the version of the model used here, graphs consist of a set of vertices joined together in a regular lattice, plus a low density of "shortcuts" which link together pairs of vertices chosen at random. We have looked at the scaling properties of small-world graphs and argued that there is only one typical length scale present other than the fundamental lattice constant, which we denote ξ and which is roughly the typical distance between the ends of shortcuts. We have shown that this length scale governs the transition of the average vertex-vertex distance on a graph from linear to logarithmic scaling with increasing system size, as well as the rate of growth of the number of vertices in a neighborhood of fixed radius about a given point. We have also shown that the value of ξ diverges on an infinite lattice as the density of shortcuts tends to zero, and therefore that the system possesses a continuous phase transition in this limit. Close to the phase transition, where ξ is large, we have shown that the average vertex-vertex distance on a finite graph obeys a simple scaling form and in any given dimension is a universal function of a single scaling variable which depends on the density of shortcuts, the system size and the average coordination number of the graph. We have calculated the form of the scaling function to fifth order in the shortcut density using a series expansion and to third order using a Padé approximant. We have defined two measures of the effective dimension D of small-world graphs and find that the value of D depends on the scale on which you look at the graph in a manner reminiscent of the behavior of multifractals. Specifically, at length scales shorter than ξ the dimension of the graph is simply that of the underlying lattice on which it is built, and for length scales larger than ξ it increases linearly, with a characteristic constant proportional to ξ. The value of D increases logarithmically with the number of vertices in the graph. We have checked all of these results by extensive numerical simulation of the model and in all cases we find good agreement between the analytic predictions and the simulation results.

In the last part of the paper we have looked at site percolation on small-world graphs as a model of the spread of information or disease in social networks. We have derived an approximate analytic expression for the percolation probability p_c at which a "giant component" of connected vertices forms on the graph and shown that this agrees well with numerical simulations. We have also performed extensive numerical measurements of the typical radius of connected clusters on the graph as a function of the percolation probability and shown by performing a scaling collapse that these obey, to a reasonable approximation, the expected scaling form in the vicinity of the percolation transition. The characteristic exponent ν takes a value close to $\frac{1}{2}$, indicating that, as far as percolation is concerned, the graph's properties are close to those of a random graph.

ACKNOWLEDGMENTS

We thank Luis Amaral, Alain Barrat, Marc Barthélémy, Roman Kotecký, Marcio de Menezes, Cris Moore, Cristian Moukarzel, Thadeu Penna, and Steve Strogatz for helpful comments and conversations, and Gilbert Strang and Henrik Eriksson for communicating to us some results from their forthcoming paper. This work was supported in part by the Santa Fe Institute and by funding from the NSF (Grant No. PHY-9600400), the DOE (Grant No. DE-FG03-94ER61951), and DARPA (Grant No. ONR N00014-95-1-0975).

---

[1] S. Milgram, Psychol. Today **2**, 60 (1967).

[2] B. Bollobás, *Random Graphs* (Academic Press, New York, 1985).

[3] D. J. Watts and S. H. Strogatz, Nature (London) **393**, 440 (1998).

[4] In previous work the letter p has been used to denote the density of random connections, rather than ϕ. We use ϕ here, however, to avoid confusion with the percolation probability introduced in Sec. VI, which is also conventionally denoted p.

[5] Watts and Strogatz used the letter k to refer to the average coordination number, but we will find it convenient to distinguish between the coordination number, which we call z, and the range of the bonds. In one dimension $z=2k$ and in general $z=2dk$ for networks based on d-dimensional lattices.

[6] D. J. Watts, Ph.D. thesis, Cornell University, 1997.

[7] D. J. Watts, *Small Worlds: The Dynamics of Networks Between Order and Randomness* (Princeton University Press, Princeton, 1999).

[8] The exact definition of ξ depends on how you measure lengths in the model. The definition given here is appropriate if ξ is measured in terms of the lattice constant of the underlying lattice. It would, however, be reasonable to measure it in terms of the number of bonds traversed between the ends of two shortcuts. Since we are measuring lattice size L in terms of the underlying lattice constant rather than number of bonds, the present definition is the more appropriate one in our case, but it would be perfectly consistent to define both ξ and L to be a factor of k smaller; all the physical results would work out the same.

[9] In a system of finite size the average distance between the ends of two shortcuts cannot be larger than $\frac{1}{2}L$, so we cannot observe this divergence once ξ is larger than this.

[10] M. Barthélémy and L. A. N. Amaral, Phys. Rev. Lett. **82**, 3180 (1999).

[11] A. Barrat cond-mat/9903323 (unpublished).

[12] M. E. J. Newman and D. J. Watts, Phys. Lett. A (in press).

[13] M. Argollo de Menezes, C. F. Moukarzel, and T. J. P. Penna, cond-mat/9903426 (unpublished).

[14] G. Strang and H. Eriksson (in preparation).

[15] D. S. Gaunt and A. J. Guttmann, in *Phase Transitions and Critical Phenomena*, edited by C. Domb and M. S. Green (Academic Press, London, 1974), Vol. 3.

[16] We are indebted to Professor S. H. Strogatz for suggesting the use of a Padé approximant in this context.

[17] We use the capital letter D to denote the dimension here, to distinguish it from the dimension d of the underlying lattice defined in Sec. II.

[18] B. B. Mandelbrot, J. Fluid Mech. **62**, 331 (1974).

[19] T. C. Halsey, M. H. Jensen, L. P. Kadanoff, I. Procaccia, and B. I. Shraiman, Phys. Rev. A **33**, 1141 (1986).

[20] A. Barrat and M. Weigt, Eur. Phys. J. B (to be published).

[21] R. Monasson, Eur. Phys. J. B (to be published).

[22] E. M. Rogers, *Diffusion of Innovations* (Free Press, New York, 1962).

[23] J. S. Coleman, E. Katz, and H. Menzel, *Medical Innovation: A Diffusion Study* (Bobbs-Merrill, Indianapolis, 1966).

[24] D. Strang, Sociol. Methods Res. **19**, 324 (1991).

[25] T. W. Valente, Soc. Networks **18**, 69 (1996).

[26] L. Sattenspiel and C. P. Simon, Math. Biosci. **90**, 367 (1988).

[27] M. Kretschmar and M. Morris, Math. Biosci. **133**, 165 (1996).

[28] I. M. Logini, Jr., Math. Biosci. **90**, 341 (1988).

[29] J. O. Kephart and S. R. White, in *Proceedings of the 1991 IEEE Computer Science Symposium on Research in Security and Privacy* (IEEE Computer Society Press, Los Alamitos, 1991).

[30] J. E. Hanson and J. O. Kephart, in *Proceedings of the National Conference on Artificial Intelligence* (MIT Press, Cambridge, MA, 1999).

[31] A closely related issue is that of disease spreading when transmission does not take place with 100% probability along every edge in the graph. This can be represented by placing random two-state variables on the *bonds* of the graph to denote whether a bond will transmit the disease. Although we will not go through the calculation in detail, an approximate figure for the point at which an epidemic occurs in this bond percolation system can be calculated by a method very similar to the one presented here for the site percolation case.

[32] D. Stauffer and A. Aharony, *Introduction to Percolation Theory*, 2nd ed. (Taylor and Francis, London, 1992).

[33] N. Alon and J. H. Spencer, *The Probabilistic Method* (Wiley, New York, 1992).

[34] J. Machta, Y. S. Choi, A. Lucke, T. Schweizer, and L. V. Chayes, Phys. Rev. Lett. **75**, 2792 (1995).

[35] G. T. Barkema and M. E. J. Newman, in *Monte Carlo Methods in Chemical Physics*, edited by D. Ferguson, J. I. Siepmann, and D. G. Truhlar (Wiley, New York, 1999).

[36] A naive recursive cluster-finding algorithm by contrast takes time proportional to L^2.

[37] M. E. J. Newman and G. T. Barkema, *Monte Carlo Methods in Statistical Physics* (Oxford University Press, Oxford, 1999).

Eur. Phys. J. B **13**, 547–560 (2000)

THE EUROPEAN PHYSICAL JOURNAL B
EDP Sciences

On the properties of small-world network models

A. Barrat[1,a] and M. Weigt[2]

[1] Laboratoire de Physique Théorique[b], bâtiment 210, Université Paris-Sud, 91405 Orsay Cedex, France
[2] CNRS-Laboratoire de Physique Théorique de l'E.N.S., 24 rue Lhomond, 75231 Paris Cedex 05, France

Received 29 March 1999 and Received in final form 21 May 1999

Abstract. We study the small-world networks recently introduced by Watts and Strogatz [Nature **393**, 440 (1998)], using analytical as well as numerical tools. We characterize the geometrical properties resulting from the coexistence of a local structure and random long-range connections, and we examine their evolution with size and disorder strength. We show that any finite value of the disorder is able to trigger a "small-world" behaviour as soon as the initial lattice is big enough, and study the crossover between a regular lattice and a "small-world" one. These results are corroborated by the investigation of an Ising model defined on the network, showing for every finite disorder fraction a crossover from a high-temperature region dominated by the underlying one-dimensional structure to a mean-field like low-temperature region. In particular there exists a finite-temperature ferromagnetic phase transition as soon as the disorder strength is finite.

PACS. 05.50.+q Lattice theory and statistics (Ising, Potts, etc.) – 64.60.Cn Order-disorder transformations; statistical mechanics of model systems – 05.70.Fh Phase transitions: general studies

1 Introduction

A recent article by Watts and Strogatz [1], showing the relevance of what they called "small-world" networks for many realistic situations, has triggered a lot of attention for these kind of networks [2–7]: this interest results from their very definition, allowing an exploration between regular and random networks.

Random networks have of course been the subject of many studies in various domains, ranging from physics to social sciences. A very important characteristic common to such lattices and for example social networks is that the length of the shortest chain connecting two vertices (or members) grows very slowly, *i.e.* in general logarithmically, with the size of the network [8]. This characteristic has important consequences for many issues, *e.g.* the speed of disease spreading [1] etc. The social psychologist Milgram [9], after realizing that the number of persons necessary to link two randomly chosen, geographically separated persons had a median number of six, has called this concept the "six degrees of separation". In addition, models defined on random networks are, due to their locally tree-like structure, of mean-field type, and can therefore be analytically more tractable than their counterparts defined on regular lattices, but, thanks to the finite connectivity of their vertices, they display however behaviours which are intrinsically not captured by the familiar infinite connectivity models [10].

However, it is well-known that many realistic networks have a local structure which is very different from random networks with finite connectivity. For example, two neighbours have many common neighbours, a property which does not hold for random networks, and which can be quantified by the introduction of the "clustering coefficient" (see Sect. 3). Such phenomena are not only found in social networks, but also *e.g.* in the connections of neural networks [1] or in the chemical bond structure of long macromolecules [11]: The one-dimensional couplings of neighbouring monomers are complemented by long-ranged interactions between monomers that are close in space although not along the chain. This interplay has been studied in fact for example in [12], but it seems that, in this case, the long-range interactions are not sufficient to really modify the properties of the one-dimensional structure of the chain[1].

The construction proposed by Watts and Strogatz [1], that we will recall in Section 2, allows to reconcile local properties of a regular network with global properties of a

[a] e-mail: Alain.Barrat@th.u-psud.fr
[b] UMR 8627

[1] For example, an Ising model defined on a self-avoiding walk with interactions between monomers neighbours in space and not only on the chain has a critical temperature $T_c = 0$, as for a one-dimensional chain [12].

random one, by introducing a certain amount of random long-range connections into an initially regular network.

The aim of this paper is to study in some detail the concepts used in [1] to characterize the "small-world" behaviour, caused by the coexistence of "short-range" and "long-range" connections. We will show that this behaviour does not appear at a finite value of the disorder p, but that, for any $p > 0$, the networks will display this behaviour as soon as their size is large enough.

This paper is organized as follows. In Section 2 we describe the procedure used to obtain small-world networks; in Section 3 we study some of their geometrical properties, *i.e.* the connectivity, the chemical distances and the "clustering" coefficient, analytically as well as numerically[2]. Section 4 contains the investigation of an Ising-model defined on a small-world lattice, where the interplay between the short- and long-range interactions leads to interesting physical effects.

2 Definition of the model(s)

The construction algorithm proposed by Watts and Strogatz for small-world networks is the following: the initial network is a one-dimensional lattice of N sites, with periodic boundary conditions (*i.e.* a ring), each vertex being connected to its $2k$ nearest neighbours. The vertices are then visited one after the other; each link connecting a vertex to one of its k nearest neighbours in the clockwise sense is left in place with probability $1-p$, and with probability p is reconnected to a randomly chosen other vertex. Long range connections are therefore introduced. Note that, even for $p = 1$, the network keeps some memory of the procedure and is not locally equivalent to a random network: each vertex has indeed *at least* k neighbours. An important consequence is that we have no isolated vertices, and the graph has usually only one component (a random graph has usually many components of various sizes)(see Fig. 1).

It is possible to obtain "small-world" networks in other ways, that yields the same physical consequences, and can be more tractable analytically. For example, the networks studied in [4,5] are obtained by *adding* long-range connections to the initial ring without diluting its one-dimensional structure; the mean connectivity then changes with the disorder. In Section 4 we will also study an initial network with multiple links between successive vertices.

3 Geometrical properties

3.1 Connectivity

For $p = 0$, each vertex has the same connectivity $2k$. On the other hand, a non-zero value of p introduces disorder into the network, in the form of a non-uniform connectivity, while maintaining a fixed average connectivity $\bar{c} = 2k$. Let us denote $P_p(c)$ the probability distribution of the connectivities.

Since k of the initial $2k$ connections of each vertex are left untouched by the construction, the connectivity of a vertex i can be written $c_i = k + n_i$, with $n_i \geq 0$. n_i can then again be divided in two parts: $n_i^1 \leq k$ links have been left in place (each one with probability $1-p$), the other $n_i^2 = n_i - n_i^1$ links have been reconnected *towards* i, each one with probability p/N. We readily obtain

$$P_1(n_i^1) = \binom{k}{n_i^1}(1-p)^{n_i^1} p^{k-n_i^1} \tag{1}$$

$$P_2(n_i^2) = \frac{(kp)^{n_i^2}}{n_i^2!}\exp(-pk) \qquad \text{for large } N \tag{2}$$

and find

$$P_p(c) = \sum_{n=0}^{\min(c-k,k)} \binom{k}{n}(1-p)^n p^{k-n} \times \frac{(kp)^{c-k-n}}{(c-k-n)!}\exp(-pk), \quad c \geq k. \tag{3}$$

We show in Figure 2 the probability distributions for $k = 3$ and various values of p: as p grows, the distribution becomes broader.

3.2 Chemical distances

We now turn to a non-local quantity of graphs: the chemical distance between its vertices, *i.e.* the minimal number of links between two vertices. We note d_{ij} the chemical distance between vertices i and j, and

$$\ell(N,p) = \frac{1}{N(N-1)}\overline{\sum_{i\neq j} d_{ij}} \tag{4}$$

the mean chemical distance, averaged over all pairs of vertices and over the disorder induced by the rewiring procedure.

Watts and Strogatz have shown that the mean distance between vertices $\ell(N,p)$ decreases very rapidly as soon as p is non-zero. They however show the curve of $\ell(N,p)$ *versus* p for only one value of N and do not study how it depends on N. For $p = 0$, we have a linear chain of sites, so that we easily find

$$\ell(N,0) = \frac{N(N+2k-2)}{4k(N-1)} \sim \frac{N}{4k}, \tag{5}$$

growing like N. On the other hand, for $p = 1$ $\ell(N,1)$ grows like $\ln(N)/\ln(2k-1)$ (inset of Fig. 3): the graph is then random. Besides, the distribution of lengths, being uniform between 1 (shortest possible distance) and $N/(2k)$ for

[2] Results in particular about the chemical distances and the onset of the small-world behaviour can also be found in [2,3,5–7].

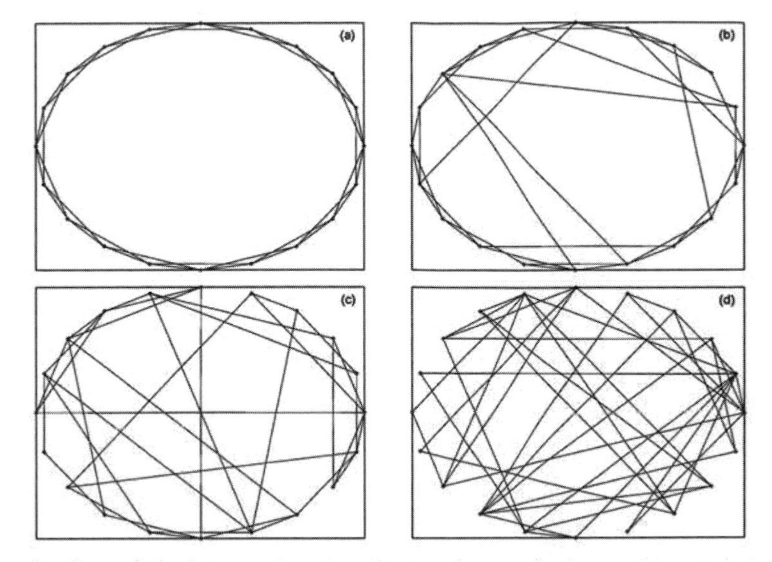

Fig. 1. Examples of networks obtained by the procedure described in the text, for $k = 2$, $N = 20$. (a): $p = 0$, regular networks; (b), (c): intermediate values of p; (d): $p = 1$.

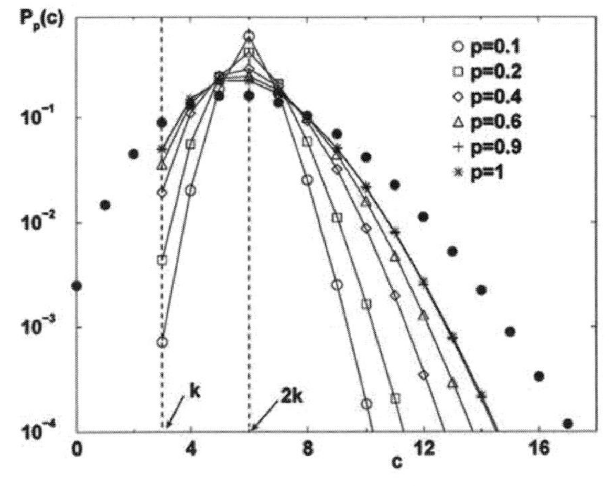

Fig. 2. Probability distributions of the connectivity c for $k = 3$ and various values of p: $c \geq k$, and the mean connectivity is $\bar{c} = 2k = 6$. The symbols are obtained by numerical simulations of small-world networks (with $N = 1000$ vertices), and the lines are a guide to the eye, joining points given by formula (3). Filled circles show the probability distribution of the connectivity c for a random network of mean connectivity is $\bar{c} = 2k = 6$ (given by $(2k)^c \exp(-2k)/c!$).

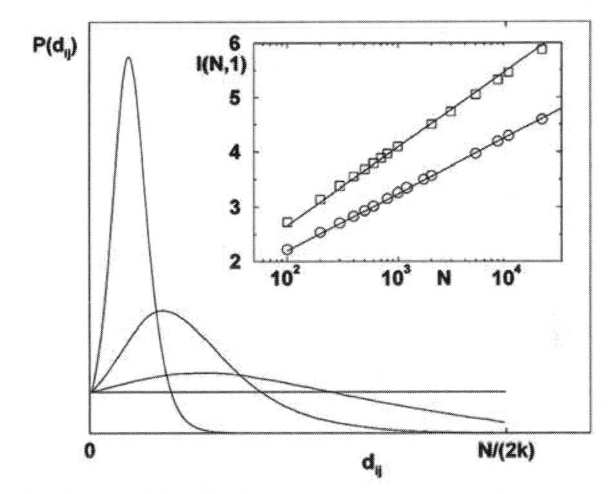

Fig. 3. Probability distribution of the distance d_{ij} between two vertices i and j of small-world graphs, for $k = 3$, $N = 2000$, $p = 2^{-20}$ (flat distribution), $p = 2^{-12}$, 2^{-10} and 2^{-8} (curves becoming more and more peaked as p grows), averaged over 500 samples for each p. The maximum value of d_{ij} is of course $N/(2k)$. Inset: $\ell(N, 1)$ *versus* N for $k = 3$ and $k = 5$, together with the $\ln(N)/\ln(2k - 1)$ straight lines.

the linear chain, becomes more and more peaked around its mean value as p grows (see Fig. 3).

It is therefore quite natural to ask if the change between these two behaviours occurs by a transition at a certain finite critical value of p or if there is a crossover phenomenon at any finite value of N, with a transition occurring only at $p = 0$. This last scenario was first proposed in [2].

We first investigate this question by numerical simulations, to study the behaviour of $\ell(N,p)$ in a systematic way, varying N and p: we use values of N from 100 to 20000, with $p = 2^a/2^{20}$, $a = 0, \cdots, 20$, and we average over 500 realizations of the disorder for each value of p. We have studied three different values of the mean connectivity: $2k = 4$, 6 and 10.

In Figure 4, we plot $\ell(N,p)/\ell(N,0)$ for various values of N and $k = 2$. It is clear that $\ell(N,p)$ decreases

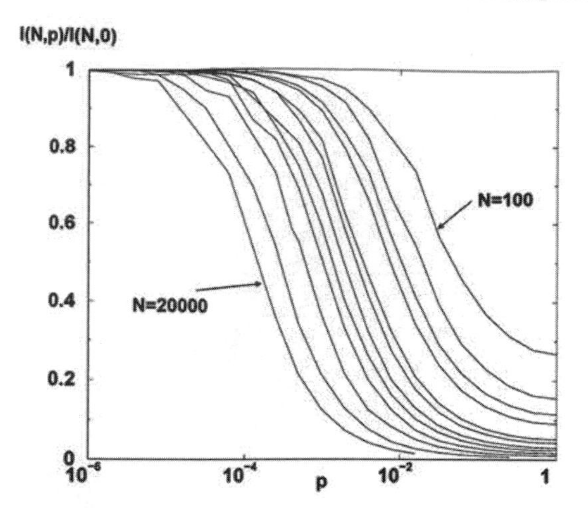

Fig. 4. Mean chemical length $\ell(N,p)$ normalized by $\ell(N,0)$, *versus* p, for $k=2$, and N from 100 to 20000: the drop in the curve occurs at lower and lower values of p as N grows.

very fast already for small p (note the logarithmic scale for p): from this point of view, the network is very soon similar to a random network. In particular, as N becomes larger, the drop in the curve occurs for smaller and smaller values of p, showing that no finite critical value of p can be determined this way: in the thermodynamic limit, $\ell(N,p)/\ell(N,0)$ goes to 0 for all $p>0$. This is a first clear indication of a crossover behaviour (as opposed to a transition at a non-zero p) that we are now going to examine in more details.

Note that the first evidence of a crossover has been given in [2] by the numerical study of system with sizes up to $N=1000$, and mean connectivities $2k=10$, 20, 30. A scaling of the form

$$\ell(N,p) \sim N^* F_k\left(\frac{N}{N^*}\right) \qquad (6)$$

was proposed, where F_k depends only on k, with $F_k(u \ll 1) \sim u$, $F_k(u \gg 1) \sim \ln u$, and $N^* \sim p^{-\tau}$ with $\tau = 2/3$ as p goes to zero. However, it can be shown [3], with a simple but rigorous argument, that τ cannot in fact be lower than 1: the mean number of rewired links is $N_r = pNk$; if $\tau < 1$, and if we take α such that $\tau < \alpha < 1$, then the scaling hypothesis implies, for large N, $\ell(N, N^{-1/\alpha}) \sim N^{\tau/\alpha}\ln(N)$ (since $N^{1-\tau/\alpha} \gg 1$ for large N); N_r however goes to zero for large N, so that the rewiring of a vanishing number of links could lead to a change in the scaling of ℓ. This obviously unphysical result shows that the hypothesis $\tau < 1$ is not valid. In addition, Newman and Watts [5], using a renormalization group analysis, have shown that $\tau = 1$ exactly. Here we will arrive at the same result, using our numerical simulations to test the scaling hypothesis, as well as analytical arguments.

To understand how strong the disorder has to be to induce a crossover, and to show that this crossover can occur, at fixed p, for $N^* \sim p^{-\tau}$, or equivalently, at fixed N, for $p^* \sim N^{-1/\tau}$, only with $\tau \geq 1$, we study the case of a finite number of rewired links, $N_r = \alpha$. This corresponds to $p = \alpha/N$. In order to show that such a value of p is not able to alter the scaling of ℓ with N, we now establish a rigorous lower bound.

For any given sample, the extremities of the α rewired links determine 2α intervals. The sum of their lengths on the ring is N, so that at least one of them has a length of order N, which is, even more precisely, larger than $N/(2\alpha)$. We call this interval $J = [i_0, j_0]$ and we consider the interval $I \subset J$, of length $N/(4\alpha) = bN$, $I = [i_0 + N/(8\alpha), i_0 + 3N/(8\alpha)]$, which has not been modified by the rewiring procedure. We now decompose the mean length between two vertices of the sample,

$$\ell = \frac{1}{N(N-1)} \sum_{i \neq j} d_{ij},$$

into two contributions: the first one comes from the pairs (i,j) with $i \in I$, $j \in I$, the second one includes all pairs (i,j) where at least one of the vertices is not an element of I. The first contribution can be estimated by formula (5), since it comes from a part of the graph which has not been modified, and at a distance big enough from any modified link:

$$\sum_{i \in I,\ j \in I} d_{ij} \geq (bN)(bN-1)\frac{bN}{4k}$$

(the inequality comes from the fact that we do not have periodic boundary conditions for this interval). We now have access to a lower bound of $\ell(N, \alpha/N)$ (which is valid for any sample, and consequently also for the average over samples):

$$\ell(N, \alpha/N) \geq \frac{1}{N(N-1)} \sum_{i \in I,\ j \in I} d_{ij} \geq \frac{b^3}{4k} N.$$

Since $\ell(N, \alpha/N)$ is smaller than $\ell(N,0) \sim N/(4k)$, this shows that

$$\ell(N, \alpha/N) = O(N). \qquad (7)$$

In other words, a *finite* number of rewired links *cannot* change the scaling at large N: $\ell(N, \alpha/N)$ is of order N for any finite α.

To complete this argument, we have computed numerically $p_{1/2}(N)$, *i.e.* the value of p such that $\ell(N, p_{1/2}(N)) = \ell(N,0)/2$. Figure 5 shows quite clearly that, for large N, $p_{1/2}(N) \sim 1/N$: a finite number of rewired links is able to divide the mean length between vertices by two[3].

Let us now go back to the scaling hypothesis of [2]. If the scaling form of equation (6) is valid, we have to compute $\ell(N,p)$ at fixed p in order to estimate $N^*(p)$. For large N, it behaves like $N^*(p)\ln(N)$ (see Fig. 6 for different values of p). For small p, $N^*(p)$ becomes bigger and bigger, so that we have to use larger and larger values of N. We show in Figure 7 that the $N^*(p)$ estimated in

[3] As shown in [3], $N^* \sim p^{-\tau}$ implies $p_{1/2}(N) \sim N^{-1/\tau}$; we thus have a clear indication that $\tau = 1$.

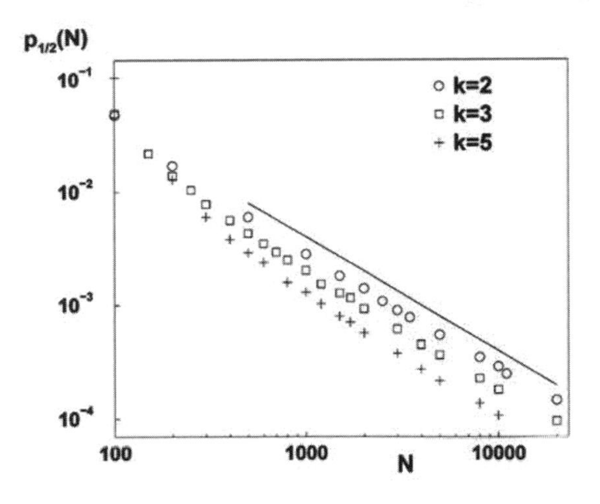

Fig. 5. $p_{1/2}(N)$ such that $\ell(N, p_{1/2}(N)) = \ell(N,0)/2$, *versus* N, for $k = 2,\ 3,\ 5$. The straight line is proportional to $1/N$.

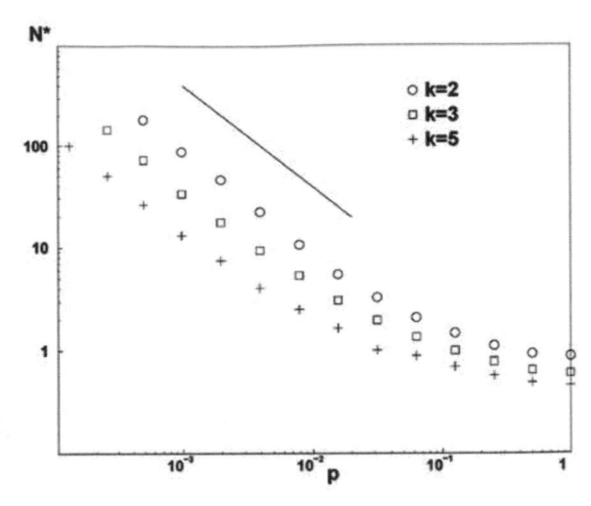

Fig. 7. $N^*(p)$ *versus* p for $k = 2,\ 3,\ 5$. The straight line is proportional to $1/p$.

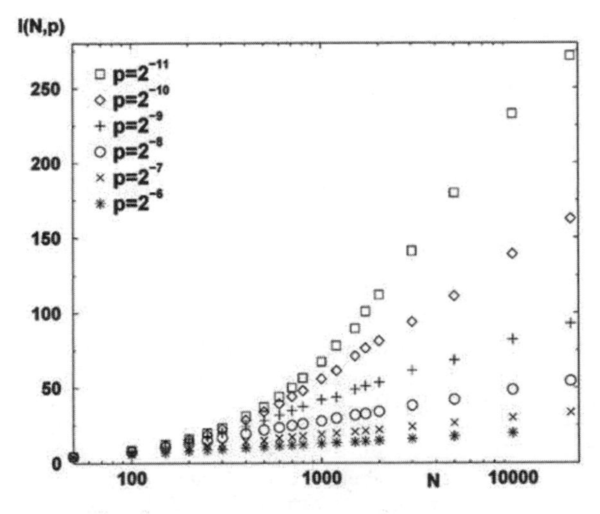

Fig. 6. $\ell(N,p)$ *versus* N, for $p = 2^{-a}$, $a = 6, \cdots, 11$, and $k = 3$. For large values of p we have a straight line in the semi-log plot, while for small values of p we observe the crossover between $\ell(N,p) \sim N$ and $\ell(N,p) \sim \ln(N)$. The value of $N^*(p)$ is given by the slope of the linear part in the semi-log plot.

this way behaves like $1/p$ for small p (and for $p \to 1$, $N^*(p) \to 1/\ln(2k-1)$, in accordance with $\ell(N,1) \sim \ln(N)/\ln(2k-1)$), giving $\tau = 1$. This is not very surprising if we consider the above discussion showing that a finite number of rewired links will change the coefficient of the scaling of ℓ with N but not the scaling itself. Moreover, $p_{1/2}(N)$ corresponds to the drop in the curves of Figure 4 and can therefore be considered as a crossover value.

Using the determined values of N^*, we plot in Figure 8 $\ell(N,p)/N^*(p)$ *versus* $N/N^*(p)$ for various values of N and p. We observe a nice collapse of the data for each value of k. Thanks to the range of values of N that we use, we are able to show the collapse over a much wider range of values than [2]. We clearly see the linear behaviour $F_k(x \ll 1) \sim x/(4k)$, and the crossover to $F_k(x \gg 1) \sim \ln(x)$. Note that, as explained in [5], we have to use values of p lower than $1/k^2$ (and of course large enough values of N, *i.e.* $N \gg k$) to obtain a clean scaling behaviour: for too large p, we are moving out of the scaling regime close to the $p = 0$-transition.

3.3 Clustering coefficient

To define the "small-world" behaviour, two ingredients are used by Watts and Strogatz [1]. The first one is the chemical length studied in the previous paragraph, which depends strongly on p and N. The second one is more local: the "clustering coefficient" $\mathcal{C}(p)$ quantifies its "cliquishness". $\mathcal{C}(p)$ is indeed defined as follows: if c_i is the number of neighbours of a vertex i, there are *a priori* $c_i(c_i-1)/2$ possible links between these neighbours. Denoting C_i the fraction of these links that are really present in the graph, $\mathcal{C}(p)$ is the average of C_i over all vertices. On a linear-log plot, $\mathcal{C}(p)/\mathcal{C}(0)$ is close to 1 for a wide range of values of p, and its drop occurs around $p \approx 0.1$. This is therefore in contrast with $\ell(N,p)$, whose drop occurs for much smaller values of p as soon as N is large enough. It is therefore an interesting question whether there is an upper threshold on p for the small-world behaviour.

We now show that a simple redefinition of $\mathcal{C}(p)$ leads to a very simple formula, without altering its physical signification, nor the shape of the curve. For $p = 0$, each vertex has $2k$ neighbours; it is easy to see that the number of links between these neighbours is $\mathcal{N}_0 = 3k(k-1)/2$. Then $\mathcal{C}(0) = \frac{3(k-1)}{2(2k-1)}$. For $p > 0$, two neighbours of i that were connected at $p = 0$ are still neighbours of i and linked together with probability $(1-p)^3$, up to terms of order $\frac{1}{N}$. The mean number of links between the neighbours of a vertex is then clearly $\mathcal{N}_0(1-p)^3 + \mathcal{O}(\frac{1}{N})$. The clustering coefficient $\mathcal{C}(p)$ is defined as the mean of the ratio $C_i = \frac{\mathcal{N}_i}{c_i(c_i-1)/2}$. If instead we define $\tilde{\mathcal{C}}(p)$ as the ratio of

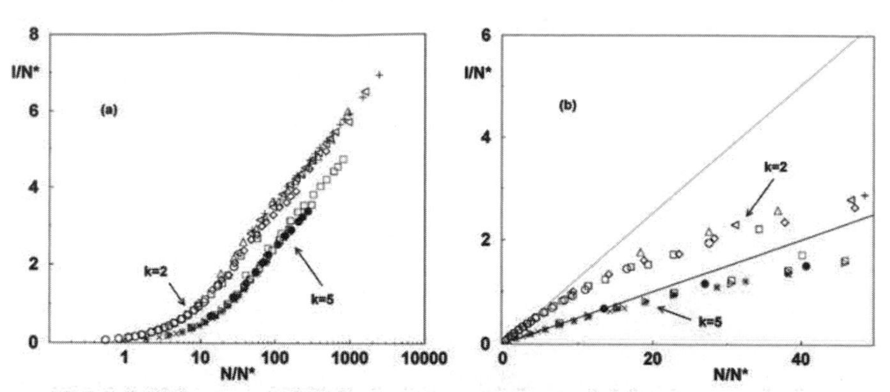

Fig. 8. Data collapse $\ell(N,p)/N^*(p)$ *versus* $N/N^*(p)$, for $k = 2$ and $k = 5$; (a): log-linear scale showing at large N/N^* the logarithmic behaviour; (b) linear scale showing at small N/N^* the linear behaviour $\ell(N,p) \sim N/(4k)$: the straight lines have slopes 1/8 and 1/20.

the mean number of links between the neighbours of a vertex and the mean number of possible links between the neighbours of a vertex, we obtain

$$\tilde{\mathcal{C}}(p) = \frac{3(k-1)}{2(2k-1)}(1-p)^3. \tag{8}$$

We check numerically, with $N = 50$ to $N = 8000$, and averaging over 5000 samples, that the two definitions lead to the same behaviour (we see in Fig. 9 that the difference between $\mathcal{C}(p)$ and $\tilde{\mathcal{C}}(p)$ is very small), and that the corrections to equation (8) are indeed of order $1/N$. The behaviour of $\mathcal{C}(p)$ is therefore very simply described by $\mathcal{C}(p) \approx \mathcal{C}(0)(1-p)^3$, and the dependence on N is very small.

To summarize this section, we have shown that the small-world behaviour – as defined by the average chemical distance and the clustering coefficient – is indeed present for any finite value of $0 < p < 1$ as soon as the network is large enough.

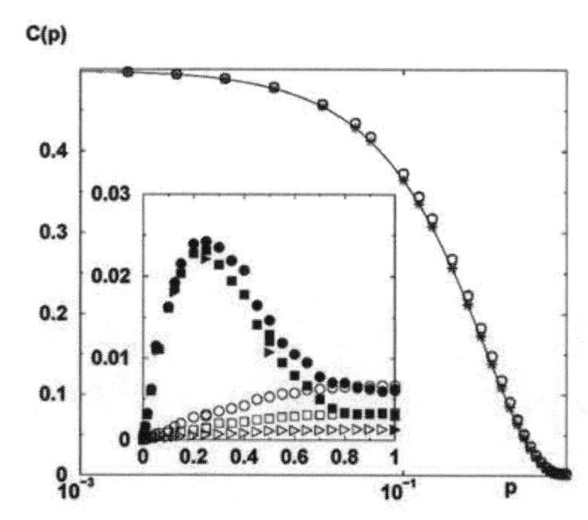

Fig. 9. $\mathcal{C}(p)$ and $\tilde{\mathcal{C}}(p)$ *versus* p, for $k = 2$ ($\mathcal{C}(0) = \tilde{\mathcal{C}}(0) = 0.5$), $N = 1000$, 2000, 5000: open symbols are for $\mathcal{C}(p)$, and the crosses are for $\tilde{\mathcal{C}}(p)$; the line is $\mathcal{C}(0)(1-p)^3$. Inset: corrections $\mathcal{C}(p) - \mathcal{C}(0)(1-p)^3$ (filled symbols) for $N = 1000$ (circles), $N = 2000$ (squares) and $N = 5000$ (triangles), and $\tilde{\mathcal{C}}(p) - \mathcal{C}(0)(1-p)^3$ (open symbols) for $N = 1000$ (circles), $N = 2000$ (squares) and $N = 5000$ (triangles). We see that the corrections go to zero as $1/N$ for $\tilde{\mathcal{C}}(p)$; the corrections for $\mathcal{C}(p)$ are larger, but anyway very small.

4 Ising model

In this section we want to investigate the consequences of the mixed geometrical structure of small-world networks on an Ising model as a prototype of statistical-mechanics models that can be defined on it. This model can be understood as a continuous interpolation of a pure one-dimensional model for $p = 0$ showing no phase transition at finite temperature to a model on a random graph[4] for $p = 1$ having a finite critical temperature $T_c(p = 1) > 0$ as long as $k \geq 2$, *cf.* [13]. In agreement with the results from Section 3, we find for every finite $p > 0$ that the low temperature behaviour of the model is of mean-field character, even if we observe a finite temperature crossover to a dominance of the one-dimensional structure. This observation confirms the value $p_c = 0$ for the onset of a non-trivial thermodynamical small-world behaviour as already found in the geometrical properties, and it shows again the crucial importance of the mixed geometrical structure, as even global quantities can be dominated by the initial ordered structure for high temperatures.

[4] As already mentioned in the introduction, every point in this model has a minimal connectivity k. So, even in the case $p = 1$, the model is not equivalent to the usual random graph where both endpoints of a link are chosen randomly.

4.1 General formalism

The system we want to study is given by its Hamiltonian

$$H(\{S_i\}) = -\sum_{i=1}^{N} S_i \sum_{j=1}^{k} S_{m(i,j)} \tag{9}$$

with N Ising spins $S_i = \pm 1$, $i = 1, ..., N$, and periodic boundary conditions, *i.e.* we identify $S_{N+1} = S_1$ etc.

in the following. The independently and identically distributed numbers $m(i,j)$ are drawn from the probability distribution

$$P(m(i,j)) = (1-p)\delta_{m(i,j),i+j} + \frac{p}{N}\sum_{l=1}^{N}\delta_{m(i,j),l}, \quad (10)$$

i.e. for $p=0$ we obtain a pure one-dimensional Ising model where every site is connected to its $2k$ nearest neighbours by ferromagnetic bonds of strength 1, whereas this structure is completely replaced by random long-range bonds for $p=1$. The number of bonds in the model is given by kN, independently of the disorder strength p. Here we consider only the case of finite probabilities $p=O(1)$, *i.e.* an extensive number of links is rewired and, according to the last section, we are therefore in the small-world regime.

In order to decide whether there exists a ferromagnetic phase transition at finite temperature or not, we have to calculate the free-energy density at inverse temperature β. Due to the existence of an extensive number as well of random as of one-dimensional links and due to the translational invariance of the distribution (10) we expect this quantity to be self-averaging, we therefore have to determine

$$\begin{aligned} f &= -\lim_{N\to\infty}\frac{1}{\beta N}\overline{\ln Z} \\ &= -\lim_{N\to\infty}\frac{1}{\beta N}\overline{\ln\sum_{\{S_i\}}\mathrm{e}^{-\beta H(\{S_i\})}}\,. \end{aligned} \quad (11)$$

The average $\overline{(\cdot)}$ over the disorder distribution $P(m(i,j))$ is achieved with the help of the replica trick

$$E12\overline{\ln Z} = \lim_{n\to 0}\partial_n\overline{Z^n} \quad (12)$$

by introducing at first a positive integer number n of replicas of the original system, averaging over the disorder and sending $n\to 0$ at the end of the calculations. Thus the replicated and disorder averaged partition function can be written as

$$\begin{aligned} \overline{Z^n} &= \overline{\sum_{\{\mathbf{S}_i\}}\exp\left\{-\beta\sum_{a=1}^{n}H(\{S_i^a\})\right\}} \\ &= \sum_{\{\mathbf{S}_i\}}\prod_{i=1}^{N}\prod_{j=1}^{k}\left((1-p)\mathrm{e}^{\beta\mathbf{S}_i\cdot\mathbf{S}_{i+j}} + \frac{p}{N}\sum_{l=1}^{N}\mathrm{e}^{\beta\mathbf{S}_i\cdot\mathbf{S}_l}\right) \end{aligned} \quad (13)$$

where we introduced the replicated Ising spins $\mathbf{S}_i = (S_i^1,...,S_i^n)$. This expression can be simplified by defining the 2^n order parameters [14]

$$c(\mathbf{S}) := \frac{1}{N}\sum_{i=1}^{N}\delta_{\mathbf{S}_i,\mathbf{S}} \quad (14)$$

giving the fraction of n-tuples in $\{\mathbf{S}_i\}$ which are equal to $\mathbf{S}\in\{-1,+1\}^n$, and their conjugates $\hat{c}(\mathbf{S})$. These order parameters have to be normalized, $\sum_{\mathbf{S}}c(\mathbf{S})=1$. After a change $\hat{c}\to i\hat{c}$ leading to real order parameters, we arrive at

$$\begin{aligned} \overline{Z^n} &= \int\prod_{\mathbf{S}}\mathrm{d}c(\mathbf{S})\;\mathrm{d}\hat{c}(\mathbf{S}) \\ &\quad\times\exp\left\{N\left(-\sum_{\mathbf{S}}c(\mathbf{S})\hat{c}(\mathbf{S}) + \frac{1}{N}\ln\operatorname{tr}\mathbf{T}^{\frac{N}{k}}\right)\right\} \\ &= \int\prod_{\mathbf{S}}\mathrm{d}c(\mathbf{S})\mathrm{d}\hat{c}(\mathbf{S})\exp\{Nf_n[c,\hat{c}]\} \end{aligned} \quad (15)$$

with an effective $2^{kn}\times 2^{kn}$-transfer matrix $\mathbf{T}$ given by its entries

$$\begin{aligned} \mathbf{T}(\mathbf{S}_1,...,\mathbf{S}_k|\mathbf{S}_{k+1},...,\mathbf{S}_{2k}) &= \prod_{i=1}^{k}\mathrm{e}^{\hat{c}(\mathbf{S}_i)} \\ \times\prod_{j=1}^{k}\left((1-p)\mathrm{e}^{\beta\mathbf{S}_i\cdot\mathbf{S}_{i+j}} + p\sum_{\mathbf{S}}c(\mathbf{S})\mathrm{e}^{\beta\mathbf{S}_i\cdot\mathbf{S}}\right). \end{aligned} \quad (16)$$

At this point we remark that the small-world Ising model offers an interesting interplay between technical concepts of mean-field theory, as represented by the global order parameters, and the theory of one-dimensional systems, here represented by the effective transfer matrix. As in the conventional transfer matrix method, the contribution of the second term in f_n can be determined by the largest eigenvalue of $\mathbf{T}$ with right (left) eigenvector $|\lambda_r\rangle$ ($\langle\lambda_l|$),

$$f_n[c,\hat{c}] = -\sum_{\mathbf{S}}c(\mathbf{S})\hat{c}(\mathbf{S}) + \ln\frac{\langle\lambda_l|\mathbf{T}|\lambda_r\rangle}{\langle\lambda_l|\lambda_r\rangle}, \quad (17)$$

but in order to calculate the integrals over the order parameters in (15) we have to use the saddle point method which implies

$$c(\mathbf{S}) = \sum_{\mathbf{S}_1,...,\mathbf{S}_{k-1}}\frac{\langle\lambda_l|\mathbf{S},\mathbf{S}_1,...,\mathbf{S}_{k-1}\rangle\langle\mathbf{S},\mathbf{S}_1,...,\mathbf{S}_{k-1}|\lambda_r\rangle}{\langle\lambda_l|\lambda_r\rangle}, \quad (18)$$

i.e. the explicit form of the transfer matrix itself depends on the eigenvectors, and the linear structure of the eigenvalue equations is destroyed.

4.2 High-temperature solution

The problem simplifies significantly in its high-temperature phase where the correct solution of the saddle point equations

$$\begin{aligned} c(\mathbf{S}) &= \frac{1}{N}\frac{\partial}{\partial\hat{c}(\mathbf{S})}\ln\operatorname{tr}\mathbf{T}^{\frac{N}{k}} \\ \hat{c}(\mathbf{S}) &= \frac{1}{N}\frac{\partial}{\partial c(\mathbf{S})}\ln\operatorname{tr}\mathbf{T}^{\frac{N}{k}} \end{aligned} \quad (19)$$

can be found without knowing the above-mentioned eigenvectors and is given by the paramagnetic values

$c_{\rm pm}(\mathbf{S}) = 1/2^n$ and $\hat{c}_{\rm pm}(\mathbf{S}) = kpa^n$. a does not depend on $\mathbf{S}$, so it can be taken out of $\ln \mathrm{tr} \mathbf{T}^{\frac{N}{k}}$ and cancels finally with $-\sum c\hat{c}$ in (15). In this phase all replicated spins $\mathbf{S}$ have the same density, and thus the average magnetization $m = \lim_{n\to 0} \sum_{\mathbf{S}} S^1 c(\mathbf{S})$ as well as the overlaps $q^{ab} = \lim_{n\to 0} \sum_{\mathbf{S}} S^a S^b c(\mathbf{S})$ vanish.

Even if this solution exists for all temperatures, it is not stable for low temperatures. The critical temperature can be determined by investigating the 2^{n+1}-dimensional fluctuation matrix

$$\begin{pmatrix} \dfrac{\partial^2 f_n[c,\hat{c}]}{\partial c\, \partial c} & \dfrac{\partial^2 f_n[c,\hat{c}]}{\partial c\, \partial \hat{c}} \\ \\ \dfrac{\partial^2 f_n[c,\hat{c}]}{\partial \hat{c}\, \partial c} & \dfrac{\partial^2 f_n[c,\hat{c}]}{\partial \hat{c}\, \partial \hat{c}} \end{pmatrix}. \tag{20}$$

The paramagnetic solution is valid as long as none of the eigenvalues of this matrix changes sign[5]. The phase transition therefore appears at the point where the first eigenvalue becomes zero and the system becomes unstable with respect to Gaussian fluctuations around the given saddle point.

4.3 Crossover from one-dimensional to mean-field behaviour

The problem in calculating these eigenvalues consists in the fact that the transfer matrix $\mathbf{T}$ is given by a sum over non-commuting matrices. So it is not clear how to obtain the eigenvectors of $\mathbf{T}$ even at the paramagnetic saddle point where the problem can be linearized again because we already know c and $\hat{c}$ and the form of the transfer matrix is fixed.

At this moment we therefore restrict to the most interesting case of small $p \ll 1$ and treat the problem by means of a first order perturbation theory in p around the pure one-dimensional model. In this case we are in principle able to calculate all the (k-dependent) eigenvectors, which are simple direct products of n eigenvectors of the pure and unreplicated transfer matrix, and hence the perturbation-theoretic corrections to their eigenvalues. The linearized transfer matrix reads

$$\mathbf{T}_{\rm lin}(\mathbf{S}_1, ..., \mathbf{S}_k | \mathbf{S}_{k+1}, ..., \mathbf{S}_{2k}) = \\ \exp\left\{ \sum_{i+1}^{k} \hat{c}(\mathbf{S}_i) + \beta \sum_{i,j=1}^{k} \mathbf{S}_i \mathbf{S}_{i+j} \right\} \\ \times \left[1 - k^2 p + p \sum_{\mathbf{S}} c(\mathbf{S}) \sum_{p,q=1}^{k} \exp\left\{ \beta \mathbf{S}_p (\mathbf{S} - \mathbf{S}_{p+q}) \right\} \right]. \tag{21}$$

As we show in some detail in Appendix A from the analysis of the entries of the fluctuation matrix (20), this perturbation expansion contains powers of a term proportional to $p\xi_0$ with ξ_0 being the correlation length of the pure system, and its first order approximation consequently breaks down when $p\xi_0$ becomes larger than $O(1)$ for increasing disorder p or decreasing temperature T. In the pure model the correlation length diverges for low temperatures as

$$\xi_0 \propto e^{k(k+1)\beta}. \tag{22}$$

Consequently, at fixed but low temperature T, we find a crossover from a weakly perturbated one-dimensional behaviour for disorder strengths $p \ll p_{\rm co}(T)$ with

$$p_{\rm co}(T) \propto \exp\left\{ \frac{-k(k+1)}{T} \right\} \tag{23}$$

to a disorder-dominated and hence mean-field like regime for larger p. This can be understood by a simple physical argument. We consider a cluster of correlated spins in the pure model which has a typical length scale $l \approx \xi_0$. Thus the number of links in this cluster is also $O(\xi_0)$ for finite k, and the average number of redirected links in this cluster at disorder strength p is approximately $p\xi_0$. For $p \ll p_{\rm co}(T)$ there are on average consequently less than one redirected link per cluster, and the system is not seriously perturbated by the disorder. The opposite holds for larger p.

This shows that an arbitrarily small, but finite fraction p of redirected links ("short cuts" in the graph) leads at sufficiently small temperature $T < T_{\rm co}(p)$,

$$T_{\rm co}(p) \propto -\frac{k(k+1)}{\log(p)}, \quad p \ll 1, \tag{24}$$

to a change of the behaviour of the model from a one-dimensional to a mean-field one, which nicely underlines the importance of both geometrical structures in the small-world lattice.

4.4 The ferromagnetic phase transition

In the low-temperature regime $T \ll T_{\rm co}(p)$ the thermodynamic behaviour is dominated by the mean-field type disorder, and we expect a finite temperature transition to a ferromagnetically ordered phase at finite temperature $T_{\rm c}(p)$ at least for sufficiently large p and $k \geq 2$. Due to the above-mentioned technical problems in diagonalizing the transfer matrix we cannot calculate this transition analytically, and we compute therefore the full line $T_{\rm c}(p)$ for $k = 2$ and $k = 3$ by means of numerical simulations. We use a cluster algorithm [15] to compute the equilibrium distribution of the magnetization, for system sizes ranging from $N = 500$ to $N = 8000$, and use Binder cumulants [16] to determine the critical point (see the inset of Fig. 10 for an example).

The important result is that we obtain a transition at a non-zero temperature for all the investigated values of p. Moreover, for small p we have, as shown in Figure 10:

$$T_{\rm c}(p) \propto -\frac{2k}{\log(p)}. \tag{25}$$

[5] Due to the common change $\hat{c} \to i\hat{c}$ one half of the eigenvalues has to be negative, the other half positive in order to insure a stable saddle point.

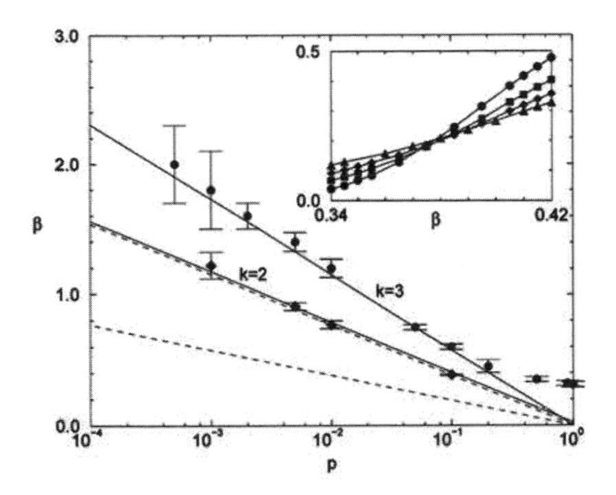

Fig. 10. Inverse critical temperature $\beta_c(p)$ for $k = 2$ (circles) and $k = 3$ (diamonds). The full lines show the asymptotic scaling (25) of $\beta_c(p \ll 1)$. The scaling (24) of the crossover between one-dimensional and mean-field behaviour is given by the dashed lines – and consistently found to be at higher temperatures as the ferromagnetic phase transition. The inset shows the β-dependence of the Binder cumulant used to determine the critical temperature for $p = 0.1$, $k = 3$ and $N = 500, 1000, 2000, 5000$ (triangles, diamonds, squares, circles).

This transition line is found to be always at smaller temperatures than the crossover temperatures, which illustrates again the mean-field character of the phase transition.

Even if the behaviour of the system is dominated by the random part of its Hamiltonian, the underlying one-dimensional structure is crucial for the existence of the phase transition and for the explicit value of the transition temperature. This becomes clear from the fact that only the existence of the short-range links leads to the existence of a macroscopic cluster for p below the percolation threshold of the random bonds, and can be supported analytically by investigating a version of the model where all one-dimensional bonds are deleted and only the random bonds for fixed p are conserved. This model shows a ferromagnetic transition only above $p_c(k) = 1 - \sqrt{(k-1)/k}$. So, even if the phase transition is induced by the presence of long-range interactions, it is based on an interplay between both structures.

4.5 A simplified model

In this subsection we present a slightly modified model where the full procedure introduced in Section 4.1 can be followed analytically, and the phase diagram can be calculated explicitly. The model has the same Hamiltonian (9), but its disorder distribution is given by

$$\tilde{P}(m(i,j)) = (1-p)\delta_{m(i,j),i+1} + \frac{p}{N}\sum_{l=1}^{N}\delta_{m(i,j),l}. \qquad (26)$$

So the underlying one-dimensional graph is changed: instead of having bonds to the next $2k$ neighbours it includes k bonds to each of the two next nearest neighbours (which, in the pure case, is equivalent to one bond of strength k). In the disordered version every of these bonds is replaced with probability p by a random bond, so the random structure of the model remains unchanged compared to the original model. Anyway, this model remains a "valid" small-world network as it consists of a mixture of a regular low-dimensional with a random long-ranged lattice. This can *e.g.* be confirmed by the fact that our simplified model also shows the scaling behaviour (6) with the same scaling exponent $\tau = 1$ as the latter depends only on the dimensionality of the regular structure, *cf.* [5]. Because of the geometrical similarity of the underlying networks we expect also a qualitatively similar thermodynamic behaviour.

Again we average the replicated partition function over the disorder and introduce the order parameters $c(\mathbf{S})$ and $\hat{c}(\mathbf{S})$. By doing this we arrive again at

$$\overline{Z^n} = \int \prod_{\mathbf{S}} \mathrm{d}c(\mathbf{S})\, \mathrm{d}\hat{c}(\mathbf{S}) \exp\{N f_n[c,\hat{c}]\} \qquad (27)$$

with a slightly changed f_n,

$$f_n[c,\hat{c}] = -\sum_{\mathbf{S}} c(\mathbf{S})\hat{c}(\mathbf{S}) + \frac{1}{N}\ln \mathrm{tr}\mathbf{T}^N \qquad (28)$$

where the effective transfer matrix is of dimension 2^n and reads

$$\mathbf{T}(\mathbf{S}_1|\mathbf{S}_2) = \mathrm{e}^{\hat{c}(\mathbf{S}_1)}\Big[(1-p)\exp\{\beta\mathbf{S}_1\cdot\mathbf{S}_2\} + p\sum_{\mathbf{S}} c(\mathbf{S})\exp\{\beta\mathbf{S}_1\cdot\mathbf{S}\}\Big]^k. \qquad (29)$$

Also in this case, the simple paramagnetic saddle point for c and $\hat{c}$ is given by $c_{\mathrm{pm}}(\mathbf{S}) = 1/2^n$, $\hat{c}_{\mathrm{pm}}(\mathbf{S}) = kpa^n$ with a β-dependent a canceling in (28), which therefore becomes

$$f_n[c_{\mathrm{pm}},\hat{c}_{\mathrm{pm}}] = \frac{1}{N}\ln \mathrm{tr}\mathbf{T}_{\mathrm{pm}}^N \qquad (30)$$

with

$$\mathbf{T}_{\mathrm{pm}}(\mathbf{S}_1|\mathbf{S}_2) = [(1-p)\exp\{\beta\mathbf{S}_1\cdot\mathbf{S}_2\} + p(\cosh\beta)^n]^k. \qquad (31)$$

This matrix can be easily diagonalized by introducing the two-dimensional orthonormalized vectors $|+\rangle = 1/\sqrt{2}\,(1,1)$ and $|-\rangle = 1/\sqrt{2}\,(1,-1)$. The eigenvectors of $\mathbf{T}_{\mathrm{pm}}$ are $|\mu\rangle = |\mu^1\rangle \otimes \cdots \otimes |\mu^n\rangle$ with $\mu^a = +,-$ for all

$$\Lambda_{cc} = k(k-1)p^2(\tanh\beta)^2 + \frac{2k^2p^2 \sum_{m=0}^{k-1} \binom{k-1}{m} p^m (1-p)^{k-m-1} \left[\tanh(k-m-1)\beta\,(\tanh\beta)^2\right]}{1 - \sum_{m=0}^{k} \binom{k}{m} p^m (1-p)^{k-m} \tanh(k-m)\beta}$$

$$\Lambda_{c\hat{c}} = -1 + \frac{kp\tanh\beta \left[1 + \sum_{m=0}^{k-1} \binom{k-1}{m} p^m (1-p)^{k-m-1} \tanh(k-m-1)\beta\right]}{1 - \sum_{m=0}^{k} \binom{k}{m} p^m (1-p)^{k-m} \tanh(k-m)\beta}$$

$$\Lambda_{\hat{c}\hat{c}} = 1 + 2\frac{\sum_{m=0}^{k} \binom{k}{m} p^m (1-p)^{k-m} \tanh(k-m)\beta}{1 - \sum_{m=0}^{k} \binom{k}{m} p^m (1-p)^{k-m} \tanh(k-m)\beta}. \tag{36}$$

$a = 1, ..., n$. With $\rho(\mu)$ being the number of factors $|+\rangle$ in $|\mu\rangle$, the eigenvalues are found to be

$$\lambda[\mu] = \lambda(\rho(\mu)) = \sum_{j=0}^{k} \binom{k}{j} (p\cosh\beta^n)^j (1-p)^{k-j} \times (2\cosh(k-j)\beta)^{\rho(\mu)} (2\sinh(k-j)\beta)^{n-\rho(\mu)}. \tag{32}$$

The behaviour of f_n in the thermodynamic limit $N \to \infty$ is completely determined by the largest eigenvalue $\lambda(n) = \lambda[+...+]$, and the paramagnetic free energy of the model reads

$$-\beta f_{\rm pm} = \lim_{n\to 0} \partial_n f_n[c_{\rm pm}, \hat{c}_{\rm pm}] = \sum_{j=0}^{k} \binom{k}{j} p^j (1-p)^{k-j} \times (j \ln\cosh\beta + \ln 2\cosh(k-j)\beta). \tag{33}$$

The second eigenvalue $\lambda(n-1) = \lambda[-+...+]$ of the transfer matrix $\mathbf{T}_{\rm pm}$ describes in the replica limit $n \to 0$ the decay of the two-point correlation function $\overline{\langle S_i S_j \rangle} \propto \lambda(n-1)^{|i-j|}$ for distances $1 \ll |i-j| \ll \ell(N,p)$, *cf.* Section 3.2, *i.e.* for points i and j whose chemical distance is given with finite probability by the one-dimensional distance $|i-j|$ and does not include random bonds. The corresponding correlation length reads

$$\xi_p = -\lim_{n\to 0} \frac{1}{\ln\lambda(n-1)} = \frac{-1}{\ln\left(\sum_{j=0}^{k} \binom{k}{j} p^j (1-p)^{k-j} \tanh(k-j)\beta\right)} \tag{34}$$

and remains finite for every non-zero temperature. So, in complete agreement with our findings for the original model in the last subsections, we can conclude that the modified model has no ferromagnetic phase transition caused by a divergence of the one-dimensional correlation length. There is nevertheless a transition due to the fact that the paramagnetic saddle point $c_{\rm pm}(\mathbf{S})$ and $\hat{c}_{\rm pm}(\mathbf{S})$ becomes unstable at a certain temperature. In order to see this we investigate again the fluctuation matrix (20) for the present model. The four blocks can be calculated (see Appendix B for details), and diagonalized simultaneously. The fluctuation mode becoming at first unstable leads to the reduced matrix

$$\begin{pmatrix} \Lambda_{cc} & \Lambda_{c\hat{c}} \\ \Lambda_{c\hat{c}} & \Lambda_{\hat{c}\hat{c}} \end{pmatrix} \tag{35}$$

with entries

see equation (36) above.

The vanishing of its determinant gives the critical temperature $T_c(p)$ which depends on p. The determinant is negative for $p = 0$ at all positive temperatures, where the paramagnetic solution is known to be correct, and positive at $T = 0$ for all $p > 0$, we thus conclude that $T_c(p > 0) > 0$. The explicit value can be calculated numerically from (35) and is shown in figure (11). The critical temperature for small disorder p behaves like

$$T_c(p) \approx -\frac{2k}{\log(2kp)}, \tag{37}$$

it consequently shows the same asymptotic p-dependence as in the original model, *cf.* (25). In addition it shows in this case the same p-dependence as the crossover temperature found from $2kp\xi_0 \propto 1$ with $\xi_0 = -1/\ln(\tanh k\beta) \propto \exp\{2k\beta\}$ for $\beta \gg 1$.

5 Summary and conclusion

In conclusion, in the first part of this work we have studied the geometrical properties of small-world networks which interpolate continuously between a one-dimensional ring

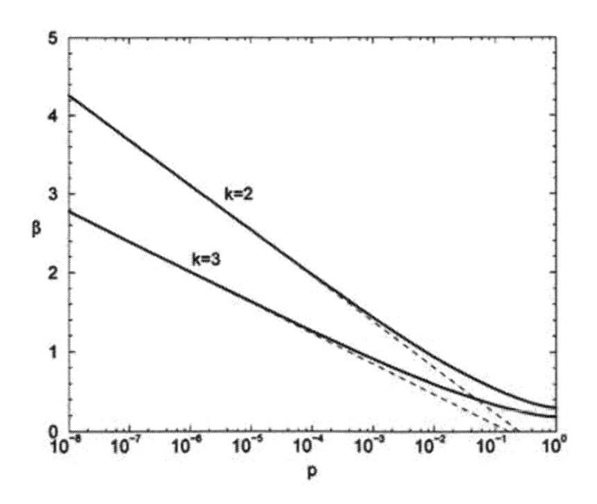

Fig. 11. The inverse phase transition temperature $\beta_c(p)$ in the simplified model for $k = 2$ and $k = 3$. The dashed lines show the asymptotic behaviour given in (37).

and a certain random graph. The coexistence of a more and more diluted local structure and of random long-ranged links leads to some very interesting features:

- Due to the local structure two neighbouring vertices have in general common neighbours, a fact which leads to a certain cliquishness. The clustering coefficient, measuring this property, was found to decrease like $(1-p)^3$ with the fraction p of randomly rewired links.

- The average length between two points characterizing global properties of the network was found to depend strongly on the amount of disorder in the network. A crossover, first proposed in [2], could be worked out: At fixed p, the average length between two vertices was found to grow linearly with the system size $N \ll O(1/p)$ for small networks, whereas it grows only logarithmically for large networks $N \gg O(1/p)$.

Therefore, the mere notion of "small-world" graph, *i.e.* the region of disorder where the local properties are still similar to those of the one-dimensional ring whereas the global properties are determined by the random short-cuts in the graph, depends on its size, and can be extended to smaller and smaller p, taking larger and larger N.

In the second part these findings where corroborated by the investigation of an Ising model defined on the small-world network. In the thermodynamic limit we found the following behaviour for fixed disorder strength p: for large temperatures, the system behaves very similarly to the pure one-dimensional system, whereas it undergoes a crossover to a mean-field like region for smaller temperatures. Finally, at low but non-zero temperature, we find a ferromagnetic phase transition. This underlines again the results of the geometrical investigations that the graph is in its small-world regime for any disorder strength at sufficiently large system sizes, *i.e.* in a region where both geometrical structures lead to interesting physical effects.

We are very grateful to G. Biroli, R. Monasson and R. Zecchina for numerous fruitful discussions. M.W. acknowledges financial support by the German Academic Exchange Service (DAAD).

Appendix A: Breakdown of the first order perturbation theory

In this appendix we want to present the first-order perturbation calculations for small disorder strengths $p \ll 1$ leading finally to the crossover phenomenon described in Section 4.3. We start from the linearized transfer matrix

$$\mathbf{T}_{\text{lin}}(\mathbf{S}_1,...,\mathbf{S}_k|\mathbf{S}_{k+1},...,\mathbf{S}_{2k}) = \exp\left\{\sum_{i+1}^{k}\hat{c}(\mathbf{S}_i) + \beta\sum_{i,j=1}^{k}\mathbf{S}_i\mathbf{S}_{i+j}\right\} \times\left[1-k^2p+p\sum_{\mathbf{S}}c(\mathbf{S})\sum_{p,q=1}^{k}\exp\left\{\beta\mathbf{S}_p(\mathbf{S}-\mathbf{S}_{p+q})\right\}\right]. \tag{A.1}$$

and calculate the elements of the fluctuation matrix (20) around the paramagnetic saddle point up to first order in p. In order to achieve this we use the 2^k (bi-)orthonormalized eigenvectors $|\lambda_\alpha\rangle$, $(\langle\lambda_\alpha|)$ $\alpha = 1,...,2^k$, of the pure and unreplicated transfer matrix

$$\mathbf{T}^{(0)}(S_1,...,S_k|S_{k+1},...,S_{2k}) = \exp\left\{\beta\sum_{i,j=1}^{k}S_iS_{i+j}\right\}. \tag{A.2}$$

We choose these eigenvectors to be ordered according to their eigenvalues. The eigenvectors of the replicated pure system are therefore given by $|\alpha\rangle = |\lambda_{\alpha^1}\rangle\otimes\cdots\otimes|\lambda_{\alpha^n}\rangle$, and the corrections of $O(p)$ can be calculated by using these vectors.

At first we realize that the second derivative of

$$f_n[c,\hat{c}] = -\sum_{\mathbf{S}}c(\mathbf{S})\hat{c}(\mathbf{S}) + \frac{1}{N}\ln\operatorname{tr}\mathbf{T}_{\text{lin}}^{\frac{N}{k}} \tag{A.3}$$

with respect to c is already of order p^2 and can therefore be neglected. The interesting entries of the fluctuation

$$\frac{\partial f_n}{\partial \hat{c}(\mathbf{S})} = -c(\mathbf{S}) + \frac{\sum_{\mathbf{S}_1,...,\mathbf{S}_{k-1}} \mathbf{T}_{\text{lin}}^{\frac{N}{k}}(\mathbf{S},\mathbf{S}_1,...,\mathbf{S}_{k-1}|\mathbf{S},\mathbf{S}_1,...,\mathbf{S}_{k-1})}{\text{tr } \mathbf{T}_{\text{lin}}^{\frac{N}{k}}}$$

$$\frac{\partial^2 f_n}{\partial \hat{c}(\mathbf{S})\partial c(\mathbf{R})} = -\delta_{\mathbf{S},\mathbf{R}} - \frac{N}{k} c_{\text{pm}}\hat{c}_{\text{pm}} + \sum_{j=0}^{N-1} \frac{\sum_{\mathbf{S}_1,...,\mathbf{S}_{k-1}} \left(\mathbf{T}_{\text{lin}}^j \frac{\partial \mathbf{T}_{\text{lin}}}{\partial c(\mathbf{R})} \mathbf{T}_{\text{lin}}^{N-j-1}\right)(\mathbf{S},\mathbf{S}_1,...,\mathbf{S}_{k-1}|\mathbf{S},\mathbf{S}_1,...,\mathbf{S}_{k-1})}{\text{tr } \mathbf{T}_{\text{lin}}^{\frac{N}{k}}}. \quad \text{(A.4)}$$

matrix consequently come from the off-diagonal blocks $\partial^2 f_n/\partial c\partial \hat{c}$. We calculate the derivatives

see equation (A.4) above.

Due to the fact that

$$\frac{\partial \mathbf{T}}{\partial c(\mathbf{R})} = p \sum_{p,q=1}^{k} \exp\left\{\beta \mathbf{S}_p(\mathbf{S}-\mathbf{S}_{p+q}) + \beta \sum_{i,j=1}^{k} \mathbf{S}_i\mathbf{S}_{i+j} + \sum_{i=1}^{k} \hat{c}(\mathbf{S}_i)\right\} \quad \text{(A.5)}$$

is already linear in p, the other $\mathbf{T}_{\text{lin}}$-factors can be replaced by the replication $(\mathbf{T}^{(0)})^{\otimes n}$ of the pure matrix. Introducing two-times the identity

$$\mathbf{1} = \sum_{\boldsymbol{\alpha}} |\boldsymbol{\alpha}\rangle\langle\boldsymbol{\alpha}| \quad \text{(A.6)}$$

where $\langle\boldsymbol{\alpha}|$ denotes the biorthogonal set of left eigenvectors into (A.4) and keeping only the exponentially dominant terms proportional to λ_1^{nN}, we can write

$$\frac{\partial^2 f_n}{\partial \hat{c}(\mathbf{S})\partial c(\mathbf{R})} = -\delta_{\mathbf{S},\mathbf{R}} - \frac{N}{k} c_{\text{pm}}\hat{c}_{\text{pm}} + \mathbf{M}_{(1...1)}(\mathbf{S},\mathbf{R}) + \sum_{\boldsymbol{\alpha}\neq(1...1)} \frac{p}{1-\lambda_{\boldsymbol{\alpha}}} \mathbf{M}_{\boldsymbol{\alpha}}(\mathbf{S},\mathbf{R}) \quad \text{(A.7)}$$

and in the limit $n \to 0$ the fluctuation modes respecting the normalization of $c(\mathbf{S})$ give rise to eigenvectors of the form

$$-1 + \frac{p}{1-\lambda_2/\lambda_1} O(p^0, (\mathrm{e}^{\beta})^0) + ... \quad \text{(A.8)}$$

with λ_1 and λ_2 being the two largest eigenvectors of $\mathbf{T}^{(0)}$. For low temperatures, where $1-\lambda_2/\lambda_1 \ll 1$, we have

$$\frac{1}{1-\lambda_2/\lambda_1} = \frac{1}{1-\exp(-\frac{1}{\xi_0})} \approx \xi_0 \quad \text{(A.9)}$$

and the correction in $O(p)$ gets arbitrarily large for low enough temperatures T. This leads directly to the crossover in the behaviour of the model for $p \propto \xi_0^{-1}$ discussed in Section 4.3.

Appendix B: Fluctuations around the paramagnetic saddle point

In this appendix we are going to present the calculations of the Gaussian fluctuation matrix at the paramagnetic saddle point solution for the modified model presented in Section 4.5 in order to determine the ferromagnetic phase transition temperature for general k and p. We start with equations (28, 29),

$$f_n[c,\hat{c}] = -\sum_{\mathbf{S}} c(\mathbf{S})\hat{c}(\mathbf{S}) + \frac{1}{N}\ln \text{tr}\mathbf{T}^N, \quad \text{(B.1)}$$

$$\mathbf{T}(\mathbf{S}_1|\mathbf{S}_2) = \mathrm{e}^{\hat{c}(\mathbf{S}_1)}\Big[(1-p)\exp\{\beta\mathbf{S}_1\cdot\mathbf{S}_2\} + p\sum_{\mathbf{S}} c(\mathbf{S})\exp\{\beta\mathbf{S}_1\cdot\mathbf{S}\}\Big]^k. \quad \text{(B.2)}$$

In the following we need the first and second derivatives of $\mathbf{T}$:

$$\frac{\partial \mathbf{T}(\mathbf{S}_1|\mathbf{S}_2)}{\partial c(\mathbf{S})} = kp\exp\{\hat{c}(\mathbf{S}_1) + \beta\mathbf{S}_1\cdot\mathbf{S}\} \times \left[(1-p)\exp\{\beta\mathbf{S}_1\cdot\mathbf{S}_2\} + p\sum_{\mathbf{S}} c(\mathbf{S})\exp\{\beta\mathbf{S}_1\cdot\mathbf{S}\}\right]^{k-1}$$

$$\frac{\partial \mathbf{T}(\mathbf{S}_1|\mathbf{S}_2)}{\partial \hat{c}(\mathbf{S})} = \mathbf{T}(\mathbf{S}_1|\mathbf{S}_2)\delta_{\mathbf{S}_1,\mathbf{S}}$$

$$\frac{\partial^2 \mathbf{T}(\mathbf{S}_1|\mathbf{S}_2)}{\partial c(\mathbf{S})\partial c(\mathbf{R})} = k(k-1)p^2\exp\{\hat{c}(\mathbf{S}_1) + \beta\mathbf{S}_1\cdot(\mathbf{S}+\mathbf{R})\} \times \left[(1-p)\exp\{\beta\mathbf{S}_1\cdot\mathbf{S}_2\} + p\sum_{\mathbf{S}} c(\mathbf{S})\exp\{\beta\mathbf{S}_1\cdot\mathbf{S}\}\right]^{k-2}$$

$$\frac{\partial^2 \mathbf{T}(\mathbf{S}_1|\mathbf{S}_2)}{\partial c(\mathbf{S})\partial \hat{c}(\mathbf{R})} = \frac{\partial \mathbf{T}(\mathbf{S}_1|\mathbf{S}_2)}{\partial c(\mathbf{S})}\delta_{\mathbf{S}_1,\mathbf{R}}$$

$$\frac{\partial^2 \mathbf{T}(\mathbf{S}_1|\mathbf{S}_2)}{\partial \hat{c}(\mathbf{S})\partial \hat{c}(\mathbf{R})} = \mathbf{T}(\mathbf{S}_1|\mathbf{S}_2)\delta_{\mathbf{S}_1,\mathbf{S}}\delta_{\mathbf{S}_1,\mathbf{R}}. \quad \text{(B.3)}$$

The resulting saddle point equations for the calculation of $\overline{Z^n}$,

$$c(\mathbf{S}) = \frac{\mathbf{T}^N(\mathbf{S}|\mathbf{S})}{\mathrm{tr}\mathbf{T}^N}$$
$$\hat{c}(\mathbf{S}) = \frac{\mathbf{T}^{N-1}\partial_{c(\mathbf{S})}\mathbf{T}}{\mathrm{tr}\mathbf{T}^N}, \tag{B.4}$$

have obviously a simple paramagnetic solution of the form $c(\mathbf{S}) = 1/2^n$ and $\hat{c}(\mathbf{S}) = 2pa(\beta)^n$, *i.e.* a solution, where every replicated spin has equal probability. Whether this is correct or not for any finite temperature depends on the eigenvalues of the Hessian matrix

$$\begin{pmatrix} \frac{\partial^2 f_n[c,\hat{c}]}{\partial c\,\partial c} & \frac{\partial^2 f_n[c,\hat{c}]}{\partial c\,\partial \hat{c}} \\ \frac{\partial^2 f_n[c,\hat{c}]}{\partial \hat{c}\,\partial c} & \frac{\partial^2 f_n[c,\hat{c}]}{\partial \hat{c}\,\partial \hat{c}} \end{pmatrix} \tag{B.5}$$

calculated at the before-mentioned saddle point. One important observation is that the structure of all four blocks in this matrix is the same, resulting in the possibility of a simultaneous diagonalization of the four blocks, so only the submatrices of 4 eigenvalues belonging to the same eigenvectors have to be considered. But at first we have to calculate the entries of (B.5), and we start with the upper left corner:

$$\frac{\partial^2 f_n}{\partial c(\mathbf{S})\partial c(\mathbf{R})} = -N\hat{c}(\mathbf{S})\hat{c}(\mathbf{R}) + \frac{\mathrm{tr}\ \mathbf{T}^{N-1}\ \partial^2\mathbf{T}/\partial c(\mathbf{S})\partial c(\mathbf{R})}{\mathrm{tr}\ \mathbf{T}^N} + \sum_{j=0}^{N-2}\frac{\partial\mathbf{T}/\partial c(\mathbf{S})\ \mathbf{T}^j\ \partial\mathbf{T}/\partial c(\mathbf{R})\ \mathbf{T}^{N-j-2}}{\mathrm{tr}\ \mathbf{T}^N}. \tag{B.6}$$

The numerator of the second term is dominated by the largest eigenvalue of $\mathbf{T}$ which, according to the notation in Section 3.5, is $|+...+\rangle$. We are only interested in the limit $n \to 0$, so we can set all n-th powers to 1 for the simplicity of our calculations.

$$\begin{aligned} \mathrm{tr}\ \mathbf{T}^{N-1}\frac{\partial^2\mathbf{T}}{\partial c(\mathbf{S})\partial c(\mathbf{R})} &= \lambda(n)^{N-1}\langle +...+|\frac{\partial^2\mathbf{T}}{\partial c(\mathbf{S})\partial c(\mathbf{R})}|+...+\rangle \\ &= k(k-1)p^2\sum_{\mathbf{S}_1,\mathbf{S}_2}\exp\{\hat{c}(\mathbf{S}_1)+\beta\mathbf{S}_1\cdot(\mathbf{S}+\mathbf{R})\} \\ &\quad\times[(1-p)\exp\{\beta\mathbf{S}_1\cdot\mathbf{S}_2\}+p]^{k-2} \\ &= k(k-1)p^2\mathrm{e}^{\hat{c}}(\cosh 2\beta)^{\frac{\mathbf{S}\cdot\mathbf{R}}{2}}. \end{aligned} \tag{B.7}$$

The last term in equation (B.6) is exponentially dominated by

$$\begin{aligned} &\mathrm{tr}\ \frac{\partial\mathbf{T}}{\partial c(\mathbf{S})}\mathbf{T}^j\frac{\partial\mathbf{T}}{\partial c(\mathbf{R})}\mathbf{T}^{N-j-2} \\ &= \lambda(n)^{N-2}\langle +...+|\frac{\partial\mathbf{T}}{\partial c(\mathbf{S})}|+...+\rangle\langle +...+|\frac{\partial\mathbf{T}}{\partial c(\mathbf{R})}|+...+\rangle \\ &+ \sum_{\boldsymbol{\mu}\neq(+...+)}\lambda[\boldsymbol{\mu}]^j\lambda(n)^{N-j-2} \\ &\times\langle +...+|\frac{\partial\mathbf{T}}{\partial c(\mathbf{S})}|\boldsymbol{\mu}\rangle\langle\boldsymbol{\mu}|\frac{\partial\mathbf{T}}{\partial c(\mathbf{R})}|+...+\rangle \\ &+ \sum_{\boldsymbol{\mu}\neq(+...+)}\lambda[\boldsymbol{\mu}]^{N-j-2}\lambda(n)^j \\ &\times\langle\boldsymbol{\mu}|\frac{\partial\mathbf{T}}{\partial c(\mathbf{S})}|+...+\rangle\langle +...+|\frac{\partial\mathbf{T}}{\partial c(\mathbf{R})}|\boldsymbol{\mu}\rangle. \end{aligned} \tag{B.8}$$

With

$$\begin{aligned} &\langle +...+|\frac{\partial\mathbf{T}}{\partial c(\mathbf{S})}|\boldsymbol{\mu}\rangle = \\ &\sum_{\mathbf{S}_1,\mathbf{S}_2} kp\mathrm{e}^{\hat{c}+\beta\mathbf{S}_1\cdot\mathbf{S}}\left((1-p)\mathrm{e}^{\beta\mathbf{S}_1\cdot\mathbf{S}_2}+p\right)^{k-1}\langle\mathbf{S}_2|\boldsymbol{\mu}\rangle \\ &= kp\mathrm{e}^{\hat{c}}\sum_{m=0}^{k-1}\binom{k-1}{m}p^m(1-p)^{k-m-1} \\ &\times[\tanh(k-m-1)\beta\ \tanh\beta]^{n-\rho(\boldsymbol{\mu})}\langle\mathbf{S}|\boldsymbol{\mu}\rangle \end{aligned}$$

$$\begin{aligned} \langle\boldsymbol{\mu}|\frac{\partial\mathbf{T}}{\partial c(\mathbf{R})}|+...+\rangle &= \sum_{\mathbf{S}_1,\mathbf{S}_2} kp\mathrm{e}^{\hat{c}+\beta\mathbf{S}_1\cdot\mathbf{S}_2} \\ &\times\left((1-p)\mathrm{e}^{\beta\mathbf{S}_1\cdot\mathbf{S}_2}+p\right)^{k-1}\langle\boldsymbol{\mu}|\mathbf{S}_1\rangle \\ &= kp\mathrm{e}^{\hat{c}}[\tanh\beta]^{n-\rho(\boldsymbol{\mu})}\langle\boldsymbol{\mu}|\mathbf{R}\rangle \end{aligned} \tag{B.9}$$

we consequently find

$$\begin{aligned} &\frac{\mathrm{tr}\ \frac{\partial\mathbf{T}}{\partial c(\mathbf{S})}\mathbf{T}^j\frac{\partial\mathbf{T}}{\partial c(\mathbf{R})}\mathbf{T}^{N-j-2}}{\mathrm{tr}\ \mathbf{T}^N} = \\ &k^2p^2\sum_{\boldsymbol{\mu}}\sum_{m=0}^{k-1}\binom{k-1}{m}p^m(1-p)^{k-m-1} \\ &\times\left[\tanh(k-m-1)\beta\ (\tanh\beta)^2\right]^{n-\rho(\boldsymbol{\mu})} \\ &\times\left(\lambda[\boldsymbol{\mu}]^j+\lambda[\boldsymbol{\mu}]^{N-j-2}-\delta_{\boldsymbol{\mu},(+...+)}\right)\langle\mathbf{S}|\boldsymbol{\mu}\rangle\ \langle\boldsymbol{\mu}|\mathbf{R}\rangle. \end{aligned} \tag{B.10}$$

It is now obvious that the matrix $\partial^2 f_n/\partial c\partial c$ has also the eigenvectors $|\mu\rangle$. The first one, $|+...+\rangle$, corresponds to fluctuations changing the normalization of $c(\mathbf{S})$ and is not allowed. So the second one, $|-+...+\rangle$ (or any other with $\rho(\boldsymbol{\mu}) = n-1$), is expected to be the dangerous one leading finally to the ferromagnetic phase transition in the Ising

$$k(k-1)p^2(\tanh\beta)^2 + \frac{2k^2p^2 \sum_{m=0}^{k-1} \binom{k-1}{m} p^m (1-p)^{k-m-1} \left[\tanh(k-m-1)\beta\,(\tanh\beta)^2\right]}{1 - \sum_{m=0}^{k} \binom{k}{m} p^m (1-p)^{k-m} \tanh(k-m)\beta}. \tag{B.11}$$

$$-1 + \frac{kp\tanh\beta \left[1 + \sum_{m=0}^{k-1} \binom{k-1}{m} p^m (1-p)^{k-m-1} \tanh(k-m-1)\beta\right]}{1 - \sum_{m=0}^{k} \binom{k}{m} p^m (1-p)^{k-m} \tanh(k-m)\beta}, \tag{B.12}$$

model. From (B.6, B.7, B.10) we obtain for this eigenvalue

see equation (B.11) above.

The calculation of the other elements of the fluctuation matrix is done analogously. Here we report only the results. The eigenvalue of $\partial^2 f_n/\partial c \partial \hat{c}$ corresponding to the eigenvector $|-+...+\rangle$ is found to be

see equation (B.12) above.

and for $\partial^2 f_n/\partial \hat{c} \partial \hat{c}$ we get the entry

$$1 + 2\frac{\sum_{m=0}^{k} \binom{k}{m} p^m (1-p)^{k-m} \tanh(k-m)\beta}{1 - \sum_{m=0}^{k} \binom{k}{m} p^m (1-p)^{k-m} \tanh(k-m)\beta} \tag{B.13}$$

leading to (35).

References

1. D.J. Watts, S.H. Strogatz, Nature **393**, 440 (1998).
2. M. Barthélémy, L.A. Nunes Amaral, Phys. Rev. Lett. **82**, 3180 (1999); Phys. Rev. Lett. **82**, 5180 (1999).
3. A. Barrat, comment on "Small-world networks: Evidence for a crossover picture" (preprint cond-mat/9903323).
4. R. Monasson, Eur. Phys. J. B **12**, 555 (1999).
5. M.E.J. Newman, D.J. Watts, "Renormalization group analysis of the small-world network model" (preprint cond-mat/9903357), Phys. Lett. A (in press).
6. M. Argollo de Menezes, C. Moukarzel, T.J.P. Penna, "First-order transition in small-world networks" (preprint cond-mat/9903426), submitted to Phys. Rev. Lett.
7. M.E.J. Newman, D.J. Watts, "Scaling and percolation in the small-world network model" (preprint cond-mat/9904419), Phys. Rev. E. (Dec. 1999).
8. B. Bollobás, *Random Graphs* (Academic Press, New-York, 1985).
9. S. Milgram, Psychology Today **2**, 60 (1967).
10. A. Barrat, R. Zecchina, Phys. Rev. E **59**, R1299 (1999).
11. P. G. de Gennes, *Scaling concepts in polymerphysics* (Cornell University Press, 1979).
12. See B.K. Chakrabarti, A.C. Maggs, R.B. Stinchcombe, J. Phys. A **18**, L373 (1985) and references therein.
13. I. Kanter, H. Sompolinsky, Phys. Rev. Lett. **58**, 164 (1987).
14. R. Monasson, J. Phys. A **31**, 513 (1998).
15. R.H. Swendsen, J.-S. Wang, Phys. Rev. Lett. **63**, 86 (1987).
16. See *e.g.* K. Binder, D.W. Heermann, *Monte-Carlo simulation in statistical physics* (Springer-Verlag, 1992).

4.3 MODELS OF SCALE-FREE NETWORKS

Several of the papers in Chapters 2 and 3 focus on the observed degree distributions of real networks, finding that for a number of systems, including citation networks, the World Wide Web, the Internet, and metabolic networks, the degree distribution approximates a power law (**Price 1965; Albert *et al.* 1999; Faloutsos *et al.* 1999; Broder *et al.* 2000**). The identification of networks with power-law degree distributions has generated a very large number of publications on such networks, *scale-free networks* as they are widely called, a term introduced by **Barabási and Albert (1999)** in the first paper reproduced in this section.

Efforts at constructing models of scale-free networks have taken network research in a new direction. Previous models, such as the random graphs and small-world models discussed in Sections 4.1 and 4.2, do not have power-law degree distributions (although the models of Section 4.1 can be adapted to study the properties of scale-free networks—see **Aiello *et al.* (2000)**). Yet there are a number of important questions concerning scale-free networks that can be illuminated by suitable models. Where do the power laws come from? How can we determine the exponent of the power law, and what affects its value? Is there any relation between power-law degree distributions and the ubiquitous power laws seen in continuous phase transitions (Stanley 1971; Binney *et al.* 1992) or in self-organized criticality (Bak 1996)? Does universality, a central concept in critical phenomena, apply here? All of these questions are tackled, to a greater or lesser extent, by papers reproduced in this section. As with the other aspects of networks discussed in this volume, it is impossible for us to do justice to all the work that has taken place in the past few years, but we will try at least to illustrate the main developments.

Barabási and Albert (1999)

The paper by **Barabási and Albert (1999)** reproduced here was largely responsible for starting the current wave of interest in scale-free networks. It makes three important contributions. First, the paper proposes that the power-law degree distribution seen in the World Wide Web is not merely a curiosity of that particular network, but that power laws could potentially be a generic property of many networks.[8] Second, the paper proposes that the properties of these networks can be explained using a model in which a network grows dynamically, rather than being a static graph, an idea that has been widely adopted in many studies since. Third, the paper proposes a specific model of a growing network that generates power-law degree distributions similar to those seen in the World Wide Web and other networks.

In support of their argument that power laws are common in real networks, Barabási and Albert offer three examples:[9] the Web itself **(Albert *et al.* 1999)**, the

---

[8]Like a number of other early papers by physicists, the paper by Barabási and Albert referred to degree as "connectivity." This usage has mostly disappeared now, and physicists by and large use the standard terminology "degree."

[9]The power transmission network of the western United States and Canada is also cited as a fourth example, but later work by **Amaral *et al.* (2000)** showed that this network is not scale-free, having a degree distribution better approximated by an exponential than by a power law.

network of film actors (**Watts and Strogatz 1998**), and citation networks (Redner 1998). While the strongest evidence comes from the World Wide Web, the movie actor network, with (at that time) about 200 000 actors, is also large enough to provide good statistics for the degree distribution, which is found to follow a power law $P(k) \sim k^{-\gamma}$ over a part of its range, with an exponent $\gamma = 2.3$. The third example, the distribution of numbers of citations that papers receive, was studied by Redner (1998), who found that the tail of the distribution followed a power law with exponent close to $\gamma = 3$.

The fact that such apparently different systems as the World Wide Web and the actor and citation networks share a scale-free degree distribution led Barabási and Albert to suggest that a common mechanism dictates the structure of all of these networks. Their suggested mechanism has two components: first, the network is growing, vertices being added continuously to it, and second, vertices gain new edges (Web links, citations, etc.) in proportion to the number they already have, a process that Barabási and Albert call *preferential attachment*. Certainly the first of these is true—each of the networks studied has grown steadily over the years. And the second appears very plausible—Web pages, for example, become well known by being linked to by many others, and it is the well-known pages that we are likely to make new links to.

Growth and preferential attachment offer a simple explanation for the emergence of scale-free networks. To demonstrate this, Barabási and Albert propose a model in which the network grows by the addition of a single new vertex at each time-step, with m edges connected to it. The other end of each edge is connected to one of the vertices already in the network, chosen at random with probability proportional to degree. Barabási and Albert analyze this model using a mean-field-like approximate argument as follows.

Let k_i be the degree of the vertex i. Then the probability that an edge belonging to a newly appearing vertex connects to i is

$$\Pi(k_i) = \frac{k_i}{\sum_j k_j}. \tag{4.26}$$

On average the degree k_i of vertex i increases at a rate proportional to k_i. If we ignore statistical fluctuations and assume that k_i always takes the mean value, its time evolution is captured by

$$\frac{\mathrm{d}k_i}{\mathrm{d}t} = m\Pi(k_i) = \frac{mk_i}{\sum_j k_j} = \frac{k_i}{2t}, \tag{4.27}$$

where we have made use of $\sum_j k_j = 2mt$. This then implies that

$$k_i(t) = m\left(\frac{t}{t_i}\right)^{1/2}, \tag{4.28}$$

where t_i is the time at which vertex i is added to the network and by definition

$k_i = m$ when $t = t_i$. Then the distribution of degrees at time t is

$$P(k) = -\left(\frac{\mathrm{d}k_i}{\mathrm{d}t_i}\right)^{-1} = \frac{2m^2 t}{k^3}. \tag{4.29}$$

That is, the degree distribution follows a power law with degree exponent $\gamma = 3$.

One problem with the model of Barabási and Albert is this exponent $\gamma = 3$, which is independent of the parameters of the model and hence cannot be varied. The exponent of the degree distribution observed in citation networks is very close to 3 (**Price 1965**; Redner 1998) but, as we have seen, for other networks the exponent differs substantially from this value. An obvious question to ask, therefore, is whether the model can be modified to yield other exponents. This question was addressed in a later paper by Albert and Barabási (2000), and also independently by **Dorogovtsev *et al.* (2000)** and by **Krapivsky (2000)** in papers reproduced later in this section.

Albert and Barabási (2000) pointed out that there are many additional processes that can shape a network, such as the addition or rewiring of edges or the removal of nodes or edges, processes that are absent from the original Barabási–Albert model. They proposed a model that incorporates the addition of new edges between existing nodes and the movement of existing edges. In this model one of three events occurs at each step:

i) With probability p, m new edges are added to the network. One end of each new edge is connected to a node selected uniformly at random from the network and the other to a node chosen using the preferential attachment process, Eq. (4.26).

ii) With probability q, m edges are "rewired," meaning that a vertex i is selected at random and one of its edges chosen at random is removed and replaced with a new edge whose other end is connected to a vertex chosen again according to the preferential attachment process.

iii) With probability $1 - p - q$, a new node is added to the network. The new node has m new edges that are connected to nodes already present in the system via preferential attachment in the normal fashion.

This model produces a degree distribution that again has a power-law tail, with an exponent γ that depends on the parameters p, q, and m, and can vary anywhere in the range from 2 to ∞.

Since the work of Albert and Barabási (and the papers by Dorogovtsev *et al.* and Krapivsky *et al.* reproduced below), a large number of other variations on the basic preferential attachment idea have been proposed, many of which also give variable degree exponents. These results show that the exponents seen in these models are not universal, and presumably the exponents seen in real networks are not universal either. There is nothing here akin to the renormalization flow fixed points that give rise to universality in the phase transitions of statistical physics.

Dorogovtsev, Mendes, and Samukhin (2000)

The solution given by **Barabási and Albert (1999)** for the degree distribution of their model is only approximate—it assumes a vertex always has the mean degree for a vertex added at that time, with no statistical fluctuations. However, it turns out to be possible to derive an exact solution for the model using the "rate equation" or "master equation" method, originally developed for problems of this kind by Simon (1955). An exact solution of this type was given by **Dorogovtsev *et al.* (2000)** in the paper reproduced here and independently by **Krapivsky *et al.* (2000)** in the following paper.

Dorogovtsev *et al.* consider a variation on the Barabási–Albert model in which preferential attachment takes place with respect only to the *incoming* edges at each vertex. Each vertex comes with m outgoing edges, which attach to other preexisting vertices with attachment probability proportional only to those vertices' in-degree. But this introduces a problem, since any vertex that has in-degree zero initially will have probability zero of gaining new incoming edges, and hence will forever have in-degree zero. To get around this problem, Dorogovtsev *et al.* add a new term to the attachment probability thus:

$$\Pi(q) \propto am + q, \tag{4.30}$$

where q is the in-degree and a is a constant. In the model of Barabási and Albert, all vertices have the same out-degree m, which never changes, and hence their total degree, in and out, is $m + q$. Thus the model of Dorogovtsev *et al.* reduces to that of Barabási and Albert if $a = 1$.

To solve their model, Dorogovtsev *et al.* consider the probability distribution of the in-degree of a vertex as a function of current time and the time at which the vertex was created. They write a difference equation for this quantity, and then sum over all vertices to get the overall degree distribution. The resulting equation is solved using a generating function method to give

$$P(q) = (1 + a)\frac{\Gamma((m + 1)a + 1)}{\Gamma(ma)} \frac{\Gamma(q + ma)}{\Gamma(q + 2 + (m + 1)a)}. \tag{4.31}$$

We can express this in terms of the total degree $k = q + m$ thus:

$$P(k) = (1 + a)\frac{\Gamma(ma + a + 1)}{\Gamma(ma)} \frac{\Gamma(k + ma - m)}{\Gamma(k + 2 + ma + a - m)}. \tag{4.32}$$

For the particular case $a = 1$ of the Barabási–Albert model, this reduces to

$$P(k) = \frac{2m(m + 1)}{k(k + 1)(k + 2)}, \tag{4.33}$$

while for the general case it is straightforward to show, using the properties of the Γ-function, that the tail of the degree distribution has the form $P(k) \sim k^{-\gamma}$, with

$\gamma = 2 + a$. Thus **Dorogovtsev *et al.* (2000)** find for this model that the exponent of the degree distribution is again nonuniversal, depending on the parameter a. This provides another possible mechanism for generating exponent values other than the value $\gamma = 3$ generated by the original model of Barabási and Albert (see above).

Interestingly, ideas and mathematical approaches similar to those discussed here, aimed at explaining the emergence of power laws in a variety of systems, have been developed independently in other fields, including economics, social science, and information science. The preferential attachment mechanism in particular has appeared in several different fields under different names. In information science it is known as "cumulative advantage" (Price 1976), in sociology as the "Matthew effect" (Merton 1968), and in economics as the "Gibrat principle" (Simon 1955).

Perhaps the most relevant historical work for our purposes is that of Price (1976), who was interested in the properties of citation networks, observing that the distribution of citations follows a power law **(Price 1965)**, as others have also done more recently (Seglen 1992; Redner 1998). He proposed a simple model to explain this observation, as well as the appearance of power laws elsewhere, such as the Lotka, Pareto, and Zipf laws. In fact, Price's model is mathematically identical to the model studied by **Dorogovtsev *et al.* (2000)** in this section. Price also found the exact solution of his model and derived the value of the exponent of the distribution. His solution was based on earlier work by Simon (1955) on the distribution of word frequencies and city sizes, which can be adapted to the case of networks, as also pointed out by several more recent authors (**Dorogovtsev *et al.* 2000**; Bornholdt and Ebel 2001).

Krapivsky, Redner, and Leyvraz (2000)

Krapivsky *et al.* (2000) independently derived the exact solution of the Barabási–Albert model discussed in the previous section, although they used a different and somewhat simpler method that avoids generating functions and special functions. Like **Dorogovtsev *et al.* (2000)**, they also proposed a variation of the model to cover additional cases of interest.

The preferential attachment hypothesis assumes that new nodes connect to existing nodes with a probability proportional to the number of edges those nodes already have, i.e., that the probability $\Pi(k)$ defined in Eq. (4.26) is linear in k. There are good reasons for supposing that it might be linear: in a citation network, for example, one could imagine that the probability of one's encountering a paper is proportional to the number of people who have cited it in the past, and hence the probability that you will also cite it is proportional to the same number. On the other hand, actions like citation and linking to Web sites are presumably subject to all sorts of psychological and sociological influences, and deviations from the linear are certainly not implausible. Krapivsky *et al.* looked at what happens when $\Pi(k)$ is not linear in k. They studied the case in which $\Pi(k)$ takes the power-law form

$$\Pi(k) \propto k^{\alpha}, \tag{4.34}$$

which includes the model studied by Barabási and Albert when $\alpha = 1$. Looking at the particular case of $m = 1$, Krapivsky *et al.* find that the resulting network is scale-free only when $\alpha = 1$; for any $\alpha \neq 1$ the degree distribution differs from a power law. For sublinear preferential attachment ($\alpha < 1$) the distribution follows a stretched exponential, while for superlinear attachment ($\alpha > 1$) there is a "winner takes all" behavior in which one node has very high degree, being connected to almost all others, with the rest of the nodes having a degree distribution with an exponential tail.

It appears, therefore, that the shape of $\Pi(k)$ plays a key role in determining the structure of the emerging network. This leads us to a further question: what in fact *is* the form of $\Pi(k)$ for real systems? To answer this question empirically, we need to study networks for which we know the order in which nodes joined the network. Data of this type are available for a number of networks, including some coauthorship networks, citation networks, the movie actor collaboration network, the Internet at the domain level, and protein interaction networks (Newman 2001d; Pastor-Satorras *et al.* 2001; Barabási *et al.* 2002; Jeong *et al.* 2003; Eisenberg and Levanon 2003), and these data have been used to test the preferential attachment hypothesis by Jeong *et al.* (2003) and by Newman (2001d). An example is given in Figure 4.6, which shows curves of $\Pi(k)$ for each of four networks, calculated by measuring the probability that a vertex gains a new edge during the course of a single year, as a function of its preexisting degree at the start of that year. As the figure shows, the results support the idea of preferential attachment, and in each case the curve appears to follow a power law, as hypothesized by Krapivsky *et al.*, Eq. (4.34). In the case of the Internet, citation networks, and biology and physics collaborations, $\Pi(k)$ depends roughly linearly on degree, $\alpha \simeq 1$, while for some other collaboration networks the dependence is sublinear.[10]

Bianconi and Barabási (2001b)

In the models discussed so far, it is assumed that current degree is the only property that affects a node's propensity to acquire new edges. As the treatment of Barabási and Albert shows (Eq. (4.28)) this results in a mean degree that increases over time as $t^{1/2}$ for all vertices, implying that the oldest nodes will always have the highest degrees in the network and the youngest nodes the lowest. As pointed out by Adamic and Huberman (2000), however, this is not normally the case in the real world. On the World Wide Web, for instance, some pages acquire a large number of edges in a very short time frame thanks to some happy combination of good content and good marketing. A prominent example is the Web search engine Google, which in just a few years grew from nothing to become one of the most highly connected nodes on the Web, displaying a growth rate that clearly surpasses that of other sites. To account for behavior of this kind, one needs a model that incorporates the idea that some nodes are intrinsically "better," or at least faster growing, than others.

---

[10]Following **Krapivsky *et al.* (2000)**, a sublinear law should lead to a stretched exponential degree distribution, but such behavior is not actually observed, at least in some of the networks. It has been suggested that this apparent contradiction may be due to the presence of another preferential attachment process governing the placement of internal edges, which in turn is linear (Jeong *et al.* 2003).

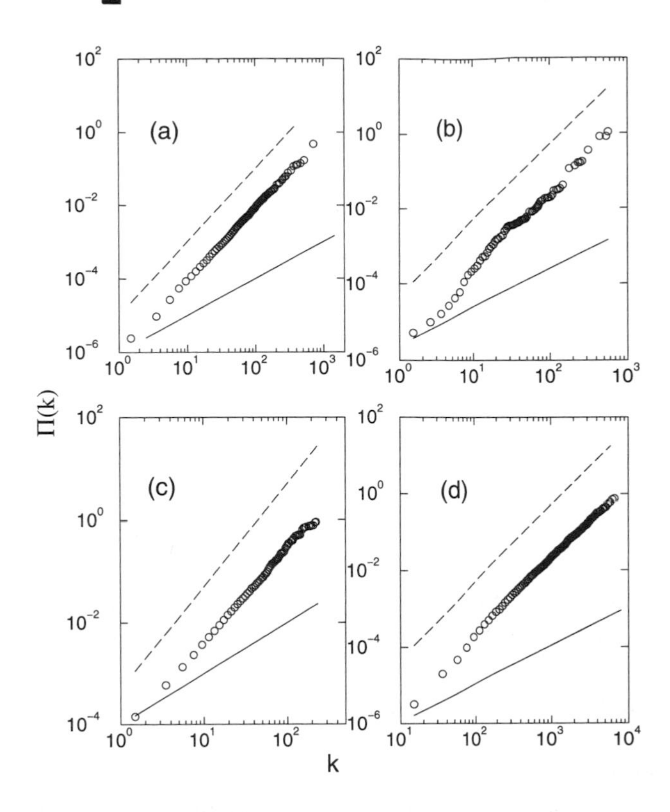

FIGURE 4.6 Measurements of the (cumulative) attachment probability $\Pi(k)$ as a function for k for (a) a citation network; (b) the Internet at the domain level; (c) a collaboration network of neuroscientists; (d) the film actor collaboration network. In all panels the dashed line corresponds to linear preferential attachment and the solid line indicates the absence of preferential attachment. After Jeong *et al.* (2003).

Such a model has been proposed by **Bianconi and Barabási (2001b)** in the paper reproduced here.

In the model of Bianconi and Barabási, each vertex in the network has a *fitness*, an intrinsic ability to compete for edges at the expense of other vertices. To represent this property, each vertex is assigned upon its first appearance a fixed fitness value η_i. The fitness of a Web page or research paper is, presumptively, determined by the utility and presentation of its content. That of an individual in a social network might be determined by the individual's social or professional skills.

At every time-step a new vertex is added to the network with fitness chosen from some distribution $\rho(\eta)$. Each such vertex connects to m others, the probability of connecting to vertex i being proportional to the product of the degree and the fitness of i:

$$\Pi = \frac{\eta_i k_i}{\sum_j \eta_j k_j}. \tag{4.35}$$

This generalized attachment mechanism means that even a relatively young vertex with few edges can acquire edges at a high rate if it has a high fitness. At the same time it is still true that a vertex of high degree is more likely to gain new connections than an otherwise identical vertex of low degree.

Analyzing this model using the approximate method of **Barabási and Albert (1999)**, Bianconi and Barabási find that the mean degree of a vertex now increases as t^{β}, where the exponent β is

$$\beta = \frac{\eta}{C}, \tag{4.36}$$

with

$$C = \int \rho(\eta) \frac{\eta}{1 - \beta(\eta)}\, \mathrm{d}\eta. \tag{4.37}$$

Thus, each vertex has its own exponent, proportional to its fitness η. Vertices with higher fitness increase their degree faster than those with lower fitness, so that vertices appearing late but having high fitness can assume a central role in the network by increasing their connectivity faster than their less fit counterparts. The fitness model has been extended to incorporate additional processes, such as internal edges—see, for example, Ergün and Rodgers (2002) and Dorogovtsev and Mendes (2003).

An extended treatment of the model has been given in another paper by Bianconi and Barabási in which they show that it can be mapped onto a noninteracting Bose gas (Bianconi and Barabási 2001a). This mapping indicates that the model should show two distinct phases, determined by the fitness distribution $\rho(\eta)$. In one phase the number of edges a vertex acquires varies with fitness but no site dominates excessively over all others, the ratio of a vertex's degree to the total number of edges in the graph tending to zero for large system sizes. All networks that develop power-law degree distributions fall in this phase. Interestingly, however, there is a second phase that corresponds to Bose–Einstein condensation in the Bose gas. In this phase the fittest node acquires a finite fraction of all the edges in the network and maintains this share of edges even as the size of the system becomes large.[11] In the presence of a Bose–Einstein condensate, the network is not scale-free, but is dominated by a single hub.

Goh, Kahng, and Kim (2001b)

The work on scale-free networks discussed so far concentrates on the power-law distribution of vertex degrees. Degree is an important property because it can be thought of as a measure of the popularity of a vertex, its "centrality" or "influence" in the parlance of network analysis. But there are other centrality measures as well, one of the most widely studied being *betweenness centrality,* which is the number of shortest paths on a network that run through the vertex of interest (Anthonisse 1971; Freeman 1977). That is, we count the shortest paths between all vertex pairs and define the betweenness of a vertex to be the number of those paths that pass

[11] This behavior is reminiscent of that seen in the model of Krapivsky *et al.* discussed earlier, although its mathematical origin is quite different.

through that vertex. In networks in which something flows between vertices, such as information of some kind, betweenness reflects the amount of traffic that passes through a vertex and hence the influence that the vertex can potentially exert over flow between other vertices.

In the paper by **Goh *et al.* (2001b)** reproduced here, the authors study the distribution of betweenness centrality in networks with power-law degree distributions. In fact, the quantity studied by Goh *et al.* is not precisely the standard betweenness centrality. In cases where there is more than one shortest path between a pair of vertices, the standard betweenness measure gives each path equal weight, as if an equal amount of flow passes along each one. Goh *et al.* use a different definition under which the flow splits in half every time a path splits. In cases where there are multiple shortest paths this can result in slightly (not greatly) different centrality scores. Goh *et al.* refer to their measure as *load,* and it seems useful to use this different name to distinguish their measure from the more traditional betweenness.

Goh *et al.* conducted their studies on growing networks similar to those discussed above and also on random graphs with power-law degree distributions of the type discussed in Section 4.1 (**Molloy and Reed 1995**, 1998; **Aiello *et al.* 2000**). The important finding of Goh *et al.* is that the load ℓ on networks of both these classes appears, from numerical calculations, to have a power-law distribution:

$$P(\ell) \sim \ell^{-\delta}. \tag{4.38}$$

Goh *et al.* also observed that the value of the exponent δ appears to be roughly constant. As pointed out earlier, the exponents for the degree distribution in preferential attachment models are not expected to have universal values, but rather depend on the details of the models. The numerical results for $P(\ell)$, however, show roughly the same value of $\delta \simeq 2.2$ for all the networks studied. Goh *et al.* also offer scaling arguments that suggest that the load should be a simple function of degree obeying

$$\ell \sim k^{(\gamma-1)/(\delta-1)}, \tag{4.39}$$

so that the best connected vertices carry the most load. (Although it is not immediately obvious, it is straightforward to show that this scaling form is consistent with Eq. (4.38).) Goh *et al.* find that their results break down if the exponent γ of the degree distribution is greater than 3, and also that there is no power-law load distribution for networks with the topology of the small-world model (see Section 4.2).

The value of the exponent δ has been subsequently revisited by several authors. Goh *et al.* (2002) in a more recent paper focusing on real-world networks concluded that protein-protein interaction networks, metabolic networks, and some coauthorship networks appear to share the exponent $\delta \simeq 2.2$. Others, however, such as the Internet and the World Wide Web, seem to have a slightly different value of $\delta \simeq 2.0$. And more recent results still (Goh *et al.* 2003) indicate that in fact the universality holds only for trees or networks with tree-like topology (such as the Barabási–Albert model with $m = 1$), for which $\delta = 2$. For networks containing a finite density of loops, the exponent δ varies depending on the network's characteristics (Barthélémy 2003).

Farkas, Derényi, Barabási, and Vicsek (2001)

The theory of random graphs has a close relative in the theory of random matrices. A graph can be represented as an adjacency matrix **A** with elements $A_{ij} = 1$ if there is an edge between vertices i and j and $A_{ij} = 0$ if there is not. Since the pioneering work of Wigner (1955, 1957, 1958), which showed that there is a deep connection between atomic spectra and the properties of random matrices, the spectral properties of random matrices have been an important topic in both physics and mathematics.

In the mathematics literature, attention in the past has focused mostly on the eigenvalues of random graphs with Poisson degree distributions. Recently, however, two papers, by **Farkas *et al.* (2001)** and Goh *et al.* (2001a), have revisited the problem, looking at graphs grown using the scale-free model of **Barabási and Albert (1999)**, which have power-law degree distributions, and also the small-world model of **Watts and Strogatz (1998)**.

The spectrum of a graph of N vertices is the set of eigenvalues of its adjacency matrix **A**. There are N eigenvalues λ_j, $j = 1, \ldots, N$, and it is useful to define the spectral density as

$$\rho(\lambda) = \frac{1}{N} \sum_{j=1}^{N} \delta(\lambda - \lambda_j), \tag{4.40}$$

which approaches a continuous function for large systems ($N \to \infty$). (Here $\delta(x)$ is the Dirac δ-function.) The spectral density is akin to the density of states in a quantum Hamiltonian system. For a random graph in which nodes are connected uniformly at random with probability p, the spectral density converges to a semicircular distribution

$$\rho(\lambda) = \begin{cases} \frac{\sqrt{4Np(1-p)-\lambda^2}}{2\pi Np(1-p)} & \text{for } |\lambda| < 2\sqrt{Np(1-p)}, \\ 0 & \text{otherwise.} \end{cases} \tag{4.41}$$

Known as Wigner's law or the semicircle law, Eq. (4.41) has many applications in quantum, statistical, and solid state physics (Mehta 1991; Crisanti *et al.* 1993; Guhr *et al.* 1998).

Farkas *et al.* (2001) and Goh *et al.* (2001a) independently found that the spectral density of scale-free networks deviates markedly from the semicircle law. Their numerical calculations reveal a spectrum with a triangle-like shape with edges decaying as a power law. These power-law tails are due to eigenvectors that are localized around the highest-degree nodes. As in the case of random graphs, the principal eigenvalue λ_1 of the adjacency matrix is clearly separated from the bulk of the spectrum. The square root of the network's highest degree k_1 can be shown to be a lower bound on the value of λ_1, and k_1 goes as $N^{1/2}$ on a scale-free network with $\gamma = 3$ (Dorogovtsev *et al.* 2001b; Moreira *et al.* 2002), so one expects that $\lambda_1 \gtrsim N^{1/4}$.

Bollobás, Riordan, Spencer, and Tusnády (2001)

The treatments of the model of **Barabási and Albert (1999)** given by, for example, **Krapivsky *et al.* (2000)** and **Dorogovtsev *et al.* (2000)** reveal the important behaviors of the model in a way that is easily understood. They are not, however, rigorous in the mathematician's sense of the word. Many details have been left out concerning the domain of validity of the solution, the typical size of stochastic deviations from the solution, and so forth. A rigorous analysis, sufficient to satisfy most mathematicians, has been given by **Bollobás *et al.* (2001)** in the paper reproduced here. They confirm that, at least within certain bounds, the results derived by the physicists are correct.

As a first step in studying the Barabási–Albert model rigorously, Bollobás *et al.* provide a precise definition of the model. Then they derive the main theorem of the paper, showing that for a network of N vertices, for degrees $k < N^{1/15}$, the degree distribution is given by

$$P(k) = \frac{2m(m+1)}{k(k+1)(k+2)}, \tag{4.42}$$

precisely as in Eq. (4.33). The restriction to small degrees, $k < N^{1/15}$, is done only to simplify the proof; as discussed by Bollobás *et al.*, this condition can be relaxed, at the cost of some effort, allowing generalization of the proof to other cases. Bollobás *et al.* also discuss growing networks without preferential attachment, sketching a proof that the degree distribution in this case decays exponentially.[12]

In some sense, the paper by Bollobás *et al.* tells us something we already know: it provides a rigorous proof of something that a physicist would consider already to have been demonstrated. More recently, however, Bollobás and collaborators have extended their analysis to reveal new results that would be quite difficult to derive using the more heuristic techniques employed by physicists. One such result is due to Bollobás and Riordan (2002), who have calculated the average vertex-vertex path length in the Barabási–Albert model. For random graphs the average vertex-vertex distance ℓ scales logarithmically with system size: $\ell \sim \log N$. In the model of Barabási and Albert, however, numerical results suggest that this is not quite correct; systematic deviations from strict logarithmic scaling can be observed for large N (Albert and Barabási 2002). Bollobás and Riordan (2002) shed light on this discrepancy, finding that ℓ has a nontrivial, doubly logarithmic correction

$$\ell \sim \frac{\log N}{\log\log N}, \tag{4.43}$$

a result that was also established independently by Cohen and Havlin (2003).

In parallel with the work of Bollobás and collaborators, a significant body of exact work on scale-free networks has been published by Chung and collaborators, who focused on random graphs with power-law degree distributions, as described in Section 4.1. They looked at a variety of different properties of their graphs,

[12] Such a network was considered previously by Barabási *et al.* (1999).

ranging from phase transition properties to path length distributions and spectral properties (**Aiello *et al.* 2000**; Aiello *et al.* 2002; Chung and Lu 2002a,b). Taken together, the work of Bollobás *et al.* and Chung *et al.*, along with work by others such as Collet and Eckmann (2002), represents the beginnings of a systematic effort to develop a rigorous body of theory dealing with more realistic network models.

Solé et al. (2002) and Vázquez et al. (2003)

For many networks the preferential attachment mechanism discussed at length in previous sections provides a plausible explanation for the origin of power-law degree distributions. But there are some other cases where power-law degree distributions are observed but where their origin is less easy to explain. For instance, as discussed in Chapter 3, a series of studies have found that biological networks, shaped by billions of years of natural selection, have power-law degree distributions. Power laws have been reported for metabolic and protein interaction networks (**Jeong *et al.* 2000**, 2001; **Wagner and Fell 2001**), for genetic regulatory networks, whose vertices are genes that are connected if they directly regulate each other's transcription (Wagner 2001; Agrawal 2002; Featherstone and Broadie 2002; Farkas *et al.* 2003), and for protein domain networks, in which vertices representing characteristic domains within proteins are connected if they cooccur in the same protein (Wuchty 2001, 2002; Apic *et al.* 2001). Biological evolution rather than the attachment of edges to vertices (preferentially or otherwise) is the dominant force shaping the structure of these networks. So where do the power laws come from?

In the two papers reproduced here, **Solé *et al.* (2002)** and **Vázquez *et al.* (2003)** argue that power laws in biological networks can be explained by the mechanism of *gene duplication*. For a cell to reproduce it needs to duplicate its genetic content and divide in two. The details of the processes involved vary between cell types, but certain steps are universal. In particular, in order to produce a genetically identical daughter cell, the DNA must be faithfully replicated. The cell's intricate copying and error-correcting mechanisms guarantee that DNA sequences are replicated with extraordinary fidelity, but there are occasional errors, including single locus mutations, insertions, and deletions, and these errors are important in creating the population diversity upon which selection acts to produce evolution. Another important though rare type of error is gene duplication. Occasionally in the replication process two copies of a stretch of the genome will be made where only one existed before, resulting in duplication of one or more genes. In some cases such mistakes kill the cell, but in other cases the duplicate copies confer evolutionary advantages on the cell and are passed on to future generations (Ohno 1970). Hemoglobin is a well-known example. Originally cells had only one hemoglobin gene, but about 500 million years ago, during the early evolution of fish species, the gene was duplicated, resulting in the two copies that many metazoan species now carry. Although the two genes were, presumably, initially the same, they have changed over the intervening interval of time and today each of them separately encodes two of the four components of the complete hemoglobin protein complex (Alberts *et al.* 2002).

The importance of gene duplication in shaping the structure of biological networks was initially highlighted by Wagner (2001). **Solé *et al.* (2002)** and **Vázquez *et al.* (2003)**, in the papers reproduced here, showed quantitatively that gene duplication is sufficient to explain the emergence of a power-law degree distribution. The fact that vertex duplication can give rise to a scale-free network had been observed previously by Kleinberg *et al.* (1999; Kumar *et al.* 2000a,b). Kleinberg *et al.* were interested in the growth of the World Wide Web and proposed a general network model that includes duplication processes, random linking, and vertex creation and deletion. In their model vertices are added to the network with a certain number of edges emanating from them, and the targets of the other ends of those edges are chosen by copying from other preexisting vertices: if a newly added vertex is to have initial degree m, then we choose other vertices at random and copy the targets of their edges until we have a total of m targets. It is straightforward to see that this copying process will give rise to a power-law degree distribution. The chance that a randomly chosen vertex i is connected to another vertex j is proportional to the degree k_j of that other vertex, and hence when we copy targets from randomly chosen vertices, we are doing so in proportion to the degree of the target. Thus we arrive once more at a process in which vertices gain new edges in proportion to the number they already have, just as in preferential attachment, and so again we get a power-law degree distribution.[13]

Solé *et al.* and Vázquez *et al.* recognized that duplication mechanisms could explain the scale-free nature of protein interaction networks as well. Both groups of authors study simple models of network evolution by gene duplication in which one starts with a few vertices (genes). At each time-step a vertex is chosen at random to undergo duplication, resulting in two identical vertices with edges linking each to the same other proteins that the duplicated protein previously interacted with. Thus the probability of a vertex gaining a new edge from one of these copied vertices is proportional to the number of edges it already has, and this mechanism, as in the preferential attachment models, produces a power-law degree distribution. With time, because of random mutations, some interactions will be lost, and this process is modeled by selecting edges at random with a certain probability and removing them from the network. Edge deletion is necessary to create diversity in the network; without it all vertices would be copies of the initial few with which the network started. It is also possible that on rare occasions proteins develop new interactions with others that they did not previously interact with. While Vázquez *et al.* ignore this possibility in their version of the model, Solé *et al.* include it by adding new edges to the system with a certain probability. This minor difference notwithstanding, both models are found in numerical simulations to give, as expected, power-law degree distributions in the limit of long times. And with appropriate choices of the model parameters it can be arranged for the emerging network to have properties similar to empirically observed biological networks.

---

[13]This idea has also been used by Krapivsky and Redner (2001), who discovered it independently, as the basis for a fast computer algorithm for growing networks with power-law degree distributions.

The most important contribution of the work by Solé *et al.* and Vázquez *et al.* is that it provides a plausible link between the scale-free form of biological networks and a well-known biological mechanism, namely gene duplication. With the increasing interest of the biological community in networks, these ideas represent only the first steps toward a greater understanding of the statistical structure of cellular networks. Further important steps in this direction have been taken by, among others, Kim *et al.* (2002) and Chung *et al.* (2003).

Emergence of Scaling in Random Networks

Albert-László Barabási* and Réka Albert

Systems as diverse as genetic networks or the World Wide Web are best described as networks with complex topology. A common property of many large networks is that the vertex connectivities follow a scale-free power-law distribution. This feature was found to be a consequence of two generic mechanisms: (i) networks expand continuously by the addition of new vertices, and (ii) new vertices attach preferentially to sites that are already well connected. A model based on these two ingredients reproduces the observed stationary scale-free distributions, which indicates that the development of large networks is governed by robust self-organizing phenomena that go beyond the particulars of the individual systems.

The inability of contemporary science to describe systems composed of nonidentical elements that have diverse and nonlocal interactions currently limits advances in many disciplines, ranging from molecular biology to computer science (*1*). The difficulty of describing these systems lies partly in their topology: Many of them form rather complex networks whose vertices are the elements of the system and whose edges represent the interactions between them. For example, liv-

Department of Physics, University of Notre Dame, Notre Dame, IN 46556, USA.

*To whom correspondence should be addressed. E-mail: alb@nd.edu

ing systems form a huge genetic network whose vertices are proteins and genes, the chemical interactions between them representing edges (2). At a different organizational level, a large network is formed by the nervous system, whose vertices are the nerve cells, connected by axons (3). But equally complex networks occur in social science, where vertices are individuals or organizations and the edges are the social interactions between them (4), or in the World Wide Web (WWW), whose vertices are HTML documents connected by links pointing from one page to another (5, 6). Because of their large size and the complexity of their interactions, the topology of these networks is largely unknown.

Traditionally, networks of complex topology have been described with the random graph theory of Erdős and Rényi (ER) (7), but in the absence of data on large networks, the predictions of the ER theory were rarely tested in the real world. However, driven by the computerization of data acquisition, such topological information is increasingly available, raising the possibility of understanding the dynamical and topological stability of large networks.

Here we report on the existence of a high degree of self-organization characterizing the large-scale properties of complex networks. Exploring several large databases describing the topology of large networks that span fields as diverse as the WWW or citation patterns in science, we show that, independent of the system and the identity of its constituents, the probability $P(k)$ that a vertex in the network interacts with k other vertices decays as a power law, following $P(k) \sim k^{-\gamma}$. This result indicates that large networks self-organize into a scale-free state, a feature unpredicted by all existing random network models. To explain the origin of this scale invariance, we show that existing network models fail to incorporate growth and preferential attachment, two key features of real networks. Using a model incorporating these two ingredients, we show that they are responsible for the power-law scaling observed in real networks. Finally, we argue that these ingredients play an easily identifiable and important role in the formation of many complex systems, which implies that our results are relevant to a large class of networks observed in nature.

Although there are many systems that form complex networks, detailed topological data is available for only a few. The collaboration graph of movie actors represents a well-documented example of a social network. Each actor is represented by a vertex, two actors being connected if they were cast together in the same movie. The probability that an actor has k links (characterizing his or her popularity) has a power-law tail for large k, following $P(k) \sim k^{-\gamma_{actor}}$, where $\gamma_{actor} = 2.3 \pm 0.1$ (Fig. 1A). A more complex network with over 800 million vertices (8) is the WWW, where a vertex is a document and the edges are the links pointing from one document to another. The topology of this graph determines the Web's connectivity and, consequently, our effectiveness in locating information on the WWW (5). Information about $P(k)$ can be obtained using robots (6), indicating that the probability that k documents point to a certain Web page follows a power law, with $\gamma_{www} = 2.1 \pm 0.1$ (Fig. 1B) (9). A network whose topology reflects the historical patterns of urban and industrial development is the electrical power grid of the western United States, the vertices being generators, transformers, and substations and the edges being to the high-voltage transmission lines between them (10). Because of the relatively modest size of the network, containing only 4941 vertices, the scaling region is less prominent but is nevertheless approximated by a power law with an exponent $\gamma_{power} \simeq 4$ (Fig. 1C). Finally, a rather large complex network is formed by the citation patterns of the scientific publications, the vertices being papers published in refereed journals and the edges being links to the articles cited in a paper. Recently Redner (11) has shown that the probability that a paper is cited k times (representing the connectivity of a paper within the network) follows a power law with exponent $\gamma_{cite} = 3$.

The above examples (12) demonstrate that many large random networks share the common feature that the distribution of their local connectivity is free of scale, following a power law for large k with an exponent γ between 2.1 and 4, which is unexpected within the framework of the existing network models. The random graph model of ER (7) assumes that we start with N vertices and connect each pair of vertices with probability p. In the model, the probability that a vertex has k edges follows a Poisson distribution $P(k) = e^{-\lambda}\lambda^k/k!$, where

$$\lambda = N\binom{N-1}{k}p^k(1-p)^{N-1-k}$$

In the small-world model recently introduced by Watts and Strogatz (WS) (10), N vertices form a one-dimensional lattice, each vertex being connected to its two nearest and next-nearest neighbors. With probability p, each edge is reconnected to a vertex chosen at random. The long-range connections generated by this process decrease the distance between the vertices, leading to a small-world phenomenon (13), often referred to as six degrees of separation (14). For $p = 0$, the probability distribution of the connectivities is $P(k) = \delta(k - z)$, where z is the coordination number in the lattice; whereas for finite p, $P(k)$ still peaks around z, but it gets broader (15). A common feature of the ER and WS models is that the probability of finding a highly connected vertex (that is, a large k) decreases exponentially with k; thus, vertices with large connectivity are practically absent. In contrast, the power-law tail characterizing $P(k)$ for the networks studied indicates that highly connected (large k) vertices have a large chance of occurring, dominating the connectivity.

There are two generic aspects of real networks that are not incorporated in these models. First, both models assume that we start with a fixed number (N) of vertices that are then randomly connected (ER model), or reconnected (WS model), without modifying N. In contrast, most real world networks are open and they form by the continuous addition of new vertices to the system, thus the number of vertices N increases throughout the lifetime of the network. For example, the actor network grows by the addition of new actors to the system, the WWW grows exponentially over time by the addition of new Web pages (8), and the research literature constantly grows by the publication of new papers. Consequently, a common feature of

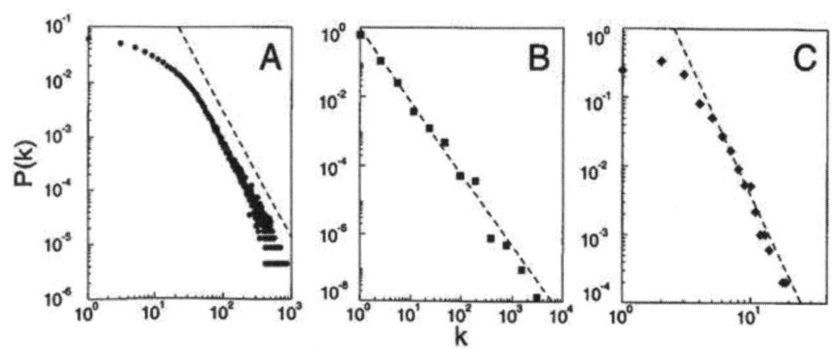

Fig. 1. The distribution function of connectivities for various large networks. **(A)** Actor collaboration graph with $N = 212{,}250$ vertices and average connectivity $\langle k \rangle = 28.78$. **(B)** WWW, $N = 325{,}729$, $\langle k \rangle = 5.46$ (6). **(C)** Power grid data, $N = 4941$, $\langle k \rangle = 2.67$. The dashed lines have slopes (A) $\gamma_{actor} = 2.3$, (B) $\gamma_{www} = 2.1$ and (C) $\gamma_{power} = 4$.

these systems is that the network continuously expands by the addition of new vertices that are connected to the vertices already present in the system.

Second, the random network models assume that the probability that two vertices are connected is random and uniform. In contrast, most real networks exhibit preferential connectivity. For example, a new actor is most likely to be cast in a supporting role with more established and better-known actors. Consequently, the probability that a new actor will be cast with an established one is much higher than that the new actor will be cast with other less-known actors. Similarly, a newly created Web page will be more likely to include links to well-known popular documents with already-high connectivity, and a new manuscript is more likely to cite a well-known and thus much-cited paper than its less-cited and consequently less-known peer. These examples indicate that the probability with which a new vertex connects to the existing vertices is not uniform; there is a higher probability that it will be linked to a vertex that already has a large number of connections.

We next show that a model based on these two ingredients naturally leads to the observed scale-invariant distribution. To incorporate the growing character of the network, starting with a small number (m_0) of vertices, at every time step we add a new vertex with $m(\leq m_0)$ edges that link the new vertex to m different vertices already present in the system. To incorporate preferential attachment, we assume that the probability Π that a new vertex will be connected to vertex i depends on the connectivity k_i of that vertex, so that $\Pi(k_i) = k_i/\Sigma_j k_j$. After t time steps, the model leads to a random network with $t + m_0$ vertices and mt edges. This network evolves into a scale-invariant state with the probability that a vertex has k edges, following a power law with an exponent $\gamma_{model} = 2.9 \pm 0.1$ (Fig. 2A). Because the power law observed for real networks describes systems of rather different sizes at different stages of their development, it is expected that a correct model should provide a distribution whose main features are independent of time. Indeed, as Fig. 2A demonstrates, $P(k)$ is independent of time (and subsequently independent of the system size $m_0 + t$), indicating that despite its continuous growth, the system organizes itself into a scale-free stationary state.

The development of the power-law scaling in the model indicates that growth and preferential attachment play an important role in network development. To verify that both ingredients are necessary, we investigated two variants of the model. Model A keeps the growing character of the network, but preferential attachment is eliminated by assuming that a new vertex is connected with equal probability to any vertex in the system [that is, $\Pi(k) = const = 1/(m_0 + t - 1)$]. Such a model (Fig. 2B) leads to $P(k) \sim \exp(-\beta k)$, indicating that the absence of preferential attachment eliminates the scale-free feature of the distribution. In model B, we start with N vertices and no edges. At each time step, we randomly select a vertex and connect it with probability $\Pi(k_i) = k_i/\Sigma_j k_j$ to vertex i in the system. Although at early times the model exhibits power-law scaling, $P(k)$ is not stationary: because N is constant and the number of edges increases with time, after $T \simeq N^2$ time steps the system reaches a state in which all vertices are connected. The failure of models A and B indicates that both ingredients—growth and preferential attachment—are needed for the development of the stationary power-law distribution observed in Fig. 1.

Because of the preferential attachment, a vertex that acquires more connections than another one will increase its connectivity at a higher rate; thus, an initial difference in the connectivity between two vertices will increase further as the network grows. The rate at which a vertex acquires edges is $\partial k_i/\partial t = k_i/2t$, which gives $k_i(t) = m(t/t_i)^{0.5}$, where t_i is the time at which vertex i was added to the system (see Fig. 2C), a scaling property that could be directly tested once time-resolved data on network connectivity becomes available. Thus older (with smaller t_i) vertices increase their connectivity at the expense of the younger (with larger t_i) ones, leading over time to some vertices that are highly connected, a "rich-get-richer" phenomenon that can be easily detected in real networks. Furthermore, this property can be used to calculate γ analytically. The probability that a vertex i has a connectivity smaller than k, $P[k_i(t) < k]$, can be written as $P(t_i > m^2t/k^2)$. Assuming that we add the vertices to the system at equal time intervals, we obtain $P(t_i > m^2t/k^2) = 1 - P(t_i \leq m^2t/k^2) = 1 - m^2t/k^2(t + m_0)$. The probability density $P(k)$ can be obtained from $P(k) = \partial P[k_i(t) < k]/\partial k$, which over long time periods leads to the stationary solution

$$P(k) = \frac{2m^2}{k^3}$$

giving $\gamma = 3$, independent of m. Although it reproduces the observed scale-free distribution, the proposed model cannot be expected to account for all aspects of the studied networks. For that, we need to model these systems in more detail. For example, in the model we assumed linear preferential attachment; that is, $\Pi(k) \sim k$. However, although in general $\Pi(k)$ could have an arbitrary nonlinear form $\Pi(k) \sim k^\alpha$, simulations indicate that scaling is present only for $\alpha = 1$. Furthermore, the exponents obtained for the different networks are scattered between 2.1 and 4. However, it is easy to modify our model to account for exponents different from $\gamma = 3$. For example, if we assume that a fraction p of the links is directed, we obtain $\gamma(p) = 3 - p$, which is supported by numerical simulations (*16*). Finally, some networks evolve not only by adding new vertices but by adding (and sometimes removing) connections between established vertices. Although these and other system-specific features could modify the exponent γ, our model offers the first successful mechanism accounting for the scale-invariant nature of real networks.

Growth and preferential attachment are mechanisms common to a number of complex systems, including business networks (*17, 18*), social networks (describing individuals or organizations), transportation networks (*19*), and so on. Consequently, we expect that the scale-invariant state observed in all systems for which detailed data has been available to us is a generic property of many complex networks, with applicability reaching far beyond the quoted examples. A better description of these systems would help in understanding other complex systems

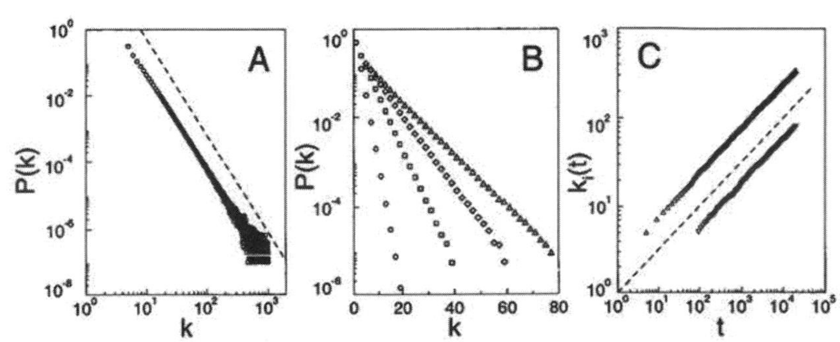

Fig. 2. **(A)** The power-law connectivity distribution at $t = 150{,}000$ (○) and $t = 200{,}000$ (□) as obtained from the model, using $m_0 = m = 5$. The slope of the dashed line is $\gamma = 2.9$. **(B)** The exponential connectivity distribution for model A, in the case of $m_0 = m = 1$ (○), $m_0 = m = 3$ (□), $m_0 = m = 5$ (◇), and $m_0 = m = 7$ (△). **(C)** Time evolution of the connectivity for two vertices added to the system at $t_1 = 5$ and $t_2 = 95$. The dashed line has slope 0.5.

as well, for which less topological information is currently available, including such important examples as genetic or signaling networks in biological systems. We often do not think of biological systems as open or growing, because their features are genetically coded. However, possible scale-free features of genetic and signaling networks could reflect the networks' evolutionary history, dominated by growth and aggregation of different constituents, leading from simple molecules to complex organisms. With the fast advances being made in mapping out genetic networks, answers to these questions might not be too far away. Similar mechanisms could explain the origin of the social and economic disparities governing competitive systems, because the scale-free inhomogeneities are the inevitable consequence of self-organization due to the local decisions made by the individual vertices, based on information that is biased toward the more visible (richer) vertices, irrespective of the nature and origin of this visibility.

References and Notes

1. R. Gallagher and T. Appenzeller, *Science* **284**, 79 (1999); R. F. Service, *ibid.*, p. 80.
2. G. Weng, U. S. Bhalla, R. Iyengar, *ibid.*, p. 92.
3. C. Koch and G. Laurent, *ibid.*, p. 96.
4. S. Wasserman and K. Faust, *Social Network Analysis* (Cambridge Univ. Press, Cambridge, 1994).
5. Members of the *Clever* project, *Sci. Am.* **280**, 54 (June 1999).
6. R. Albert, H. Jeong, A.-L. Barabási, *Nature* **401**, 130 (1999); A.-L. Barabási, R. Albert, H. Jeong, *Physica A* **272**, 173 (1999); see also http://www.nd.edu/~networks.
7. P. Erdős and A. Rényi, *Publ. Math. Inst. Hung. Acad. Sci.* **5**, 17 (1960); B. Bollobás, *Random Graphs* (Academic Press, London, 1985).
8. S. Lawrence and C. L. Giles, *Science* **280**, 98 (1998); *Nature* **400**, 107 (1999).
9. In addition to the distribution of incoming links, the WWW displays a number of other scale-free features characterizing the organization of the Web pages within a domain [B. A. Huberman and L. A. Adamic, *Nature* **401**, 131 (1999)], the distribution of searches [B. A. Huberman, P. L. T. Pirolli, J. E. Pitkow, R. J. Lukose, *Science* **280**, 95 (1998)], or the number of links per Web page (*6*).
10. D. J. Watts and S. H. Strogatz, *Nature* **393**, 440 (1998).
11. S. Redner, *Eur. Phys. J. B* **4**, 131 (1998).
12. We also studied the neural network of the worm *Caenorhabditis elegans* (*3, 10*) and the benchmark diagram of a computer chip (see http://vlsicad.cs.ucla.edu/~cheese/ispd98.html). We found that $P(k)$ for both was consistent with power-law tails, despite the fact that for *C. elegans* the relatively small size of the system (306 vertices) severely limits the data quality, whereas for the wiring diagram of the chips, vertices with over 200 edges have been eliminated from the database.
13. S. Milgram, *Psychol. Today* **2**, 60 (1967); M. Kochen, ed., *The Small World* (Ablex, Norwood, NJ, 1989).
14. J. Guare, *Six Degrees of Separation: A Play* (Vintage Books, New York, 1990).
15. M. Barthélémy and L. A. N. Amaral, *Phys. Rev. Lett.* **82**, 15 (1999).
16. For most networks, the connectivity m of the newly added vertices is not constant. However, choosing m randomly will not change the exponent γ (Y. Tu, personal communication).
17. W. B. Arthur, *Science* **284**, 107 (1999).
18. Preferential attachment was also used to model evolving networks (L. A. N. Amaral and M. Barthélémy, personal communication).
19. J. R. Banavar, A. Maritan, A. Rinaldo, *Nature* **399**, 130 (1999).

20. We thank D. J. Watts for providing the *C. elegans* and power grid data, B. C. Tjaden for supplying the actor data, H. Jeong for collecting the data on the WWW, and L. A. N. Amaral for helpful discussions. This work was partially supported by NSF Career Award DMR-9710998.

24 June 1999; accepted 2 September 1999

VOLUME 85, NUMBER 21 PHYSICAL REVIEW LETTERS 20 NOVEMBER 2000

Structure of Growing Networks with Preferential Linking

S. N. Dorogovtsev,[1,2,*] J. F. F. Mendes,[1,†] and A. N. Samukhin[2,‡]
[1]Departamento de Física and Centro de Física do Porto, Faculdade de Ciências, Universidade do Porto, Rua do Campo Alegre 687, 4169-007 Porto, Portugal
[2]A. F. Ioffe Physico-Technical Institute, 194021 St. Petersburg, Russia
(Received 10 April 2000)

The model of growing networks with the preferential attachment of new links is generalized to include initial attractiveness of sites. We find the exact form of the stationary distribution of the number of incoming links of sites in the limit of long times, $P(q)$, and the long-time limit of the average connectivity $\overline{q}(s,t)$ of a site s at time t (one site is added per unit of time). At long times, $P(q) \sim q^{-\gamma}$ at $q \to \infty$ and $\overline{q}(s,t) \sim (s/t)^{-\beta}$ at $s/t \to 0$, where the exponent γ varies from 2 to ∞ depending on the initial attractiveness of sites. We show that the relation $\beta(\gamma - 1) = 1$ between the exponents is universal.

PACS numbers: 84.35.+i, 05.40.–a, 05.50.+q, 87.18.Sn

It was observed recently that the distributions of several quantities in various growing networks have a power-law form. This scaling behavior was observed in the World Wide Web, in neural and social networks, in nets of citations of scientific papers, etc.; see [1–14], and references therein. These observations challenge us to find the general reasons of such behavior. It is only recently that scientists became aware of the ever increasing impact of various evolving networks on everyone's life. Earlier studies [15–20] concentrated on simple random networks, and it was recently discovered that many complex networks are hierarchically organized [5,6,21].

Mostly, the interest is concentrated on the distribution of shortest paths between the different sites of a network [1] and on the distribution of the number of connections with a site [2–6]. The second quantity is obviously simpler to obtain than the first one but even for it, in the case of the networks with scaling behavior, no exact results are known.

The only known mechanism of self-organization of a growing network into a free-scale structure is preferential linking [7–9], i.e., new links are preferentially attached to sites with high numbers of connections. A simple model of a growing network with preferential linking was proposed by Barabási and Albert [7] (BA model). At each time step a new site is added. It connects with old sites by a fixed number of links. The probability of an old site to get a new link is proportional to the total number of connections with this site. It was found in [7,8] that the distribution of the number of links has a power-law form at long times. The value of the corresponding scaling exponent, γ, obtained using a mean-field approach, equals 3. This value is close to that one observed in the network of citations [3], but other examples of evolving networks show different values of γ. Introduction of the aging of sites changes γ [22] and may even break the scaling behavior [13,22].

In the present Letter, we generalize the BA model and find the exact form of the distribution of *incoming* links of sites in the limit of large sizes of the growing network. We derive a scaling relation connecting the scaling exponent of the distribution of incoming links and the exponent of the temporal behavior of the average connectivity, and demonstrate that it applies for a large class of evolving networks.

The model.—At each time step a new site appears (see Fig. 1). Simultaneously, m new *directed* links coming out from nonspecified sites are introduced. Let the connectivity q_s be the number of incoming links to a site s, i.e., to a site added at time s. The new links are distributed between sites according to the following rule. The probability that a new link points to a given site s is proportional to the following characteristic of the site:

$$A_s = A + q_s , \qquad (1)$$

thereafter called its *attractiveness*. All sites are born with some initial attractiveness $A \geq 0$, but afterwards it increases because of the q_s term. The introduced parameter A, the initial attractiveness, governs the probability for "young" sites to get new links.

We emphasize that we do not specify sites from which the new links come out. They may come out from the new site, from old sites, or even from outside of the network. Our results do not depend on that. Therefore, the model describes also the particular case when every new site is the source of all the m new links like in the BA model. In this case, every site has m outgoing links and the total

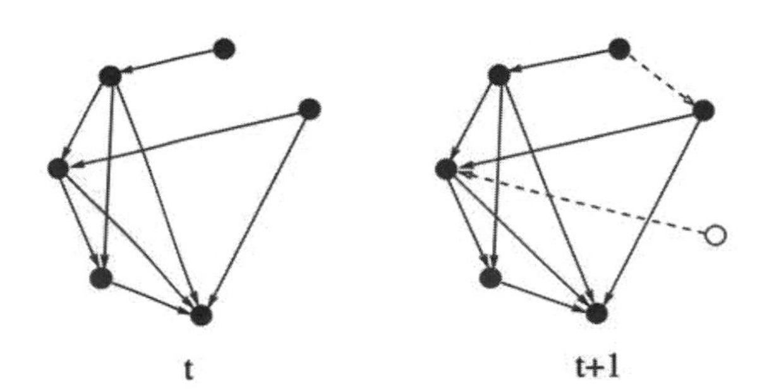

FIG. 1. Illustration of the growing network under consideration. Each instant a new site (open circle) and m (here, $m = 2$) new directed links (dashed arrows) are added. These links are distributed between the sites according to the rule introduced in the text.

0031-9007/00/85(21)/4633(4)$15.00

number of its connections equals $q_s + m$. This number coincides with the attractiveness of the site, Eq. (1), if one sets $A = m$. In this case, our rule for the distribution of new links among sites coincides with the corresponding rule of the BA model. Hence, the model that we consider here is equivalent to the BA model in the particular case of the initial attractiveness equal to m.

In fact, our model may be mapped to the following general problem. Each instant, m new particles (i.e., incoming links) have to be distributed between an *increasing* number (one per time step) of boxes (i.e., sites) according to the introduced rule.

The master equation.—Let us derive the equation for the distribution $p(q,s,t)$ of the connectivity q of the site s. At time t ($t = 1,2,\ldots$) the network consists of t sites connected by $m(t-1)$ directed links (it is convenient to assume that initially ($t = 1$) we have one site with m incoming links—the resulting behavior at long times is independent of the initial condition). The total attractiveness of the network at time t is $A_\Sigma = (m + A)t = (1 + a)m$, where $a \equiv A/m$. Note that one may allow multiple links, i.e., the connectivity of a given site may increase simultaneously by more than one. That is not essential for long times, when probability to receive simultaneously more than one link of the m is vanishing. The probability that a new link is connected with the site s equals A_s/A_Σ. The probability for the site s to receive exactly l new links of the m injected is $\mathcal{P}_s^{(ml)} = \binom{m}{l}(A_s/A_\Sigma)^l(1 - A_s/A_\Sigma)^{m-l}$. Hence, the connectivity distribution of a site obeys the following master equation:

$$p(q,s,t+1) = \sum_{l=0}^{m} \mathcal{P}_s^{(ml)} p(q-l,s,t) = \sum_{l=0}^{m} \binom{m}{l} \left[\frac{q-l+am}{(1+a)mt}\right]^l \left[1 - \frac{q-l+am}{(1+a)mt}\right]^{m-l} p(q-l,s,t). \tag{2}$$

Equation (2) is supplied with the initial condition $p(q,s,s) = \delta(q)$, which means that sites are born with zero connectivity (i.e., without incoming links in our definition).

The connectivity distribution of the entire network is $P(q,t) = \sum_{u=1}^{t} p(q,u,t)/t$. Summing up Eq. (2) over s from 1 to t, one gets

$$(t+1)P(q,t+1) - p(q,t+1,t+1) = \left(t - \frac{q+am}{1+a}\right)P(q,t) + \frac{q-1+am}{1+a}P(q-1,t) + \mathcal{O}\left(\frac{P}{t}\right). \tag{3}$$

At long times, we obtain

$$(1+a)t\frac{\partial P}{\partial t}(q,t) + (1+a)P(q,t) + (q+am)P(q,t) - (q-1+am)P(q-1,t) = (1+a)\delta(q). \tag{4}$$

Finally, assuming that the limit $P(q) = P(q, t \to \infty)$ exists, we get the following equation for the stationary connectivity distribution:

$$(1+a)P(q) + (q+ma)P(q) - (q-1+ma)P(q-1) = (1+a)\delta(q). \tag{5}$$

The stationary distribution.—To solve Eq. (5) one may use the Z transform of the distribution function:

$$\Phi(z) = \sum_{q=0}^{\infty} P(q)z^q. \tag{6}$$

Then one gets from Eq. (5)

$$z(1-z)\frac{d\Phi}{dz} + ma(1-z)\Phi + (1+a)\Phi = 1 + a. \tag{7}$$

The solution of Eq. (7) that is analytic at $z = 0$ has the following form:

$$\Phi(z) = (1+a)z^{-1-(m+1)a}(1-z)^{1+a}\int_0^z dx\,\frac{x^{(m+1)a}}{(1-x)^{2+a}} = \frac{1+a}{1+(m+1)a}\,{}_2F_1[1,ma;2+(m+1)a;z], \tag{8}$$

where ${}_2F_1[\,]$ is the hypergeometric function. Using its expansion [23] in z, we obtain, comparing with Eq. (6),

$$P(q) = (1+a)\frac{\Gamma[(m+1)a+1]}{\Gamma(ma)}\frac{\Gamma(q+ma)}{\Gamma[q+2+(m+1)a]}, \tag{9}$$

that is our main result (see Fig. 2). In particular, when $a = 1$, that corresponds to the case $A_s = m + q_s$, studied in [7,8], we get

$$P(q) = \frac{2m(m+1)}{(q+m)(q+m+1)(q+m+2)}. \tag{10}$$

This expression in the limit $q \to \infty$ approaches the corresponding result of [7,8] obtained in the frames of an approximate scheme, but the prefactors are different. In fact, the "mean field" approach, used in [8], is equivalent to the continuous-q approximation in our discrete-difference equations. Indeed, if we replace the finite difference with a derivative over q, we get the expression obtained in [7,8].

For $ma + q \gg 1$, the distribution function (9) takes the form

$$P(q) \cong (1 + a)\frac{\Gamma[(m+1)a+1]}{\Gamma(ma)}(q + ma)^{-(2+a)}. \tag{11}$$

Therefore, we find the scaling exponent γ of the distribution function:

$$\gamma = 2 + a = 2 + A/m, \tag{12}$$

where A is the initial attractiveness of a site.

The distribution $p(q,s,t)$.—Let us find the connectivity distribution $p(q,s,t)$ for the site s. At long times $t \gg 1$, keeping only two leading terms in $1/t$ in Eq. (2), one gets

$$p(q,s,t+1) = \left[1 - \frac{q+am}{(1+a)t}\right]p(q,s,t) + \frac{q-1+am}{(1+a)t}\,p(q-1,s,t) + \mathcal{O}\left(\frac{p}{t^2}\right). \tag{13}$$

Assuming that the scale of time variation is much larger than 1, we can replace the finite t difference with a derivative

$$(1+a)t\frac{\partial p}{\partial t}(q,s,t) = (q-1+am)p(q-1,s,t) - (q+am)p(q,s,t). \tag{14}$$

Finally, using the Z transform in the similar way as before, we obtain the solution of Eq. (14), i.e., the connectivity distribution of individual sites:

$$p(q,s,t) = \frac{\Gamma(am+q)}{\Gamma(am)q!}\left(\frac{s}{t}\right)^{am/(1+a)}\left[1 - \left(\frac{s}{t}\right)^{1/(1+a)}\right]^q. \tag{15}$$

Hence, this distribution has an exponential tail. Now one may get also the expression for the average connectivity of a given site:

$$\overline{q}(s,t) = \sum_{q=0}^{\infty} qp(q,s,t) = am\left[\left(\frac{s}{t}\right)^{-1/(1+a)} - 1\right]. \tag{16}$$

Thus, at a fixed time t the average connectivity of an old site $s \ll t$ depends upon its age as $\sim s^{-\beta}$, where the exponent $\beta = 1/(1+a)$. Therefore, we have the following relation between the exponents of the considered network:

$$\beta(\gamma - 1) = 1, \tag{17}$$

that was previously obtained in the continuous approximation [22].

We can show that Eq. (17) is universal and may be obtained from the most general considerations. In fact, we assume only that the averaged connectivity $\overline{q}(s,t)$ and the connectivity distribution $P(q)$ show scaling behavior. Then, in the scaling region, the quantity of interest, i.e., the probability $p(q,s,t)$, has to be of the following form: $p(q,s,t) = (s/t)^{\Delta_1} f[q^{\Delta_2}(s/t)^{\Delta_3}]$. Obviously, one can set $\Delta_2 = 1$. $\Delta_1 = \Delta_3$ because of the normalization condition for $p(q,s,t)$ at a fixed s, $\sum_{q=0}^{\infty} p(q,s,t) = 1$. Then, the relation $\overline{q}(s,t) \propto (s/t)^{-\beta}$ leads to $\Delta_1 = \Delta_3 = \beta$ [we use the definition (16)], and finally, inserting $p(q,s,t)$ in such a form into the relation $P(q) \propto q^{-\gamma}$ at large q and t, one gets the relation (17).

The network that we consider here belongs to the class of scale-free growing networks (we use the classification of growing networks presented in [13]). For the known real networks of this class (see the most complete description [13]), no data for the variation of the average connectivity of a site with its age are available yet. These data are necessary to obtain the exponent β. It would be intriguing to study this quantity in the real scale-free networks and to check Eq. (17). One should note that, for the network with aging of sites, the relation (17) was already confirmed by the simulation [22].

The particular form of the scaling function $f(\xi)$, $\xi \equiv q(s/t)^{-\beta}$, depends on the specific model of the growing network. In the case under consideration, it follows from Eq. (15) that

$$f(\xi) = \frac{1}{\Gamma(am)}\xi^{am-1}\exp(-\xi) \tag{18}$$

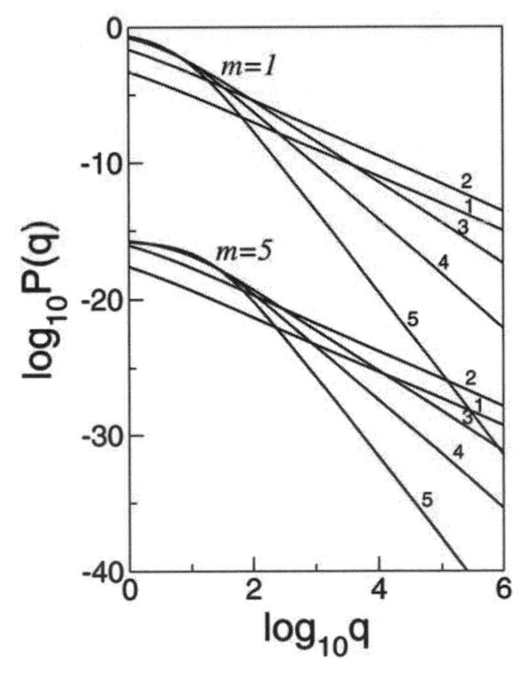

FIG. 2. Log-log plot of the distribution of the incoming links of sites for $m = 1$ and $m = 5$ (the curves for $m = 5$ are displaced down by 15). (1) $a = 0.001$, (2) $a = 0.05$, (3) $a = 1.0$ (BA model), (4) $a = 2.0$, (5) $a = 4.0$.

for $s/t \to 0$, $q \to \infty$ and the fixed $q(s/t)^{\beta}$. (Here, we use the asymptote: $\Gamma(am + q)/q! \to q^{am-1}$ at $q \to \infty$.)

Discussion.—In the limit of zero initial attractiveness of sites ($a = 0$) all the links lead to the first site since all others have no chance to get a new incoming link. In this case, Eqs. (12) and (17) give $\gamma = 2$ and $\beta = 1$. For $a = 1$, i.e., for the BA model, $\gamma = 3$ and $\beta = 1/2$ [7,8]. Finally, when $a \to \infty$, i.e., all sites have equal attractivity all the time, and the scaling breaks, one sees from Eqs. (12) and (17) that $\gamma \to \infty$ and $\beta \to 0$. In the last case, our system appears to be out of a class of networks with preferential linking. Note that the ranges of variation of γ, $2 < \gamma < \infty$, and β, $1 < \beta < 0$, are the same as for the scale-free network with aging of sites [22]. Note also that the observed value of the scaling exponent of the distribution of incoming links in the World Wide Web, 2.1 [5,9], is in this range.

We see that the approach of [7,8] based on the continuous-q approximation gives the proper values for the critical exponents (see also [22] for the network with aging of sites). Thus, this approximation is effective for calculation of the exponents of the scale-free networks.

A two-parameter fitting was proposed in [24] to describe the observed distribution for the citations of scientific papers [3]. One sees that the connectivity distribution of the considered growing networks is of quite different form. It seems that the difference occurs because we study the *growing* structure unlike the approach [24].

In conclusion, in the limit of long times (large sizes of the growing networks), we have found the exact solution of the master equation for the distribution of incoming links. This solution demonstrates the existence of the scaling region in a class of the growing networks with the preferential linking.

The considered growing networks are self-organized into free-scale structures. The input flow of the new links is distributed between the increasing number of sites. The scaling exponents are determined by the value of the initial attractiveness ascribed to every new site. Depending on this quantity, the scaling exponent γ of the connectivity distribution takes values from 2 to ∞. We have shown that the relation (17) between the scaling exponents γ and β is valid for a wide class of growing networks.

S. N. D. thanks PRAXIS XXI (Portugal) for Research Grant No. PRAXIS XXI/BCC/16418/98. J. F. F. M. was partially supported by FCT Sapiens Project No. 33141/99. We also thank E. J. S. Lage for reading the manuscript and A. V. Goltsev and Yu. G. Pogorelov for many useful discussions.

Note added.—After submission of this manuscript we learned of Krapivsky and co-workers' work [25] which overlaps some of our results.

*Email address: sdorogov@fc.up.pt
†Email address: jfmendes@fc.up.pt
‡Email address: alnis@samaln.ioffe.rssi.ru

[1] D. J. Watts and S. H. Strogatz, Nature (London) **393**, 440 (1998).
[2] J. Laherrere and D. Sornette, Eur. Phys. J. B **2**, 525 (1998).
[3] S. Redner, Eur. Phys. J. B **4**, 131 (1998).
[4] B. A. Huberman, P. L. T. Pirolli, J. E. Pitkow, and R. J. Lukose, Science **280**, 95 (1998).
[5] R. Albert, H. Jeong, and A.-L. Barabási, Nature (London) **401**, 130 (1999).
[6] B. A. Huberman and L. A. Adamic, Nature (London) **401**, 131 (1999).
[7] A.-L. Barabási and R. Albert, Science **286**, 509 (1999).
[8] A.-L. Barabási, R. Albert, and H. Jeong, Physica (Amsterdam) **272A**, 173 (1999).
[9] R. Kumar, P. Raghavan, S. Rajagopalan, and A. Tomkins, in *Proceedings of the 25th VLDB Conference, Edinburgh, 1999* (Morgan Kaufmann, Orlando, FL, 1999), pp. 639–650.
[10] M. Barthélémy and L. A. N. Amaral, Phys. Rev. Lett. **82**, 3180 (1999); Phys. Rev. Lett. **82**, 5180(E) (1999).
[11] M. E. J. Newman and D. J. Watts, Phys. Lett. A **263**, 341 (1999).
[12] A. Barrat and M. Weigt, Eur. Phys. J. B **13**, 547 (2000).
[13] L. A. N. Amaral, A. Scala, M. Barthelemy, and H. E. Stanley, Proc. Natl. Acad. Sci. U.S.A. **97**, 11 149 (2000).
[14] S. N. Dorogovtsev and J. F. F. Mendes, Europhys. Lett. **50**, 1 (2000).
[15] P. Erdös and A. Renýi, Publ. Math. Inst. Hung. Acad. Sci. **5**, 17 (1960).
[16] B. Bollobás, *Random Graphs* (Academic Press, London, 1985).
[17] G. Parisi, J. Phys. A **19**, L675 (1986).
[18] B. Derrida and Y. Pomeau, Europhys. Lett. **1**, 45 (1986).
[19] B. Derrida and D. Stauffer, Europhys. Lett. **2**, 739 (1987).
[20] R. Monasson and R. Zecchina, Phys. Rev. Lett. **75**, 2432 (1995).
[21] D. J. Watts, *Small Worlds* (Princeton University Press, Princeton, New Jersey, 1999).
[22] S. N. Dorogovtsev and J. F. F. Mendes, Phys. Rev. E **62**, 1842 (2000).
[23] H. Bateman and A. Erdélyi, *Higher Transcendental Functions* (McGraw-Hill, New York, 1953), Vol. 1.
[24] C. Tsallis and M. P. de Albuquerque, Eur. Phys. J. B **13**, 777 (2000).
[25] P. L. Krapivsky, S. Redner, and F. Leyvraz, preceding Letter, Phys. Rev. Lett. **85**, 4629 (2000).

VOLUME 85, NUMBER 21 PHYSICAL REVIEW LETTERS 20 NOVEMBER 2000

Connectivity of Growing Random Networks

P. L. Krapivsky,[1,2] S. Redner,[1] and F. Leyvraz[3]
[1]Center for BioDynamics, Center for Polymer Studies, and Department of Physics, Boston University, Boston, Massachusetts 02215
[2]CNRS, IRSAMC, Laboratoire de Physique Quantique, Université Paul Sabatier, 31062 Toulouse, France
[3]Centro Internacional de Ciencias, Cuernavaca, Morelos, Mexico
(Received 8 May 2000)

A solution for the time- and age-dependent connectivity distribution of a growing random network is presented. The network is built by adding sites that link to earlier sites with a probability A_k which depends on the number of preexisting links k to that site. For homogeneous connection kernels, $A_k \sim k^\gamma$, different behaviors arise for $\gamma < 1$, $\gamma > 1$, and $\gamma = 1$. For $\gamma < 1$, the number of sites with k links, N_k, varies as a stretched exponential. For $\gamma > 1$, a single site connects to nearly all other sites. In the borderline case $A_k \sim k$, the power law $N_k \sim k^{-\nu}$ is found, where the exponent ν can be tuned to any value in the range $2 < \nu < \infty$.

PACS numbers: 84.35.+i, 05.40.–a, 05.50.+q, 87.18.Sn

Random networks play an important role in epidemiology, ecology (food webs), and many other fields. The geometry of such fixed topology networks have been extensively investigated [1–7]. However, networks based on human interactions, such as transportation systems, electrical distribution systems, biological systems, and the Internet, are open and continuously growing, and new approaches are rapidly developing to understand their structure and time evolution [8–12].

In this Letter, we apply a rate equation approach to solve the growing random network (GN) model, a special case of which was introduced in [13] to account for the distribution of citations and other growing networks [13–18]. Our approach is ideally suited for the GN and is much simpler than the standard probabilistic [1] or generating function [2] techniques. The rate equation formulation can be adapted to study more general evolving graph systems, such as networks with site deletion and link rearrangement.

The GN model is defined as follows. At each time step, a new site is added and a directed link to one of the earlier sites is created. In terms of citations, we may interpret the sites in Fig. 1 as publications, and the directed link from one paper to another as a citation to the earlier publication. This growing network has a directed tree graph topology where the basic elements are sites which are connected by directed links. The structure of this graph is determined by the connection kernel A_k, which is the probability that a newly introduced site links to an existing site with k links ($k - 1$ incoming and 1 outgoing). We will solve for the connectivity distribution $N_k(t)$, defined as the average number of sites with k links as a function of the connection kernel A_k.

We focus on a class of homogeneous connection kernels, $A_k = k^\gamma$, with $\gamma \geq 0$ reflecting the tendency of preferential linking to popular sites. As we shall show, the connectivity distribution crucially depends on whether γ is smaller than, larger than, or equal to unity. For $\gamma < 1$, the connectivity distribution decreases as a stretched exponential in k. The case $\gamma > 1$ leads to phenomenon akin to gelation [19] in which a single "gel" site connects to nearly every other site of the graph. For $\gamma > 2$, this phenomenon is so extreme that the number of connections between other sites is finite in an infinite graph. A power law distribution $N_k \sim k^{-\nu}$ arises *only* for $\gamma = 1$. In this case, finer details of the dependence of the connection kernel on k affect the exponent ν. Hence we consider a more general class of *asymptotically* linear connection kernels, $A_k \sim k$ as $k \to \infty$. We show that ν is tunable to any value in the range $2 < \nu < \infty$. In particular, we can naturally generate values of ν between 2 and 3, as observed in the web graph [10–12] and in movie actor collaboration networks [13].

The rate equations for the time evolution of the connectivity distribution $N_k(t)$ are

$$\frac{dN_k}{dt} = \frac{1}{M_\gamma}[(k-1)^\gamma N_{k-1} - k^\gamma N_k] + \delta_{k1}. \quad (1)$$

The first term accounts for the process in which a site with $k - 1$ links is connected to the new site, leading to a gain in the number of sites with k links. This happens with probability $(k-1)^\gamma/M_\gamma$, where $M_\gamma(t) = \sum j^\gamma N_j(t)$ provides the proper normalization. A corresponding role is played by the second (loss) term on the right-hand side of Eq. (1). The last term accounts for the continuous introduction of new sites with no incoming links.

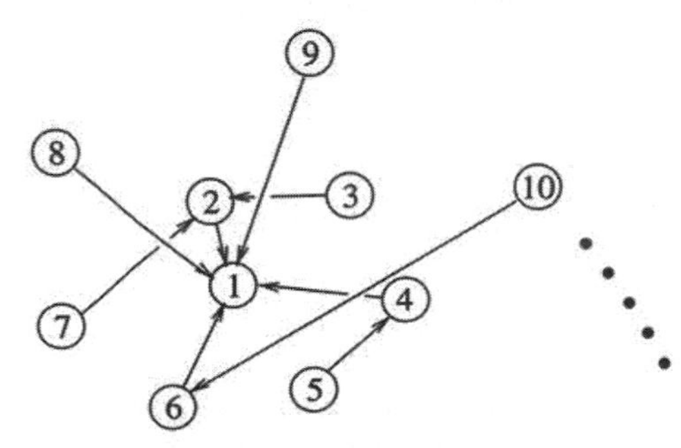

FIG. 1. Schematic illustration of the evolution of the growing random network. Sites are added sequentially and a single link joins the new site to an earlier site.

We start by finding the low-order moments $M_n(t)$ of the connectivity distribution. Summing Eqs. (1) over all k gives the rate equation for the total number of sites, $\dot{M}_0 = 1$, whose solution is $M_0(t) = M_0(0) + t$. The first moment (the total number of bond end points) obeys $\dot{M}_1 = 2$, which gives $M_1(t) = M_1(0) + 2t$. The first two moments are therefore *independent* of γ, while higher moments and the connectivity distribution itself do depend on γ.

For the linear connection kernel, Eqs. (1) can be solved for an arbitrary initial condition. We limit ourselves to the most interesting asymptotic regime ($t \to \infty$) where the initial condition is irrelevant. Using $M_1 = 2t$, we solve the first few of Eqs. (1) and obtain $N_1 = 2t/3, N_2 = t/6$, etc., which implies that the N_k grow linearly with time. Accordingly, we substitute $N_k(t) = tn_k$ in Eqs. (1) to yield the recursion relation $n_k = n_{k-1}(k-1)/(k+2)$. Solving for n_k then gives

$$n_k = \frac{4}{k(k+1)(k+2)}. \tag{2}$$

To solve the model with a sublinear connection kernel, $0 < \gamma < 1$, notice that M_γ satisfies the obvious inequalities $M_0 \le M_\gamma \le M_1$. Consequently, in the long-time limit

$$M_\gamma = \mu t, \qquad 1 \le \mu \le 2, \tag{3}$$

with a yet undetermined prefactor $\mu = \mu(\gamma)$. Now substituting $N_k(t) = tn_k$ and $M_\gamma = \mu t$ into Eqs. (1) and again solving for n_k we obtain

$$n_k = \frac{\mu}{k^\gamma} \prod_{j=1}^{k} \left(1 + \frac{\mu}{j^\gamma}\right)^{-1}, \tag{4}$$

whose asymptotic behavior is

$$n_k \sim \begin{cases} k^{-\gamma} \exp[-\mu(\frac{k^{1-\gamma}-2^{1-\gamma}}{1-\gamma})] & \frac{1}{2} < \gamma < 1, \\ k^{(\mu^2-1/2)} \exp[-2\mu\sqrt{k}] & \gamma = \frac{1}{2}, \\ k^{-\gamma} \exp[-\mu\frac{k^{1-\gamma}}{1-\gamma} + \frac{\mu^2}{2}\frac{k^{1-2\gamma}}{1-2\gamma}] & \frac{1}{3} < \gamma < \frac{1}{2}, \end{cases} \tag{5}$$

etc. This pattern in (5) continues *ad infinitum*: Whenever γ decreases below $1/m$, with m a positive integer, an additional term in the exponential arises from the now relevant contribution of the next higher-order term in the expansion of the product in Eq. (4).

To complete the solution for the n_k, we need to establish the dependence of the amplitude μ on γ. Using the defining relation $M_\gamma/t = \mu = \sum_{k\ge1} k^\gamma n_k$, together with Eq. (4), we obtain the implicit relation for $\mu(\gamma)$,

$$\mu = \sum_{k=2}^{\infty} \prod_{j=2}^{k} \left(1 + \frac{\mu}{j^\gamma}\right)^{-1}. \tag{6}$$

Despite the simplicity of this exact expression, it is not easy to extract explicit information except for the limiting cases $\gamma = 0$ and $\gamma = 1$, where $\mu = 1$ and $\mu = 2$, respectively, and the corresponding connectivity distributions are given by $n_k = 2^{-k}$ and by Eq. (2). However, numerical evaluation shows that μ varies smoothly between 1 and 2 as γ increases from 0 to 1 (Fig. 2). This result, together with Eq. (4), provides a comprehensive description of the connectivity distribution in the regime $0 \le \gamma \le 1$. It is worth emphasizing that for $0.8 \lesssim \gamma \le 1$, n_k depends weakly on γ for $1 \le k \le 1000$. Thus, it is difficult to discriminate between different γ's and even to distinguish a power law from a stretched exponential in the GN model. This subtlety was already encountered in the analysis of the citation distribution [15,16].

A striking feature of the GN model is that we can "tune" the exponent ν by augmenting the linear connection kernel to the asymptotically linear connection kernel, with $A_k \to a_\infty k$ as $k \to \infty$, but otherwise *arbitrary*. For this asymptotically linear kernel, by repeating the steps leading to Eq. (4) we find

$$n_k = \frac{\mu}{A_k} \prod_{j=1}^{k} \left(1 + \frac{\mu}{A_j}\right)^{-1}. \tag{7}$$

Expanding the product in Eq. (7) leads to $n_k \sim k^{-\nu}$ with $\nu = 1 + \mu/a_\infty$, while the amplitude μ is found from

$$\mu = A_1 \sum_{k=2}^{\infty} \prod_{j=2}^{k} \left(1 + \frac{\mu}{A_j}\right)^{-1}. \tag{8}$$

As an explicit example, consider the connection kernel $A_1 = 1$ and $A_k = a_\infty k$ for $k \ge 2$. In this case, we can reduce Eq. (8) to a quadratic equation from which we obtain $\nu = (3 + \sqrt{1 + 8/a_\infty})/2$ which can indeed be tuned to *any* value larger than 2.

The GN model with superlinear connection kernels, $\gamma > 1$, exhibits a "winner takes all" phenomenon, namely, the emergence of a single dominant gel site which is linked to almost every other site. A particularly singular behavior occurs for $\gamma > 2$, where there is a nonzero probability that the initial site is connected to every other site of the graph. To determine this probability, it is convenient to consider a discrete time version process where one site is introduced at each step which always links to the initial site. After N steps, the probability that the new site will link to the

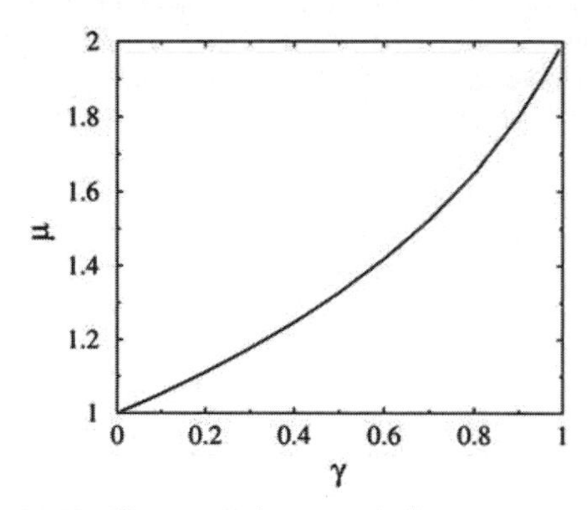

FIG. 2. The amplitude μ in $M_\gamma(t) = \mu t$ vs γ.

initial site is $N^\gamma/(N + N^\gamma)$. This pattern continues indefinitely with probability

$$\mathcal{P} = \prod_{N=1}^{\infty} \frac{1}{1 + N^{1-\gamma}}. \tag{9}$$

Clearly, $\mathcal{P} = 0$ when $\gamma \leq 2$ but $\mathcal{P} > 0$ when $\gamma > 2$. Thus for $\gamma > 2$ there is a nonzero probability that the initial site connects to all other sites.

To determine the behavior for general $\gamma > 1$, we need the asymptotic time dependence of M_γ. To this end, it is useful to consider the discretized version of the master equations Eq. (1), where the time t is limited to integer values. Then $N_k(t) = 0$ whenever $k > t$ and the rate equation for $N_k(k)$ immediately leads to

$$N_k(k) = \frac{(k-1)^\gamma N_{k-1}(k-1)}{M_\gamma(k-1)} = N_2(2) \prod_{j=2}^{k-1} \frac{j^\gamma}{M_\gamma(j)}. \tag{10}$$

From this and the obvious fact that $N_k(k)$ must be less than unity, it follows that $M_\gamma(t)$ cannot grow more slowly than t^γ. On the other hand, $M_\gamma(t)$ cannot grow faster than t^γ as follows from the estimate:

$$M_\gamma(t) = \sum_{k=1}^{t} k^\gamma N_k(t) \leq t^{\gamma-1} \sum_{k=1}^{t} k N_k(t) = t^{\gamma-1} M_1(t). \tag{11}$$

Thus $M_\gamma \propto t^\gamma$. In fact, the amplitude of t^γ is unity as will be derived self-consistently after solving for the N_k's.

We now use $M_\gamma \sim t^\gamma$ in the rate equations to solve recursively for each N_k. Starting with the equation $\dot{N}_1 = 1 - N_1/M_\gamma$, the second term on the right-hand side is subdominant; neglecting this term gives $N_1 = t$. Continuing this same line of reasoning for each successive rate equation gives the leading behavior of N_k,

$$N_k = J_k t^{k-(k-1)\gamma} \quad \text{for } k \geq 1, \tag{12}$$

with $J_k = \prod_{j=1}^{k-1} j^\gamma/[1 + j(1-\gamma)]$. This pattern of behavior for N_k continues as long as its exponent $k - (k-1)\gamma$ remains positive, or $k < \gamma/(\gamma - 1)$. The full behavior of the N_k may be determined straightforwardly by keeping the next correction terms in the rate equations. For example, $N_1 = t - t^{2-\gamma}/(2-\gamma) + \ldots$.

For $k > \gamma/(\gamma - 1)$, each N_k has a finite limiting value in the long-time limit. Since the total number of connections equals $2t$ and t of them are associated with N_1, the remaining t links must all connect to a single site which has t connections (up to corrections which grow no faster than sublinearly with time). Consequently, the amplitude of M_γ equals unity, as argued above.

Thus for superlinear kernels, the GN undergoes an infinite sequence of connectivity transitions as a function of γ. For $\gamma > 2$ all but a finite number of sites are linked to the gel site which has the rest of the links of the network. This is the "winner takes all" situation. For $3/2 < \gamma < 2$, the number of sites with two links grows as $t^{2-\gamma}$, while the number of sites with more than two links is again finite. For $4/3 < \gamma < 3/2$, the number of sites with three links grows as $t^{3-2\gamma}$ and the number with more than three is finite. Generally, for $\frac{m+1}{m} < \gamma < \frac{m}{m-1}$, the number of sites with more than m links is finite, while $N_k \sim t^{k-(k-1)\gamma}$ for $k \leq m$. Logarithmic corrections also arise at the transition points.

The connectivity distribution leads to an amusing consequence for the most popular site. Its connectivity $k_{\max}$ is determined by $\sum_{k>k_{\max}} N_k = 1$; that is, there is one site whose connectivity lies in the range $(k_{\max}, \infty)$. This criterion gives

$$k_{\max} \sim \begin{cases} (\ln t)^{1/(1-\gamma)} & 0 \leq \gamma < 1; \\ t^{1/(\nu-1)} & \text{asymptotically linear;} \\ t & \text{superlinear.} \end{cases} \tag{13}$$

Since t also equals the total number of sites, we can compare this prediction about the most popular site with available data from the Institute of Scientific Information based on 783 339 papers with 6 716 198 total citations (details in Ref. [16]). Here the most cited paper had 8904 citations. This accords with the first line of Eq. (13) for $\gamma \approx 0.86$, and also with the second when $\nu \approx 2.5$.

In addition to the connectivity of a site, we also may ask about its *age*. Within the GN model, older sites should clearly be more highly connected. We quantify this feature and also determine how the connection kernel affects the combined age and connectivity distribution. Note that our model does *not* have explicit aging where the connection kernel depends on the age of each site; this feature is treated in Ref. [17].

Let $c_k(t,a)$ be the average number of sites of age a which have $k - 1$ incoming links at time t. Here age a means that the site was introduced at time $t - a$. The quantity $c_k(t,a)$ evolves according to

$$\frac{\partial c_k}{\partial t} + \frac{\partial c_k}{\partial a} = \frac{1}{M_\gamma}[(k-1)^\gamma c_{k-1} - k^\gamma c_k] + \delta_{k1}\delta(a). \tag{14}$$

The second term on the left-hand side accounts for the aging of sites, while the right-hand side accounts for the (age independent) connection changing processes. Consider first the linear kernel, $A_k = k$. Let us focus again on the most interesting limit, namely, asymptotic behavior. Then we can disregard the initial condition and write $M_1(t) = 2t$. This transforms Eq. (14) into

$$\left(\frac{\partial}{\partial t} + \frac{\partial}{\partial a}\right) c_k = \frac{(k-1)c_{k-1} - k c_k}{2t} + \delta_{k1}\delta(a). \tag{15}$$

The homogeneous form of this equation suggests that solution should be self-similar. Specifically, one can seek a solution as a function of the *single* variable a/t rather than two separate variables, $c_k(t,a) = f_k(a/t)$. This simplifies

VOLUME 85, NUMBER 21 PHYSICAL REVIEW LETTERS 20 NOVEMBER 2000

the partial differential equation (15) into an ordinary differential equation for $f_k(x)$ which can easily be solved. In terms of the original variables of a and t, we find

$$c_k(t,a) = \sqrt{1 - \frac{a}{t}}\left[1 - \sqrt{1 - \frac{a}{t}}\right]^{k-1}. \quad (16)$$

Notice that this age distribution satisfies the normalization requirement, $N_k(t) = \int_0^t da\, c_k(t,a)$. As expected, young sites (those with $a/t \to 0$) typically have a small connectivity while old sites have large connectivity. Further, old sites have a broad distribution of connectivities up to a characteristic number which asymptotically grows as $\langle k \rangle \sim (1 - a/t)^{-1/2}$ as $a \to t$. These properties and related issues may be worthwhile to investigate in citation and other information networks.

Similarly, we can obtain $c_k(t,a)$ for the GN model with an arbitrary homogeneous connection kernel [20] which grows slower than linearly in k. Assuming a self-similar solution $c_k(t,a) = f_k(a/t)$, applying a Laplace transform, we find a recursion relation for $\hat{f}_k$ whose solution is identical in structure to Eq. (4). Although it appears impossible to perform the inverse Laplace transform in explicit form for arbitrary k, we can compute $c_k(t,a)$ for small k; for example, we find $c_1 = (1 - a/t)^{1/\mu}$. The behavior also simplifies in the large-k limit. Here we find that the age of sites with k links is peaked about the value a_k which satisfies

$$\frac{a_k}{t} \simeq \begin{cases} 1 - \exp(-\mu \frac{k^{1-\gamma}}{1-\gamma}) & \gamma < 1; \\ 1 - \frac{12}{(k+3)(k+4)} & \gamma = 1. \end{cases} \quad (17)$$

This shows how old sites are better connected.

In summary, we solved for both the connectivity distribution and the age-dependent structure of the growing random network. The most interesting connectivity arises in a network with an asymptotically linear connection kernel. Here the number of sites with k connections has the power-law form $N_k \sim k^{-\nu}$, with ν tunable to any value in the range $2 < \nu < \infty$. This accords with the connectivity distributions observed in various contemporary examples of growing networks.

We are grateful for Grants No. NSF INT9600232, No. NSF DMR9978902, and No. DGAPA IN112998 for financial support.

Note added.—While writing this manuscript we learned of Ref. [21] which overlaps some of our results. We thank J. Mendes for informing us of this work.

[1] B. Bollobás, *Random Graphs* (Academic Press, London, 1985).
[2] S. Janson, T. Luczak, and A. Rucinski, *Random Graphs* (Wiley, New York, 2000).
[3] S. A. Kauffman, *The Origin of Order: Self-Organization and Selection in Evolution* (Oxford University Press, London, 1993).
[4] S. Wasserman and K. Faust, *Social Network Analysis* (Cambridge University Press, Cambridge, 1994).
[5] B. Derrida and H. Flyvbjerg, J. Phys. (Paris) **48**, 971 (1987).
[6] H. Flyvbjerg and N. J. Kjaer, J. Phys. A **21**, 1695 (1988).
[7] U. Bastolla and G. Parisi, Physica (Amsterdam) **98D**, 1 (1996).
[8] R. V. Solé and S. C. Manrubia, Phys. Rev. E **54**, R42 (1996).
[9] S. Jain and S. Krishna, Phys. Rev. Lett. **81**, 5684 (1998).
[10] J. Kleinberg, R. Kumar, P. Raphavan, S. Rajagopalan, and A. Tomkins, in *Proceedings of the International Conference on Combinatorics and Computing,* Lecture Notes in Computer Science Vol. 1627 (Springer-Verlag, Berlin, 1999).
[11] S. R. Kumar, P. Raphavan, S. Rajagopalan, and A. Tomkins, in *Proceedings of the 25th Very Large Databases Conference, Edinburgh, Scotland, 1999* (Morgan Kaufmann, Orlando, FL, 1999).
[12] A. Broder, R. Kumar, F. Maghoul, P. Raphavan, S. Rajagopalan, R. Stata, A. Tomkins, and J. Wiener, Comput. Netw. **33**, 309 (2000).
[13] A. L. Barabási and R. Albert, Science **286**, 509 (1999).
[14] A. J. Lotka, J. Wash. Acad. Sci. **16**, 317 (1926); W. Shockley, Proc. IRE **45**, 279 (1957).
[15] J. Laherrere and D. Sornette, Eur. Phys. J. B **2**, 525 (1998).
[16] S. Redner, Eur. Phys. J. B **4**, 131 (1998).
[17] S. N. Dorogovtsev and J. F. F. Mendes, Phys. Rev. E **62**, 1842 (2000).
[18] B. A. Huberman, P. L. T. Pirolli, J. E. Pitkow, and R. Lukose, Science **280**, 95 (1998); S. M. Maurer and B. A. Huberman, nlin.CD/0003041.
[19] M. H. Ernst, in *Fundamental Problems in Stat. Physics VI,* edited by E. G. D. Cohen (Elsevier, New York, 1985).
[20] P. L. Krapivsky and S. Redner (to be published).
[21] S. N. Dorogovtsev, J. F. F. Mendes, and A. N. Samukhin, cond-mat/0004434.

EUROPHYSICS LETTERS 15 May 2001
Europhys. Lett., **54** (4), pp. 436–442 (2001)

Competition and multiscaling in evolving networks

G. BIANCONI[1] and A.-L. BARABÁSI[1,2]

[1] *Department of Physics, University of Notre Dame - Notre Dame, IN 46556, USA*
[2] *Institute for Advanced Studies, Collegium Budapest*
Szentháromság utca 2, H-1014 Budapest, Hungary

(received 3 November 2000; accepted in final form 2 March 2001)

PACS. 05.65.+b – Self-organized systems.
PACS. 89.75.-k – Complex systems.
PACS. 89.75.Hc – Networks and genealogical trees.

Abstract. – The rate at which nodes in a network increase their connectivity depends on their fitness to compete for links. For example, in social networks some individuals acquire more social links than others, or on the www some webpages attract considerably more links than others. We find that this competition for links translates into multiscaling, *i.e.* a fitness-dependent dynamic exponent, allowing fitter nodes to overcome the more connected but less fit ones. Uncovering this fitter-gets-richer phenomenon can help us understand in quantitative terms the evolution of many competitive systems in nature and society.

The complexity of many systems can be attributed to the interwoven web in which their constituents interact with each other. For example, the society is organized in a social web, whose nodes are individuals and links represent various social interactions, or the www forms a complex web whose nodes are documents and links are URLs. While for a long time these networks have been modeled as completely random [1,2], recently there is increasing evidence that they in fact have a number of generic non-random characteristics, obeying various scaling laws or displaying short length-scale clustering [3–16].

A generic property of these complex systems is that they constantly evolve in time. This implies that the underlying networks are not static, but continuously change through the addition and/or removal of new nodes and links. Consequently, we have to uncover the dynamical forces that act at the level of individual nodes, whose cumulative effect determines the system's large-scale topology. A step in this direction was the scale-free model [8], that incorporates the fact that network evolution is driven by at least two coexisting mechanisms: 1) growth, implying that networks continuously expand by the addition of new nodes; 2) preferential attachment, mimicking the fact that a new node links with higher probability to nodes that already have a large number of links. With these two ingredients the scale-free model predicts the emergence of a power law connectivity distribution, observed in many systems [3, 8–10], ranging from the Internet to citation networks. Furthermore, extensions of this model, including rewiring [11] or aging [12,13], have been able to account for more realistic aspects of the network evolution, such as the existence of various scaling exponents or cutoffs in the connectivity distribution.

The scale-free model neglects an important aspect of competitive systems: not all nodes are equally successful in acquiring links [17]. The model predicts that all nodes increase their connectivity in time as $k_i(t) = (t/t_i)^\beta$, where $\beta = 1/2$ and t_i is the time at which node i has been added into the system. Consequently, the oldest nodes will have the highest number of links, since they had the longest timeframe to acquire them.

On the other hand, numerous examples convincingly indicate that in real systems a node's connectivity and growth rate does not depend on its age alone. For example, in social systems some individuals are better in turning a random meeting into a lasting social link than others. On the www some documents through a combination of good content and marketing acquire a large number of links in a very short time, easily overtaking older websites. Finally, some research papers in a short timeframe acquire a very large number of citations. We tend to associate these differences with some intrinsic quality of the nodes, such as the social skills of an individual, the content of a web page, or the content of a scientific article. We will call this the node's fitness, describing its ability to compete for links at the expense of other nodes.

In this paper we propose a simple model that allows us to investigate this competitive aspect of real networks in quantitative terms. Assuming that the existence of a fitness modifies the preferential attachment to compete for links, we find that different fitness translates into multiscaling in the dynamical evolution: the time dependence of a node's connectivity depends on the fitness of the node. We develop the continuum model for this competitive evolving network, allowing us to calculate β analytically and derive a general expression for the connectivity distribution. We find that the analytical predictions are in excellent agreement with the results obtained from numerical simulations.

The fitness model. – The examples discussed above indicate that nodes have different ability (fitness) to compete for links. To account for these differences we introduce a fitness parameter, η_i, that we assign to each node, and assume that it is unchanged in time (*i.e.* η_i represents a quenched noise) [18]. Starting with a small number of nodes, at every timestep we add a new node i with fitness η_i, where η is chosen from the distribution $\rho(\eta)$. Each new node i has m links that are connected to the nodes already present in the system. We assume that the probability Π_i that a new node will connect to a node i already present in the network depends on the connectivity k_i and on the fitness η_i of that node, such that

$$\Pi_i = \frac{\eta_i k_i}{\sum_j \eta_j k_j}. \tag{1}$$

This generalized preferential attachment [8] incorporates in the simplest possible way that fitness and connectivity jointly determine the rate at which new links are added to a given node, *i.e.* even a relatively young node with a few links can acquire links at a high rate if it has a large fitness parameter. To address the scaling properties of this model we first develop a continuum theory, allowing us to predict the connectivity distribution [8, 11, 12]. A node i will increase its connectivity k_i at a rate that is proportional to the probability (1) that a new node will attach to it, giving

$$\frac{\partial k_i}{\partial t} = m \frac{\eta_i k_i}{\sum_j k_j \eta_j}. \tag{2}$$

The factor m accounts for the fact that each new node adds m links to the system. If $\rho(\eta) = \delta(\eta - 1)$, *i.e.* all fitness are equal, (2) reduces to the scale-free model, which predicts that $k_i(t) \sim t^{1/2}$ [8]. In order to solve (2) we assume that similarly to the scale-free model the time evolution of k_i follows a power law, but there is multiscaling in the system, *i.e.* the dynamic exponent depends on the fitness η_i,

$$k_{\eta_i}(t, t_0) = m\left(\frac{t}{t_0}\right)^{\beta(\eta_i)}, \tag{3}$$

where t_0 is the time at which the node i was born. The dynamic exponent $\beta(\eta)$ is bounded, *i.e.* $0 < \beta(\eta) < 1$ because a node always increases the number of links in time ($\beta(\eta) > 0$) and $k_i(t)$ cannot increases faster than t ($\beta(\eta) < 1$). We first calculate the mean of the sum $\sum_j \eta_j k_j$ over all possible realizations of the quenched noise $\{\eta\}$. Since each node is born at a different time t_0, the sum over j can be written as an integral over t_0:

$$\begin{aligned}\left\langle \sum_j \eta_j k_j \right\rangle &= \int \mathrm{d}\eta\rho(\eta)\,\eta \int_1^t \mathrm{d}t_0\; k_\eta(t, t_0) \\ &= \int \mathrm{d}\eta\; \eta\rho(\eta) m\,\frac{(t - t^{\beta(\eta)})}{1 - \beta(\eta)}.\end{aligned} \tag{4}$$

Since $\beta(\eta) < 1$, in the $t \to \infty$ limit $t^{\beta(\eta)}$ can be neglected compared to t, thus we obtain

$$\left\langle \sum_j \eta_j k_j \right\rangle \overset{t\to\infty}{=} Cmt(1 + \mathrm{O}(t^{-\epsilon}), \tag{5}$$

where

$$\begin{aligned}\epsilon &= (1 - \max_\eta \beta(\eta)) > 0, \\ C &= \int \mathrm{d}\eta\rho(\eta)\frac{\eta}{1 - \beta(\eta)}.\end{aligned} \tag{6}$$

Using (5), and the notation $k_\eta = k_{\eta_i}(t, t_0)$, the dynamic equation (2) can be written as

$$\frac{\partial k_\eta}{\partial t} = \frac{\eta k_\eta}{Ct}, \tag{7}$$

which has a solution of form (3), given that

$$\beta(\eta) = \frac{\eta}{C}, \tag{8}$$

thereby confirming the self-consistent nature of the assumption (3). To complete the calculation we need to determine C from (6) after substituting $\beta(\eta)$ with η/C,

$$1 = \int_0^{\eta_{\max}} \mathrm{d}\eta\rho(\eta)\frac{1}{\frac{C}{\eta} - 1}, \tag{9}$$

where $\eta_{\max}$ is the maximum possible fitness in the system [19]. Apparently, (9) is a singular integral. However, since $\beta(\eta) = \eta/C < 1$ for every value of η, we have $C > \eta_{\max}$, thus the integration limit never reaches the singularity. Note also that, since $\sum_j \eta_j k_j \le \eta_{\max} \sum_j k_j = 2mt\eta_{\max}$, we have, using (5), that $C \le 2\eta_{\max}$.

Finally, we can calculate the connectivity distribution $P(k)$, which gives the probability that a node has k links. If there is a single dynamic exponent β, the connectivity distribution follows the power law $P(k) \sim k^\gamma$, where the connectivity exponent is given by $\gamma = 1/\beta + 1$.

However, in this model we have a spectrum of dynamic exponents $\beta(\eta)$, thus $P(k)$ is given by a weighted sum over different power laws. To find $P(k)$ we need to calculate the cumulative probability that for a certain node $k_\eta(t) > k$,

$$P(k_\eta(t) > k) = P\left(t_0 < t\left(\frac{m}{k}\right)^{C/\eta}\right) = t\left(\frac{m}{k}\right)^{\frac{C}{\eta}}. \tag{10}$$

Thus the connectivity distribution, *i.e.* the probability that a node has k links, is given by the integral

$$P(k) = \int_0^{\eta_{\max}} \mathrm{d}\eta \frac{\partial P(k_\eta(t) > k)}{\partial t} \propto \int \mathrm{d}\eta \rho(\eta) \frac{C}{\eta}\left(\frac{m}{k}\right)^{\frac{C}{\eta}+1}. \tag{11}$$

Scale-free model. - Given the fitness distribution $\rho(\eta)$, the continuum theory allows us to predict both the dynamics, described by the dynamic exponent $\beta(\eta)$ (eqs. (8) and (9)), and the topology, characterized by the connectivity distribution $P(k)$ (eq. (11)). To demonstrate the validity of our predictions, in the following we calculate these quantities for two different $\rho(\eta)$ functions. As a first application, let us consider the simplest case, corresponding to the scale-free model, when all fitnesses are equal. Thus we have $\rho(\eta) = \delta(\eta - 1)$, which, inserted in (9), gives $C = 2$, which represents the largest possible value of C. Using (8) we obtain $\beta = 1/2$ and from (11) we get $P(k) \propto k^{-3}$, the known scaling of the scale-free model. Thus the scale-free model represents an extreme case of the fitness model considered here, the connectivity exponent taking up the largest possible value of γ.

Uniform fitness distribution. - The behavior of the system is far more interesting, however, when nodes with different fitness compete for links. The simplest such case, which already offers nontrivial multiscaling, is obtained when $\rho(\eta)$ is chosen uniformly from the interval $[0, 1]$. The constant C can be determined again from (9), which gives

$$\exp[-2/C] = 1 - 1/C, \tag{12}$$

whose solution is $C^* = 1.255$. Thus, according to (8), each node will have a different dynamic exponent, given by $\beta(\eta) \sim \frac{\eta}{C^*}$. Using (11) we obtain

$$P(k) \propto \int_0^1 \mathrm{d}\eta \frac{C^*}{\eta} \frac{1}{k^{1+C^*/\eta}} \sim \frac{k^{-(1+C^*)}}{\log(k)}, \tag{13}$$

i.e. the connectivity distribution follows a generalized power law, with an inverse logarithmic correction.

To check the predictions of the continuum theory we performed numerical simulations of the discrete fitness model, choosing fitness with equal probability from the interval $[0, 1]$. Most important is to test the validity of the ansatz (3), for which we recorded the time evolution of nodes with different fitness η. As fig. 1 shows, we find that $k_i(t)$ follows a power law for all η, and the scaling exponent, $\beta(\eta)$, depends on η, being larger for nodes with larger fitness. Equation (6) predicts that the sum $\langle \sum_i \eta_i k_i \rangle / mt \to C^*$ in the $t \to \infty$ limit, where C^* is given by (12) as $C^* = 1.255$. Indeed as the inset in fig. 1 shows, the discrete network model indicates that this sum converges to the analytically predicted value. Figure 1 allows us to determine

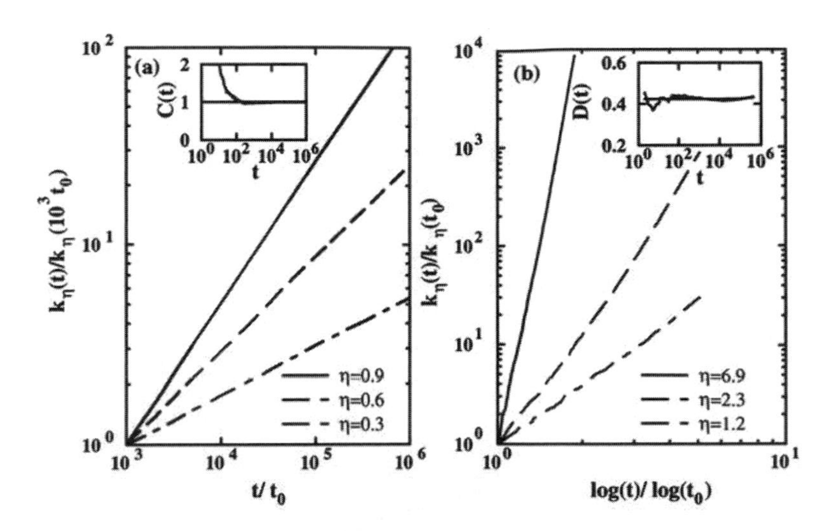

Fig. 1 – (a) Time dependence of the connectivity, $k_\eta(t)$, for nodes with fitness $\eta = 0.3$, 0.6 and 0.9. Note that $k_\eta(t)$ follows a power law in each case and the dynamic exponent $\beta(\eta)$, given by the slope of $k(t)$, increases with η. While in the simulation the fitness of the nodes has been drawn uniformly, between [0, 1], in the figure we show only the connectivity of three nodes with selected fitness. In the simulation we used $m = 2$ and the shown curves represent averages over 20 runs. Inset: Asymptotic convergence of $C(t) = (\sum_{i=1}^{t} \eta_i k_i)/t$ to the analytically predicted limit $C^* = 1.255$, shown as a horizontal line (see eq. (12)). (b) The same as (a) for the exponential fitness distribution demonstrating that $k(t)$ follows (14). The inset shows the convergence of $D(t) = (\sum_{i=1}^{t} \eta_i k_i)/(mtln(t))$ to $D^* = 0.45$.

numerically the exponent $\beta(\eta)$, and compare it to the prediction (8). As the inset in fig. 2 indicates, we obtain excellent agreement between the numerically determined exponents and the prediction of the continuum theory. Finally, in fig. 2 we show the agreement between the prediction (13) and the numerical results for the connectivity distribution $P(k)$.

An interesting feature of the numerically determined connectivity distribution (fig. 2) is the appearance of a few nodes that have higher number of links than predicted by the connectivity distribution. Such highly connected hubs, appearing as a horizontal line with large k on the log-log plot, are present in many systems, including the www [3] or the metabolic network of a cell [20], clearly visible if we do not use logarithmic binning. This indicates that the appearance of a few "super hubs", *i.e.* nodes that have connections in excess to that predicted by a power law, is a generic feature of competitive systems.

Exponential fitness distribution. – If the $\rho(\eta)$ distribution has an infinite support, the integral (9) contains a singularity at $\eta = C$, and the self-consistent calculation cannot be applied. To recover the behavior of such systems, we studied numerically the case $\rho(\eta) = e^{-\eta}$. In a system with a finite support for which there is a $\eta_{\max}$ such that $\rho(\eta_{\max}) \neq 0$, the system will reach $\eta_{\max}$ within a finite time. That is, within a short timeframe a η' will appear that is infinitely close to $\eta_{\max}$, and the likelihood of finding an $\eta > \eta'$ goes to zero. This is not the case for an infinite support: at any time there is a finite probability that an $\eta > \eta_{\max}$ will appear, as, strictly speaking, $\eta_{\max} = \infty$. The average time required for a large η to appear scales as $\tau(\eta) \sim 1/p(\eta) \sim e^{\eta}$, indicating that $\eta_{\max}$ scales as $\eta_{\max} \sim \ln(t)$, and $\Sigma_i \eta_i k_i(t) \leq D \ln(t) t$.

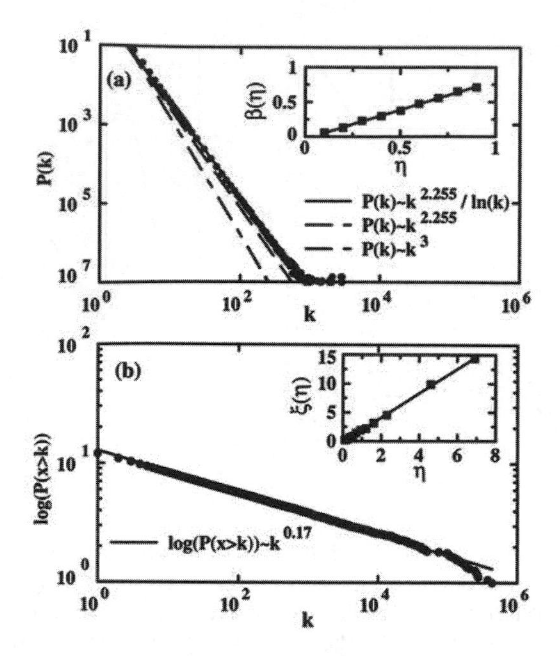

Fig. 2 – (a) Connectivity distribution in the fitness model, obtained for a network with $m = 2$ and $N = 10^6$ nodes and uniform fitness distribution. The upper solid line that goes along the circles provided by the numerical simulations corresponds to the theoretical prediction (13), with $\gamma = 2.25$. The dashed line corresponds to a simple fit $P(k) \sim k^{-2.255}$ without the logarithmic correction, while the long-dashed curve correspond to $P(k) \sim k^{-3}$, as predicted by the scale-free model [8], in which all fitnesses are equal. Inset: The dependence of the dynamic exponent $\beta(\eta)$ on the fitness parameter η in the case of a uniform $\rho(\eta)$ distribution. The squares were obtained from the numerical simulations while the solid line corresponds to the analytical prediction $\beta(\eta) = \eta/1.255$. (b) Same as (a) but for exponential fitness distribution. To decrease statistical fluctuations we show the cumulative distribution $P(x > k)$ [13], which follows a stretched exponential. The inset shows $\xi(\eta)$ as determined form fig. 1b using eq. (14) and demonstrate that $\xi(\eta)$ is linear in η, in line with the theoretical predictions.

Assuming that $\Sigma_i \eta_i k_i(t)/D \ln(t)t \to Dm$ in the $t \to \infty$ limit, using (2) we obtain that

$$k(t) = k(t_0)\left(\frac{\ln(t)}{\ln(t_0)}\right)^{\xi(\eta)}, \tag{14}$$

where $\xi(\eta) = D/\eta$. As fig. 1b shows, we find that indeed $k(t)$ scales as a power of $\ln(t)$, while fig. 2b (inset) shows that the power depends linearly on η. Interestingly, numerical simulations indicate that in this case $P(k)$ follows a stretched exponential (fig. 2b).

Discussion – The fitness model investigated in this paper reflects the basic properties of many real systems in which the nodes compete for links with other nodes, thus a node can acquire links only at the expense of the other nodes. The competitive nature of the model is guaranteed by the fact that nodes that are already in the system have to compete with a linearly increasing number of other nodes for a link. We find that allowing for different fitness, multiscaling emerges and the time dependence of a node's connectivity depends on the fitness parameter, η. This allows nodes with a higher fitness to enter the system at a later time and overcome nodes that have been in the system for a much longer timeframe. Our results indicate, however, that not all $\rho(\eta)$ distributions will result in a power law time dependence and connectivity distribution. If $\rho(\eta)$ decays exponentially, we find that $P(k)$ follows a stretched exponential and $k(t)$ follows a complex combination of logarithmic and power law behavior. This indicates that $P(k)$ is not robust against changes in the functional form of the fitness distribution: with an appropriate choice of $\rho(\eta)$ one can obtain a non–power-law distribution. As many real networks display $P(k)$ that are best approximated with a power law, this implies that there are some restrictions regarding the nature of the $\rho(\eta)$ distribution. For example, an exponential is clearly not appropriate. Understanding the restrictions on the classes of $\rho(\eta)$ which support a power law $P(k)$ is a formidable challenge for further work.

* * *

We wish to acknowledge useful discussions with R. Albert, I. Derényi, H. Jeong, E. Szathmáry and T. Vicsek. This research was partially supported by NSF Career Award DMR-9710998.

REFERENCES

[1] Erdös P. and Rényi A., *Publ. Math. Inst. Acad. Sci.*, **4** (1960) 17.
[2] Bollobás B., *Random Graphs* (Academic Press, London) 1985.
[3] Albert R., Jeong H. and Barabási A.-L., *Nature*, **401** (1999) 130.
[4] Newman M. E. J., *J. Stat. Phys.*, **101** (2000) 819.
[5] Caldarelli G., Marchetti R. and Pietronero L., *Europhys. Lett.*, **52** (2000) 386.
[6] Kleinberg J., Kumar S. R., Raghavan P., Rajagopalan S. and Tomkins A., *International Conference on Combinatorics and Computing* (1999).
[7] Huberman B. A. and Adamic L. A., *Nature*, **401** (1999) 131.
[8] Barabási A.-L. and Albert R., *Science*, **286** (1999) 509.
[9] Faloutsos M., Faloutsos P. and Faloutsos C., *Comput. Commun. Rev.*, **29** (1999) 251.
[10] Redner S., *Eur. Phys. J. B*, **4** (1998) 131.
[11] Albert R. and Barabási A.-L., *Phys. Rev. Lett.*, **85** (2000) 5234.
[12] Dorogovtsev S. N. and Mendes J. F. F., *Phys. Rev. E*, **62** (2000) 1842.
[13] Amaral L. A. N., Scala A., Barthèlèmy M. and Stanley H. E., *Proc. Natl. Acad. Sci. USA*, **97** (2000) 11149.
[14] Watts D. J. and Strogatz S. H., *Nature*, **393** (1998) 440.
[15] Barthélémy M. and Amaral L. A. N., *Phys. Rev. Lett.*, **82** (1999) 3180.
[16] Banavar J. R. and Maritan A. and Rinaldo A., *Nature*, **399** (1999) 130.
[17] Adamic L. A. and Huberman B. A., *Science*, **287** (2000) 2115.
[18] Note that in some real systems the fitness can change with time, for example a research field can slowly close down or an actor can suspend acting diminishing the ability of the corresponding nodes to compete for links [12, 13].
[19] Equation (9) can also be derived from the normalization condition $2k_0t = \sum_{j\epsilon N(t)} k_j$, a "mass conservation" law, giving the total number of links in the network at time t.
[20] Jeong H., Tombor B., Albert R., Oltvai Z. and Barabási, A.-L., *Nature*, **407** (2000) 651.

VOLUME 87, NUMBER 27 PHYSICAL REVIEW LETTERS 31 DECEMBER 2001

Universal Behavior of Load Distribution in Scale-Free Networks

K.-I. Goh, B. Kahng, and D. Kim

School of Physics and Center for Theoretical Physics, Seoul National University, Seoul 151-747, Korea

(Received 26 June 2001; published 12 December 2001)

We study a problem of data packet transport in scale-free networks whose degree distribution follows a power law with the exponent γ. Load, or "betweenness centrality," of a vertex is the accumulated total number of data packets passing through that vertex when every pair of vertices sends and receives a data packet along the shortest path connecting the pair. It is found that the load distribution follows a power law with the exponent $\delta \approx 2.2(1)$, insensitive to different values of γ in the range, $2 < \gamma \leq 3$, and different mean degrees, which is valid for both undirected and directed cases. Thus, we conjecture that the load exponent is a universal quantity to characterize scale-free networks.

DOI: 10.1103/PhysRevLett.87.278701 PACS numbers: 89.75.Hc, 05.10.-a, 89.70.+c, 89.75.Da

Complex systems consist of many constituents such as individuals, substrates, and companies in social, biological, and economic systems, respectively, showing cooperative phenomena between constituents through diverse interactions and adaptations to the pattern they create [1,2]. Interactions may be described in terms of graphs, consisting of vertices and edges, where vertices (edges) represent the constituents (their interactions). This approach was initiated by Erdös and Rényi (ER) [3]. In the ER model, the number of vertices is fixed, while edges connecting one vertex to another occur randomly with certain probability. However, the ER model is too random to describe real complex systems. Recently, Watts and Strogatz (WS) [4] introduced a small-world network, where a fraction of edges on a regular lattice is rewired with probability p_{WS} to other vertices. More recently, Barabási and Albert (BA) [5–7] introduced an evolving network where the number of vertices N increases linearly with time rather than fixed, and a newly introduced vertex is connected to m already existing vertices, following the so-called preferential attachment (PA) rule. When the number of edges k incident upon a vertex is called the degree of the vertex, the PA rule means that the probability for the new vertex to connect to an already existing vertex is proportional to the degree k of the selected vertex. Then the degree distribution $P_D(k)$ follows a power law $P_D(k) \sim k^{-\gamma}$ with $\gamma = 3$ for the BA model, while for the ER and WS models, it follows a Poisson distribution. Networks whose degree distribution follows a power law, called scale-free (SF) networks [8], are ubiquitous in real-world networks such as the World Wide Web [9–11], the Internet [12–14], the citation network [15] and the author collaboration network of scientific papers [16–18], and the metabolic networks in biological organisms [19]. On the other hand, there also exist random networks such as the actor network whose degree distribution follows a power law but has a sharp cutoff in its tail [20]. Thus, it has been proposed that the degree distribution can be used to classify a variety of diverse real-world networks [20]. In SF networks, one may wonder if the exponent γ is universal in analogy with the theory of critical phenomena; however, the exponent γ turns out to be sensitive to the detail of network structure. Thus, a universal quantity for SF networks is yet to be found. From a theoretical viewpoint, it is important to find a universal quantity for SF networks, which is the purpose of this Letter.

A common feature between the WS and SF networks would be the small-world property that the mean separation between two vertices, averaged over all pairs of vertices (called the diameter hereafter), is shorter than that of a regular lattice. The small-world property in SF networks results from the presence of a few vertices with high degree. In particular, the hub, the vertex whose degree is the largest, plays a dominant role in reducing the diameter of the system. Diameters of many complex networks in the real world are small, allowing objects transmitted through the network such as neural spikes on neural network, or data packets on the Internet, to travel from one vertex to another quickly along the shortest path. The shortest paths are indeed of relevance to network transport properties. When a data packet is sent from one vertex to another through SF networks such as the Internet, it is efficient to take a road along the shortest paths between the two. Then vertices with higher degrees should be heavily loaded and jammed by lots of data packets passing along the shortest paths. To prevent such Internet traffic congestions and allow data packets to travel in a free-flow state, one has to enhance the capacity, the rate of data transmission, of each vertex to the extent that the capacity of each vertex is large enough to handle appropriately defined "load."

In this Letter, we define and study such a quantity, which we simply call load, to characterize the transport dynamics in SF networks. In fact, this quantity turns out to be equivalent to "betweenness centrality" which was introduced in a social network to quantify how much power is centralized to people in social networks [17,21]. While it has been noted that the betweenness centrality has a long tail [22], here we focus our attention on its probability distribution for various SF networks with different degree exponents. Thus knowing the distribution of such a quantity enables us to not only estimate the capacity of each vertex needed for a free-flow state, but also to understand the power distribution in social networks, which is another purpose of this Letter.

To be specific, we suppose that a data packet is sent from a vertex i to j, for every ordered pair of vertices (i, j). For a given pair (i, j), it is transmitted along the shortest path between them. If there exist more than one shortest paths, the data packet would encounter one or more branching points. In this case, we assume that the data packet is divided evenly by the number of branches at each branching point as it travels. Then we define the load ℓ_k at a vertex k as the total amount of data packets passing through that vertex k when all pairs of vertices send and receive one unit of data packet between them. Here, we do not take into account the time delay of data transfer at each vertex or edge, so that all data are delivered in a unit time, regardless of the distance between any two vertices.

We find numerically that the load distribution $P_L(\ell)$ follows a power law $P_L(\ell) \sim \ell^{-\delta}$. Moreover, the exponent $\delta \approx 2.2$ we obtained is insensitive to the detail of the SF network structure as long as the degree exponent is in the range $2 < \gamma \leq 3$. The SF networks we used do not permit the rewiring process, and the number of vertices is linearly proportional to that of edges. When $\gamma > 3$, δ increases as γ increases, however. The universal behavior is valid for directed networks as well, when $2 < \{\gamma_{\rm in}, \gamma_{\rm out}\} \leq 3$. Since the degree exponents in most of the real-world SF networks satisfy $2 < \gamma \leq 3$, the universal behavior is interesting.

We construct a couple of classes of undirected SF networks both in the static and evolving ways. Each class of networks includes a control parameter, according to which the degree exponent is determined. First, we deal with the static case. There are N vertices in the system from the beginning, which are indexed by an integer i $(i = 1, \ldots, N)$. We assign the weight $p_i = i^{-\alpha}$ to each vertex, where α is a control parameter in $[0, 1)$. Next, we select two different vertices (i, j) with probabilities equal to the normalized weights, $p_i/\sum_k p_k$ and $p_j/\sum_k p_k$, respectively, and add an edge between them unless one exists already. This process is repeated until mN edges are made in the system. Then the mean degree is $2m$. Since edges are connected to a vertex with frequency proportional to the weight of that vertex, the degree at that vertex is given as

$$\frac{k_i}{\sum_j k_j} \approx \frac{(1-\alpha)}{N^{1-\alpha} i^{\alpha}}, \tag{1}$$

where $\sum_j k_j = 2mN$. Then it follows that the degree distribution follows the power law, $P_D(k) \sim k^{-\gamma}$, where γ is given by

$$\gamma = (1+\alpha)/\alpha. \tag{2}$$

Thus, adjusting the parameter α in [0,1), we can obtain various values of the exponent γ in the range $2 < \gamma < \infty$.

Once a SF network is constructed, we select an ordered pair of vertices (i, j) on the network and identify the shortest path(s) between them and measure the load on each vertex along the shortest path using the modified version of the breath-first search algorithm introduced by Newman [17]. It is found numerically that the load ℓ_i at vertex i follows the formula

$$\frac{\ell_i}{\sum_j \ell_j} \sim \frac{1}{N^{1-\beta} i^{\beta}}, \tag{3}$$

with $\beta = 0.80(5)$. This value of β is insensitive to different values of the exponent γ in the range $2 < \gamma \leq 3$, as shown in the inset in Fig. 1. The total load $\sum_j \ell_j$ scales as $\sim N^2 \log N$. This is because there are N^2 pairs of vertices in the system and the sum of the load contributed by each pair of vertices is equal to the distance between the two vertices, which is proportional to $\log N$. Therefore, the load ℓ_i at a vertex i is given as

$$\ell_i \sim (N \log N)(N/i)^{\beta}. \tag{4}$$

From Eq. (4), it follows that the load distribution scales as $P_L(\ell) \sim \ell^{-\delta}$, with $\delta = 1 + 1/\beta \approx 2.2(1)$, independent of γ in the range $2 < \gamma \leq 3$. A direct measure of $P_L(\ell)$ also gives $\delta \approx 2.2(1)$ as shown in Fig. 1. We also check δ for different mean degrees $m = 2$, 4, and 6, but we obtain the same value, $\delta \approx 2.2(1)$. Thus, we conclude that the exponent δ is a generic quantity for this network. Note that Eqs. (1) and (4) combined give a scaling relation between the load and the degree for this network as

$$\ell \sim k^{(\gamma-1)/(\delta-1)}. \tag{5}$$

Thus, when and only when $\gamma = \delta$, the load at each vertex is directly proportional to its degree. Otherwise, it scales nonlinearly. On the other hand, for $\gamma > 3$, the exponent δ depends on the exponent γ in a way that it increases as γ increases. Eventually, the load distribution decays

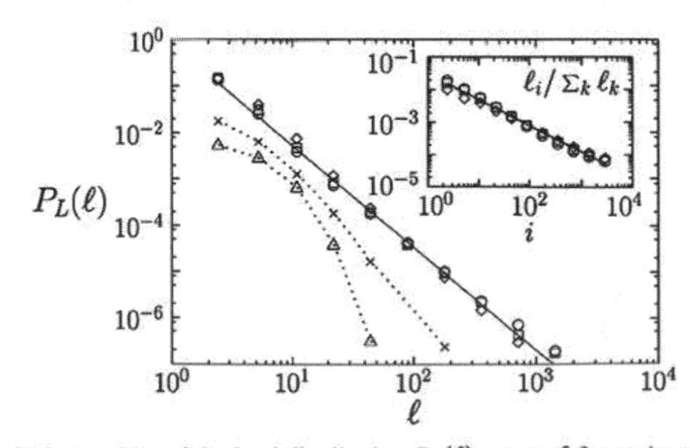

FIG. 1. Plot of the load distribution $P_L(\ell)$ versus ℓ for various $\gamma = 2.2$ (○), 2.5 (□), 3.0 (◇), 4.0 (×), and ∞ (△) in double logarithmic scales. The linear fit (solid line) has a slope -2.2. Data for $\gamma > 3.0$ are shifted vertically for clearance. Dotted lines are guides to the eye. Simulations are performed for $N = 10\,000$ and $m = 2$ and all data points are log-binned, averaged over ten configurations. Inset: Plot of the normalized load $\ell_i/\sum_k \ell_k$ versus vertex index i in double logarithmic scales for various $\gamma = 2.2$ (○), 2.5 (□), and 3.0 (◇).

exponentially for $\gamma = \infty$ as shown in Fig. 1. Thus, the transport properties of the SF networks with $\gamma > 3$ are fundamentally different from those with $2 < \gamma \leq 3$. This is probably due to the fact that for $\gamma > 3$, the second moment of $P_D(k)$ exists, while for $\gamma \leq 3$, it does not.

We examine the system-size dependent behavior of the load at the hub, ℓ_h, for the static model. According to Eq. (4), ℓ_h behaves as $\ell_h \sim N^{1.8} \log N$ in the range $2 < \gamma \leq 3$, while for $\gamma > 3$, ℓ_h increases with N but at a much slower rate than that for $2 < \gamma \leq 3$ as shown in Fig. 2. That implies that the shortest pathways between two vertices become diversified, and they do not necessarily pass through the hub for $\gamma > 3$. That may be related to the result that epidemic threshold is null in the range $2 < \gamma \leq 3$, while it is finite for $\gamma > 3$ in SF networks, because there exist many other shortest paths not passing through the hub for $\gamma > 3$, so that the infection of the hub does not always lead to the infection of the entire system. Thus, epidemic threshold is finite for $\gamma > 3$ [24].

Next, we generate other SF networks in an evolving way, using the methods proposed by Kumar *et al.* [23] and by Dorogovtsev *et al.* [7]. In these cases, we also find the same results as in the case of static models.

We also consider the case of directed SF network. The directed SF networks are generated following the static rule. In this case, we assign two weights $p_i = i^{-\alpha_{\rm out}}$ and $q_i = i^{-\alpha_{\rm in}}$ $(i = 1, \ldots, N)$ to each vertex for outgoing and incoming edges, respectively. Both control parameters $\alpha_{\rm out}$ and $\alpha_{\rm in}$ are in the interval $[0, 1)$. Then two different vertices (i, j) are selected with probabilities, $p_i / \sum_k p_k$ and $q_j / \sum_k q_k$, respectively, and an edge from the vertex i to j is created with an arrow, $i \rightarrow j$. The SF networks generated in this way show the power law in both outgoing and incoming degree distributions with the exponents $\gamma_{\rm out}$ and $\gamma_{\rm in}$, respectively. They are given as $\gamma_{\rm out} = (1 + \alpha_{\rm out})/\alpha_{\rm out}$ and $\gamma_{\rm in} = (1 + \alpha_{\rm in})/\alpha_{\rm in}$. Thus, choosing various values of $\alpha_{\rm out}$ and $\alpha_{\rm in}$, we can determine different exponents $\gamma_{\rm out}$ and $\gamma_{\rm in}$. Following the same steps as for the undirected case, we obtain the load distribution on the directed SF networks. The load exponent δ obtained is $\approx 2.3(1)$ as shown in Fig. 3, consistent with the one for the undirected case, also being independent of $\gamma_{\rm out}$ and $\gamma_{\rm in}$ in $2 < \{\gamma_{\rm out}, \gamma_{\rm in}\} \leq 3$. Therefore, we conjecture that the load exponent is a universal value for both the undirected and directed cases.

To see if such universal value of δ appears in the real-world network, we analyzed the coauthorship network, where nodes represent scientists and they are connected if they wrote a paper together. The data are collected in the field of the neuroscience, published in the period 1991–1998 [18]. This network is appropriate to test the load, i.e., the betweenness centrality distribution, because it does not include a rewiring process as it evolves, and its degree exponent $\gamma \approx 2.2$ lies in the range $2 < \gamma \leq 3$. As shown in Fig. 4, the load distribution follows a power law with the exponent $\delta \approx 2.2$, in good agreement with the value obtained in the previous models.

We also check the load distribution for the case when data travel with constant speed, so that the time delay of data transfer is proportional to the distance between two vertices. We find that the time delay effect does not change the load distribution and the conclusion of this work. The reason of this result is that when the time delay is accounted, load at each vertex is reduced roughly by a factor $\log N$, proportional to the diameter, which is negligible compared with the load without the time delay estimated to be $\sim N^{1.8} \log N$ in Eq. (4). Because of this small-world property, the universal behavior remains unchanged under the time delay of data transmission.

Finally, we mention the load distribution of the small-world network of WS which is not scale-free. It is found that its load distribution does not obey a power law but shows a combined behavior of two Poisson-type decays resulting from short-ranged and long-ranged connections, respectively, as shown in Fig. 5. We also find the average

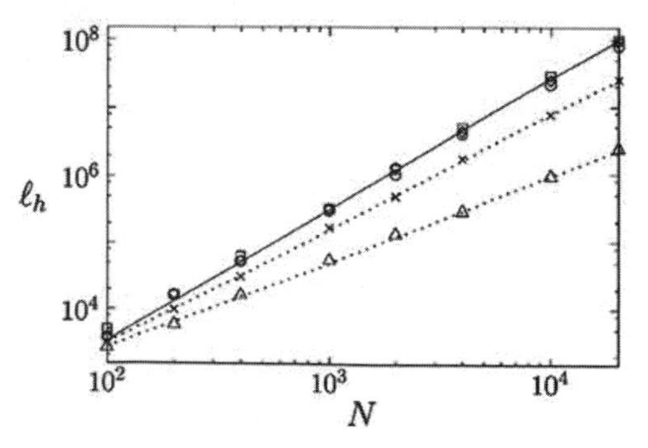

FIG. 2. Plot of the system-size dependence of the load at the hub versus system size N for various $\gamma = 2.2$ (○), 2.5 (□), 3.0 (◇), 4.0 (×), and ∞ (△). The solid line is $N^{1.8} \log N$ and dotted lines have slopes 1.70 and 1.25, respectively, from top to bottom. Simulations are performed for $m = 2$ and all data points are averaged over ten configurations.

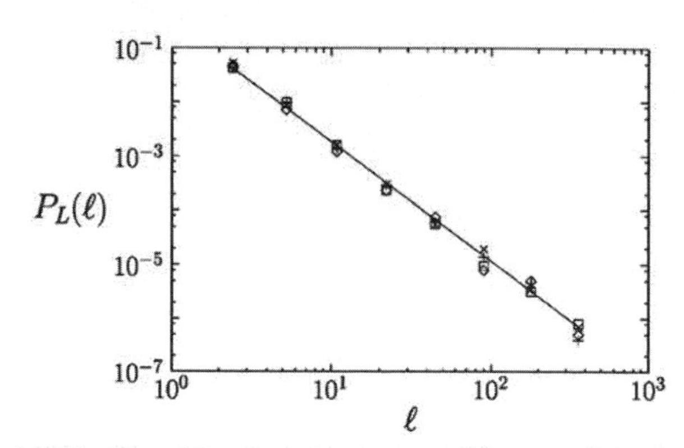

FIG. 3. Plot of the load distribution $P_L(\ell)$ versus ℓ for the directed case. The data are obtained for $(\gamma_{\rm in}, \gamma_{\rm out}) = (2.1, 2.3)$ (◇), (2.1, 2.7) (+), (2.5, 2.7) (□), and (2.5, 2.2) (×). The fitted line has a slope -2.3. All data points are log-binned.

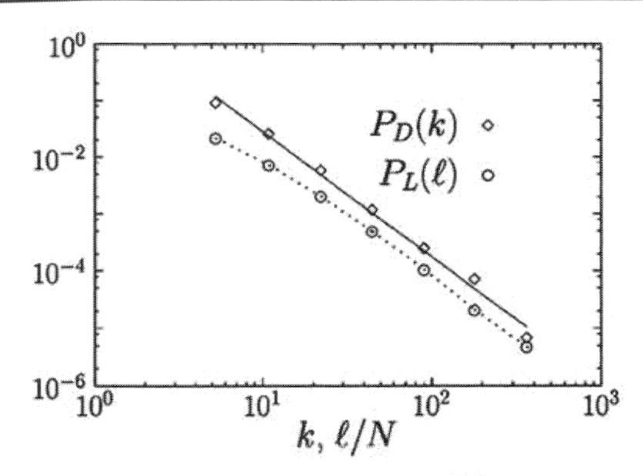

FIG. 4. Plot of the degree distribution $P_D(k)$ (◇) and the load distribution $P_L(\ell)$ (○) for a real-world network, the coauthorship network. The number of vertices (different authors) are 205 202. Least-squares fit (solid line) has a slope −2.2. All data points are log-binned.

load, $\bar{\ell}(p_{WS}) \equiv (1/N)\sum_i \ell_i(p_{WS})$, as a function of the rewiring probability p_{WS} decays rapidly with increasing p_{WS}, behaving similar to the diameter in the WS model, as shown in the inset of Fig. 5.

In conclusion, we have considered a problem of data packet transport on scale-free networks generated according to preferential attachment rules and introduced a physical quantity, load $\{\ell_i\}$ at each vertex. We found that the load distribution follows a power law, $P_L(\ell) \sim \ell^{-\delta}$, with the exponent $\delta \approx 2.2(1)$, which turns out to be insensitive to the degree exponent in the range $(2, 3]$ when the rewiring process is not included and the networks are of unaccelerated growth. Moreover, it is also the same for both directed and undirected cases within our numerical uncertainties. Therefore, we conjecture that the load exponent is a generic quantity to characterize scale-free networks. The universal behavior we found may have interesting implications to the interplay of SF network structure and dynamics. For $\gamma > 3$, however, the load exponent δ increases as the degree exponent γ increases, and eventually the load distribution decays exponentially as $\gamma \to \infty$. It would be interesting to examine the robustness of the universal behavior of the load distribution under some modifications of generating rules for SF networks such as the rewiring process and acceleration growth, which, however, is beyond the scope of the current study.

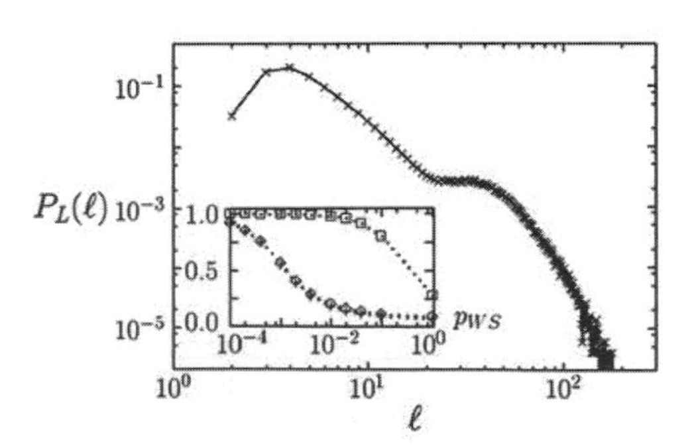

FIG. 5. Plot of the load distribution $P_L(\ell)$ versus load ℓ for the small-world network. Simulations are performed for system size $N = 1000$, and average degree $\langle k \rangle = 10$, and the rewiring probability $p_{WS} = 0.01$, averaged over 500 configurations. Inset: Plot of the average load (◇), diameter (+), clustering coefficient (□) versus the rewiring probability p_{WS}. All the data are normalized by the corresponding values at $p_{WS} = 0$. Dotted lines are guides to the eye.

We thank M. E. J. Newman for introducing Refs. [17, 21,22] and H. Jeong for providing the data for the coauthorship network. This work is supported by Grant No. 2000-2-11200-002-3 from the BRP program of the KOSEF.

[1] S. H. Strogatz, Nature (London) **410**, 268 (2001).
[2] N. Goldenfeld and L. P. Kadanoff, Science **284**, 87 (1999).
[3] P. Erdös and A. Rényi, Publ. Math. Inst. Hung. Acad. Sci. Ser. A **5**, 17 (1960).
[4] D. J. Watts and S. H. Strogatz, Nature (London) **393**, 440 (1998).
[5] A.-L. Barabási and R. Albert, Science **286**, 509 (1999).
[6] P. L. Krapivsky, S. Redner, and F. Leyvraz, Phys. Rev. Lett. **85**, 4629 (2000); P. L. Krapivsky and S. Redner, Phys. Rev. E **63**, 066123 (2001).
[7] S. N. Dorogovtsev, J. F. F. Mendes, and A. N. Samukhin, Phys. Rev. Lett. **85**, 4633 (2000).
[8] A.-L. Barabási, R. Albert, and H. Jeong, Physica (Amsterdam) **272A**, 173 (1999).
[9] R. Albert, H. Jeong, and A.-L. Barabási, Nature (London) **401**, 130 (1999).
[10] D. Butler, Nature (London) **405**, 112 (2000).
[11] A. Broder *et al.*, Comput. Networks **33**, 309 (2000).
[12] E. W. Zegura, K. L. Calvert, and M. J. Donahoo, IEEE/ACM Trans. Network **5**, 770 (1997).
[13] M. Faloutsos, P. Faloutsos, and C. Faloutsos, Comput. Commun. Rev. **29**, 251 (1999).
[14] R. Pastor-Satorras, A. Vázquez, and A. Vespignani, cond-mat/0105161.
[15] S. Redner, Eur. Phys. J. B **4**, 131 (1998).
[16] M. E. J. Newman, Proc. Natl. Acad. Sci. U.S.A. **98**, 404 (2001).
[17] M. E. J. Newman, Phys. Rev. E **64**, 016131 (2001); **64**, 016132 (2001).
[18] A.-L. Barabási, H. Jeong, Z. Neda, E. Ravasz, A. Schubert, and T. Vicsek, cond-mat/0104162.
[19] H. Jeong, B. Tombor, R. Albert, Z. N. Oltvani, and A.-L. Barabási, Nature (London) **407**, 651 (2000).
[20] L. A. Amaral, A. Scala, M. Barthélémy, and H. E. Stanley, Proc. Natl. Acad. Sci. U.S.A. **97**, 11 149 (2000).
[21] L. C. Freeman, Sociometry **40**, 35 (1977).
[22] M. E. J. Newman (private communication).
[23] R. Kumar *et al.*, in *Proceedings of the 41st Annual Symposium on Foundations of Computer Science, Redondo Beach, CA, 2000* (IEEE Computer Society, Los Alamitos, CA, 2000).
[24] R. Pastor-Satorras and A. Vespignani, Phys. Rev. E **63**, 066117 (2001).

PHYSICAL REVIEW E, VOLUME 64, 026704

Spectra of "real-world" graphs: Beyond the semicircle law

Illés J. Farkas,[1,*] Imre Derényi,[2,3,†] Albert-László Barabási,[2,4,‡] and Tamás Vicsek[1,2,§]

[1]*Department of Biological Physics, Eötvös University, Pázmány Péter Sétány 1A, H-1117 Budapest, Hungary*

[2]*Collegium Budapest, Institute for Advanced Study, Szentháromság utca 2, H-1014 Budapest, Hungary*

[3]*Institut Curie, UMR 168, 26 rue d'Ulm, F-75248 Paris 05, France*

[4]*Department of Physics, University of Notre Dame, Notre Dame, Indiana 46556*

(Received 19 February 2001; published 20 July 2001)

Many natural and social systems develop complex networks that are usually modeled as random graphs. The eigenvalue spectrum of these graphs provides information about their structural properties. While the semicircle law is known to describe the spectral densities of uncorrelated random graphs, much less is known about the spectra of real-world graphs, describing such complex systems as the Internet, metabolic pathways, networks of power stations, scientific collaborations, or movie actors, which are inherently correlated and usually very sparse. An important limitation in addressing the spectra of these systems is that the numerical determination of the spectra for systems with more than a few thousand nodes is prohibitively time and memory consuming. Making use of recent advances in algorithms for spectral characterization, here we develop methods to determine the eigenvalues of networks comparable in size to real systems, obtaining several surprising results on the spectra of adjacency matrices corresponding to models of real-world graphs. We find that when the number of links grows as the number of nodes, the spectral density of uncorrelated random matrices does not converge to the semicircle law. Furthermore, the spectra of real-world graphs have specific features, depending on the details of the corresponding models. In particular, scale-free graphs develop a trianglelike spectral density with a power-law tail, while small-world graphs have a complex spectral density consisting of several sharp peaks. These and further results indicate that the spectra of correlated graphs represent a practical tool for graph classification and can provide useful insight into the relevant structural properties of real networks.

DOI: 10.1103/PhysRevE.64.026704 PACS number(s): 02.60.−x, 68.55.−a, 68.65.−k, 05.45.−a

I. INTRODUCTION

Random graphs [1,2] have long been used for modeling the evolution and topology of systems made up of large assemblies of similar units. The uncorrelated random graph model—which assumes each pair of the graph's vertices to be connected with equal and independent probabilities—treats a network as an assembly of equivalent units. This model, introduced by the mathematicians Paul Erdős and Alfréd Rényi [1], has been much investigated in the mathematical literature [2]. However, the increasing availability of large maps of real-life networks has indicated that real networks are fundamentally correlated systems, and in many respects their topology deviates from the uncorrelated random graph model. Consequently, the attention has shifted towards more advanced graph models which are designed to generate topologies in line with the existing empirical results [3–14]. Examples of real networks, that serve as a benchmark for the current modeling efforts, include the Internet [6,15–17], the World-Wide Web [8,18], networks of collaborating movie actors and those of collaborating scientists [13,14], the power grid [4,5], and the metabolic network of numerous living organisms [9,19].

These are the systems that we will call *"real-world" networks* or *graphs*. Several converging reasons explain the enhanced current interest in such real graphs. First, the amount of topological data available on such large structures has increased dramatically during the past few years thanks to the computerization of data collection in various fields, from sociology to biology. Second, the hitherto unseen speed of growth of some of these complex networks—e.g., the Internet—and their pervasiveness in affecting many aspects of our lives has created the need to understand the topology, origin, and evolution of such structures. Finally, the increased computational power available on almost every desktop has allowed us to study such systems in unprecedented detail.

The proliferation of data has lead to a flurry of activity towards understanding the general properties of real networks. These efforts have resulted in the introduction of two classes of models, commonly called *small-world graphs* [4,5] and the *scale-free networks* [10,11]. The first aims to capture the clustering observed in real graphs, while the second reproduces the power-law degree distribution present in many real networks. However, until now, most analyses of these models and data sets have been confined to real-space characteristics, which capture their static structural properties e.g., degree sequences, shortest connecting paths, and clustering coefficients. In contrast, there is extensive literature demonstrating that the properties of graphs and the associated adjacency matrices are well characterized by spectral methods, that provide global measures of the network properties [20,21]. In this paper we offer a detailed analysis of the

*Email address: fij@elte.hu

†Email address: derenyi@angel.elte.hu

‡Email address: alb@nd.edu

§Email address: vicsek@angel.elte.hu

most studied network models using algebraic tools intrinsic to large random graphs.

The paper is organized as follows. Section II introduces the main random graph models used for the topological description of large assemblies of connected units. Section III lists the—analytical and numerical—tools that we used and developed to convert the topological features of graphs into algebraic invariants. Section IV contains our results concerning the spectra and special eigenvalues of the three main types of random graph models: sparse uncorrelated random graphs in Sec. IV A, small-world graphs in Sec. IV B, and scale-free networks in Sec. IV C. Section IV D gives simple algorithms for testing the graph's structure, and Sec. IV E investigates the variance of structure within single random graph models.

II. MODELS OF RANDOM GRAPHS

A. The uncorrelated random graph model and the semicircle law

1. Definitions

Throughout this paper we will use the term "*graph*" for a set of points (vertices) connected by undirected lines (edges); no multiple edges and no loops connecting a vertex to itself are allowed. We will call two vertices of the graph "*neighbors*," if they are connected by an edge. Based on Ref. [1], we shall use the term "*uncorrelated random graph*" for a graph if (i) the probability for any pair of the graph's vertices being connected is the same, p; (ii) these probabilities are independent variables.

Any graph G can be represented by its *adjacency matrix* $A(G)$, which is a real symmetric matrix: $A_{ij}=A_{ji}=1$, if vertices i and j are connected, or 0, if these two vertices are not connected. The main algebraic tool that we will use for the analysis of graphs will be the spectrum—i.e., the set of eigenvalues—of the graph's adjacency matrix. The spectrum of the graph's adjacency matrix is also called the *spectrum of the graph.*

2. Applying the semicircle law for the spectrum of the uncorrelated random graph

A general form of the semicircle law for real symmetric matrices is the following [20,22,23]. If A is a real symmetric $N\times N$ uncorrelated random matrix, $\langle A_{ij}\rangle\equiv 0$ and $\langle A_{ij}^2\rangle=\sigma^2$ for every $i\neq j$, and with increasing N each moment of each $|A_{ij}|$ remains finite, then in the $N\to\infty$ limit the spectral density—i.e., the density of eigenvalues—of $A/\sqrt{N}$ converges to the semicircular distribution

$$\rho(\lambda)=\begin{cases}(2\pi\sigma^2)^{-1}\sqrt{4\sigma^2-\lambda^2} & \text{if } |\lambda|<2\sigma\\ 0 & \text{otherwise.}\end{cases} \tag{1}$$

This theorem is also known as *Wigner's law* [22], and its extensions to further matrix ensembles have long been used for the stochastic treatment of complex quantum-mechanical systems lying far beyond the reach of exact methods [24,25].

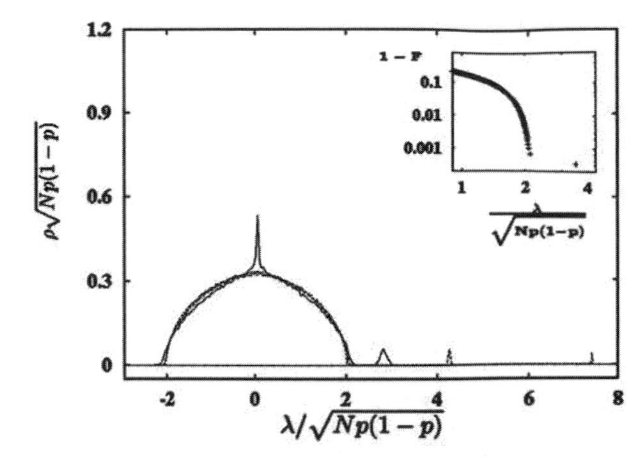

FIG. 1. If $N\to\infty$ and $p=$const, the average spectral density of an uncorrelated random graph converges to a semicircle, the first eigenvalue grows as N, and the second is proportional to $\sqrt{N}$ (see Sec. II A). *Main panel:* The spectral density is shown for $p=0.05$ and three different system sizes: $N=100$ (—), $N=300$ (– –), and $N=1000$ (- - -). In all three cases, the complete spectrum of 1000 graphs was computed and averaged. *Inset:* At the edge of the semicircle, i.e., in the $\lambda\approx\pm 2\sqrt{Np(1-p)}$ regions, the spectral density decays exponentially, and with $N\to\infty$, the decay rate diverges [20,29]. Here, $F(\lambda)=N^{-1}\Sigma_{\lambda_i<\lambda}1$ is the cumulative spectral distribution function, and $1-F$ is shown for a graph with $N=3000$ vertices and 15 000 edges.

Later, the semicircle law was found to have many applications in statistical physics and solid-state physics as well [20,21,26].

Note, that for the adjacency matrix of the uncorrelated random graph many of the semicircle law's conditions do not hold, e.g., the expectation value of the entries is a nonzero constant: $p\neq 0$. Nevertheless, in the $N\to\infty$ limit, the rescaled spectral density of the uncorrelated random graph converges to the semicircle law of Eq. (1) [27]. An illustration of the convergence of the average spectral density to the semicircular distribution can be seen on Fig. 1. It is necessary to make a comment concerning figures here. In order to keep figures simple, for the spectral density plots we have chosen to show the spectral density of the original matrix A and to rescale the horizontal (λ) and vertical (ρ) axes by $\sigma^{-1}N^{-1/2}=[Np(1-p)]^{-1/2}$ and $\sigma N^{1/2}=[Np(1-p)]^{1/2}$.

Some further results on the behavior of the uncorrelated random graph's eigenvalues, relevant for the analysis of real-world graphs as well, include the following: The principal eigenvalue (the largest eigenvalue λ_1) grows much faster than the second eigenvalue: $\lim_{N\to\infty}(\lambda_1/N)=p$ with probability 1, whereas for every $\epsilon>1/2$, $\lim_{N\to\infty}(\lambda_2/N^\epsilon)=0$ (see Refs. [27,28] and Fig. 1). A similar relation holds for the smallest eigenvalue λ_N: for every $\epsilon>1/2$, $\lim_{N\to\infty}(\lambda_N/N^\epsilon)=0$. In other words, if $\langle k_i\rangle$ denotes the average number of connections of a vertex in the graph, then λ_1 scales as $pN\approx\langle k_i\rangle$, and the width of the "bulk" part of the spectrum, the set of the eigenvalues $\{\lambda_2,\ldots,\lambda_N\}$, scales as $\sigma\sqrt{N}$. Lastly, the semicircular distribution's edges are known to decay exponentially, and the number of eigenvalues in the $\lambda > O(\sqrt{N})$ tail has been shown to be of the order of 1 [20,29].

B. Real-world graphs

The two main models proposed to describe real-world graphs are the *small-world model* and the *scale-free model.*

1. Small-world graphs

The small-world graph [4,5,30] is created by randomly rewiring some of the edges of a regular [31] ring graph. The regular ring graph is created as follows. First draw the vertices $1,2,\ldots,N$ on a circle in ascending order. Then, for every i, connect vertex i to the vertices lying closest to it on the circle: vertices $i-k/2,\ldots,i-1,i+1,\ldots,i+k/2$, where every number should be understood modulo N (k is an even number). Figure 9 will show later that this algorithm creates a regular graph indeed, because the degree [31] of any vertex is the same number k. Next, starting from vertex 1 and proceeding towards N, perform the *rewiring step*. For vertex 1, consider the first "forward connection," i.e., the connection to vertex 2. With probability p_r, reconnect vertex 1 to another vertex chosen uniformly at random and without allowing multiple edges. Proceed toward the remaining forward connections of vertex 1, and then perform this step for the remaining $N-1$ vertices also. For the rewiring, use equal and independent probabilities. Note that in the small-world model the density of edges is $p=\langle k_i\rangle/(N-1)\approx k/N$. Throughout this paper, we will use only $k>2$.

If we use $p_r=0$ in the small-world model, the original regular graph is preserved, and for $p_r=1$, one obtains a random graph that differs from the uncorrelated random graph only slightly: every vertex has a minimum degree of $k/2$. Next, we will need two definitions. The *separation* between vertices i and j, denoted by L_{ij}, is the number of edges in the shortest path connecting them. The *clustering coefficient* at vertex i, denoted by C_i, is the number of existing edges among the neighbors of vertex i divided by the number of all possible connections between them. In the small-world model, both L_{ij} and C_i are functions of the rewiring probability p_r. Based on the above definitions of $L_{ij}(p_r)$ and $C_i(p_r)$, the characteristics of the small-world phenomenon, which occurs for intermediate values of p_r, can be given as follows [4,5]: (i) the average separation between two vertices, $L(p_r)$, drops dramatically below $L(p_r=0)$, whereas (ii) the average clustering coefficient $C(p_r)$ remains high, close to $C(p_r=0)$. Note that the rewiring procedure is carried out independently for every edge; therefore, the degree sequence and also other distributions in the system, e.g., path length and loop size, decay exponentially.

2. The scale-free model

The scale-free model assumes a random graph to be a *growing* set of vertices and edges, where the location of new edges is determined by a *preferential attachment rule* [10,11]. Starting from an initial set of m_0 isolated vertices, one adds one new vertex and m new edges at every time step t. (Throughout this paper, we will use $m=m_0$.) The m new edges connect the new vertex and m different vertices chosen from the N old vertices. The ith old vertex is chosen with probability $k_i/\Sigma_{j=1,N}k_j$, where k_i is the degree of vertex i. [The density of edges in a scale-free graph is $p=\langle k_i\rangle/(N-1)\approx 2m/N$.] In contrast to the small-world model, the distribution of degrees in a scale-free graph converges to a power law when $N\rightarrow\infty$, which has been shown to be a combined effect of growth and the preferential attachment [11]. Thus, in the infinite time or size limit, the scale-free model has no characteristic scale in the degree size [14,32–37].

3. Related models

Lately, numerous other models have been suggested for a *unified description* of real-world graphs [14,32–35,37–40]. Models of growing networks with aging vertices were found to display both heavy tailed and exponentially decaying degree sequences [34–36] as a function of the speed of aging. Generalized preferential attachment rules have helped us better understand the origin of the exponents and correlations emerging in these systems [32,33]. Also, investigations of more complex network models—using aging or an additional fixed cost of edges [12] or preferential growth and random rewiring [37]—have shown, that in the "frequent rewiring, fast aging, high cost" limiting case, one obtains a graph with an exponentially decaying degree sequence, whereas in the "no rewiring, no aging, zero cost" limiting case the degree sequence will decay as a power law. According to studies of scientific collaboration networks [13,14] and further social and biological structures [12,19,41], a significant proportion of large networks lies between the two extremes. In such cases, the characterization of the system using a small number of algebraic constants could facilitate the classification of real-world networks.

III. TOOLS

A. Analytical

1. The spectrum of the graph

The spectrum of a graph is the set of eigenvalues of the graph's adjacency matrix. The physical meaning of a graph's eigenpair (an eigenvector and its eigenvalue) can be illustrated by the following example. Write each component of a vector $\vec{v}$ on the corresponding vertex of the graph: v_i on vertex i. Next, on every vertex write the sum of the numbers found on the neighbors of vertex i. If the resulting vector is a multiple of $\vec{v}$, then $\vec{v}$ is an eigenvector, and the multiplier is the corresponding eigenvalue of the graph.

The spectral density of a graph is the density of the eigenvalues of its adjacency matrix. For a finite system, this can be written as a sum of δ functions

$$\rho(\lambda):=\frac{1}{N}\sum_{j=1}^{N}\delta(\lambda-\lambda_j), \tag{2}$$

which converges to a continuous function with $N\rightarrow\infty$ (λ_j is the jth largest eigenvalue of the graph's adjacency matrix).

The spectral density of a graph can be directly related to the graph's topological features: the kth moment M_k of $\rho(\lambda)$ can be written as

$$M_k = \frac{1}{N}\sum_{j=1}^{N}(\lambda_j)^k = \frac{1}{N}\mathrm{Tr}(A^k)$$
$$= \frac{1}{N}\sum_{i_1,i_2,\ldots,i_k} A_{i_1,i_2}A_{i_2,i_3}\cdots A_{i_k,i_1}. \quad (3)$$

From the topological point of view, $D_k = NM_k$ is the number of directed paths (loops) of the underlying—undirected—graph, that return to their starting vertex after k steps. On a tree, the length of any such path can be an even number only, because these paths contain any edge an even number of times: once such a path has left its starting point by choosing a starting edge, no alternative route for returning to the starting point is available. However, if the graph contains loops of odd length, the path length can be an odd number, as well.

2. *Extremal eigenvalues*

In an uncorrelated random graph the principal eigenvalue λ_1 shows the density of edges and λ_2 can be related to the conductance of the graph as a network of resistances [42]. An important property of all graphs is the following: the principal eigenvector $\vec{e}_1$ of the adjacency matrix is a non-negative vector (all components are non-negative), and if the graph has no isolated vertices, $\vec{e}_1$ is a positive vector [43]. All other eigenvectors are orthogonal to $\vec{e}_1$, therefore they all have entries with mixed signs.

3. *The inverse participation ratios of eigenvectors*

The inverse participation ratio of the normalized jth eigenvector $\vec{e}_j$ is defined as [26]

$$I_j = \sum_{k=1}^{N}[(e_j)_k]^4. \quad (4)$$

If the components of an eigenvector are identical, $(e_j)_i = 1/\sqrt{N}$ for every i, then $I_j = 1/N$. For an eigenvector with one single nonzero component, $(e_j)_i = \delta_{i,i'}$, the inverse participation ratio is 1. The comparison of these two extremal cases illustrates that with the help of the inverse participation ratio, one can tell whether only $O(1)$ or as many as $O(N)$ components of an eigenvector differ significantly from 0, i.e., whether an eigenvector is localized or nonlocalized.

B. Numerical

1. *General real symmetric eigenvalue solver*

To compute the eigenpairs of graphs below *the size* $N = 5000$, we used the general real symmetric eigenvalue solver of Ref. [44]. This algorithm requires the allocation of memory space to all entries of the matrix, thus to compute the spectrum of a graph of size $N = 20\,000$ ($N = 1\,000\,000$) using this general method with double precision floating point arithmetic, one would need 3.2 GB (8 TB) memory space and the execution of approximately $30N^2 = 1.2\times10^{10}$ (3×10^{13}) floating point operations [44]. Consequently, we need to develop more efficient algorithms to investigate the properties of graphs with sizes comparable to real-world networks.

2. *Iterative eigenvalue solver based on the thick-restart Lanczos algorithm*

The spectrum of a real-world graph is the spectrum of a sparse real symmetric matrix; therefore, the most efficient algorithms that can give a handful of the top n_d eigenvalues—and the corresponding eigenvectors—of a large graph are iterative methods [45]. These methods allow the matrix to be stored in any compact format, as long as matrix-vector multiplication can be carried out at a high speed. *Iterative methods use little memory*: only the nonzero entries of the matrix and a few vectors of size N need to be stored. The price for computational speed lies in the number of the obtained eigenvalues: iterative methods compute only a handful of the largest (or smallest) eigenvalues of a matrix. To compute the eigenvalues of graphs above the size $N = 5\,000$, we have developed algorithms using a specially modified version of the thick-restart Lanczos algorithm [46,47]. The modifications and some of the main technical parameters of our software are explained in the following paragraphs.

Even though iterative eigenvalue methods are mostly used to obtain the top eigenvalues of a matrix, after minor modifications the internal eigenvalues in the vicinity of a fixed $\lambda = \lambda_0$ point can be computed as well. For this, extremely sparse matrices are usually "shift-inverted," i.e., to find those eigenvalues of A that are closest to λ_0, the highest and lowest eigenvalues of $(A-\lambda_0 I)^{-1}$ are searched for. However, because of the extremely high cost of matrix inversion in our case, for the computation of internal eigenvalues we suggest using the *"shift-square" method* with the matrix

$$B = [\lambda^*/2 - (A-\lambda_0 I)^2]^{2n+1}. \quad (5)$$

Here λ^* is the largest eigenvalue of $(A-\lambda_0 I)^2$, I is the identity matrix, and n is a positive integer. Transforming the matrix A into B transforms the spectrum of A in the following manner. First, the spectrum is shifted to the left by λ_0. Then, the spectrum is "folded" (and squared) at the origin such that all eigenvalues will be negative. Next, the spectrum is linearly rescaled and shifted to the right, with the following effect: (i) the whole spectrum will lie in the symmetric interval $[-\lambda^*/2, \lambda^*/2]$ and (ii) those eigenvalues that were closest to λ_0 in the spectrum of A will be the largest now, i.e., they will be the eigenvalues closest to $\lambda^*/2$. Now, raising all eigenvalues to the $(2n+1)$st power increases the relative difference, $1-\lambda_i/\lambda_j$, between the top eigenvalues λ_i and λ_j by a factor of $2n+1$. This allows the iterative method to find the top eigenvalues of B more quickly. One can compute the corresponding eigenvalues (those being closest to λ_0) of the original matrix, A: if $\vec{b}_1, \vec{b}_2, \ldots, \vec{b}_{n_d}$ are the normalized eigenvectors of the n_d largest eigenvalues of

B, then for A the n_d eigenvalues closest to λ_0 will be, not necessarily in ascending order, $\vec{b}_1 A \vec{b}_1, \vec{b}_2 A \vec{b}_2, \ldots, \vec{b}_{n_d} A \vec{b}_{n_d}$.

The thick-restart Lanczos method uses memory space for the nonzero entries of the $N \times N$ large adjacency matrix, and n_g+1 vectors of length N, where n_g ($n_g > n_d$) is usually between 10 and 100. Besides the relatively small size of required memory, we could also exploit the fact that the nonzero entries of a graph's adjacency matrix are all 1's: during matrix-vector multiplication—which is usually the most time-consuming step of an iterative method—only additions had to be carried out instead of multiplications.

The numerical spectral density functions of large graphs ($N \geq 5000$) of this paper were obtained using the following steps. To compute the spectral density of the adjacency matrix A at an internal $\lambda = \lambda_0$ location, first the n_d eigenvalues closest to λ_0 were searched for. Next, the distance between the smallest and the largest of the obtained eigenvalues was computed. Finally, to obtain $\rho(\lambda_0)$ this distance was multiplied by $N/(n_d-1)$, and was averaged using n_{av} different graphs. We used double precision floating point arithmetic, and the iterations were stopped if (i) at least n_{it} iterations had been carried out and (ii) the lengths of the residual vectors belonging to the n_d selected eigenpairs were all below $\varepsilon = 10^{-12}$ [46].

IV. RESULTS

A. Sparse uncorrelated random graphs: The semicircle law is not universal

In the uncorrelated random graph model of Erdős and Rényi, the total number of edges grows quadratically with the number of vertices: $N_{edge} = N\langle k_i \rangle = Np(N-1) \approx pN^2$. However, in many real-world graphs edges are "expensive," and the growth rate of the number of connections remains well below this rate. For this reason, we also investigated the spectra of such uncorrelated networks, for which the probability of any two vertices being connected changes with the size of the system using $pN^\alpha = c = \text{const}$. Two special cases are $\alpha=0$ (the Erdős-Rényi model) and $\alpha=1$. In the second case, $pN \to \text{const}$ as $N \to \infty$, i.e., the average degree remains constant.

For $\alpha<1$ *and* $N \to \infty$, there exists an infinite cluster of connected vertices (in fact, it exists for every $\alpha \leq 1$ [2]). Moreover, the expectation value of any k_i converges to infinity, thus any vertex is almost surely connected to the infinite cluster. The spectral density function converges to the semicircular distribution of Eq. (1) because the total weight of isolated subgraphs decreases exponentially with growing system size. (A detailed analysis of this issue is available in Ref. [48].)

For $\alpha=1$ *and* $N \to \infty$ (see Fig. 2), the probability for a vertex to belong to a cluster of any finite size remains also finite [49]. Therefore, the limiting spectral density contains the weighted sum of the spectral densities of all finite graphs [50]. The most striking deviation from the semicircle law in this case is the elevated central part of the spectral density. The probability for a vertex to belong to an isolated cluster of size s decreases exponentially with s [49]; therefore, the number of large isolated clusters is low. The eigenvalues of a graph with s vertices are bounded by $-\sqrt{s-1}$ and $\sqrt{s-1}$. For these two reasons, the amplitudes of δ functions decay exponentially, as the absolute value of their locations, $|\lambda|$, increases.

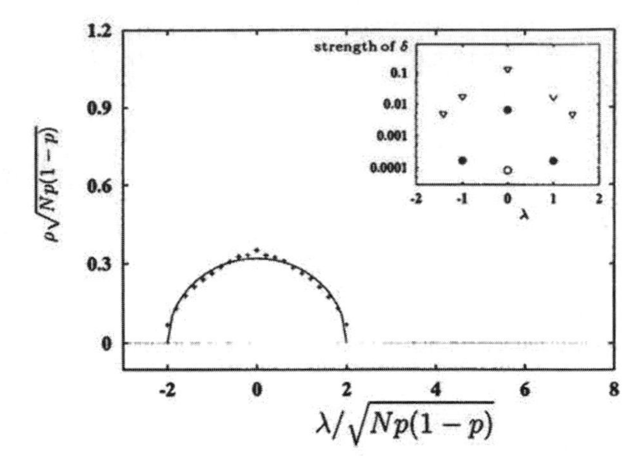

FIG. 2. If $N \to \infty$ and $pN = \text{const}$, the spectral density of the uncorrelated random graph does not converge to a semicircle. *Main panel:* Symbols show the spectrum of an uncorrelated random graph (20 000 vertices and 100 000 edges) measured with the iterative method using $n_{av}=1$, $n_d=101$, and $n_g=250$. A solid line shows the semicircular distribution for comparison. (Note that the principal eigenvalue λ_1 is not shown here because here at any λ_0 point the average first-neighbor distance among $n_d=101$ eigenvalues was used to measure the spectral density.) *Inset:* Strength of δ functions in $\rho(\lambda)$ "caused" by isolated clusters of sizes 1, 2, and 3 in uncorrelated random graphs (see Ref. [50] for a detailed explanation). Symbols are for graphs with 20 000 vertices and 20 000 edges (▽), 50 000 edges (●), and 100 000 edges (○). Results were averaged for three different graphs everywhere.

The principal eigenvalue of this graph converges to a constant: $\lim_{N\to\infty}(\lambda_1) = pN = c$, and $\rho(\lambda)$ will be symmetric in the $N \to \infty$ limit. Therefore, in the limit, all odd moments (M_{2k+1}), and thus the number of all loops with odd length (D_{2k+1}), disappear. This is a salient feature of graphs with tree structure (because on a tree every edge must be used an even number of times in order to return to the initial vertex), indicating that the structure of a sparse uncorrelated random graph becomes more and more treelike. This can also be understood by considering that the typical distance (length of the shortest path) between two vertices on both a sparse uncorrelated random graph and a regular tree with the same number of edges scales as $\ln(N)$. So except for a few shortcuts a sparse uncorrelated random graph looks like a tree.

B. The small-world graph

Triangles are abundant in the graph

For $p_r=0$ the small-world graph is *regular and also periodical.* Because of the highly ordered structure, $\rho(\lambda)$ contains numerous singularities, which are listed in Sec. VI A (see also Fig. 3). Note that $\rho(\lambda)$ has a high third moment. (Remember, that we use only $k>2$.)

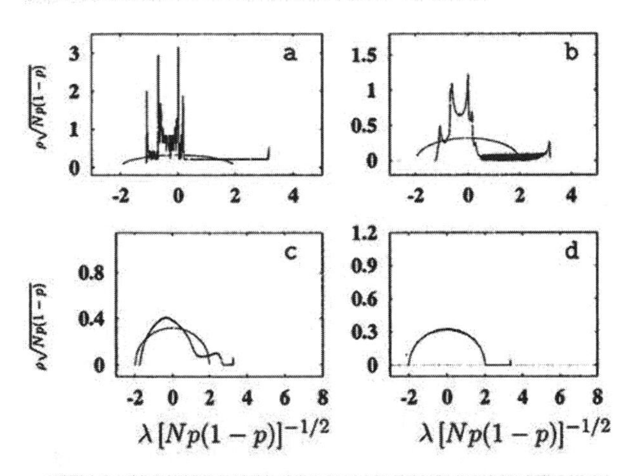

FIG. 3. Spectral densities of small-world graphs using the complete spectra. The solid line shows the semicircular distribution for comparison. (a) Spectral density of the regular ring graph created from the small-world model with $p_r=0$, $k=10$, and $N=1000$. (b) For $p_r=0.01$, the average spectral density of small-world graphs contains sharp maxima, which are the "blurred" remnants of the singularities of the $p_r=0$ case. Topologically, this means that the graph is still almost regular, but it contains a small number of impurities. In other words, after a small perturbation, the system is no longer degenerate. (c) The average spectral density computed for the $p_r=0.3$ case shows that the third moment of $\rho(\lambda)$ is preserved even for very high values of p_r, where there is already no sign of any blurred singularity (i.e., regular structure). This means that even though all remaining regular islands have been destroyed already, *triangles are still dominant.* (d) If $p_r=1$, then the spectral density of the small-world graph converges to a semicircle. In (b), (c), and (d), 1000 different graphs with $N=1000$ and $k=10$ were used for averaging.

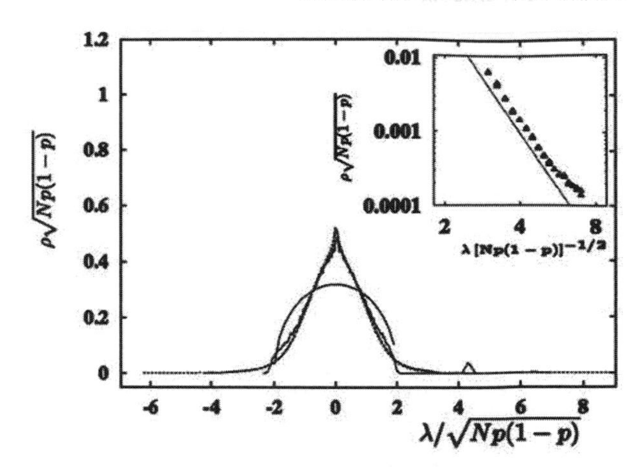

FIG. 4. *Main panel:* The average spectral densities of scale-free graphs with $m=m_0=5$ and $N=100$ (—), $N=1000$ (– –), and $N=7000$ (- - -) vertices. (In all three cases, the complete spectrum of 1000 graphs was used.) Another continuous line shows the semicircular distribution for comparison. Observe that (i) the central part of the scale-free graph's spectral density is trianglelike, not semicircular and (ii) the edges show a power-law decay, whereas the semicircular distribution's edges decay exponentially, i.e., it decays exponentially at the edges [20]. *Inset:* The upper edge of the spectral density for scale-free graphs with $N=40\,000$ vertices, the average degree of a vertex being $\langle k_i \rangle = 2m = 10$ as before. Note that both axes are logarithmic, indicating that $\rho(\lambda)$ has a power-law tail. Here we used the iterative eigenvalue solver of Sec. III B 2 with $n_d=21$, $n_{av}=3$, and $n_g=60$. The line with the slope -5 in this figure is a guide to the eye.

If we increase p_r such that the small-world region is reached, i.e., the periodical structure of the graph is perturbed, then singularities become blurred and are transformed into high local maxima, but $\rho(\lambda)$ retains a strong skewness (see Fig. 3). This is in good agreement with the results of Refs. [30,51], where it has been shown that the local structure of the small-world graph is ordered; however, already a very small number of shortcuts can drastically change the graph's global structure.

In the $p_r=1$ case the small-world model becomes very similar to the uncorrelated random graph: the only difference is that here, the minimum degree of any vertex is a positive constant $k/2$, whereas in an uncorrelated random graph the degree of a vertex can be any non-negative number. Accordingly, $\rho(\lambda)$ becomes a semicircle for $p_r=1$ (Fig. 3). Nevertheless, it should be noted that as p_r converges to 1, a high value of M_3 is preserved even for p_r close to 1, where all local maxima have already vanished. The third moment of $\rho(\lambda)$ gives the number of triangles in the graph (see Sec. III A 1); the lack of high local maxima, i.e., the remnants of singularities, shows the absence of an ordered structure. From the above we conclude, that—from the spectrum's point of view—the high number of triangles is one of the most basic properties of the small-world model, and it is preserved much longer than regularity or periodicity if the level of randomness p_r is increased. This is in good agreement with the results of Ref. [19] where the high number of small cycles is found to be a fundamental property of small-world networks. As an application, the high number of small cycles results in special diffusion on small-world graphs [61].

C. The scale-free graph

For $m=m_0=1$, the scale-free graph is a tree by definition and its spectrum is symmetric [43]. In the $m>1$ case $\rho(\lambda)$ consists of several well distinguishable parts (see Fig. 4). The "bulk" part of the spectral density—the set of the eigenvalues $\{\lambda_2, \ldots, \lambda_N\}$—converges to a symmetric continuous function which has a trianglelike shape for the normalized λ values up to 1.5 and has power-law tails.

The central part of the spectral density lies well above the semicircle. Since the scale-free graph is fully connected by definition, the increased number of eigenvalues with small magnitudes cannot be accounted to isolated clusters, as before in the case of the sparse uncorrelated random graph. As an explanation, we suggest, that the eigenvectors of these eigenvalues are localized on a small subset of the graph's vertices. (This idea is supported by the high inverse participation ratios of these eigenvectors, see Fig. 7.)

1. The spectral density of the scale-free graph decays as a power law

The inset of Fig. 4 shows the tail of the bulk part of the spectral density for a graph with $N=40\,000$ vertices and

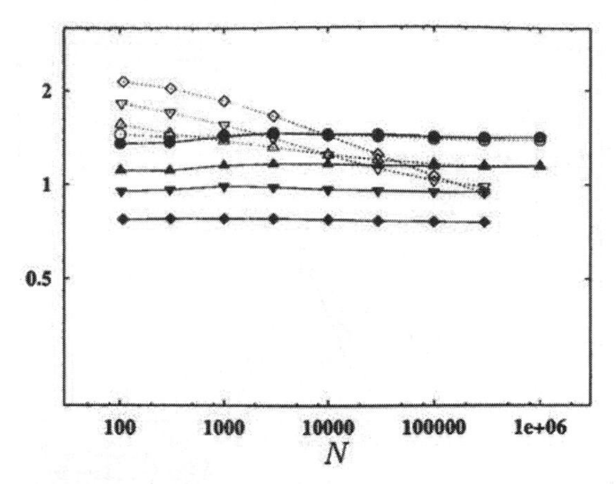

FIG. 5. Comparison of the length of the longest row vector $\sqrt{k_1}$ and the principal eigenvalue λ_1 in scale-free graphs. Open symbols show $\lambda_1/(\sqrt{m}N^{1/4})$, closed symbols show $\sqrt{k_1}/(\sqrt{m}N^{1/4})$. The parameter values are $m=1$ (○), $m=2$ (△), $m=4$ (▽), and $m=8$ (◇). Each data point is an average for nine graphs. For the reader's convenience, data points are connected. If $m>1$ and the network is small, the principal eigenvalue λ_1 of a scale-free graph is determined by the largest row vectors jointly: the largest eigenvalue is above $\sqrt{k_1}$ and the growth rate of λ_1 stays below the maximum possible growth rate, which is $\lambda_1 \propto N^{1/4}$. If $m=1$, or the network is large, the effect of row vectors other than the longest on λ_1 vanishes: the principal eigenvalue converges to the length of the longest row vector, and it grows as $\lambda_1 \propto N^{1/4}$. Our results show a crossover in the growth rate of the scale-free model's principal eigenvalue.

200 000 edges (i.e., $pN=10$). Comparing this to the inset of Fig. 1, where the number of vertices and edges is the same as here, one can observe the *power-law decay* at the edge of the bulk part of $\rho(\lambda)$. As shown later, in Sec. IV D, the power-law decay in this region is caused by localized eigenvectors; these eigenvectors are localized on vertices with the highest degrees. The power-law decay of the degree sequence, i.e., the existence of very high degrees, is, in turn, due to the preferential attachment rule of the scale-free model.

2. The growth rate of the principal eigenvalue shows a crossover in the level of correlations

Since the adjacency matrix of a graph is a non-negative symmetric matrix, the graph's largest eigenvalue λ_1 is also the largest in magnitude (see, e.g., Theorem 0.2 of Ref. [43]). Considering the effect of the adjacency matrix on the base vectors $(b_i)_j=\delta_{ij}$ $(i=1,2,\ldots,N)$, it can be shown that a lower bound for λ_1 is given by the length of the longest row vector of the adjacency matrix, which is the square root of the graph's largest degree k_1. Knowing that the largest degree of a scale-free graph grows as $\sqrt{N}$ [11], one expects λ_1 to grow as $N^{1/4}$ for large enough systems.

Figure 5 shows a rescaled plot of the scale-free graph's largest eigenvalue for different values of m. In this figure, λ_1 is compared to the length of the longest row vector $\sqrt{k_1}$ on the "natural scale" of these values, which is $\sqrt{m}N^{1/4}$ [11]. It is clear that if $m>1$ and the system is small, then through several decades (a) λ_1 is larger than $\sqrt{k_1}$ and (b) the growth rate of λ_1 is well below the expected rate of $N^{1/4}$. In the $m=1$ case, and for large systems, (a) the difference between λ_1 and $\sqrt{k_1}$ vanishes and (b) the growth rate of the principal eigenvalue will be maximal, too. This crossover in the behavior of the scale-free graph's principal eigenvalue is a specific property of sparse growing correlated graphs, and it is a result of the changing level of correlations between the longest row vectors (see Sec. VI B).

3. Comparing the role of the principal eigenvalue in the scale-free graph and the α=1 uncorrelated random graph: A comparison of structures

Now we will compare the role of the principal eigenvalue in the $m>1$ scale-free graph and the $\alpha=1$ uncorrelated random graph through its effect on the moments of the spectral density. On Figs. 4 and 5 one can observe that (i) the principal eigenvalue of the scale-free graph is detached from the rest of the spectrum, and (ii) as $N\to\infty$, it grows as $N^{1/4}$ (see also Secs. IV C 2 and VI B). It can be also seen that in the limit, the bulk part will be symmetric, and its width will be constant (Fig. 4 rescales this constant width merely by another constant, namely $[Np(1-p)]^{-1/2}$. Because of the symmetry of the bulk part, in the $N\to\infty$ limit, the third moment of $\rho(\lambda)$ is determined exclusively by the contribution of the principal eigenvalue, which is $N^{-1}(\lambda_1)^3 \propto N^{-1/4}$. For each moment above the third (e.g., for the lth moment), with growing N, the contribution of the bulk part to this moment will scale as $\mathcal{O}(1)$, and the contribution of the principal eigenvalue will scale as $N^{-1+l/4}$. In summary, in the $N\to\infty$ limit, the scale-free graph's first eigenvalue has a significant contribution to the fourth moment; the fifth and all higher moments are determined exclusively by λ_1: the lth moment will scale as $N^{-1+l/4}$.

In contrast to the above, the principal eigenvalue of the $\alpha=1$ uncorrelated random graph converges to the constant $pN=c$ in the $N\to\infty$ limit, and the width of the bulk part also remains constant (see Fig. 2). Given a fixed number l the contribution of the principal eigenvalue to the lth moment of the spectral density will change as $N^{-1}c^l$ in the $N\to\infty$ limit. The contribution of the bulk part will scale as $\mathcal{O}(1)$, therefore all even moments of the spectral density will scale as $\mathcal{O}(1)$ in the $N\to\infty$ limit, and all odd moments will converge to 0.

The difference between the growth rate of the moments of $\rho(\lambda)$ in the above two models (scale-free graph and $\alpha=1$ uncorrelated random graph model) can be interpreted as a sign of different structure (see Sec. III A 1). In the $N\to\infty$ limit, the average degree of a vertex converges to a constant in both models: $\lim_{N\to\infty}\langle k_i\rangle=pN=c=2m$. (Both graphs will have the same number of edges per vertex.) On the other hand, in the limit all moments of the $\alpha=1$ uncorrelated random graph's spectral density converge to a constant, whereas the moments M_l $(l=5,6,\ldots)$ of the scale-free graph's $\rho(\lambda)$ will diverge as $N^{-1+l/4}$. In other words: the number of loops of length l in the $\alpha=1$ uncorrelated random graph will grow as $D_l=NM_l=\mathcal{O}(N)$, whereas for the scale-free graph for every $l\geq 3$, the number of these loops will grow as

$D_l = NM_l = \mathcal{O}(N^{l/4})$. From this we conclude that in the limit, the role of loops is negligible in the $\alpha=1$ uncorrelated random graph, whereas it is large in the scale-free graph. In fact, the growth rate of the number of loops in the scale-free graph exceeds all polynomial growth rates: the longer the loop size (l) investigated, the higher the growth rate of the number of these loops ($N^{l/4}$) will be. Note that the relative number of triangles (i.e., the third moment of the spectral density, M_l/N) will disappear in the scale-free graph, if $N \to \infty$.

In summary, the spectrum of the scale-free model converges to a trianglelike shape in the center, and the edges of the bulk part decay slowly. The first eigenvalue is detached from the rest of the spectrum, and it shows an anomalous growth rate. Eigenvalues with large magnitudes belong to eigenvectors localized on vertices with many neighbors. In the present context, the absence of triangles, the high number of loops with length above $l=3$, and the buildup of correlations are the basic properties of the scale-free model.

D. Testing the structure of a "real-world" graph

To analyze the structure of a large sparse random graph (correlated or not), here we suggest several tests that can be performed within $\mathcal{O}(N)$ CPU time, use $\mathcal{O}(N)$ floating point operations, and can clearly differentiate between the three "pure" types of random graph models treated in Sec. IV. Furthermore, these tests allow one to quantify the relation between any real-world graph and the three basic types of random graphs.

1. Extremal eigenvalues

In Sec. III A 2 we have already mentioned that the extremal eigenvalues contain useful information on the structure of the graph. As the spectra of uncorrelated random graphs (Fig. 1) and scale-free networks (Fig. 4) show, the principal eigenvalue of random graphs is often detached from the rest of the spectrum. For these two network types, the remaining bulk part of the spectrum, i.e., the set $\{\lambda_2, \ldots, \lambda_N\}$, converges to a symmetric distribution, thus the quantity

$$R := \frac{\lambda_1 - \lambda_2}{\lambda_2 - \lambda_N} \tag{6}$$

measures the distance of the first eigenvalue from the main part of $\rho(\lambda)$ normalized by the extension of the main part. (R can be connected to the chromatic number of the graph [52].)

Note that in the $N \to \infty$ limit the $\alpha=0$ sparse uncorrelated random graph's principal eigenvalue will scale as $\langle k_i \rangle$, whereas both λ_2 and $|-\lambda_N|$ will scale as $2\sqrt{\langle k_i \rangle}$. Therefore, if $\langle k_i \rangle > 4$, the principal eigenvalue will be detached from the bulk part of the spectrum and R will scale as $(\sqrt{\langle k_i \rangle} - 2)/4$. If, however, $\langle k_i \rangle \leq 4$, λ_1 will not be detached from the bulk part, it will converge to 0.

The above explanation and Fig. 6 show that in the $\langle k_i \rangle > 4$ sparse uncorrelated random graph model and the scale-free network, λ_1 and the rest of the spectrum are well separated, which gives similarly high values for R in small systems. In large systems, R of the sparse uncorrelated random graph converges to a constant, while R in the scale-free model decays as a power-law function of N. The reason for this drop is the increasing denominator on the right-hand side of Eq. (6): λ_2 and λ_N are the extremal eigenvalues in the lower and upper long tails of $\rho(\lambda)$, therefore, as N increases, the expectation values of λ_2 and $-\lambda_N$ grow as quickly as that of λ_1. On the other hand, the small-world network shows much lower values of R already for small systems: here, λ_1 is not detached from the rest of the spectrum, which is a consequence of the almost periodical structure of the graph.

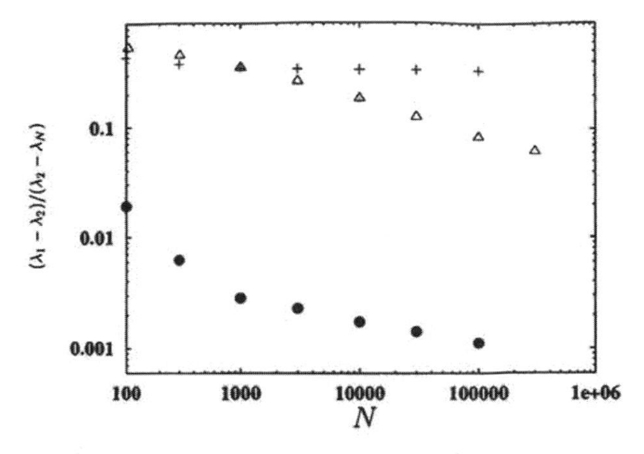

FIG. 6. The ratio $R = (\lambda_1 - \lambda_2)/(\lambda_2 - \lambda_N)$ for sparse uncorrelated random graphs (+), small-world graphs with $p_r = 0.01$ (●), and scale-free networks (△). All graphs have an average degree of $\langle k_i \rangle = 10$, and at each data point, the number of graphs used for averaging was 9. Observe, that for the uncorrelated random graph, R converges to a constant (see Sec. III A 2), whereas it decays rapidly for the two other types of networks, as $N \to \infty$. On the other hand, the latter two network types (small-world and scale-free) differ significantly in their magnitudes of R.

On Fig. 6 graphs with the same number of vertices and edges are compared. For large ($N \geq 10\,000$) systems and for sparse uncorrelated random graphs R converges to a constant, whereas for scale-free graphs and small-world networks it decays as a power law. The latter two networks significantly differ in the magnitude of R. In summary, the suggested quantity R has been shown to be appropriate for distinguishing between the following graph structures: (i) periodical or almost periodical (small world), (ii) uncorrelated nonperiodical, and (iii) strongly correlated nonperiodical (scale free).

2. Inverse participation ratios of extremal eigenpairs

Figure 7 shows the inverse participation ratios of the eigenvectors of an uncorrelated random graph, a small-world graph with $p_r = 0.01$, and a scale-free graph. Even though all

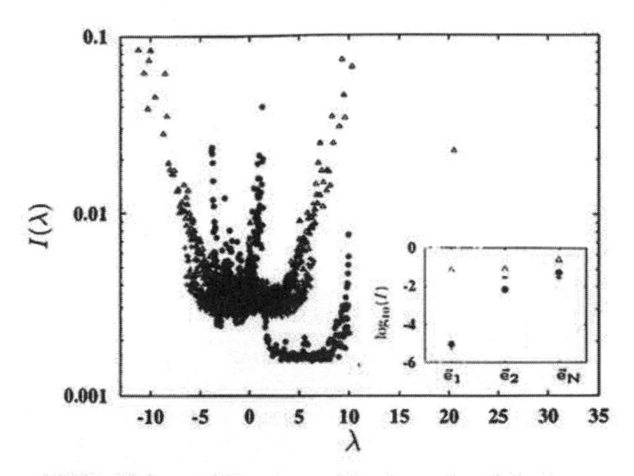

FIG. 7. *Main panel:* Inverse participation ratios of the eigenvectors of three graphs shown as a function of the corresponding eigenvalues: uncorrelated random graph (+), small-world graph with $p_r=0.01$ (●), and scale-free graph (△). All three graphs have $N=1000$ vertices, and the average degree of a vertex is $\langle k_i \rangle = 10$. Observe that the eigenvectors of the sparse uncorrelated random graph and the small-world network are usually nonlocalized [$I(\lambda)$ is close to $1/N$]. On the contrary, eigenvectors belonging to the scale-free graph's extremal eigenvalues are highly localized with $I(\lambda)$ approaching 0.1. Note also that for $\lambda \approx 0$, the scale-free graph's $I(\lambda)$ has a significant "spike" indicating again the localization of eigenvectors. *Inset:* Inverse participation ratios of the first, second, and Nth eigenvectors of an uncorrelated random graph (+), a small-world graph with $p_r=0.01$ (●), and a scale-free graph (△). For each data point, the number of vertices was $N=300\,000$ and the number of edges was 1 500 000. Clearly, the principal eigenvector of the scale-free graph is localized, while the principal eigenvector of the other two systems (the uncorrelated models) is not. Note also that the inverse participation ratios of the second and Nth eigenvectors clearly differ in the small-world graph—the spectrum of this graph has already been shown to be strongly asymmetric—whereas in the uncorrelated random graph the inverse participation ratios of $\vec{e}_2$ and $\vec{e}_N$ are approximately the same. Thus, with the help of the inverse participation ratios of $\vec{e}_1$, $\vec{e}_2$, and $\vec{e}_N$, one can identify the three main types of random graphs used here.

three graphs have the same number of vertices ($N=1000$) and edges (5000), one can observe rather specific features (see also the inset of Fig. 7).

The uncorrelated random graph's eigenvectors show very little difference in their level of localization, except for the principal eigenvector, which is much less localized than the other eigenvectors; $I(\lambda_2)$ and $I(\lambda_N)$ are almost equal. For the small-world graph's eigenvectors, $I(\lambda)$ has many different plateaus and spikes; the principal eigenvector is not localized, and the second and Nth eigenvectors have high, but different, $I(\lambda)$ values. The eigenvectors belonging to the scale-free graph's largest and smallest eigenvalues are localized on the "largest" vertices. The long tails of the bulk part of $\rho(\lambda)$ are due to these vertices. All three investigated eigenvectors ($\vec{e}_1$, $\vec{e}_2$, and $\vec{e}_N$) of the scale-free graph are highly localized. Consequently, the inverse participation ratios of the eigenvectors $\vec{e}_1$, $\vec{e}_2$, and $\vec{e}_N$ are handy for the identification of the three basic types of random graph models used.

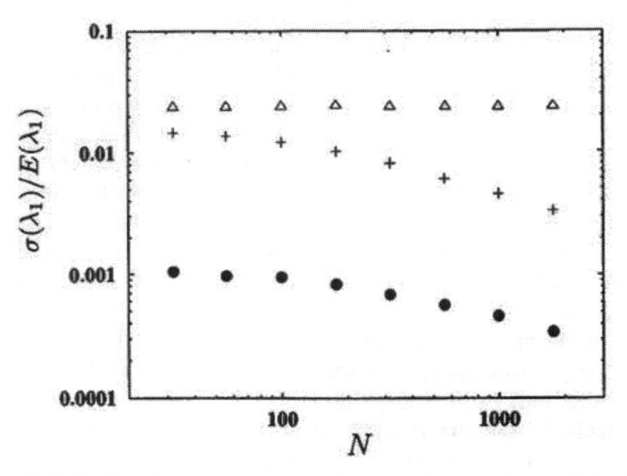

FIG. 8. Size dependence of the relative variance of the principal eigenvalue, i.e., $\sigma(\lambda_1)/E(\lambda_1)$, for sparse uncorrelated random graphs (+), small-world graphs with $p_r=0.01$ (●), and scale-free graphs (△). The average degree of a vertex is $\langle k_i \rangle = 10$, and 1000 graphs were used for averaging at every point. Observe that in the uncorrelated random graph and the small-world model $\sigma(\lambda_1)/E(\lambda_1)$ decays with increasing system size; however, for scale-free graphs with the same number of edges and vertices, it remains constant.

E. Structural variances

Relative variance of the principal eigenvalue for different types of networks: The scale-free graph and self-similarity

Figure 8 shows the relative variance of the principal eigenvalue, i.e., $\sigma(\lambda_1)/E(\lambda_1)$, for the three basic random graph types.

For nonsparse uncorrelated random graphs ($N \to \infty$ and $p=$const) this quantity is known to decay at a rate which is faster than exponential [28,53]. Comparing sparse graphs with the same number of vertices and edges, one can see that in the sparse uncorrelated random graph and the small-world model the relative variance of the principal eigenvalue drops quickly with growing system size. In the scale-free model, however, the relative variance of the principal eigenvalue's distribution remains constant with an increasing number of vertices.

In fractals, fluctuations do not disappear as the size of the system is increased, while in the scale-free graph, the relative variance of the principal eigenvalue is independent of system size. In this sense, the scale-free graph resembles self-similar systems.

V. CONCLUSIONS

We have performed a detailed analysis of the complete spectra, eigenvalues, and the eigenvectors' inverse participation ratios in three types of sparse random graphs: the sparse uncorrelated random graph, the small-world model, and the scale-free network. Connecting the topological features of these graphs to algebraic quantities, we have demonstrated that (i) the semi circle law is not universal, not even for the uncorrelated random graph model; (ii) the small-world graph is inherently noncorrelated and contains a high number of

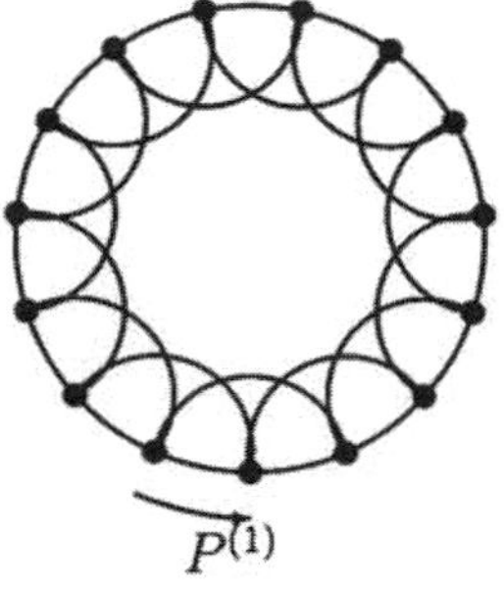

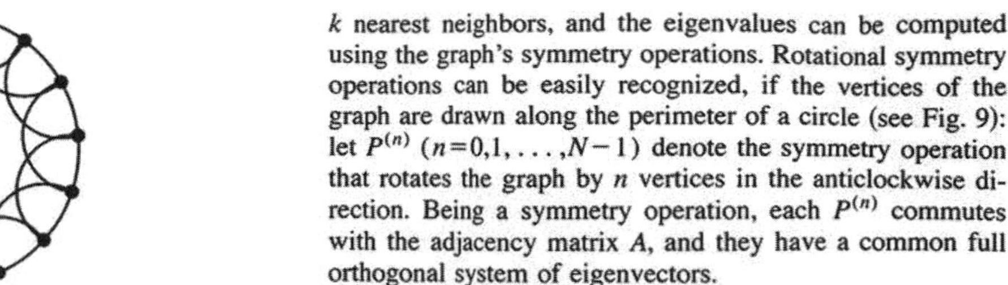

FIG. 9. The regular ring graph obtained from the small-world model in the $p_r=0$ case: rotations ($P^{(n)}$ for every $n=0,1,\ldots,N-1$) are symmetry operations of the graph. The $P^{(n)}$ operators (there are N of them) can be used to create a full orthogonal basis of the adjacency matrix A: taking any $P^{(n)}$, it commutes with A, therefore they have a common full orthogonal system of eigenvectors. (For a clear illustration of symmetries, this figure shows a graph with only $N=15$ vertices and $k=4$ connections per vertex.)

triangles; (iii) the spectral density of the scale-free graph is made up of three, well distinguishable parts (center, tails of bulk, first eigenvalue), and as $N\rightarrow\infty$, triangles become negligible and the level of correlations changes.

We have presented practical tools for the identification of the above-mentioned basic types of random graphs and further, for the classification of real-world graphs. The robust eigenvector techniques and observations outlined in this paper combined with previous studies are likely to improve our understanding of large sparse correlated random structures. Examples for algebraic techniques already in use for large sparse correlated random structures are analyses of the Internet [6,18] and search engines [54,62] and mappings [55,56] of the World-Wide Web. Besides the improvement of these techniques, the present work may turn out to be useful for analyzing the correlation structure of the transactions between a very high number of economical and financial units, which has already been started in, e.g., Refs. [57–59]. Lastly, we hope to have provided quantitative tools for the classification of further "real-world" networks, e.g., social and biological networks.

Note added in proof. Recently, we were made aware of a manuscript by Goh, Kahng, and Kim [63] investigating the spectral properties of scale-free networks. Also, our attention has been drawn to a recent publication of Bauer and Golinelli [64] on the spectral properties of uncorrelated random graphs.

ACKNOWLEDGMENTS

We thank D. Petz, G. Stoyan, G. Tusnády, K. Wu, and B. Kahng for helpful discussions and suggestions. This research was partially supported by a HNSF Grant No. OTKA T033104 and NSF Grant No. PHY-9988674.

APPENDIX A: THE SPECTRUM OF A SMALL-WORLD GRAPH FOR $p_r=0$ REWIRING PROBABILITY

1. Derivation of the spectral density

If the rewiring probability of a small-world graph is $p_r=0$, then the graph is regular, each vertex is connected to its k nearest neighbors, and the eigenvalues can be computed using the graph's symmetry operations. Rotational symmetry operations can be easily recognized, if the vertices of the graph are drawn along the perimeter of a circle (see Fig. 9): let $P^{(n)}$ ($n=0,1,\ldots,N-1$) denote the symmetry operation that rotates the graph by n vertices in the anticlockwise direction. Being a symmetry operation, each $P^{(n)}$ commutes with the adjacency matrix A, and they have a common full orthogonal system of eigenvectors.

Now, we will create a full orthogonal basis of A. (We will treat only the case when N is an even number; odd N's can be treated similarly.) It is known that the eigenvalues of A are real; however, to simplify calculations, we will use complex numbers first. The eigenvectors of every $P^{(n)}$ are $\vec{e}_1,\vec{e}_2,\ldots,\vec{e}_N$,

$$(e_l)_j=\exp\left(2\pi i\frac{jl}{N}\right), \tag{A1}$$

where $l=0,2,\ldots,N-1$ and $i=\sqrt{-1}$. The eigenvalue of $P^{(n)}$ on $\vec{e}_l$ is

$$s_l^{(n)}=\exp\left(2\pi i\frac{nl}{N}\right). \tag{A2}$$

By adding these values pairwise, one can obtain the N eigenvalues of the graph

$$\lambda_l=2\sum_{j=1}^{k/2}\cos\left(2\pi\frac{jl}{N}\right). \tag{A3}$$

In the previous exponential form the right-hand side is a summation for a geometrical series; therefore,

$$\lambda_l=\frac{\sin[(k+1)l\pi/N]}{\sin(l\pi/N)}-1. \tag{A4}$$

In the $N\rightarrow\infty$ limit, this converges to

$$\lambda(x)=\frac{\sin[(k+1)x]}{\sin(x)}-1, \tag{A5}$$

where x is evenly distributed in the interval $[0,\pi]$.

2. Singularities of the spectral density

The spectral density is singular in $\lambda=\lambda(x)$, if and only if $(d\lambda/dx)(x)=0$, which is equivalent to

$$(k+1)\tan(x)=\tan[(k+1)x]. \tag{A6}$$

Since k is an even number, both this equation and Eq. (A5) are invariant under the transformation $x\mapsto\pi-x$, therefore only the $x\epsilon[0,\pi/2]$ solutions will give different λ values. If $k=10$ (see Fig. 3), Eq. (A6) has $k/2+1=6$ solutions in $[0,\pi/2]$, which are $x=0$, 0.410, 0.704, 0.994, 1.28, and $\pi/2$. Therefore, according to Eq. (A5), in the $N\rightarrow\infty$ limit the spectral density will be singular in the following points:

$$\lambda_i=-3.46,-2.19,-2,0.043,0.536,\text{ and } k=10. \tag{A7}$$

APPENDIX B: CROSSOVER IN THE GROWTH RATE OF THE SCALE-FREE GRAPH'S PRINCIPAL EIGENVALUE

The largest eigenvalue is influenced only by the longest row vector if and only if the two longest row vectors are almost orthogonal:

$$\vec{v}_1\vec{v}_2 \ll |\vec{v}_1||\vec{v}_2|. \tag{B1}$$

For $m>1$, the left-hand side (lhs) of Eq. (B1) is the number of simultaneous 1's in the two longest row vectors, and the rhs can be approximated with $|\vec{v}_1|^2=k_1$, the largest degree of the graph. It is known [11] that for large j ($j>i$), the jth vertex will be connected to vertex i with probability $P_{ij}=m/(2\sqrt{ij})$. Thus, we can write Eq. (B1) in the following forms:

$$\sum_{t=1}^{N} P_{1t}^2 \ll \sum_{t=1}^{N} P_{1t} \tag{B2}$$

or

$$\frac{\sqrt{N_c}}{\ln N_c} \gg \frac{m}{4}, \tag{B3}$$

where N_c is the critical system size.

[1] P. Erdős and A. Rényi, Publ Math **6**, 290 (1959); Publ. Math. Inst. Hung. Acad. Sci. **5**, 17 (1960); **5**, 290 (1959); Acta Math. Acad. Sci. Hung. **12**, 261 (1961).
[2] B. Bollobás, *Random Graphs* (Academic, London, 1985).
[3] S. Redner, Eur. Phys. J. B **4**, 131 (1998).
[4] D.J. Watts and S.H. Strogatz, Nature (London) **393**, 440 (1998).
[5] D.J. Watts, *Small Worlds: The Dynamics of Networks Between Order and Randomness (Princeton Reviews in Complexity)* (Princeton University Press, Princeton, NJ, 1999).
[6] M. Faloutsos, P. Faloutsos, and C. Faloutsos, Comput. Commun. Rev. **29**, 251 (1999).
[7] L.A. Adamic and B.A. Huberman, Nature (London) **401**, 131 (1999).
[8] R. Albert, H. Jeong, and A.-L. Barabási, Nature (London) **401**, 130 (1999).
[9] H. Jeong, B. Tombor, R. Albert, Z.N. Oltvai, and A.-L. Barabási, Nature (London) **407**, 651 (2000).
[10] A.-L. Barabási and R. Albert, Science **286**, 509 (1999).
[11] A.-L. Barabási, R. Albert, and H. Jeong, Physica A **272**, 173 (1999).
[12] L.A.N. Amaral, A. Scala, M. Barthélémy, and H.E. Stanley, e-print cond-mat/0001458.
[13] A.L. Barabási, H. Jeong, E. Ravasz, Z. Néda, T. Vicsek, and A. Schubert (unpublished).
[14] M.E.J. Newman, Proc. Natl. Acad. Sci. USA **98**, 404 (2001);e-print cond-mat/0011144.
[15] A. Medina, I. Matta, and J. Byers, Comput Commun. Rev. **30**, 18 (2000).
[16] R. Cohen, K. Erez, D. ben-Avraham and S. Havlin, Phys. Rev. Lett. **85**, 4626 (2000).
[17] R. Cohen, K. Erez, D. ben-Avraham, and S. Havlin, Phys. Rev. Lett. **86**, 3682 (2001).
[18] A. Broder, R. Kumar, F. Maghoul, P. Raghavan, S. Rajalopagan, R. Stata, A. Tomkins, and J. Wiener, in Proceeding of the 9th International World-Wide Web Conference, 2000 (un published), see (http://www.almaden.ibm.com/cs/k53/www9.final/).
[19] Petra M. Gleiss, Peter F. Stadler, Andreas Wagner, and David A. Fell (unpublished).
[20] M.L. Mehta, *Random Matrices*, 2nd ed. (Academic, New York, 1991).
[21] A. Crisanti, G. Paladin, and A. Vulpiani, *Products of Random Matrices in Statistical Physics*, Springer Series in Solid-State Sciences Vol. 104 (Springer, Berlin, 1993).
[22] E.P. Wigner, Ann. Math. **62**, 548 (1955); **65**, 203 (1957); **67**, 325 (1958).
[23] F. Hiai and D. Petz *The Semicircle Law, Free Random Variables and Entropy* (American Mathematical Society, Providence, 2000), Section 4.1.
[24] F.J. Dyson, J. Math. Phys. **3**, 140 (1962).
[25] E.P. Wigner, SIAM Rev. **9**, 1 (1967).
[26] T. Guhr, A. Müller-Groeling, and H.A. Weidenmüller, Phys. Rep. **299**, 189 (1998).
[27] F. Juhász, in *Algebraic Methods in Graph Theory* (North Holland, Amsterdam, 1981), pp. 313–316.
[28] D. Cvetkovic and P. Rowlinson, Linear Multilinear Algebra **28**, 3 (1990).
[29] B.V. Bronx, J. Math. Phys. **5**, 215 (1964).
[30] M.E.J. Newman, C. Moore, and D.J. Watts, Phys. Rev. Lett. **84**, 3201 (2000).
[31] The number of edges meeting at one vertex of a graph is called the degree of that vertex, and a graph is called regular if every vertex has the same degree.
[32] P.L. Krapivsky and S. Redner, e-print cond-mat/0011094.
[33] P.L. Krapivsky, G.J. Rodgers, and S. Redner, e-print cond-mat/0012181.
[34] S.N. Dorogovtsev and J.F.F. Mendes, Phys. Rev. E **62**, 1842 (2000).
[35] S.N. Dorogovtsev, J.F.F. Mendes, and A.N. Samukhin, e-print cond-mat/0011077.
[36] S.N. Dorogovtsev and J.F.F. Mendes, Phys. Rev. E **63**, 056125 (2001).
[37] R. Albert and A.-L. Barabási, Phys. Rev. Lett. **85**, 5234 (2000).
[38] G. Bianconi and A.-L. Barabási, e-print cond-mat/0011224.
[39] G. Bianconi, and A.-L. Barabási, e-print cond-mat/0011029.
[40] A. Vazquez, e-print cond-mat/0006132.
[41] J.M. Montoya and R.V. Solé (unpublished); R.V. Solé and J. M. Montoya (unpublished).
[42] *Handbook of Combinatorics*, edited by R.L. Graham, M. Grötschel, and L. Lovász (North-Holland, Amsterdam, 1995).
[43] D. M. Cvetković, M. Doob, and H. Sachs, *Spectra of Graphs* (VEB Deutscher Verlag der Wissenschaften, Berlin, 1980).
[44] W.H. Press, S.A. Teukolsky, W.T. Vetterling, and B.P. Flan-

nery, *Numerical Recipes in C*, 2nd ed. (Cambridge University Press, Cambridge, 1995), Secs. 11.2 and 11.3.

[45] B.N. Parlett, *The Symmetric Eigenvalue Problem* (SIAM, Philadelphia, PA, 1998).

[46] K. Wu, A. Canning, H.D. Simon, and L.-W. Wang, J. Comput. Phys. **154**, 156 (1999).

[47] K. Wu and H. Simon, Lawrence Berkeley National Laboratory Report No. 41412, 1998.

[48] B. Bollobás and A.G. Thomason, Random Graphs '83 (Poznan, Poland, 1983).

[49] If $n_{s,e}$ denotes the number of isolated clusters with s vertices and e edges, then the number of isolated clusters of size s can be written as $n_s = \Sigma_{e=s-1}^{s(s-1)/2} n_{s,e}$. The $e = s-1$ case corresponds to stars of size s, and $n_{s,s-1}$ can be given as $\binom{N}{s} s p^{s-1} (1-p)^{s(N-s)+[s(s-1)/2]-(s-1)}$. If $N \to \infty$ and $pN \to c$, then for any finite s this converges to $Nf(s)$, where $f(s) = (2\pi)^{-1/2} s^{-s+1/2} c^{s-1} \exp(s - sc)$, which is an exponentially decreasing function of s, if $c > 1$. Similarly to this case, in the $N \to \infty$ and $pN \to c$ limit, for all values of e, $n_{s,e}$ is an exponentially decaying function of s, if $c > 1$. Since the number of possible e values for a fixed s is a polynomial function of s, n_s will also decay exponentially, if $c > 1$, $N \to \infty$ and $pN \to c$.

[50] In a graph with N vertices, the contribution of one isolated vertex to the spectral density is $N^{-1} \delta(\lambda)$, and the contribution of an isolated cluster with two vertices is $N^{-1}[\delta(\lambda+1) + \delta(\lambda-1)]$. An isolated cluster with three vertices and two edges will give $N^{-1}[\delta(\lambda+\sqrt{2}) + \delta(\lambda) + \delta(\lambda-\sqrt{2})]$, and one with three vertices and three edges adds $N^{-1}[2\delta(\lambda+1) + \delta(\lambda-2)]$ to the spectral density.

[51] M. Barthélémy and L.A.N. Amaral, Phys. Rev. Lett. **82**, 3180 (1999); **82**, 5180(E) (1999).

[52] A graph is said to be properly colored if any two vertices connected by an edge have different colors, and the *chromatic number* $\chi(G)$ of a graph G is the smallest such natural number k for which the graph can be properly colored using k colors. On a complete graph (a graph where any two vertices are connected) with N vertices, $\chi = N$, while on a tree, $\chi = 2$. It should be noted that, in general, the chromatic number of a graph is not determined by its spectrum; in fact, there exist graphs with identical spectra and different chromatic numbers [43]. A well-known relation [60] connecting the *chromatic number and the extremal eigenvalues* of any graph is the following: $\chi(G) \geq 1 + \lambda_1/(-\lambda_N)$. If the bulk part of the spectrum is symmetrical and the first eigenvalue is separated—as is the case for uncorrelated random graphs and scale-free networks—then R can be written as $[1 + \lambda_1/(-\lambda_N)]/z - 1 \leq \chi/z - 1$.

[53] C. McDiarmid, *Surveys in Combinatorics, 1989 (Norwich, 1989)*, London Math. Soc. Lecture Note Series Vol. 141 (Cambridge University Press, Cambridge, 1989), pp. 148–188.

[54] J. Kleinberg, S.R. Kumar, P. Raghavan, S. Rajagopalan, and A. Tomkins, Proceedings of the International Conference on Combinations and Computers, 1999 (unpublished); see also the homepage of the "CLEVER" project at ⟨http://www.almaden.ibm.com/cs/k53/clever.html⟩.

[55] The CAIDA Plankton project: Visualizing NLANR's Web Cache Hierarchy ⟨http://www.caida.org/tools/visualization/plankton/⟩.

[56] Y. Shavitt, X. Sun, A. Wool, and B. Yener, Lucent Technologies Technical Report No. 10009674-000214-01TM, 2000.

[57] L. Laloux, P. Cizeau, J.-P. Bouchaud, and M. Potters, Phys. Rev. Lett. **83**, 1467 (1999).

[58] V. Plerou, P. Gopikrishnan, B. Rosenow, L.A.N. Amaral, and H.E. Stanley, Phys. Rev. Lett. **83**, 1471 (1999).

[59] R.N. Mantegna, e-print cond-mat/9802256.

[60] N. Biggs, *Algebraic Graph Theory* (Cambridge University Press, Cambridge, 1974).

[61] S. Jespersen, I.M. Sokolov, and A. Blumen, Phys. Rev. E **62**, 4405 (2000).

[62] The "citation graph analysis" feature of the NECI Scientific Literature Digital Library [formerly CiteSeer (http://citeseer.nj.nec.com/)] is based on algebraic search techniques from Ref. [54].

[63] K.-I. Goh, B. Kahng, and D. Kim, e-print cond-mat/0103337.

[64] M. Bauer and O. Golinelli, J. Stat. Phys. **103**, 301 (2001).

The Degree Sequence of a Scale-Free Random Graph Process

Béla Bollobás,[1,2] Oliver Riordan,[2] Joel Spencer,[3] Gábor Tusnády[4]
[1]*Department of Mathematical Sciences, University of Memphis, Memphis, Tennessee 38152*
[2]*Trinity College, Cambridge CB2 1TQ, United Kingdom*
[3]*Courant Institute of Mathematical Sciences, New York University, New York, New York, 10003*
[4]*Rényi Institute, Budapest, Hungary*

Received 29 August 2000; accepted 23 January 2001

ABSTRACT: Recently, Barabási and Albert [2] suggested modeling complex real-world networks such as the worldwide web as follows: consider a random graph process in which vertices are added to the graph one at a time and joined to a fixed number of earlier vertices, selected with probabilities proportional to their degrees. In [2] and, with Jeong, in [3], Barabási and Albert suggested that after many steps the proportion $P(d)$ of vertices with degree d should obey a power law $P(d) \alpha d^{-\gamma}$. They obtained $\gamma = 2.9 \pm 0.1$ by experiment and gave a simple heuristic argument suggesting that $\gamma = 3$. Here we obtain $P(d)$ asymptotically for all $d \leq n^{1/15}$, where n is the number of vertices, proving as a consequence that $\gamma = 3$.
 Random Struct. Alg., 18, 279–290, 2001

1. INTRODUCTION

Recently there has been considerable interest in using random graphs to model complex real-world networks to gain an insight into their properties. One of the

Correspondence to: Oliver Riordan; e-mail: omr10@dpmms.cam.ac.uk
Contract grant sponsor: NSF.
Contract grant number: DSM9971788.

most basic properties of a graph or network is its degree sequence. For the standard random graph model $\mathcal{G}(n, m)$ of all graphs with m edges on a fixed set of n vertices, introduced by Erdős and Rényi in [8] and studied in detail in [9], there is a "characteristic" degree $2m/n$: the vertex degrees have approximately a Poisson or normal distribution with mean $2m/n$. The same applies to the closely related model $\mathcal{G}(n, p)$ introduced by Gilbert [10], where vertices are joined independently with probability p. In contrast, Barabási and Albert [2], as well as several other groups (see [4, 14] and the references therein), noticed that in many real-world examples the degree sequence has a "scale-free" power law distribution: the fraction $P(d)$ of vertices with degree d is proportional over a large range to $d^{-\gamma}$, where γ is a constant independent of the size of the network. To explain this phenomenon, Barabási and Albert [2] suggested the following random graph process as a model.

> ... starting with a small number (m_0) of vertices, at every time step we add a new vertex with $m(\leq m_0)$ edges that link the new vertex to m different vertices already present in the system. To incorporate preferential attachment, we assume that the probability Π that a new vertex will be connected to a vertex i depends on the connectivity k_i of that vertex, so that $\Pi(k_i) = k_i / \sum_j k_j$. After t steps the model leads to a random network with $t + m_0$ vertices and mt edges.

This process is intended as a highly simplified model of the growth of the world-wide web, for example, the vertices representing sites or web pages, and the edges links from sites to earlier sites. The preferential attachment assumption is based on the idea that a new site is more likely to link to existing sites which are "popular" at the time the site is added. For $m = 1$ this process is very similar to the nonuniform random recursive tree process considered in [15, 17, 18]. An alternative model, replacing the preferential attachment assumption by a notion of "link copying" is given in [12, 14]. We shall discuss these models briefly in the final section.

In [2, 3] it is stated that computer experiments for the process above suggest that $P(d) \alpha d^{-\gamma}$ with $\gamma = 2.9 \pm 0.1$. In [3], the following heuristic argument is given to suggest that $\gamma = 3$: consider the degree d_i of the ith new vertex v_i at time t, i.e., when there are $t + m_0$ vertices and mt edges. When a new vertex is added, the probability that it is joined to v_i is md_i over the sum of the degrees, i.e., over $2mt$. This suggests the "mean-field theory"

$$\frac{\mathrm{d}\, d_i}{\mathrm{d}t} = \frac{d_i}{2t}.$$

With the initial condition that $d_i = m$ when $t = i$ this gives $d_i = m(t/i)^{1/2}$, which yields $\gamma = 3$.

Here we show how one can calculate the exact distribution of d_i at time t and obtain an asymptotic formula for $P(d)$, $d \leq t^{1/15}$, which gives $\gamma = 3$ as a simple consequence. The first step is to give an exact definition of a random graph process that fits the rather vague description given above.

2. THE MODEL

The description of the random graph process quoted above is rather imprecise. First, as the degrees are initially zero, it is not clear how the process is supposed to get started. More seriously, the expected number of edges linking a new vertex v to earlier vertices is $\sum_i \Pi(k_i) = 1$, rather than m. Also, when choosing in one go a set S of m earlier vertices as the neighbors of v, the distribution of S is *not* specified by giving the marginal probability that each vertex lies in S. For a trivial example, suppose that $m = 2$ and that the first four vertices form a four-cycle. Then for any $0 \leq \alpha \leq 1/4$ we could join the fifth vertex to each adjacent pair with probability α and to each nonadjacent pair with probability $1/2 - 2\alpha$. This suggests that for $m > 1$ we should choose the neighbors of v one at a time. Once doing so, it is very natural to allow some of these neighbors to be the same, creating multiple edges in the graph. Here we shall consider the precise model introduced in [6], which turns out to be particularly pleasant to work with. This model fits the description above except that it allows multiple edges and also loops—in terms of the interpretation there is no reason to exclude these. Once the process gets started there will in any case not be many loops or multiple edges, so they should have little effect overall. The following definition is essentially as given in [6]; we write $d_G(v)$ for the total (in plus out) degree of the vertex v in the graph G.

We start with the case $m = 1$. Consider a fixed sequence of vertices $v_1, v_2, \ldots$. We shall inductively define a random graph process $(G_1^t)_{t\geq 0}$ so that G_1^t is a directed graph on $\{v_i : 1 \leq i \leq t\}$, as follows. Start with G_1^0 the "graph" with no vertices, or with G_1^1 the graph with one vertex and one loop. Given G_1^{t-1}, form G_1^t by adding the vertex v_t together with a single edge directed from v_t to v_i, where i is chosen randomly with

$$\mathbb{P}(i = s) = \begin{cases} d_{G_1^{t-1}(v_s)}/(2t-1) & 1 \leq s \leq t-1 \\ 1/(2t-1) & s = t. \end{cases} \tag{1}$$

In other words, we send an edge e from v_t to a random vertex v_i, where the probability that a vertex is chosen as v_i is proportional to its (total) degree at the time, counting e as already contributing one to the degree of v_t. For $m > 1$ we add m edges from v_t one at a time, counting the previous edges as well as the "outward half" of the edge being added as already contributing to the degrees. Equivalently, we define the process $(G_m^t)_{t\geq 0}$ by running the process (G_1^t) on a sequence $v_1', v_2', \ldots$; the graph G_m^t is formed from G_1^{mt} by identifying the vertices $v_1', v_2', \ldots, v_m'$ to form v_1, identifying $v_{m+1}', v_{m+2}', \ldots v_{2m}'$ to form v_2, and so on.

We shall write $\mathcal{G}_m^n$ for the probability space of directed graphs on n vertices $v_1, v_2, \ldots, v_n$, where a random $G_m^n \in \mathcal{G}_m^n$ has the distribution derived from the process above. As G_m^n is defined in terms of G_1^{mn}, for most of the time we shall consider the case $m = 1$. As noted in [6], there is an alternative description of the distribution of G_1^n in terms of pairings.

An *n-pairing* $\mathcal{P}$ is a partition of the set $\{1, 2, \ldots, 2n\}$ into pairs, so there are $(2n-1)!! = (2n)!/(n!2^n)$ n-pairings. Thinking of the elements $1, 2, \ldots, 2n$ of the ground set as points on the x axis, and the pairs as chords joining them, we shall speak of the left and right endpoint of each pair.

We form a directed graph $\phi(\mathcal{P})$ from an n-pairing $\mathcal{P}$ as follows: starting from the left, merge all endpoints up to and including the first right endpoint reached to form the vertex v_1. Then merge all further endpoints up to the next right endpoint to form v_2, and so on to v_n. For the edges, replace each pair by a directed edge from the vertex corresponding to its right endpoint to that corresponding to its left endpoint. As noted in [6], if $\mathcal{P}$ is chosen uniformly at random from all $(2n-1)!!$ n-pairings, then $\phi(\mathcal{P})$ has the same distribution as a random $G_1^n \in \mathcal{G}_1^n$. This statement is easy to prove by induction on n: thinking in terms of pairings of distinct points on the x axis, one can obtain a random $(n-1)$-pairing from a random n-pairing by deleting the pair containing the rightmost point. The reverse process, starting from an $(n-1)$-pairing $\mathcal{P}$, is to add a new pair with its right endpoint to the right of everything in $\mathcal{P}$ and its left endpoint in one of the $2n-1$ possible places. Now a vertex of degree d in $\phi(\mathcal{P})$ corresponds to d intervals between endpoints in $\mathcal{P}$. The effect of adding a new pair to $\mathcal{P}$ as described is thus to add a new vertex to $\phi(\mathcal{P})$ together with a new edge to a vertex chosen according to (1), with $t = n$.

The advantage of this description from pairings is that it gives us a simple non-recursive definition of the distribution of G_m^n, enabling us to calculate properties of G_m^n directly. We now use this to study the degrees of G_m^n.

3. THE DEGREES OF G_m^n

In [2] it was suggested that the fraction of vertices of G_m^n having degree d should fall off as d^{-3} as $d \to \infty$. We shall prove the following precise version of this statement, writing $\#_m^n(d)$ for the number of vertices of G_m^n with *indegree* equal to d, i.e., with (total) degree $m + d$.

Theorem 1. *Let $m \geq 1$ be fixed, and let $(G_m^n)_{n\geq 0}$ be the random graph process defined in Section 2. Let*

$$\alpha_{m,d} = \frac{2m(m+1)}{(d+m)(d+m+1)(d+m+2)},$$

and let $\epsilon > 0$ be fixed. Then with probability tending to 1 as $n \to \infty$ we have

$$(1-\epsilon)\alpha_{m,d} \leq \frac{\#_m^n(d)}{n} \leq (1+\epsilon)\alpha_{m,d}$$

for every d in the range $0 \leq d \leq n^{1/15}$.

In turns out that we only need to calculate the expectation of $\#_m^n(d)$; the concentration result is then given by applying the following standard inequality due to Azuma [1] and Hoeffding [11] (see also [5]).

Lemma 2 (Azuma–Hoeffding inequality). *Let $(X_t)_{t=0}^n$ be a martingale with $|X_{t+1} - X_t| \leq c$ for $t = 0, \ldots, n-1$. Then*

$$\mathbb{P}(|X_n - X_0| \geq x) \leq \exp\left(-\frac{x^2}{2c^2 n}\right).$$

The strategy of the proof is as follows. First, as mentioned earlier, the results for general m will follow from those for $m = 1$. We shall use the pairing model to find explicitly the distribution of D_k, the sum of the first k degrees, in this case, and also the distribution of the next degree, $d_{G_1^n}(v_{k+1})$, given D_k. One could combine these formulae to give a rather unilluminating expression for the distribution of $d_{G_1^n}(v_{k+1})$; instead we show that D_k is concentrated about a certain value and hence find approximately the probability that $d_{G_1^n}(v_{k+1}) = d$. Summing over k gives us the expectation of $\#_1^n(d)$, and concentration follows from Lemma 2.

Before turning to the distributions of the (total) degrees for $m = 1$, we note that their expectations are easy to calculate exactly:

$$\mathbb{E}(d_{G_1^t}(v_t)) = 1 + \frac{1}{2t-1}.$$

Also, for $s < t$,

$$\mathbb{E}\big(d_{G_1^t}(v_s) \,|\, d_{G_1^{t-1}}(v_s)\big) = d_{G_1^{t-1}}(v_s) + \frac{d_{G_1^{t-1}}(v_s)}{2t-1},$$

which implies that

$$\mathbb{E}\big(d_{G_1^t}(v_s)\big) = \frac{2t}{2t-1}\,\mathbb{E}\big(d_{G_1^{t-1}}(v_s)\big).$$

Thus, for $1 \le s \le n$,

$$\mathbb{E}\big(d_{G_1^n}(v_s)\big) = \prod_{i=s}^{n} \frac{2i}{2i-1} = \frac{4^{n-s+1} n!^2 (2s-2)!}{(2n)!(s-1)!^2} = \sqrt{n/s}\,(1 + O(1/s)),$$

using Stirling's formula.

If every degree of G_1^n were equal to its expectation this would give the proposed distribution, but in fact the degrees can be far from their expectations. Indeed we shall see that for almost all vertices the most likely degree is 1!

Let us write d_i for $d_{G_1^n}(v_i)$, i.e., for the (total) degree of the vertex v_i in the graph G_1^n. Our aim is to describe the distributions of the individual d_i. To do this it turns out to be useful to consider their sums $D_k = \sum_{i=1}^k d_i$.

Consider first the event $\{D_k - 2k = s\}$, where $0 \le s \le n - k$. This is the event that the last $n - k$ vertices of G_1^n send exactly s edges to the first k vertices. This event corresponds to pairings $\mathcal{P}$ in which the kth right endpoint is $2k + s$. Consider any pairing $\mathcal{P}$ with this property. We shall split $\mathcal{P}$ into two partial pairings, the *left partial pairing* $\mathcal{L}$ and the *right partial pairing* $\mathcal{R}$, each consisting of some number of pairs together with some unpaired elements. For $\mathcal{L}$ we take the partial pairing on $\{1, \ldots, 2k + s\}$ induced by $\mathcal{P}$, for $\mathcal{R}$ that on $\{2k + s + 1, \ldots, 2n\}$. From the restriction on $\mathcal{P}$, in $\mathcal{L}$ the element $2k + s$ must be paired with one of $\{1, \ldots, 2k + s - 1\}$, precisely s of the remaining $2k + s - 2$ elements must be unpaired, and the other $2(k - 1)$ elements must be paired off somehow. Any of the

$$(2k+s-1)\binom{2k+s-2}{s}\frac{(2k-2)!}{2^{k-1}(k-1)!}$$

partial pairings obtained in this way may arise as $\mathcal{L}$. Similarly, for $\mathcal{R}$ there are

$$\binom{2n-2k-s}{s}\frac{(2n-2k-2s)!}{2^{n-k-s}(n-k-s)!}$$

possibilities. Any possible $\mathcal{L}$ may be combined with any possible $\mathcal{R}$ to form $\mathcal{P}$ by pairing off the unpaired elements of $\mathcal{L}$ with those of $\mathcal{R}$ in any of $s!$ ways. Multiplying together and dividing by the total number $(2n)!/(2^n n!)$ of n-pairings we see that for $1 \le k \le n$ and $0 \le s \le n-k$,

$$\mathbb{P}(D_k - 2k = s) = \frac{(2k+s-1)!(2n-2k-s)!n!2^{s+1}}{s!(k-1)!(n-k-s)!(2n)!}. \tag{2}$$

From the expression above it is easy to deduce a concentration result for D_k. For k with $1 \le k \le n$ let us write $p_s = p_{s,k}$ for the probability above, and let

$$r_s = \frac{p_{s+1}}{p_s} = 2\frac{(2k+s)(n-k-s)}{(s+1)(2n-2k-s)}.$$

Note that r_s is a decreasing function of s. Allowing s to be a real number for the moment, the unique positive solution to $r_s = 1$ is given by

$$s = -2k + \sqrt{4kn - 2n + 1/4} + 1/2.$$

Thus $s_0 = \lceil -2k + \sqrt{4kn - 2n + 1/4} + 1/2 \rceil$ is one of the at most two most likely values of $D_k - 2k$. Also, for n larger than some constant we have

$$\begin{aligned}\frac{r_{s+1}}{r_s} &= \left(1 - \frac{2k-1}{(s+2)(2k+s)}\right)\left(1 - \frac{n-k}{(2n-2k-s-1)(n-k-s)}\right)\\ &\le \left(1 - \frac{2k-1}{2n^2}\right)\left(1 - \frac{n-k}{2n^2}\right)\\ &\le \exp\left(-\frac{2k-1}{2n^2}\right)\exp\left(-\frac{n-k}{2n^2}\right) \le \exp\left(-\frac{1}{2n}\right).\end{aligned}$$

As $r_{s_0} \le 1$ it follows that $r_{s_0+x} \le \exp(-x/(2n))$ for $x > 0$ and hence that $p_{s_0+x} \le \exp(-x(x-1)/(4n))$. A similar bound on p_{s_0-x} shows that

$$\mathbb{P}\left(|D_k - (2k+s_0)| \ge 3\sqrt{n\log n}\right) = o(n^{-1}).$$

In fact, as $|s_0 - (2\sqrt{kn} - 2k)| \le 2\sqrt{n}$ for each k, we obtain

$$\mathbb{P}\left(|D_k - 2\sqrt{kn}| \ge 4\sqrt{n\log n}\right) = o(n^{-1}). \tag{3}$$

We now turn to the probability that $d_{k+1} = d+1$, i.e., that the indegree of v_{k+1} is d, given D_k. Suppose that $1 \le k \le n-1$ and $0 \le s \le n-k$, and consider a left partial pairing $\mathcal{L}$ as above. We have already seen that each such $\mathcal{L}$ has

$$s!\binom{2n-2k-s}{s}(2n-2k-2s-1)!!$$

extensions to an n-pairing. Such an extension corresponds to a graph with $d_{k+1} = d+1$ if and only if $2k+s+d+1$ is a right endpoint, and each of $2k+s+1, \ldots, 2k+s+d$ is a left endpoint. Noting that the element paired with $2k+s+d+1$ must be either one of the s unpaired elements in $\mathcal{L}$ or one of $2k+s+1, \ldots, 2k+s+d$, and that $s-1+d$ pairs start before $2k+s+d+1$ and end after this point, each $\mathcal{L}$ has exactly

$$(s+d)(s+d-1)!\binom{2n-2k-s-d-1}{s+d-1}(2n-2k-2s-2d-1)!!$$

such extensions, and for $0 \le d \le n-k-s$ we have, writing $(a)_b$ for $a!/(a-b)!$,

$$\mathbb{P}(d_{k+1} = d+1 \mid D_k - 2k = s) = (s+d)2^d \frac{(n-k-s)_d}{(2n-2k-s)_{d+1}}. \tag{4}$$

It is easy to see that (4) also applies when $k = s = 0$, when we obtain $\mathbb{P}(d_1 = d+1) = d2^d(n)_d/(2n)_{d+1}$. For $k \ge 1$ we can of course combine (2) and (4) to give a rather unilluminating expression for $\mathbb{P}(d_{k+1} = d+1)$. Instead, we shall use (3) and (4) to estimate the expectation of $\#_1^n(d)$, the number of vertices of G_1^n with indegree d. Above and in what follows the functions implied by $o(.)$ or $\sim$ notation are to be interpreted as depending on n only, not on d or k. Also, the constant implied by $O(.)$ notation is absolute.

Let $M = \lfloor n^{4/5}/\log n \rfloor$, let $k = k(n)$ be any function satisfying $M \le k \le n-M$, and let $d = d(n)$ be any function satisfying $0 \le d \le n^{1/15}$. For any D with $|D - 2\sqrt{kn}| \le 4\sqrt{n\log n}$ we can use (4) to write $\mathbb{P}(d_{k+1} = d+1 \mid D_k = D)$ as

$$(2\sqrt{kn} - 2k + O(\sqrt{n\log n}))2^d \frac{(n+k-2\sqrt{kn}+O(\sqrt{n\log n}))^d}{(2n-2\sqrt{kn}+O(\sqrt{n\log n}))^{d+1}}.$$

Using the bounds on d and k we find that the ratio of $n+k-2\sqrt{kn} = (\sqrt{n}-\sqrt{k})^2$ to $d\sqrt{n\log n}$ tends to infinity as $n \to \infty$, as does $(2n-2\sqrt{kn})/(d\sqrt{n\log n})$. Also, $\sqrt{n\log n} = o(2\sqrt{kn}-2k)$, so the probability above is equal to

$$(1+o(1))\frac{2\sqrt{kn}-2k}{2n-2\sqrt{kn}}\left(\frac{2(\sqrt{n}-\sqrt{k})^2}{2(n-\sqrt{kn})}\right)^d \sim \sqrt{\kappa}\left(1-\sqrt{\kappa}\right)^d,$$

where $\kappa = k/n$. As this estimate applies uniformly to $\mathbb{P}(d_{k+1} = d+1 \mid D_k = D)$ for all D with $|D - 2\sqrt{kn}| \le 4\sqrt{n\log n}$, we see from (3) that

$$\mathbb{P}(d_{k+1} = d+1) = o(n^{-1}) + (1+o(1))\sqrt{\kappa}(1-\sqrt{\kappa})^d.$$

In particular, although it is not relevant for the proof, we note that for almost every vertex the most likely indegree is zero.

Keeping n and d fixed and varying k in the range $M \le k \le n-M$, as the estimate above is uniform in k we find that the expected number of vertices v_{k+1}, $M \le k \le n-M$, with degree equal to $d+1$ can be written as

$$o(1) + \sum_{k=M}^{n-M}(1+o(1))\sqrt{k/n}(1-\sqrt{k/n})^d.$$

Thus, as all terms in the sum are positive, we have

$$\mathbb{E}(\#_1^n(d)) = O(M) + o(1) + (1+o(1)) \sum_{k=M}^{n-M} \sqrt{k/n}(1-\sqrt{k/n})^d. \tag{5}$$

Writing $f = \sqrt{\kappa}(1-\sqrt{\kappa})^d$, we have

$$\frac{1}{f}\frac{\mathrm{d}f}{\mathrm{d}\kappa} = \frac{\kappa^{-1}}{2} - \frac{d}{2}\frac{\kappa^{-1/2}}{1-\kappa^{1/2}}.$$

Provided $n\kappa$ and $n(1-\kappa)$ tend to infinity, the proportional change in f as κ changes by $1/n$ is thus $o(1)$ uniformly in κ. It follows that the sum in (5) can be written as

$$(1+o(1))n \int_{(M+1)/n}^{1-M/n} \sqrt{\kappa}(1-\sqrt{\kappa})^d \,\mathrm{d}\kappa \sim n \int_0^1 \sqrt{\kappa}(1-\sqrt{\kappa})^d \,\mathrm{d}\kappa.$$

It is easy to evaluate this integral by substituting $\kappa = (1-u)^2$, and we obtain

$$\mathbb{E}(\#_1^n(d)) = O(M) + (1+o(1))\frac{4n}{(d+1)(d+2)(d+3)} \sim \frac{4n}{(d+1)(d+2)(d+3)},$$

which is the required form of the distribution.

At this point, let us return to the general case $m \geq 1$. Suppose that m is a constant fixed once and for all, and let d'_k be the degree of v_k in the graph G_m^n. We shall estimate $\mathbb{P}(d'_{k+1} = d+m)$, keeping the notation d_K for degrees in the graph G_1^{mn} from which G_m^n is obtained. For the estimate we look at the distributions of $d_{K+1}, \ldots, d_{K+m}$ in G_1^N, where $K = mk$ and $N = mn$. The argument giving the conditional probability estimate (4) actually applies to the conditional probability given the entire sequence of earlier degrees. For $M \leq k \leq n-M$ and $d \leq n^{1/15}$ our earlier estimates show that, provided no $|D_{K'} - 2\sqrt{K'N}|$ is too large,

$$\begin{aligned}\mathbb{P}(d_{K+j+1} = d+1 \,|\, d_1, d_2, \ldots, d_{K+j}) &\sim \sqrt{(K+j)/N}\left(1-\sqrt{(K+j)/N}\right)^d \\ &\sim \sqrt{\kappa}(1-\sqrt{\kappa})^d,\end{aligned}$$

with $\kappa = k/n = K/N$. Thus, using (3),

$$\begin{aligned}\mathbb{P}(d'_{k+1} = d+m) &= o(n^{-1}) + (1+o(1)) \sum_{a_1+\cdots+a_m=d} \prod_{j=1}^m \sqrt{\kappa}(1-\sqrt{\kappa})^{a_j} \\ &= o(n^{-1}) + (1+o(1))\binom{d+m-1}{m-1}\kappa^{m/2}(1-\sqrt{\kappa})^d.\end{aligned}$$

Proceeding as before we can express the expectation of the number $\#_m^n(d)$ of vertices of G_m^n with indegree d in terms of

$$\int_0^1 \kappa^{m/2}(1-\sqrt{\kappa})^d \,\mathrm{d}\kappa = 2\int_0^1 (1-u)^{m+1}u^d \,\mathrm{d}u = 2\frac{(m+1)!d!}{(d+m+2)!},$$

where we have again substituted $\kappa = (1-u)^2$. We find that for $0 \le d \le n^{1/15}$,

$$\mathbb{E}(\#_m^n(d)) \sim \frac{2m(m+1)n}{(d+m)(d+m+1)(d+m+2)}, \tag{6}$$

uniformly in d. We are now ready to prove Theorem 1.

Proof of Theorem 1. We return to considering the graph G_m^n as one graph from the process $(G_m^t)_{t\ge 0}$. Fix $m \ge 1$, $n \ge 1$ and $0 \le d \le n^{1/15}$, and consider the martingale $X_t = \mathbb{E}(\#_m^n(d) \mid G_m^t)$ for $0 \le t \le n$. We have $X_n = \#_m^n(d)$, while $X_0 = \mathbb{E}(\#_m^n(d))$. We claim that the differences $|X_{t+1} - X_t|$ are bounded by two. To see this note that whether at stage t we join v_t to v_i or v_j does not affect the degrees at later times of vertices v_k, $k \notin \{i, j\}$. More precisely, the joint distribution of all other degrees is the same in either case. Since we are just counting vertices with a particular degree, no matter how much the degrees of v_i and v_j are changed in G_m^n, this changes $\#_m^n(d)$ by at most two.

An alternative way of seeing this is to say that at stage t we add a half edge h_{2t-1} directed from $v_{\lceil t/m \rceil}$ paired with a half edge h_{2t} directed to some other vertex, and to consider h_{2t} not as attached to a random vertex, but rather as associated with equal probability to any of $h_1, \ldots, h_{2t-1}$. In the final graph a half edge h_{2t} is attached to $v_{\lceil t/m \rceil}$, while a half edge h_{2t-1} is attached to the vertex the half edge it is associated to is attached to. If we change the choice made at stage t, the effect on the final graph is to move the half edge h_{2t-1} and all later half edges associated directly or indirectly to h_{2t-1} together. This operation only affects two degrees.

Applying Lemma 2, the Azuma–Hoeffding inequality, we find that for each d with $0 \le d \le n^{1/15}$ we have

$$\mathbb{P}\left(\left|\#_m^n(d) - \mathbb{E}(\#_m^n(d))\right| \ge \sqrt{n \log n}\right) \le e^{-\log n/8} = o(n^{-1/15}).$$

Noting from (6) that in this range $\mathbb{E}(\#_m^n(d)) \sim \frac{2m(m+1)n}{(d+m)(d+m+1)(d+m+2)}$ and that this is much larger than $\sqrt{n \log n}$, the result follows. ■

It is natural to ask how far Theorem 1 can be extended to degrees $d > n^{1/15}$. The bound $d \le n^{1/15}$ was chosen to make the proof as simple as possible, and can certainly be weakened considerably, by choosing a suitable cutoff and considering "early" and "late" vertices separately. For large d, Eq. (6) suggests that the expected number of vertices with degree at least d should be roughly $m(m+1)n/d^2$ and hence that the maximum degree should be $\Theta(\sqrt{n})$. It turns out that this is indeed the case, as could be proved using, for example, the analysis of the pairing model given in [6].

4. UNIFORM ATTACHMENT

In [2, 3] it is stated that the preferential attachment assumption of the model is needed to obtain a power-law degree distribution; experimental and heuristic results are given suggesting that with uniform attachment the degrees with be geometrically distributed. It is easy to prove a precise result for this case along the lines of Theorem 1. As the argument is similar to but much simpler than that given above, we only give a rough outline.

Consider a random process in which vertices are added one at a time, starting from any given finite graph G. Suppose that when the vertex v_i is added, it is joined to m earlier vertices, in such a way that the expected number of edges from v_i to v_k is the same for all $k < i$. (It does not matter whether we allow multiple edges or not.) Then the expected indegree of v_k when n vertices have been added is exactly $m\sum_{i=k+1}^{n}\frac{1}{i}$. For $\kappa = k/n$ bounded away from 0 and 1, it is easy to see the degree of v_k has asymptotically a Poisson distribution with mean $\lambda \sim -m\log(\kappa)$. Thus, arguing as before, for any fixed d the proportion of vertices with indegree d is asymptotically

$$\int_0^1 \frac{\lambda^d}{d!}e^{-\lambda}\,\mathrm{d}\kappa = \frac{m^d}{d!}\int_0^1 (-\log\kappa)^d\kappa^m\,\mathrm{d}\kappa = \frac{m^d}{(m+1)^{d+1}},$$

giving the expected geometric distribution.

5. CONCLUDING REMARKS

It is presumably possible to prove a weaker version of Theorem 1 using the following continuous model, which is much more precise than the "mean-field theory" of [3].

Consider a vertex v born at time t_0 uniformly distributed in $[0, 1]$. When born it has weight (degree) m. If at time t the vertex has weight i then it gets a "hit" in the infinitesimal time interval $[t, t+\mathrm{d}t]$ with probability $m\frac{i}{2mtn}n\,\mathrm{d}t = \frac{i}{2t}\,\mathrm{d}t$. If it does get a hit its weight is incremented by one. The connection is that if vertices are born at time intervals of $1/n$ then at time t the sum of all degrees is $2mnt$, and in an interval of length $\mathrm{d}t$ there are $n\,\mathrm{d}t$ vertices born, each having m chances to send an edge to v. The differential equations that arise can easily be solved explicitly; one finds that, conditional on t_0, the probability that v has (total) degree $i \geq m$ at time $t > t_0$ is given by

$$\binom{i-1}{m-1}\rho^{m/2}(1-\sqrt{\rho})^{i-m},$$

where $\rho = t_0/t$. At time 1, writing κ for t_0 and d for $i-m$ as before, this reduces to the expression

$$\binom{d+m-1}{m-1}\kappa^{m/2}(1-\sqrt{\kappa})^{d}$$

seen in Section 3. For constant d, integrating over κ as before suggests the bounds on $\#_m^n(d)$ given in Theorem 1.

Recall that when defining the process (G_1^t) above, it was convenient to start from the "graph" with no vertices, or from the graph with one vertex and no loops. As far as our results are concerned, however, it makes no difference where we start. Given any finite graph G one can define a similar process to (G_1^t), or (G_m^t), starting from G. Now the joint distribution of the degrees of $v_{s+1}, \ldots, v_t$ in G_m^t is independent of G_m^s. If G has s edges then, speaking loosely, for $m = 1$ we can just "pretend" that G has s vertices as well. As far as all new vertices are concerned, the process starting from G is then indistinguishable from the process $(G_1^t)_{t>s}$, so asymptotic results such as Theorem 1 are unaffected by the starting graph G.

We finish by comparing the model considered here with one much older random process and two new ones. Graphs in which each vertex (apart from the first few) is joined to a fixed number m of randomly chosen earlier vertices are known in the literature as *random recursive dags*, or *random recursive trees* if $m = 1$ (see, e.g., [7]). For $m > 1$ only uniform random recursive dags have been studied significantly. For $m = 1$, however, nonuniform random recursive trees with attachment probabilities proportional to the degrees (also known as random plane-oriented recursive trees) have been studied; see [7, 16, 18], for example. These objects are very close to the random graph G_1^n considered here. The only differences are that loops are not allowed and that the root vertex is sometimes treated in a slightly different way. The expected number of vertices of degree $d = d(n)$ in these objects was found to within an additive constant by Szymański [18]; a concentration result for d fixed was given by Lu and Feng [15]. For a survey of results on random recursive trees see [17].

Finally, a rather different model for the worldwide web graph was introduced in [12]. Again, vertices are born one at a time, but instead of preferential attachment, each new vertex picks an old vertex to copy from and copies a randomly selected part of its neighborhood, as well as choosing (uniformly) new neighbors of its own. In [13, 14] it is shown that such models also give power-law degree distributions, as well as explaining the high number of dense bipartite subgraphs found in the web graph.

REFERENCES

[1] K. Azuma, Weighted sums of certain dependent variables, Tôhoku Math J 3 (1967), 357–367.

[2] A.-L. Barabási and R. Albert, Emergence of scaling in random networks, Science 286 (1999), 509–512.

[3] A.-L. Barabási, R. Albert, and H. Jeong, Mean-field theory for scale-free random networks, Physica A 272 (1999), 173–187.

[4] A.-L. Barabási, R. Albert, and H. Jeong, Scale-free characteristics of random networks: the topology of the world-wide web, Physica A 281 (2000), 69–77.

[5] B. Bollobás, Martingales, isoperimetric inequalities and random graphs, in Combinatorics (Eger, 1987), Colloq. Math. Soc. János Bolyai, Vol. 52, North-Holland, Amsterdam, 1988, 113–139.

[6] B. Bollobás and O.M. Riordan, The diameter of a scale-free random graph, submitted for publication.

[7] L. Devroye and J. Lu, The strong convergence of maximal degrees in uniform random recursive trees and dags, Random Structures Algorithms 7 (1995), 1–14.

[8] P. Erdős and A. Rényi, On random graphs. I, Publ Math Debrecen 6 (1959), 290–297.

[9] P. Erdős and A. Rényi, On the evolution of random graphs, Magyar Tud Akad Mat Kutató Int Kőzl 5 (1960), 17–61.

[10] E.N. Gilbert, Random graphs, Ann Math Statist 30 (1959), 1141–1144.

[11] W. Hoeffding, Probability inequalities for sums of bounded random variables, J Amer Statist Assoc 58 (1963), 13–30.

[12] J. Kleinberg, R. Kumar, P. Raghavan, S. Rajagopalan, and A. Tomkins, The web as a graph: measurements, models, and methods, COCOON 1999.

[13] R. Kumar, P. Raghavan, S. Rajagopalan, and A. Tomkins, Extracting large scale knowledge bases from the web, VLDB 1999.

[14] R. Kumar, P. Raghavan, S. Rajagopalan, D. Sivakumar, A. Tomkins, and E. Upfal, Stochastic models for the web graph, FOCS 2000.

[15] J. Lu and Q. Feng, Strong consistency of the number of vertices of given degrees in nonuniform random recursive trees, Yokohama Math J 45 (1998), 61–69.

[16] H.M. Mahmoud, R.T. Smythe, and J. Szymański, On the structure of random plane-oriented recursive trees and their branches, Random Structures Algorithms 4 (1993), 151–176.

[17] H.M. Mahmoud and R.T. Smythe, A survey of recursive trees, Theory Probability Math Statist 51 (1995), 1–27.

[18] J. Szymański, On a nonuniform random recursive tree, Annals Discrete Math 33 (1987), 297–306.

Advances in Complex Systems

A MODEL OF LARGE-SCALE PROTEOME EVOLUTION

Ricard V. Solé[1,2,3], Romualdo Pastor-Satorras[1], Eric Smith[2], and Thomas B. Kepler[2]

[1] *ICREA-Complex Systems Research Group, FEN*
Universitat Politècnica de Catalunya, Campus Nord B4, 08034 Barcelona, Spain

[2] *Santa Fe Institute, 1399 Hyde Park Road, New Mexico 87501, USA*

[3] *NASA-associated Astrobiology Institute, INTA/CSIC, Carr. del Ajalvir Km4, Madrid, Spain*

The next step in the understanding of the genome organization, after the determination of complete sequences, involves proteomics. The proteome includes the whole set of protein-protein interactions, and two recent independent studies have shown that its topology displays a number of surprising features shared by other complex networks, both natural and artificial. In order to understand the origins of this topology and its evolutionary implications, we present a simple model of proteome evolution that is able to reproduce many of the observed statistical regularities reported from the analysis of the yeast proteome. Our results suggest that the observed patterns can be explained by a process of gene duplication and diversification that would evolve proteome networks under a selection pressure, favoring robustness against failure of its individual components.

Keywords: Genomics, proteomics, gene duplication, small-world, networks

1. Introduction

The genome is one of the most fascinating examples of the importance of emergence from network interactions. The recent sequencing of the human genome [23,38] revealed some unexpected features and confirmed that *"the sequence is only the first level of understanding of the genome"* [38]. The next fundamental step beyond the determination of the genome sequence involves the study of the properties of the proteins the genes encode, as well as their interactions [12]. Protein interactions play a key role at many different levels and its failure can lead to cell malfunction or even apoptosis, in some cases triggering neoplastic transformation. This is the case, for example, of the feedback loop between two well-known proteins, MDM2 and p53: in some types of cancers, amplification of the first (an oncoprotein) leads to the inactivation of p53, a tumor-suppressor gene that is central in the control of the cell cycle and death [47].

Understanding the specific details of protein-protein interactions is an essential part of our understanding of the proteome, but a complementary approach is provided by the observation that network-like effects play also a key role. Using again p53 as an example, this gene is actually involved in a large number of interaction pathways dealing with cell signaling, the maintenance of genetic stability, or the induction of cellular differentiation [39]. The failure in p53, as when a highly connected node in the Internet breaks [1], has severe consequences.

Additional insight is provided by the observation that in many cases the total suppression of a given gene in a given organism leads to a small phenotypic effect or even no effect at all [32,41]. These observations support the idea that, although some genes might play a key role and their suppression is lethal, many others can be replaced in their function by some redundancy implicit in the network of interacting proteins.

Protein-protein interaction maps have been studied, at different levels, in a variety of organisms including viruses [5,13,25], prokaryotes [31], yeast [18], and multicellular organisms such as *C. elegans* [44]. Most previous studies have used the so-called two-hybrid assay [14] based on the properties of site-specific transcriptional activators. Although differences exist between different two-hybrid projects [16] the statistical patterns used in our study are robust.

Recent studies have revealed a surprising result: the protein-protein interaction networks in the yeast *Saccharomyces cerevisiae* share some universal features with other complex networks [35]. These studies actually offer the first global view of the proteome map. These are very heterogeneous networks: The probability $P(k)$ that a given protein interacts with other k proteins is given by a power law, i.e. $P(k) \sim k^{-\gamma}$ with $\gamma \approx 2.5$ (see figure 1), with a sharp cut-off for large k. This distribution is thus very different from the Poissonian shape expected from a simple (Erdos-Renyi) random graph [6,22]. Additionally, these maps also display the so-called small-world (SW) effect: they are highly clustered (i.e. each node has a well-defined neighborhood of "close" nodes) but the minimum distance between any two randomly chosen nodes in the graph is short, a characteristic feature of random graphs [45].

As shown in previous studies [1] this type of networks is extremely robust against random node removal but also very fragile when removal is performed selectively on the most connected nodes. SW networks appear to be present in a wide range of systems, including artificial ones [4,2,10,29] and also in neural networks [45,34], metabolic pathways [8,20,43] (see also [28]), even in human language organization [9]. The implications of these topologies are enormous also for our understanding of epidemics [30,24].

The experimental observations on the proteome map can be summarized as follows:

(1) The proteome map is a sparse graph, with a small average number of links per protein. In [42] an average connectivity $\bar{K} \sim 1.9 - 2.3$ was reported for the proteome map of *S. cerevisiae*. This observation is also consistent with the study of the global organization of the *E. coli* gene network from available information on transcriptional regulation [36].

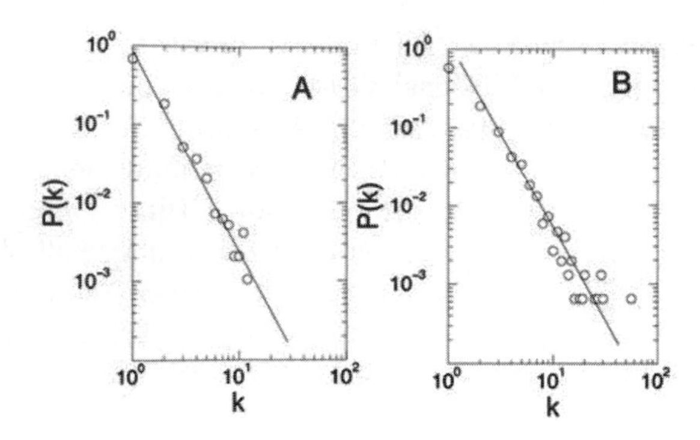

Fig. 1. Degree distributions for two different data sets from the Yeast proteome: A: Ref. [42]; B: Ref. [19]. Both distributions display scaling behavior in their degree distribution $P(k)$, i.e. $P(k) \sim k^{-\gamma}$, a sharp cut-off for large k and very small average connectivities: $\bar{K}_A = 1.83$ (total graph) and $\bar{K}_B = 2.3$ (giant component), respectively. The slopes are $\gamma_A \approx 2.5 \pm 0.15$ and $\gamma_B \approx 2.4 \pm 0.21$.

(2) It exhibits a SW pattern, different from the properties displayed by purely random (Poissonian) graphs.

(3) The degree distribution of links follows a power-law with a well-defined cut-off. To be more precise, Jeong *et al.* [19] reported a functional form for the degree distribution of *S. cerevisiae*

$$P(k) \simeq (k_0 + k)^{-\gamma} e^{-k/k_c}. \tag{1.1}$$

A best fit of the real data to this form yields a degree exponent $\gamma \approx 2.5$ and a cut-off $k_c \approx 20$. This could have adaptive significance as a source of robustness against mutations.

In this paper we present a model of proteome evolution aimed at capturing the main properties exhibited by protein networks. The basic ingredients of the model are gene duplication plus re-wiring of the protein interactions, two elements known to be the essential driving forces in genome evolution [27]. The model does not include functionality or dynamics of the proteins involved, but it is a topologically-based approximation to the overall features of the proteome graph and intends to capture some of the generic features of proteome evolution.

During the completion of this work we became aware of a paper by Vázquez et al., Ref. [37], in which a related model of proteome evolution, showing multifractal connectivity properties, is described and analyzed.

2. Proteome growth model

Here we restrict our rules to single-gene duplications, which occur in most cases due to unequal crossover [27], plus re-wiring. Multiple duplications should be considered in future extensions of these models: molecular evidence shows that even whole-genome duplications have actually occurred in *S. cerevisiae* [46] (see also Ref. [40]).

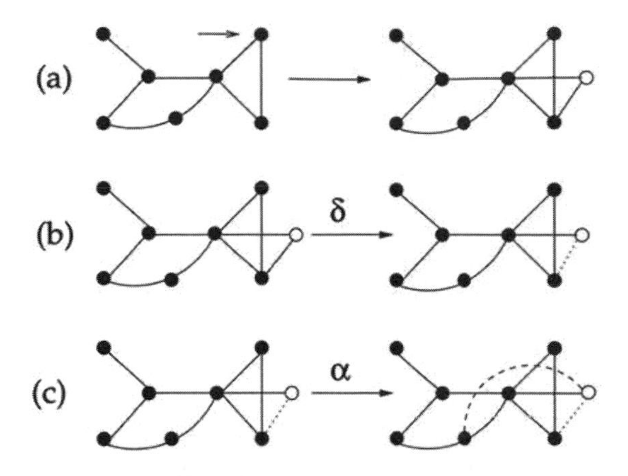

Fig. 2. Growing network by duplication of nodes. First (a) duplication occurs after randomly selecting a node (arrow). The links from the newly created node (white) now can experience deletion (b) and new links can be created (c); these events occur with probabilities δ and α, respectively.

Re-wiring has also been used in dynamical models of the evolution of robustness in complex organisms [7].

It is worth mentioning that the study of metabolic networks provides some support to the rule of *preferential attachment* [4] as a candidate mechanism to explain the origins of the scale-free topology. Scale-free graphs are easily obtained in a growing network provided that the links to new nodes are made preferentially from nodes that already have many links. A direct consequence is that vertices with many connections are those that have been incorporated early. This seems to be plausible in the early history of metabolic nets, and this view is supported by some available evidence [43]. A similar argument can be made with proteome maps, since there are strong connections between the evolution of metabolic pathways and genome evolution, and other scenarios have also been proposed, including optimization [11]. Here we do not consider preferential attachment rules, although future studies should explore the possible contributions of different mechanisms to the evolution of network biocomplexity. In this context, new integrated analyses of cellular pathways using microarrays and quantitative proteomics [17] will help to obtain a more detailed picture of how these networks are organized.

The proteome graph at any given step t (i.e. after t duplications) will be indicated as $\Omega_p(t)$. The rules of the model, summarized in figure 2, are implemented as follows. Each time step: (a) one node in the graph is randomly chosen and duplicated; (b) the links emerging from the new generated node are removed with probability δ; (c) finally, new links (not previously present) can be created between the new node and all the rest of the nodes with probability α. Step (a) implements gene duplication, in which both the original and the replicated proteins retain the same structural properties and, consequently, the same set of interactions. The rewiring steps (b) and (c) implement the possible mutations of the replicated gene, which translate into the deletion and addition of interactions, with different probabilities.

Since we have two free parameters, we should first constrain their possible values by using the available empirical data. As a first step, we can estimate the asymptotic average connectivity exhibited by the model in a mean-field approximation (see also Ref. [37]). Let us indicate by $\bar{K}_N$ the average connectivity of the system when it is composed by N nodes. It is not difficult to see that the increase in the average connectivity after one iteration step of the model is proportional to

$$\frac{d\bar{K}_N}{dN} \simeq \bar{K}_{N+1} - \bar{K}_N = \frac{1}{N}\left[\bar{K}_N - 2\delta\bar{K}_N + 2\alpha(N - \bar{K}_N)\right]. \tag{2.1}$$

The first term accounts for the duplication of one node, the second represents the average elimination of $\delta\bar{K}_N$ links emanating from the new node, and the last term represents the addition of $\alpha(N - \bar{K}_N)$ new connections pointing to the new node. Eq. (2.1) is a linear equation which easily solved, yielding

$$\bar{K}_N = \frac{\alpha N}{\alpha + \delta} + \left(\bar{K}_1 - \frac{\alpha}{\alpha + \delta}\right) N^{\Gamma}, \tag{2.2}$$

where $\Gamma = 1 - 2\alpha - 2\delta$ and $\bar{K}_1$ is the initial average connectivity of the system. This solution leads to an increasing connectivity through time. In order to have a finite $\bar{K}$ in the limit of large N, we must impose the condition $\alpha = \beta/N$, where β is a constant. That is, the rate of addition of new links (the establishment of new viable interactions between proteins) is inversely proportional to the network size, and thus much smaller than the deletion rate δ, in agreement with the rates observed in [42]. In this case, for large N, we get

$$\frac{d\bar{K}_N}{dN} = \frac{1}{N}(1 - 2\delta)\bar{K}_N + \frac{2\beta}{N}. \tag{2.3}$$

The solution of this equation is

$$\bar{K}_N = \frac{2\beta}{2\delta - 1} + \left(\bar{K}_1 - \frac{2\beta}{2\delta - 1}\right) N^{1-2\delta}. \tag{2.4}$$

For $\delta > 1/2$ a finite connectivity is reached,

$$\bar{K} \equiv \bar{K}_\infty = \frac{2\beta}{2\delta - 1}. \tag{2.5}$$

The previous expression imposes the boundary condition $\delta > 1/2$, necessary in order to obtain a well-defined limiting average connectivity. Eq. (2.5), together with the experimental estimates of $\bar{K} \sim 1.9 - 2.3$, allows to set a first restriction to the parameters β and δ. Imposing $\bar{K} = 2$, we are led to the relation

$$\beta = 2\delta - 1. \tag{2.6}$$

Moreover, estimations of addition and deletion rates α and δ from yeast [42] give a ratio $\alpha/\delta \leq 10^{-3}$. For proteomes of size $N \sim 10^3$, as in the case of the yeast,

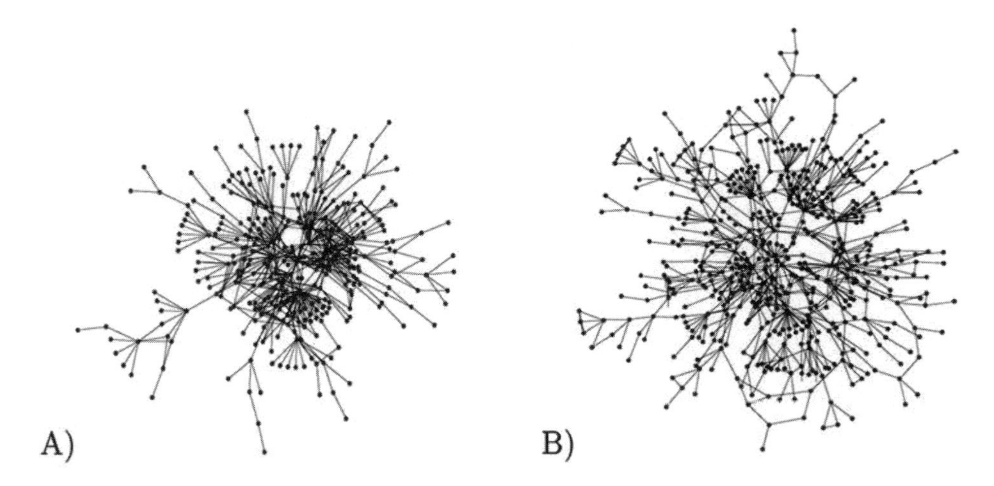

Fig. 3. A) An example of a small proteome interaction map (giant component, Ω_∞) generated by the model with $N = 10^3$, $\delta = 0.58$, and $\beta = 0.16$. B) Real yeast proteome map obtained from the MIPS database [26]. We can observe the close similitude between the real map and the output of the model.

this leads to $\beta/\delta \le 10^{-3}N \sim 1$. Using the safe approximation $\beta/\delta \approx 0.25$, together with the constraint (2.5), we obtain the approximate values

$$\delta = 0.58, \qquad \beta = 0.16. \tag{2.7}$$

which will be used through the rest of the paper.

Simulations of the model start form a connected ring of $N_0 = 5$ nodes, and proceed by iterating the rules until the desired network size is achieved.

3. Results

Computer simulations of the proposed model reproduce many of the regularities observed in the real proteome data. As an example of the output of the model, in figure 3A we show an example of the giant component Ω_∞ (the largest cluster of connected proteins) of a realization of the model with $N = 10^3$ nodes. This figure clearly resembles the giant component of real yeast networks, as we can see comparing with figure 3B[a], and we can appreciate the presence of a few highly connected hubs plus many nodes with a relatively small number of connections. The size of the giant component for $N = 10^3$, averaged of 10^4 networks, is $|\Omega_\infty| = 472 \pm 87$, in good agreement with Wagner's data $|\Omega_\infty^W| = 466$ for a yeast with a similar total number of proteins (the high variance in our result is due to the large fluctuations in the model for such small network size N). On the other hand, in figure 4 we plot the connectivity $P(k)$ obtained for networks of size $N = 10^3$. In this figure we observe that the resulting connectivity distribution can be fitted

[a]Figure kindly provided by W. Basalaj (see http://www.cl.cam.uk~wb204/GD99/#Mewes).

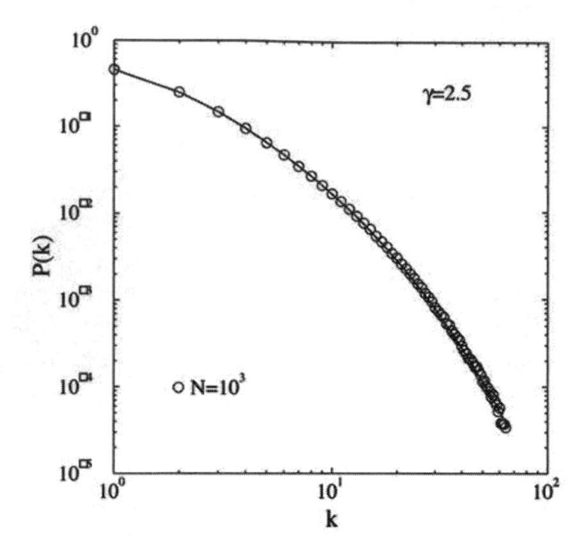

Fig. 4. Degree distribution $P(k)$ for the model, averaged over 10^4 networks of size $N = 10^3$. The distribution shows a characteristic power law behavior, with exponent $\gamma = 2.5 \pm 0.1$ and an exponential cut-off $k_c \simeq 28$.

to a power-law win an exponential cut-off, of the form given by Eq. (1.1), with parameters $\gamma = 2.5 \pm 0.1$ and $k_c \simeq 28$, in good agreement with the measurements reported in Refs [42] and [19].

An additional observation from Wagner's study of the yeast proteome is the presence of SW properties. We have found also similar topological features in our model, using the considered set of parameters. The proteome graph is defined by a pair $\Omega_p = (W_p, E_p)$, where $W_p = \{p_i\}, (i = 1, ..., N)$ is the set of N proteins and $E_p = \{\{p_i, p_j\}\}$ is the set of edges/connections between proteins. The *adjacency matrix* ξ_{ij} indicates that an interaction exists between proteins $p_i, p_j \in \Omega_p$ ($\xi_{ij} = 1$) or that the interaction is absent ($\xi_{ij} = 0$). Two connected proteins are thus called *adjacent* and the *degree* of a given protein is the number of edges that connect it with other proteins.

The SW pattern can be detected from the analysis of two basic statistical quantities: the *clustering coefficient* C_v and the *average path length* L. Let us consider the adjacency matrix and indicate by $\Gamma_i = \{p_i \mid \xi_{ij} = 1\}$ the set of nearest neighbors of a protein $p_i \in W_p$. The clustering coefficient for this protein is defined as the number of connections between the proteins $p_j \in \Gamma_i$ [45]. Denoting

$$\mathcal{L}_i = \sum_{j=1}^{N} \xi_{ij} \left[\sum_{k \in \Gamma_i} \xi_{jk} \right], \tag{3.1}$$

we define the clustering coefficient of the i-th protein as

$$c_v(i) = \frac{2\mathcal{L}_i}{k_i(k_i - 1)}, \tag{3.2}$$

Table 1. Comparison between the observed regularities in the yeast proteome [42], the model predictions with $N = 10^3$, $\delta = 0.58$ and $\beta = 0.16$, and a random network with the same size and average connectivity as the model. The quantities X represent averages over the whole graph; X^g represent averages over the giant component.

	Yeast proteome	Network model	Random network
$\bar{K}$	1.83	2.2 ± 0.5	2.00 ± 0.06
$\bar{K}^g$	2.3	4.3 ± 0.5	2.41 ± 0.05
γ	2.5	2.5 ± 0.1	—
$\|\Omega_\infty\|$	466	472 ± 87	795 ± 22
C_v^g	2.2×10^{-2}	1.0×10^{-2}	1.5×10^{-3}
L^g	7.14	5.1 ± 0.5	9.0 ± 0.4

where k_i is the connectivity of the i-th protein. The clustering coefficient is defined as the average of $c_v(i)$ over all the proteins,

$$C_v = \frac{1}{N}\sum_{i=1}^{N} c_v(i), \tag{3.3}$$

and it provides a measure of the average fraction of pairs of neighbors of a node that are also neighbors of each other.

The average path length L is defined as follows: Given two proteins $p_i, p_j \in W_p$, let $L_{min}(i,j)$ be the minimum path length connecting these two proteins in Ω_p. The average path length L will be:

$$L = \frac{2}{N(N-1)}\sum_{i<j}^{N} L_{min}(i,j) \tag{3.4}$$

Random graphs, where nodes are randomly connected with a given probability p [6], have a clustering coefficient inversely proportional to the network size, $C_v^{\text{rand}} \approx \bar{K}/N$, and an average path length proportional to the logarithm of the network size, $L^{\text{rand}} \approx \log N / \log \bar{K}$. At the other extreme, regular lattices with only nearest-neighbor connections among units are typically clustered and exhibit long average paths. Graphs with SW structure are characterized by a high clustering with $C_v \gg C_v^{rand}$, while possessing an average path comparable with a random graph with the same connectivity and number of nodes.

In Table 1 we report the values of $\bar{K}$, γ, $|\Omega_\infty|$, C_v, and L for our model, compared with the values reported for the yeast *S. cerevisiae* [19,42], and the values corresponding to a random graph with size and connectivity comparable with both the model and the real data. Except the average connectivity of the giant component, which is slightly larger for the model, all the magnitudes for the model compare quite well with the values measured for the yeast. On the other hand, the values obtained for a random graph support the conjecture of the SW properties of the protein network put forward in Ref. [42].

4. Discussion

The analysis of complex biological networks in terms of random graphs is not new. Early work suggested that the understanding of some general principles of genome organization might be the result of emergent properties within random networks of interacting units [21,22]. An important difference emerges, however, from the new results about highly heterogeneous networks: the topological organization of metabolic and protein graphs is very different from the one expected under totally random wiring and as a result of their heterogeneity, new qualitative phenomena emerge (such as the robustness against mutation). This supports the view that cellular functions are carried out by networks made up by many species of interacting molecules and that networks of interactions might be at least as important as the units themselves [15,33].

Our study has shown that the macroscopic features exhibited by the proteome are also present in our simple model. This is surprising, since it is obvious that different proteins and protein interactions play different roles and operate under very different time scales and our model lacks such specific properties, dynamics or explicit functionality. Using estimated rates of addition and deletion of protein interactions as well as the average connectivity of the yeast proteome, we accurately reproduce the available statistical regularities exhibited by the real proteome. In this context, although data from yeast might involve several sources of bias, it has been shown that the same type of distribution is observable in other organisms, such as the protein interaction map of the human gastric pathogen *Helicobacter pylori* or in the p53 network (Jeong and Barabási, personal communication).

These results suggest that the global organization of protein interaction maps can be explained by means of a simple process of gene duplication plus diversification. These are indeed the mechanisms known to be operating in genome evolution (although the magnitude of the duplication event can be different). One important point to be explored by further extensions of this model is the origin of the specific parameters used. The use of evolutionary algorithms and optimization procedures might provide a consistent explanation of the particular values observed and their relevance in terms of functionality. A different source of validation of our model might be the study of proteome maps resulting from the evolution of resident genomes [3]: the genomes of endosymbionts and cellular organelles display an evolutionary degradation that somehow describe an inverse rule of proteome reduction. Reductive evolution can be almost extreme, and available data of resident proteomes might help to understand how proteome maps get simplified under the environmental conditions defined by the host genome. If highly connected nodes play a relevant role here, perhaps resident genomes shrink by loosing weakly connected nodes first.

Most of the classic literature within this area deal with the phylogenetic consequences of duplication and do not consider the underlying dynamics of interactions between genes. We can see, however, that the final topology has nontrivial consequences: this type of scale-free network will display an extraordinary robustness against random removal of nodes [1] and thus it can have a selective role. But an open question arises: is the scale-free organization observed in real proteomes a byproduct of the pattern of duplication plus rewiring (perhaps under a low-cost constraint in wiring) and thus we have "robustness for free"? The alternative is of course a fine-tuning of the process in which selection for robustness has been obtained by accepting or rejecting single changes. Further model approximations and molecular data might provide answers to these fundamental questions.

Acknowledgements

The authors thank J. Mittenthal, R. Ferrer, J. Montoya, S. Kauffman and A. Wuensche for useful discussions. This work has been supported by a grant PB97-0693 and by the Santa Fe Institute (RVS). RPS acknowledges financial support from the Ministerio de Ciencia y Tecnología (Spain).

References

[1] R. A. Albert, H. Jeong, and A.-L. Barabási. Error and attack tolerance of complex networks. *Nature*, 406:378–382, 2000.

[2] L. A. N. Amaral, A. Scala, M. Barthélémy, and H. E. Stanley. Classes of small-world networks. *Proc. Natl. Acad. Sci. USA*, 97:11149–11152, 2000.

[3] S. G. E. Andersson and C. Kurland. Reductive evolution of resident genomes. *Trends Microbiol.* 6: 263-268, 1998.

[4] A.-L. Barabási and R. Albert. Emergence of scaling in random networks. *Science*, 286:509–511, 1999.

[5] P. L. Bartel, J. A. Roecklein, D. SenGupta, and S. A. Fields. A protein linkage map of *Escherichia coli* bacteriohage t7. *Nature Genet.*, 12:72–77, 1996.

[6] B. Bollobás. *Random Graphs.* Academic Press, London, 1985.

[7] S. Bornholdt and K. Sneppen. Robustness as an evolutionary principle. *Proc. Roy. Soc. Lond. B*, 267:2281–2286, 2000.

[8] D. Fell and A. Wagner. The small world of metabolism. *Nature Biotech.*, 18:1121, 2000.

[9] R. Ferrer i Cancho, C. Janssen, and R. V. Solé. The small world of human language. *Procs. Roy. Soc. London B*, 268:2261–2266, 2001.

[10] R. Ferrer i Cancho, C. Janssen, and R. V. Solé. The topology of technology graphs: small world pattern in electronic circuits. *Phys. Rev. E*, 63:32767, 2001.

[11] R. Ferrer i Cancho and R. V. Solé. Optimization in complex networks. *Phys. Rev. Lett.* (submitted, 2000).

[12] S. Fields. Proteomics in genomeland. *Science*, 409:861–921, 2001.

[13] M. Flajolet, G. Rotondo, L. Daviet, F. Bergametti, G. Inchauspe, P. Tiollais, C. Transy, and P. Legrain. A genomic approach to the Hepatitis C virus generates a protein interaction map. *Gene*, 242:369–379, 2000.

[14] M. Fromont-Racine, J. C. Rain, and P. Legrain. Towards a functional analysis of the yeast genome through exhaustive two-hybrid screens. *Nature Genet.*, 16:277–282, 1997.

[15] L. H. Hartwell, J. J. Hopfield, S. Leibler, and A. W. Murray. From molecular to modular cell biology. *Nature*, 402:C47–C52, 1999.

[16] T. R. Hazbun and S. Fields. Networking proteins in yeast. *Proc. Natl. Acad. Sci. USA*, 98:4277–4278, 2001.
[17] T. Ideker, V. Thorsson, J. A. Ranish et al. Integrated genomic and proteomic analyses of a systematically perturbed metabolic network. *Science* 292: 929-934, 2001.
[18] T. Ito, K. Tashiro, S. Muta, R. Ozawa, T. Chiba, M. Nishizawa, K. Yamamoto, S. Kuhara, and Y. Sakaki. Toward a protein-protein interaction map of the budding yeast: A comprehensive system to examine two-hybrid interactions in all possible combinations between the yeast proteins. *Proc. Natl. Acad. Sci. USA*, 97:1143–1147, 2000.
[19] H. Jeong, S. Mason, A. L. Barabási, and Z. N. Oltvai. Lethality and centrality in protein networks. *Nature*, 411:41, 2001.
[20] H. Jeong, B. Tombor, R. Albert, Z. N.Oltvai, and A.-L. Barabasi. The large-scale organization of metabolic networks. *Nature*, 407:651–654, 2001.
[21] S. A. Kauffman. Metabolic stability and epigenesis in randomly connected nets. *J. Theor. Biol.*, 22:437–467, 1962.
[22] S. A. Kauffman. *Origins of Order*. Oxford, New York, 1993.
[23] E. S. Lander and *et al.* Initial sequencing and analysis of the human genome. *Nature*, 409:861–921, 2001.
[24] A. L. Lloyd and R. M. May. How viruses spread among computers and people. *Science*, 292:1316–1317, 2001.
[25] S. McCraith, T. Holtzman, B. Moss, and S. Fields. Genome-wide analysis of vaccinia virus protein-protein interactions. *Proc. Natl. Acad. Sci. USA*, 97:4879–4884, 2000.
[26] H. W. Mewes, K. Heumann, A. Kaps, K. Mayer, F. Pfeiffer, S. Stocker, and D. Frishman. Mips: a database for genomes and protein sequences. *Nucleic Acids Res.*, 27:44–48, 1999.
[27] S. Ohno. *Evolution by gene duplication.* Springer, Berlin, 1970.
[28] C. A. Onzonnis and P. D. Karp. Global properties of the metabolic map of Escherichia coli. *Genome Res.* 10: 568-576, 2000.
[29] R. Pastor-Satorras, A. Vázquez, and A. Vespignani. Dynamical and correlation properties of the internet. *Phys. Rev. Lett.*, 87:258701, 2001.
[30] R. Pastor-Satorras and A. Vespignani. Epidemic spreading in scale-free networks. *Phys. Rev. Lett.*, 86:3200–3203, 2001.
[31] J. C. Rain, L. Selig, H. De Reuse, V. Battaglia, C. Reverdy, S. Simon, G. Lenzen, F. Petel, J. Wojcik, V. Schachter, Y. Chemama, A. S. Labigne, and P. Legrain. The protein-protein interaction map of Helicobacter pylori. *Nature*, 409:743, 2001.
[32] P. Ross-Macdonald, P. S. R. Coelho, T. Roemer, S. Agarwal, A. Kumar, R. Jansen, K. H. Cheung, A. Sheehan, D. Symoniatis, L. Umansky, M. Heldtman, F. K. Nelson, H. Iwasaki, K. Hager, M. Gerstein, P. Miller, G. S. Roeder, and M. Snyder. Large-scale analysis of the yeast genome by transposon tagging and gene disruption. *Nature*, 402:413–418, 1999.
[33] R. V. Solé, I. Salazar-Ciudad, and S. A. Newman. Gene network dynamics and the evolution of development. *Trends Ecol. Evol.*, 15:479–480, 2000.
[34] K. E. Stephan, C-C. Hilgetag, G. A. P. C. Burns, M. A. O'Neill, M. P. Young and R. Kötter. Computational analysis of functional connectivity between areas of primate cerebral cortex. *Phil. Trans. Roy. Soc. Lond. B* 355: 111-126, 2000.
[35] S. H. Strogatz. Exploring complex networks. *Nature*, 410:268–276, 2001.
[36] D. Thieffry, A. M. Huerta, E. Pérez-Rueda, and J. Collado-Vives. From specific gene regulation to genomic networks: a global analysis of transcriptional regulation in *Escherichia coli. BioEssays*, 20:433–440, 1998.
[37] A. Vázquez, A. Flammini, A. Maritan, and A. Vespignani. Modelling of protein

interaction networks, 2001. cond-mat/0108043.

[38] J. C. Venter and *et al.* The sequence of the human genome. *Science*, 291:1305, 2001.

[39] B. Vogelstein, D. Lane, and A. J. Levine. Surfing the p53 network. *Nature*, 408:307–310, 2000.

[40] A. Wagner. Evolution of gene networks by gene duplications: A mathematical model and its implications on genome organization. *Proc. Natl. Acad. Sci. USA*, 91:4387–4391, 1994.

[41] A. Wagner. Robustness against mutations in genetic networks of yeast. *Nature Genet.*, 24: 355-361, 2000.

[42] A. Wagner. The yeast protein interaction network evolves rapidly and contains few redundant duplicate genes. *Mol. Biol. Evol.*, 18:1283–1292, 2001.

[43] A. Wagner. and D. A. Fell. The small world inside large metabolic networks. *Proc. Roy. Soc. London B* 268: 1803-1810, 2001.

[44] A. J. M. Walhout, R. Sordella, X. W. Lu, J. L. Hartley, G. F. Temple, M. A. Brasch, N. Thierry-Mieg, and M. Vidal. Protein interaction mapping in c. elegans using proteins involved in vulval development. *Science*, 287:116–122, 2000.

[45] D. J. Watts and S. H. Strogatz. Colective dynamics of 'small-world' networks. *Nature*, 393:440–442, 1998.

[46] K. H. Wolfe and D. C. Shields. Molecular evidence for an ancient duplication of the entire yeast genome. *Nature*, 387:708–713, 1997.

[47] X. Wu, J. H. Bayle, D. Olson, and A. J. Levine. The P53 MDM-2 autoregulatory feedback loop. *Gen. Dev.*, 7:1126, 1993.

ComPlexUs
ORIGINAL RESEARCH PAPER
ComPlexUs 2003;1:38–44; DOI:10.1159/000067642
Received: March 25, 2002
Accepted after revision: October 22, 2002

Modeling of Protein Interaction Networks

Alexei Vázquez[a] Alessandro Flammini[a]
Amos Maritan[a,b] Alessandro Vespignani[b]

[a] International School for Advanced Studies and INFM and
[b] The Abdus Salam International Centre for Theoretical Physics, Trieste, Italy

Key Words
Protein interaction network · Duplication · Divergence · Multifractal ·
PACS No. 89.75.-k · 87.15.Kg · 87.23.Kg

Abstract
We introduce a graph-generating model aimed at representing the evolution of protein interaction networks. The model is based on the hypothesis of evolution by duplication and divergence of the genes which produce proteins. The obtained graphs have multifractal properties recovering the absence of a characteristic connectivity as found in real data of protein interaction networks. The error tolerance of the model to random or targeted damage is in very good agreement with the behavior obtained in real protein network analyses. The proposed model is a first step in the identification of the evolutionary dynamics leading to the development of protein functions and interactions.

Fax +41 61 306 12 34
E-Mail karger@karger.ch
www.karger.com
KARGER

1424–8492/03/0011–0038
$19.50/0
Accessible online at:
www.karger.com/cpu

Amos Maritan
SISSA, Via Beirut 2–4
I-34014 Trieste (Italy)
Tel. + 39 040 2240462, Fax + 39 040 3787528
E-Mail maritan@sissa.it

Synopsis

Proteins interact with one another. Biology is complex. Hence, one might expect that even for a simple organism, the detailed map of the interactions between its various proteins would be a tangled web of overwhelming complexity, more or less random in its appearance. Yet as researchers have recently learned, networks of protein-protein interactions are very far from being random; instead, they possess subtle but significant order. In pioneering experiments, Uetz et al. [1] have mapped out 2,238 pair-wise interactions between 1,825 proteins in the single-celled yeast *Saccharomyces cerevisae* – also known as baker's or brewer's yeast. As it turns out, this network reveals noteworthy mathematical regularities and shares important topological features with other complex networks, including webs of social interaction, the World Wide Web and the Internet.

Like these other networks, the yeast protein network has the 'small world' property – following along links in the network, it requires only a handful of steps to go from any one protein to any other. Another similarity is the manner in which the links are shared out among the various proteins: empirically, the probability that a protein interacts with k other proteins follows a power-law distribution, $P(k) \sim k^{-\gamma}$, with $\gamma \cong 2.5$. Networks of this sort are called 'scale-free', and the average number of links $\langle k \rangle$ offers little insight into the real network topology, which is highly heterogeneous. The 'connectivity' of the various nodes varies considerably: while many nodes have only 1 or a few links, a handful of 'super-connected' hubs have very many.

What is the origin of this architecture? Since scale-free networks arise in diverse settings, it seems likely that some general process might be at work, and this appears to be the case. A crucial insight is that real-world networks have grown to be what they are. Suppose a network starts out small and grows through the progressive addition of new elements, and that each new element, upon entering the network, establishes links at random to a few of the older elements. As Barabási and Albert [5] showed a few years ago, the scale-free architecture

The complete genome sequencing gives for the first time the means to analyze organisms on a genomic scale. This implies the understanding of the role of a huge number of gene products and their interactions. For instance, it becomes a fundamental task to assign functions to uncharacterized proteins, traditionally identified on the basis of their biochemical role. The functional assignment has progressed considerably by partnering proteins of similar functions, leading eventually to the drawing of protein interaction networks (PINs). This has been accomplished in the case of the yeast *Saccharomyces cerevisiae*, where two hybrid analyses and biochemical protein interaction data have been used to generate a web-like view of the protein-protein interaction network [1]. The topology of the obtained graph has been recently studied in order to identify and characterize its intricate architecture [2, 3]. Surprisingly, the PIN resulted in a very highly heterogeneous network with scale-free (SF) connectivity properties.

SF networks differ from regular (local or non-local) networks in the small average link distance between any two nodes (small-world property) [4] and a statistically significant probability that nodes possess a large number of connections k compared to the average connectivity $\langle k \rangle$ [5]. This reflects in a power-law connectivity distribution $P(k) \sim k^{-\gamma}$ for the probability that a node has k links pointing to other nodes. These topological properties, which appear to be realized in many natural and technological networks [6], are due to the interplay of the network growth and the 'preferential attachment' rule. In other words, new appearing nodes have a higher probability to get connected to another node already counting a large number of links. These ingredients encompassed in the Barabási and Albert model [5] are present in all the successive SF network models [7–9] and are fundamental for the spontaneous evolution of networks in an SF architecture. From this perspective it is natural to ask about the microscopic process that drives the placement of nodes and links in the case of the protein network.

In the present article we study a growing network model whose microscopic mechanisms are inspired by the duplication and the functional complementation of genes [2, 10, 11]. In this evolutive model all proteins in a family evolved from a common ancestor through gene duplications and mutations (divergence), and the protein network is the blueprint of the entire history of the genome evolution. This duplication-divergence (DD) model is analyzed by analytical calculations and numerical simulations in order to characterize the topological and large-scale properties of the generated networks. We find that the networks show the absence of any characteristic connectivity and exhibit a connectivity distribution with multifractal properties. The n-th moment of the distribution behaves as a power law of the network size N with an exponent that has a nonlinear dependency with n. In addition, the evolution of all moments with N crosses from a divergent behavior to a finite asymptotic value at a given value of the mutation parameters. The generated networks can be directly compared with the *S. cerevisiae* PIN data reported on the Biotech website [12], composed of 1,825 proteins (nodes) connected by 2,238 identified interactions (links). In agreement with the PIN analysis we find that the connectivity distribution can be approximately fitted by a single apparent power-law behavior, and the model's parameter can be optimized in order to quantitatively reproduce magnitudes such as the average connectivity or the clustering coefficient. Further, we compare the simulated tolerance of the DD network to random deletion of individual nodes with that obtained by deleting the most connected ones. While the DD network proves to be fragile in the latter case, it is extremely resistant to random damages, in agreement with the behavior recently observed for the PIN [3]. The analogous topological properties found in the proposed model and the experimental PIN represent a first step towards a possible understanding of protein-protein interactions in terms of gene evolution.

Proteins are divided in families according to their sequence and functional similarities [13, 14]. The existence of these families can be explained using the evolutive

Synopsis

emerges naturally from this algorithm if the placement of links follows a rule of 'preferential attachment', that is, if the probability of linking to a node grows in direct proportion to the number of links that node has already.

For a network such as the World Wide Web it is easy to see how this mechanism might come into play. The creator of a new web page naturally provides some links to other pages, and the links chosen will to some extent reflect the visibility of these sites; one has to at least know about a site in order to link to it. People tend to provide links to popular sites that already have a large number of links pointing to them (such as Yahoo, Amazon or Google), rather than to more obscure sites. Hence, preferential attachment may well explain the scale-free character of the World Wide Web.

What about protein networks? It is less clear how preferential attachment might play a role here, and consequently, the origin of the scale-free architecture is more mysterious. But cellular biochemistry has emerged through a long history of biological evolution, and Vázquez et al. show how evolution can produce scale-free networks. They explore a model for the evolution of protein networks that accurately reproduces the topological features seen in the yeast *S. cerevisae.*

As Vázquez et al. point out, proteins fall into families according to similarities in their amino-acid sequences and functions, and it is natural to suppose that such proteins have all evolved from a common ancestor. A favored hypothesis views such evolution as taking place through a sequence of gene duplications – a relatively frequent occurrence during cell reproduction. Following each duplication, the two resulting genes are identical for the moment, and naturally lead to the production of identical proteins. But between duplications, random genetic mutations will lead to a slow divergence of the genes and their associated proteins. This repetitive, two-stage process can be captured in a relatively simple model for the growth of a protein interaction network – the 'duplication-divergence model'.

hypothesis that all proteins in a family evolved from a common ancestor [15]. This evolution is thought to take place through gene or entire genome duplications, resulting in redundant genes. After the duplication, redundant genes diverge and evolve to perform different biological functions. According to the classic model [15], after duplication the duplicate genes have fully overlapping functions. Later on, one of the copies may either become nonfunctional due to degenerative mutations or it can acquire a novel beneficial function and become preserved by natural selection. In a more recent framework [10, 11] it is proposed that both duplicate genes are subject to degenerative mutations and lose some functions, but jointly retain the full set of functions present in the ancestral gene. The outcome of this evolution results in complex PINs with physical, genetic, and biochemical interactions among them. In this article we will restrict our analysis to the physical interactions, and from now on use the term 'interaction' to denote a physical interaction. There is a large number of transient protein-protein interactions that controls and regulates almost all cellular processes. These transient interactions can be detected using the two-hybrid experiments [16]. Two-hybrid experiments are known to be prone to false positives, and it is hard to establish to what extent protein-protein interaction networks can be considered complete and error-free [17]. Nevertheless, the statistical properties of the PIN obtained by different groups are essentially identical, as observed by Yook et al. [18].

The evolution of the PIN can be translated into a growing network model. We can consider each node of the network as the protein expressed by a gene. After gene duplication, both expressed proteins will have the same interactions. This corresponds to the addition of a new node in the network with links pointing to the neighbors of its ancestor. Moreover, if the ancestor is a self-interacting protein, the copy will have also an interaction with it [2]. Eventually, some of the common links will be removed because of the divergence process. We can formalize this process by defining an evolving network in which, at each time step, a node is added according to the following rules:

Duplication: A node i is selected at random. A new node i', with a link to all the neighbors of i, is created. With probability p a link between i and i' is established.

Divergence: For each of the nodes j linked to i and i' we choose randomly one of the two links (i, j) or (i', j) and remove it with probability q.

p is a parameter that models the creation of an interaction between the duplicates of a self-interacting protein and its possible loss due to the divergence of the duplicates. The other parameter q represents the loss of interactions between the duplicates and their neighbors due to the divergence of the duplicates. Our purpose is to provide a minimal model that captures the main effects of the duplication and divergence mechanisms in the PIN topology. Thus p and q take into account fine details in some effective way.

For practical purposes, the DD algorithm starts with two connected nodes and repeats the duplication-divergence rules N times. Since genome evolution analysis [2, 19] supports the idea that the divergence of duplicate genes takes place shortly after the duplication, we can assume that the divergence process always occurs before any new duplication takes place; i.e. there is a time scale separation between duplication and mutation rates. This allows us to consider the number of nodes in the network, N, as a measure of time (in arbitrary units). It is worth remarking that the algorithm does not include the creations of new links; i.e. the developing of new interactions between gene products. This process has been argued to have a probability relatively smaller than the divergence one [2]. However, we found that introducing a probability in the DD algorithm to develop new random connections does not change the network topology.

Another assumption is the one-to-one relation between genes and proteins. It is possible that a gene expresses more than 1 protein. This would mean that a set of proteins, those expressed by the same gene, will have a correlated evolution. In any case, the number of proteins expressed by one

Synopsis

In this model, each node in the network represents a protein that is expressed by a gene, and the network grows as follows. At each time step, one selects a random node – call it node i – and carries out two steps in sequence. First comes duplication. In association with node i, a new node i' enters the network and is linked to the same nodes to which i is linked. This reflects the idea that the new protein, the result of duplication, is identical to the old protein; hence, it interacts with other proteins in the very same way. Also, with some probability p, a link is added between i and i' to account for the possibility that these two (identical) proteins also interact. Next comes divergence. Mutations in the genes associated with i and i' will gradually produce differences in these proteins, altering their interactions. The model accounts for this divergence by considering in turn all the proteins j to which i and i' are linked, selecting one of these at random, and removing it with probability q.

As Vázquez et al. readily acknowledge, this model leaves aside many finer details of the genetic evolution that lies behind the duplication and divergence process. The model aims only to capture the most basic factors affecting the topological evolution of the network. New proteins enter the network following duplication and, because of subsequent mutations, carry with them some but not all of the interaction links of the protein from which they sprouted. Starting with a small seed network (such as two proteins linked to one another) and running the growth procedure many times, one produces a large, complex network. The important question: Do such networks resemble the real protein interaction networks of biology?

One quantity to explore is the distribution of nodes according to their connectivity. Using a computer, Vázquez et al. ran the model 100 times, on each occasion stopping when the number of nodes reached $N = 1,825$ – the number of proteins in the available data for the yeast *S. cerevisae.* (As the model follows a process of random growth, the network turns out different in each run; hence, it takes many runs to reveal the average behavior behind the statis-

gene is not comparable with the network size, hence it is expected that a more general model allowing for this correlated evolution will exhibit the same qualitative features.

In order to provide a general analytical understanding of the model, we use a mean-field approach for the moments distribution behavior. Let us define $\langle k \rangle_N$ as the average connectivity of the network with N nodes. After a duplication event $N \to N+1$ the average connectivity is given by

$$\langle k \rangle_{N+1} = \frac{\langle N \rangle \langle k \rangle_N + 2p + (1-2q) \langle k \rangle_N}{N+1} \quad (1)$$

On average, there will be a gain proportional to $2p$ because of the interaction between duplicates, to $\langle k \rangle$ because of duplication, and a loss proportional to $2q \langle k \rangle_N$ due to the divergence process. For large N, taking the continuum limit, we obtain a differential equation for $\langle k \rangle$. For $q > 1/2$, $\langle k \rangle$ grows with N but saturates to the stationary value $k_\infty = 2p/(2q-1) + \mathcal{O}(N^{1-2q})$. By contrast, for $q < 1/2$, $\langle k \rangle$ grows with N as N^{1-2q}. At $q = q_1 = 1/2$ there is a dramatic change of behavior in the large scale connectivity properties. Analogous equations can be written for higher order moments $\langle k^n \rangle$, and for all n we find a value q_n at which the moments cross from a divergent behavior to a finite value for $N \to \infty$. More interestingly, it is possible to write the generalized exponents $\sigma_n(q)$ characterizing the moments divergence as $\langle k^n \rangle \sim N^{\sigma n(q)}$. The lengthy calculation that will be reported elsewhere [23] gives the mean-field estimate.

$$\sigma_n(q) = n(1-q) - 2 + 2(1-q/2)^n. \quad (2)$$

The nonlinear behavior with n is indicative of a multifractal connectivity distribution. In order to support the analytical calculations, we performed numerical simulations generating DD networks with a size ranging from $N = 10^3$ to $N = 10^6$. In figure 1 we show the generalized exponents $\sigma_n(q)$ as a function of the divergence parameter q. As predicted by the analytical calculations, $\sigma_n = 0$ at a critical value q_n. The general phase diagrams obtained is in good agreement with the mean-field predictions and the multifractal picture.

Noticeably, multifractal features are present also in a recently introduced model of growing networks [20] where, in analogy with the duplication process, newly added nodes inherit the network connectivity properties from parent nodes. Thus multifractality appears related to local inheritance mechanisms. Multifractal distributions have a rich scaling structure, where the SF behavior is characterized by a continuum of exponents. This behavior is, however, contrary to the usual exponentially bounded distribution and it is interesting to understand why, in the DD model, we generate the large connectivity fluctuation needed for SF behavior. A given node with connectivity k receives, with probability k/N, a new link if one of its neighbors is chosen in the duplication processes. In this case the newly added node will establish a new link to it with a probability of $1 - q$. Hence, the probability that the degree of a node increases by one is

$$\omega_{kN} \sim (1-q)k/N, \quad (3)$$

where we have neglected the constant contribution given by the self-interaction probability. This shows that even if evolutionary rules of the DD model are local, they introduce an effective linear preferential attachment known to be at the origin of SF connectivity distribution [21, 24, 25]. However, because the link deletion of duplicate nodes introduces additional heterogeneity to the problem we obtain a multifractal behavior.

The peculiarities of the duplication and divergence process manifest quantitatively in other features characterizing the topology of the network, e.g. the tendency to generate biconnected triplets and quadruples of nodes. These are sets of nodes connected by a simple cycle of links, thus forming a triangle or a square. In the DD model, triangle formation is a pronounced effect since with probability $p(1-q)$, the duplicating genes and any neighbor of the parent gene will form a new triangle. Analogously, duplicating genes and any couple of neighbors of the parent gene will form a new square with the probability $(1-q)^2$. An indication of triangle formation in networks

tical fluctuations.) The model has two adjustable parameters p and q, and, as figure 2 shows, it generates networks that closely resemble the real network in yeast if one chooses $p = 0.1$ and $q = 0.7$.

For both the model and for the real data, the figure shows Zipf plots – the logarithm of the number of links k in a node versus the logarithm of the node's rank r (the most highly-connected node having rank 1, the next most highly-connected having rank 2, and so on). For intermediate values of k, both curves are approximately linear, reflecting a power-law or scale-free relationship between connectivity k and rank r. This is not quite the probability distribution of nodes according to their connectivity, but it is easy to show that these two quantities are closely related – and if one is a power-law relationship, then so is the other. Hence, the Zipf plots imply that the distribution of nodes by connectivity is also scale-free – the model captures this aspect of the real network. It also turns out that any single network generated by the model also follows this scale-free pattern, with the exponent $\gamma \simeq 2.5$, as was found empirically for the yeast protein network.

These results appear to reflect a mechanism of preferential attachment that is effective in the evolution of protein networks. In the duplication-divergence model, as Vázquez et al. point out, the node to be duplicated at each time step is chosen at random over the full network; hence highly linked nodes have a better chance of being a neighbor of the selected node than do less connected nodes. Indeed, the chance of a node being the neighbor of the node selected for duplication increases in direct proportion to its connectivity k, since highly connected nodes have more neighbors. Consequently, more highly connected nodes have a greater chance of receiving one of the new links created in the duplication process, and preferential attachment sneaks into the model in a subtle way.

But the duplication-divergence model also has some richer features that distinguish it from most of the earlier models of network growth based on the preferential attachment idea. Power-law distributions are closely associated with fractals – highly

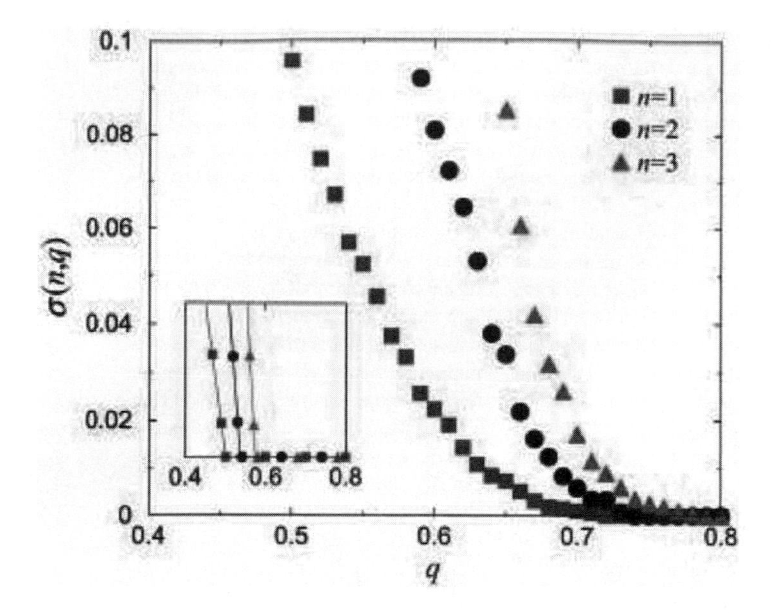

Fig. 1. The exponent $\sigma_n(q)$ as a function of q for different values of n. The symbols were obtained from numerical simulations of the model. The moments $\langle k^n \rangle$ were computed as a function of N in networks with a size ranging from $N = 10^3$ to $N = 10^6$. The exponents $\sigma_n(q)$ are obtained from the power-law fit of the plot $\langle k^n \rangle$ vs. N. In the inset we show the corresponding mean-field behavior, as obtained by equation 2, which is in qualitative agreement with the numerical results.

is given by the clustering coefficient $C_\Delta = 3N_\Delta/N_\Lambda$ [26], where N_Δ is the number of biconnected triplets (triangles) and N_Λ is the total number of simply connected triplets. Similarly, it is possible to define the square coefficient $C_\Box = 4N_\Box/N_{\Pi}$, with $N_\Box$ the number of squares in the network and N_{Π} the number of simply connected quadruples. By measuring these quantities in the yeast *S. cerevisiae* in the Biotech PIN [12], we obtained $C_\Delta = 0.23$ and $C_\Box = 0.11$. These values are one order of magnitude larger than those obtained for a SF random graph and other growing network models, for which it has been shown that the clustering coefficient is algebraically decaying with the network size [21]. By contrast, the DD model shows clustering coefficients saturating at a finite value, and it is possible to tune the parameters p and q in order to recover the real data estimates, keeping the average degree as that of the PIN $\langle k \rangle \simeq 2.4$. A reasonable agreement with the values obtained for the real PIN is found when $p \simeq 0.1$ and $q \simeq 0.7$, which yields networks with $C_\Delta = 0.10(5)$ and $C_\Box = 0.10(2)$. The value of p obtained in this way is close to the fraction of self-interacting proteins reported for the PIN (0.04) [1]. Thus, considering that p is an effective parameter that takes into account self-interactions but that may also include other effects, the agreement is very good. Noticeably, for these values of the parameters the DD model generates networks where other quantities are in good agreement with those obtained from experimental data. A pictorial representation of this agreement is provided in figure 2, where we compare the Zipf plot of the connectivity obtained from 10^3 realizations of the DD model with optimized p and q and that of the yeast *S. cerevisiae* PIN. The DD networks are composed of $N = 1{,}825$ nodes, as for the yeast PIN. The agreement is very good, considering the relatively large statistical fluctuations we have for this network size. Error bars on the DD model refer to statistical fluctuations on single realizations. It is worth noticing that despite the evident multifractal nature

Synopsis

irregular fractured surfaces, coastlines and other objects that have details over a broad range of scales, and exhibit some kind of self-similarity. Fractals can also be considered as geometrical objects of non-integer dimension. The mass of a steel cube of side L grows in proportion to L^3, reflecting the fact that the cube is a three-dimensional object; in contrast, the mass of a fractal of linear size L grows as L^D, D being the fractal dimension.

A scale-free network can also be considered as a fractal, and its character can be explored through the connectivity distribution. For their network model, Vázquez et al. have analyzed the behavior of the average connectivity $\langle k \rangle$ and of higher moments $\langle k^n \rangle$ as N becomes very large, finding that each of these approaches a finite value for some values of q, but grows without bound for others. For these values, they report the analytical result $\langle k^n \rangle \sim N^{\sigma_n(q)}$, with the exponent $\sigma_n(q)$ given in their equation 2 (simulations confirm this result). This result is significant. Mathematically, if the network were a pure fractal, described by a single fractal dimension, then $\sigma_n(q)$ would depend linearly on n. The nonlinear dependence on n in equation 2 implies that a single fractal dimension does not suffice – the duplication divergence model generates more complex networks having so-called 'multi-fractal' properties. Such networks can be thought of as a statistical mixture of many fractals of different dimensions.

Vázquez et al. have investigated their model further, showing that it also produces other network features that compare favorably to the yeast protein network. This network reveals a number of proteins linked in cyclic fashion into triangles or squares. The duplication-divergence model generates such cycles readily, and the authors compare the networks it generates to the real protein data by considering 'clustering coefficients'. In the case of triangular cycles, one can consider the ratio of the number of triangles to the number of protein triplets that are connected by only two links. This ratio offers a rough measure of the tendency for triangle formation, and a similar approach can be used for squares or higher cycles. With $p = 0.1$ and $q = 0.7$,

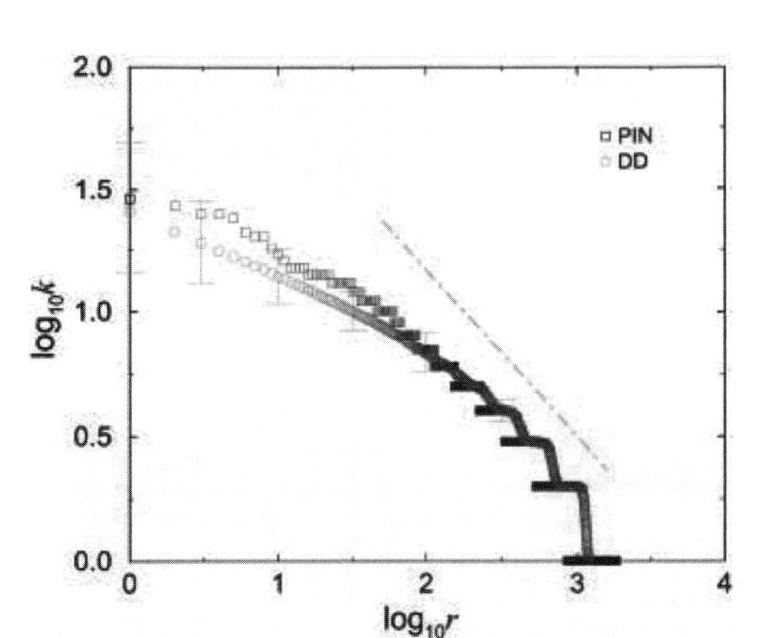

Fig. 2. Zipf plot for the PIN and the DD model with $p = 0.1$, $q = 0.7$ with $N = 1{,}825$. k is the connectivity of a node and r is its rank in decreasing order of k. Error bars represent standard deviation on a single network realization. The straight line is a power law with exponent $1/(1-\gamma)$, with $\gamma = 2.5$, which will correspond to a power-law connectivity distribution $p(k) \sim k^{-\gamma}$.

of the DD model, for a single realization of size consistent with that of the PIN, the intermediate k behavior can be approximated by an effective algebraic decay with exponent $1/(\gamma - 1)$ with $\gamma \simeq 2.5$, as found by Jeong et al. [3]. However, the plot in figure 2 shows a curvature that deviates from the algebraic behavior, evidencing the multifractal nature of the connectivity distribution.

Finally, we examined the bahavior of the DD model under random and selective deletion of nodes and compared it with those obtained for the yeast PIN [3]. Resilience to damage is indeed considered an extremely relevant property of a network. From an applicative point of view it gives a measure of how robust a network is against disruptive modifications, and how far one can go in altering it without destroying its connectivity and therefore functionality. In the random deletion process of a fraction f of nodes and the relevant links, we observed that the network fragments into several disconnected components, with the largest connected component a size of $N(f)$. For random graphs it is known that the fraction of nodes $P(f) = N(f)/N$ belonging to the largest remaining network undergoes an inverse percolation transition [27–29]. In the thermodynamic limit ($N \rightarrow \infty$), above a fraction f_c of deleted nodes, the density P drops to zero; i.e. no dominant network (giant component) is left. On the contrary, in SF networks the density of the largest cluster drops to zero only in the limit $f_c \rightarrow 1$, denoting a high resilience to random damages. Associated with this property, we observed that SF networks are very fragile with respect to targeted removal of the highest connected nodes. In this case a small fraction of removed site fragments completely destroyed the network ($P \rightarrow 0$). A similar behavior has been observed also in the yeast PIN [3]. Figure 3 shows the density of the largest remaining network versus the fraction of removed nodes both for the PIN and the DD model, and for random and selective nodes removal. The latter case consists in systematically removing nodes with the highest degree. The DD network tolerance to damage is determined by the SF nature of its multifractal distribution, and the obtained curves are in very good agreement with the corresponding ones for the yeast PIN. It is worth noting again that the parameters used for the DD network have not been independently estimated, but are those obtained from the previous optimization of the clustering coefficients. The striking analogies in the tolerance behavior are an

Synopsis

these clustering coefficients match well with the data, whereas earlier network models based on preferential attachment produce clustering coefficients about ten times too small for networks of size $N = 1{,}825$.

As one final test, Vázquez et al. study the topological resilience of the networks it generates to the removal of nodes. As they note, this is an important property since it reflects on the network's ability to function in the face of accidental damage or attack. A noteworthy characteristic of scale-free networks is their ability to 'hang together' when sites are removed at random. Even when a large fraction of the nodes have been removed, the network will remain as one more or less fully connected whole. Conversely, scale-free networks are highly vulnerable to intelligent attack, i.e. to a targeted removal of nodes beginning with the most highly connected. Vázquez et al. show that networks created through the duplication-divergence process also work in this fashion, falling apart gracefully in the face of random damage, and collapsing quickly under directed attack (fig. 3). Earlier studies have revealed closely similar behavior in the protein network in yeast.

The results reported here represent an impressive step forward, as they tie basic algorithms of network growth to a real biological basis, naturally reproducing the scale-free network architecture observed in protein networks. At the same time, these results also reflect back on the emerging science of complex networks more generally, as they point the way to networks having richer multi-fractal structures. Complex networks in other settings might well reveal similar characteristics.

Mark Buchanan

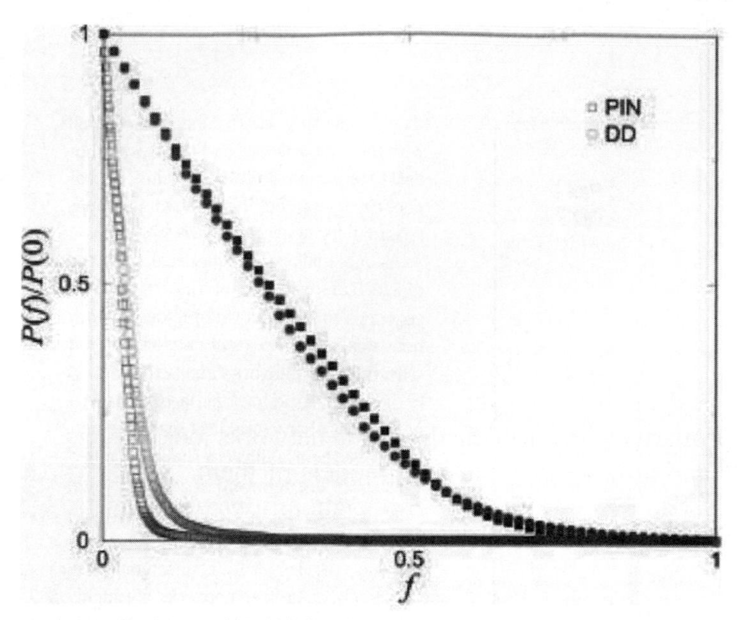

Fig. 3. Fraction of nodes $P(f) = N(f)/N$ in the largest network after a fraction f of the nodes has been removed for the PIN and the DD model. $N(f)$ is defined as the size of the largest network of connected sites. Two different removal strategies have been used, random (filled symbols) and selective (open symbols). In both cases, the DD model curves were obtained after an average of over 100 network realizations.

important test to assess the efficacy of the DD model in reproducing the PIN topology.

In conclusion, we presented a physically motivated dynamic model for the network of protein interactions in biological systems. The model is based on a simple process of gene duplication and differentiation, which is believed to be the main mechanism beyond the evolution of PINs. Although the resulting networks share common features with other SF networks, they present novel and intriguing properties, both in the degree distribution (multifractality) and in their topology. The model reproduces with noticeable accuracy the topological properties of the real PIN of the yeast *S. cerevisiae.*

After submission we became aware of a work treating a model similar to the one introduced here [30].

Acknowledgments

A.M. and A.F. acknowledge funding from Murst Gfin'99. A.V. has been partially supported by the European Network Contract No. ERBFMRXCT980183.

References

1 Uetz PL, et al: A comprehensive analysis of protein-protein interactions in Saccharomyces cerevisiae. Nature 2000;403:623.
2 Wagner A: The yeast protein interaction network evolves rapidly and contains few redundant duplicate genes. Mol Biol Evol 2001;18:1283–1292.
3 Jeong H, Mason SP, Barabási A-L, Oltvai ZN: Lethality and centrality in protein networks. Nature 2001; 411:41.
4 Watts DJ, Strogatz SH: Collective dynamics of 'small-world' networks. Nature 1998;393:440.
5 Barabási A-L, Albert R: Emergence of scaling in random networks. Science 1999;286:509.
6 Strogatz SH: Exploring complex networks. Nature 2001;410:268.
7 Albert R, Barabási A-L: Topology of evolving networks: Local events and universality. Phys Rev Lett 2000;85:5234.
8 Krapivsky PL, Redner S, Leyvraz F: Connectivity of growing random networks. Phys Rev Lett 2000;85: 4629.
9 Dorogovtsev SN, Mendes JFF: Scaling behaviour of developing and decaying networks. Europhys Lett 2000;52:33.
10 Force A, Lynch M, Pickett, FB, Amores A, Yan Y-I, Postlethwait J: The preservation of duplicate genes by complementary degenerative mutations. Genetics 1999;151:1531.
11 Lynch M, Force A: The probability of duplicate-gene preservation by subfunctionalization. Genetics 2000; 154:459.
12 http://www.biotech.nature.com/web_extras.
13 Henikoff S, Greene EA, Pietrokovski S, Bork P, Attwood TK, Hood L: Gene families: The taxonomy of protein paralogs and chimeras. Science 1997;278:609.
14 Tatusov RL, Koonin EV, Lipman DJ: A genomic perspective on protein families. Science 1997;278:631.
15 Ohno S: Evolution by Gene Duplication. Berlin, Springer, 1970.
16 Phizicky EM, Fields S: Protein-protein interactions: Methods for detection and analysis. Microbiol Rev 1995;59:94.
17 von Mering C, Krause R, Snel B, Cornell M, Oliver SG, Fields S, Bork P: Comparative assessment of large-scale data sets of protein-protein interactions. Nature 2002;417:399.
18 Yook S-H, Oltvai ZN, Barabási A-L: Functional and topological characterization of protein interaction networks. Preprint.
19 Huynen MA, Bork P: Measuring genome evolution. Proc Natl Acad Sci USA 1998;95:5849.
20 Dorogovtsev SN, Mendes JFF, Samukhin AN: Multifractal properties of growing networks. Europhys Lett 2002;57:334.
21 Albert R, Barabási A-L: Statistical mechanics of complex networks. Rev Mod Phys 2001;74:47.
22 Dorogovtsev SN, Mendes JFF: Evolution of networks. Adv Phys 2002;51:1079.
23 Vázquez A, Flammini A, Maritan A, Vespignani A: Modeling of protein interaction networks. In preparation.
24 Dorogovtsev SN, Mendes JFF: Evolution of networks with aging of sites. Phys Rev E Stat Phys Plasmas Fluids Relat Interdiscip Topics 2000;62:1842.
25 Krapivsky PL, Redner S: Organization of growing random networks. Phys Rev E Stat Phys Plasmas Fluids Relat Interdiscip Topics 2001;63:066123.
26 Newman MEJ, Strogatz SH, Watts DJ: Random graphs with arbitrary degree distributions and their applications. Phys Rev E Stat Phys Plasmas Fluids Relat Interdiscip Topics 2001;64:026118.
27 Albert RA, Jeong H, Barabási A-L: Error and attack tolerance of complex networks. Nature 2000;406:378.
28 Callaway DS, Newman MEJ, Strogatz SH, Watts DJ: Network robustness and fragility: Percolation on random graphs. Phys Rev Lett 2000;85:5468.
29 Cohen R, Erez K, ben-Avraham D, Havlin S: Breakdown of the internet under intentional attack. Phys Rev Lett 2001;86:3682.
30 Solé RV, Pastor-Satorras R, Smith ED, Kepler T: A model of large-scale proteome evolution. Adv Comp Syst 2002;5:43.

Chapter Five
Applications

In this chapter, we consider some applications of recent discoveries concerning networks to questions of practical significance. The subject matter of the previous chapters has been concerned primarily with understanding what networks look like, how they got to be that way, and how we can construct models of them. All of this, however, should be regarded as just a first step toward the ultimate goal of understanding the function of systems built on networks: the reason for studying the structure of the Internet is to help us understand how computers interact, the reason for studying social networks is to help us understand society, and so forth. Unfortunately, developing an understanding of a complete system is a much harder task than simply looking at the underlying network. Our studies of the function of networked systems are much less well developed than our studies of network structure. There is, however, one particular area in which a number of important breakthroughs have been made, and that is in the study of the propagation of things over networks, such as disease or information or rumors. Broadly construed, this is the topic of all of the papers reproduced in this chapter.

5.1 EPIDEMICS AND RUMORS

There is more than one way in which something could spread across a network. For example, we could have a conserved fluid of some kind propagating through a network of pipes or wires, so that the total amount of fluid, summed over the whole network, would be constant. This type of propagation might be described by a diffusion equation or some other mass-conserving equation of motion. Diffusive propagation in the Watts–Strogatz small-world model, for instance, has been studied by Monasson (1999) and by Pandit and Amritkar (2001). However, conservative flow is not usually appropriate for modeling information propagation because most kinds of information are not conserved. Ideas are not like apples. If you have an apple and you give it to a friend, you no longer have an apple. If you have an idea and you give it to a friend, you still have the idea. Diseases are the same: if you give the flu to someone, that doesn't mean you no longer have it.

One of the simplest models of the spread of nonconserved information on a network is one in which an idea or a disease starts at a single vertex and spreads first to all the neighboring vertices of that one. Then from those vertices it spreads to all of their neighbors, and so forth, until there are no accessible vertices left that have not been "infected." This process is known in the computer science literature as breadth-first search. Numerical algorithms based on it are used, among other things, for finding the vertices in a connected component of a graph: one simply chooses a starting vertex then performs breadth-first search until there are no more uninfected vertices accessible. The final set of infected vertices is precisely the component to which the starting vertex belongs. Breadth-first search can be implemented in a very few lines of computer code using a first-in/first-out buffer, or queue, and runs to completion in time proportional to the number of edges in the component.

In Chapters 3 and 4 we saw that typical networks are composed of a large number of small components plus, optionally, a single giant component that fills an extensive portion of the network. In many models there is a phase transition with some parameter or parameters of the system at which the giant component appears. If we simulate the spread of a rumor or disease on such a network using breadth-first search, either we will see a small outbreak corresponding to one of the small components, or, if there is a giant component and we happen to start our breadth-first search within it, we will see a giant outbreak that fills a large portion of the system. The latter is precisely what we call an "epidemic" when we are talking about disease, and the phase transition at which the giant component appears in a graph is also a phase transition between a regime in which epidemics are not possible and a regime in which they are. This epidemic transition has been well studied by mathematical epidemiologists for many decades (Bailey 1975; Anderson and May 1991; Hethcote 2000). Note that being "above" the epidemic transition, in the region where epidemics can occur, does not guarantee that one will occur. Even above the transition there are typically still many disconnected small components in a graph, and an outbreak starting with one of these will give rise to only a few cases of the disease before the component is exhausted and the disease fizzles out. In fact, if the disease starts its breadth-first search at a randomly chosen vertex in the network, then the probability of seeing an epidemic is precisely equal to the fraction of the graph occupied by the giant component.

It is believed, however, that the social network of the world does not have any small components to speak of. Virtually everyone in the world, with very few exceptions if any, is connected to the giant component of the world's network. Does this mean that every disease outbreak will affect everyone in the world? Clearly not. The reason is that exhaustive breadth-first search is not a very good model for the spread of disease (or rumors or news or computer viruses either). It is not the case that a disease spreads with 100% certainty to all the neighbors of an infectious vertex. In general, there is only some finite chance of any given contact been two people resulting in infection. Furthermore, some people may be more susceptible to infection than others, some may be more effective at passing the disease on than others, and these things may vary with time or be correlated between one individual and another. Any serious epidemiological model will take at least some of these things into account.

Traditional models in epidemiology have been particularly good at taking time variation into account, along with such things as age structure in populations, population turnover, or reinfection (ability to catch a disease more than once). However, traditional epidemiological models almost all make use of the so-called "fully mixed" approximation, which is the approximation that transmission is equally likely between any pair of individuals in a population or subpopulation. This is equivalent to saying that within a population the pattern of contacts is a fully connected graph—there is an edge between every vertex and every other. More sophisticated models divide the population into subgroups of some kind (e.g., by age) and stipulate different degrees of infectivity within and between groups. Within each group, however, the fully mixed approximation is still assumed. Given what we have seen of real-world networks in the last few chapters, this is obviously a problematic assumption, and it is this assumption that the papers in this section attempt to move away from. By studying, either analytically or numerically, the behavior of various epidemiological models on networks, the authors hope to get a better idea of how real diseases will behave, and in some cases they have found entirely new behaviors that had not previously been observed in epidemiological models.

Ball, Mollison, and Scalia-Tomba (1997)

Our first paper on this subject deals with the spread of disease on networks, and is notable as one of the first to treat such problems mathematically with precision. In this paper, Frank Ball and collaborators consider the spread of disease on a certain limited class of networks, but their ideas and techniques have much broader applicability, and have been applied to more general network models by others. The networks considered by **Ball *et al.* (1997)** are networks with "two levels of mixing," meaning that each vertex in the network belongs both to the network as a whole and to one of a specified set of subgroups within the network. In epidemiological terms, one could think of the network as being divided, for example, into families, with each person belonging to just one family. Disease transmission within families is assumed to take place much more easily (i.e., with higher probability) than transmission between families.

Disease spreading is modeled using the susceptible/infective/removed (SIR) model. This is one of the most fundamental of epidemic models, applicable to diseases that occur primarily in brief outbreaks and that confer upon their hosts at least temporary immunity to catching the same disease again. Many common bacterial or viral infections fall into this class. In the model, people can be in one of three states: susceptible (S), meaning they can catch the disease but haven't yet; infective (I), meaning they have caught the disease and can pass it on to others; and removed (R), meaning they have recovered from the disease and can neither pass it on nor catch it again, or they have died.[1] The only allowed transitions in the model are from S to I and from I to R, the first taking place with a certain probability per

[1] In common parlance we might refer to the I stage as "infectious" rather than the more technical "infective." Many people also refer to the R stage as "recovered," although "removed" is more general, encompassing both recovered and deceased individuals. SIR models are also similar in some ways to models of excitable media, and the R stage is sometimes called "refractory" in acknowledgment of this similarity.

unit time if a susceptible individual has a network neighbor who is infective, and the second taking place at a stochastically constant rate once one becomes infected, regardless of the states of one's neighbors. The number of susceptible individuals in the network decreases monotonically in the model—there is no mechanism for making new susceptible individuals—so the disease always dies out in the end (although the number of susceptible individuals does not necessarily tend to zero).

Normally, the SIR model is studied in the fully mixed approximation, which, as discussed above, assumes equal probability of contact between all members of a population group (Bailey 1975). Epidemiologists certainly appreciated the importance of network structure in the propagation of disease before the work of Ball *et al.* (see, for instance, Sattenspiel and Simon (1988) or Longini (1988)), but previous attempts to incorporate it into calculations mostly took the form of "patching up" the standard fully mixed models, rather than employing a fundamentally network-based approach. For example, one might divide the population into groups of some kind—perhaps representing families—but then have the disease spread in a fully mixed fashion both within and between those groups, so that the progress of the disease can be represented using differential equations. The work of Ball *et al.* departs from this by abandoning differential equations in favor of a framework based on probability generating functions similar to those we saw in Section 4.1. Probability generating functions had been employed in epidemiological models before (see Kretzschmar and Morris (1996), for instance), but the paper by Ball *et al.* represents a particularly lucid and effective example of their use.

The primary results of Ball *et al.*'s paper can be summarized as follows. They find that the rapid spread of a disease within groups such as families can lead to epidemic outbreaks in the population as a whole, even when the probability of interfamily communication of the disease is low enough that epidemic outbreaks normally would not be possible. The reason for this is the following. If transmission between family members takes place readily enough that most members of a family will contract the disease once one of them does, then we can regard the disease as spreading on a "supernetwork" in which vertices are families, not individuals. Roughly speaking, spread of the disease between families will take place with n^2 times the normal person-to-person probability, where n is the number of people in a family. This increase in the probability of transmission can be enough to push an otherwise nonepidemic disease into the epidemic regime. Using generating functions, the authors show exactly where the threshold for transmission of the disease falls, and also how many people are expected to be infected if an epidemic does occur, for arbitrary distributions of the sizes of family groups.

The paper of Ball *et al.* is also notable in that the authors introduce and analyze a model, which they call the "great circle model," that is remarkably similar to the small-world model introduced by **Watts and Strogatz (1998)** the following year and discussed in Section 4.2. Many of the analytic results of Ball *et al.* concerning their great circle model have been independently rediscovered and extended in the physics literature, although unfortunately Ball *et al.* rarely receive their due credit.

Keeling (1999)

The work of Ball *et al.* effectively provides a solution for the behavior of SIR-type epidemics in the small-world model, which, as described in detail in Section 4.2, is a model of a social network that introduces clustering (or network transitivity) into the random graph in a controlled way. An alternative approach to calculating the effect of clustering on SIR epidemics has been presented by **Keeling (1999)**. (Keeling does not refer to the clustering as "clustering," although he cites Watts and Strogatz, who introduced the term. He does, however, define a parameter ϕ that is precisely the clustering coefficient, in the form used by **Newman *et al.* (2001)**.) In his paper Keeling considers a more traditional differential-equation-based approach to the question, which in fact ignores most of the network features of the problem. What Keeling's method does is to include, in approximate form, the effect of the short-scale structure of the network—the clustering—but treat everything else using a standard fully mixed approximation. Thus things like the effect of degree distributions are absent from his calculations. But the effect of clustering is made very clear.

The standard fully mixed SIR epidemic model assumes that the rate at which the fraction S of individuals in the population that are in the susceptible state decreases is proportional to the number I of infective individuals from whom the susceptibles can catch the disease. Thus

$$\frac{\mathrm{d}S}{\mathrm{d}t} = -\beta SI, \tag{5.1}$$

where β is a constant representing the time rate of formation of new contacts capable of transmitting the disease between individuals. Similar equations can be written down for $\mathrm{d}I/\mathrm{d}t$ and $\mathrm{d}R/\mathrm{d}t$. Keeling takes this idea one step further. He denotes by $[AB]$ the density of pairs of individuals who are joined by an edge in the network and who have states A and B, where $A, B \in \{S, I, R\}$. Then these densities obey differential equations such as

$$\frac{\mathrm{d}[SS]}{\mathrm{d}t} = -2\tau[SSI]. \tag{5.2}$$

Allowing for symmetries and conservation laws, five such equations are necessary to specify the dynamics of the entire system. All the equations involve densities of connected triples $[ABC]$ on the right-hand side, and these densities of triples depend on the clustering of the network. Keeling shows that one can approximate these densities in terms of the pair densities $[AB]$ if one also knows the clustering coefficient, and so one can derive a closed set of approximate differential equations for the time evolution of the pair densities as functions of various rate parameters and the clustering coefficient. Among other results, Keeling uses his equations to show that as the clustering coefficient is increased, we move below the epidemic threshold of the model into the nonepidemic regime. The common-sense explanation for this is that a high clustering coefficient means many short loops in the network. As a disease spreads in such a network it repeatedly encounters individuals who already have the disease and so cannot catch it again. Thus the rate at which the disease infects new victims is lower than if all individuals encountered were susceptible, and this lowering of the rate can push us into the nonepidemic regime. (A similar result has recently been shown for disease spread in a true network model by Newman (2003c).) Similarly Keeling finds that a lower fraction of the population need be vaccinated against a disease to prevent an epidemic if the clustering coefficient is high.

Kuperman and Abramson (2001)

This paper and the following one **(Pastor-Satorras and Vespignani 2001)** address the behavior of endemic diseases spreading on networks with topology defined by two recently proposed network models—the Watts–Strogatz small-world model (see Section 4.2) and the Barabási–Albert scale-free model (see Section 4.3).

Real diseases do not, as in the SIR model, simply die out after a single outbreak. If they did, they wouldn't exist any more. Real diseases persist, for one of two reasons. Either they rely on population turnover (births and deaths) to provide a continual supply of new victims to infect, or they rely on past victims losing their acquired immunity after a certain amount of time, so that they can be infected again.[2] Many childhood diseases, such as chicken pox and measles, fall into the first category, as does HIV. The second category, which is the one considered by Kuperman and Abramson, includes infections that undergo rapid evolution in order to evade the immune system. Influenza and the common cold are two obvious examples. The standard model for diseases of this latter type is the SIRS model, which, like the SIR model discussed above, divides hosts into susceptible, infective, and removed categories. The difference is that individuals in the removed category move back into the susceptible category after a certain amount of time, so that the population is "recycled" by the disease:

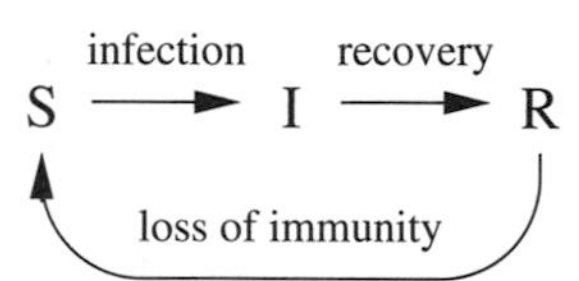

This model is similar in its general behavior to reaction-diffusion processes such as the Belousov–Zhabotinsky reaction (Zhabotinsky 1991; Strogatz 1994; Winfree 2000). In those processes, systems also pass from an initial "susceptible" state into an "infected" one, before relaxing into a refractory state from which they cannot be infected. The refractory state ends, however, after some fixed time and the process cycles around. On regular lattices, reaction-diffusion processes can give rise to beautiful patterns often characterized by spiral waves (Winfree 1974, 2000). Epidemiological studies are more often carried out in fully mixed or random graph settings, where spiral waves are not seen, although there may be systemic oscillations in some regimes (Cooke *et al.* 1977).

[2]A third, rare possibility is that a disease is always (or mostly) cured by medical intervention so that its victims do not acquire immunity and can be infected again immediately. Gonorrhea is an example of this type of infection.

Like **Keeling (1999)**, **Kuperman and Abramson (2001)** were interested in the effect of clustering on the propagation of disease and performed simulation studies of the SIRS model on networks with the topology defined by the small-world model of Section 4.2, which has strong clustering. (They used the original version of the model, in which links are "rewired," rather than the version in which extra shortcuts are added.) Their principal finding was that the model shows two distinct classes of behavior as the parameter p denoting the shortcut density is varied. For low values of p, there is a persistent low-level endemic infection, with minor statistical fluctuations in the fraction of individuals carrying the disease. For higher p the model develops clear almost-periodic oscillations in the number of infectives similar to those seen in SIRS models in a fully mixed environment (Cooke *et al.* 1977; Girvan *et al.* 2002). The interesting thing is that there appears to be a sharp transition between the two regimes in the limit of large system size, which occurs at a finite value of p. (The sharpness cannot be proved, since the results are only numerical, but the simulations look convincing, with the sharpness of the transition clearly increasing as system size gets larger.) Recall from Section 4.2 that most properties of the small-world model show a phase transition at $p = 0$, but nothing special at finite values.

The reason for the sudden appearance of synchronized oscillations as p increases is not clear. Kuperman and Abramson suggest that it may be associated with the decrease in the clustering coefficient as p gets larger, which may reduce the number of independently synchronized regions in the lattice, allowing the lattice as a whole to synchronize. For the moment, however, this must be regarded as conjecture; we do not yet have a good theoretical explanation for the observations of Kuperman and Abramson.

Pastor-Satorras and Vespignani (2001)

In this paper, **Pastor-Satorras and Vespignani (2001)** address the behavior of an endemic disease model on networks with scale-free degree distributions, of the type discussed in Section 4.3. Their motivation for the work was an interest in the dynamics of computer virus infections, rather than diseases of humans or animals, which is why they look at scale-free networks; computer viruses spread over the Internet and the Internet has a scale-free form, as demonstrated by **Faloutsos *et al.* (1999)**.[3]

Pastor-Satorras and Vespignani use a simplified version of the SIRS model of the previous section to model their computer viruses—the SIS model, in which there is no refractory period during which a computer cannot contract the virus. Computers in this model pass straight from the infective state back to the susceptible state. In their work Pastor-Satorras and Vespignani grow networks according to the

[3] It has been pointed out (Lloyd and May 2001; Balthrop *et al.* 2004) that the effective network that many viruses spread over is not the Internet but is instead a social network of friends and acquaintances who use the Internet. For example, Pastor-Satorras and Vespignani consider macro viruses, which spread in email attachments, often by searching the email address books of users of the computers that they infect. The structure of the network formed by the addresses in an address book is essentially unconnected with the physical structure of the Internet. However, some such networks appear to be scale-free, so the work of Pastor-Satorras and Vespignani is of interest nonetheless.

scale-free model of **Barabási and Albert (1999)** described in Section 4.3 and then simulate the SIS model on this network, starting with some fixed initial number of infective computers.

In contrast with the results of Kuperman and Abramson discussed above, Pastor-Satorras and Vespignani do not find oscillations in the number of infected individuals for any value of the independent parameter of the model. (There is only one important parameter, which measures the ratio of how long computers stay in the infective state to how likely they are to infect others while they are there.) However, they do observe another striking effect. No matter what value the parameter takes, the system is always in the epidemic regime; there is no epidemic threshold in this system. No matter how short a time computers spend in the infective state or how little they pass on the virus, the virus remains endemic. They also provide a mean-field treatment of their model that confirms the same finding, showing that the average fraction of the population infected at any one time decreases exponentially with the infectiousness of the disease, but never reaches zero. In a commentary on this paper, Lloyd and May (2001) later pointed out that in general there is expected to be no epidemic threshold on a network whose degree distribution has a divergent second moment, as the degree distribution in the Barabási–Albert networks does.

One of the nice features of the paper by Pastor-Satorras and Vespignani is that in addition to their theoretical arguments they look at actual data for real computer viruses. Drawing on data for viruses of three different types (boot, file, and macro viruses) over a 50-month period, they calculate survivorship as a function of time for the starting cohort of different viruses—that is, they calculate the fraction of the cohort that remains in circulation after a given interval. They find that this number decays exponentially, but with quite a long time constant (7 months for file viruses and about 14 months for others), so that even after two or three years a moderate number of viruses survive. Since this is much longer than the supposed time-scale for the spread of viruses from one computer to another, they conclude that some viruses must, by some mechanism, be remaining in the system at a low endemic level of infection for long periods of time. As they point out, this is an unlikely occurrence in a system with a normal epidemic threshold. It would imply that the system was coincidentally tuned to a point just fractionally above the threshold. In a system with no epidemic threshold, however, it is not unlikely at all: it simply means that the viruses in question have a low intrinsic infectiousness.

Watts (2002)

A slightly different slant on the spread of influences across networks is given by **Watts (2002)**, who has looked at the behavior of cascading processes. Unlike disease, the spread of some kinds of information, such as rumors, fashions, or opinions, depends not simply on "susceptible" individuals having contact with "infective" ones, but on their having contact with such individuals in sufficient numbers to persuade them to change their position or beliefs on an issue. The rise from obscurity to wild popularity of books by unknown authors or sleeper movies or teenage fashions seems to be driven by a dynamics of this sort. People have a threshold for adoption of trends. When they see a sufficient number of others wearing a particular style, or hear from enough of their friends that a particular book is good, they will jump on the bandwagon themselves. Watts proposes a model of this process in which individuals are represented by the vertices of a network and social contacts of one kind or another between individuals are represented by edges. Each individual has a threshold ϕ for adoption of the trend being modeled, which is chosen at random from a specified distribution. When the fraction of a person's contacts who have themselves adopted the trend rises above this threshold, they adopt it also. This gives rise to cascading behavior in the network: once a sufficient number of people have adopted the trend it is virtually guaranteed that most of the rest will also, giving a kind of "avalanche" of adoption. Similar behavior is seen in some physical models also, such as bootstrap percolation (Adler 1991) and the zero-temperature random-field Ising model (Sethna *et al.* 1993).

Watts gives an exact solution for his model on random graphs for the case where initially a low density of vertices has adopted the trend. The solution depends crucially on the presence in the network of individuals who have very low thresholds ϕ. In particular, there must exist in the network a sufficient density of individuals whose thresholds are so low that they will adopt the trend if only a single one of their neighbors does. For these people, the trend really does act like a disease: they can catch it from only a single infective neighbor. As Watts argues, the trend will only propagate and cause a cascade if the density of these individuals is high enough that they form a percolating subgraph in the network. Then, using generating function methods similar to those of **Callaway *et al.* (2000)**, he derives the parameter values at which this happens. The fundamental result of this analysis is that, as a function of the average degree z of a vertex in the graph, there are two different thresholds for cascading spread of information. Below $z = 1$, no cascades happen because the network itself has no giant component, as we described in our discussion of random graphs in Section 4.1. In some sense, this threshold is trivial, since it is a property solely of the graph and not of the cascade process. The other threshold, however, is nontrivial. Cascades also cease occurring when z is large enough, the exact value depending on the chosen distribution of the variables ϕ. The reason for this upper threshold is that as z becomes large, the value of ϕ that a vertex needs to have if it is to adopt the trend when only a single one of its neighbors does becomes small, and hence there are fewer such vertices. For large enough z these vertices fail to percolate and so cascades stop happening.

Interestingly, there is evidence that the cascading spread of trends really is driven primarily by people with low thresholds for adoption. In the literature on the subject these people are called *early adopters* (Gladwell 2000). We all know a few of these people (or maybe you are one yourself). These are the people who latch onto the new fashions months or years before the rest of us; they are the ones who have the newest gadgets in their homes, the ones who tell us about the hot new books or music before anyone else. The rest of us belong to the *early majority* or the *late majority*, those in the middle of Watts's distribution of ϕ, or in its tail. Watts's model therefore reflects in a mathematical way some home truths of sales and advertising that company executives and political scientists have been pondering for many years.

5.2 ROBUSTNESS OF NETWORKS

A question that has been the focus of considerable recent research is the following. If we have information or disease propagating through a network, how robust is the propagation to failure or removal of vertices? The Internet, for example, is a highly robust network because there are many different paths by which information can get from any vertex to any other. Even if an Internet router fails (routers are the high-speed computers that form the "backbone" of the Internet), the system is capable of rerouting data around the blockage and making sure it gets to its destination. This kind of robustness to failure was the subject of one of the earliest theorems in graph theory. In 1927 Karl Menger proved a remarkable result. Two different paths through a network that connect the same two vertices are described as *node-independent* if they have no vertices in common other than the vertices at their ends. Menger showed that the number of node-independent paths between two vertices is always exactly equal to the minimum number of other vertices in the network that must fail in order for those two vertices to become disconnected from one another (Menger 1927). Although at first sight this may appear to be a trivial result, it is actually quite tricky to prove. The problem is that the node-independent paths are not in general unique—there can be more than one way of choosing them so that they connect the two vertices of interest and have no other vertices in common. A reasonably straightforward proof of Menger's result is given by Harary (1995).

When we look at a network like the Internet, however, we are not usually concerned about the connection between any two particular vertices. You personally may scream your lungs out if you are unable to connect to your favorite Web site today, but the world at large probably won't care very much. In fact, it turns out that a small percentage of the routers on the Internet are always nonfunctional at any moment. What's important in the big picture is that the Internet as a whole keep on functioning despite this endemic level of failure. As a guide to this kind of average behavior of the network, Menger's theorem is not very useful. Recent work has therefore focused instead on the following question: if a certain fraction of the vertices in a network fail, what effect will that have on the connection between a typical pair of vertices in the remaining network? The question can also be rephrased in terms of disease. If a certain fraction of all the people in a network are removed in some way from a network—by immunization against disease, for instance—what effect will this have on the spread of the disease? As we will see, if vertices are removed at random the effect is usually very slight; random removal is normally a poor way of immunizing a population against a disease. If, however, the removed vertices are chosen carefully, rather than at random, quite the reverse may be true.

The next three papers deal with the question of the resilience of networks to the removal of their vertices. The first of the three, by Reka Albert and collaborators, was the first to discuss the idea. The latter two present analytic theories of network resilience that confirm the numerical results of Albert *et al.*

Albert, Jeong, and Barabási (2000)

Albert *et al.* discuss network resilience for two specific types of model networks, random graphs (see Section 4.1) and scale-free networks, meaning graphs with power-law degree distributions, specifically those grown using the model previously proposed by Barabási and Albert (see Section 4.3). The principal conclusion of the paper is that the scale-free networks are substantially more robust to the random deletion of vertices than Erdős–Rényi random graphs, but substantially less robust to deletion that specifically targets those vertices with the highest degrees.

Albert *et al.* study two specific properties of graphs in this paper: mean vertex-vertex distance and size of the largest component. (The former they refer to loosely as the "diameter" of the network, although strictly the diameter is the *maximum* vertex-vertex distance, rather than the mean.) They study each of these separately for Erdős–Rényi random graphs and scale-free networks as vertices are randomly deleted from each. The calculations are numerical and show clearly that the scale-free network is more resilient. The mean vertex-vertex distance in the scale-free network increases as vertices are deleted, but does so much more slowly than in the random graph. Thus if vertex-vertex distance is regarded as a measure of the average quality of the connection between vertices, then most people will have better connections on average in a scale-free network, if vertices fail at random. Similarly, the size of the largest component goes down substantially more slowly as vertices are deleted in the scale-free network than in the random graph. If the size of the largest component is equated with the number of people in the network who can use the network effectively (for communication, say), then again this implies that the scale-free topology is more desirable. Note, however, that the paper examines resilience to node removal for scale-free and random networks only, leaving open the question of the resilience of other network structures. Indeed, more recent work has shown that there exist networks with even higher resilience than scale-free networks (Costa 2004; Rozenfeld and ben-Avraham 2004; Tanizawa *et al.* 2005).

When we turn to removal of the vertices with the highest degree in the network, however, the picture is reversed. Albert *et al.* find that the scale-free network proves vulnerable to deletion of high-degree vertices, by contrast with the random graph, whose behavior is almost identical whether one deletes the highest-degree vertices or vertices chosen at random. Thus the scale-free networks are highly robust to random failure, but highly fragile to targeted attacks.

Taking their simulations one step further, Albert *et al.* also study the behavior of actual real-world networks under random or targeted deletion of vertices. Starting with a map of the Internet at the level of "autonomous systems" (i.e., groups of computers, see Chapter 3), and a map of a small section of the World Wide Web, they again simulate the deletion of vertices and find results very similar to those for the model systems. Both networks are known to have power-law degree distributions (**Faloutsos *et al.* 1999**; **Albert *et al.* 1999**; Kleinberg *et al.* 1999; **Broder *et al.* 2000**; Chen *et al.* 2002) and their behavior closely matches that of the scale-free model networks, with typical vertex-vertex distances and largest component sizes that vary little upon random deletion of nodes, but greatly upon targeted deletion.

Albert and coworkers have gone on to apply their ideas about network robustness to a number of other systems, particularly metabolic networks (**Jeong *et al.* 2000**) and protein networks (Jeong *et al.* 2001).

Cohen, Erez, ben-Avraham, and Havlin (2000)

Our second and third papers on the robustness question extend the ideas of **Albert *et al.* (2000)** discussed above by providing mathematical proofs of the results that they discovered by computer simulation. **Cohen *et al.* (2000)** study the case of random (not targeted) deletion of vertices on graphs with arbitrary degree distributions. They study random graphs with arbitrary degree distributions, of the kind discussed in Section 4.1. This differs slightly from the approach of Albert *et al.*, who studied graphs grown according to the growth model of **Barabási and Albert (1999)**. Random graphs and grown graphs are known to have measurable statistical differences, even though they may have the same degree distributions (Callaway *et al.* 2001). As we will see, however, as far as random vertex deletion goes, the properties of the two appear very similar.

The calculation of Cohen *et al.* is exact, and relies on the observation that if we delete a fraction p of the vertices from a random graph, chosen uniformly at random, then the resulting graph is still a random graph: if vertices were connected together at random before the deletion, then the remaining vertices are still connected at random after it. The degree distribution may change however. The authors show that if the probability of having degree k before the deletion is $P(k)$, then afterward it will be

$$P'(k) = \sum_{k_0=k}^{\infty} P(k_0) \binom{k_0}{k} (1-p)^k p^{k_0-k}. \tag{5.3}$$

Combining this result with Molloy and Reed's expression, Eq. (4.4), for the position of the phase transition in a random graph with arbitrary degree distribution (**Molloy and Reed 1995**), Cohen *et al.* thereby show that the transition on their graphs takes place when the fraction p of deleted vertices takes the critical value

$$p_c = 1 - \frac{1}{\langle k_0^2 \rangle / \langle k_0 \rangle - 1}, \tag{5.4}$$

where $\langle k_0 \rangle$ and $\langle k_0^2 \rangle$ are the mean and mean square degrees in the degree distribution of the original network.

The phase transition in this model corresponds to the point at which so many vertices have been deleted from the network that no giant component exists any more. If we consider the case of a communications network in which those vertices not in the giant component are cut off from communication with the bulk of the network, then the phase transition is also the point at which effective communication is no longer possible for anyone. Certainly this is a point which we would not want our communications networks to reach.

Cohen *et al.* test their results using a power-law degree distribution $P(k) \sim k^{-\alpha}$. Their findings confirm the resilience seen in power-law networks by Albert *et al.* For values of the exponent α above 3, they find that there is a phase transition for $p < 1$, but for $\alpha \leq 3$ the giant component never disappears from the network no matter how many vertices one deletes, at least in the limit of large network size. They also confirm this by numerical simulations of the same system in which they explicitly measure the size of the largest component and show that as system size becomes large it does indeed persist as p approaches 1.

Callaway, Newman, Strogatz, and Watts (2000)

Callaway *et al.* (2000) extend the calculations of **Cohen *et al.* (2000)** for random graphs with arbitrary degree distributions, showing that it is possible to calculate exactly not only the threshold for the phase transition but also the size of the giant component and the average size of non-giant components. They also generalize these calculations to the case where probability of deletion of a vertex is any arbitrary function of vertex degree, studying in particular the case where deletion probability is 1 above some maximum degree and 0 below it, which is equivalent to the deletion of the highest degree vertices. Thus their calculations provide exact solutions for both the random and targeted deletion of **Albert *et al.* (2000)**.[4]

Callaway *et al.* viewed the resilience question as a form of percolation model in which vertices on a network are occupied or not to represent whether they are functional or failing, and solved for the resulting model's properties using a generating function formalism similar to that introduced by Moore and Newman (2000b) for percolation in the small-world model. They defined two fundamental generating functions

$$F_0(x) = \sum_{k=0}^{\infty} p_k q_k x^k, \qquad F_1(x) = \frac{\sum_k k p_k q_k x^{k-1}}{\sum_k k p_k}, \tag{5.5}$$

where p_k is the probability that a vertex has degree k, and q_k is the probability that a vertex with degree k is occupied. They then showed that the average component size, the giant component size, and the position of the phase transition threshold can all be expressed simply in terms of these functions. For example, the size S of

[4]The calculations of Callaway *et al.* and Albert *et al.* are not exactly equivalent for two reasons. First, like those of Cohen *et al.*, the calculations of Callaway *et al.* are for random graphs with arbitrary degree distributions, whereas those of Albert *et al.* are for the growing graph model of **Barabási and Albert (1999)**. Second, the deletion of vertices was done in slightly different ways in the two papers. Albert *et al.* performed an iterative deletion in which the highest-degree vertex in the graph is found and deleted, and then the degrees of vertices are recalculated, allowing for edges that have disappeared because of the deletion. The calculation of Callaway *et al.* on the other hand is equivalent to the deletion of those vertices that have the highest degrees in the initial graph. As we will see, however, neither of these differences seems to affect the qualitative outcome in any noticeable way.

the giant component is a solution of the equations

$$S = F_0(1) - F_0(u), \qquad u = 1 - F_1(1) + F_1(u). \tag{5.6}$$

For the special case of random deletion of vertices, where $q_k = q$ is independent of k, this leads to a value q_c for the critical fraction of vertices that need to be deleted to destroy the giant component which is precisely equivalent to the result of **Cohen *et al.* (2000)**, Eq. (5.4). Callaway *et al.* also looked at the question of targeted attack, in which high degree vertices are removed first. Setting $q_k = \theta(k_{\max} - k)$, so that all vertices with degree $k > k_{\max}$ are deleted, they calculated the size of the resulting giant component exactly in the limit of large graph size. As Cohen *et al.* did, they also performed explicit simulations that confirm their results.

Their findings were in good agreement with the results from the numerical studies of **Albert *et al.* (2000)**, showing that for graphs with power-law degree distributions, the giant component size dwindles very rapidly as the highest-degree vertices are removed. For a power-law distribution with exponent 2.7—a typical value for the Web graph—they found that only about 1% of the highest degree vertices need be deleted to destroy the giant component completely, equivalent to destroying all long-range communication on a communication network with the same topology.

5.3 SEARCHING NETWORKS

The final issue addressed in this chapter is the issue of search on networks. Many networks contain information stored at their vertices. The World Wide Web is a good example—its pages are vertices in a network but they are also stores of information, which is their primary reason for existing in the first place. Citation networks of academic papers (**Price 1965**; Egghe and Rousseau 1990; Redner 1998) have a similar structure, with information-bearing papers being linked to one another via bibliographical references. Some computer networks, particularly distributed databases such as digital libraries, and peer-to-peer (P2P) networks such as file-sharing networks, also take the form of graphs with information stored at their nodes. Algorithms for finding information in networks such as these are of obvious technological importance, but initial attempts to create such algorithms tended to rely on rather old-fashioned techniques that ignored the network nature of the problem. The earliest Web search engines, for example, simply constructed catalogs of every Web page they could find, and then performed searches using standard text matching and relevance criteria. Similarly, the early methods for searching peer-to-peer networks either used centralized indexes of content, or used brute-force network traversal (breadth-first search, for example) to scan all vertices in real time, a technique that scales poorly with system size. In recent work, more intelligent search schemes have been proposed that take into account the structure of the network, and may give much better performance. The three papers in this section all deal with schemes of this nature.

Kleinberg (1999)

The archetypal example of search on a network is the problem of searching for information in the World Wide Web (Lawrence and Giles 1999). At the time of writing, the largest catalogs of the Web record more than 8 billion separate pages of information, and the number is growing all the time. Searching such a large and distributed dataset is a formidable task, but a worthwhile one. The Web would be virtually useless without search engines.

The first step in creating a Web search engine is to "crawl" the Web using a computer program called, naturally, a crawler. A crawler starts from a given Web page and iteratively follows links from that page to find some or all of the pages that can be reached from it. Essentially the crawler performs a breadth- or depth-first search on the Web graph (see Section 5.1). Each page encountered is then indexed: the text it contains is split into words, and lists are created of which words appear in which pages. Then when one wants to know which pages contain a given word or set of words, one simply retrieves the relevant index (or intersection of indexes for a group of words). Ranking pages by frequency or prominence of the words of interest provides a guide to which pages are most likely to yield useful information. Of course there are many practical details that make real search engines far more complicated than what we describe here, but in essence this is the method used by most of the early engines.

As mentioned above, however, this method is unsatisfactory. For example, it fails badly when the pages you most want to find do not contain the words you are looking for, or fail to feature them prominently or frequently. To borrow an example from **Kleinberg (1999)**, the chances are that the home pages of Honda or Toyota do not prominently feature the words "automobile manufacturer," so that a search for these words is unlikely to turn up these pages. And even if the correct pages are found by the simple word search, the ranking of results according to word frequency or prominence may well place the best pages far down the list after pages on which the words occur more often but which otherwise contain little information of use.

Search engine performance can, however, be dramatically improved if we recall that the Web is not just a set of pages—it is a network. There is considerable information stored in the structure of the hyperlinks between pages on the World Wide Web which can tell us which pages are most likely to contain useful information. At the simplest level, it is reasonable to assume that people will create links to pages they find useful more often than to useless ones. Thus the number of links that point to a page should give us a simple measure of the page's importance. However, we can do better than this. Not all links are of equal value. A link from a page that is itself important probably carries more weight than a link from a page that is useless and irrelevant. This idea forms the basis for a Web search scheme proposed by Brin and Page (1998), which goes like this.

Given a set of n pages on a given topic, denoted by $i = 1, \ldots, n$, we give each one a nonnegative weight x_i. We wish to choose these weights so that pages have high weight if they are linked to by many other high-weight pages, and low weight if they are linked to by few others or by pages with low weight. To achieve this, we make the weight for each page proportional to the sum of the weights of the pages that link to it. If we define the adjacency matrix **A** for the (directed) link graph of the pages with elements

$$A_{ij} = \begin{cases} 1 & \text{if there is link from } j \text{ to } i, \\ 0 & \text{otherwise,} \end{cases} \tag{5.7}$$

then $x_i = \lambda^{-1} \sum_j A_{ij} x_j$ for some $\lambda > 0$, or in matrix form

$$\lambda \mathbf{x} = \mathbf{A} \cdot \mathbf{x}, \tag{5.8}$$

where $\mathbf{x}$ is the vector whose elements are the x_i. The weights defined in this way are called the *eigenvector centralities* of the vertices (Wasserman and Faust 1994; Scott 2000), since $\mathbf{x}$ is an eigenvector of the adjacency matrix with eigenvalue λ. Provided the link graph of our set of pages is connected (there are no separate components), there is only one eigenvector that makes all the weights nonnegative and this is the unique eigenvector corresponding to the largest eigenvalue,[5] which can be found trivially by repeated multiplication of the adjacency matrix into any initial trial vector that has some component in the direction of the leading eigenvector. The popular Web search engine Google uses a scheme called "PageRank" which is similar to this (although with significantly more bells and whistles) and appears to work well.[6]

The main question we have not addressed is how one picks the initial starting set of pages. One wants a set that includes most of the pages that might be relevant to the topic of interest and excludes most others, in the interests of keeping down the size of the set. A reasonable set size to work with, taking the power of current computing resources and the sparsity of the Web into account, is about 10 000. One could simply take the 10 000 highest-ranked results returned by a normal text-based Web search for the words of interest. However, as pointed out above, the best pages matching our query may not even contain those words. So instead the usual strategy is to take a smaller number of the highest-ranked results from a text search—typically on the order of 100 or so—plus all the pages that those pages point to. Optionally one may also include some of the pages that point to the 100. The resulting set normally seems to include most of the pages of interest. Even if the words "automobile manufacturer" do not appear anywhere on the Honda home page, the chances are good that they will appear on another page that points to Honda's. Another way to create the starting set of pages is to give an initial page that one is interested in, such as `www.honda.com`. Then one can add to that page all other pages that it points to or that point to it (and possibly farther out than that, depending on the kind of set required), and then proceed as before.

[5]This is a result of the Perron–Frobenius theorem (MacCluer 2000).

[6]Two important differences in the scheme used by Google are that links are given a weight that goes inversely as the out-degree of the referring vertex, so that vertices that point to a lot of others count for less than those that point to only a few, and each vertex is given a certain base weight regardless of its incoming links, so that even vertices with no links have a nonzero score (Brin and Page 1998).

The first paper reproduced in this section, by Jon Kleinberg, gives a clear presentation of the ideas behind graph-based Web search algorithms of this kind, and also takes the ideas of Brin and Page one step farther (**Kleinberg 1999**). Kleinberg notes that, while being pointed to by another important page will certainly lend a page importance itself, it is also the case that pointing to an important page can lend you importance. If you point to a lot of pages that are themselves highly ranked, it is an indicator that you know what you are talking about and therefore that people should take you seriously. In fact, there are really two kinds of importance for Web pages. There are pages that are themselves likely to be relevant as results in your search—Kleinberg calls these pages "authorities"—and there are ones that may not contain the information you are looking for but that reliably point to pages that do—Kleinberg calls these "hubs."

To allow for these two kinds of importance, Kleinberg extends the scheme of Brin and Page (1998) by giving each page two weights. One, the *authority weight* x_i, indicates as before how informative the page is estimated to be for your particular search. The other, the *hub weight* y_i, indicates how good the page is at pointing to useful pages. The Honda home page, for example, probably contains a lot of useful information about Honda motor cars, but it is highly unlikely that it contains links to the home pages of competing automobile manufacturers. On the other hand, there are many excellent hub pages on the Web maintained by car enthusiasts, which contain less specific information about the cars, but which reliably point to the best sources for such information. Pages of these two types should get high authority and hub weights, respectively. It is also possible in appropriate circumstances for pages to have both high authority weight and high hub weight, or, of course, to have neither.

The fundamental definition of a good authority, then, is that it is pointed to by many good hubs. And the fundamental definition of a good hub is that it points to many good authorities. In matrix form, we can express this as

$$\lambda \mathbf{x} = \mathbf{A} \cdot \mathbf{y}, \qquad \mu \mathbf{y} = \mathbf{A}^T \cdot \mathbf{x}, \tag{5.9}$$

where $\mathbf{A}^T$ is the transpose of $\mathbf{A}$. If we want only the authority weights x_i and not the hub weights, we can eliminate $\mathbf{y}$ and hence show that $\mathbf{x}$ is the leading eigenvector of the symmetric matrix $\mathbf{AA}^T$, with eigenvalue $\lambda\mu$:

$$\lambda\mu \, \mathbf{x} = \mathbf{AA}^T \cdot \mathbf{x}. \tag{5.10}$$

Kleinberg illustrates this idea with a number of examples, and the results are uncannily accurate. Taking, for example, a base set built around the home page `www.honda.com` of the Honda motor company, as described above, he finds that the top ten authorities returned by the algorithm are, respectively, the home pages of Toyota, Honda, Ford, BMW, Volvo, Saturn, Nissan, Audi, Dodge, and Chrysler. (Only one result from any one Web site was returned, which ensures that the top ten are not all, say, from Toyota.) This result is doubly remarkable when one considers, as noted above, that it is highly unlikely that these pages link to one another in any prominent fashion.

Kleinberg's Web search ideas have been implemented in the commercial search engine Teoma, which provides the muscle behind the Ask Jeeves Web site, among others. Some smaller-scale test implementations have also been created (Bharat and Henzinger 1998; Chakrabarti *et al.* 1998a,b). Kleinberg also discusses possible applications to bibliographic search: the network formed by citations between academic papers is similar in some respects to the link graph of the Web, and certainly papers exist in most fields that play the roles of hub and authority. It may well be that a version of Kleinberg's method could yield an effective way of searching that much older repository of accumulated wisdom, the printed word.

Adamic, Lukose, Puniyani, and Huberman (2001)

The Web search methods discussed above rely on one's ability to construct a centralized index of Web pages, which can then be searched rapidly for information of interest. However some networks have no such central index, and the construction of one may be difficult or undesirable. Peer-to-peer networks are one example. These are ad hoc networks of computers, each storing some subset of a particular collection of data and connected to one another via a virtual network of computer-computer communication links. A typical use for such networks is the sharing of data or data files. Examples of generic peer-to-peer networks include FastTrack and Gnutella.[7]

Because of their ephemeral and distributed nature, these networks have no central index, so how are we to search them for content? A simple method, which is implemented in Gnutella,[8] is for a computer first to query all of its network neighbors for the required data item. We assume that this item is unique and identifiable—a particular file, for example—so that the queried computers can state categorically either that they possess the item in question or that they do not. If any of them possesses the item, the search ends. If not, then each of the queried computers queries *its* neighbors, excluding any that have already been queried. And so the process repeats until a computer is reached that has the desired item.

Fundamentally, this search strategy is equivalent to the breadth-first search of Section 5.1. As with breadth-first search, the number of nodes in the network that need to be queried to find a given item is on average $O(n)$, that is, linear in the total volume n of the graph. But if one supposes that all n vertices are presenting queries at some constant average rate r per unit time, then the total rate at which individual nodes are receiving queries is $O(rn^2)$. This n^2 dependence on the network size is a bad thing: it means that the amount of work that each computer in the network has to do increases as the size of the network increases, and this places a limit on the size of the network. Once the rate at which work is demanded of each computerexceeds

---

[7]Local area file-sharing networks such as SMB or AppleShare count as peer-to-peer networks also, although search problems on such networks usually raise no special issues because the networks are small enough for the simplest strategies to work well.

[8]Gnutella's search strategy is more sophisticated than what we describe here, but the basic principle, and the resulting scaling, are essentially the same.

the power of the computers to perform that work, the network can grow no larger and still function (Ritter 2000).

In the paper reproduced here, **Adamic *et al.* (2001)** show that this problem can be circumvented, at least in some cases, by appealing to its network nature. Like many of the networks discussed in Chapter 3, the pattern of connections between computers in peer-to-peer networks appears to be right-skewed, and possibly power-law in form. Adamic *et al.* propose to exploit this using a modified search algorithm as follows. The computer initiating the search once again starts by querying each of its network neighbors to see if they have the data item of interest. And as before, if they do the search finishes. The crucial difference occurs if none of them have the desired item. In this case the query passes to only one of the neighbors, the one with highest degree. The job of taking care of the query can be thought of as a baton or torch passed from one computer to another, but only in the possession of one at a time. The new computer that receives the torch repeats the process of querying all of *its* neighbors (except those that have been queried before), and if none of them has the desired item, the torch is passed on once again to the neighbor with the highest degree that has not previously held it.

Adamic *et al.* show both analytically and numerically for the case of power-law distributed networks that this algorithm takes an average of $O(\log^2 n)$ steps to find a given item, and hence the total amount of work being done in the whole network is $O(rn \log^2 n)$ if all n computers are performing queries at an average rate r. This is much better than the previous $O(rn^2)$ behavior. Although it is not quite linear in the volume of the network, the logarithm increases sufficiently slowly for large n that for practical purposes it may as well be, and so the average rate at which each computer must perform work is roughly constant as the network grows.

While this is a nice result, some caveats are in order. First, we note that the number of queries that the average computer *receives* in the algorithm of Adamic *et al.* is still $O(n)$. After all, in order to find an item, one must still query the computer on which it lives, and one will probably have to query a large fraction of them before finding it. Adamic *et al.* suggest a possible solution to this, that all computers should maintain a local index describing the data stored by their neighbors. There is still no global index in the network, but each computer stores a small subindex of its local environment. This means that when the "torch" is passed to a new computer that computer does not actually need to query its neighbors; it can simply look in its local index to see if any of them has the item required and contact the relevant neighbor if one of them does.

A more serious problem with the algorithm is that most of the queries are handled by a small number of computers. Since the torch is passed preferentially to high-degree vertices, it tends to find its way rapidly to the few with the highest degrees, and those few are repeatedly entrusted with most of the work for most of the searches. The amount of work required of these "supernodes" is much greater than the average and hence the algorithm again fails to scale well with system size unless those computers have sufficient power to perform the required operations. However, only the few high-degree computers need this power, rather than all computers on the network, and it is easier to increase the power of a small number of computers than to increase the power of them all. Thus the algorithm of Adamic *et al.* could be seen as a way of redistributing the computational burden of performing searches onto a few nodes of the network, which can be given increased computational power to handle the task.[9] In this way the algorithm achieves some of the same goals as search using a centralized index, but without need of the index itself.

Kleinberg (2000a)

Another aspect of the search problem has been considered by **Kleinberg (2000a)** in the last paper of this chapter (see also Kleinberg (2000b)). In Chapter 2 we have reprinted the influential paper by **Travers and Milgram (1969)** in which the authors describe Milgram's small-world experiments involving the passing of letters from person to person in an attempt to reach a specified target person. Travers and Milgram found that the typical number of steps to the target individual from a source chosen roughly speaking at random was quite small—about six on average—and so concluded that randomly chosen pairs of individuals tend to be separated by only a short chain of intermediate acquaintances.

As Kleinberg points out, however, there is another conclusion that can be drawn from the small-world experiments: Milgram's results demonstrate not only that there exist short paths between individuals in a social network, but also that ordinary people are good at finding those short paths. This is, upon reflection, perhaps an even more surprising result than the existence of the paths in the first place. The participants in Milgram's study had no special knowledge of the network connecting them to the target person. Most people know only who their friends are and perhaps a few of their friends' friends. Nonetheless, it proved possible to get a message to a distant target in a small number of steps. This indicates that there is something quite special about the structure of the network. On a random graph, for instance, as Kleinberg points out, short paths between vertices exist, but no one would be able to find them given only the kind of information that people have in realistic situations.

Kleinberg proposes a simple model to illustrate this point. The model is an extension of the small-world model discussed in Section 4.2. In the normal small-world model, vertices are connected together on a regular lattice, and a low density of long-range "shortcuts" are added between randomly chosen vertices, either by introducing extra edges or by "rewiring" those that already exist. Kleinberg considers this model on a two-dimensional square lattice, but adds the twist that the shortcuts have a power-law length distribution, with the probability of a shortcut falling between two vertices a distance d apart on the lattice going as $d^{-\alpha}$, where α is a constant. He then considers the expected efficiency of a Milgram-type experiment on such a network, in which a "letter" is passed from vertex to vertex in an

[9] Some peer-to-peer networks, such as FastTrack, already use a variant of this idea in which the most powerful computers are explicitly designated as supernodes and entrusted with most of the work of performing searches.

attempt to reach a randomly chosen target. At each step the vertex currently holding the letter passes it to whichever of its neighbors is closest, in lattice distance, to the target vertex. If there were no shortcuts in the network then this algorithm would reach the target in a time that increases linearly with the linear dimension of the lattice for a letter that starts at a random position—the letter would simply get passed straight across the lattice until it reached the target.

When shortcuts are introduced, however, the picture is more complicated. Kleinberg derives analytically a lower bound on the time (i.e., number of steps) needed for this "greedy" algorithm to find a random target from a random starting point as a function of the exponent α. His main finding is that the amount of time needed is still polynomial in the lattice size for almost all values of α. That is, a short path may exist to the target—typically one with length only logarithmic in the lattice size—but the algorithm with only local information about network structure cannot find it. Kleinberg shows that for one special value only, $\alpha = 2$, does the greedy algorithm manage on average to find the target in time (poly)logarithmic in the lattice size. He points out that the value $\alpha = 2$ corresponds to the case in which each vertex has equal numbers of network neighbors at all Euclidean distances (other than at distance one, for which the square lattice itself contributes connections). There are more potential neighbors at greater distances from a vertex on a two-dimensional lattice, but less chance of knowing each one, and for $\alpha = 2$ these two effects exactly cancel out. Kleinberg argues that this result should extend to higher dimensions also, with $\alpha = d$ being the special value that allows efficient navigation of the network for lattices of any dimension d. The physical insight behind this result is that for $\alpha < d$ there are too many long shortcuts and not enough short ones, making the network too random, so that short paths to the target exist but are hard to find; for $\alpha > d$ there are not enough long shortcuts to get to the target quickly, so that short paths through the network don't even exist.

Returning to Milgram's experiments, what conclusions can one draw from Kleinberg's work? The principal conclusion seems to be that the ability to find a target quickly using only local information, as the participants in the small-world experiments did, is not something that should be taken for granted. It is easy to construct networks on which such tasks are difficult or impossible, and indeed it appears that, at least within some classes of model networks, and quite likely more generally too, *most* networks present significant challenges to local search algorithms. Thus the fact that Milgram's participants were able to find their target tells us that at least some social networks have quite an unusual structure. What precisely that structure might be has been the subject of several more recent papers by Kleinberg and others (Kleinberg 2000b, 2002; Watts *et al.* 2002).

The Annals of Applied Probability
1997, Vol. 7, No. 1, 46–89

EPIDEMICS WITH TWO LEVELS OF MIXING

BY FRANK BALL, DENIS MOLLISON AND GIANPAOLO SCALIA-TOMBA

University of Nottingham, Heriot-Watt University and Universita La Sapienza

We consider epidemics with removal (SIR epidemics) in populations that mix at two levels: *global* and *local*. We develop a general modelling framework for such processes, which allows us to analyze the conditions under which a large outbreak is possible, the size of such outbreaks when they can occur and the implications for vaccination strategies, in each case comparing our results with the simpler homogeneous mixing case.

More precisely, we consider models in which each infectious individual i has a global probability p_G for infecting each other individual in the population and a local probability p_L, typically much larger, of infecting each other individual among a set of neighbors $\mathscr{N}(i)$. Our main concern is the case where the population is partitioned into local groups or households, but our approach also applies to cases where neighborhoods do not form a partition, for instance, to spatial models with a mixture of local (e.g., nearest-neighbor) and global contacts.

We use a variety of theoretical approaches: a random graph framework for the initial exposition of the simple case where an individual's contacts are independent; branching process approximations for the general threshold result; and an embedding representation for rigorous results on the final size of outbreaks.

From the applied viewpoint the key result is that, compared with the homogeneous mixing model in which individuals make contacts simply with probability p_G, the local infectious contacts have an "amplification" effect. The basic reproductive ratio of the epidemic is increased from its individual-to-individual value R_G in the absence of local infections to a group-to-group value $R_* = \mu R_G$, where μ is the mean size of an outbreak, started by a randomly chosen individual, in which only local infections count. Where the groups are large and the within-group epidemics are above threshold, this amplification can permit an outbreak in the whole population at very low levels of p_G, for instance, for $p_G = O(1/Nn)$ in a population of N divided into groups of size n.

The implication of these results for control strategies is that vaccination should be directed preferentially toward reducing μ; we discuss the conditions under which the *equalizing strategy*, aimed at leaving unvaccinated sets of neighbors of equal sizes, is optimal. We also discuss the estimation of our threshold parameter R_* from data on epidemics among households.

1. Introduction.

1.1. *Mixing at two levels.* In the spread of infectious disease, heterogeneities in population behavior often play a key role in determining whether a major epidemic outbreak occurs and, if it does, its rate of spread and the

Received September 1995; revised July 1996.

AMS 1991 *subject classifications.* Primary 92D30, 60K35; secondary 05C80, 60J80.

Key words and phrases. Epidemic, household, vaccination, estimation, final outcome, random graphs, Reed–Frost, branching processes, Gontcharoff polynomials, threshold parameters, asymptotic distribution, embedding, R_0.

final size of the epidemic. Here we shall analyze in depth one of the simplest and most basic kinds of heterogeneity, in which each infectious individual i has not only a global probability p_G for infecting each other individual in the population, but also a local probability p_L, typically much larger, of infecting each other individual among a set of neighbors $\mathcal{N}(i)$. We shall call these *global* and *local* infections, respectively.

This kind of model is of application to a wide variety of epidemic situations, especially in infectious diseases of humans, where local mixing in social groups such as households and schools can play a crucial role in facilitating the spread of infection. Such models are also of considerable interest in ecology: see, for instance, the reviews by Kareiva (1990) of "patch dynamics," especially the references to "island" and "meta-population" models, and by Hanski and Gilpin (1991).

Perhaps the simplest such model is that of a population partitioned into equal sized groups. That is to say, we have m groups, each of n individuals, giving a total population of size $N = mn$; two individuals are neighbors if and only if they belong to the same group. We shall analyze two cases, that of *households* where n takes a fixed, typically fairly small, value, and the case of *large groups*, where we consider what happens as $n \to \infty$. In either case, we can generalize the model to allow for unequal group sizes $\{n_i\colon 1 \le i \le m$, with $\sum_{i=1}^{m} n_i = N\}$; this level of generality is of course vital in applications (see Section 5).

Our basic model is a generalization of the standard SIR epidemic: we consider individuals of just one type, who once infected make contacts in Poisson processes during an infectious period T_I, where the T_I's for different individuals are independent and identically distributed (see Section 3.1). There is no difficulty of principle, however, to extending the same treatment to models with different within-group contact structures, for instance, the models of Gertsbakh (1977) (see Section 5.2.2) or de Koeijer, Diekmann and Reijnders (1995), or with several types of individual with different contact probabilities, which might for instance represent children and adults (see Section 3.6).

Our basic model assumes that global contacts are equally likely to be with any other individual. Because of this, it is natural to describe the variability of group size through its *size-biased* distribution $\{\pi_k\}$, with π_k being defined as the probability that a randomly chosen *individual* lives in a group of size k (see Section 2.1). Much of our analysis would apply equally well if global contacts were chosen on some other basis, for instance, according to the ordinary group size distribution $\{h_k\}$, where h_k is defined as the probability that a randomly chosen *group* is of size k; notable exceptions to this are our results on optimal vaccination strategies, which do depend on how global contacts are chosen (see Section 5.2.1).

Returning to simple models, another basic case is where individuals are arranged in space (e.g., equally spaced around a circle) and the neighbors of an individual are defined as those within a certain distance, in the simplest case as just an individual's nearest neighbors. Note that in this case the sets of neighbors will overlap, rather than partitioning the population. From the theoretical point of view, this model can be regarded as a kind of limit of a dispersal model with local and long distance interactions, in which the distribution of the latter degenerates into the uniform distribution [for dispersal distributions with extreme behavior, such as "great leaps forward," or with infinite velocity, see Mollison (1972) and Mollison and Levin 1995)]. Possible applications include the spread of infection between pigs in a line of stalls (M. de Jong, personal communication) and the recent North Sea seal epidemic [Bolker et al. (1995)]. Note that the latter requires a model with three levels—long distance and local contacts between seal colonies, perhaps following the great circle model (see Section 2.3), plus within-colony contacts, perhaps following the model of de Koeijer, Diekmann and Reijnders (1995)—but there is no difficulty in principle in extending our analysis to deal with mixing at three levels.

1.2. *Contents.* We shall restrict attention to the SIR model, that is, where there are just three possible states for an individual, *susceptible* (S), *infected and infectious* (I) and *removed* (R), and the only possible transitions are S → I and I → R. We shall also assume that the sets of contacts made by different individuals are independent of each other. In general, infections made by the same individual will be correlated, if only because of the dependence induced by the variability of the length of the infectious period.

In Section 2 we describe how a random graph framework can be used to analyze the non-time-dependent aspects of epidemics. This is especially helpful in the "independent links" (or Reed–Frost) case where infections made by the same individual are independent, since we can then use an undirected graph. Further, the technique of first considering only local infections allows us to partition the population, whether or not it was in separate groups initially, so that global contacts can be considered using a "clumped Reed–Frost" model: essentially simple homogeneous mixing, but with individuals replaced by the clumps formed by local contacts.

We are thus able to derive explicit expressions for thresholds and the final size of epidemics. Recall that in an epidemic with a single level of mixing, the condition for a large outbreak to be possible is that the reproductive ratio R_0 should be greater than 1, where R_0 can be loosely defined as the expected number of potentially infectious contacts of a single infectious individual [see, e.g., Diekmann, Heesterbeek and Metz (1990) and Dietz (1993)].

For epidemics with two levels of mixing, we find that the condition for a large outbreak to be possible is given by $R_* > 1$, where $R_* = R_G \mu$, the product of the reproductive ratio R_G for global contacts and the mean size μ of outbreaks utilizing only local contacts.

The threshold parameter R_* thus provides a natural generalization of R_0 to two levels of mixing, but we must emphasize that R_* is a group-to-group—or more precisely clump-to-clump—reproductive ratio: it is the expected number of clumps contacted by all individuals in the clump of a random individual. While it reduces to R_0 when only local contacts matter (and the clump size is therefore equal to 1), it can differ substantially when group sizes are large (see example of Section 2.5).

The case of large subgroups (Section 2.4) is of particular interest, because it is then meaningful to ask whether individual subgroups are above their local threshold. If so, then the "amplification factor" μ is $O(n)$, so that a global epidemic can occur when p_G is only $O(1/(Nn))$.

An important practical implication of this threshold result is the difference it makes for a vaccination strategy according to whether it succeeds in reducing individual groups below their local threshold. This is explored informally in Section 2.5 in an example showing how dramatic the difference can be; more generally, optimal vaccination strategies are considered later in Section 5.2.

In Section 3 we consider a more general model for the spread of an epidemic among a population consisting of m groups or households. We start with the case where the households are of equal size n. Infectious individuals make local contacts at rate $n\lambda_L$ and global contacts at rate λ_G during an infectious period T_I that follows any arbitrary but specified distribution. (The independent links case of Section 2 corresponds to the special case when T_I is constant.) The individual contacted by a local (global) contact is chosen uniformly at random from the n ($N = mn$) individuals in the infective's group (whole population). Thus the individual-to-individual local and global infection rates are λ_L and λ_G/N, respectively.

In order to derive explicit expressions relating to the threshold behavior of our model, a number of properties of single population SIR stochastic epidemics are required. For convenience, these are collected together in Section 3.2. In Section 3.3.1 we show that the early stages of our epidemic can be approximated by a branching process, whose individuals are single group epidemic processes. Moreover, this approximation can be made precise by considering a sequence of epidemics in which the number of groups $m \to \infty$. This enables us to determine a threshold parameter R_* for our epidemic, such that, in the limit as $m \to \infty$, global epidemics occur with nonzero probability if and only if $R_* > 1$. Here, a global epidemic is one which affects infinitely many groups as $m \to \infty$. We also determine the probability that a global epidemic occurs and various properties of nonglobal epidemics. In Section 3.3.2 we discuss the threshold parameter R_*. In particular, we compare it with the classical basic reproductive ratio R_G that applies when only global contacts are considered, and we show that our model displays a similar amplification effect to that described in Section 2.3 for the independent links case.

In Section 3.4 we use a heuristic argument to determine the distribution of the total size within a typical group in the event of a global epidemic occurring. In Section 3.5, we extend our results to the situation in which the group sizes are not all equal. In Section 3.6 we derive a threshold parameter for the proliferation of infectious individuals for our model and discuss the relationship of our methodology and results to recent papers of Becker and Dietz (1995) and Becker and Hall (1996).

In Section 4 we show that the embedding approach of Scalia-Tomba (1985, 1990) can be extended to epidemics with two levels of mixing and we make use of this to provide a formal derivation of the asymptotic distribution for the final size of a global epidemic.

In Section 5 we consider applications of our results. In Section 5.1 we describe a method of estimating the threshold parameter R_* from household final size data, using data on the spread of an influenza epidemic as an illustration. Finally, in Section 5.2 we consider vaccination programs, whose aim must be to reduce R_* to below unity, and for a variety of models we examine conditions under which the *equalizing strategy*, which leaves the numbers of susceptibles in each group as nearly equal as possible, is optimal.

1.3. *Related work.* Papers on epidemics with two levels of mixing go back at least as far as Rushton and Mautner (1955) on deterministic simple epidemics and Bartlett (1957) and Daley (1967) on the stochastic side, but the earliest works relevant to the present approach are two independent papers dating from 1972. Watson (1972) considered a model with a fixed number of large groups, giving deterministic results and some heuristic stochastic approximations supported by simulations. [A more detailed analysis of the threshold behavior of Watson's deterministic model is given in Daley and Gani (1994).]

A more abstract paper, but perhaps representing a greater advance, was that of Bartoszyński (1972), who considered a group epidemic model which corresponds to the limiting branching process that we use in Section 3.1 and derived a threshold condition for it which is essentially the same as our $R_G\mu > 1$. However, his model was described in rather general terms and hence his results are not so explicit as ours. Indeed, he does not in general say how the probability of choosing a contact in a group of size k is to be specified, leaving it open that it might be either $\{\pi_k\}$ or $\{h_k\}$ (in the notation of Section 1.2), though in one specific example he makes it clear that he means the latter.

May and Anderson (1984), who acknowledged earlier work by Hethcote (1978) and Post, DeAngelis and Travis (1983), who considered the vaccination problem for a deterministic epidemic among a finite number of (large) groups, showing that the equalizing strategy is optimal (see Section 5.2.3).

Work on outbreaks within households in the presence of community infection—but without considering the dynamics of the latter—has a long history: see, for example, the discussion in Longini and Koopman (1982) and, for more recent work, Becker (1989) and Addy, Longini and Haber (1991).

The present work is the elaboration of a "back-of-envelope" answer to a question raised by Klaus Dietz at the Newton Institute in 1993, as to whether whole-population models could be found that would justify such applied work on household models and provide a framework for their development and extension. We are grateful to Klaus Dietz and Niels Becker for exchanging preprints as our work has proceeded in parallel: we comment on their papers [Becker and Dietz (1995) and Becker and Hall (1996)] in Section 3.6.

2. Random graphs and independent contacts.

2.1. *Introduction: the random graph framework.* We will often not be interested in the time course of the epidemic, but only in which individuals become infected (indeed, we may only be interested in their total number, the *final size* of the epidemic). In that case we can make good use of the representation of the spread of the epidemic by a directed graph, in which the individuals are represented by the nodes of the graph, and we draw an arrow from one individual to another to indicate that the first, if infected, will make an infectious contact with the second [see, e.g., Barbour and Mollison (1989)].

As already mentioned (Section 1.2), in general infections made by the same individual will be correlated, because of the dependence induced by the variability of the length of the infectious period. We shall later prove results on thresholds and final outbreak size for such more general models. However, the special case where they are independent—the generalization to two levels of mixing of the well-known Reed–Frost model—is well worth considering first, as then we can use an *undirected graph*, with links rather than arrows [see Barbour and Mollison (1989); also von Bahr and Martin-Löf (1980), and Ball (1983a)], and analysis is much clearer and simpler. The key idea here is that if an individual's contacts are independent of each other and if the probability that i infects j is the same as the probability that j infects i, then we can represent both the latter events by the *same*, undirected link in the contact graph.

An undirected graph can be partitioned into connected components, and the set of those infected during the epidemic will consist precisely of the connected component(s) to which those initially infected belong. In the simple case of homogeneous mixing with contact probability p—the basic Reed–Frost model—the corresponding graph is the simple random graph on a set of N nodes (i.e., a population of size N) $G(N, p)$ [Barbour and Mollison (1989)]. For large N, this graph has a single "giant" component if and only if $R_0 > 1$, where $R_0 = Np$; it then contains a proportion z of the population, where z is the largest root of $z = 1 - \exp(-R_0 z)$ [Bollobas (1985)]. Thus, if there is a large outbreak it will affect approximately this proportion z of the population. Further, if the initial number of infected $I(0) = 1$, the probability of a large outbreak ζ is simply the probability that the initially infected individual lies in the giant component and is therefore equal to z. Note that the equality of ζ to z is quite special to this case. In general (see Section 3.3.1) the probability $\zeta(I(0))$ of a large outbreak depends sensitively both on $I(0)$ [$\zeta(I(0)) = 1 - (1 - \zeta)^{I(0)}$] and on the assumption that the contacts of an individual are independent.

In the remainder of this section, we extend the use of undirected random graphs to find the threshold conditions and final size for epidemics in a large population with two levels of mixing, with particular emphasis on the case of large local groups (Section 2.4) and on the implications for vaccination strategies (Section 2.5).

2.2. *Local contacts and the clumped Reed–Frost model.* We first describe in detail two models in which each individual has a small number of local contacts, and show how they can both be considered as special cases of a clumped Reed–Frost model.

The first of these is the households model described above, where we fix the size or size distribution of households. Then, if we consider only local contacts, these partition each household into a number of connected components. Once we have done this, these connected components summarize the local interactions—whether two separate components originate from the same household or not is irrelevant when we complete our model by adding the global contacts, since the probability of such contacts is to be the same, independently, for each pair of individuals.

Note that this deconstruction relies on our model allowing individuals in the same group to have both a local probability p_L *and* (independently) a global probability p_G of contacting each other. Our model is essentially identical to one in which we allow them only a local probability $p'_L = 1-(1-p_L)(1-p_G)$. (We of course require $p_G \leq p'_L$ here, which is no problem as our main interest is in the case $p_G \ll p_L$, when $p'_L \approx p_L$.)

The probability, π_k say, that an individual chosen at random from the whole population belongs to a component of size k can be calculated—in principle at least—from the distribution of household size and standard methods for the Reed–Frost model (see Section 3.2). The component sizes will not be exactly independent of each other, because of dependence of sizes within households, but this effect will be negligible provided the number of households is large.

Our second model, the "great circle," is one where the population is not partitioned into households. Instead, we have individuals located in one-dimensional space. For simplicity we shall just consider the case where each individual has two neighbors—one on each side; to avoid boundary problems it is convenient to take the space to be the circumference of a circle. We allow infectious local links, of probability p_L, between each pair of neighbors, and global links, as usual, of equal probability p_G for each pair in the population. When we consider first the local contacts, these again partition the whole population into connected components. In this case the probability π_k of belonging to a component of size k is given by the double geometric distribution of parameter p_L [$\pi_k = kp_L^{k-1}(1-p_L)^2$, $k = 1, 2, \ldots$]. Again the component sizes are not exactly independent, but we can neglect their dependence in what follows provided that the population is large relative to the mean component size.

We note that this "neighbors plus global links" model could be generalized to other isotropic spatial structures, such as a regular toroidal lattice or a tessellation of a Poisson process on a sphere, and that we could allow further than nearest-neighbor contacts, provided that the local contacts give only relatively small connected components.

2.3. *Threshold and final size for the clumped Reed–Frost model.* Both the models introduced above are special cases of the following, which we shall call the clumped Reed–Frost model. In this, the population consists of clumps, the ith clump having weight w_i. We then run a Reed–Frost type epidemic (that is, with independent and symmetric contacts) in this population, with probability $1-\exp(-cw_iw_j)$ for a contact between clumps i and j. We relate this to our original models by taking a locally connected component containing k individuals to be a clump of weight k and by taking $p_G = 1 - \exp(-c)$.

We shall use π_k to denote the probability that a randomly chosen individual belongs to a clump of size k and use μ to denote the mean clump size $\sum_k k\pi_k$. We shall assume that the mean clump size is finite. For both the household and nearest-neighbor models, in the limit of large total population the clump weights will be chosen independently from the distribution $\{\pi_k\}$. Note that $\{\pi_k\}$ is the *size-biased* distribution for clump size, as distinct from the formulation in which we define the probability h_k that a randomly chosen clump is of size k. The distributions $\{\pi_k\}$ and $\{h_k\}$ are simply related, with $\pi_k = kh_k/\sum_j jh_j$. Note also that if μ_h and σ_h^2 are, respectively, the mean and variance of the distribution $\{h_k\}$, then $\mu = \mu_h + \sigma_h^2/\mu_h$.

Assuming that the number of clumps is large, the probability of a large outbreak can be found by considering the branching process which approximates its early stages (see Section 3.3.1), in which individuals correspond to clumps in the epidemic process and the offspring of a given clump are the clumps that it directly tries to infect in the clumped Reed–Frost epidemic. The approximation (which can be made fully rigorous; see Section 3.3.1) assumes that each new clump contacted in the epidemic process is still susceptible. For large N, the number of clumps contacted by an individual in the epidemic process is Poisson(Np_G), with probability generating function (p.g.f.) $\exp(Np_G(s-1))$, and the clump size distribution is $\{\pi_k\}$, with p.g.f. $G_\pi(s)$ say. Thus the number of clumps contacted by a given clump, that is, the offspring distribution for the approximating branching process, has p.g.f. $G_\pi(\exp(Np_G(s-1)))$. It follows that the probability of a large outbreak is the largest solution ζ (≤ 1) of $1-\zeta = G_\pi(\exp(-Np_G\zeta))$, that is, of $1-\zeta = \sum_{k=1}^{\infty} \pi_k \exp(-kNp_G\zeta)$. Further, ζ will be greater than 0 if and only if the mean number of offspring from a clump $Np_G\mu$ is greater than 1.

Thus the basic reproductive ratio for the epidemic among clumps is $R_* = \mu Np_G = \mu R_G$, where $R_G = Np_G$ is the basic reproductive ratio for the ordinary Reed–Frost epidemic—that is, where we only have global contacts so that all clumps are of size 1. [More strictly, we should perhaps use $R_G = (N-1)p_G$, but if we are interested in values of N sufficiently small that this matters, we should be worrying about the correct definition of thresholds in finite populations; see Nåsell (1995).]

The probability ζ here is that of a large outbreak started by a random individual. If we know that the initial infection(s) is (are) in a clump of size k, consideration of the first generation of contacts shows that the probability of a large outbreak ζ_k is related to ζ by $\zeta_k = 1-\exp(-kNp_G\zeta)$ (see Section 3.3.1 for more detail).

In either case, just as for the simple Reed–Frost model, we can argue that the probability of a large outbreak is the same as the probability that an individual belongs to the giant connected component of the contact graph, which in turn is the same as the (proportional) final size of the epidemic conditional on a large outbreak. Thus the final size $\approx \zeta N$, and the probability that an individual in a clump of size k (or equivalently the whole of that clump) is infected during the epidemic is ζ_k. We may check the consistency of these results: $\zeta = \sum_k \pi_k \zeta_k = \sum \pi_k - \sum \pi_k \exp(-\zeta k N p_G) = 1 - G_\pi(\exp(-\zeta N p_G))$. Also, thinking of $1 - \zeta_k$ as the probability that a clump of size k escapes infection, we note that the number of links from each individual in the clump to the giant component is Poisson$(\zeta N p_G)$, so that the probability of having no such links should indeed be $\exp(-kNp_G\zeta)$.

Note that for the great circle model with independent links there are only two parameters, p_L and p_G, and it is easy to show that $G_\pi(s) = (1-p_L)^2 s/(1-p_L s)^2$ and $R_* = R_G\mu = Np_G(1+p_L)/(1-p_L)$.

2.4. *Epidemics among giants.* We consider here the simple case of a large number (m) of large households of equal sizes n. For large households the idea of a threshold for local contacts makes sense. When we consider only these local contacts, each household has its own local simple Reed–Frost epidemic, with population size n and basic reproductive ratio $R_L = np_L$. It is well known that the behavior of these single-group models goes through a phase transition at around the value $R_L = 1$ [e.g., Whittle (1955), von Bahr and Martin-Löf (1980), Nåsell (1995) and Ball and Nåsell (1994)], and it is interesting to examine the implications of this for the present two level model.

If $R_L \leq 1$, the epidemics in individual households are below threshold and the contact graph within each household consists of components all small compared with n, in fact of size $O(1)$. Then the analysis of the previous section applies, with R_* being greater than $R_G = Np_G$ by the factor μ equal to the mean size of these components; but μ is only $O(1)$, so we still require $p_G = O(1/N)$ to get a global epidemic—that is, a large outbreak at the interhousehold level.

The situation is more interesting when $R_L > 1$, so that the within-household epidemics are above threshold. Then each has its own giant connected component, of size nz_h say, and it is easy to see that the epidemic among the meta-population of giants has $R_* = nNp_G z_h^2$. Further, it is not too difficult to see that the members of households out with the giants do not significantly affect the probability or size of the overall outbreak.

Since $z_h = O(1)$ in this case, it only requires p_G to be $O(1/Nn)$ for R_* to be greater than 1, and thus make possible a large outbreak among the giants; that is, R_G need only be $O(1/n)$. Then the proportion of giants forming the "meta-giant" component of those involved in this large-scale Reed–Frost epidemic is given by $1 - z_g = \exp(-R_* z_g)$, so that the final proportion of the whole population affected is $z_h z_g$, and, as usual in the independent links case, this is also the probability of a large outbreak arising from an initial infected individual: here z_h is represents the probability that the individual belongs to its local giant, and z_g is the probability that this giant belongs to the meta-giant.

In all cases the local contacts have an amplifying effect on the global epidemic. But, for large households this amplification undergoes a significant change (we might call this a phase transition) from $O(1)$ to $O(n)$ as we reach the local threshold ($R_L = 1$) at which the households go through their individual phase transitions [von Bahr and Martin-Löf (1980)].

Finally here, we note the consistency of these essentially asymptotic results with those of the previous section. If in the clumped Reed–Frost model we let the clump distribution tend to that concentrated on nz_h, with probability z_h, and 0, with probability $1 - z_h$, then $\mu = nz_h^2$ so that both models agree that $R_* = nNp_G z_h^2$ and $G_\pi(s) = (1 - z_h) + z_h s^{nz_h}$. So that the equation for the final size becomes $1 - z = (1 - z_h)1 + z_h \exp(-Np_G z n z_h)$, which, if we write $z_g = z/z_h$, boils down to $1 - z_g = \exp(-R_* z_g)$ as obtained above for the giant epidemic.

2.5. *Vaccination strategies in relation to local thresholds.* In a homogeneously mixing population, the minimum proportion v that we need to vaccinate to render the remaining susceptible population sub-threshold is given by $R'_G = (1 - v)R_G = 1$; that is, we require $v \leq 1 - 1/R_G$.

With our two levels of mixing, we have found that the basic reproductive ratio is $R_* = \mu R_G$. For a population divided into large groups, R_* can take large values, since μ will be a significant proportion of group size if groups are above their individual thresholds ($R_L > 1$). (Recall, from Section 1.2, that R_* is a parameter describing group-to-group infection and is therefore not directly comparable with individual-to-individual reproductive ratios such as R_G and R_L.)

Now vaccination of a proportion v of the population will still simply reduce R_G *pro rata* to $R'_G = (1 - v)R_G$, but the effect on μ will depend on the distribution of vaccination among the population. We shall consider the question of optimal vaccination strategies in more detail and generality in Section 5.2; here we simply indicate the practical importance of this question.

For the groups or households model, one strategy is to vaccinate whole groups. Let us assume for simplicity that if they are of different sizes, we choose groups at random, that is, according to the distribution $\{\pi_k\}$. Then μ will be unchanged, so that the overall reproductive ratio will simply become $R'_* = (1 - v)R_*$. However, a strategy in which we vaccinate a proportion of those in each group—for instance, the strategy in which we simply vaccinate members of the overall population chosen at random—can also reduce μ and thus reduce R_* further. In the case where vaccination changes groups from being above to below their local threshold, the difference can be dramatic, as the following simple numerical example illustrates.

Suppose that our population is divided into groups of size $n = 1000$ (perhaps schools or local communities) and that the reproductive ratio R_G for global contacts is 1 (the exact value is not important for what follows). Suppose also that $p_L = 0.003$, so that the reproductive ratio for local contacts is $R_L = (n - 1)p_L \approx 3$. Then $z_h \approx 0.95$, whence (in the notation of the last section) $\mu \approx nz_h^2 \approx 900$, and hence the overall reproductive ratio is $R_* = \mu R_G \approx 900$. (In contrast, the individual-to-individual reproductive ratio here is $R_0 \approx R_L + R_G \approx 4$.)

We now consider two alternative strategies for vaccinating 80% of the population. First note that with any such strategy, R'_G will be $(1 - 0.8)R_G = 0.2$. If we have a "patchy" vaccination program that vaccinates whole groups, we will have $R'_* = \mu R'_G \approx 900 \times 0.2 = 180$, still far above threshold. However, if we have a uniform vaccination program, in which approximately 80% of each group are vaccinated, the local reproductive ratio will be brought down to $R'_L \approx (1 - 0.8) \times 3 = 0.6$. The groups will thus be below their local thresholds and their new mean clump size is easily calculated (from an approximating branching process, as in Section 3.3.2) to be $\mu' \approx 1/(1 - 0.6) = 2.5$. Thus in this case we will have $R'_* = \mu' R'_G \approx 2.5 \times 0.2 = 0.5$, so that vaccination will succeed in bringing the infection below threshold.

We can go further: from the practical point of view it is interesting to consider a program aimed at uniform coverage, but which is inadequate in some groups, meaning that in them there are still enough susceptibles left for the group to be above its local threshold. We find that where the initial value of R_* is large, a quite small proportion of groups with inadequate coverage suffices to leave the population as a whole above threshold, that is, $R'_* > 1$. Extending our example of groups of size 1000 with $R_G = 1$, $R_L = 3$, if we have a program which generally vaccinates 80% within each group, the program will fail ($R'_* > 1$) if there is just 1% of the groups in which vaccination coverage is only 50%.

3. The model with a general infectious period.

3.1. *The basic model.* We now consider a generalization of the households model of Section 2, in which the infectious period may follow any arbitrary but specified distribution. Let the population consist of N individuals, subdivided into m groups each of size n. (We shall treat the case of unequal group sizes in Section 3.5.) The infectious periods of different infectives are independently and identically distributed according to a random variable T_I. Throughout its infectious period a given infective makes contact with each other susceptible in the population at the points of a homogeneous Poisson process having rate λ_G/N and, additionally, with each susceptible in its own group at the points of a homogeneous Poisson process having rate λ_L. All the Poisson processes describing infectious contacts (whether or not either or both of the individuals involved are the same), as well the random variables describing infectious periods, are assumed to be mutually independent.

Note that we have chosen here to formulate our model so that an individual i can make both local and global contacts with a susceptible in its own group, which may seem slightly unnatural. However, for the groups model of this section it facilitates our analysis by putting all individuals on an equal footing with respect to global contacts, and it is essential for cases such as the great circle model where the population is not partitioned into groups (see Section 2.3).

The alternative would be to treat all within-group contacts as local, at rate $\lambda'_L = \lambda_L + \lambda_G/N$. It follows from the superposition and splitting (or "coloring") properties of the Poisson process [see, e.g., Kingman (1993)] that for the groups model (provided $\lambda'_L \geq \lambda_G/N$) the two formulations are exactly equivalent; and of course $\lambda'_L \to \lambda_L$ as $N \to \infty$.

For ease of exposition we shall assume that there is no latent period. However, all our results can be generalized to a model that incorporates a latent period. In particular, the final outcome of the epidemic is invariant to very general assumptions concerning a latent period. This can be seen by considering the random graph associated with the epidemic, in which for any two nodes, i, j say, a directed arc from i to j is present if and only if i will infect j if i is an infective and j is a susceptible.

The epidemic is initiated by a number of individuals becoming infected at time $t = 0$. We shall consider the spread of the epidemic in the asymptotic situation where the number of groups m tends to infinity while the group size n is held fixed.

If in this model we let T_I take a constant value t_I, then the epidemic has the same final outcome as the Reed–Frost model of Section 2 with $p_L = 1-\exp(-\lambda_L t_I)$ and $p_G = 1-\exp(-\lambda_G t_I/N)$. If instead we let T_I follow an exponential distribution, then our model reduces to the "equivalent classes" model of Watson (1972). Watson studied the deterministic version of the equivalent classes model, and also the branching process approximation as the group size n tends to infinity with the number of groups m fixed and finite. This contrasts sharply with our asymptotic regime outlined above.

3.2. *Final outcome of a single population SIR stochastic epidemic.* Consider a closed homogeneously mixing population consisting initially of n susceptibles and a infectives, who have just been infected. Suppose, as above, that the infectious period is distributed according to a random variable T_I and that throughout its infectious period a given infective infects a given susceptible at rate λ_L. The epidemic ceases as soon as there are no infectives present in the population. Let T be the final size of the epidemic, that is, the total number of initial susceptibles that are ultimately infected by the epidemic. Let T_A be the severity of the epidemic, that is, the sum of the infectious periods of all individuals infected during the course of the epidemic, including the a initial infectives. Note that T_A is equal to the area under the trajectory of infectives; see, for example, Downton (1972). The joint distribution of (T, T_A) is studied in Ball (1986). More recently, a general framework for analyzing the final size and severity of SIR stochastic epidemics has been developed in a series of papers by Lefèvre and Picard; see, for example, Picard and Lefèvre (1990). A key tool in their framework is a nonstandard family of polynomials, first introduced by Gontcharoff (1937), which we now outline.

Let $U = u_0, u_1, \ldots$ be a given sequence of real numbers. Then the Gontcharoff polynomials attached to U, $G_0(x|U), G_1(x|U), \ldots$, are defined recursively by the triangular system of equations

$$\sum_{j=0}^{i} \frac{u_j^{i-j}}{(i-j)!} G_j(x|U) = \frac{x^i}{i!}, \qquad i = 0, 1, \ldots. \tag{3.1}$$

For $i = 1, 2, \ldots$, the polynomial $G_i(x|U)$ admits the integral representation

$$G_i(x|U) = \int_{u_0}^{x} \int_{u_1}^{\xi_0} \int_{u_2}^{\xi_1} \cdots \int_{u_{i-1}}^{\xi_{i-2}} d\xi_0\, d\xi_1\, d\xi_2 \cdots d\xi_{i-1}; \tag{3.2}$$

see, for example, Lefèvre and Picard [(1990), (2.5)]. Another property of Gontcharoff polynomials [see (2.7) of Lefèvre and Picard (1990)] that we shall require is

$$G_i^{(j)}(x|U) = G_{i-j}(x|E^jU), \qquad 0 \le j \le i, \tag{3.3}$$

where E^jU is the sequence $u_j, u_{j+1}, \ldots$ and $G_i^{(j)}(x|U)$ is the jth derivative of $G_i(x|U)$. Note that $G_i^{(j)}(x|U) = 0$ if $j > i$.

For the single population epidemic model, let $\phi(\theta) = \mathrm{E}[\exp(-\theta T_I)]$, $\theta \ge 0$, be the moment generating function of T_I and let

$$\phi_{n,a}(s, \theta) = \mathrm{E}[s^{n-T}\exp(-\theta T_A)], \qquad \theta \ge 0. \tag{3.4}$$

Then it follows from Proposition 3.3 of Picard and Lefèvre (1990) [see also Ball and Clancy (1993)] that

$$\phi_{n,a}(s, \theta) = \sum_{i=0}^{n} \frac{n!}{(n-i)!}\phi(\theta + \lambda_L i)^{n+a-i}G_i(s|U), \tag{3.5}$$

where the sequence U is given by $u_i = \phi(\theta + \lambda_L i)$, $i = 0, 1, \ldots$.

Let $\mu_{n,a} = \mathrm{E}[T]$ be the mean final size of the above epidemic. Then by differentiating (3.5) with respect to s and setting $s = 1$ and $\theta = 0$, it follows using (3.3) that

$$\mu_{n,a} = n - \sum_{i=1}^{n} \frac{n!}{(n-i)!} q_i^{n+a-i}\alpha_i, \tag{3.6}$$

where $q_i = \phi(\lambda_L i)$ and $\alpha_i = G_{i-1}(1|V)$. Here the sequence V is given by $v_i = \phi(\lambda_L(i+1)) = q_{i+1}$ (for $i = 0, 1, \ldots$). We may call the q_i's the escape probabilities, since $q_i = \mathrm{E}[\exp(-i\lambda_L T_I)]$ is the probability that a set of i individuals exposed to a single infective in the same group all escape infection by it. From this interpretation it is immediate that the q_i's, and hence the v_i's, are monotone nonincreasing:

$$q_i \ge q_{i+1} \quad \text{for all } i \ge 0 \text{ (note that } q_0 = 1). \tag{3.7}$$

Note that it is straightforward to compute $\alpha_1, \alpha_2, \ldots$ numerically using the recursive definition of the Gontcharoff family of polynomials given in (3.1).

In Sections 3.4 and 3.5 we shall require the fact that $\alpha_i > 0$, $i = 1, 2, \ldots$, which we now prove. The integral definition of $G_i(x|U)$ given in (3.2) implies that $G_i(x|U) > 0$ for $i = 0, 1, \ldots$, provided that $x > u_0 \ge u_1 \ge \cdots \ge 0$. [This gives a new and elegant proof of a result proved in Gani and Shanbhag (1974).] The strict positivity of the α_i's follows immediately from (3.7) (remembering that $v_i = q_{i+1}$).

We shall need the moment generating function of T_A,

$$\psi_{n,a}(\theta) = \mathrm{E}[\exp(-\theta T_A)]$$

say, which can be obtained by setting $s = 1$ in (3.5).

Consider now an extension of the single population epidemic model, in which susceptibles can also be infected from outside the population. Specifically, suppose that each of the n initial susceptibles has probability π of avoiding infection from outside the population during the course of the epidemic, independently of other susceptibles in the population. This extended model has been considered by Addy, Longini and Haber (1991), who derived recursive expressions for the probability generating function of T and the moment generating function of T_A. The final outcome of the extended model with outside infection has the same distribution as that of the single population model with initial numbers of infectives and susceptibles $a+Y$ and $n-Y$, respectively, where Y is a realization of a binomial random variable with parameters n and $1-\pi$. (This follows by considering the random graph associated with the epidemic.)

Let $\tilde{\phi}_{n,a}(s,\theta)=\mathrm{E}[s^{n-T}\exp(-\theta T_A)]$, $\theta \geq 0$, be the joint generating function of (T, T_A) for the model with outside infection. Then conditioning on the value of Y and using (3.5) yields

$$(3.8)\quad \tilde{\phi}_{n,a}(s,\theta)=\sum_{k=0}^{n}\binom{n}{k}\pi^k(1-\pi)^{n-k}\sum_{i=0}^{k}\frac{k!}{(k-i)!}\phi(\theta+\lambda_L i)^{n+a-i}G_i(s|U),$$

which on changing the order of summation gives, after a little algebra,

$$(3.9)\qquad \tilde{\phi}_{n,a}(s,\theta)=\sum_{i=0}^{n}\frac{n!}{(n-i)!}\phi(\theta+\lambda_L i)^{n+a-i}\pi^i G_i(s|U).$$

Let $\tilde{\mu}_{n,a}=\mathrm{E}[T]$ be the mean final size for the epidemic with outside infection. Then arguing as in the derivation of (3.6) yields

$$(3.10)\qquad \tilde{\mu}_{n,a}=n-\sum_{i=1}^{n}\frac{n!}{(n-i)!}q_i^{n+a-i}\pi^i\alpha_i.$$

We now give expressions for the final size distribution of the single population epidemic model with outside infection. Let $\tilde{P}_k^n=\Pr\{T=k\}$, $k=0, 1,\ldots,n$. Then setting $\theta=0$ in (3.9), differentiating $n-k$ times with respect to s and using (3.3) yields

$$(3.11)\ \tilde{P}_k^n=\frac{1}{(n-k)!}\sum_{i=n-k}^{n}\frac{n!}{(n-i)!}q_i^{n+a-i}\pi^i G_{i-n+k}(0|E^{n-k}U),\quad k=0,1,\ldots,n,$$

where the sequence U is given by $u_i=q_i=\phi(\lambda_L i)$, $i=0,1,\ldots$.

Addy, Longini and Haber (1991) gave a similar expression to (3.11), but not using Gontcharoff polynomials. They also showed that the final size probabilities can be determined from the triangular system of linear equations

$$(3.12)\qquad \sum_{i=0}^{k}\binom{n-i}{k-i}\tilde{P}_i^n/\{q_{n-k}^{a+i}\pi^{n-k}\}=\binom{n}{k},\qquad k=0,1,\ldots,n.$$

Setting $\pi=1$ in (3.12) yields a set of linear equations governing the final size distribution of the epidemic without outside infection; see Ball (1986).

The above systems of equations are in principle straightforward to solve numerically, because of their triangular structure. Numerical problems due to rounding errors can occur even for moderate values of n, perhaps $n = 50$ or 100. However, in many applications, n will correspond to group or household size and will typically be small, say $n \leq 5$ or 10, permitting the required properties to be calculated accurately.

3.3. *Initial stages of a multigroup epidemic.*

3.3.1. *Branching process approximation.* Suppose that the number of groups m, and hence the total population N, is large. Then during the early stages of the epidemic, every time a between group infection occurs the contacted individual is likely to be in a previously uninfected group. Thus the initial stages of the epidemic can be approximated by a branching process, in which the units are single group epidemic processes and the offspring of a given unit are those groups that are directly infected by infectives in that unit.

The approximation can be made precise by considering a sequence of epidemics with fixed group size n, indexed by the number of groups m, and using the coupling argument of Ball (1983b) and Ball and Donnelly (1995). Specifically, the epidemic processes and the approximating branching process can be constructed on the same probability space $(\Omega, \mathscr{F}, P)$ such that if $A \subseteq \Omega$ denotes the set on which the branching process goes extinct, then (i) for P-almost all $\omega \in A$ the process of infectives in the epidemic process and the branching process agree over the time interval $[0, \infty)$ for all sufficiently large m and (ii) for P-almost all $\omega \in \Omega \setminus A$ the epidemic process and the branching process agree over $[0, c \log m]$ for all sufficiently large m, for any $c < (2\alpha)^{-1}$, where α is the Malthusian parameter of the branching process. The result in (ii) is the best possible in the sense that if $c > (2\alpha)^{-1}$, then, for all sufficiently large m, the epidemic process and the branching process disagree over part of the interval $[0, c \log m]$, and the maximum difference tends to infinity as $m \to \infty$. The Malthusian parameter α can be obtained as follows. For $t \geq 0$, let $Y(t)$ denote the number of infectives at time t in the single group epidemic model of Section 3.2, when initially there are one infective and $n - 1$ susceptibles. Then, provided that the branching process is supercritical, α is the unique solution in $(0, \infty)$ of the equation

$$\int_0^\infty \lambda_G \mathrm{E}[Y(t)] \exp(-\alpha t)\, dt = 1 \tag{3.13}$$

[see Ball (1996) for details].

The final size of the approximating branching process can be obtained by considering its embedded Galton–Watson process, whose offspring distribution can be derived as follows. A typical unit in the branching process commences with one of the susceptibles in the group being infected from outside. That infective will start an epidemic within its own group. Each infective in this single-group epidemic independently makes infections outside the group at rate λ_G throughout its infectious period. Hence the total number of outside infections emanating from the group under consideration follows a Poisson distribution with random mean $\lambda_G T_A$, where T_A is the severity of the single-group epidemic. Further, in the branching process approximation, all of these outside infections are with susceptibles in distinct groups, so the offspring distribution, R say, of the embedded Galton–Watson process is also Poisson with random mean $\lambda_G T_A$. Let $R_* = \mathrm{E}[R]$. Then, letting T be the final size of the single group epidemic and using the Wald identity for epidemics proved in Ball (1986), we obtain

$$\begin{aligned} R_* &= \lambda_G \mathrm{E}[T_A] \\ &= \lambda_G(1+\mathrm{E}[T])\mathrm{E}[T_I] \\ &= \lambda_G(1+\mu_{n-1,1})\mathrm{E}[T_I]. \end{aligned} \tag{3.14}$$

Note that, as in the clumped Reed–Frost model, R_* is of the form $R_* = \mu R_G$, where $R_G = \lambda_G \mathrm{E}[T_I]$ is the basic reproductive ratio for the model in which all the groups are of size 1, that is, $n = 1$, and $\mu = 1 + \mu_{n-1,1}$ is the mean clump size.

To obtain a threshold theorem for the multigroup epidemic process, we say that a global epidemic occurs if in the limit as $m \to \infty$ the epidemic infects infinitely many groups. By standard branching process theory [see, e.g., Jagers (1975)], global epidemics can occur if and only if $R_* > 1$, so R_* may be viewed as the threshold parameter for the multigroup epidemic. Note that for any given set of parameter values, R_* can be computed using (3.6). Indeed, for small values of the group size n, explicit expressions for α_i, and hence for R_*, can easily be obtained.

The probability of a global epidemic depends on the number and configuration of initial infectives. Consider first the case in which the epidemic is initiated by just one of the susceptibles becoming infected. Then, again by standard branching process theory, the probability of a global epidemic is $\zeta = 1-\tau$, where τ is the smallest root in $[0, 1]$ of the equation $f(s) = s$. Here $f(s)$ is the probability generating function of R, which, conditioning on the value of T_A, is given by

$$\begin{aligned} f(s) &= \mathrm{E}[s^R] \\ &= \mathrm{E}[\mathrm{E}[s^R|T_A]] \\ &= \mathrm{E}[\exp(-\lambda_G T_A(1-s))] \\ &= \psi_{n-1,1}(\lambda_G(1-s)), \qquad 0 \le s \le 1. \end{aligned} \tag{3.15}$$

For $i = 1, 2, \ldots, n$, let τ_i be the probability of a nonglobal epidemic when initially there is one infectious group containing i infectives and $n-i$ susceptibles, so $\tau_1 = \tau$. Let Z be the size of the first generation in the embedded Galton–Watson process; that is, Z is the total number of outside infections emanating from the initial single-group epidemic. Then, again conditioning on the value of T_A,

$$\begin{aligned}\tau_i &= \mathrm{E}[\tau^Z] \\ &= \mathrm{E}[\mathrm{E}[\tau^Z|T_A]] \\ &= \mathrm{E}[\exp(-\lambda_G T_A(1-\tau))] \\ &= \psi_{n-i,i}(\lambda_G(1-\tau)).\end{aligned} \tag{3.16}$$

Finally, if initially there are a_i infectious groups with i infectives and $n-i$ susceptibles, $i = 1, 2, \ldots, n$, then

$$\Pr\{\text{global epidemic}\} = 1 - \prod_{i=1}^{n} \tau_i^{a_i}. \tag{3.17}$$

Note that $\psi_{n-i,i}(\theta)$, and hence τ_i, $i = 1, 2, \ldots, n$, are straightforward to compute by setting $s = 1$ in (3.5) and using the recursive definition (3.1) for the quantities $G_0(1|U), G_1(1|U), \ldots, G_{n-1}(1|U)$.

Other properties of the approximating branching process are straightforward to determine. Suppose that initially there is one infectious group containing just one infective. Let $\tilde{N}$ and $\tilde{G}$ be, respectively, the total number of individuals and total number of groups infected by the epidemic, where now the initial infective and the initial infectious group are included. As before, let T and T_A be, respectively, the final size and severity of the single group epidemic in the initial infectious group. Let $h(s_1, s_2) = \mathrm{E}[s_1^{\tilde{N}} s_2^{\tilde{G}}]$ be the joint probability generating function of $(\tilde{N}, \tilde{G})$ under the branching process approximation. Then, conditioning on (T, T_A),

$$\begin{aligned}h(s_1, s_2) &= \mathrm{E}\left[\mathrm{E}\left[s_1^{\tilde{N}} s_2^{\tilde{G}}|T, T_A\right]\right] \\ &= \mathrm{E}\left[\mathrm{E}\left[s_1^{1+T+\sum_{i=1}^{Z}\tilde{N}_i} s_2^{1+\sum_{i=1}^{Z}\tilde{G}_i}|T, T_A\right]\right],\end{aligned} \tag{3.18}$$

where, as above, Z is the size of the first generation in the embedded Galton–Watson process and $(\tilde{N}_1, \tilde{G}_1), (\tilde{N}_2, \tilde{G}_2), \ldots, (\tilde{N}_Z, \tilde{G}_Z)$ are independent and identically distributed copies of $(\tilde{N}, \tilde{G})$. Now Z is Poisson with mean $\lambda_G T_A$, so

$$\begin{aligned}h(s_1, s_2) &= s_1 s_2 \mathrm{E}[s_1^T \mathrm{E}[h(s_1, s_2)^Z|T, T_A]] \\ &= s_1 s_2 \mathrm{E}[s_1^T \exp(-\lambda_G T_A(1 - h(s_1, s_2)))] \\ &= s_1 s_2 \hat{\phi}_{n-1,1}(s_1, \lambda_G(1 - h(s_1, s_2))),\end{aligned} \tag{3.19}$$

where

$$\begin{aligned}\hat{\phi}_{n-1,1}(s, \theta) &= \mathrm{E}[s^T \exp(-\theta T_A)] \\ &= s^{n-1}\phi_{n-1,1}(s^{-1}, \theta)\end{aligned} \tag{3.20}$$

is the joint generating function of (T, T_A). Thus $h(s_1, s_2)$ satisfies the functional equation

$$h(s_1, s_2) = s_1^n s_2 \phi_{n-1,1}(s_1^{-1}, \lambda_G(1 - h(s_1, s_2))). \tag{3.21}$$

Appropriate differentiation of (3.21) yields expressions for the moments of $\tilde{N}$ and $\tilde{G}$, such as $\mathrm{E}[\tilde{N}]$, $\mathrm{E}[\tilde{G}]$, $\mathrm{var}(\tilde{N})$, $\mathrm{var}(\tilde{G})$ and $\mathrm{cov}(\tilde{N}, \tilde{G})$. Note that (3.3) and the recursive definition (3.1) of Gontcharoff polynomials enables the derivatives of $\phi_{n-1,1}(s, \theta)$ and hence the above moments to be calculated. When $R_* \geq 1$, the above moments are all infinite. However, if $R_* > 1$, then moments conditional upon the occurrence of a nonglobal epidemic can be derived from (3.21).

An alternative approach to determining the limiting properties of $(\tilde{N}, \tilde{G})$ as the number of groups $m \to \infty$ is via the two-type branching process described in Section 3.6.

3.3.2. *Discussion of the threshold parameter* R_*. We now discuss the relationship of our threshold parameter R_* to the classical reproductive ratio R_0 [see, e.g., Diekmann, Heesterbeek and Metz (1990)] for the multigroup epidemic. For definiteness of argument, suppose that the infectious period T_I follows an exponential distribution with mean γ^{-1}, so that our model becomes a multigroup generalization of the general stochastic epidemic [see, e.g., Bailey (1975), Chapter 6]. The deterministic version of our model is then expressed by the differential equations

$$\frac{dx_i}{dt} = -\Big(\lambda_L y_i + N^{-1}\lambda_G \sum_{j\neq i} y_j\Big)x_i,$$

$$(3.22)\qquad \frac{dy_i}{dt} = \Big(\lambda_L y_i + N^{-1}\lambda_G \sum_{j\neq i} y_j\Big)x_i - \gamma y_i, \qquad i = 1, 2, \ldots, m,$$

where the groups are labelled $1, 2, \ldots, m$ and $x_i(t)$ and $y_i(t)$ are, respectively, the numbers of susceptibles and infectives in the ith group at time t.

The reproductive ratio for the above deterministic model, usually defined informally (in a stochastic sense!) as the expected number of infectious contacts made by a single initial infective in an otherwise susceptible population, is $R_0 = \{(n-1)\lambda_L + \lambda_G\}/\gamma$. In the deterministic setting, a major epidemic occurs if and only if $R_0 > 1$. However, as we shall see soon, $R_0 > 1$ does not generally provide a good indication as to whether a global epidemic can occur in our stochastic model. This is because a deterministic model can only be a good approximation to the more realistic stochastic model if all the population sizes are large [cf. the convergence theorems of Kurtz (1970, 1981)], but in the multigroup epidemic the group size n is often small. Thus the deterministic model (3.22) will not generally provide an adequate description of the multigroup epidemic. Indeed, a more appropriate deterministic model is one described by a system of differential equations for $x_{i,j}(t)$, $0 \leq i, j \leq n$, where $x_{i,j}(t)$ is the number of groups with i susceptibles and j infectives at time t.

It is now convenient to assume that the multigroup epidemic model is parameterized so that the within-group infection rate is $(n-1)^{-1}\lambda_L$ and, for the purpose of illustration, that the time axis is linearly rescaled so that $\gamma = 1$. Under these assumptions, $R_0 = \lambda_L + \lambda_G$ independently of the group size n. The threshold parameter R_* can be calculated using (3.14). Figure 1 shows for various group sizes n the graph of critical values of (λ_L, λ_G) so that $R_* = 1$.

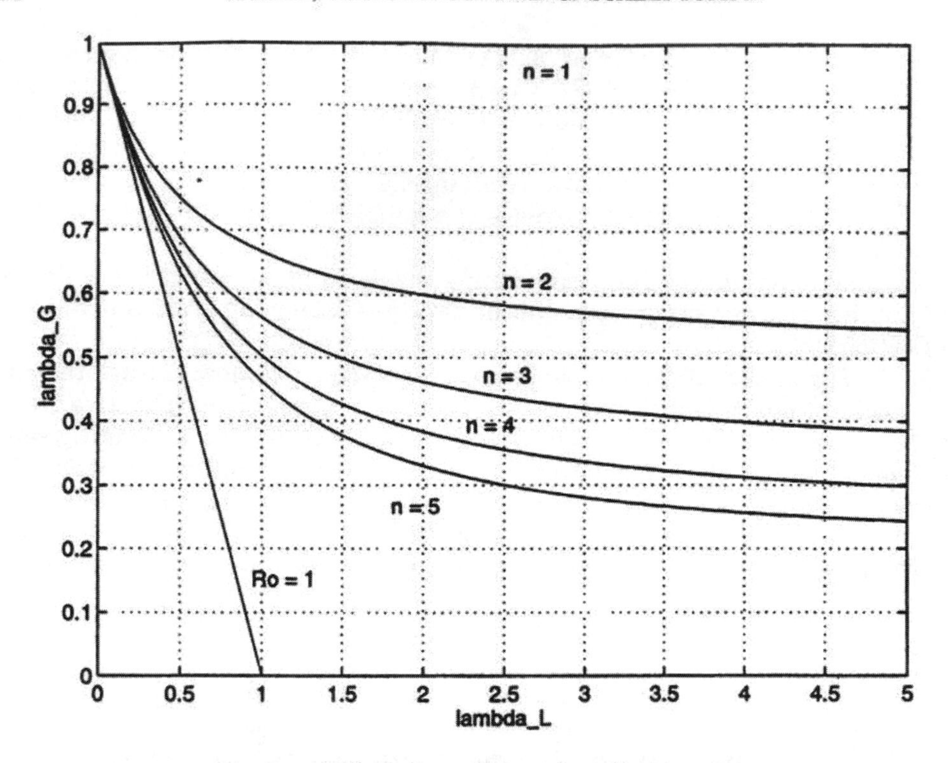

FIG. 1. *Critical values of* (λ_L, λ_G) *so that* $R_* = 1$.

The corresponding graph for $R_0 = 1$ is also shown. When the group size $n = 1$, the graph corresponding to $R_* = 1$ is constant at $\lambda_G = 1$, since then there can be no within-group spread of infection and the value of λ_L is irrelevant. For $n = 1, 2, \dots$, the $R_* = 1$ graph for $n + 1$ lies below that for n and it is shown below that the $R_* = 1$ graph converges to the $R_0 = 1$ graph as $n \to \infty$. As noted above, $R_0 = 1$ does not provide a good indicator as to whether a global epidemic can occur when the group size n is small.

We now return to the model with general T_I and examine the asymptotic behavior of R_* as $n \to \infty$. We still assume without loss of generality that $\mathrm{E}[T_I] = 1$ and that the within-group infection rate is $(n-1)^{-1}\lambda_L$. Thus from (3.14), $R_* = \lambda_G(1 + \mu_{n-1,1})$, so we are interested in the asymptotic behavior of $\mu_{n-1,1}$ as $n \to \infty$. For large n, the early stages of the single-group epidemic process can be approximated by a branching process and Ball (1983b) shows how to make the approximation precise in the limit as $n \to \infty$. Specifically, a sequence of epidemic processes indexed by n and the approximating branching process can be constructed on the same probability space so that, as $n \to \infty$, if the branching process goes extinct, then the final size of the epidemic process converges almost surely to the final size of the branching process, and if the branching process does not go extinct, then the final size of the epidemic process converges almost surely to ∞. Moreover, in the latter case von Bahr and Martin-Löf (1980) show that $n^{1/2}(T/n - \rho)$ converges in distribution to a normal random variable with zero mean. Here T is the final size of the single population epidemic and ρ is the largest root in $(0, 1)$ of the equation $1 - x = \exp(-\lambda_L x)$. Note that ρ is the proportion of initial susceptibles that are ultimately infected in the limiting deterministic epidemic as $n \to \infty$.

Suppose first that $\lambda_L < 1$, so that the single-group epidemic is below threshold. The mean final size of the approximating branching process is $\lambda_L/(1-\lambda_L)$, so by the dominated convergence theorem $\lim_{n\to\infty} \mu_{n-1,1} = \lambda_L/(1-\lambda_L)$. Thus R_* converges up to $\lambda_G(1+(1-\lambda_L)^{-1}\lambda_L) = \lambda_G/(1-\lambda_L)$ as n tends to ∞. Hence, if $\lambda_G < 1 - \lambda_L$ only nonglobal epidemics can occur, however large the group size n is, while if $\lambda_G > 1 - \lambda_L$ global epidemics can occur provided that n is sufficiently large. Note that as $n \to \infty$ the equation $R_* = 1$ converges to $\lambda_L + \lambda_G = 1$, that is, $R_0 = 1$.

Now suppose that $\lambda_L > 1$ so that the single-group epidemic is above threshold. Let q be the probability that the approximating branching process (to the single-group epidemic) goes extinct. Recall from Waugh (1958) and Daly (1979) that, conditional upon extinction, a supercritical Galton–Watson process with offspring probability generating function $g(s)$ behaves as a (subcritical) Galton–Watson process with offspring probability generating function $q^{-1}g(qs)$. The number of contacts made by the initial infective in the single population epidemic is Poisson with (random) mean $\lambda_L T_I$, so the offspring probability generating function of the Galton–Watson process embedded in the approximating branching process is $g(s) = \phi(\lambda_L(1 - s))$, where $\phi(\theta) = \mathrm{E}[\exp(-\theta T_I)]$. Thus q is the unique solution in $(0, 1)$ of the equation $\phi(\lambda_L(1-s)) = s$ and the offspring mean for the embedded Galton–Watson process conditioned upon extinction, $\tilde{m}$ say, is given by $\tilde{m} = -\lambda_L \phi^{(1)}(\lambda_L(1 - q))$. Further, conditional upon extinction, the mean final size of the approximating branching process is $\tilde{m}/(1 - \tilde{m})$. Combining all this with the above von Bahr and Martin-Löf limit theorem and recalling that $R_* = \lambda_G(1 + \mu_{n-1,1})$ yields

$$R_* \sim \lambda_G\left\{1 + \frac{q\tilde{m}}{1-\tilde{m}} + (1-q)\rho(n-1)\right\} \quad \text{as } n \to \infty. \tag{3.23}$$

Thus, in contrast to the situation when $\lambda_L < 1$, for $\lambda_L > 1$ global epidemics can always occur if n is sufficiently large, whatever the value of λ_G (provided it is not zero).

In the critical case $\lambda_L = 1$, the mean final size of the approximating branching process is infinite, so again global epidemics can always occur provided that n is sufficiently large.

Another way of viewing the above is to assume that λ_L is fixed and examine the behavior, as $n \to \infty$, of the critical value, λ_G^{crit} say, of λ_G for global epidemics to be possible. It follows from the preceding arguments that if $\lambda_L < 1$, then $\lambda_G^{\text{crit}} = O(1)$ as $n \to \infty$, whilst if $\lambda_L > 1$, then $\lambda_G^{\text{crit}} = O(n^{-1})$ as $n \to \infty$. This corresponds to the amplification effect discussed for the multigroup Reed–Frost epidemic in Section 2.5.

We can use our model to study the efficacy of various vaccination strategies. For example, as in Section 2.5, consider two vaccination policies: a local one in which a fixed proportion, θ say, of groups is completely vaccinated and a global one in which a proportion θ of susceptibles in every group is vaccinated. For convenience we suppose that θn is an integer. Under both policies the rate at which a given infective makes outside infections is $(1-\theta)\lambda_G$. However, in the local policy such an infection is with a group having n susceptibles, so $R_* = (1-\theta)\lambda_G(1+\mu_{n-1,1})$, but in the global policy it is with a group having $(1-\theta)n$ susceptibles, so $R_* = (1-\theta)\lambda_G(1+\mu_{(1-\theta)n-1,1})$. Clearly $\mu_{n-1,1} > \mu_{(1-\theta)n-1,1}$ so the global policy will be more effective in preventing the spread of a global epidemic.

3.4. *Final outcome of a multigroup epidemic.* In this subsection we consider the final outcome of the multigroup epidemic as m, the number of groups, becomes large. In Section 3.3.1 we examined the final outcome of a nonglobal epidemic; here we shall be concerned with what happens in the event of a global epidemic. Our argument will be heuristic, a formal proof being delayed until Section 4.2.

Let z be the expected proportion of initial susceptibles that are infected by a global epidemic. Thus z can be interpreted as the probability that a given initial susceptible, who is not in one of the initially infectious groups, is ultimately infected by the epidemic.

Fix attention on a single group that initially contained no infectives. We can decompose the ultimate spread of infection within that group by first determining which of the initial susceptibles are infected from outside the group, and then letting these individuals initiate a single population epidemic among the remaining susceptibles in the group.

This decomposition justifies the approach taken by a number of applied authors, who treated the probability π of an external (primary) contact as independent of the internal (secondary) process that they analyzed (see Sections 1.3, 3.2 and 5.1).

Let $\tilde{T}$ be the total person time units of infection present in the population at large over the whole course of the epidemic. Then for large m, $\tilde{T} \sim Nz\mathrm{E}[T_I]$. At any time a given susceptible in the group under consideration is being infected from outside the group with intensity $N^{-1}\lambda_G$ per outside infective. Thus, as $m \to \infty$, each given susceptible in the group independently avoids infection from outside with probability $\pi = \exp(-\lambda_G z\mathrm{E}[T_I])$. It follows that the ultimate spread of infection within the group has the same distribution as that of the extended model of Addy, Longini and Haber (1991) described in Section 3.2. Hence, the mean final size of the epidemic within the group is given by setting $a = 0$ in (3.10). However, the mean final size also equals zn, since z is the expected proportion of susceptibles that are ultimately infected. Thus we can deduce the equation

$$nz = n - \sum_{i=1}^{n} \frac{n!}{(n-i)!} q_i^{n-i} \pi^i \alpha_i, \tag{3.24}$$

which, since $\pi = \exp(-\lambda_G z\mathrm{E}[T_I])$, is an implicit equation for z. Clearly $z = 0$ is always a solution of (3.24). We now show that there is a (unique) second solution in $(0, 1)$ if and only if $R_* > 1$.

It is convenient to rearrange (3.24) into

$$n(1-z) = \sum_{i=1}^{n} \frac{n!}{(n-i)!} q_i^{n-i} \exp(-\lambda_G z \mathrm{E}[T_I] i)\alpha_i. \tag{3.25}$$

We proved in Section 3.2 that $\alpha_i > 0$, $i = 1, 2, \ldots$, so the right-hand side of (3.25) is a convex function of z. Thus (3.25) has at most two solutions since its left-hand side is linear in z. Further, by examining the values at $z = 0$ of the derivatives with respect to z of the two sides of (3.25), we see that there is a second solution if and only if

$$\lambda_G \mathrm{E}[T_I] \sum_{i=1}^{n} \frac{(n-1)!}{(n-i)!} q_i^{n-i} i\alpha_i > 1. \tag{3.26}$$

Now

$$\begin{aligned} \sum_{i=1}^{n} \frac{(n-1)!}{(n-i)!} q_i^{n-i} i\alpha_i &= \sum_{i=1}^{n} \frac{(n-1)!}{(n-i)!} q_i^{n-i} \alpha_i (n-(n-i)) \\ &= \sum_{i=1}^{n} \frac{n!}{(n-i)!} q_i^{n-i} \alpha_i - \sum_{i=1}^{n-1} \frac{(n-1)!}{(n-i-1)!} q_i^{n-i} \alpha_i. \end{aligned} \tag{3.27}$$

From (3.6), the second sum on the right-hand side of (3.27) is $n - 1 - \mu_{n-1,1}$. The first sum can be evaluated by recalling that $\alpha_i = G_{i-1}(1|V)$, where the sequence V is given by $v_i = q_{i+1} = \phi(\lambda_L(i+1))$, $i = 0, 1, \ldots$. We obtain

$$\begin{aligned} \sum_{i=1}^{n} \frac{n!}{(n-i)!} q_i^{n-i} \alpha_i &= \sum_{i=1}^{n} \frac{n!}{(n-i)!} v_{i-1}^{n-i} G_{i-1}(1|V) \\ &= n \sum_{i=0}^{n-1} \frac{(n-1)!}{(n-1-i)!} v_i^{n-1-i} G_i(1|V) \\ &= n, \end{aligned} \tag{3.28}$$

where in the last step we have used the recursive definition (3.1) of the Gontcharoff polynomials $G_i(x|V)$, $i = 0, 1, \ldots$. Putting all this together, we obtain from (3.27) that

$$\sum_{i=1}^{n} \frac{(n-1)!}{(n-i)!} q_i^{n-i} i\alpha_i = 1 + \mu_{n-1,1}. \tag{3.29}$$

Hence from (3.26) and the expression for R_* given in (3.14), (3.24) has a solution in $(0, 1)$ if and only if $R_* > 1$. When $R_* > 1$ the solution of (3.24) in $(0, 1)$ gives the expected proportion of initial susceptibles ultimately infected by a global epidemic.

As noted earlier, in the event of a global epidemic the final size in a group that did not have initial infectives is distributed as the final size of the extended model of Addy, Longini and Haber (1991) with $\pi = \exp(-\lambda_G z \mathrm{E}[T_I])$. This distribution may be calculated by using (3.11) or (3.12). Figure 2 illustrates, for various values of λ_L and λ_G, the final size distribution in a group

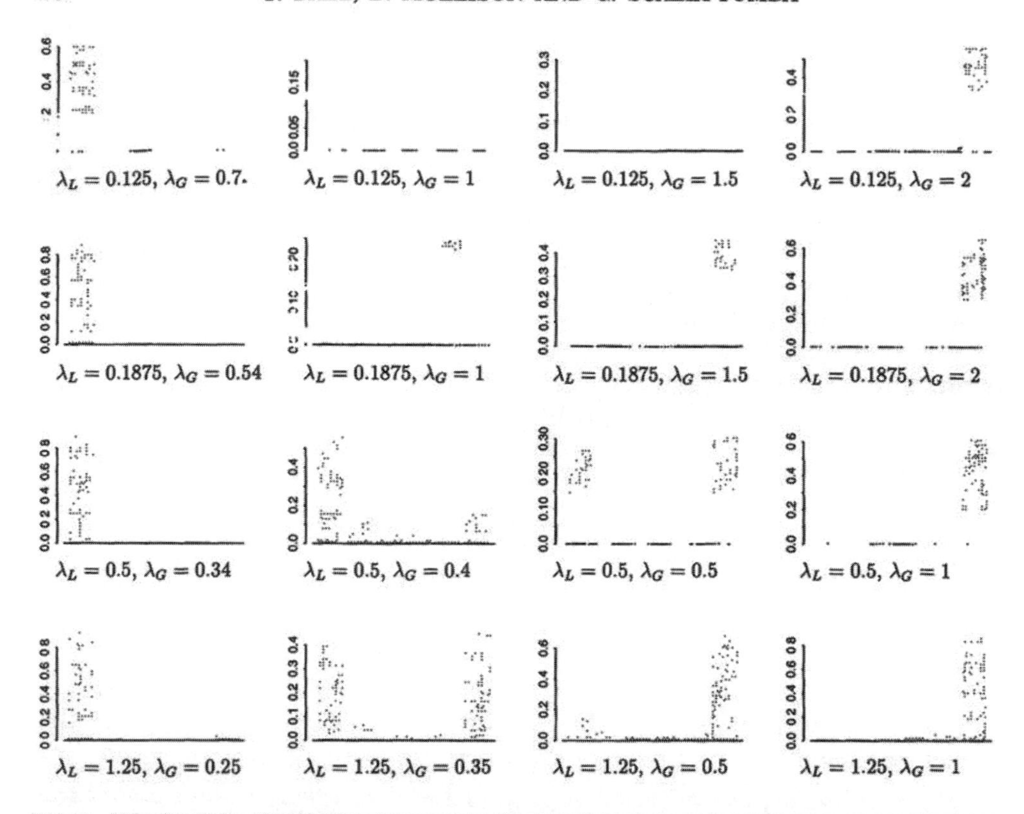

FIG. 2. *The final size distribution in a group when the infectious period T_I follows an exponential distribution with mean 1 and the group size $n = 5$. Note that the critical values of λ_G are (for each row, starting at the top)* $\lambda_G^{\text{crit}} = 0.6336,\ 0.5321,\ 0.3296,\ 0.2429$.

when the infectious period T_I follows an exponential distribution with mean 1 and the group size $n = 5$. Figure 2 also gives, for each choice of λ_L, the critical value λ_G^{crit} of λ_G for global epidemics to be possible. Notice the difference in the shape of the distribution according to whether the local reproductive ratio $R_L = (n-1)\lambda_L$ is less than or greater than 1, that is, according to whether the within-group epidemic is below or above its threshold. When $R_L < 1$ the distribution is unimodal for all values of $\lambda_G > \lambda_G^{\text{crit}}$, with the mode increasing from 0 for values of λ_G just greater than λ_G^{crit} to n (= 5) for sufficiently large values of λ_G. When $R_L > 1$ the distribution is initially bimodal as λ_G is increased from λ_G^{crit}, but becomes unimodal with the mode either at or close to n for sufficiently large values of λ_G. (In our examples the mode is always at $n = 5$ but this is unlikely to be the case in general.) The shape of the group final size distribution can be explained in terms of the threshold behavior of the single-group epidemic. When $R_L < 1$, only minor epidemics will occur within a group, but as λ_G increases so does the number of group members infected from the population at large, and hence also the size of the epidemic within the group. When $R_L > 1$, the within-group epidemic is above threshold, so major epidemics can occur as soon as $\lambda_G > \lambda_G^{\text{crit}}$. Thus the distribution is bimodal, being a mixture of two components, one corresponding to a minor epidemic and the other to a major epidemic. Again, as λ_G increases, so does the number of outside infections, and eventually the minor epidemic component will disappear.

3.5. *Unequal group sizes.* We now consider the situation in which the group sizes are not all equal. For $n = 1, 2, \ldots$, let m_n be the number of groups of size n. Let $m = \sum_{n=1}^{\infty} m_n$ be the total number of groups and $N = \sum_{n=1}^{\infty} nm_n$ be the total number of individuals. As before, the infectious periods of different infectives are independently and identically distributed according to a random variable T_I, and throughout its infectious period a given infective makes contact with each other susceptible at rate λ_G/N, and additionally with each susceptible in its own group at rate λ_L. We examine the asymptotic situation in which the number of groups m tends to infinity in such a way that $m_n/m \to h_n$, $n = 1, 2, \ldots$, and $\sum_{n=1}^{\infty} h_n = 1$. Thus, for $n = 1, 2, \ldots$, h_n is the asymptotic proportion of groups of size n. Let $\mu_h = \sum_{n=1}^{\infty} nh_n$ be the asymptotic mean group size and assume that $\mu_h < \infty$.

The initial stages of the multigroup epidemic can be approximated by a multitype branching process, in which the units are single-group epidemic processes, the offspring of a given unit are those groups that are directly infected by infectives in that unit and type corresponds to group size. Again the approximation can be made precise in the limit as $m \to \infty$ by using the coupling argument of Ball (1983b) and Ball and Donnelly (1995). Label the types $1, 2, \ldots$, according to group size and let $\Lambda = [\lambda_{ij}]$ be the offspring mean matrix of the embedded multitype Galton–Watson process. Thus λ_{ij} is the expected number of type j groups infected by infectives from a type i group single population epidemic.

Let $T^{(i)}$ and $T_A^{(i)}$ be, respectively, the final size and severity of a single population epidemic in which initially there are 1 infective and $i-1$ susceptibles. As in Section 3.3.1, the total number of outside infections emanating from a type i group follows a Poisson distribution with random mean $\lambda_G T_A^{(i)}$. The probability that a given outside infection is with an individual in a group of size j is $jm_j/N = jh_j/\mu_h$. Hence

$$\begin{aligned} \lambda_{ij} &= \lambda_G \mathrm{E}[T_A^{(i)}] jh_j/\mu_h \\ &= \lambda_G(1 + \mathrm{E}[T^{(i)}])\mathrm{E}[T_I] jh_j/\mu_h \\ &= \lambda_G(1 + \mu_{i-1,1})\mathrm{E}[T_I] jh_j/\mu_h, \end{aligned} \tag{3.30}$$

using the Wald identity for epidemics.

The multiplicative structure of the matrix Λ given by (3.30) implies that its maximal eigenvalue is

$$R_* = \lambda_G \mathrm{E}[T_I]\mu_h^{-1} \sum_{n=1}^{\infty} (1 + \mu_{n-1,1}) nh_n. \tag{3.31}$$

By standard branching process theory a global epidemic (corresponding to nonextinction of the approximating multitype branching process) has nonzero probability of occurring if and only if $R_* > 1$. Formulae implicitly giving the probability of a global epidemic and properties of a nonglobal epidemic can be derived as in Section 3.3.1.

Note that again R_* is of the form $R_* = \mu R_G$, where $R_G = \lambda_G \mathrm{E}[T_I]$ is the basic reproductive ratio for the model in which all the groups are of size 1, and $\mu = \mu_h^{-1} \sum_{n=1}^{\infty}(1+\mu_{n-1,1})nh_n$ is the size-biased mean clump size. The formula for μ uses the fact that if π_i (as in Sections 1 and 2) is the probability that an individual chosen at random from the population is in a group of size i, then

$$\pi_i = \mu_h^{-1} i h_i, \qquad i = 1, 2, \ldots \tag{3.32}$$

[thus $\mu = \sum_{n=1}^{\infty}(1+\mu_{n-1,1})\pi_n$]. Indeed, using the size-biased sampling, the initial stages of the epidemic can be approximated by a single-type branching process (in which the units are single-group epidemic processes) whose offspring distribution is Poisson with random mean, which is a mixture of $T_A^{(1)}, T_A^{(2)}, \ldots$ with respective mixing probabilities $\pi_1, \pi_2, \ldots$. Note that this second, single-type approximation avoids any difficulties caused by the possibility of there being infinitely many types in the multitype approximation.

We now turn to the final outcome of a global epidemic. Let z be the probability that a randomly chosen initial susceptible is ultimately infected by the epidemic and, for $n = 1, 2, \ldots$, let z_n be the same probability for a randomly chosen initial susceptible in a group of size n. The size-biased sampling implies that

$$z = \mu_h^{-1} \sum_{n=1}^{\infty} n z_n h_n. \tag{3.33}$$

Fix attention on a group of size n that did not contain any initial infectives. Arguing as in Section 3.4, the probability that a given susceptible in that group avoids infection from outside is $\pi = \exp(-\lambda_G z \mathrm{E}[T_I])$, and the expected final size of the epidemic within that group is $\tilde{\mu}_{n,0}$. Thus, using (3.10),

$$n z_n = n - \sum_{i=1}^{n} \frac{n!}{(n-i)!} q_i^{n-i} \pi^i \alpha_i, \qquad n = 1, 2, \ldots. \tag{3.34}$$

Summing (3.34) over n and using (3.33) yields

$$z = 1 - \sum_{n=1}^{\infty} \mu_h^{-1} h_n \sum_{i=1}^{n} \frac{n!}{(n-i)!} q_i^{n-i} \pi^i \alpha_i, \tag{3.35}$$

which, since $\pi = \exp(-\lambda_G z \mathrm{E}[T_I])$, is an implicit equation for z. Clearly, $z = 0$ is always a solution of (3.35) and similar arguments to those used in Section 3.4 show that there is a (unique) second solution in $(0, 1)$ if and only if $R_* > 1$. When $R_* > 1$, the root of (3.35) in $(0, 1)$ gives the expected proportion of initial susceptibles that are ultimately infected by a global epidemic. As in Section 3.5, the total spread of infection within a group not having initial infectives has the same distribution as in the extended model of Addy, Longini and Haber (1991), with $\pi = \exp(-\lambda_G z \mathrm{E}[T_I])$.

3.6. *Threshold parameter for the proliferation of infectious individuals.* As noted in Section 1.3, Becker and Dietz (1995) consider a model for highly infectious diseases in which it is assumed that once one individual in a group (household) is infected then so is everyone else in that group. This assumption corresponds to setting $\lambda_L = \infty$ in our model. Becker and Dietz (1995) derive two threshold parameters for their model: one, which they call R_0, for the proliferation of infectious individuals and another, which they call R_{H0}, for the proliferation of infected households. Clearly, R_{H0} is the same as our R_* and it is easily checked that setting $\lambda_L = \infty$ in (3.31) yields Becker and Dietz's formula for R_{H0}. We now determine a threshold parameter for the proliferation of infectious individuals for our model, which, to avoid confusion with our earlier notation, we denote by R_I.

Consider the branching process approximation of Section 3.3.1 and call an initial infective in a group a *primary* case and all subsequent infectives in that group *secondary* cases. Thus both primary and secondary cases can give rise to further primary cases, but only primary cases can give rise to secondary cases. We now consider a two-type branching process, in which types 1 and 2 denote primary and secondary cases, respectively. Then, using the size-biased sampling, it is easily seen that the mean matrix for the above two-type branching process is given, in the notation of Section 3.5, by

$$M = \begin{bmatrix} R_G & \mu - 1 \\ R_G & 0 \end{bmatrix}. \tag{3.36}$$

The threshold parameter R_I is given by the maximal eigenvalue of M and a simple calculation shows that

$$R_I = \frac{R_G}{2}\left(1 + \sqrt{1 + 4(\mu - 1)/R_G}\right). \tag{3.37}$$

Again, it is easily checked that setting $\lambda = \infty$ in (3.37) yields the corresponding expression for R_0 given in Becker and Dietz (1995). Note that we only had to find the maximal eigenvalue of a 2×2 matrix, whereas Becker and Dietz's approach required the maximal eigenvalue of an $N \times N$ matrix (where N denoted their largest group size), though, of course, there is a lot of structure in their $N \times N$ matrix, which is owing to our size-biased sampling. The same comment also applies to our derivation of R_*. Note also that $R_I = 1$ if and only if $R_* = 1$, so the critical values of $R_* = 1$ shown in Figure 1 also apply to $R_I = 1$.

For homogeneously mixing epidemic models it is well known that the critical fraction, v^* say, of susceptibles that have to be vaccinated to make a supercritical epidemic critical is

$$v^* = 1 - 1/R_0. \tag{3.38}$$

Becker and Dietz (1995) show that (3.38) still holds for their model if individuals are vaccinated independently with probability v^* and, in our notation, R_0 is replaced by R_I. However, this is *not* the case for our model. Equation

(3.38) holds for Becker and Dietz's model since vaccinating individuals independently with probability v reduces both R_G and $\mu - 1$ *pro rata* to $(1-v)R_G$ and $(1-v)(\mu-1)$, respectively. The second of these reductions does not hold for our model since the mean size of a single population epidemic is not linear in the initial number of susceptibles.

In a later paper, Becker and Hall (1996) consider the household threshold parameter R_{H0} and associated vaccination strategies for the spread of an epidemic among a population of households made up of individuals of p different types, labelled $1, \ldots, p$. The methodology of Section 3.5 can be extended to encompass this situation, using appropriate size-biased sampling. To determine the household threshold parameter R_{H0}, we consider a p-type branching process, where p is the number of types of individuals present in the population, whilst to determine the individual threshold parameter R_I we consider a $2p$-type branching process, where again we distinguish between primary and secondary cases. In order to derive explicit expressions for the threshold parameters we need results for multitype epidemics analogous to those given for single population epidemics in Section 3.2. These can be found, for example, in Ball (1986), Picard and Lefèvre (1990) or Ball and Clancy (1993). The details are rather involved and will be published separately. Becker and Hall (1996) sidestep these complications by labelling households according to the type of epidemic that occurs in them (so that the type space, $\mathscr{T}$ say, for households soon gets rather large) and then defining their model in terms of the mean number of type τ households generated by a type i individual, for $\tau \in \mathscr{T}$ and $i = 1, \ldots, p$. They then get more explicit results by again making the very special assumption that once one individual in a household is infected, then the whole household becomes infected. Thus the extension of our methodology to the multitype setting is concerned with models that are described at a more basic level (i.e., in terms of individual infectious periods and individual-to-individual infection rates) than those considered by Becker and Hall (1996).

4. Embedding representations of the final size of the epidemic.

4.1. *Embedding and the asymptotic distribution of final size.* Showing that the epidemic process and its final size in a population can be constructed by sampling an appropriate embedding process at suitably defined stopping times may be of interest in itself since it yields an alternative way of constructing some aspects of the epidemic process, but it also turns out to be an efficient tool for studying the distribution of the final size and related quantities, in particular in asymptotic situations. In the present case, the construction and methods of Scalia-Tomba (1985, 1990) can be used, with minor modifications, to show that the final size of the epidemic, in a large population, is either small, with probabilities related to the approximating branching process, or large, with an approximately normal distribution around the mean expected from the deterministic approximation.

The general idea of the construction is to create a process describing the number of individuals in the population who would become infected, with infection being considered as coming from outside the population, and then creating the epidemic within the population by letting the infectious individuals in the population define the amount of infection to which the remaining susceptibles will be exposed. The final size of the epidemic will then typically be characterized by a balance equation stating that the epidemic stops when the total "infection pressure" generated by those infected in the population (including initial infectives) becomes equal to the infection pressure needed to infect the same individuals. One then proceeds to show that the embedding process is asymptotically Gaussian and that the balance equation translates into a first-crossing problem for the embedding process. Finally, some further calculations are needed to clarify the "either small or large" character of the epidemic process.

4.2. *The case of distributed infectious period and households.* We will first carry out the construction for the situation considered in Section 3.1, in which the population is composed of a priori defined groups or households with given sizes.

4.2.1. *The basic household process* $(R(t), A(t))$. Let $(R(t), A(t))$ describe what has happened to a household of size n, with no initial infectives, after having been subjected to global infection pressure $t \geq 0$, that is, when it has been exposed to t time units of global infection. Here R is the number of household members having had the disease [in the terminology of Section 3.2, R is the final size when there are 0 initial infectives and $\pi = \exp(-\lambda_G t/N)$] and A is the cumulative sum of the infectious periods of these individuals. We assume that, as far as the process $(R(t), A(t))$ is concerned, local (within-household) infections are instantaneous. Thus R and A are constant except for a finite number of (simultaneous) jumps, corresponding to infection from outside of a susceptible individual; they are nondecreasing with $0 \leq R \leq n$ and $0 \leq A \leq \Sigma$, where Σ is the sum of n independent copies of T_I. The R and A components are strongly correlated (as T and T_A in Section 3.2) and may even be equal if T_I takes a constant value t_I (Reed–Frost case).

It may be useful to have a more concrete construction of (R, A). Label the individuals in the household $1, 2, \ldots, n$. For $k = 1, 2, \ldots, n$, let individual k be endowed with random variables $(Q_G^{(k)}, Q_L^{(k)}, T_I^{(k)})$, where Q_G and Q_L are the thresholds for global and local infections, respectively. Thus Q_G is the total time units of global infection that has to be present before a given individual is globally infected, so Q_G follows a negative exponential distribution with rate λ_G/N. Similarly, Q_L is the total time units of local infection that has to be present before a given individual is locally infected, so Q_L follows a negative exponential distribution with rate λ_L. All these random variables, whether for the same or different individuals, are assumed to be independent.

To construct the associated realization of (R, A), first, the n Q_G-values are marked on the t-axis. The first jump of (R, A) occurs at the smallest of these, that is, at the least amount of global infection necessary to infect an individual in our previously completely susceptible household. This infected individual will initiate an epidemic among the remaining $n-1$ susceptibles that is determined by the values of $\{Q_T^{(k)}, T_I^{(k)}\}$ [see, e.g., the construction of Sellke (1983) as described in Ball (1986)]. Let T and T_A be, respectively, the total size and severity of this epidemic, where T includes the initial infective. Then the size of the first jump of (R, A) is (T, T_A). Next, the $T-1$ marks corresponding to the individuals who are no longer susceptible should be deleted from the t-axis. The next jump of the process will then occur at the smallest remaining mark, at which point one of the individuals not infected by the epidemic corresponding to the first mark will be globally infected. This individual will initiate an epidemic among the other remaining susceptibles. The total size and severity of this second epidemic is the size of the second jump of (R, A) and so on. This view of (R, A) yields, for instance, easy estimates of the increments of the process, since these depend on finding at least one of the original marks in the time interval considered.

The results in Section 3.2 [in particular (3.9) and (3.10)] are directly interpretable in terms of (R, A). We have $\mathrm{E}[R(t)] = \tilde{\mu}_{n,0}$, with $\pi = \exp(-\lambda_G t/N)$ [we will use the notation $\tilde{\mu}_{n,0}(\pi)$ in the sequel, to make the dependence on π explicit], and $\mathrm{E}[A(t)] = \mathrm{E}[T_I]\mathrm{E}[R(t)]$, since Wald's identity for epidemics applies for any fixed number of initial infectives. Let us further, for $t, s \geq 0$, denote $\mathrm{cov}(R(t), R(s))$ by $c_n^R(t, s)$, $\mathrm{cov}(A(t), A(s))$ by $c_n^A(t, s)$ and $\mathrm{cov}(R(t), A(s))$ by $c_n^B(t, s)$. For $t = s$, the covariances can be derived from (3.9). For $t \leq s$, say, one may, at least in theory, use the properties of the exponential distributions to derive the covariances: given $(t, R(t), A(t))$, $R(s) - R(t)$ and $A(s) - A(t)$ will have the same distribution as $(R(s-t), A(s-t))$ in a household starting with $n - R(t)$ susceptibles. However, it will be seen in the sequel that explicit determination of the covariance functions is not essential for the derivation of the main results; they will only be needed explicitly for the case $t = s$, for a particular choice of t.

4.2.2. *Embedding the epidemic process.* Assume that each household in the population has a process $(R_i(t), A_i(t))$, $i = 1, \ldots, m$, of the type described above. Let $R_\bullet(t) = \sum R_i(t)$ and $A_\bullet(t) = \sum A_i(t)$. Assume also that an initial amount T_0 of infectious time is applied to the initially totally susceptible population. We can now define a sequence of stochastic times in which to consider $(R_\bullet(t), A_\bullet(t))$ (these correspond roughly to a description of the epidemic by cumulated generations, with anticipated local or within household infections):

$$\begin{aligned}
T_0 &\to R_\bullet(T_0), A_\bullet(T_0),\\
T_1 &= T_0 + A_\bullet(T_0) \to R_\bullet(T_1), A_\bullet(T_1),\\
&\vdots\\
T_{k+1} &= T_0 + A_\bullet(T_k) \to R_\bullet(T_{k+1}), A_\bullet(T_{k+1}),\\
&\vdots
\end{aligned}$$

Thus T_1 is the total amount of infection that has been generated in the population after the local household epidemics initiated by the initial T_0 units of infectious time have occurred. These T_1 units of infection may create further global infections which may in turn give rise to further local infections, after which there will have been a total of T_2 units of infectious time generated in the population. The process continues until the additional infectious time created by a set of local infections is not enough to give rise to further global infections. Consequently, the above sequence stops at $T_\infty := \min\{t \geq 0\colon t = T_0 + A_\bullet(t)\}$ (see Figure 3). Then $R_\bullet(T_\infty)$ represents the final size of the epidemic in the population and $T_\infty = A_\bullet(T_\infty) + T_0$ represents its severity.

4.2.3. *Asymptotic distribution of the embedding process and of the final size of the epidemic.* For $n = 1, 2, \ldots$, let m_n be the number of households of size n, m the total number of households, $N = \sum nm_n$ the total number of individuals in the population, $\theta_n = m_n/m$ the proportions of households of size n and $\tilde{m}_1 = \sum n\theta_n < \infty$. Then $\mathrm{E}[R_\bullet(t)] = \sum m_n \tilde{\mu}_{n,0}(\exp(-\lambda_G t/N))$. In order to handle the bivariate character of $(R_\bullet, A_\bullet)$, let us use the Cramér–Wold device and define, for $(\alpha, \beta) \in \mathbb{R}^2$,

$$(4.1)\quad Z_m^{(\alpha,\beta)}(t) = \frac{1}{\sqrt{m}}\Big(\alpha\big(R_\bullet(Nt) - \mathrm{E}[R_\bullet(Nt)]\big) + \beta\big(A_\bullet(Nt) - \mathrm{E}[A_\bullet(Nt)]\big)\Big).$$

Proceeding as in Scalia-Tomba (1990), under the further condition that $\sum n^2\theta_n < \infty$, it can be shown that, as $m \to \infty$, $Z_m^{(\alpha,\beta)}$ converges in distribution, on $D[0,\infty)$ with the Skohorod topology, to a Gaussian process with mean 0 and covariance function

$$(4.2)\quad \gamma^{(\alpha,\beta)}(t,s) = \sum \theta_n\big(\alpha^2 c_n^R(t,s) + \alpha\beta(c_n^B(t,s) + c_n^B(s,t)) + \beta^2 c_n^A(t,s)\big),$$

where π, at times t and s, now equals $\exp(-\lambda_G t)$ and $\exp(-\lambda_G s)$, respectively.

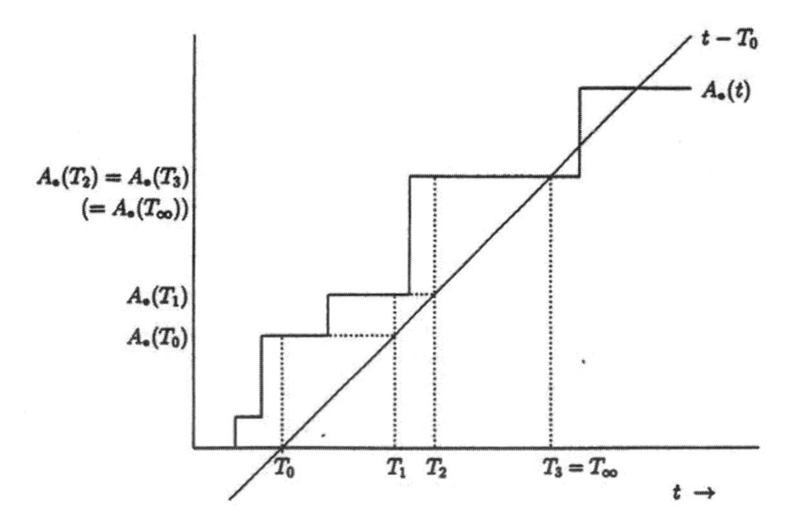

FIG. 3. *Determination of the severity of the epidemic from the embedded process $A_\bullet(t)$.*

Now let $r(t) = \sum \theta_n \tilde{\mu}_{n,0}(\exp(-\lambda_G t))$, $a(t) = \mathrm{E}[T_I]r(t)$, $\tilde{r}(t) = r(t)/\tilde{m}_1$ and $\tilde{a}(t) = a(t)/\tilde{m}_1$. Then, letting $(\alpha, \beta) = (0, 1)$, we have that, as $m \to \infty$,

$$\tilde{A}_m(t) = \frac{1}{\sqrt{m}}(A_\bullet(Nt) - ma(t)) \tag{4.3}$$

converges weakly to a Gaussian process with mean 0 and covariance function $\gamma^{(0,1)}$. Assume now that $T_0/N \to \mu_0 > 0$ as $m \to \infty$. Then

$$\frac{T_\infty}{N} = \min\left\{t\colon t = \frac{T_0}{N} + \tilde{a}(t) + \frac{\sqrt{m}}{N}\tilde{A}_m(t)\right\}, \tag{4.4}$$

and, since $(\sqrt{m}/N)\tilde{A}_m(t)$ converges uniformly to 0 on any compact subset of $[0, \infty)$, we have that $T_\infty/N \to \tau(\mu_0) := \min\{t\colon t = \mu_0 + \tilde{a}(t)\}$. We may then conclude that $Z_m^{(\alpha,\beta)}(T_\infty/N)$ converges in distribution to $Z_m^{(\alpha,\beta)}(\tau(\mu_0))$ for all (α, β), which means that the vector

$$\frac{N}{\sqrt{m}}\left(\frac{R_\bullet(T_\infty)}{N} - \tilde{r}\left(\frac{T_\infty}{N}\right), \frac{A_\bullet(T_\infty)}{N} - \tilde{a}\left(\frac{T_\infty}{N}\right)\right) \tag{4.5}$$

converges to a bivariate normal distribution with mean 0 and covariance matrix $M(\mu_0)$ with elements $M_{11} = \sum \theta_n c_n^R(\tau(\mu_0), \tau(\mu_0); \pi = \exp(-\lambda_G \tau(\mu_0)))$, and $M_{12} = M_{21}$ and M_{22} of similar form, with c_n^B and c_n^A replacing c_n^R. By using the identities satisfied by T_∞ and by $\tau(\mu_0)$ [see Scalia-Tomba (1990)], this result can be recast into the convergence in distribution of the vector

$$\sqrt{m}\left(\frac{R_\bullet(T_\infty)}{N} - \tilde{r}(\tau(\mu_0)), \frac{A_\bullet(T_\infty)}{N} - \tilde{a}(\tau(\mu_0))\right) \tag{4.6}$$

to a bivariate normal distribution with mean 0 and covariance matrix $\tilde{m}_1^{-2}(1 - \tilde{a}'(\tau(\mu_0)))^{-2} AM(\mu_0)A^T$, where $A_{11} = 1 - \tilde{a}'(\tau(\mu_0))$, $A_{12} = \tilde{r}'(\tau(\mu_0))$, $A_{21} = 0$ and $A_{22} = 1$.

This is the basic result on asymptotic normality of the final size of the epidemic, around the value predicted by "deterministic" considerations, when the epidemic is started by a large amount of initial infection [some algebra will show that the definitions of $\tau(\mu_0)$ and, consequently, of $\tilde{r}(\tau(\mu_0))$ and $\tilde{a}(\tau(\mu_0))$ agree with (3.35), when $\mu_0 = 0$].

The most interesting case to study is, however, when T_0 remains fixed as $m \to \infty$, corresponding to few initial infectives. One must then combine the branching process approximations of Sections 3.3.1 and 3.5 with the asymptotic normality results shown above. Once again, the strategy in Scalia-Tomba (1985, 1990), of studying the final size distribution in different ranges of values, may be followed. Let us, for simplicity, denote the final size of the epidemic in a population with m households by T_m and assume that the epidemic is above threshold (otherwise, the results in Sections 3.3.1 and 3.5 account for the whole asymptotic distribution). The branching process approximations of Sections 3.3.1 and 3.5 then show that $\Pr\{T_m = k\} \to p(k)$, for $k = 0, 1, \ldots,$ where $p(\cdot)$ is the distribution of the total size in the approximating branching process. This distribution has total mass $\tau < 1$, say, corresponding to the event of extinction of the approximating branching process. It therefore remains to show that the remaining probability mass $1-\tau$ is concentrated around the deterministic solution for a large epidemic [see (3.35); the solution corresponds to $\rho = \tilde{r}(\tau(0))$, with the convention that the non-zero solution should be taken when $\mu_0 = 0$]. One then starts by studying $\Pr\{k < T_m < a_m\}$, where $\{a_m\}$ satisfies $a_m \to \infty$ but $a_m/m \to 0$ as $m \to \infty$, with the aim of showing that

$$\lim_{k\to\infty} \lim_{m\to\infty} \Pr\{k < T_m < a_m\} = 0. \tag{4.7}$$

The coupling construction of an approximating branching process by Ball and Donnelly (1995) (see Section 3.3.1), combined with the introduction of a lower bounding branching process [cf. Whittle (1955), Ball and Clancy (1992) and Andersson (1993)], can be used for this purpose. The approximating branching process $B_U(t)$, say, is always larger than the infectives process $I_m(t)$, since every contact is considered as a new individual in B_U, but some contacts do not yield new infectives in I_m, since contacts may occur with already infected or removed individuals. This mechanism amounts to a thinning of the branching process, with thinning probabilities depending on the total progeny up to the time point considered for the contact. To make things simpler, one can therefore apply thinning to each contact in the branching process, with fixed probability $\varepsilon > 0$, which will overestimate the "true" thinning probabilities as long as the total number of individuals ever having been infected is less than εN, thus constructing a second branching process $B_L(t)$, for which we will have $B_L(t) \leq I_m(t) \leq B_U(t)$, at least as long as the total epidemic is less than εN. If we denote the distribution functions of final size (total progeny) by F_L, F_m, and F_U, respectively, we will then have $F_U(i) \leq F_m(i) \leq F_L(i)$, for all $i \leq \varepsilon N$. Thus, $0 \leq F_m(a_m) - F_m(k) \leq F_L(a_m) - F_U(k) \leq \tau(\varepsilon) - F_U(k)$, where $\tau(\varepsilon)$ is the extinction probability in the ε-thinned process. However, since $\tau(\varepsilon) \to \tau$ as $\varepsilon \to 0$ and $F_U(k) \to \tau$ as $k \to \infty$, one obtains the desired result.

The remaining range, as long as $\{a_m\}$ is taken so that $a_m/\sqrt{m} \to \infty$, can be studied using the Gaussian process approximation [see Scalia-Tomba (1985)] to show that the crossing condition, equivalent to achieving the final size, can only be fulfilled in a $O(\sqrt{m})$ neighborhood of the "deterministic" value. Having thus accounted for the whole asymptotic probability mass, one now proceeds by showing that $\Pr\{(T_m - \rho N)/\sqrt{N} \in K\} = \Pr\{(T_m - \rho N)/\sqrt{N} \in K \mid T_m > a_m\}\Pr\{T_m > a_m\} \approx \Pr\{(T_m - \rho N)/\sqrt{N} \in K \mid T_m > a_m\}(1-\tau)$, for large N and $K \subset R$ bounded. The final step consists in showing that the epidemic process, conditioned on $T_m > a_m$, that is, on having a large epidemic, again follows the Gaussian approximation derived above, now with $\mu_0 = 0$. However, the conditioning event involves members from at most a_m households and times, used as arguments in the $Z_m^{(\alpha,\beta)}$ process, of order $O(a_m/m)$. The effect on the (conditional) limit law of $Z_m^{(\alpha,\beta)}$ will be vanishingly small and the limit law will be unchanged, at least as long as the removal of any set of a_m households from the total set of m households does not affect asymptotic proportions or means. This last requirement is equivalent to the uniform integrability of the sequence of household size proportion distributions, indexed by m.

4.3. *The case of fixed infection probabilities.* In the case studied in Sections 2.1 and 2.2, in which the infection probabilities are fixed and independent, it is possible to construct an embedding process directly based on the clumps formed by the local infection process. Let $\{C_k\}$ denote the total number of local components of size $k = 1, \dots, N$ that have been formed by local infection, in a population of size N. Closely following Scalia-Tomba (1985, 1990), we now construct an epidemic between components, with susceptibility and infectivity proportional to size. To each component of size k, we attach a threshold variable with geometric distribution with "success probability" $= 1-(1-p_G)^k$, representing the number of individual infection attempts necessary to infect the component. We denote these variables by $\{Q_{kj}\}$, $1 \le j \le C_k$, $k = 1, \dots, N$. We now define processes $X_{kj}(t) = 1_{\{Q_{kj} \le t\}}$ and $X_k(t) = \sum_j X_{kj}(t)$, which represent the numbers of k-components that have been infected after t infection attempts on the population. Finally, we define $X(t) = \sum_k kX_k(t)$, the total number of individuals infected after t infection attempts. We now construct the generations of the epidemic process by considering $X(t)$ at suitably defined random times. Assuming that the epidemic is started by m_0 initial infectives global to the population, we set $T_1 = m_0$, $T_2 = m_0 + X(T_1)$ and, in general, $T_{k+1} = m_0 + X(T_k)$. These times form an increasing sequence which stops at $T_\infty = \min\{t\colon t = m_0 + X(t)\}$. The final size of the epidemic in the population is then $X(T_\infty)$.

We would now want to consider the asymptotic situation $N \to \infty$, $p_G \approx \lambda_G/N$, local infection probabilities fixed and m_0 either fixed or increasing with N. Except for the randomness of $\{C_k\}$, the problem is similar to the situations studied in Scalia-Tomba (1985, 1990). It can therefore be expected that similar results will be valid, modified only by the additional randomness generated by $\{C_k\}$. However, in models like the great circle (Section 2.1) or the epidemic among giants (Section 2.3), there will potentially be an infinite number of types (sizes) of local components. Work is in progress on how best to resolve the technical problems arising in these situations; the results will be published separately.

5. Applications.

5.1. *Estimating R_* from household total size data.* A number of authors [e.g., Longini and Koopman (1982), Becker (1989) and Addy, Longini and Haber (1991)] have previously considered household infection data, using a secondary attack rate that corresponds to our λ_L. They do not explicitly model the spread of the infection through the wider population, simply assuming that each individual is exposed to the same probability of external infection (via a primary attack rate).

The present paper provides a framework for modelling the spread through the whole population that is consistent with these previous household models (see Section 3.4). We can thus estimate parameters of the internal process similarly to those authors and additionally relate the probability of external infection to our global infection rate parameter λ_G, and thus estimate the overall group-to-group reproductive ratio R_*.

In this subsection we describe a method for estimating the threshold parameter R_* when the available data are the total number of individuals in each group that are ultimately infected by the epidemic. We shall assume that the number of groups m is large, that a global epidemic has occurred and that the distribution of the infectious period is known. Our method is based on Addy, Longini and Haber (1991); see Becker (1989) for alternative approaches.

Consider first the extended model of Addy, Longini and Haber (1991) and suppose that the total size of such epidemics in a number of independent groups is known. Let λ_L denote the individual-to-individual local infection rate and let λ_G denote the individual-to-population global infection rate as defined in Section 3.1. Addy, Longini and Haber (1991) give an algorithm for obtaining maximum likelihood estimates of the local infection rate λ_L and the probability π that a random individual escapes infection. [The likelihood is straightforward to compute numerically using (3.12).] In our situation the epidemic total sizes in different groups are not mutually independent, but if the number of groups is large, the total sizes will be approximately independent in the event of a global epidemic. Thus estimates for λ_L and π can be derived using the method of Addy, Longini and Haber (1991). An estimate for z can then be obtained using (3.35), allowing λ_G to be estimated from the equation $\pi = \exp(-\lambda_G z \mathrm{E}[T_I])$. An estimate for R_* can then be obtained from (3.31).

As a simple example, we consider data on the spread of an influenza epidemic in Tecumseh, Michigan, analyzed in Addy, Longini and Haber (1991). The data do not exactly fit our situation since (a) they are combined data over two separate epidemics, (b) only 10% of households are included, (c) households of more than five individuals are omitted and (d) it is likely that some susceptibles were in fact immune or at least highly resistant to the strain involved in the outbreak (Klaus Dietz and Jim Koopman, personal communication), which will lead to underestimation of R_* [see Dietz (1993)]. The present analysis should therefore be viewed as simply illustrative of our methodology. The data are shown in Table 1.

Addy, Longini and Haber (1991) considered two possible distributions for the infectious period, namely, $T_I \equiv 4.1$ days and T_I follows a gamma distribution with probability density function $f(t) = c^2 t \exp(-ct)$, $t > 0$, where $c = 2/4.1 \approx 0.49$. For the model with a constant infectious period, Addy, Longini and Haber (1991) obtained the estimates $\hat{\lambda}_L = 0.0423$ and $\hat{\pi} = 0.8677$, from which $\hat{p}_L = 0.1592$, $\hat{\lambda}_G = 0.1950$ and $\hat{z} = 0.1775$. From these we can calculate $\hat{R}_G = 0.7995$, $\hat{\mu} = 1.4145$ and hence $\hat{R}_* = \mu R_G = 1.1309$. For the model with a gamma distributed infectious period, Addy, Longini and Haber (1991) obtained the estimates $\hat{\lambda}_L = 0.0446$ and $\hat{\pi} = 0.8674$, from which $\hat{p}_L = 0.1605$, $\hat{\lambda}_G = 0.1955$ and $\hat{z} = 0.1775$; from which $\hat{R}_G = 0.8006$, $\hat{\mu} = 1.4102$ and $\hat{R}_* = 1.1303$. Note that the two models give very similar estimates of R_*

TABLE 1

Observed distribution of influenza A(H3N2) *infections in* 1977–1978 *and* 1980–1981 *combined epidemics in Tecumseh, Michigan*

	No. of susceptibles* per household				
No. infected	1	2	3	4	5
0	110	149	72	60	13
1	23	27	23	20	9
2		13	6	16	5
3			7	8	2
4				2	1
5					1
Total	133	189	108	106	31

*The criterion for classifying individuals as susceptible is a preseason hemagglutination inhibition test detecting no antibody in a dilution of 1 in 128 or less. Households with more than five susceptibles are deleted from all analyses. [From Addy, Longini and Haber (1991).]

and other parameters. Also, the observed proportion of initial susceptibles ultimately infected by the disease is $250/1414 = 0.1768$, which is in close agreement with the estimate $\hat{z} = 0.1775$ fitted from both models.

The moral appears to be that the data are inadequate to discriminate between rival models for within-group contacts, but the parameters that determine the overall spread of the epidemic—μ and R_G, and hence $R_* = \mu R_G$—are not sensitive to this inadequacy. Similar conclusions can be drawn from the comparison by Islam, O'Shaughnessy and Smith (1995) of their fit of the somewhat extreme inverse Gertsbakh model (see Section 5.2.2) with the results of Haber, Longini and Cotsonis (1988) for some similar sets of household epidemic data.

The method of Addy, Longini and Haber (1991) also yields approximate confidence sets for (λ_L, π). Thus an approximate confidence interval for R_* could be obtained, since R_* is a function of λ_L and π. The above method will always yield an estimate of R_* that is larger than 1. This is because (3.35) is essentially deterministic, and in a deterministic model an initial trace of infection can only lead to a nonzero proportion of the population ultimately being infected if the model is above threshold. Thus the above method of estimating R_* should only be used if there is a good reason to believe that a global epidemic has occurred.

5.2. *Vaccination: the equalizing strategy.* The fundamental aim of a vaccination program must be to reduce the basic reproductive ratio R_* to below unity. In Section 2.5 we examined the implications of this for a simple example of large groups of equal sizes. Now, having seen (Section 3) that the relation

$R_* = \mu R_G$ holds for a wider set of models, we return to examine the question of optimal vaccination strategies in more generality. We shall compare different strategies that vaccinate a fixed proportion v of the population.

As noted in Section 2.5, any vaccination strategy will reduce R_G simply *pro rata*, so the difference between strategies will lie in how they affect the (size-biased) mean component size μ. In practice, of course, we do not know the component sizes at the time of vaccination, and even for small group sizes the evaluation of the distribution and mean of the component size in general requires quite complicated iterative calculation (see Section 3.2).

When we consider more general within-group distributions, one straightforward strategy that suggests itself is to leave the numbers of susceptibles in each group as nearly equal as possible. We call this the *equalizing strategy*, and conjecture that it is optimal for the groups model for any infectious period distribution.

5.2.1. *The equalizing strategy for groups or households.* It is easy to see that the equalizing strategy will be optimal if and only if, for all n, two groups of n susceptibles contribute less to μ than a pair of groups of sizes $n-1$ and $n+1$. Because the probability of a global infection hitting a group with n susceptibles is proportional to n, this condition is equivalent to the sequence $(n\mu_n)$ being convex; here $\mu_n = \mu_{1,n-1}$ is the mean size of an outbreak in a group of n susceptibles which is started by just one of them becoming infected. The condition that $(n\mu_n)$ be convex is in turn equivalent to the requirement that the second difference $D_n \equiv n\mu_n - 2(n-1)\mu_{n-1} + (n-2)\mu_{n-2}$ be greater than or equal to 0 for all n.

It is easily shown that the equalizing strategy is optimal for the simple "all-or-none case" where within-group outbreaks are either of size 1, with probability q_1, or of size n (this can be thought of as arising when the infectious period is either of length 0 or ∞, with respective probabilities q_1 and $1-q_1$). For this case, $\mu_n = q_1 + n(1-q_1)$, whence $D_n = 2 - 2q_1 \geq 0$.

Note that the model of Becker and Dietz (1995) is the even more special "all" case ($q_1 = 0$, and therefore $D_n = 2$), where within-group infectivity is so high that no one escapes. [The details of their calculations are a little complicated because they work in terms of the non-size-biased distribution $\{h_k\}$ rather than $\{\pi_k\}$ (see Section 2.3), but it is not difficult to check that their conclusions are consistent with the optimal strategy being to minimize μ, and that this is achieved by using the equalizing strategy.]

The all-or-none case is one in which the within-group infections by an individual are maximally correlated. The other extreme, for models with a general infectious period, is the independent links or Reed–Frost case considered in Section 2. For this, we have calculated D_n for $n = 2, \ldots, 15$ using Maple (see Figure 4), and the conjecture appears to hold for all these values, with a pattern suggesting that it is likely to hold for all n.

Indeed, on the basis of this, and a similar plot for the epidemic with exponentially distributed infectious period (for $n = 2, \ldots, 8$), we conjecture that in fact $D_n \geq 2 - 2q_1$ for all n, so that the simple all-or-none case is the lower bound.

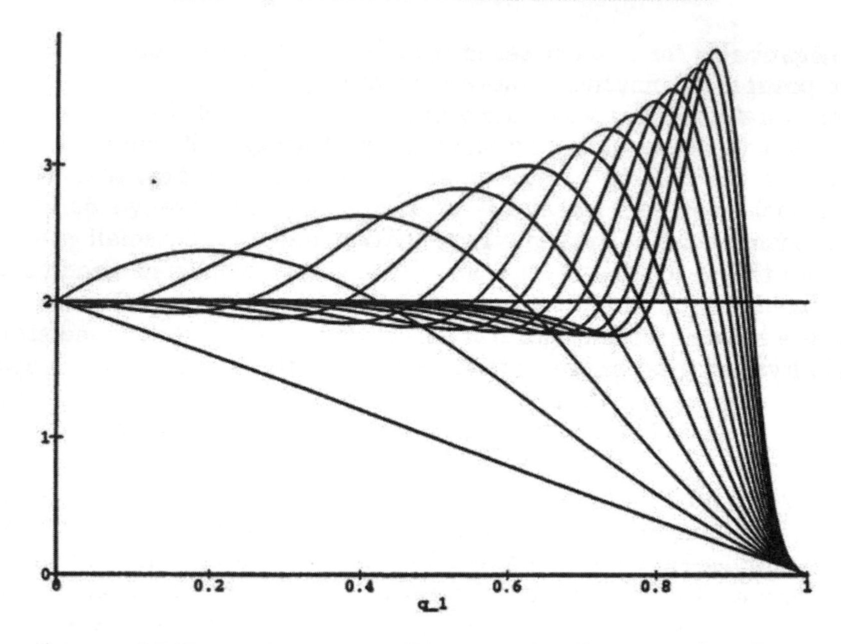

FIG. 4. *The second difference D_n, conjectured to be greater than or equal to $2 - 2q_1$, plotted against q_1 for $n = 2, \ldots, 15$ for the independent links (Reed–Frost) case.*

Our calculations for the independent links and exponential infectious period models used the techniques of Section 3.2, particularly (3.6). For small values of n, we can work out explicit expressions for D_n for a general infectious period in terms of the sequence (q_k), where q_k is as before [see following (3.6)] the probability that an infective will infect none of a set of k susceptibles in the same group. Let $d_k \equiv q_k - q_{k+1}$ be greater than or equal to 0 because (q_k) is a monotone nonincreasing sequence. We find $D_2 = 2d_0$, $D_3 = 2d_0 + 6d_0d_1$ and $D_4 = 2d_0 + 12d_0(q_1d_1 + d_0d_2 + 2d_1d_2)$, in each case greater than or equal to our conjectured minimum of $2d_0$.

In the remainder of this subsection, we prove that the equalizing strategy is optimal for a variant of the groups model in which individual contacts are *negatively* correlated (Section 5.2.2), for the case of large groups (Section 5.2.3), and, in Section 5.2.4, for the great circle model introduced in Section 2.3.

Note that for the different class of models where global infections choose groups with equal probabilities, instead of individuals with equal probabilities, it is easy to show that the equalizing strategy is *not* optimal. For this case, the appropriate second difference is $D_n = \mu_n + \mu_{n-1} - 2\mu_{n-2}$, and it is quite easy to show that $D_2 = 1 - 2q_1$, which is less than 0 for $q_1 > 1/2$, and that, in general, $D_n = -1/n < 0$ at $q_1 = 1$.

5.2.2. *The equalizing strategy for the Gertsbakh model and its inverse.* Gertsbakh (1977) introduced a model in which each individual makes exactly one (potentially infectious) contact. This does not seem realistic in the context of epidemics, but is of some theoretical interest in that it provides an example with negative correlation between the contacts made by an individual, which is not possible for our basic general infectious period model as defined in Section 3.1. Gertsbakh (1977) also considered the inverse of this model, in which there is exactly one contact *to* each individual, and this inverse model has recently been fitted to data by Islam, O'Shaughnessy and Smith (1995). Both models can be generalized, replacing "exactly one contact" by "one contact with probability $(n-1)p$, otherwise none" [we choose this parameterization so that p is the probability of contacting any one specific individual; $p \leq 1/(n-1)$].

For the inverse model, with one initial infective in a group of size n, the probability that the total size of the group epidemic will be k is

$$p'_k = \binom{n-1}{k-1} p(kp)^{k-2}(1-kp)^{n-k} \qquad (1 \leq k \leq n) \tag{5.1}$$

[Islam, O'Shaughnessy and Smith (1995)]. It is not easy to calculate the mean outbreak size μ_n from (5.1). However, it follows from a simple and general result (see Appendix) that μ_n is the same as for the forward model, for which calculations turn out to be easier. We can write down the corresponding probability that the total size of the group epidemic will be k for the forward model,

$$p_k = \frac{(n-1)!}{(n-k)!}(p^{k-1} - (n-k)p^k) \qquad (1 \leq k \leq n). \tag{5.2}$$

From (5.2) it is straightforward to show that (for both the Gertsbakh model and its inverse)

$$\mu_n = \sum_{k=0}^{n-1} \frac{(n-1)!}{(n-k-1)!} p^k. \tag{5.3}$$

From this, matching powers of p in $D_n \equiv n\mu_n + (n-2)\mu_{n-2} - 2(n-1)\mu_{n-1}$, we have that

$$D_n = \sum_{k=1}^{n-1} \frac{(n-2)!}{(n-k-1)!} k(k+1)p^k. \tag{5.4}$$

This is a sum of nonnegative terms and is therefore greater than or equal to its first term, $2p = 2 - 2q_1$, so our conjecture holds for both the Gertsbakh model and its inverse.

5.2.3. *The equalizing strategy for large groups.* We next consider the vaccination problem in the limiting case where groups are large (see Sections 2.4 and 2.5). Let $r_i = (n_i - 1)p_L \approx n_i p_L$ denote the local reproductive ratio in group i, whose size is n_i, and let $R_* = R_G \mu$ be the overall reproductive ratio as before. Then we can distinguish three cases:

1. *Safe*: $R_* < 1$ and $r_i < 1$ for all groups; only minor local outbreaks can occur.
2. *Restricted*: $R_* < 1$, but $r_i > 1$ for some groups; the overall (between-group) epidemic is below threshold, but an outbreak can affect significant proportions $[O(n_i)]$ within a small number of groups.

3. *Generalized*: $R_* > 1$: a global outbreak affecting a significant proportion of the whole population $[O(N)]$ can occur.

This classification follows that of Watson (1972) except that we have changed the name for case (1): Watson called this localized, which given our use of "local" in this paper would more appropriately be applied to case (2). We could further subdivide (3) according to whether $r_i > 1$ or $r_i < 1$ for most groups: this determines whether the outbreaks in each group are driven by local or global contacts.

In case (2), it is arguable whether we need to vaccinate. If we wish to, then the requirement is to reduce each group below its local threshold, which requires that we vaccinate a proportion $v_i \geq 1 - 1/r_i$ in group i. Note that in case (2), $R_* = R_G\mu < 1$ although some groups are above their local threshold and therefore $\mu \gg 1$; thus we must have $R_G = o(1)$ in this case.

We can also have $R_G = o(1)$ in case (3), provided that $r_i > 1$ for some groups. This is the interesting case already referred to in Sections 2.4 and 3.3.2, where the large amplification from local infections in such groups puts the population as a whole above threshold although the global infection probability p_G is extremely small. In this case, it is easy to see that the equalizing strategy will, as we increase the proportion vaccinated v, take us to case (2) before we reach the fully satisfactory situation of case (1).

To keep matters simple, we shall make the stronger requirement that our vaccination strategy must take us to case (1), where not even a restricted outbreak can occur. We can then prove that the equalizing strategy is optimal, as follows.

We assume that we have m groups of sizes n_i, $i = 1, \ldots, m$, giving a total population $N = \sum_i n_i$. We consider the asymptotic regime where the sizes tend to infinity with the relative sizes fixed, that is, $n_i = f_i N$, where $\{f_i\}$ is a fixed distribution, and we scale p_G and p_L so that the global reproductive ratio $R_G = Np_G$ and mean within-group reproductive ratio $\bar{R}_L = Np_L/m$ (note that for once this is not a size-biased mean) remain fixed. We could also let the number of groups $m \to \infty$, provided we keep the distribution of group sizes fixed.

As usual, we consider a strategy that vaccinates a proportion v of the population, asking what distribution of the vaccine among groups will achieve $R_* < 1$ for minimal v. We shall use primes to denote modified values of parameters after vaccination, including s'_i to denote the number of susceptibles in group i after vaccination.

Since we require $r'_i < 1$ for all i, we can use the branching process approximation for within-group contacts to give us $\mu'_i = 1/(1 - r'_i)$. Hence μ', the size-biased mean of the μ'_i's, is $\sum_i \mu'_i(s'_i / \sum_j s'_j)$. Now $r'_i = s'_i p_L$ and $R'_G = p_G \sum_j s'_j$, whence $R'_* = R'_G\mu' = c\sum_i \mu'_i r'_i = c\sum_i r'_i/(1 - r'_i)$, where $c = p_G/p_L$. Also, $\sum_i r'_i = p_L \sum_i s'_i = (1 - v)Np_L$, that is, $\sum r'_i \propto 1 - v$.

Thus our vaccination problem is equivalent to maximizing $\sum r'_i$ subject to keeping $\sum r'_i/(1 - r'_i)$ fixed and subject to the constraints $r'_i \leq r_i$ (i.e., $s'_i \leq n_i$; violating this constraint would require a negative number of vaccinations in group i). Since $f(x) = x/(1-x)$ is convex over the range of interest (0,1), the solution to this is to take all the r_i''s equal, as far as possible, and since $s_i' \propto r_i'$, this *is* the equalizing strategy. By "as far as possible" we mean that if the constraints $s_i' \leq n_i$ prevent us from making the s_i''s equal, we leave $s_i' = n_i$ in all groups up to a certain size n_v and take $s_i' = n_v$ wherever $n_i > n_v$.

To find the value of v required for the straightforward case where we can make the s_i''s equal, we note that their value will then be $(1-v)N/m$ and hence the r_i''s will all equal $(1-v)\bar{R}_L$. Also $R_G' = (1-v)R_G$ and so $R_*' = R_o'\mu' = (1-v)R_G/(1-(1-v)\bar{R}_L)$, whence the condition $R_*' < 1$ yields

$$v > 1 - 1/R_0, \tag{5.5}$$

where $R_0 \equiv R_G + \bar{R}_L$, the mean reproductive ratio for an individual from a randomly chosen household.

In conclusion, we note that our analysis only required consideration of the branching process regime at the start of a potential outbreak. One consequence of this [see, e.g., Mollison (1995), Section 2.3] is that our results will hold for the deterministic version of our large group model; this gives an indirect way of recovering the deterministic result of May and Anderson (1984) [see also Hethcote and Van Ark (1986)].

5.2.4. *The equalizing strategy for the great circle model.* We conclude by proving that the equalizing strategy is optimal for the great circle model: that is, the optimal policy is to spread the vaccinations around the circle as evenly as possible.

We consider then a population of N individuals spaced equally around a circle. If we vaccinate a fixed number m of individuals, so that $v = m/N$, then the (*non*-size-biased) mean length of the intervals of susceptibles between these will be $\tau \equiv (N-m)/m = (1/v) - 1$. Now consider choosing a susceptible at random and then looking at the numbers of susceptibles T_+ and T_-, respectively, to its right and left between it and the next vaccinated individual. Let $T = 1 + T_+ + T_-$. Then $\mathbf{E}[T]$ is the size-biased mean for a group (interval) of susceptibles.

Consider first the case where local contacts always infect, so that $\mathbf{E}[T]$ will also be the mean clump size for the epidemic. That $\mathbf{E}[T]$ is minimal when T is as near constant as possible, that is, it is $= \tau$ when τ is an integer, and has a distribution concentrated on the two integers either side of τ otherwise, is a well-known result for renewal processes: it can be thought of as saying that waiting times for buses will be minimal if they are scheduled at equal intervals.

What we prove is a generalization of this result that takes account of our actual local infection process. It turns out that we can allow a more general local infection process than the basic great circle model in which infections by different individuals are independent.

Thus, secondly, consider a model in which the local outbreak caused by an individual i, in the absence of vaccination, has an arbitrary distribution on intervals containing that individual. Suppose that it consists of C_+ individuals to the right of i and C_- individuals to the left, so that its total size is $C = 1 + C_+ + C_-$. Let $p_r = \mathrm{P}(C_+ \geq r)$.

We are now ready to put the vaccination process and the infection process together; we need of course to assume that these are independent. When we include the information on vaccinated intervals, the local outbreak caused by i becomes $D = 1 + R + L$, where $R = \min(T_+, C_+)$ and $L = \min(T_-, C_-)$. Our target is to find the distribution of T that minimizes $\mu = \mathbf{E}[D]$.

Now

$$\mathbf{E}[R] = \sum_{r=1}^{\infty} \mathrm{P}(R \geq r) = \sum_{r=1}^{\infty} \mathrm{P}(C_+ \geq r)\mathrm{P}(T_+ \geq r) = \sum_{r=1}^{\infty} p_r \mathrm{P}(T_+ \geq r).$$

Next comes the crucial step: $\mathrm{P}(T_+ < r) = \sum_{j=1}^{r} \mathrm{P}(T_+ = j-1)$. But, $T_+ = j-1$ only if the individual at j (i.e., j steps to the right of individual i) is vaccinated, and this has probability $1/\tau = m/(N-m)$. Hence $\mathrm{P}(T_+ = j-1) \leq 1/\tau$ and therefore $\mathrm{P}(T_+ \geq r) \geq 1 - \sum_{j=1}^{r} 1/\tau = 1 - r/\tau$. Of course we also have $\mathrm{P}(T_+ \geq r) \geq 0$, so that $\mathbf{E}[R] = \sum_{r=1}^{\infty} p_r \mathrm{P}(T_+ \geq r)$ will be minimized by taking $\mathrm{P}(T_+ \geq r) = \max(1 - r/\tau, 0)$. A mirror argument for $\mathbf{E}[L]$ leads to the corresponding condition $\mathrm{P}(T_- \geq r) = \max(1 - r/\tau, 0)$.

To see that these minima are uniquely attained when the distribution of T is concentrated on $[\tau]$ (the integer part of τ) and $[\tau] + 1$, note that equality in the argument of the last paragraph [turning "only if" into "if and only if" and hence giving $\mathrm{P}(T_+ = j-1) = 1/\tau$] requires that it is impossible to have two vaccinated individuals within the range $j = 0$ to $[\tau] - 1$; hence $T \geq [\tau]$. And $T \leq [\tau] + 1$ because otherwise it would be possible for T_+ (and T_-) to be $= [\tau] + 1$, which would contradict $\mathrm{P}(T_+ \geq r) = \max(1 - r/\tau, 0)$.

We have thus shown that the equalizing vaccination strategy is optimal for the generalized great circle model, in which the local outbreak caused by an individual takes an arbitrary distribution on the intervals containing the individual.

APPENDIX

Mean size of inverse epidemics. We prove here a general lemma, used in Section 5.2.1, which tells us that the mean size of an epidemic started by a randomly chosen individual is the same as for the model where we invert the contact structure.

LEMMA 1. *Consider any directed graph Γ on a finite number of nodes N. Let $\mu(\Gamma)$ be the mean size of the set of points that can be reached from a randomly chosen initial node of Γ by following links of the graph. Let Γ' denote the inverse of Γ, in which the direction of every link is reversed. Then $\mu(\Gamma') = \mu(\Gamma)$.*

REMARK. Γ can either be a specific given graph or, as in our epidemic models here, a random graph, with any desired distribution as long as it is (a.s.) finite.

PROOF OF LEMMA 1. Suppose first that Γ is a specific given graph. For any pair of nodes i, j of Γ, let $I_{ij} = 1$ if j can be reached from i, $=0$ otherwise. Then the size of the set of points that can be reached from i is $\sum_j I_{ij}$ and hence the mean size if i is chosen randomly, that is, from the uniform distribution over the set of nodes, is $\mu(\Gamma) = (1/N)\sum_i \sum_j I_{ij}$.

Similarly, for the inverse graph, $\mu(\Gamma') = (1/N)\sum_j \sum_i I'_{ji}$, but i can be reached from j in Γ' if and only if j can be reached from i in Γ. Therefore $I'_{ji} = I_{ij}$ for all i and j, whence $\mu(\Gamma') = \mu(\Gamma)$.

Finally, if Γ is a random graph, we simply take expectations of $\mu(\Gamma)$ and $\mu(\Gamma')$ with respect to its distribution. □

Acknowledgments. We are grateful to Klaus Dietz and Niels Becker for stimulating discussions, for comments on an earlier draft of this paper and for showing us preprints of their related work (see Sections 1.3 and 3.6), and to the Isaac Newton Institute for Mathematical Sciences in Cambridge, during whose Epidemic Models program in 1993 this work began.

REFERENCES

ADDY, C. L., LONGINI, I. M. and HABER, M. (1991). A generalized stochastic model for the analysis of infectious disease final size data. *Biometrics* **47** 961–974.

ANDERSSON, M. (1993). The final size of a multitype chain-binomial epidemic process. M.Sc. dissertation, Technical Report 1993:31, Dept. Mathematics, Chalmers Univ. Technology, Göteborg, Sweden.

BAILEY, N. T. J. (1975). *The Mathematical Theory of Infectious Diseases and Its Applications*. Griffin, London.

BALL, F. G. (1983a). A threshold theorem for the Reed–Frost chain-binomial epidemic. *J. Appl. Probab.* **20** 153–157.

BALL, F. G. (1983b). The threshold behavior of epidemic models. *J. Appl. Probab.* **20** 227–241.

BALL, F. G. (1986). A unified approach to the distribution of total size and total area under the trajectory of infectives in epidemic models. *Adv. in Appl. Probab.* **18** 289–310.

BALL, F. G. (1996). Threshold behaviour in stochastic epidemics among households. *Applied Probability 1. Lecture Notes in Statist.* **114** 253–266. Springer, Berlin.

BALL, F. G. and CLANCY, D. (1992). The final outcome of a generalised stochastic multitype epidemic model. Technical Report 92-4, Nottingham Statistics Group.

BALL, F. G. and CLANCY, D. (1993). The final size and severity of a generalised stochastic multitype epidemic model. *Adv. in Appl. Probab.* **25** 721–736.

BALL, F. G. and DONNELLY, P. J. (1995). Strong approximations for epidemic models. *Stochastic Process Appl.* **55** 1–21.

BALL, F. G. and NÅSELL, I. (1994). The shape of the size distribution of an epidemic in a finite population. *Math. Biosci.* **123** 167–181.

BARBOUR, A. D. and MOLLISON, D. (1989). Epidemics and random graphs. In *Stochastic Processes in Epidemic Theory. Lecture Notes in Biomath.* **86** 86–89. Springer, Berlin.

BARTLETT, M. S. (1957). Measles periodicity and community size. *J. Roy. Statist. Soc. Ser. A* **120** 48–70.

BARTOSZYŃSKI, R. (1972). On a certain model of an epidemic. *Applicationes Mathematicae* **13** 139–151.

BECKER, N. G. (1989). *Analysis of Infectious Disease Data*. Chapman and Hall, London.

BECKER, N. G. and DIETZ, K. (1995). The effect of the household distribution on transmission and control of highly infectious diseases. *Math. Biosci.* **127** 207–219.

BECKER, N. G. and HALL, R. (1996). Immunization levels for preventing epidemics in a community of households made up of individuals of different types. *Math. Biosci.* **132** 205–216.

BOLKER, B. M., ALTMANN, M., AUBERT, M., BALL, F. G., BARLOW, N. D., BOWERS, R. G., DOBSON, A. P., ELKINGTON, J. S., GARNETT, G. P., GILLIGAN, C. A., HASSELL, M. P., ISHAM, V., JACQUEZ, J. A., KLECZKOWSKI, A., LEVIN, S. A., MAY, R. M., METZ, J. A. J., MOLLISON, D., MORRIS, M., REAL, L. A., SATTENSPIEL, L., SWINTON, J., WHITE, P. and WILLIAMS, B. G. (1995). Group report: spatial dynamics of infectious diseases in natural populations. In *Ecology of Infectious Diseases in Natural Populations* (B. T. Grenfell and A. P. Dobson, eds.) 399–420. Cambridge Univ. Press.

BOLLOBÁS, B. (1985). *Random Graphs*. Academic Press, London.

DALEY, D. J. (1967). Some aspects of Markov chains in queueing theory and epidemiology. Ph.D. thesis, Cambridge Univ.

DALEY, D. J. and GANI, J. (1994). A deterministic general epidemic model in a stratified population. In *Probability, Statistics and Optimisation* (F. P. Kelly, ed.) 117–132. Wiley, Chichester.

DALY, F. (1979). Collapsing supercritical branching processes. *J. Appl. Probab.* **16** 732–739.

DE KOEIJER, A. A., DIEKMANN, O. and REIJNDERS, P. J. H. (1995). A mechanistic model to describe the spread of phocid distemper. Report AM-R9514, CWI, Amsterdam.

DIEKMANN, O., HEESTERBEEK, J. A. P. and METZ, J. A. J. (1990). On the definition and the computation of the basic reproduction ratio R_0 in models for infectious diseases in heterogeneous populations. *J. Math. Biol.* **28** 365–382.

DIETZ, K. (1993). The estimation of the basic reproduction number for infectious diseases. *Statistical Methods in Medical Research* **2** 23–41.

DOWNTON, F. (1972). A correction to "The area under the infectives trajectory of the general stochastic epidemic." *J. Appl. Probab.* **9** 873–876.

GANI, J. and SHANBHAG, D. N. (1974). An extension of Raikov's theorem derivable from a result in epidemic theory. *Z. Wahrsch. Verw. Gebiete* **29** 33–37.

GERTSBAKH, I. B. (1977). Epidemic process on a random graph: some preliminary results. *J. Appl. Probab.* **14** 427–438.

GONTCHAROFF, W. (1937). *Détermination des Fonctions Entières par Interpolation*. Hermann, Paris.

HABER, M., LONGINI, I. M. and COTSONIS, G. A. (1988). Models for the statistical analysis of infectious disease data. *Biometrics* **44** 163–173.

HANSKI, I. and GILPIN, M. (1991). Metapopulation dynamics: brief history and conceptual domain. *Biology Journal of the Linnean Society* **42** 3–16.

HETHCOTE, H. W. (1978). An immunization model for a heterogeneous population. *Theoret. Population Biol.* **14** 338–349.

HETHCOTE, H. W. and VAN ARK, J. W. (1986). Epidemiological models for heterogeneous populations: proportionate mixing, parameter estimation and immunization programs. *Math. Biosci.* **84** 84–118.

ISLAM, M. N., O'SHAUGHNESSY, C. D. and SMITH, B. (1996). A random graph model for the final-size distribution of household infections. *Statistics in Medicine* **15** 837–843.

JAGERS, P. (1975). *Branching Processes with Biological Applications*. Wiley, New York.

KAREIVA, P. (1990). Population dynamics in spatially complex environments: theory and data. *Philos. Trans. Roy. Soc. London Ser. B* **330** 175–190.

KINGMAN, J. F. C. (1993). *Poisson Processes*. Clarendon Press, Oxford.

KURTZ, T. G. (1970). Limit theorems for sequences of jump Markov processes approximating ordinary differential processes. *J. Appl. Probab.* **8** 344–356.

KURTZ, T. G. (1981). *Approximation of Population Processes*. SIAM, Philadelphia.

LEFÈVRE, C. and PICARD, PH. (1990). A non-standard family of polynomials and the final size distribution of Reed–Frost epidemic processes. *Adv. in Appl. Probab.* **22** 25–48.

LONGINI, I. M. and KOOPMAN, J. S. (1982). Household and community transmission parameters from final distributions of infections in households. *Biometrics* **38** 115–126.

MAY, R. M. and ANDERSON, R. M. (1984). Spatial heterogeneity and the design of immunization programs. *Math. Biosci.* **72** 83–111.

MOLLISON, D. (1972). The rate of propagation of simple epidemics. *Proc. Sixth Berkeley Symp. Math. Statist. Probab.* **3** 579–614. Univ. California Press, Berkeley.

MOLLISON, D. (1995). The structure of epidemic models. In *Epidemic Models: Their Structure and Relation to Data* (D. Mollison, ed.) 17–33. Cambridge Univ. Press.

MOLLISON, D. and LEVIN, S. A. (1995). Spatial dynamics of parasitism. In *Ecology of Infectious Diseases in Natural Populations* (B. T. Grenfell and A. Dobson, eds.) 384–398. Cambridge Univ. Press.

NÅSELL, I. (1995). The threshold concept in stochastic epidemic and endemic models. In *Epidemic Models: Their Structure and Relation to Data* (D. Mollison, ed.) 71–83. Cambridge Univ. Press.

PICARD, PH. and LEFÈVRE, C. (1990). A unified analysis of the final size and severity distribution in collective Reed–Frost epidemic processes. *Adv. in Appl. Probab.* **22** 269–294.

POST, W. M., DEANGELIS, D. L. and TRAVIS, C. C. (1983). Endemic disease in environments with spatially heterogeneous host populations. *Math. Biosci.* **63** 289–302.

RUSHTON, S. and MAUTNER, A. J. (1955). The deterministic model of a simple epidemic for more than one community. *Biometrika* **42** 126–132.

SCALIA-TOMBA, G. (1985). Asymptotic final size distribution for some chain-binomial processes. *Adv. in Appl. Probab.* **17** 477–495.

SCALIA-TOMBA, G. (1990). On the asymptotic final size distribution of epidemics in heterogeneous populations. In *Stochastic Processes in Epidemic Theory. Lecture Notes in Biomath.* **86** 189–196. Springer, Berlin.

SELLKE, T. (1983). On the asymptotic distribution of the size of a stochastic epidemic. *J. Appl. Probab.* **20** 390–394.

VON BAHR, B. and MARTIN-LÖF, A. (1980). Threshold limit theorems for some epidemic processes. *Adv. in Appl. Probab.* **12** 319–349.

WATSON, R. K. (1972). On an epidemic in a stratified population. *J. Appl. Probab.* **9** 659–666.

WAUGH, W. A. O'N. (1958). Conditioned Markov processes. *Biometrika* **45** 241–249.

WHITTLE, P. (1955). The outcome of a stochastic epidemic—a note on Bailey's paper. *Biometrika* **42** 116–122.

FRANK BALL
DEPARTMENT OF MATHEMATICS
UNIVERSITY PARK
NOTTINGHAM NG7 2RD
ENGLAND
E-MAIL: fgb@maths.nott.ac.uk

DENIS MOLLISON
DEPARTMENT OF ACTUARIAL MATHEMATICS AND STATISTICS
RICCARTON
EDINBURGH EH14 4AS
SCOTLAND
E-MAIL: denis@ma.hw.ac.uk

GIANPAOLO SCALIA-TOMBA
DEPARTMENT OF MATHEMATICAL METHODS AND MODELS
UNIVERSITA LA SAPIENZA
VIA A SCARPA 10
00161 ROMA
ITALY
E-MAIL: gianpi@dmmm.uniroma1.it

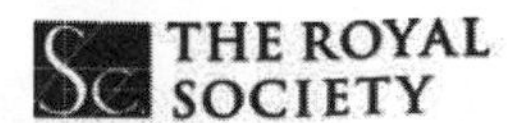

The effects of local spatial structure on epidemiological invasions

M. J. Keeling

Zoology Department, University of Cambridge, Downing Street, Cambridge CB2 3EJ, UK (*matt@zoo.cam.ac.uk*)

Predicting the likely success of invasions is vitally important in ecology and especially epidemiology. Whether an organism can successfully invade and persist in the short-term is highly dependent on the spatial correlations that develop in the early stages of invasion. By modelling the correlations between individuals, we are able to understand the role of spatial heterogeneity in invasion dynamics without the need for large-scale computer simulations. Here, a natural methodology is developed for modelling the behaviour of individuals in a fixed network. This formulation is applied to the spread of a disease through a structured network to determine invasion thresholds and some statistical properties of a single epidemic.

Keywords: correlation equations; contact networks; spatial heterogeneity; basic reproductive ratio; final size; invasion

1. INTRODUCTION

Invasion is one of the most fundamental concepts in ecology and epidemiology. It is only with a firm quantitative understanding of this ubiquitous phenomenon that we can accurately determine vaccination thresholds and evolutionary selection as well as more common forms of invasion (Kornberg & Williamson 1987). Invading organisms are initial highly aggregated, with only limited spatial spread, and therefore suffer from far more intraspecific competition than non-spatial models would predict. As infectious diseases provide the best documented and most accurately modelled problems in ecology, this paper shall concentrate on analytical results for the invasion of an infection into a spatially distributed host population.

Spatial models, from meta-population models to partial differential equations (PDEs), have become increasingly popular in both ecology and epidemiology. It has become obvious that in many situations spatial patterns and correlations play a vital role; this is especially true for invasions. Using the correlations between individuals to capture the essential spatial characteristics is not new to ecology (Hassell & May 1974), although only recently have these correlations been treated as dynamic variables (Dickman 1986; Matsuda 1987; Matsuda *et al.* 1992; Sato *et al.* 1994; Levin & Durrett 1996; Keeling & Rand 1999).

Correlation models, and in particular pair-wise models, have been primarily used to describe the behaviour of simple spatial models (such as probabilistic cellular automata) in terms of a set of ordinary differential equations (ODEs). However, these correlation models can be used in their own right (Dietz & Hadeler 1988; Altmann 1995; Keeling *et al.* 1997) and can provide a more general framework and neighbourhood structure than is feasible in traditional spatial models. Correlation models are of most use when the interactions between individuals (or sites) can be considered as occurring on a network—this is the case for communicable diseases.

When considering the spread of an epidemic, it is the contact structure between individuals that determines the progress of the disease through the population (Barbour & Mollison 1990). One of the simplest models which captures the fundamental features of infection dynamics is the SIR (susceptible–infectious–recovered) model (Anderson & May 1992; Mollison 1995; Grenfell & Dobson 1995).

The first section examines the structure of a network and the behaviour of various quantities when the network is specified by a graph or contact matrix. In § 3, a method for closing the system of ODEs at the level of pairs is formulated, (i.e. we express the number of triples in terms of the number of pairs). Section 4 combines the theoretical arguments to derive the equations for an SIR disease spreading across a network. Sections 5 and 6 examine two fundamental properties of the SIR model the basic reproductive ratio R_0 and the final size of the epidemic. The final section considers how these two properties are changed by vaccination, and predicts vaccination thresholds.

2. GENERAL THEORY

One of the simplest assumptions for disease transmission is that the contact structure forms a network of links between individuals (or nodes), with all links being of equal strength. Such a network is often referred to as a graph. We can describe a network involving N individuals by a matrix $G \in \{0,1\}^{N^2}$,

$$G_{ij} = \begin{cases} 1 & \text{if } i \text{ and } j \text{ are connected} \\ 0 & \text{otherwise} \end{cases}$$

Proc. R. Soc. Lond. B (1999) **266**, 859–867
Received 8 December 1998 *Accepted* 25 January 1999

As all links are bidirectional and self-contact is not allowed, this places two constraints upon the matrix: $G = G^T$ and $G_{ii} = 0$. From this matrix, we can calculate the number of connected pairs and triples in the graph,

$$\text{number of pairs} = \|G\| = nN,$$
$$\text{number of triples} = \|G^2\| - \text{trace}(G^2).$$

Here, $\|G\| = \sum_{i,j} G_{ij}$ is the sum of all the elements in the matrix and n is therefore the average number of neighbours per node. The number of triples is calculated as the number of nodes which are joined by two connections, given that the nodes are distinct.

It should be noted that there is ambiguity in the precise form of a triple; for three connected nodes, it is possible to form triangular loops as well as linear arrangements (Keeling *et al.* 1997; Morris 1997; Rand 1999; Van Baalen 1999). These loops are very important in the spread of a disease, as we will show later. Let ϕ be defined as the ratio of triangles to triples, this is a simple measure of how interconnected the local neighbourhoods are. When ϕ is large, the members of a connected pair will be connected to many common nodes, your neighbour's neighbours are also your neighbours (figure 1*a*), whereas when ϕ is small there are few common nodes, and long-range connections dominate (figure 1*b*). As triangles are three linked nodes with the same start and end point, ϕ can be expressed in terms of the graph

$$\phi = \frac{\text{number of triangles}}{\text{number of triples}} = \frac{\text{trace}(G^3)}{\|G^2\| - \text{trace}(G^2)}. \quad (1)$$

For most communicable diseases, we are likely to find that each individual experiences only a small proportion of the population ($n \ll N$) and yet ϕ is generally large as there tends to be complete interaction within small social groups (cf. Watts & Strogatz 1998). We believe that in general, much of the underlying structure of the network can be characterized in terms of the average number of neighbours (n) and the interconnectedness (ϕ). Throughout this paper, we will assume that both n and ϕ are fixed for all sites, i.e. the network is homogeneous.

To begin to consider the dynamics of individuals, it is necessary to define a set of functions which inform us about the state of each node. Let A_i be equal to one if the individual at node i is of type A or zero otherwise. This allows us to define rigorously the number of single, pairs and triples of each type,

$$\text{singles of type A} = [A] = \sum_i A_i,$$
$$\text{pairs of type A–B} = [AB] = \sum_{i,j} A_i B_j G_{ij},$$
$$\text{triples of type A–B–C} = [ABC] = \sum_{i,j,k} A_i B_j C_k G_{ij} G_{jk}. \quad (2)$$

This method of counting means that pairs are counted once in each direction so that $[AB] = [BA]$ and that $[AA]$ is even. From equations (2), we can recover the natural rules for summing singles, pairs and triples,

(*a*)

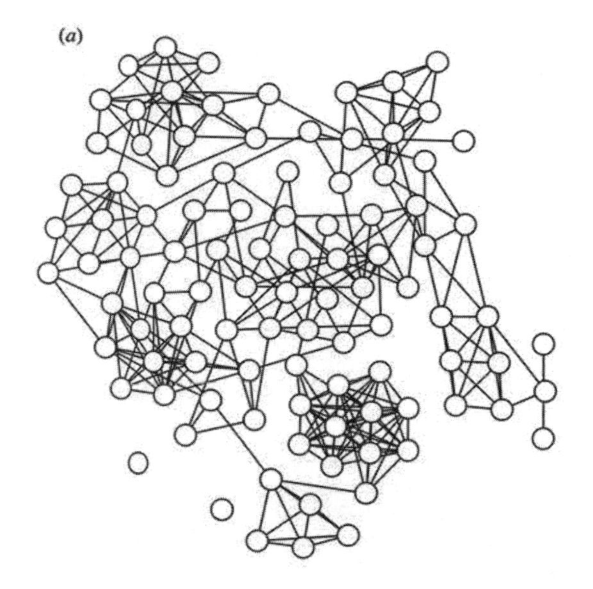

(*b*)

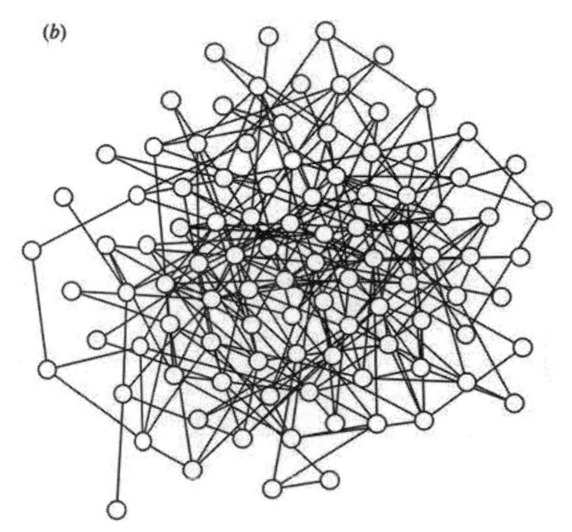

Figure 1. Examples of a network of 100 nodes, with an average of five connections per node ($n = 5$). In (*a*) $\phi = 0.7$ and triangles are common, whereas in (*b*) $\phi = 0.2$ and there is less obvious structure. These graphs were obtained by placing nodes randomly in two dimensions and weighting the probability of a connection between nodes by the distance.

$$\sum_A [A] = \sum_A \sum_i A_i = \sum_i \left(\sum_A A_i \right) = N, \quad (3)$$

$$\sum_B [AB] = \sum_B \sum_{i \neq j} A_i B_j G_{ij} = \sum_{i,j} A_i G_{ij} = n[A], \quad (4)$$

$$\sum_C [ABC] = \sum_C \sum_{j, i \neq k} A_i B_j C_k G_{ij} G_{jk} = \frac{n(n-1)}{N} [A][B]. \quad (5)$$

Note that, in general, there is no corresponding formula for $\sum_B [ABC]$.

As well as the number of pairs, it is often useful to consider the multiplicative correlation between connected nodes of various types. We shall define C_{AB} to be the correlation between nodes of type A and B,

$$C_{AB} = N^2 \frac{\sum_{i\neq j} A_i B_j G_{ij}}{\sum_{i\neq j} A_i B_j \sum_{i\neq j} G_{ij}} = \frac{N}{n}\frac{[AB]}{[A][B]}. \qquad (6)$$

From this, it can be seen that $C_{AB} \in [0,N]$. When $C_{AB} = 1$, then A and B are uncorrelated so their placement with respect to each other is random.

3. FORMULATING A MODEL BY CLOSING THE SYSTEM

Let f be any function of the states of nodes in the network. If f can be assumed to be continuous in the limit of large populations, then its behaviour can be captured by the differential equation (Morris 1997; Rand 1999)

$$\frac{df}{dt} = \sum_{\text{events}} \text{rate of event} \times \text{change in } f \text{ due to event} \qquad (7)$$

Normally, f is considered to be either the number of nodes of a given type ($[A]$) or the number of pairs of a given type ($[AB]$), allowing the formulation of equations for the behaviour of the system.

In the vast majority of ecological and epidemiological systems, any change in the behaviour or state of an individual will be dependent on the state of its neighbours. For example, a susceptible individual surrounded by infectious neighbours is likely to become infected. Let $Q_i(B|A)$ be the number of B neighbours surrounding node i, given that node i is state A. Q_i is considered to be comprised two parts: the expected value $\overline{Q}$, together with some associated error σ which is not necessarily small.

$$Q_i(B|A) = \overline{Q}(B|A) + \sigma_i(B|A) = \frac{[AB]}{[A]} + \sigma_i(B|A).$$

One of the main features of individual-based models with a small neighbourhood, is that individuals often experience large fluctuations in their environment, so σ may be large compared to $\overline{Q}$ (Keeling & Rand 1999). By assuming a distribution for the Q_i terms and, hence, a distribution for the errors σ_i, one can produce expressions for the expected number of triples and higher-order connections in terms of pairs (Rand 1999). For a fixed number of neighbours per site, the most likely form for the errors is multinomial, in which case

$$[ABC] = \sum_{j,B_j=1} Q_j(A|B)Q_j(C|B), \qquad (8)$$

$$= \sum_{j,B_j=1} \left(\frac{[AB]}{[B]} + \sigma_j(A|B)\right)\left(\frac{[BC]}{[B]} + \sigma_j(C|B)\right),$$

$$= \frac{[AB][BC]}{[B]} + \sum_{j,B_j=1} \sigma_j(A|B)\sigma_j(C|B),$$

$$= \left(\frac{n-1}{n}\right)\frac{[AB][BC]}{[B]}. \qquad (9)$$

A similar technique can be applied if other nonlinear arrangements of the Q_i terms arise or if other distributions for σ are plausible. In the SIR example given below, only linear combinations of pairs and triples will appear, and therefore equation (9) will be all that is needed. In the remainder of the paper, for convenience of notation, we shall set $\zeta = (n-1)/n$.

In the above calculation of the number of triples, no consideration has been given to the structure of the network; in particular the number of triangular and higher order loops has been ignored. If A–B–C form a triangle, we must also allow for the fact that the correlation between A and C is also important. The most natural way to include this correlation is

$$[ABC] \approx \zeta\frac{[AB][BC]}{[B]}C_{AC} = \frac{\zeta N}{n}\frac{[AB][BC][AC]}{[A][B][C]}.$$

Hence, bringing in the full network structure by including the ratio between triangles and triples,

$$[ABC] \approx \zeta\frac{[AB][BC]}{[B]}\left((1-\phi) + \phi\frac{N}{n}\frac{[AC]}{[A][C]}\right). \qquad (10)$$

It can be hoped that in choosing ϕ for any particular applications, it may capture the effects due to all loops of three nodes and higher. It should be expected that the effect from four node loops is far less than that from three node loops etc., and so the ϕ which describes the system best should be slightly larger than the exact ratio of triangles to triples.

4. THE SIR MODEL

With these tools, we are now in a position to consider the spread of a disease through a network of nodes; in particular, a network with n neighbours per site and with an interconnectedness of ϕ. Throughout this work, the mean field limit ($n \to N \to \infty$, $\phi \to 1$) will be calculated as a check and a comparison.

The most basic epidemic model with a recovered status is the simple epidemic, described by the SIR model without the demographic processes of birth and death (Kermack & McKendrick 1927; Anderson & May 1992; Mollison 1995). Such a model exhibits a single epidemic. Each individual can be in one of three states: susceptible to the disease, infectious when they can spread the disease to susceptibles, and recovered when they have been infectious but can no longer spread or catch the disease (cf. the rapid spread of influenza). This framework forms the basis of almost all epidemiological models (Grenfell *et al.* 1992; Grenfell & Dobson 1995; Keeling & Grenfell 1997). Denoting the number of susceptible, infectious and recovered individuals by S, I and R respectively, the mean field equations are

$$\dot{S} = -\beta \frac{S}{N} I,$$
$$\dot{I} = \beta \frac{S}{N} I - gI, \qquad (11)$$
$$\dot{R} = gI.$$

Here, β is the contact parameter, and $1/g$ is the infectious period. This model has been analysed and applied to numerous situations (Anderson & May 1992; Mollison 1995 and references therein), the main results are as follows.

(i) An epidemic can only occur if $R_0 = \beta/g > 1$.
(ii) S is monotonically decreasing, R is monotonically increasing and I is unimodal.
(iii) The epidemic eventually dies out with some proportion of susceptibles, S_∞, remaining

$$S_\infty = \exp((S_\infty - 1)R_0).$$

For the correlation model, the equations describe the behaviour of A–B pairs instead of the behaviour of individuals. As we have moved from a global to a more individual approach, we will define τ the transmission rate across a connection to be β/n; hence, the potential for spreading infection is equal between the two approaches.

There exist nine distinct types of pairs; however, due to symmetries ($[AB] = [BA]$) and the fact that the sum over all pairs remains constant, only five differential equations are necessary.

$$[\dot{SS}] = -2\tau[SSI],$$
$$[\dot{SI}] = \tau([SSI] - [ISI] - [SI]) - g[SI],$$
$$[\dot{SR}] = -\tau[RSI] + g[SI], \qquad (12)$$
$$[\dot{II}] = 2\tau([ISI] + [SI]) - 2g[II],$$
$$[\dot{IR}] = \tau[RSI] + g([II] - [IR]),$$

Using equation (10), the system can be closed at the level of pairs, assuming a multinomial distribution of neighbours. Figure 2 shows a comparison between the results of the correlation equations (12) and the average of astochastic simulation modelling the spread of a disease across a network. There is good quantitative agreement between the equations and the full model. It is the characteristic shape of the S–I correlation (figure 2*b*) with its rapid initial decline that leads to the major difference between the correlation and mean field equations.

When $\phi = 0$, the ODEs can be uncoupled in a natural way by expressing the system in terms of $\overline{Q}(S|S) = [SS]/[S]$, $\overline{Q}(I|S) = [SI]/[S]$ and the number of new cases $C = \tau[SI]$,

$$\dot{C} = C\left(\tau\frac{n-1}{n}\overline{Q}(S|S) - \frac{n-1}{n}\tau\overline{Q}(I|S) - \tau - g\right),$$
$$\overline{Q}(\dot{S}|S)) = -\tau\frac{n-2}{n}\overline{Q}(S|S)\overline{Q}(I|S), \qquad (13)$$
$$\overline{Q}(\dot{I}|S) = \tau\frac{n-1}{n}\overline{Q}(S|S)\overline{Q}(I|S) - (\tau+g)\overline{Q}(I|S) + \frac{\tau}{n}\overline{Q}(I|S)^2.$$

Therefore, when there are no triangular connections, the pairwise ODEs can be reduced to three dimensions, only one more than the mean-field model. The extra dimension accounts for the correlation between susceptibles and infectious individuals.

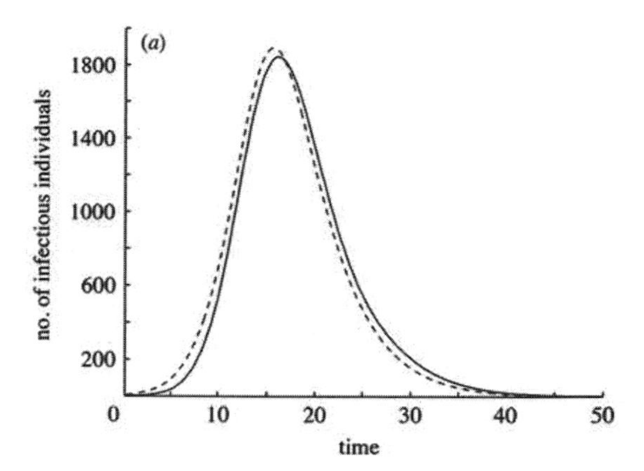

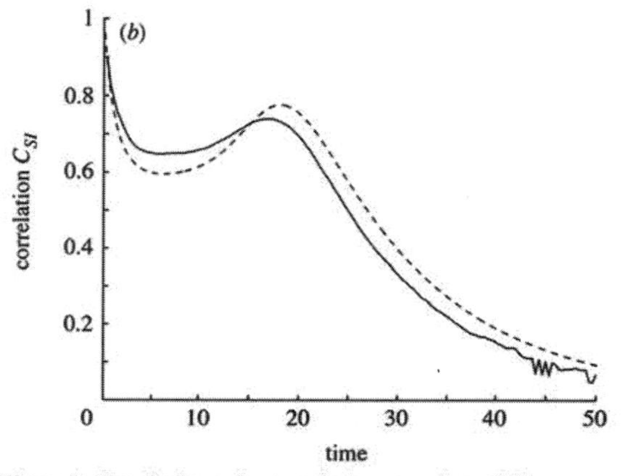

Figure 2. Results from the correlation equations (12) (solid line) and from the average of 100 stochastic simulations of an epidemic on a network (dashed line). The network contained $N = 6000$ nodes, with each individual having $n = 6$ neighbours and an interconnectedness $\phi = 0.2$. (*a*) shows the number of infectious individuals over time, (*b*) shows the correlation between infectious and susceptibles C_{SI}. The discrepancies between the two results may in part be due to the imprecise synchronization of epidemics across the many realizations (because of the stochastic nature of the system), leading to a smoothing-out of the simulation results.

5. THE BASIC REPRODUCTIVE RATIO, R_0

The basic reproductive ratio, R_0, is the most fundamental quantity in epidemiology (Diekmann *et al.* 1990; Anderson & May 1992; de Jong *et al.* 1994). R_0 is defined as the average number of secondary cases produced by an infectious individual in a totally susceptible population. It informs us whether a disease can ever invade a population, and is useful in the calculation of many other quantities. For the simple epidemic,

$$R_0 = \frac{\beta}{g}.$$

Let us consider the initial phase of an infection invading a total susceptible population.

$$[\dot{I}] = \tau[SI] - g[I],$$
$$= \left(\beta\frac{[S]}{N}C_{SI} - g\right)[I],$$

Because $[S]$ is assumed to be equal to N initially, this gives $R_0 = C_{SI}\beta/g$. As a single infectious individual is placed in a sea of susceptibles, initially we find that susceptibles and infectious nodes must be uncorrelated ($C_{SI} = 1$). It would therefore appear that R_0 is the same for the mean-field and the network model. However, this approach is flawed; consider the early behaviour of C_{SI}

$$\dot{C_{SI}} = \frac{N}{n}\frac{\mathrm{d}}{\mathrm{d}t}\left(\frac{[SI]}{[S][I]}\right) \rightarrow$$
$$-\tau\left(C_{SI} + C_{SI}^2 - n\zeta(C_{SI} - C_{SI}^2)(1-\phi) + n\zeta C_{SI}^2\phi\frac{[I]C_{II}}{N}\right) \quad (14)$$

as $[S]/N \rightarrow 1$, $C_{SS} \rightarrow 1$ and $[I]/N \rightarrow 0$. The initial growth in the proportion of infectious nodes is small, however equation (14) shows that the correlation between S and I decays at order one. This means that in the region of the network that has been invaded, there is rapid development of the spatial structure as captured by the local correlations. In the early development of the epidemic C_{SI} converges to a quasi-equilibrium C_{SI}^*. It is therefore advisable to measure R_0 once the local spatial pattern has formed and C_{SI} has equilibrated. As C_{SI}^* will be less than one, the value of R_0 will be similarly reduced.

From equation (14), it is clear that, in general, the quasi-equilibrium value also depends on the value of $C_{II}[I]/N$. The correlations between infectious nodes grow fast enough that $[I]C_{II}/N$ (which is interpreted as the probability that a neighbour of an infectious individual is also infectious) is of order one, even when the density of infectious individuals is small.

$$\frac{\mathrm{d}}{\mathrm{d}t}\frac{[I]C_{II}}{N} = \frac{1}{n}\frac{\mathrm{d}}{\mathrm{d}t}\left(\frac{[II]}{[I]}\right) \Rightarrow \left(\frac{IC_{II}}{N}\right) \rightarrow \frac{2\tau C_{SI}}{g + \beta C_{SI} - 2\zeta\beta C_{SI}^2\phi} \quad (15)$$

Therefore, from equations (14) and (15), it is found that C_{SI}^* satisfies

$$n\zeta(1-C_{SI}^*)(1-\phi) - \frac{2\zeta\beta\phi C_{SI}^{*2}}{g + \beta C_{SI}^* - 2\zeta\beta C_{SI}^{*2}\phi} - C_{SI}^* = 1 \quad (16)$$

The values of C_{SI} and hence R_0 can be seen to depend only of the values of n, ϕ and β/g. Figure 3 shows the behaviour of C_{SI} (and hence R_0) and the associated values of $[I]C_{II}/N$ for a range of n and ϕ.

As R_0 is proportional to the S–I correlation, it is clear that the network models have a lower basic reproductive ratio than their mean-field counterparts. Therefore, there will exist conditions when the mean-field equations will predict that an epidemic will take off, but the network model will show that this is not the case. We can examine this result analytically in the limiting case when there are no triangular loops, $\phi = 0$.

$$n\zeta(1 - C_{SI}^*) - C_{SI}^* = 1 \Rightarrow C_{SI}^* = 1 - \frac{2}{n}$$
$$\Rightarrow R_0 = \left(1 - \frac{2}{n}\right)\frac{\beta}{g}$$

Similarly, by considering the limiting case when β is large, we find that the disease always dies out if

$$\phi > \frac{n-2}{n-1}.$$

This limiting value of ϕ corresponds to a highly interconnected system, where all but one of the neighbours is within a completely interconnected group.

It should be clear that whether a disease can invade a totally susceptible population (whether $R_0 > 1$) is dependent upon the network structure as well as the individual disease parameters. We find that the potential to invade is reduced by having few neighbours or a highly interconnected network structure, with the effects of ϕ being reduced as the number of neighbours increases.

6. THE FINAL SIZE OF A SINGLE EPIDEMIC

One of the main characteristics of an SIR epidemic in a population without births is the final size of the epidemic. This is the total number of individuals that become infected during the course of the epidemic and is equal to the final number of recovered individuals $R_\infty = 1 - S_\infty$. The final size of simple epidemics has been much studied in deterministic and stochastic models (Kermack & McKendrick 1927; Ball & Nasell 1994; Islam *et al.* 1996). It should be noted that when the birth rate is low compared to the epidemic time (e.g. influenza), this formulation still gives an accurate prediction of the number infected.

Given a fixed number of neighbours n per site and $\phi = 0$, analytical results have been developed (Diekmann *et al.* 1998) to show how, for a network of connected nodes, R_∞ is smaller than predicted by mean-field assumptions. For the network model

$$R_\infty = 1 - \left(1 - \frac{\tau}{g+\tau} + \theta\frac{\tau}{g+\tau}\right)^n, \quad (17)$$

where

$$\theta = \left(1 - \frac{\tau}{g+\tau} + \theta\frac{\tau}{g+\tau}\right)^{n-1}.$$

When $n \rightarrow \infty$, and keeping $n\tau = \beta$, this returns to the standard final size result of Kermack & McKendrick (1927) calculated as the long-term limit of the SIR equations. Unfortunately, the long-term limit of equations (12) cannot be solved as easily. However, as shown in Appendix A, for the case where $\phi = 0$, using the reduced form (13) the final size can be obtained and agrees with the value calculated by Diekmann *et al.* (1998). There is no obvious means of extracting the final size equations when $\phi \neq 0$, instead numerical results from integration of the network equations will be used to demonstrate the behaviour (figure 4).

The correlation model predicts a lower final size than the mean-field equations, and as ϕ is increased the final

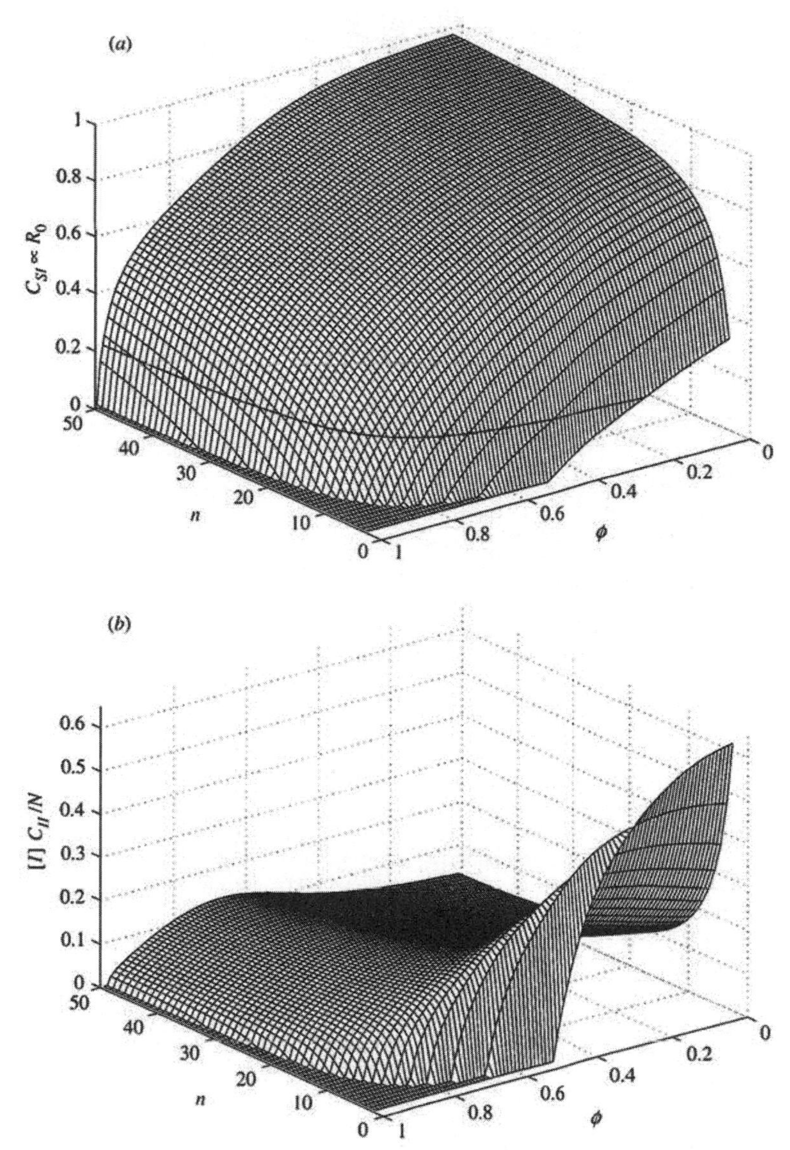

Figure 3. From equations (14) and (15), we can calculate the correlation between susceptibles and infectious nodes in the initial stages of invasion (*a*) and the probability that a neighbour of an infectious node is also infectious (*b*). For both these models, we have taken $g = 0.1$ and $\beta = 0.5$. The contour on graph *a* represents the line where $R_0 = 1$, above this line epidemics can start; for the comparative mean-field model, $R_0 = 5$.

size is further reduced. In all cases, it is true that $R_\infty \to 1$ as $\beta \to \infty$. Figure 4 also illustrates the effect ϕ has on the invasion threshold ($R_0 = 1$) which can be seen to occur at increasing transmissibilities for increasing ϕ. Therefore, not only does the proportion of triangular contacts (ϕ) reduce the initial spread of an epidemic, it also limits the final proportion of the population that the epidemic reaches.

7. VACCINATION

One of the main aims of epidemiology is to understand the role of vaccination and hence predict the level of vaccination necessary to eradicate a disease. If we assume that vaccination confers lifelong immunity to the disease, then we can amalgamate the vaccination and recovered individuals into a single non-susceptible class. For the simple SIR model without births or deaths, the effects of vaccination can be captured by starting the population with a fraction $V = 1 - S$ in the non-susceptible class, so equations (12) and (13) still hold.

When starting with only a proportion S of the population being susceptible, the invasion is characterized by the effective reproductive ratio R. For the mean-field models, there is a simple relationship between R and R_0,

$$R = SR_0 = \frac{\beta S}{g}.$$

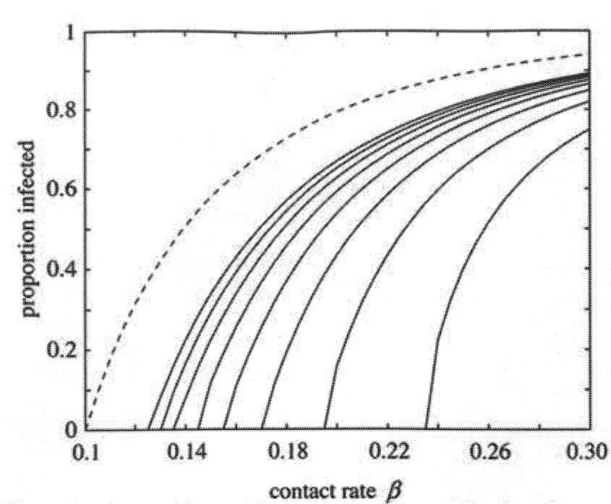

Figure 4. Comparison of theoretical and numerical results for $n = 10$ and $g = 0.1$. The dashed line is the theoretical proportion infected in the mean-field model. The black line is the theoretical result (and numerical solution of the ODEs (12)) when $\phi = 0$. The grey lines are for increasing ϕ (0.1, 0.2, 0.3, 0.4, 0.5, 0.6, 0.7, 0.8) and show how on a more interconnected graph we should expect smaller epidemics. The points where the curves meet the horizontal axis correspond to the invasion threshold when $R_0 = 1$. Again, $g = 0.1$.

Hence, using this approximation, the disease can only invade if the proportion susceptible is greater than some threshold value $S_T = g/\beta$. When considering the invasion on a network, we must take into account the initial spatial distribution of the susceptibles as well as their overall density. In general, we should expect susceptibles to be aggregated so that $C_{SS} > 1$. As shown in § 5, local spatial structure develops over a short time-scale; the effect this has on R can be captured by examining the correlation C_{SI}.

$$C_{SI} \rightarrow \frac{(n-1)[S]C_{SS}(1-\phi) - N}{n[S] + (n-1)[I]C_{II}\phi - (n-1)[S]C_{SS}\phi},$$

$$[I]C_{II} \rightarrow \frac{2\tau N[S]C_{SI}}{gN + n\tau[S]C_{SI} - 2(n-1))\tau[S]C_{SI}^2\phi}.$$

When, $\phi = 0$, so that there is no spatial structure present in the network, we find

$$C_{SI} \rightarrow \frac{(n-1)[S]C_{SS} - [N]}{n[S]},$$
$$\Rightarrow R = \frac{\tau}{g}\left[(n-1)\frac{[S]}{N}C_{SS} - 1\right]. \tag{18}$$

Hence, it is both the proportion of susceptibles, $[S]/N$, and the correlation between susceptibles, C_{SS} which controls the initial spread of the disease. The disease can only invade if

$$\frac{[S]}{N} > \frac{1}{C_{SS}}\frac{g+\tau}{\tau(n-1)},$$
$$\Rightarrow \frac{[S]}{N} > \frac{1}{(n-1)C_{SS}}\left[\frac{1}{R_0}(n-2) + 1\right] > \frac{S_T}{C_{SS}}.$$

In general, irrespective of the contact rate β, a disease cannot invade if

$$\frac{[S]}{N} < \frac{1}{(n-1)(1-\phi)C_{SS}},$$

which gives us a vaccination threshold that takes into account the network structure as well as the aggregation of the susceptibles. The aggregation of susceptible individuals is a monotonic function of the aggregation of vaccinated individuals. We therefore find that the vaccination threshold necessary for disease eradication increases with the number of neighbours and the aggregation of vaccination, but decreases with increasing ϕ.

As detailed in § 6, it is also possible to formulate equations for the final size of an epidemic when $\phi = 0$. Using the same approach, starting with a proportion S_0 of susceptibles with correlation C_{SS}, we find

$$S_\infty = S_0\left(1 - \frac{\tau}{\tau+g}C_{SS}S_0 + \frac{\tau}{\tau+g}C_{SS}S_0^{1/n}S_\infty^{1-1/n}\right)^n. \tag{19}$$

Final proportion infected $= S_0 - S_\infty$.

From which we find that both larger initial densities and larger initial aggregations will lead to a greater proportion of the population being infected.

8. DISCUSSION

Using the simple epidemiological SIR model, we have shown that the spatial structure which develops during the early stages of invasion can determine its success or failure. The limited spread of invading organisms means that they suffer far greater intraspecific competition than the homogeneously mixed, mean-field equations would suggest. This greater intraspecific competition leads to reduced success at invasion. This paper has demonstrated, primarily by numerical integration, that intraspecific competition, and therefore departure from the mean-field model, is greatest when the neighbourhood is small and there are many local connections (ϕ is large).

We have defined the reproductive ratio (R_0) as being the average number of individuals produced by an invading organism once the local spatial structure has reached equilibrium. Thus, this is a measure of whether an organism can enter the population and persist in the short-term; the standard mean-field result informs us whether the organism can simply enter the population. Therefore, for some parameter values, an organism may be able to enter a population, but cannot survive in the environment that rapidly develops around it. However, such situations are likely to be indistinguishable from chance events. It should be realized that invasion is a highly stochastic phenomenon and many invasions will fail simply by 'bad luck'. We believe that, in general, by the time the local spatial structure has developed, there should be sufficiently many cases that stochastic extinctions can be ignored. Thus, invasion is best characterized by a short stochastic entry phase followed by a more deterministic growth phase governed by the value of R_0 calculated above.

For communicable diseases, assuming the infection is spread via a network of contacts is the natural way to model the effects of spatial correlations. We have shown that departure from the standard mean-field results is greatest when n is small and ϕ is large. This is often the case for sexually transmitted diseases where most of the partners come from a small social group. For other human diseases, such as childhood epidemics, there are many more contacts, but they are still highly interconnected. Therefore we should expect that for many common communicable diseases the fact that a network consists of a limited number of connected pairs will play an important role in the dynamics. If we allow the network to be dynamic, such that connections can break and reform (cf. Dietz & Hadeler 1988), then the behaviour becomes more like the mean-field model, with the correlations decaying to one as the movement of individuals becomes large.

For these correlation models to prove useful in practical epidemiological problems requires greater parameterization at the level of an individual, therefore what is needed is a method of extracting local parameters from global results. However, this method forms a valuable link between mathematically simplistic mean-field equations and computationally intensive individual-based models, and provides insights into the role of individuals and spatial correlations in ecological invasions.

This research was supported by the Wellcome Trust and the Royal Society. I wish to thank David Rand for his many helpful comments and insights as well as Minus van Baalen and Odo Diekmann.

APPENDIX A

(a) *Long-term limiting behaviour of the correlation model*

Equation (13) cannot be solved for the final size of the epidemic so easily as the standard SIR model. It is necessary to consider the behaviour of some new combinations of parameters; let us define

$$\zeta=\frac{n-1}{n},\quad P=\frac{[SI]}{[S]^{\zeta}},\quad R=\frac{[SS]}{[S]^{\zeta}},\quad T=\frac{[SS]}{\exp(n[S]^{1/n})S^{2\zeta}}.$$

Now, from equation (12)

$$\begin{aligned}\dot{P}&=\tau\zeta\frac{[SS]}{[S]}P-(\tau+g)P,\\ \dot{R}&=-2\tau\zeta\frac{[SS]}{[S]}P+[SS]\frac{\zeta}{[S]^{\zeta+1}}\tau[SI],\\ &=-\tau\zeta\frac{[SS]}{[S]}P,\\ \dot{T}&=\tau TP,\end{aligned}\qquad\text{(A1)}$$

and from equation (13),

$$\overline{Q}(\dot{S}|S)=\frac{\mathrm{d}}{\mathrm{d}t}\frac{[SS]}{[S]}=-\tau\frac{n-2}{n}\overline{Q}(S|S)\overline{Q}(I|S),$$

and

$$\frac{\mathrm{d}[S]}{\mathrm{d}t}=-\tau[S]\overline{Q}(I|S)\Rightarrow[SS]=n[S]^{2\zeta}.$$

In particular, $[SS]_\infty=n[S]_\infty^{2\zeta}$.

Integrating the three equations (A1) from $t=0$ to ∞,

$$P(\infty)-P(0)=0=\tau\zeta\int_0^\infty\frac{[SS]}{[S]}P\mathrm{d}t-(\tau+g)\int_0^\infty P\mathrm{d}t,\quad\text{(A2)}$$

$$R(\infty)-R(0)=\frac{[SS]_\infty}{[S]_\infty^{\zeta}}-n=-\tau\zeta\int_0^\infty\frac{[SS]}{[S]}P\mathrm{d}t,\qquad\text{(A3)}$$

$$\ln(T(\infty))-\ln(T(0))=n-n[S]_\infty^{1/n}=\tau\int_0^\infty P\mathrm{d}t.\qquad\text{(A4)}$$

Therefore, from equation (A2) and substituting the expressions from equations (A3) and (A4),

$$-\frac{(\tau+g)}{\tau}(n-n[S]_\infty^{1/n})=\frac{[SS]_\infty}{[S]_\infty^{\zeta}}-n.$$

So setting $S_\infty=[S]_\infty/N$, the equation for the final size of an epidemic can be recovered.

$$\begin{aligned}-(\tau+g)(1-S_\infty^{1/n})&=\tau(S_\infty^{\zeta}-1),\\ S_\infty^{1/n}&=1+\frac{\tau}{\tau+g}(S_\infty^{\zeta}-1),\\ S_\infty&=\left(1-\frac{\tau}{\tau+g}+S_\infty^{\zeta}\frac{\tau}{\tau+g}\right)^n.\end{aligned}$$

This is the same solution as equation (17) because $\theta=S_\infty^{\zeta}$. It is doubtful whether a similar technique could be applied to the situation when $\phi\neq0$.

REFERENCES

Altmann, M. 1995 Susceptible–infectious–recovered epidemic models with dynamic partnerships *J. Math. Biol.* **33**, 661–675.
Anderson, R. M. & May, R. M. 1992 *Infectious diseases of humans.* Oxford University Press.
Ball, F. & Nasell, I. 1994 The shape of the size distribution of an epidemic in a finite population. *Math. Biosci.* **123**, 167–181.
Barbour, A. & Mollison, D. 1990 Epidemics and random graphs. In *Stochastic processes in epidemic theory* (ed. J. P. Gabriel, C. Lefevre & P. Picard), pp. 86–89. New York: Springer.
de Jong, M. C. M., Diekmann, O. & Heesterbeek, J. A. P. 1994 The computation of R_0 for discrete-time epidemic models with dynamic heterogeneity. *Math. Biosci.* **119**, 97–114.
Dickman, R. 1986 Kinetic phase transitions in a surface reaction model: mean field theory. *Phys. Rev.* A **34**, 4246–4250.
Diekmann, O., Heesterbeek, J. A. P. & Metz, J. A. J. 1990 On the definition and the computation of the basic reproduction ratio R_0, in models for infectious diseases in heterogeneous populations. *J. Math. Biol.* **28**, 365–382.
Diekmann, O., de Jong, M. C. M. & Metz, J. A. J. 1998 A deterministic epidemic model taking account of repeated contacts between the same individuals. *J. Appl. Prob.* **35**, 448–462.
Dietz, K. & Hadeler, K. P. 1988 Epidemiological models for sexually transmitted diseases *J. Math. Biol.* **26**, 1–25.
Grenfell, B. T. & Dobson, A. P. 1995 *Ecology of infectious diseases in natural populations.* Cambridge University Press.
Grenfell, B. T., Lonergan, M. E. & Harwood, J. 1992 Quantitative investigations of the epidemiology of phocine distemper virus (PDV) in European common seal populations. *Sci. Total Environ.* **115**, 15–29.
Hassell, M. P. & May, R. M. 1974 Aggregation in predators and insect parasites and its effects on stability. *J. Anim. Ecol.* **43**, 567–594.
Islam, M. N., Oshaughnessy, C. D. & Smith, B. 1996 A random graph model for the final-size distribution of household infections. *Statist. Med.* **15**, 837–843.
Keeling, M. J. & Grenfell, B. T. 1997 Disease extinction and community size: modeling the persistence of measles. *Science* **275**, 65–67.

Keeling, M. J. & Rand, D. A. 1999 Spatial correlations and local fluctuations in host–parasite systems. In *From finite to infinite dimensional dynamical systems* (ed. P. A. Glendinning). Amsterdam: Kluwer.

Keeling, M. J., Rand, D. A. & Morris, A. J. 1997 Correlation models for childhood epidemics. *Proc. R. Soc. Lond.* B **264**, 1149–1156.

Kermack, W. O. & Mc.Kendrick, A. G. 1927 Contributions to the mathematical theory of epidemics. 1. *Proc. R. Soc. Edinb.* A **115**, 700–721 (reprinted in *Bull. Math. Biol.* **53**, 33–55).

Kornberg, H. & Williamson, M. H. (eds) 1987 *Quantitative aspects of the ecology of biological invasions.* London: The Royal Society.

Levin, S. A. & Durrett, R. 1996 From individuals to epidemics. *Proc. R. Soc. Lond.* B **351**, 1615–1621.

Matsuda, H. 1987 Conditions for the evolution of altruism. In *Animal societies: theories and facts* (ed. Y. Ito, J. P. Brown & J. Kikkawa), pp. 67–80. Tokyo: Japanese Scientific Society Press.

Matsuda, H., Ogita, N., Sasaki, A. & Sato, K. 1992 Statistical mechanics of population: the lattice Lotka–Volterra model. *Prog. Theor. Phys.* **88**, 1035–1049.

Mollison, D. 1995 *Epidemic models: their structure and relation to data.* Cambridge University Press.

Morris, A. J. 1997 Representing spatial interactions in simple ecological models. PhD thesis, University of Warwick, UK.

Rand, D. A. 1999 Correlation equations for spatial ecologies. In *Advanced ecological theory* (ed. J. McGlade), pp. 99–143. London: Blackwell Scientific Publishing.

Sato, K., Matsuda, H. & Sasaki, A. 1994 Pathogen invasion and host extinction in lattice structured populations. *J. Math. Biol.* **32**, 251–268.

Van Baalen, M. 1999 Pair dynamics and configuration approximations. In *The geometry of ecological interactions: simplifying spatial complexity* (ed. U. Dieckmann, & J. Metz). Cambridge University Press. (In the press.)

Watts, D. J. & Strogatz, S. H. 1998 Collective dynamics of 'small-world' networks. *Nature* **393**, 440–442.

As this paper exceeds the maximum length normally permitted, the author has agreed to contribute to production costs.

VOLUME 86, NUMBER 13 PHYSICAL REVIEW LETTERS 26 MARCH 2001

Small World Effect in an Epidemiological Model

Marcelo Kuperman[1,*] and Guillermo Abramson[1,2,†]

[1]*Centro Atómico Bariloche and Instituto Balseiro, 8400 S. C. de Bariloche, Argentina*
[2]*Consejo Nacional de Investigaciones Científicas y Técnicas, Argentina*

(Received 5 October 2000)

A model for the spread of an infection is analyzed for different population structures. The interactions within the population are described by small world networks, ranging from ordered lattices to random graphs. For the more ordered systems, there is a fluctuating endemic state of low infection. At a finite value of the disorder of the network, we find a transition to self-sustained oscillations in the size of the infected subpopulation.

DOI: 10.1103/PhysRevLett.86.2909 PACS numbers: 89.75.Hc, 05.65.+b, 87.19.Xx, 87.23.Ge

I. Introduction.—How does the dynamics of an infectious disease depend on the *structure* of a population? A great amount of work has been done on the phenomenological description of particular epidemic situations [1–4]. A classical mathematical approach to these problems deals with well mixed populations, where the subpopulations involved (typically susceptible, infected, and removed) interact in proportion to their sizes. With these zero dimensional models it has been possible to study, among other epidemic features, the existence of threshold values for the spread of an infection [5], the asymptotic solution for the density of infected people [6–8], and the effect of stochastic fluctuations on the modulation of an epidemic situation [9]. A second classical approach describes spatially extended subpopulations, such as elements in a lattice. In this, the geographic spread of an epidemic can be analyzed as a reaction-diffusion process [10–13], bearing close similarity to paradigmatic reactions such as Belousov-Zhabotinskii's.

Real populations rarely fall into either of these categories, being neither well mixed nor lattices. Recently introduced by Watts and Strogatz [14], small world networks attempt to translate, into an abstract model, the complex topology of social interactions. Small worlds may play an important role in the study of the influence of the network structure upon the dynamics of many social processes, such as disease spreading, formation of public opinion, distribution of wealth, transmission of cultural traits, etc. [15]. In relation to epidemiological models, it has been shown that small world networks present a much faster epidemic propagation than reaction-diffusion models or discrete models based on regular lattices of a social network [16].

In the original model of small worlds a single parameter p, running from 0 to 1, characterizes the degree of disorder of the network, respectively, ranging from a regular lattice to a completely random graph. It has been shown that geometrical properties, as well as certain statistical mechanics properties, show a transition at $p_c = 0$ in the limit of large systems, $N \to \infty$ [17]. That is, any finite value of the disorder induces the small world behavior. In this Letter we show that a sharp transition in the behavior of an infection dynamics exists at a finite value of p.

II. Epidemic model.—We analyze a simple model of the spread of an infectious disease. We want, mainly, to point to the role played by the network structure on the temporal dynamics of the epidemic. The disease has three stages: susceptible (S), infected (I), and refractory (R). An element of the population is described by a single dynamical variable adopting one of these three values. Susceptible elements can pass to the infected state through contagion by an infected one. Infected elements pass to the refractory state after an infection time τ_I. Refractory elements return to the susceptible state after a recovery time τ_R. This kind of system is usually called *SIRS*, for the cycle that a single element goes over. The contagion is possible only during the S phase, and only by an I element. During the R phase, the elements are immune and do not infect. *SIRS* models are excitable systems, known to display relaxation oscillations in mean field or well-mixed approaches. In spatially extended versions space-time oscillations can occur, due to the interaction between neighboring elements. Both kinds of behavior are analogous to reaction-diffusion systems such as the Belousov-Zhabotinskii reaction [6,10]. An *SIR* system on a one-dimensional lattice with local and global interactions has already been studied before [18]. In this work, a socially sensible network was used to study the spread of an infection, well before the introduction of small-world networks.

The interactions between the elements of the population are described by a small world network. The links represent the contact between subjects, and infection can proceed only through them. As in the Watts and Strogatz model, the small worlds we study are random networks built upon a topological ring with N vertices and coordination number $2K$. Each link connecting a vertex to a neighbor in the clockwise sense is then rewired at random, with probability p, to any vertex of the system. With probability $(1 - p)$ the original link is preserved. Self-connections and multiple connections are prohibited. With this procedure, we have a regular lattice at $p = 0$, and progressively random graphs for $p > 0$. The long range links that appear at any $p > 0$ trigger the small world phenomenon. At $p = 1$ all the links have been rewired, and the result is similar to (though not exactly) a completely random network. This algorithm should be used with caution, since it can produce disconnected graphs. We have used only connected ones for our analysis.

Time proceeds by discrete steps. Each element is characterized by a time counter $\tau_i(t) = 0, 1, \ldots, \tau_I + \tau_R \equiv \tau_0$, describing its phase in the cycle of the disease. The epidemiological state π_i of the element (S, I, or R) depends on this phase in the following way:

$$\begin{aligned} \pi_i(t) &= S \quad \text{if } \tau_i(t) = 0, \\ \pi_i(t) &= I \quad \text{if } \tau_i(t) \in (1, \tau_I), \\ \pi_i(t) &= R \quad \text{if } \tau_i(t) \in (\tau_I + 1, \tau_0). \end{aligned} \tag{1}$$

The state of an element in the next step depends on its current phase in the cycle, and the state of its neighbors in the network. The rules of evolution are the following:

$$\begin{aligned} \tau_i(t+1) &= 0 && \text{if } \tau_i(t) = 0 \text{ and no infection occurs}, \\ \tau_i(t+1) &= 1 && \text{if } \tau_i(t) = 0 \text{ and } i \text{ becomes infected}, \\ \tau_i(t+1) &= \tau_i(t) + 1 && \text{if } 1 \le \tau_i(t) < \tau_0, \\ \tau_i(t+1) &= 0 && \text{if } \tau_i(t) = \tau_0. \end{aligned} \tag{2}$$

That is, a susceptible element stays as such, at $\tau = 0$, until it becomes infected. Once infected, it goes (deterministically) over a cycle that lasts τ_0 time steps. During the first τ_I time steps, it is infected and can potentially transmit the disease to a susceptible neighbor. During the last τ_R time steps of the cycle, it remains in state R, immune but not contagious. After the cycle is complete, it returns to the susceptible state.

The contagion of a susceptible element by an infected one, and the subsequent excitation of the disease cycle in the new infected, occur stochastically at a local level. Say that the element i is susceptible, and that it has k_i neighbors, of which k_{inf} are infected. Then, i will become infected with probability k_{inf}/k_i. Observe that i will become infected with probability 1 if all its neighbors are infected. Besides this parameter-free mechanism, there may be other reasonable choices. For example, if the susceptible had a probability q of contagion with each infected neighbor, we would have a probability of infection $[1 - (1 - q)^{k_{\text{inf}}}]$. We have tested that both these criteria give qualitatively the same results for values of $q \lesssim 0.2$. For other values of q, the behaviors are outlined at the end of Section III.

III. Numerical results.—We have performed extensive numerical simulations of the described model. Networks with $N = 10^3$ to 10^6 vertices have been explored, with $K = 3$ to 10. A typical realization starts with the generation of the random network and the initialization of the state of the elements. An initial fraction of 0.1 infected, and the rest susceptible, was used in all the results shown here. Other initial conditions have been explored as well, and no changes have been observed in the behavior. After a transient a stationary state is achieved, and the computations are followed for several thousand time steps to perform statistical averages.

We show in Fig. 1 part of three time series displaying the fraction of infected elements in the system, $n_{\text{inf}}(t)$. The three curves correspond to systems with different values of the disorder parameter: $p = 0.01$ (top), 0.2 (middle), and 0.9 (bottom). The three systems have $N = 10^4$ and $K = 3$, and infection cycles with $\tau_I = 4$ and $\tau_R = 9$. The initial state is random with $n_{\text{inf}}(0) = 0.1$. The 400 time steps shown are representative of the stationary state. We can see clearly a transition from an endemic situation to an oscillatory one. At $p = 0.01$ (top), where the network is nearly a regular lattice, the stationary state is a fixed point, with fluctuations. The situation corresponds to that of an endemic infection, with a low and persistent fraction of infected individuals. At high values of p—like the case with $p = 0.9$ shown in the figure (bottom)—large amplitude, self-sustained oscillations develop. The situation is almost periodic, with a very well defined period and small fluctuations in amplitude. The period is slightly longer than τ_0, since it includes the average time that a susceptible individual remains at state S, before being infected. Epidemiologically, the situation resembles the periodic epidemic patterns typical of large populations [3]. A mean field model of the system, expected to resemble the behavior at $p = 1$, can easily be shown to exhibit these oscillations. The transition between both behaviors is apparent—in this relatively small system—at the intermediate value of

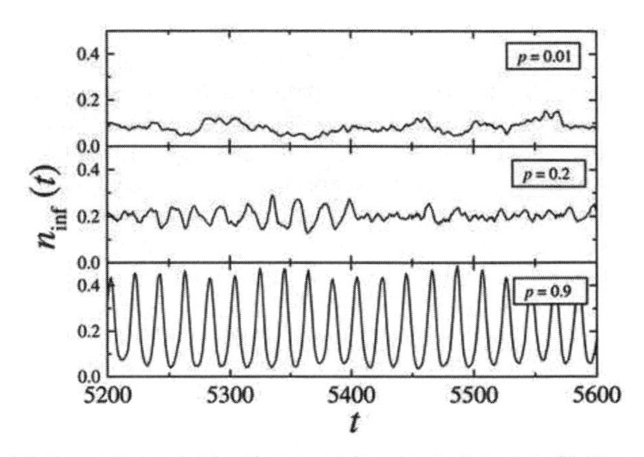

FIG. 1. Fraction of infected elements as a function of time. Three time series are shown, corresponding to different values of the disorder parameter p, as shown in the legends. Other parameters are $N = 10^4$, $K = 3$, $\tau_I = 4$, $\tau_R = 9$, $N_{\text{inf}}(0) = 0.1$.

disorder $p = 0.2$, shown in the middle curve. Here a low-amplitude periodic pattern can be seen, appearing and disappearing again in a background of strong fluctuations. Moreover, the mean value of infection is seen to grow with p.

The formation of persistent oscillations corresponds to a spontaneous synchronization of a significant fraction of the elements in the system. Their phases $\tau_i(t)$ in the epidemic cycle become synchronized, and they go over the disease process together, becoming ill at the same time, and recovering at the same time. We can characterize this behavior with a synchronization parameter [19], defined as

$$\sigma(t) = \left| \frac{1}{N} \sum_{j=1}^{N} e^{i\phi_j(t)} \right|, \tag{3}$$

where $\phi_j = 2\pi(\tau_j - 1)/\tau_0$ is a geometrical phase corresponding to τ_j. We have chosen to let the states $\tau = 0$ out of the sum in (3), and take into account only the deterministic part of the cycles.

When the system is not synchronized, the phases are widely spread in the cycle and the complex numbers $e^{i\phi}$ are correspondingly spread in the unit circle. In this situation σ is small. On the other hand, when a significant part of the elements are synchronized in the cycle, σ is large. The synchronization would be strictly $\sigma = 1$ if all the elements were at the same state at the same time. However, such a state would end up in $N_{\rm inf} = 0$ after τ_0 time steps, and the epidemic would end since the system is closed and no spontaneous infection of susceptibles is being taken into account in the model.

In Fig. 2 we show the synchronization parameter σ, obtained as a time average of $\sigma(t)$ over 2000 time steps and a subsequent average over realizations of the system. Several curves are shown, corresponding to system sizes $N = 10^3$, 10^4, 10^5, and 10^6. A transition in the synchronization can be seen as p runs from 0 to 1. The transition becomes sharper for large systems, at a value of the disorder parameter $p_c \approx 0.4$. It is worthwhile to note that the transition to synchronization occurs as a function of the structure of the network, contrasting the phenomenon of synchronization as a function of the strength of the interaction, as in other systems of coupled oscillators [20]. This behavior is observed for a wide range of values of τ_I and τ_R. The amplest oscillations take place around $\tau_I/\tau_R = 1$. They disappear when τ_I is significantly greater or smaller than τ_R.

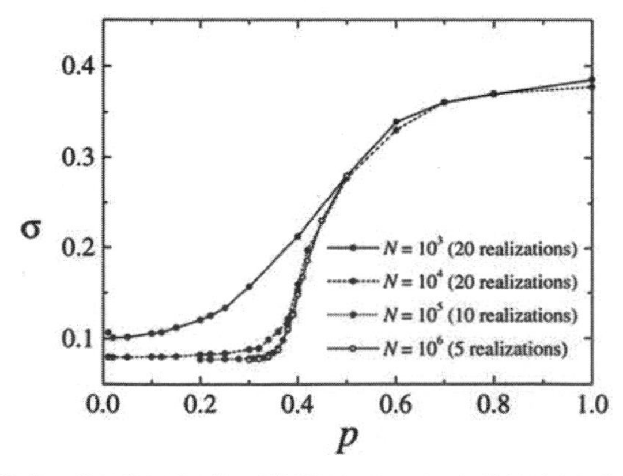

FIG. 2. Synchronization of the system as a function of the disorder parameter p. Three curves are shown, corresponding to different system sizes $N = 10^3$, $N = 10^4$, $N = 10^5$ and 10^6, as shown in the legend. Each point corresponds to a time average of 2000 time steps, and a subsequent average over a number of realizations of the networks and the initial condition, as shown in the legend. Other parameters are $K = 3$, $\tau_I = 4$, $\tau_R = 9$, $N_{\rm inf}(0) = 0.1$.

All the systems shown in Figs. 1 and 2 have $K = 3$. We have explored higher values of K as well. The picture is qualitatively the same, with a sharp transition from a quasifixed point to a quasilimit cycle at a finite value of p. The critical value p_c shifts toward lower values for growing K. This is reasonable, since higher values of K approach the system to a globally coupled one (at $K = N$ all the elements interact with every other one, even at $p = 0$). So, the mean field behavior (the oscillations) can be expected to occur at lower values of p. In Fig. 3 this effect can be seen for a system with $N = 10^4$ and growing values of $K = 3$, 5, and 10.

The alternative mechanism of contagion, introduced at the end of Section II, produces essentially the same behaviors at an intermediate range of values of q, around $q \approx 0.2$. For lower values of q, the regular oscillations are not observed. Instead, the irregular oscillatory behavior (like that in Fig. 1, center) extends up to $p = 1$. As q approaches 0, the infection is less virulent and, eventually, no self-sustained patterns are possible. For values of $q \gtrsim 0.3$, the infection is virulent enough to display the self-sustained oscillations even at low values of p—we observed irregular oscillations down to $p \approx 0.01$. Regular oscillations appear at $p \approx 0.2$. Their amplitude is great enough—at $p \gtrsim 0.5$—to produce the simultaneous infection of almost the whole population, followed by its immunity and

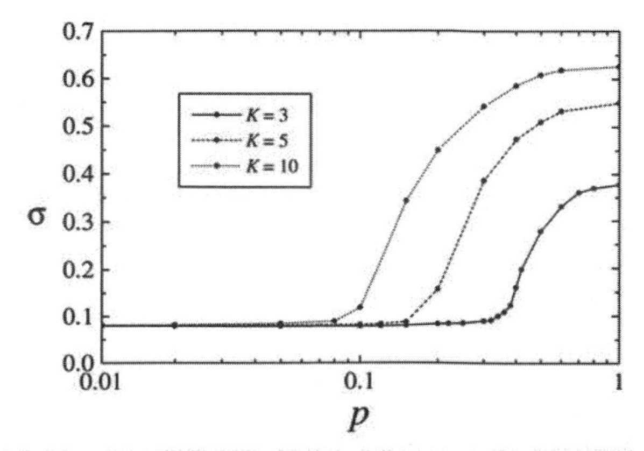

FIG. 3. Synchronization of the system as a function of the disorder parameter p. Three systems with different values of the coordination number K are shown. All of them have $N = 10^4$, $\tau_I = 4$, $\tau_R = 9$, $N_{\rm inf}(0) = 0.1$.

VOLUME 86, NUMBER 13 PHYSICAL REVIEW LETTERS 26 MARCH 2001

subsequent extinction of the disease. Additionally, in this case, for all values of p, there is an increase of the mean value of the infected fraction of the population.

IV. Discussion.—Why does this transition to synchronization take place? Unfortunately, we do not have analytical arguments to describe this yet. As mentioned before, a mean field model can be shown to predict the oscillations, but this can be expected to describe only the network at $p = 1$, and it does not shed light on the nature of the transition at lower values of disorder. We can advance only some conjectures, based on our observation of the dynamical behavior of the system in real time on the computer screen.

An explanation involving $L(p)$, the typical distance between pairs of elements (defined as the size of the minimum path connecting two elements) is not plausible, since L is known to behave critically with $p_c = 0$, and we observe the transition at $p > 0$. Complementarily to L, small world networks can be described by the degree of clusterization $C(p)$ [21]. At low values of p, the networks are rather regular and highly clustered. As p approaches 1, C decreases. The crossover from high to low clusterization occurs at a higher value of p, compared to that observed in the decay of L. For systems with $K = 3$, such as those mainly studied here, we have that this crossover is mostly concentrated between $p = 0.1$ and $p = 0.5$, precisely where the onset of oscillations occurs. Moreover, the change in the average clusterization $C(p)$ is accompanied by a corresponding one in the distribution of the clusterization at the element level, $c_i(p)$. Highly ordered networks, at low p, show not only a large value of the average clusterization $C(p)$ but also a small dispersion around it. On the other extreme, highly disordered networks, with $p > 0.5$, exhibit a low average clusterization and also a low dispersion around it. There is an intermediate range of p, between 0.1 and 0.5, where the average clusterization shifts from high to low, but the distributions are wide. This indicates that, in this range, the system is a mixture of highly ordered, highly clustered regions and random, lowly clustered ones. If we consider that the clustering structure determines a partition of the whole system into smaller, interacting, subsystems, the global behavior of the system could be interpreted as a superposition of the subsystems' dynamics. When p is small, the existence of large clusters (essentially one-dimensional in our model) inhibits the oscillatory behavior because, once the infection breaks into such a region, it remains restricted to it a long time until all the individuals have completed the cycle, orderly and deterministically. As p increases, the number of big, ordered regions decreases. Elements within small regions with some long range links go through the infection cycle and become infected again before long. Some degree of synchronization can be seen here (Fig. 1, center, corresponds to $p = 0.2$). When a critical value of p is reached, and the system is essentially composed of enough small regions of low clusterization and a similar local dynamics, the synchronized, periodic global behavior establishes spontaneously.

In summary, we have observed a transition at a finite value of the disorder in a small world model. The dynamical behavior of an *SIRS* epidemiological model changes from an irregular, low-amplitude evolution at small p to a spontaneous state of wide amplitude oscillations at large p. This may be related to observed patterns in real epidemics [3], where an effect of the social structure is observed in the dynamics of the disease.

The authors thank D. H. Zanette for fruitful discussions.

---

*Email address: kuperman@cab.cnea.gov.ar
†Email address: abramson@cab.cnea.gov.ar

[1] C. McEvedy, Sci. Am. **258**, No. 2, 74 (1988).
[2] M. M. Kaplan and R. G. Webster, Sci. Am. **237**, No. 6, 88 (1977).
[3] A. Cliff and P. Haggett, Sci. Am. **250**, No. 5, 110 (1984).
[4] N. Hirschhorn and W. Greenough III, Sci. Am. **225**, No. 2, 15 (1971).
[5] R. M. Anderson and R. M. May, Science **215**, 1053 (1982); Nature (London) **318**, 323 (1982).
[6] J. D. Murray, *Mathematical Biology* (Springer-Verlag, Berlin, 1993).
[7] N. T. Bailey, *The Mathematical Theory of Infectious Diseases* (Griffin, London, 1975).
[8] F. C. Hoppensteadt, *Mathematical Theories of Populations: Demographics, Genetics and Epidemics,* CBMS Lectures Vol. 20 (SIAM Publications, Philadelphia, 1975).
[9] P. Landa and A. Zaikin, in *Applied Nonlinear Dynamics and Stochastic Systems Near the Millennium,* edited by J. Kadkte and A. Bulsara (AIP, New York, 1997), p. 321.
[10] A. Mikhailov, *Foundations of Synergetic I* (Springer-Verlag, Berlin, 1994).
[11] J. Murray, E. Stanley, and D. Brown, Proc. R. Soc. London B **229**, 111 (1986).
[12] A. Källen, P. Arcury, and J. Murray, J. Theor. Biol. **116**, 377 (1985).
[13] M. Fuentes and M. N. Kuperman, Physica (Amsterdam) **267A**, 471 (1999).
[14] D. J. Watts and S. H. Strogatz, Nature (London) **393**, 440 (1998).
[15] D. J. Watts, *Small Worlds* (Princeton University Press, Princeton, NJ, 1999).
[16] C. Moore and M. E. J. Newman, Phys. Rev. E **61**, 5678 (2000).
[17] A. Barrat and M. Weigt, Eur. Phys. J. B **13**, 547 (2000).
[18] F. Ball, D. Mollison, and G. Scalia-Tomba, Ann. Appl. Prob. **7**, 46 (1997).
[19] Y. Kuramoto, *Chemical Oscillations, Waves, and Turbulence* (Springer, Berlin, 1984).
[20] J. F. Heagy, T. L. Carroll, and L. M. Pecora, Phys. Rev. E **50**, 1874 (1994).
[21] $C(p)$ is an average, over the system, of the local clusterization $c_i(p)$. This is defined as the number of neighbors, of element i, that are neighbors among themselves, normalized to the value that this would have if all of them were connected to one another—namely, $k_i(k_i - 1)/2$.

VOLUME 86, NUMBER 14 PHYSICAL REVIEW LETTERS 2 APRIL 2001

Epidemic Spreading in Scale-Free Networks

Romualdo Pastor-Satorras[1] and Alessandro Vespignani[2]

[1]*Departament de Física i Enginyeria Nuclear, Universitat Politècnica de Catalunya, Campus Nord, Mòdul B4, 08034 Barcelona, Spain*

[2]*The Abdus Salam International Centre for Theoretical Physics (ICTP), P.O. Box 586, 34100 Trieste, Italy*

(Received 20 October 2000)

The Internet has a very complex connectivity recently modeled by the class of scale-free networks. This feature, which appears to be very efficient for a communications network, favors at the same time the spreading of computer viruses. We analyze real data from computer virus infections and find the average lifetime and persistence of viral strains on the Internet. We define a dynamical model for the spreading of infections on scale-free networks, finding the absence of an epidemic threshold and its associated critical behavior. This new epidemiological framework rationalizes data of computer viruses and could help in the understanding of other spreading phenomena on communication and social networks.

DOI: 10.1103/PhysRevLett.86.3200 PACS numbers: 89.75.Hc, 05.50.+q, 05.70.Ln

Many social, biological, and communication systems can be properly described by complex networks whose nodes represent individuals or organizations, and links mimic the interactions among them [1,2]. Particularly interesting examples are the Internet [3,4] and the World Wide Web [5], which have been extensively studied because of their technological and economical relevance. These studies have revealed, among other facts, that the probability that a node of these networks has k connections follows a scale-free distribution $P(k) \sim k^{-\gamma}$, with an exponent γ that ranges between 2 and 3. The presence of nodes with a very large number of connections (local clustering) is indeed the key ingredient in the modeling of these networks with the recent introduction of scale-free (SF) graphs [6].

In view of the wide occurrence of complex networks in nature it is of great interest to inspect the effect of their features on epidemic and disease spreading [7], and more in general in the context of the nonequilibrium phase transitions typical of these phenomena [8]. The study of epidemics on these networks finds an immediate practical application in the understanding of computer virus spreading [9,10], and could also be relevant to the fields of epidemiology [11] and pollution control [12].

In this Letter, we analyze data from real computer virus epidemics, providing a statistical characterization that points out the importance of incorporating the peculiar topology of scale-free networks in the theoretical description of these infections. With this aim, we study by large scale simulations and analytical methods the susceptible-infected-susceptible (SIS) [11] model on SF graphs. We find the absence of an epidemic threshold and its associated critical behavior, which implies that SF networks are prone to the spreading and the persistence of infections at whatever spreading rate the epidemic agents possess. The absence of the epidemic threshold—a standard element in mathematical epidemiology [11]—radically changes many of the standard conclusions drawn in epidemic modeling. The present results are also relevant in the field of absorbing-state phase transitions and catalytic reactions [8].

The analysis of computer viruses has been the subject of a continuous interest in the computer science community [10,13–15], mainly following approaches borrowed from biological epidemiology [11]. The standard model used in the study of computer virus infections is the SIS epidemiological model. Each node of the network represents an individual and each link is a connection along which the infection can spread to other systems. Individuals exist only in two discrete states, "healthy" or "infected." At each time step, each susceptible (healthy) node is infected with rate ν if it is connected to one or more infected nodes. At the same time, infected nodes are cured and become again susceptible with rate δ, defining an effective spreading rate $\lambda = \nu/\delta$ [16]. Without lack of generality, we can set $\delta = 1$. This model implicitly considers the presence of antivirus software, since all infected individuals eventually return to the susceptible state, and represents the case in which computer users do not become more alert with respect to viral infection once they have cleaned their computers which can again become infected [15]. The updating can be performed with both parallel and sequential dynamics [8]. In models with local connectivity (Euclidean lattices) and random graphs, the most significant result is the general prediction of a nonzero epidemic threshold λ_c [8,11]. If the value of λ is above the threshold, $\lambda \geq \lambda_c$, the infection spreads and becomes persistent. Below it, $\lambda < \lambda_c$, the infection dies out exponentially fast. The epidemic threshold is actually equivalent to a critical point in a nonequilibrium phase transition. In this case, the critical point separates an active phase with a stationary density of infected nodes from a phase with only healthy nodes and null activity. In particular, it is easy to recognize that the SIS model is a generalization of the contact process model that has been extensively studied in the context of absorbing-state phase transitions [8]. Statistical observations of virus incidents in the wild, on the other hand, indicate that all surviving viruses saturate to a very

 0031-9007/01/86(14)/3200(4)$15.00

low level of persistence, affecting just a tiny fraction of the total number of computers [10]. This fact is in striking contradiction with the theoretical predictions unless in the very unlikely chance that *all* computer viruses have an effective spreading rate tuned just infinitesimally above the threshold. This points out that the view obtained so far with the modeling of computer virus epidemics is very instructive but not completely adequate to represent the real phenomenon.

In order to gain further insight into the spreading properties of viruses in the wild, we have analyzed the prevalence data reported by the Virus Bulletin [17] from February 1996 to March 2000, covering a time window of 50 months. We have analyzed in particular the *surviving probability* of homogeneous groups of viruses, classified according to their infection mechanism [9]. We consider the total number of viruses of a given strain that are born and died within our observation window. Hence, we calculate the surviving probability $P_s(t)$ of the strain as the fraction of viruses still alive at time t after their birth. Figure 1 shows that the surviving probability suffers a sharp drop in the first two months of a virus' life. This is a well-known feature [10,13] indicating that statistically only a small percentage of viruses gives rise to a significant outbreak in the computer community. Figure 1, on the other hand, shows for larger times a clean exponential tail, $P_s(t) \sim \exp(-t/\tau)$, where τ represents the characteristic lifetime of the virus strain [18]. The numerical fit of the data yields $\tau \simeq 14$ months for boot and macroviruses and $\tau \simeq 6$–9 months for file viruses. The values of τ are relatively independent of the observation window considered, i.e., the analysis of the viruses that are born and die in a time range of less than 50 months yields results compatible with the full data

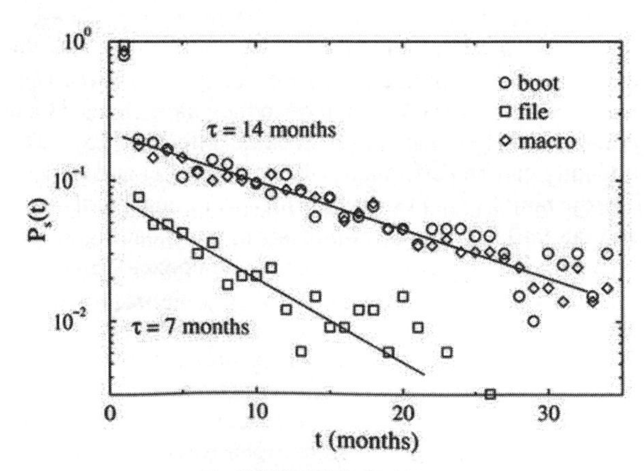

FIG. 1. Surviving probability for viruses in the wild. The 814 different viruses analyzed have been grouped in three main strains [9]: file viruses infect a computer when running an infected application; boot viruses also spread via infected applications, but copy themselves into the boot sector of the hard drive and are thus immune to a computer reboot; macroviruses infect data files and are thus platform independent. The presence of an exponential decay is evident in the plot, with characteristic time τ.

set, with larger fluctuations, however, due to the smaller statistics. These characteristic times are impressively large if compared with the interval in which antivirus software is available on the market (usually within days or weeks after the first incident report) and corresponds to the occurrence of metastable endemic states. Such a long lifetime on the scale of the typical spread/recovery rates would suggest an effective spreading rate much larger than the epidemic threshold. On the other hand, this is again discordant with the always low prevalence levels of computer viruses.

The key point in understanding the puzzling properties exhibited by computer viruses resides in the capacity of many of them to propagate via data exchange with communication protocols (FTP, emails, etc.) [10]. Viruses will spread preferentially to computers which are highly connected to the outer world and thus are proportionally exchanging more data and information. It is thus rather intuitive to consider the Internet topology as the effective one on which the spreading occurs. The scale-free connectivity of the Internet implies that each node has a statistically significant probability of having a very large number of connections compared to the average connectivity $\langle k \rangle$ of the network. That opposes conventional random networks (local or nonlocal) in which each node has approximately the same number of links $k \simeq \langle k \rangle$ [19]. It is then natural to foresee that scale-free properties should be included in a theory of epidemic spreading of computer viruses.

To address the effects of scale-free connectivity in epidemic spreading we study the SIS model on SF networks. As a prototypical example, we consider the graph generated by using the algorithm devised in Ref. [6]. We start from a small number m_0 of disconnected nodes; every time step a new node is added, with m links that are connected to an old node i with k_i links according to the probability $k_i/\sum_j k_j$. After iterating this scheme a sufficient number of times, we obtain a network composed by N nodes with connectivity distribution $P(k) \sim k^{-3}$ and average connectivity $\langle k \rangle = 2m$. In this work we take $m = 3$. We have performed numerical simulations on graphs with the number of nodes ranging from $N = 10^3$ to $N = 8.5 \times 10^6$ and studied the variation in time and the stationary properties of the density of infected nodes ρ in surviving infections; i.e., the virus prevalence. Initially we infect half of the nodes in the network, and iterate the rules of the SIS model with parallel updating. After an initial transient regime, the system stabilizes in a steady state with a constant average density of infected nodes. In this steady state, nodes are infected recurrently, without apparent periodicity. The prevalence is computed averaging over at least 100 different starting configurations, performed on at least 10 different realizations of the random networks.

The first arresting evidence from simulations is the *absence* of an epidemic threshold, i.e., $\lambda_c = 0$. In Fig. 2 we show the virus prevalence in the steady state that decays with decreasing λ as $\rho \sim \exp(-C/\lambda)$, where C is a constant. This implies that for any finite value of λ the virus

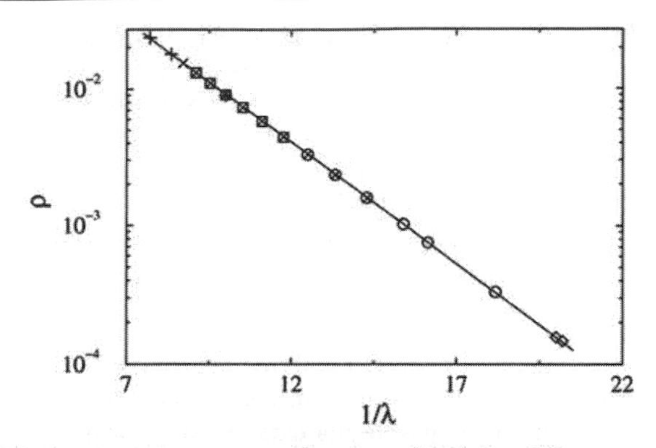

FIG. 2. Persistence ρ as a function of $1/\lambda$ for different network sizes: $N = 10^5$ (+), $N = 5 \times 10^5$ (□), $N = 10^6$ (×), $N = 5 \times 10^6$ (○), and $N = 8.5 \times 10^6$ (◇). The linear behavior on the semilogarithmic scale proves the stretched exponential behavior predicted for ρ. The full line is a fit to the form $\rho \sim \exp(-C/\lambda)$.

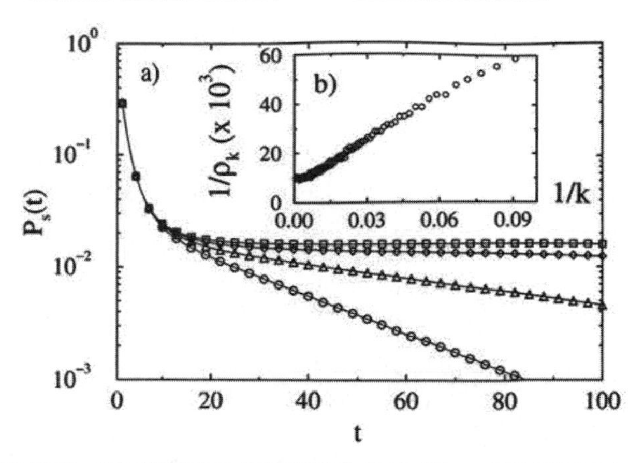

FIG. 3. (a) Surviving probability $P_s(t)$ for a spreading rate $\lambda = 0.065$ in scale-free networks of size $N = 5 \times 10^5$ (□), $N = 2.5 \times 10^4$ (◇), $N = 1.25 \times 10^4$ (△), and $N = 6.25 \times 10^3$ (○). The exponential behavior, following a sharp initial drop, is compatible with the data analysis of Fig. 1. (b) Relative density ρ_k versus k^{-1} in a SF network of size $N = 5 \times 10^5$ and spreading rate $\lambda = 0.1$. The plot recovers the form predicted in Eq. (2).

can pervade the system with a finite prevalence, in sufficiently large networks. In all networks with bounded connectivity the steady state prevalence is always null below the epidemic threshold; i.e., all infections die out. Further evidence to our results is given by the total absence of scaling of ρ with the number of nodes that is, on the contrary, typical of epidemic transitions in the proximity of a finite threshold [8]. This allows us to exclude the presence of any spurious results due to network finite size effects. The present result can be intuitively understood by noticing that for usual lattices, the higher the node's connectivity, the smaller the epidemic threshold. In a SF network the unbounded fluctuations in connectivity ($\langle k^2 \rangle = \infty$) play the role of an infinite connectivity, annulling thus the threshold.

Finally, we analyze the spreading of infections starting from a localized virus source. We observe that the spreading growth in time has an algebraic form that is in agreement with real data that never found an exponential increase of a virus in the wild. Noteworthy, by applying the definition of surviving probability $P_s(t)$ used to analyze real data, we recover in our model the same exponential behavior in time (see Fig. 3a). The characteristic lifetime depends on the spreading rate and the network sizes, allowing us to relate the average lifetime of a viral strain with an effective spreading rate and the Internet size [20]. At the same time, the divergence of lifetimes for larger networks points out that viruses live longer if the Internet expands.

We can also approach the system analytically by writing the single-site equation governing the time evolution of $\rho(t)$. In order to take into account connectivity fluctuations, we consider the relative density $\rho_k(t)$ of infected nodes with given connectivity k; i.e., the probability that a node with k links is infected. The dynamical mean-field (MF) reaction rate equations can be written as [8,21]

$$\partial_t \rho_k(t) = -\rho_k(t) + \lambda k[1 - \rho_k(t)]\Theta(\lambda). \quad (1)$$

The creation term considers the probability that a node with k links is healthy $[1 - \rho_k(t)]$ and gets the infection via a connected node. The probability of this event is proportional to the infection rate, the number of connections, and the probability $\Theta(\lambda)$ that any given link points to an infected node. The MF character of this equation stems from the fact that we have neglected the density correlations among the different nodes. However, we have relaxed the homogeneity assumption on the node's connectivity usually implemented in regular networks. By imposing stationarity $[\partial_t \rho_k(t) = 0]$ we find the stationary densities

$$\rho_k = \frac{k\lambda\Theta(\lambda)}{1 + k\lambda\Theta(\lambda)}, \quad (2)$$

denoting that the higher the node connectivity, the higher the probability to be infected. This inhomogeneity must be taken into account in the self-consistent calculation of $\Theta(\lambda)$. Indeed, the probability that a link points to a node with s links is proportional to $sP(s)$. In other words, a randomly chosen link is more likely to be connected to a node with high connectivity, yielding

$$\Theta(\lambda) = \sum_k \frac{kP(k)\rho_k}{\sum_s sP(s)}. \quad (3)$$

Since ρ_k is on its turn function of $\Theta(\lambda)$, we obtain a consistency equation that allows us to find $\Theta(\lambda)$ and ρ_k. Finally we can calculate the order parameter by evaluating the relation $\rho = \sum_k P(k)\rho_k$ that expresses the average density of infected nodes in the system. In the SF model considered here, we have a connectivity distribution $P(k) = 2m^2/k^{-3}$, where k is approximated as a continuous variable [6]. In this case, integration of Eq. (3) allows one to write $\Theta(\lambda) = e^{-1/m\lambda}/\lambda m$, at lowest order in λ. Averaging over ρ_k, this finally gives

$$\rho \simeq 2e^{-1/m\lambda}. \quad (4)$$

This very intuitive calculation recovers the numerical findings and confirms the surprising absence of any epidemic threshold or critical point in the model; i.e., $\lambda_c = 0$. Finally, as a further check of our analytical results, we have numerically computed in our model the relative densities ρ_k, recovering the predicted dependence upon k of Eq. (2) (see Fig. 3b). It is also worth remarking that the present framework can be generalized to networks with $2 < \gamma \leq 3$, recovering qualitatively the same results. Only for $\gamma > 4$, epidemics on SF networks have the same properties as on random networks. A detailed analysis of the various cases will be presented elsewhere [22].

The emerging picture for epidemic spreading in complex networks emphasizes the role of topology in epidemic modeling. In particular, the absence of epidemic threshold and critical behavior in a wide range of scale-free network provide an unexpected result that changes radically many standard conclusions on epidemic spreading. This indicates that infections can proliferate on these scale-free networks whatever spreading rates they may have. This very bad news is, however, balanced by the exponentially small prevalence for a wide range of spreading rates ($\lambda \ll 1$). This point appears to be particularly relevant in the case of technological networks such as the Internet [4] that show scale-free connectivity with exponents $\gamma \simeq 2.5$. For instance, the present picture qualitatively fits the observation from real data of computer virus spreading, and could solve the long-standing problem of the generalized low prevalence of computer viruses without assuming any global tuning of the spreading rates.

This work has been partially supported by the European Network Contract No. ERBFMRXCT980183. R.P.-S. also acknowledges support from the Grant No. CICYT PB97-0693. We thank S. Franz, M.-C. Miguel, R.V. Solé, M. Vergassola, S. Visintin, S. Zapperi, and R. Zecchina for helpful comments and discussions.

[1] See the special section on Complex systems [Science **284**, 79 (1999)]; S. Wasserman and K. Faust, *Social Network Analysis* (Cambridge University Press, Cambridge, 1994).

[2] L. A. N. Amaral, A. Scala, M. Barthélémy, and H. E. Stanley, Proc. Natl. Acad. Sci. U.S.A. **97**, 11 149 (2000).

[3] M. Faloutsos, P. Faloutsos, and C. Faloutsos, ACM SIGCOMM '99, Comput. Commun. Rev. **29**, 251 (1999).

[4] A. Medina, I. Matt, and J. Byers, Comput. Commun. Rev. **30**, 18 (2000); G. Caldarelli, R. Marchetti, and L. Pietronero, Europhys. Lett. **52**, 386 (2000).

[5] R. Albert, H. Jeong, and A.-L. Barabási, Nature (London) **401**, 130 (1999).

[6] A.-L. Barabási and R. Albert, Science **286**, 509 (1999); A.-L. Barabási, R. Albert, and H. Jeong, Physica (Amsterdam) **272A**, 173 (1999).

[7] C. Moore and M. E. J. Newman, Phys. Rev. E **61**, 5678 (2000).

[8] J. Marro and R. Dickman, *Nonequilibrium Phase Transitions in Lattice Models* (Cambridge University Press, Cambridge, 1999).

[9] F. B. Cohen, *A Short Course on Computer Viruses* (John Wiley & Sons, New York, 1994).

[10] J. O. Kephart, G. B. Sorkin, D. M. Chess, and S. R. White, Sci. Am. **277**, No. 5, 56 (1997); S. R. White, in *Proceedings of the Virus Bulletin Conference, Munich, 1998.* Available on-line at http://www.research.ibm.com/antivirus/SciPapers.htm.

[11] N. T. J. Bailey, *The Mathematical Theory of Infectious Diseases* (Griffin, London, 1975), 2nd ed.; J. D. Murray, *Mathematical Biology* (Springer-Verlag, Berlin, 1993).

[12] M. K. Hill, *Understanding Environmental Pollution* (Cambridge University Press, Cambridge, 1997).

[13] J. O. Kephart, S. R. White, and D. M. Chess, IEEE Spectr. **30**, 20 (1993).

[14] W. H. Murray, Comput. Sec. **7**, 130 (1988).

[15] J. O. Kephart and S. R. White, in *Proceedings of the 1991 IEEE Computer Society Symposium on Research in Security and Privacy (SSP '91)* (IEEE, Washington, 1991), p. 343.

[16] It is also possible to define models in which the infection rate is proportional to the number of infected nearest neighbors [8]. In the small prevalence regime we are interested in, both prescriptions yield exactly the same behavior.

[17] Virus prevalence data publicly available at the web site http://www.virusbtn.com/Prevalence/.

[18] This is the usual way in which it is determined the survival probability in numerical simulations of spreading models; see Ref. [8].

[19] P. Erdös and P. Rényi, Publ. Math. Inst. Hung. Acad. Sci. **5**, 17 (1960); D. J. Watts and S. H. Strogatz, Nature (London) **393**, 440 (1998); A. Barrat and M. Weigt, Eur. Phys. J. B **13**, 547 (2000).

[20] This characteristic scaling is often encountered at absorbing-state phase transitions in finite size systems [8]. In general, $P_s(\infty)$ is finite only for infinite size networks.

[21] G. Szabó, Phys. Rev. E **62**, 7474 (2000).

[22] R. Pastor-Satorras and A. Vespignani (unpublished).

A simple model of global cascades on random networks

Duncan J. Watts*

Department of Sociology, Columbia University New York, NY 10027

Communicated by Murray Gell-Mann, Santa Fe Institute, Santa Fe, NM, February 14, 2002 (received for review May 29, 2001)

The origin of large but rare cascades that are triggered by small initial shocks is a phenomenon that manifests itself as diversely as cultural fads, collective action, the diffusion of norms and innovations, and cascading failures in infrastructure and organizational networks. This paper presents a possible explanation of this phenomenon in terms of a sparse, random network of interacting agents whose decisions are determined by the actions of their neighbors according to a simple threshold rule. Two regimes are identified in which the network is susceptible to very large cascades—herein called global cascades—that occur very rarely. When cascade propagation is limited by the connectivity of the network, a power law distribution of cascade sizes is observed, analogous to the cluster size distribution in standard percolation theory and avalanches in self-organized criticality. But when the network is highly connected, cascade propagation is limited instead by the local stability of the nodes themselves, and the size distribution of cascades is bimodal, implying a more extreme kind of instability that is correspondingly harder to anticipate. In the first regime, where the distribution of network neighbors is highly skewed, it is found that the most connected nodes are far more likely than average nodes to trigger cascades, but not in the second regime. Finally, it is shown that heterogeneity plays an ambiguous role in determining a system's stability: increasingly heterogeneous thresholds make the system more vulnerable to global cascades; but an increasingly heterogeneous degree distribution makes it less vulnerable.

How is it that small initial shocks can cascade to affect or disrupt large systems that have proven stable with respect to similar disturbances in the past? Why do some books, movies, and albums emerge out of obscurity, and with small marketing budgets, to become popular hits (1), when many *a priori* indistinguishable efforts fail to rise above the noise? Why does the stock market exhibit occasional large fluctuations that cannot be traced to the arrival of any correspondingly significant piece of information (2)? How do large, grassroots social movements start in the absence of centralized control or public communication (3)?

These phenomena are all examples of what economists call *information cascades* (ref. 4; but which are herein called simply *cascades*), during which individuals in a population exhibit herd-like behavior because they are making decisions based on the actions of other individuals rather than relying on their own information about the problem. Although they are generated by quite different mechanisms, cascades in social and economic systems (3–6) are similar to cascading failures in physical infrastructure networks (7, 8) and complex organizations (9) in that initial failures increase the likelihood of subsequent failures, leading to eventual outcomes that, like the August 10, 1996 cascading failure in the western United States power transmission grid (8), are extremely difficult to predict, even when the properties of the individual components are well understood. Not as newsworthy, but just as important as the cascades themselves, is that the very same systems routinely display great stability in the presence of continual small failures and shocks that are at least as large as the shocks that ultimately generate a cascade. Cascades can therefore be regarded as a specific manifestation of the *robust yet fragile* nature of many complex systems (10): a system may appear stable for long periods of time and withstand many external shocks (robust), then suddenly and apparently inexplicably exhibit a large cascade (fragile).

Although the social, economic, and physical mechanisms responsible for the occurrence of cascades are complex and may vary widely across systems and even between particular cascades in the same system, it is proposed in this paper that some generic features of cascades can be explained in terms of the connectivity of the network by which influence is transmitted between individuals. Specifically, this paper addresses the set of qualitative observations that (*i*) global (i.e., very large) cascades can be triggered by exogenous events (shocks) that are very small relative to the system size, and (*ii*) global cascades occur rarely relative to the number of shocks that the system receives, and may be triggered by shocks that are *a priori* indistinguishable from shocks that do not.

Model Motivation: Binary Decisions with Externalities

This model is motivated by considering a population of individuals each of whom must decide between two alternative actions, and whose decisions depend explicitly on the actions of other members of the population. In social and economic systems, decision makers often pay attention to each other either because they have limited information about the problem itself or limited ability to process even the information that is available (6). When deciding which movie (11) or restaurant (12) to visit, we often have little information with which to evaluate the alternatives, so frequently we rely on the recommendation of friends, or simply pick the movie or restaurant to which most people are going. Even when we have access to plentiful information, such as when we evaluate new technologies, risky financial assets, or job candidates, we often lack the ability to make sense of it; hence, again we rely on the advice of trusted friends, colleagues, or advisors. In other decision making scenarios, such as in collective action problems (3) or social dilemmas (13), an individual's payoff is an explicit function of the actions of others. And in other problems still, involving say the diffusion of a new technology (14), the utility of a single additional unit—a fax machine for example—may depend on the number of units that have already been sold. In all these problems, therefore, regardless of the details, individual decision makers have an incentive to pay attention to the decisions of others.

In economic terms, this entire class of problems is known generically as *binary decisions with externalities* (6). As simplistic as it appears, a binary decision framework is relevant to surprisingly complex problems. To take an extreme example, the creation of a political coalition or an international treaty is unquestionably a complex, multifaceted process with many potential outcomes. But once the coalition exists or the treaty has been drafted, the decision of whether or not to join is essential a binary one. Similar reasoning applies to a firm's choice between two technologies, or an individual's choice between two neighborhood restaurants—the factors involved in the decision may be many, but the decision itself can be regarded as binary.

Both the detailed mechanisms involved in binary decision problems, and also the origins of the externalities can vary widely across specific problems. Nevertheless, in many applications that have been examined in the economics and sociology literature—for

*E-mail: djw24@columbia.edu.

 www.pnas.org/cgi/doi/10.1073/pnas.082090499

example, fads (1, 4, 5), riots (15), crime (16), competing technologies (14), and the spread of innovations (17, 18), conventions (6), and cooperation (13)—the decision itself can be considered a function solely of the *relative* number of other agents who are observed to choose one alternative over the other (6). Because many decisions are inherently costly, requiring commitment of time or resources, the relevant decision function frequently exhibits a strong *threshold* nature: agents display inertia in switching states, but once their personal threshold has been reached, the action of even a single neighbor can tip them from one state to another.

Model Specification

A particularly simple binary decision rule with externalities that captures the essential features outlined above is the following: An individual agent observes the current states (either 0 or 1) of k other agents, which we call its *neighbors*, and adopts state 1 if at least a threshold fraction ϕ of its k neighbors are in state 1, else it adopts state 0.

To account for variations in knowledge, preferences, and observational capabilities across the population of decision-making agents, both individual thresholds and also the number of neighbors k are allowed to be heterogeneous. First, each agent is assigned a threshold ϕ drawn at random from a distribution $f(\phi)$ that is defined on the unit interval and normalized such that $\int_0^1 f(\phi)d\phi = 1$, but which is otherwise arbitrary. Next, we construct a network of n agents, in which each agent is connected to k neighbors with probability p_k and the average number of neighbors is $\langle k \rangle = z$. Although we shall continue to speak of an agent's *neighbors*, we should think of them simply as the set of incoming signals that are relevant to the problem at hand. More formally, we say that agents are represented by *vertices* (or *nodes*) in a graph; neighboring vertices are joined by *edges*; p_k is the *degree distribution* of the graph; and z is the *average degree* (in physics, z is usually called the *coordination number*). To model the dynamics of cascades, the population is initially all-off (state 0) and is perturbed at time $t = 0$ by a small fraction $\Phi_0 \ll 1$ of vertices that are switched on (state 1). The population then evolves at successive time steps with all vertices updating their states in random, asynchronous order according to the threshold rule above. Once a vertex has switched on, it remains on (*active*) for the duration of the dynamics.

In the social science literature, decision rules of this kind are usually derived either from the payoff structure of noncooperative games such as the prisoner's dilemma (3, 6), or from stochastic sampling procedures (18). But when regarded more generally as a change of state—not just a decision—the model belongs to a larger class of contagion problems that includes models of failures in engineered systems such as power transmission networks (8) or the internet (19, 20), epidemiological (21) and percolation (22, 23) models of disease spreading, and a multiplicity of cellular-automata models including random-field Ising models (24), bootstrap percolation (25, 26), majority voting (27, 28), spreading activation (29), and self-organized criticality (8, 29).

The model, however, differs from these other contagion models in some important respects. (*i*) Unlike epidemiological models, where contagion events between pairs of individuals are *independent*, the threshold rule effectively introduces *local dependencies*; that is, the effect that a single infected neighbor will have on a given node depends critically on the states of the node's other neighbors. (*ii*) Unlike bootstrap percolation, and self-organized criticality models (which also exhibit local dependencies), the threshold is not expressed in terms of the absolute number of a node's neighbors choosing a given alternative, but the corresponding fraction of the neighborhood. This is a natural condition to impose for decision making problems, because the more signals a decision maker receives, the less significant any one signal becomes. (*iii*) Unlike random-field Ising and majority vote models, which are typically modeled on regular lattices, here we are concerned with heterogeneous networks; that is, networks in which individuals have different numbers of neighbors. All these features—*local dependencies*, *fractional thresholds*, and *heterogeneity*—are essential to the dynamics of cascades. Furthermore, although they are clearly related by the threshold condition, network heterogeneity and threshold heterogeneity turn out not to be equivalent, and therefore need to be considered separately.

Exact Solution on an Arbitrary Random Graph

The main objective of this paper is to explore how the vulnerability of interconnected systems to global cascades depends on the network of interpersonal influences governing the information that individuals have about the world, and therefore their decisions. Because building relationships and gathering information are both costly exercises, interaction and influence networks tend to be very sparse (17)—a characteristic that appears to be true of real networks in general (30)—so we consider only the properties of networks with $z \ll n$. In the absence of any known geometry for the problem, a natural first choice for a sparse interaction network is an undirected random graph (31), with n vertices and specified degree distribution p_k. Although random graphs are not considered to be highly realistic models of most real-world networks (30), they are often used as first approximations (19, 20, 32) because of their relative tractability, and this tradition is followed here. Our approach concentrates on two quantities: (*i*) the probability that a *global cascade* will be triggered by a single node (or small seed of nodes), where we define a global cascade formally as cascade that occupies a finite fraction of an infinite network; and (*ii*) the expected size of a global cascade once it is triggered. When describing our results, the term *cascade* therefore refers to an event of any size triggered by an initial seed, whereas *global cascade* is reserved for *sufficiently large* cascades (in practice, this means more than a fixed fraction of large, but finite network).

In any sufficiently large random graph with $z < c \ln n$ (where c is some constant) and $\Phi_0 \ll 1$ (i.e., sparsely connected with a small initial seed), we can assume that the local neighborhood of a small seed will not contain any short cycles; hence, no vertex neighboring the initial seed will be adjacent to more than one seed member. This approximation becomes exact in the case of an infinite network, with finite z, or a seed consisting of a single vertex. Under this condition, the only way in which the seed can grow is if at least one of its immediate neighbors has a threshold such that $\phi \leq 1/k$, or equivalently has degree $k \leq K = \lfloor 1/\phi \rfloor$. We call vertices that are unstable in this one-step sense, *vulnerable*, and those that are not, *stable*, noting that the distinction only applies when the seed in question is *small* (numerical simulations suggest that seeds that are three orders of magnitude less than the system size are sufficiently small). The case of large seeds will be discussed later.

Although the vulnerability condition is quite general, for concreteness we use the language of the diffusion of innovations (17), in which the initial seed plays the role of the *innovators*, and vulnerable vertices correspond to *early adopters*. Unless the innovators are connected to a community of early adopters, no cascade is possible. In fact, as we show below, the success or failure of an innovation may depend less on the number and characteristics of the innovators themselves than on the structure of the community of early adopters. Clearly, the more early adopters exist in the network, the more likely it is that an innovation will spread. But the extent of its growth—and hence the susceptibility of the network as a whole—depends not only on the number of early adopters, but on how connected they are to one another, and also to the much larger community consisting of the *early and late majority*, who do not tend to respond to the innovators directly, but who can be influenced indirectly if exposed to multiple early adopters. In the context of this model, we conjecture that the required condition for a global cascade is that the subnetwork of vulnerable vertices must *percolate* (22) throughout the network as a whole, which is to say that the largest, connected vulnerable cluster must occupy a finite fraction of an infinite network. Regardless of how connected the network as a whole might be, the claim here is that only if the largest *vulnerable* cluster percolates are global cascades possible.

This condition, which we call the *cascade condition* (see Eq. **5** below), has the considerable advantage of reducing a complex dynamics problem to a static, percolation problem that can be solved using a generating function approach. A similar technique has been used elsewhere (20, 32) to study the connectivity properties of random graphs; here the basic approach is modified (described in detail in ref. 32) to focus on vulnerable vertices. By construction, every vertex has degree k with probability p_k, and by the vulnerability condition above, a vertex with degree k is vulnerable with probability $\rho_k = P[\phi \leq 1/k]$. Hence, the probability of a vertex u having degree k and being vulnerable is $\rho_k p_k$, and the corresponding generating function of vulnerable vertex degree is:

$$G_0(x) = \textstyle\sum_k \rho_k p_k x^k, \qquad \text{[1a]}$$

$$\text{where } \rho_k = \begin{cases} 1 & k=0 \\ F(1/k) & k>0 \end{cases} \qquad \text{[1b]}$$

and $F(\phi) = \int_0^\phi f(\varphi)d\varphi$. By incorporating all of the information contained in the degree distribution and the threshold distribution, $G_0(x)$ generates all of the moments of the degree distribution solely of *vulnerable* vertices, where the relevant moments can be extracted by evaluating the derivatives of $G_0(x)$ at $x = 1$. For the purposes of this paper, the two most important quantities are (*i*) the vulnerable fraction of the population $P_v = G_0(1)$, and (*ii*) the average degree of vulnerable vertices $z_v = G_0'(1)$. Because we are interested in the propagation of cascades from one vertex to another, we also require the degree distribution of a vulnerable vertex v that is a random neighbor of our initially chosen vertex u. The larger the degree of v, the more likely it is to be a neighbor of u; hence, the probability of choosing v is proportional to kp_k, and the correctly normalized generating function $G_1(x)$ corresponding to a neighbor of u is:

$$G_1(x) = \frac{\sum_k k\rho_k p_k x^{k-1}}{\sum_k kp_k} = \frac{G_0'(x)}{z}. \qquad \text{[2]}$$

To calculate the properties of clusters of vulnerable vertices (the community structure of the early adopters), we introduce the analogous generating functions

$$H_0(x) = \textstyle\sum_n q_n x^n \text{ and } H_1(x) = \sum_n r_n x^n,$$

where q_n is the probability that a randomly chosen vertex will belong to a vulnerable cluster of size n, and r_n is the corresponding probability for a neighbor of an initially chosen vertex. Any *finite* cluster of size n that we arrive at by following a random edge can be regarded as composed of smaller such clusters, whose cumulative sizes must sum to n. Because a sufficiently large random graph below percolation can be regarded as a pure branching structure, we can therefore ignore the possibility that the subclusters will be connected in cycles, so each subcluster can be treated independently of the others. (The presence of an infinite cluster above percolation will be dealt with below.) Hence, the probability of a finite cluster of size n is simply the product of the probabilities of its (also finite) subclusters. It follows from the properties of generating functions (20, 32) that $H_1(x)$ satisfies the following self-consistency equation:

$$H_1(x) = [1 - G_1(1)] + xG_1(H_1(x)), \qquad \text{[3a]}$$

from which $H_0(x)$ can be computed according to

$$H_0(x) = [1 - G_0(1)] + xG_0(H_1(x)). \qquad \text{[3b]}$$

where the first term in both Eqs. **3a** and **3b** corresponds to the probability that the vertex chosen is not vulnerable, and the second term accounts for the size distribution of vulnerable clusters attached to a vertex that is, itself, vulnerable. $H_0(x)$ therefore generates all moments of the distribution of vulnerable cluster sizes, the most important of which, for our current purpose, is the *average vulnerable cluster size* $\langle n \rangle = H_0'(1)$, because this is the quantity that diverges at percolation. Substituting the expressions for $H_0(x)$ and $H_1(x)$ above, we find that

$$\langle n \rangle = G_0(1) + (G_0'(1))^2/(z - G_0''(1)) = P + z_v^2/(z - G_0''(1)), \qquad \text{[4]}$$

which diverges when

$$G_0''(1) = \textstyle\sum_k k(k-1)\rho_k p_k = z. \qquad \text{[5]}$$

Eq. **5**—the *cascade condition*—is interpreted as follows: When $G_0''(1) < z$, all vulnerable clusters in the network are small; hence, the early adopters are isolated from each other and will be unable to generate the momentum necessary for a cascade to become global. But when $G_0''(1) > z$, the typical size of vulnerable clusters is infinite, implying the presence of a percolating vulnerable cluster, in which case random initial shocks should trigger global cascades with finite probability. Because Eq. **5** marks the transition between these two regimes, or *phases*, at which the average cluster size diverges and global cascades first commence, it is called a *phase transition* (31–33). The conditions necessary to generate global cascades can, in other words, be determined by locating the position and nature of the relevant phase transition. Note, however, that the $k(k-1)$ term in Eq. **5** is monotonically increasing in k, but ρ_k is monotonically decreasing. Thus we would expect that Eq. **5** will have either two solutions (resulting in two phase transitions), or none at all, in contrast with the usual percolation model, which exhibits a single phase transition in z for all finite values of the occupation probability. Furthermore, in the case where we have two solutions, we should observe a continuous interval in z, inside which cascades occur.

Results and Discussion

Although the cascade condition (Eq. **5**) applies to random graphs with arbitrary degree distributions p_k and threshold distributions $f(\phi)$ (expressed through the weighting function ρ_k), we can illustrate its main features for the special case of a uniform random graph (in which any pair of vertices is connected with probability $p = z/n$), and where all vertices have the same threshold ϕ; that is, $f(\phi) = \delta(\phi - \phi_*)$. A characteristic of uniform random graphs is that $p_k = e^{-z}z^k/k!$ (the Poisson distribution), in which case Eq. **5** reduces to $zQ(K_* - 1, z) = 1$, where $K_* = \lfloor 1/\phi_* \rfloor$ and $Q(a,x)$ is the incomplete gamma function. Fig. 1 expresses the cascade condition graphically as a boundary in the (ϕ_*, z) phase diagram (dashed line) and compares it to the region (outlined by solid circles) in which cascades are observed over 1,000 realizations of the dynamics (each realization consists of a randomly constructed network of 10,000 vertices, in which a single vertex is switched on at $t = 0$). Because the simulated system is finite, the predicted and actual boundaries of the cascade window do not agree perfectly, but they are very similar. In particular, as predicted above, both display a lower and an upper boundary as a function of the average degree z, at which the characteristic time scale of the dynamics diverges (see Fig. 2*a*).

To understand the nature of the phase transitions that define the boundaries of the cascade window, we solve exactly for the fractional size S_v of the vulnerable cluster inside the cascade window. Because the generation function approach requires the largest vulnerable cluster to be a pure branching structure, and because the vulnerable cluster will, in general, contain cycles above percolation, Eq. **4** only applies below percolation, which is why Eq. **5** can only specify the boundary of the cascade window. However, we can still solve for S_v above the phase transition, as well as below it, by evaluating $H_0(1)$ exclusively over the set of finite clusters; that is, by

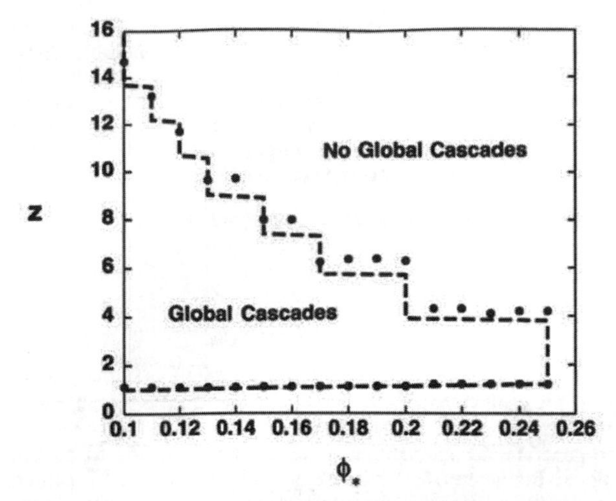

Fig. 1. Cascade windows for the threshold model. The dashed line encloses the region of the (ϕ_*, z) plane in which the cascade condition (Eq. 5) is satisfied for a uniform random graph with a homogenous threshold distribution $f(\phi) = \delta(\phi - \phi_*)$. The solid circles outline the region in which global cascades occur for the same parameter settings in the full dynamical model for n = 10,000 (averaged over 100 random single-node perturbations).

explicitly excluding the percolating cluster (when it exists) from the sum $\Sigma_n q_n x^n$. Using Eq. **3b**, it follows that $S_v = 1 - H_0(1) = P - G_0(H_1(1))$, where $H_1(1)$ satisfies Eq. **3a**. Outside the cascade window, the only solution to Eq. **3a** is $H_1(1) = 1$, which yields $S_v = 0$ (and therefore no cascades) as expected. But inside the cascade window, where the percolating vulnerable cluster exists, Eq. **3a** has an additional solution that corresponds to a non-zero value of S_v. In the special case of a uniform random graph with homogeneous thresholds, we obtain $S_v = Q(K_* + 1, z) - e^{z(H_1-1)}Q(K_* + 1, zH_1)$, in which H_1 satisfies $H_1 = 1 - Q(K_*, z) + e^{z(H_1-1)}Q(K_*, zH_1)$. We contrast this expression with that for the size of the entire connected component of the graph, $S = 1 - e^{-zS}$ (32), which is equivalent to allowing $K_* \to \infty$ (or $\phi_* \to 0$). In Fig. 2*b* we show the exact solutions for both S_v (long-dashed line) and S (solid line) for the case of $\phi_* = 0.18$, and compare these quantities with the frequency and size of global cascades observed in the full dynamical simulation of 10,000 nodes averaged over 1,000 random realizations of the network and the initial condition. (The corresponding numerical values for S_v and S are indistinguishable from the analytical curves, except near the upper boundary of the window.)

The frequency of global cascades (open circles)—that is, cascades that are "successful"—is obviously related to the size of the vulnerable component: the larger is S_v, the more likely a randomly chosen initial site is to be a part of it. In particular, if S_v does not percolate, then global cascades are impossible. Fig. 2*b* clearly supports this intuition, but it is equally clear that, within the cascade window, S_v seriously underestimates the likelihood of a global cascade. The reason is that, according to our original decision rule, an individual's choice of state depends only on the states of its neighbors; hence, even stable vertices, although they do not participate in the initial stages of a global cascade, can still trigger them as long as they are directly *adjacent* to the vulnerable cluster. The true likelihood of a global cascade is therefore determined by the size of what we call the *extended vulnerable cluster* S_e, consisting of the vulnerable cluster itself, and any stable vertices immediately adjacent to it. We have not solved for S_e exactly (although this may be possible), but it is relatively simple to determine numerically, and as the corresponding (dotted) curve in Fig. 2*b* demonstrates, the average value of S_e is an excellent approximation to the observed frequency of global cascades.

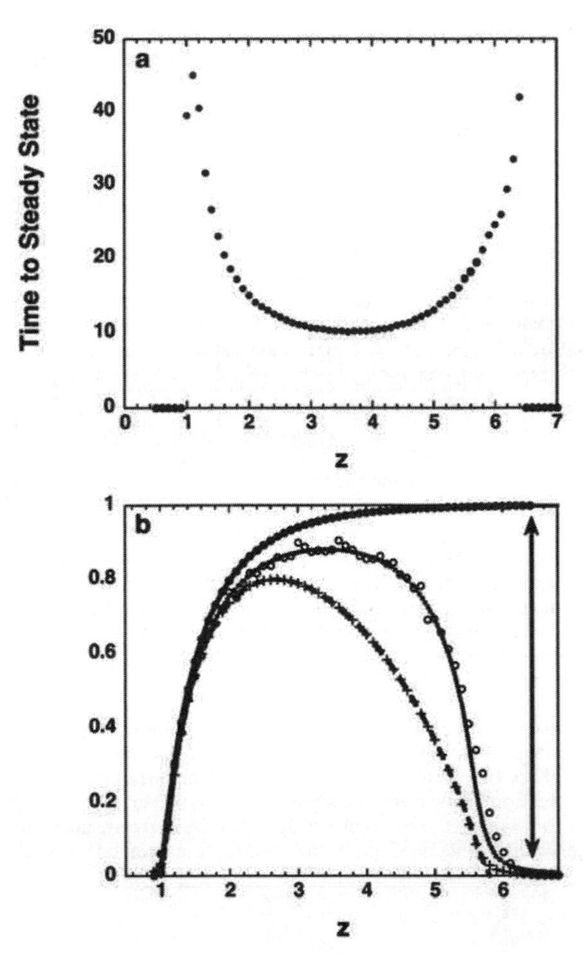

Fig. 2. Cross section of the cascade window from Fig. 1, at $\phi_* = 0.18$. (*a*) The average time required for a cascade to terminate diverges at both the lower and upper boundaries of the cascade window, indicating two phase transitions. (*b*) Comparison between connected components of the network and the properties of global cascades. The frequency of global cascades in the numerical model (open circles) is well approximated by the fractional size of the *extended vulnerable cluster* (short dashes). For comparison, the size of the *vulnerable cluster* is also shown, both the exact solution derived in the text (long dashes) and the average over 1,000 realizations of a random graph (crosses). The exact and numerical solutions agree everywhere except at the upper phase transition, where the finite size of the network (n = 10,000) affects the numerical results. Finally, the average size of global cascades is shown (solid circles) and compared with the exact solution for the largest connected component (solid line).

The average size of global cascades (solid circles) is clearly not governed either by the size of the vulnerable cluster S_v, or by S_e, but by S, the connectivity of the network as a whole. This is a surprising result, the reason for which is not entirely clear, but a plausible explanation is as follows. If a global cascade is triggered by an initially small seed striking the extended vulnerable cluster, it is guaranteed to occupy the entire vulnerable cluster, and therefore a finite fraction of even an infinite network. At this stage, the small-seed condition no longer holds, and so nodes that are still in the off state can now have multiple (early-adopting) neighbors in the on state. Hence, even individuals that were originally classified as stable (the early and late majority) can now be toppled, allowing the cascade to occupy not just the vulnerable component that allowed the cascade to spread initially, but the entire connected component of the graph. That the activation of a percolating

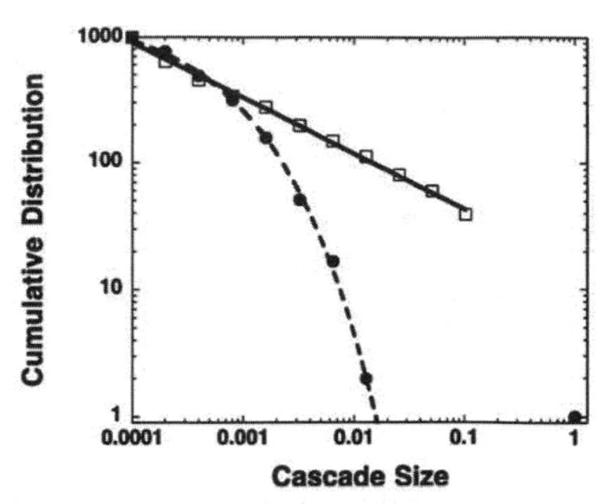

Fig. 3. Cumulative distributions of cascade sizes at the lower and upper critical points, for $n = 1{,}000$ and $z = 1.05$ (open squares) and $z = 6.14$ (solid circles), respectively. The straight line on the double logarithmic scale indicates that cascades at the lower critical point are power-law distributed, with slope 3/2 (the cumulative distribution has slope 1/2). By contrast, the distribution at the upper critical point is bimodal, with an exponential tail at small cascade size, and a second peak at the size of the entire system corresponding to a single global cascade. Above the upper boundary, the global cascade disappears and large cascades are always exponentially unlikely.

vulnerable cluster should always be sufficient to activate the entire connected component, even when the former is a very small fraction of the latter, is not an obvious result, but it appears to hold consistently, at least within the class of random graphs. Whether or not it turns out to hold for networks more general than random graphs is a matter of current investigation.

As Figs. 1 and 2 suggest, the onset of global cascades can occur in two distinct regimes—a low connectivity regime and a high connectivity regime—corresponding to the lower and upper phase transitions respectively. The nature of the phase transitions at the two boundaries is different, and this has important consequences for the apparent stability of the systems involved. As Fig. 3 (open squares) demonstrates, the cumulative distribution of cascades at the lower boundary of the cascade window follows a power law, analogous to the distribution of avalanches in models of self-organized criticality (29) or the cluster size distribution at criticality for standard percolation (22). In fact, the slope of the cascade size distribution is indistinguishable from the known critical exponent $\alpha = 3/2$ for the cluster size distribution of random graphs at percolation (32). This result is expected because, when $z \simeq 1$, most vertices satisfy the vulnerability condition, so the propagation of cascades is constrained principally by the connectivity of the network, which for random graphs is known to undergo a second-order phase transition at $z = 1$ (31).

The upper boundary, however, is different. Here, the propagation of cascades is limited not by the connectivity of the network, but by the local stability of the vertices. Most vertices in this regime have so many neighbors that they cannot be toppled by a single neighbor perturbation; hence, most initial shocks immediately encounter stable vertices. Most cascades therefore die out before spreading very far, giving the appearance that large cascades are exponentially unlikely. A percolating vulnerable cluster, however, still exists, so very rarely a cascade will be triggered in which case the high connectivity of the network ensures that it will be extremely large, typically much larger than cascades at the lower phase transition. The result is a distribution of cascade sizes that is bimodal rather than a power law (see Fig. 3, solid circles). As the upper phase transition is approached from below, global cascades become

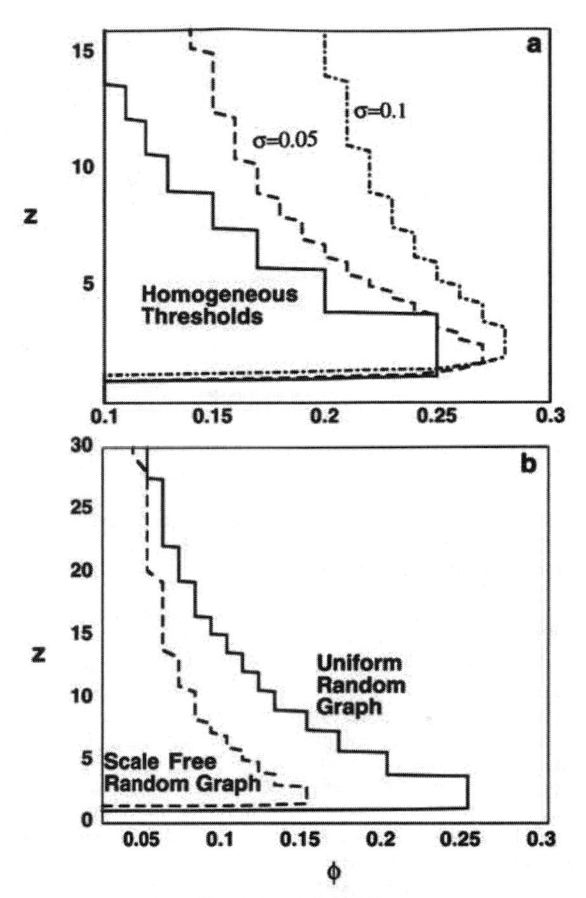

Fig. 4. Analytically derived cascade windows for heterogeneous networks. The solid lines are the same as Fig. 1. (*a*) The dashed lines represent cascade windows for uniform random graphs, but where the threshold distributions (ϕ) are normally distributed with mean ϕ and SD $\sigma = 0.05$ and $\sigma = 0.1$. (*b*) The dashed line represents the cascade window for a random graph with a degree distribution that is a power law with exponent τ and exponential cut-off κ_0, where τ has been fixed at $\tau = 2.5$ and κ_0 has been adjusted to generate graphs with variable z.

larger, but increasingly rare, until they disappear altogether, implying a discontinuous (i.e., first-order) phase transition in the size of successful cascades (see Fig. 2*b*, solid circles). The main consequence of the first-order phase transition is that just inside the boundary of the window, where global cascades occur very rarely (Fig. 3 shows only a single cascade occurring in 1,000 random trials), the system will in general be indistinguishable from one that is highly stable, exhibiting only tiny cascades for many initial shocks before generating a massive, global cascade in response to a shock that is *a priori* indistinguishable from any other.

These qualitative results are quite general within the class of random networks, applying to arbitrary distributions both of thresholds $f(\phi)$ and degree p_k. Variations in either distribution, however, can affect the quantitative results—and thus the effective vulnerability of the system—considerably, as is demonstrated in Fig. 4 *a* and *b*. Fig. 4*a* shows the original cascade window for homogeneous thresholds (solid line) and also two windows (dashed lines) derived by the same generating function method, but corresponding to threshold distributions $f(\phi)$ that are normally distributed with mean ϕ_* and increasing standard deviation σ. Numerical results (not shown) correspond to the analytically derived windows. Clearly, increased heterogeneity of thresholds causes the system to be less stable, yielding cascades over a greater range of both ϕ_* and z. Fig. 4*b*, however, presents a different view of heterogeneity. Now the threshold distribution is held fixed, with all vertices exhibiting the same threshold, but the distribution of degree p_k is given by $p_k = Ck^{-\tau}e^{-k/\kappa}$ $(k > 0)$, where C, τ, and κ are constants that can be adjusted such that we retain $\langle k \rangle = z$. This class of power-law random graphs has attracted much recent interest (19, 20, 32) as a model of many real networks, including the internet. Unlike the Poisson distribution of a uniform random graph, which is sharply peaked around a well defined mean, power law distributions are highly skewed with long tails, corresponding to increased network heterogeneity. Fig. 4*b* implies that random graphs with power law degree distributions tend to be much less vulnerable to random shocks than uniform random graphs with the same z, a point observed elsewhere (19, 20) with respect to network connectivity. Although this distinction between threshold and network heterogeneity is slightly surprising (because both kinds of heterogeneity are related by the fractional threshold condition), it is understandable. Nodes that are vulnerable because of a low threshold can still be well connected to the network, making them ideal early adopters. But nodes that are vulnerable to small perturbations because they have very few neighbors are therefore also poorly connected; hence, they have difficulty propagating any influence.

Network heterogeneity has an additional, complicating effect. Although networks with highly skewed degree distributions are more stable overall, within the cascade window they display increased susceptibility with respect to initial shocks that explicitly target high-degree nodes (19), even though such nodes are unlikely to be vulnerable themselves. If instead of choosing an initial node at random, we deliberately target a node with degree k, then the probability of at least one of its neighbors being a part of the largest vulnerable cluster, and therefore the probability of triggering a cascade, is $P_k = 1 - (1 - S_v)^k$, where S_v is the strength of the vulnerable cluster—a prediction that is well fit by numerical data (not shown) for uniform random graphs. Near the boundaries of the cascade window, where S_v is small, $P_k \simeq kS_v$, implying that the ratio between the probability of a global cascade being triggered by the most connected node in the network (with $k = k_{max}$) and an average node (with $k = z$) is approximately k_{max}/z, which is a rough measure of the skewness of the degree distribution p_k. Networks with highly skewed p_k (such as uniform random graphs near the lower cascade boundary in Fig. 1, or those with power-law degree distributions) should therefore exhibit the property that their most connected nodes are disproportionately likely to trigger global cascades when chosen as initial sites. By contrast, networks in which p_k is sharply peaked, with rapidly decaying tails (such as near the upper boundary of Fig. 1) will not display this property. Numerical results for uniform random graphs (not shown) support this conclusion. Hence, the value of deliberately targeting highly connected initial nodes depends significantly on the global degree distribution, and therefore, in the case of uniform random graphs, whether the system is in its high-connectivity or low-connectivity regime.

Conclusions

Global cascades in social and economic systems, as well as cascading failures in engineered networks, display two striking qualitative features: they occur rarely, but by definition are large when they do. This general observation, however, presents an empirical mystery. Both power-law and bimodal distributions of cascades would satisfy the claim of infrequent, large events, but these distributions are otherwise quite different, and might require quite different explanations. Unfortunately a lack of empirical data detailing cascade size distributions prevents us from determining which distribution (if either) correctly describes which systems. Here we have motivated and analyzed a simple, binary-decision model that, under different conditions, exhibits both kinds of behaviors and thus sets up some testable predictions about cascades in real systems. When the network of interpersonal influences is sufficiently sparse, the propagation of cascades is limited by the global connectivity of the network; and when it is sufficiently dense, cascade propagation is limited by the stability of the individual nodes. In the first case, cascades exhibit a power-law distribution at the corresponding critical point, and the most highly connected nodes are critical in triggering cascades. In the second case, the distribution of cascades is bimodal, and nodes with average connectivity, by virtue of their greater frequency, are much more likely to serve as triggers. In the latter regime, the system displays a more dramatic kind of robust-yet-fragile quality than in the former, remaining almost completely stable throughout many shocks before exhibiting a sudden and giant cascade—a feature that would make global cascades exceptionally hard to anticipate. Finally, systemic heterogeneity has mixed effects on systemic stability. On the one hand, increased heterogeneity of individual thresholds appears to increase the likelihood of global cascades; but on the other hand, increased heterogeneity of vertex degree appears to reduce it. It is hoped that the introduction of this simple framework will stimulate theoretical and empirical efforts to analyze more realistic network models (incorporating social structure, for example) and obtain comprehensive data on the frequency, size, and time scales of global cascades in real networked systems.

This paper benefited from conversations with D. Callaway, A. Lo, M. Newman, and especially S. Strogatz. The research reported was conducted at the Massachusetts Institute of Technology Sloan School of Management under the sponsorship of A. Lo.

1. Gladwell, M. (2000) *The Tipping Point: How Little Things Can Make a Big Difference* (Little, Brown, New York).
2. Shiller, R. J. (1995) *Am. Econ. Rev.* **85,** 181–185.
3. Lohmann, S. (1994) *World Politics* **47,** 42–101.
4. Bikhchandani, S., Hirshleifer, D. & Welch, I. (1992) *J. Pol. Econ.* **100,** 992–1026.
5. Aguirre, B. E., Quarantelli, E. L. & Mendoza, J. L. (1988) *Am. Soc. Rev.* **53,** 569–584.
6. Schelling, T. C. (1973) *J. Conflict Resolution* **17,** 381–428.
7. Kosterev, D. N., Taylor, C. W. & Mittelstadt, W. A. (1999) *IEEE Trans. Power Systems* **14,** 967–979.
8. Sachtjen, M. L., Carreras, B. A. & Lynch, V. E. (2000) *Phys. Rev. E* **61,** 4877–4882.
9. Perrow, C. (1984) *Normal Accidents: Living with High-Risk Technologies* (Basic Books, New York).
10. Carlson, J. M. & Doyle, J. (1999) *Phys. Rev. E* **60,** 1412–1427.
11. De Vany, A. S. & Walls, W. D. (1996) *Econ. J.* **106,** 1493–1514.
12. Banerjee, A. V. (1992) *Quart. J. Econ.* **107,** 797–817.
13. Glance, N. S. & Huberman, B. A. (1993) *J. Math. Soc.* **17,** 281–302.
14. Arthur, W. B. (1989) *Econ. J.* **99,** 116–131.
15. Granovetter, M. (1978) *Am. J. Soc.* **83,** 1420–1443.
16. Glaeser, E. L., Sacerdote, B. & Schheinkman, J. A. (1996) *Quart. J. Econ.* **111,** 507–548.
17. Valente, T. W. (1995) *Network Models of the Diffusion of Innovations* (Hampton Press, Cresskill, NJ).
18. Arthur, W. B. & Lane, D. A. (1993) *Structural Change and Economic Dynamics* **4,** 81–103.
19. Albert, R., Jeong, H. & Barabasi, A. L. (2000) *Nature (London)* **406,** 378–382.
20. Callaway, D. S., Newman, M. E. J., Strogatz, S. H. & Watts, D. J. (2000) *Phys. Rev. Lett.* **85,** 5468–5471.
21. Keeling, M. J. (1999) *Proc. R. Soc. London B* **266,** 859–867.
22. Stauffer, D. & Aharony, A. (1991) *Introduction to Percolation Theory* (Taylor and Francis, London).
23. Newman, M. E. J. & Watts, D. J. (1999) *Phys. Rev. E* **60,** 7332–7342.
24. Sethna, J. P., Dahmen, K., Kartha, S., Krumhansl, J. A., Roberts, B. W. & Shore, J. D. (1993) *Phys. Rev. Lett.* **70,** 3347–3350.
25. Adler, J. (1991) *Physics A* **171,** 453–470.
26. Solomon, S., Weisbuch, G., de Arcangelis, L., Jan, N. & Stauffer, D. (2000) *Physica A* **277,** 239–247.
27. Watts, D. J. (1999) *Small Worlds: The Dynamics of Networks Between Order and Randomness* (Princeton Univ. Press, Princeton).
28. Shrager, J., Hogg, T. & Huberman, B. A. (1987) *Science* **236,** 1092–1094.
29. Bak, P., Tang, C. & Wiesenfeld, K. (1987) *Phys. Rev. Lett.* **59,** 381–384.
30. Strogatz, S. H. (2001) *Nature (London)* **410,** 268–276.
31. Bollobas, B. (1985) *Random Graphs* (Academic, London).
32. Newman, M. E. J., Strogatz, S. H. & Watts, D. J. (2001) *Phys. Rev. E* **64,** 02611.8.
33. Stanley, H. E. (1971) *Introduction to Phase Transitions and Critical Phenomena* (Oxford Univ. Press, Oxford).

Error and attack tolerance of complex networks

Réka Albert, Hawoong Jeong & Albert-László Barabási

Department of Physics, 225 Nieuwland Science Hall, University of Notre Dame, Notre Dame, Indiana 46556, USA

Many complex systems display a surprising degree of tolerance against errors. For example, relatively simple organisms grow, persist and reproduce despite drastic pharmaceutical or environmental interventions, an error tolerance attributed to the robustness of the underlying metabolic network[1]. Complex communication networks[2] display a surprising degree of robustness: although key components regularly malfunction, local failures rarely lead to the loss of the global information-carrying ability of the network. The stability of these and other complex systems is often attributed to the redundant wiring of the functional web defined by the systems' components. Here we demonstrate that error tolerance is not shared by all redundant systems: it is displayed only by a class of inhomogeneously wired networks,

called scale-free networks, which include the World-Wide Web[3–5], the Internet[6], social networks[7] and cells[8]. We find that such networks display an unexpected degree of robustness, the ability of their nodes to communicate being unaffected even by unrealistically high failure rates. However, error tolerance comes at a high price in that these networks are extremely vulnerable to attacks (that is, to the selection and removal of a few nodes that play a vital role in maintaining the network's connectivity). Such error tolerance and attack vulnerability are generic properties of communication networks.

The increasing availability of topological data on large networks, aided by the computerization of data acquisition, had led to great advances in our understanding of the generic aspects of network structure and development[9–16]. The existing empirical and theoretical results indicate that complex networks can be divided into two major classes based on their connectivity distribution $P(k)$, giving the probability that a node in the network is connected to k other nodes. The first class of networks is characterized by a $P(k)$ that peaks at an average $\langle k \rangle$ and decays exponentially for large k. The most investigated examples of such exponential networks are the random graph model of Erdös and Rényi[9,10] and the small-world model of Watts and Strogatz[11], both leading to a fairly homogeneous network, in which each node has approximately the same number of links, $k \simeq \langle k \rangle$. In contrast, results on the World-Wide Web (WWW)[3–5], the Internet[6] and other large networks[17–19] indicate that many systems belong to a class of inhomogeneous networks, called scale-free networks, for which $P(k)$ decays as a power-law, that is $P(k) \sim k^{-\gamma}$, free of a characteristic scale. Whereas the probability that a node has a very large number of connections ($k \gg \langle k \rangle$) is practically prohibited in exponential networks, highly connected nodes are statistically significant in scale-free networks (Fig. 1).

We start by investigating the robustness of the two basic connectivity distribution models, the Erdös–Rényi (ER) model[9,10] that produces a network with an exponential tail, and the scale-free model[17] with a power-law tail. In the ER model we first define the N nodes, and then connect each pair of nodes with probability p. This algorithm generates a homogeneous network (Fig. 1), whose connectivity follows a Poisson distribution peaked at $\langle k \rangle$ and decaying exponentially for $k \gg \langle k \rangle$.

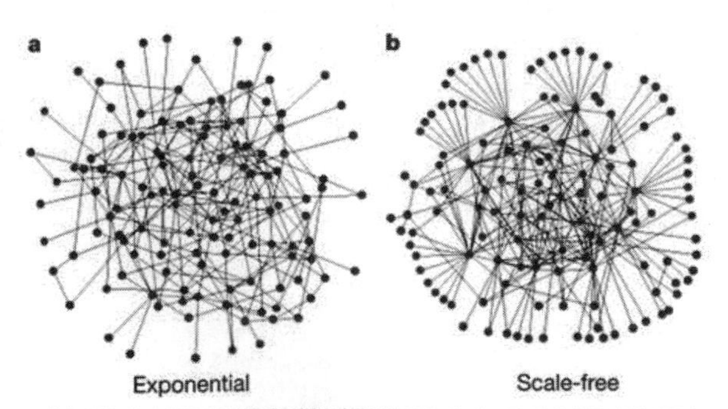

Figure 1 Visual illustration of the difference between an exponential and a scale-free network. **a**, The exponential network is homogeneous: most nodes have approximately the same number of links. **b**, The scale-free network is inhomogeneous: the majority of the nodes have one or two links but a few nodes have a large number of links, guaranteeing that the system is fully connected. Red, the five nodes with the highest number of links; green, their first neighbours. Although in the exponential network only 27% of the nodes are reached by the five most connected nodes, in the scale-free network more than 60% are reached, demonstrating the importance of the connected nodes in the scale-free network Both networks contain 130 nodes and 215 links ($\langle k \rangle = 3.3$). The network visualization was done using the Pajek program for large network analysis: ⟨http://vlado.fmf.uni-lj.si/pub/networks/pajek/pajekman.htm⟩.

The inhomogeneous connectivity distribution of many real networks is reproduced by the scale-free model[17,18] that incorporates two ingredients common to real networks: growth and preferential attachment. The model starts with m_0 nodes. At every time step t a new node is introduced, which is connected to m of the already-existing nodes. The probability Π_i that the new node is connected to node i depends on the connectivity k_i of node i such that $\Pi_i = k_i/\Sigma_j k_j$. For large t the connectivity distribution is a power-law following $P(k) = 2m^2/k^3$.

The interconnectedness of a network is described by its diameter d, defined as the average length of the shortest paths between any two nodes in the network. The diameter characterizes the ability of two nodes to communicate with each other: the smaller d is, the shorter is the expected path between them. Networks with a very large number of nodes can have quite a small diameter; for example, the diameter of the WWW, with over 800 million nodes[20], is around 19 (ref. 3), whereas social networks with over six billion individuals

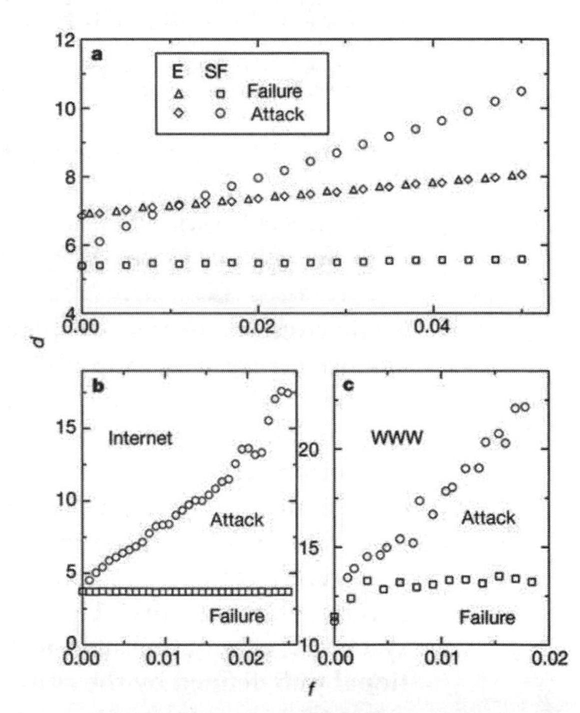

Figure 2 Changes in the diameter d of the network as a function of the fraction f of the removed nodes. **a**, Comparison between the exponential (E) and scale-free (SF) network models, each containing $N = 10{,}000$ nodes and 20,000 links (that is, $\langle k \rangle = 4$). The blue symbols correspond to the diameter of the exponential (triangles) and the scale-free (squares) networks when a fraction f of the nodes are removed randomly (error tolerance). Red symbols show the response of the exponential (diamonds) and the scale-free (circles) networks to attacks, when the most connected nodes are removed. We determined the f dependence of the diameter for different system sizes (N = 1,000; 5,000; 20,000) and found that the obtained curves, apart from a logarithmic size correction, overlap with those shown in **a**, indicating that the results are independent of the size of the system. We note that the diameter of the unperturbed ($f = 0$) scale-free network is smaller than that of the exponential network, indicating that scale-free networks use the links available to them more efficiently, generating a more interconnected web. **b**, The changes in the diameter of the Internet under random failures (squares) or attacks (circles). We used the topological map of the Internet, containing 6,209 nodes and 12,200 links ($\langle k \rangle = 3.4$), collected by the National Laboratory for Applied Network Research ⟨http://moat.nlanr.net/Routing/rawdata/⟩. **c**, Error (squares) and attack (circles) survivability of the World-Wide Web, measured on a sample containing 325,729 nodes and 1,498,353 links[3], such that $\langle k \rangle = 4.59$.

are believed to have a diameter of around six[21]. To compare the two network models properly, we generated networks that have the same number of nodes and links, such that $P(k)$ follows a Poisson distribution for the exponential network, and a power law for the scale-free network.

To address the error tolerance of the networks, we study the changes in diameter when a small fraction f of the nodes is removed. The malfunctioning (absence) of any node in general increases the distance between the remaining nodes, as it can eliminate some paths that contribute to the system's interconnectedness. Indeed, for the exponential network the diameter increases monotonically with f (Fig. 2a); thus, despite its redundant wiring (Fig. 1), it is increasingly difficult for the remaining nodes to communicate with each other. This behaviour is rooted in the homogeneity of the network: since all nodes have approximately the same number of links, they all contribute equally to the network's diameter, thus the removal of each node causes the same amount of damage. In contrast, we observe a drastically different and surprising behaviour for the scale-free network (Fig. 2a): the diameter remains unchanged under an increasing level of errors. Thus even when as many as 5% of the nodes fail, the communication between the remaining nodes in the network is unaffected. This robustness of scale-free networks is rooted in their extremely inhomogeneous connectivity distribution: because the power-law distribution implies that the majority of nodes have only a few links, nodes with small connectivity will be selected with much higher probability. The removal of these 'small' nodes does not alter the path structure of the remaining nodes, and thus has no impact on the overall network topology.

An informed agent that attempts to deliberately damage a network will not eliminate the nodes randomly, but will preferentially target the most connected nodes. To simulate an attack we first remove the most connected node, and continue selecting and removing nodes in decreasing order of their connectivity k. Measuring the diameter of an exponential network under attack, we find that, owing to the homogeneity of the network, there is no substantial difference whether the nodes are selected randomly or in decreasing order of connectivity (Fig. 2a). On the other hand, a drastically different behaviour is observed for scale-free networks. When the most connected nodes are eliminated, the diameter of the scale-free network increases rapidly, doubling its original value if 5% of the nodes are removed. This vulnerability to attacks is rooted in the inhomogeneity of the connectivity distribution: the connectivity is maintained by a few highly connected nodes (Fig. 1b), whose removal drastically alters the network's topology, and

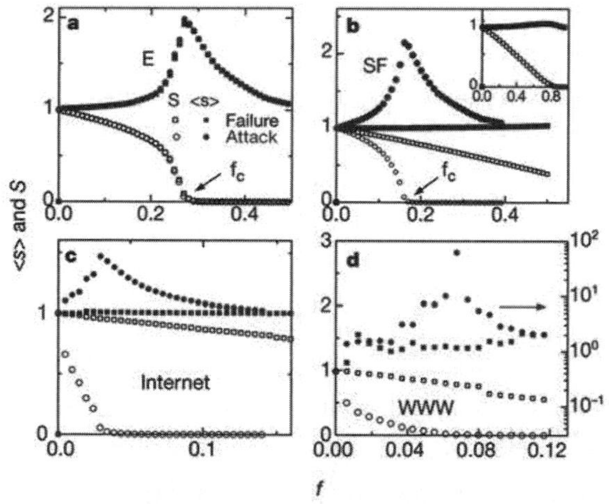

Figure 3 Network fragmentation under random failures and attacks. The relative size of the largest cluster S (open symbols) and the average size of the isolated clusters $\langle s \rangle$ (filled symbols) as a function of the fraction of removed nodes f for the same systems as in Fig. 2. The size S is defined as the fraction of nodes contained in the largest cluster (that is, $S = 1$ for $f = 0$). **a**, Fragmentation of the exponential network under random failures (squares) and attacks (circles). **b**, Fragmentation of the scale-free network under random failures (blue squares) and attacks (red circles). The inset shows the error tolerance curves for the whole range of f, indicating that the main cluster falls apart only after it has been completely deflated. We note that the behaviour of the scale-free network under errors is consistent with an extremely delayed percolation transition: at unrealistically high error rates ($f_{max} \simeq 0.75$) we do observe a very small peak in $\langle s \rangle$ ($\langle s_{max} \rangle \simeq 1.06$) even in the case of random failures, indicating the existence of a critical point. For **a** and **b** we repeated the analysis for systems of sizes $N = 1{,}000$, 5,000 and 20,000, finding that the obtained S and $\langle s \rangle$ curves overlap with the one shown here, indicating that the overall clustering scenario and the value of the critical point is independent of the size of the system. **c**, **d**, Fragmentation of the Internet (**c**) and WWW (**d**), using the topological data described in Fig. 2. The symbols are the same as in **b**. $\langle s \rangle$ in **d** in the case of attack is shown on a different scale, drawn in the right side of the frame. Whereas for small f we have $\langle s \rangle \simeq 1.5$, at $f_c^w = 0.067$ the average fragment size abruptly increases, peaking at $\langle s_{max} \rangle \simeq 60$, then decays rapidly. For the attack curve in **d** we ordered the nodes as a function of the number of outgoing links, k_{out}. We note that while the three studied networks, the scale-free model, the Internet and the WWW have different γ, $\langle k \rangle$ and clustering coefficient[11], their response to attacks and errors is identical. Indeed, we find that the difference between these quantities changes only f_c and the magnitude of d, S and $\langle s \rangle$, but not the nature of the response of these networks to perturbations.

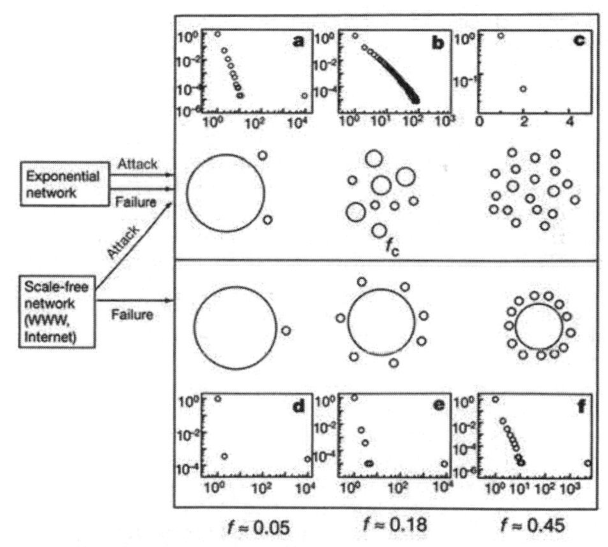

Figure 4 Summary of the response of a network to failures or attacks. **a–f**, The cluster size distribution for various values of f when a scale-free network of parameters given in Fig. 3b is subject to random failures (**a–c**) or attacks (**d–f**). Upper panels, exponential networks under random failures and attacks and scale-free networks under attacks behave similarly. For small f, clusters of different sizes break down, although there is still a large cluster. This is supported by the cluster size distribution: although we see a few fragments of sizes between 1 and 16, there is a large cluster of size 9,000 (the size of the original system being 10,000). At a critical f_c (see Fig. 3) the network breaks into small fragments between sizes 1 and 100 (**b**) and the large cluster disappears. At even higher f (**c**) the clusters are further fragmented into single nodes or clusters of size two. Lower panels, scale-free networks follow a different scenario under random failures: the size of the largest cluster decreases slowly as first single nodes, then small clusters break off. Indeed, at $f = 0.05$ only single and double nodes break off (**d**). At $f = 0.18$, the network is fragmented (**b**) under attack, but under failures the large cluster of size 8,000 coexists with isolated clusters of sizes 1 to 5 (**e**). Even for an unrealistically high error rate of $f = 0.45$ the large cluster persists, the size of the broken-off fragments not exceeding 11 (**f**).

decreases the ability of the remaining nodes to communicate with each other.

When nodes are removed from a network, clusters of nodes whose links to the system disappear may be cut off (fragmented) from the main cluster. To better understand the impact of failures and attacks on the network structure, we next investigate this fragmentation process. We measure the size of the largest cluster, S, shown as a fraction of the total system size, when a fraction f of the nodes are removed either randomly or in an attack mode. We find that for the exponential network, as we increase f, S displays a threshold-like behaviour such that for $f > f_c^e \simeq 0.28$ we have $S \simeq 0$. Similar behaviour is observed when we monitor the average size $\langle s \rangle$ of the isolated clusters (that is, all the clusters except the largest one), finding that $\langle s \rangle$ increases rapidly until $\langle s \rangle \simeq 2$ at f_c^e, after which it decreases to $\langle s \rangle = 1$. These results indicate the following breakdown scenario (Fig. 3a). For small f, only single nodes break apart, $\langle s \rangle \simeq 1$, but as f increases, the size of the fragments that fall off the main cluster increases, displaying unusual behaviour at f_c^e. At f_c^e the system falls apart; the main cluster breaks into small pieces, leading to $S \simeq 0$, and the size of the fragments, $\langle s \rangle$, peaks. As we continue to remove nodes ($f > f_c^e$), we fragment these isolated clusters, leading to a decreasing $\langle s \rangle$. Because the ER model is equivalent to infinite dimensional percolation[22], the observed threshold behaviour is qualitatively similar to the percolation critical point.

However, the response of a scale-free network to attacks and failures is rather different (Fig. 3b). For random failures no threshold for fragmentation is observed; instead, the size of the largest cluster slowly decreases. The fact that $\langle s \rangle \approx 1$ for most f values indicates that the network is deflated by nodes breaking off one by one, the increasing error level leading to the isolation of single nodes only, not clusters of nodes. Thus, in contrast with the catastrophic fragmentation of the exponential network at f_c^e, the scale-free network stays together as a large cluster for very high values of f, providing additional evidence of the topological stability of these networks under random failures. This behaviour is consistent with the existence of an extremely delayed critical point (Fig. 3) where the network falls apart only after the main cluster has been completely deflated. On the other hand, the response to attack of the scale-free network is similar (but swifter) to the response to attack and failure of the exponential network (Fig. 3b): at a critical threshold $f_c^{sf} \simeq 0.18$, smaller than the value $f_c^e \simeq 0.28$ observed for the exponential network, the system breaks apart, forming many isolated clusters (Fig. 4).

Although great efforts are being made to design error-tolerant and low-yield components for communication systems, little is known about the effect of errors and attacks on the large-scale connectivity of the network. Next, we investigate the error and attack tolerance of two networks of increasing economic and strategic importance: the Internet and the WWW.

Faloutsos *et al.*[6] investigated the topological properties of the Internet at the router and inter-domain level, finding that the connectivity distribution follows a power-law, $P(k) \sim k^{-2.48}$. Consequently, we expect that it should display the error tolerance and attack vulnerability predicted by our study. To test this, we used the latest survey of the Internet topology, giving the network at the inter-domain (autonomous system) level. Indeed, we find that the diameter of the Internet is unaffected by the random removal of as high as 2.5% of the nodes (an order of magnitude larger than the failure rate (0.33%) of the Internet routers[23]), whereas if the same percentage of the most connected nodes are eliminated (attack), d more than triples (Fig. 2b). Similarly, the large connected cluster persists for high rates of random node removal, but if nodes are removed in the attack mode, the size of the fragments that break off increases rapidly, the critical point appearing at $f_c^I \simeq 0.03$ (Fig. 3b).

The WWW forms a huge directed graph whose nodes are documents and edges are the URL hyperlinks that point from one document to another, its topology determining the search engines' ability to locate information on it. The WWW is also a scale-free network: the probabilities $P_{out}(k)$ and $P_{in}(k)$ that a document has k outgoing and incoming links follow a power-law over several orders of magnitude, that is, $P(k) \sim k^{-\gamma}$, with $\gamma_{in} = 2.1$ and $\gamma_{out} = 2.45$[3,4,24]. Since no complete topological map of the WWW is available, we limited our study to a subset of the web containing 325,729 nodes and 1,469,680 links ($\langle k \rangle = 4.59$) (ref. 3). Despite the directedness of the links, the response of the system is similar to the undirected networks we investigated earlier: after a slight initial increase, d remains constant in the case of random failures and increases for attacks (Fig. 2c). The network survives as a large cluster under high rates of failure, but the behaviour of $\langle s \rangle$ indicates that under attack the system abruptly falls apart at $f_c^w = 0.067$ (Fig. 3c).

In summary, we find that scale-free networks display a surprisingly high degree of tolerance against random failures, a property not shared by their exponential counterparts. This robustness is probably the basis of the error tolerance of many complex systems, ranging from cells[8] to distributed communication systems. It also explains why, despite frequent router problems[23], we rarely experience global network outages or, despite the temporary unavailability of many web pages, our ability to surf and locate information on the web is unaffected. However, the error tolerance comes at the expense of attack survivability: the diameter of these networks increases rapidly and they break into many isolated fragments when the most connected nodes are targeted. Such decreased attack survivability is useful for drug design[8], but it is less encouraging for communication systems, such as the Internet or the WWW. Although it is generally thought that attacks on networks with distributed resource management are less successful, our results indicate otherwise. The topological weaknesses of the current communication networks, rooted in their inhomogeneous connectivity distribution, seriously reduce their attack survivability. This could be exploited by those seeking to damage these systems. □

Received 14 February; accepted 7 June 2000.

1. Hartwell, L. H., Hopfield, J. J., Leibler, S. & Murray, A. W. From molecular to modular cell biology. *Nature* **402,** 47–52 (1999).
2. Claffy, K., Monk, T. E. *et al.* Internet tomography. *Nature Web Matters* [online] (7 Jan. 99) ⟨http://helix.nature.com/webmatters/tomog/tomog.html⟩ (1999).
3. Albert, R., Jeong, H. & Barabási, A.-L. Diameter of the World-Wide Web. *Nature* **401,** 130–131 (1999).
4. Kumar, R., Raghavan, P., Rajalopagan, S. & Tomkins, A. in *Proc. 9th ACM Symp. on Principles of Database Systems* 1–10 (Association for Computing Machinery, New York, 2000).
5. Huberman, B. A. & Adamic, L. A. Growth dynamics of the World-Wide Web. *Nature* **401,** 131 (1999).
6. Faloutsos, M., Faloutsos, P. & Faloutsos, C. On power-law relationships of the internet topology, ACM SIGCOMM '99. *Comput. Commun. Rev.* **29,** 251–263 (1999).
7. Wasserman, S. & Faust, K. *Social Network Analysis* (Cambridge Univ. Press, Cambridge, 1994).
8. Jeong, H., Tombor, B., Albert, R., Oltvai, Z. & Barabási, A.-L. The large-scale organization of metabolic networks. *Nature* (in the press).
9. Erdős, P. & Rényi, A. On the evolution of random graphs. *Publ. Math. Inst. Hung. Acad. Sci.* **5,** 17–60 (1960).
10. Bollobás, B. *Random Graphs* (Academic, London, 1985).
11. Watts, D. J. & Strogatz, S. H. Collective dynamics of 'small-world' networks. *Nature* **393,** 440–442 (1998).
12. Zegura, E. W., Calvert, K. L. & Donahoo, M. J. A quantitative comparison of graph-based models for internet topology. *IEEE/ACM Trans. Network.* **5,** 770–787 (1997).
13. Williams, R. J. & Martinez, N. D. Simple rules yield complex food webs. *Nature* **404,** 180–183 (2000).
14. Maritan, A., Colaiori, F., Flammini, A., Cieplak, M. & Banavar, J. Universality classes of optimal channel networks. *Science* **272,** 984–986 (1996).
15. Banavar, J. R., Maritan, A. & Rinaldo, A. Size and form in efficient transportation networks. *Nature* **399,** 130–132 (1999).
16. Barthélémy, M. & Amaral, L. A. N. Small-world networks: evidence for a crossover picture. *Phys. Rev. Lett.* **82,** 3180–3183 (1999).
17. Barabási, A.-L. & Albert, R. Emergence of scaling in random networks. *Science* **286,** 509–511 (1999).
18. Barabási, A.-L., Albert, R. & Jeong, H. Mean-field theory for scale-free random networks. *Physica A* **272,** 173–187 (1999).
19. Redner, S. How popular is your paper? An empirical study of the citation distribution. *Euro. Phys. J. B* **4,** 131–134 (1998).
20. Lawrence, S. & Giles, C. L. Accessibility of information on the web. *Nature* **400,** 107–109 (1999).
21. Milgram, S. The small-world problem. *Psychol. Today* **2,** 60–67 (1967).
22. Bunde, A. & Havlin, S. (eds) *Fractals and Disordered Systems* (Springer, New York, 1996).
23. Paxson, V. End-to-end routing behavior in the internet. *IEEE/ACM Trans. Network.* **5,** 601–618 (1997).
24. Adamic, L. A. The small world web. *Lect. Notes Comput. Sci.* **1696,** 443–452 (1999).

VOLUME 85, NUMBER 21 PHYSICAL REVIEW LETTERS 20 NOVEMBER 2000

Resilience of the Internet to Random Breakdowns

Reuven Cohen,[1,*] Keren Erez,[1] Daniel ben-Avraham,[2] and Shlomo Havlin[1]
[1]*Minerva Center and Department of Physics, Bar-Ilan University, Ramat-Gan 52900, Israel*
[2]*Physics Department and Center for Statistical Physics (CISP), Clarkson University, Potsdam, New York 13699-5820*
(Received 11 July 2000; revised manuscript received 31 August 2000)

A common property of many large networks, including the Internet, is that the connectivity of the various nodes follows a scale-free power-law distribution, $P(k) = ck^{-\alpha}$. We study the stability of such networks with respect to crashes, such as random removal of sites. Our approach, based on percolation theory, leads to a general condition for the critical fraction of nodes, p_c, that needs to be removed before the network disintegrates. We show analytically and numerically that for $\alpha \leq 3$ the transition never takes place, unless the network is finite. In the special case of the physical structure of the Internet ($\alpha \approx 2.5$), we find that it is impressively robust, with $p_c > 0.99$.

PACS numbers: 84.35.+i, 02.50.Cw, 05.50.+q, 64.60.Ak

Recently there has been increasing interest in the formation of random networks and in the connectivity of these networks, especially in the context of the Internet [1–9]. When such networks are subject to random breakdowns—a fraction p of the nodes and their connections are removed randomly—their integrity might be compromised: when p exceeds a certain threshold, $p > p_c$, the network disintegrates into smaller, disconnected parts. Below that critical threshold, there still exists a connected cluster that spans the entire system (its size is proportional to that of the entire system). Random breakdown in networks can be seen as a case of infinite-dimensional percolation. Two cases that have been solved exactly are Cayley trees [10] and Erdős-Rényi (ER) random graphs [11], where the networks collapse at known thresholds p_c. Percolation on small-world networks (i.e., networks where every node is connected to its neighbors, plus some random long-range connections [12]) has also been studied by Moore and Newman [13]. Albert *et al.* have raised the question of random failures and intentional attack on networks [1]. Here we consider random breakdown in the Internet (and similar networks) and introduce an analytical approach to finding the critical point. The site connectivity of the physical structure of the Internet, where each communication node is considered as a site, is power law, to a good approximation [14]. We introduce a new general criterion for the percolation critical threshold of randomly connected networks. Using this criterion, we show analytically that the Internet undergoes no transition under random breakdown of its nodes. In other words, a connected cluster of sites that spans the Internet survives even for arbitrarily large fractions of crashed sites.

We consider networks whose nodes are connected randomly to each other, so that the probability for any two nodes to be connected depends solely on their respective connectivity (the number of connections emanating from a node). We argue that, for randomly connected networks with connectivity distribution $P(k)$, the critical breakdown threshold may be found by the following criterion: if loops of connected nodes may be neglected, the percolation transition takes place when a node (i), connected to a node (j) in the spanning cluster, is also connected to at least one other node—otherwise the spanning cluster is fragmented. This may be written as

$$\langle k_i \,|\, i \leftrightarrow j \rangle = \sum_{k_i} k_i P(k_i \,|\, i \leftrightarrow j) = 2\,, \qquad (1)$$

where the angular brackets denote an ensemble average, k_i is the connectivity of node i, and $P(k_i \,|\, i \leftrightarrow j)$ is the conditional probability that node i has connectivity k_i, given that it is connected to node j. But, by Bayes rule for conditional probabilities $P(k_i \,|\, i \leftrightarrow j) = P(k_i, i \leftrightarrow j)/P(i \leftrightarrow j) = P(i \leftrightarrow j \,|\, k_i)P(k_i)/P(i \leftrightarrow j)$, where $P(k_i, i \leftrightarrow j)$ is the *joint* probability that node i has connectivity k_i and that it is connected to node j. For randomly connected networks (neglecting loops) $P(i \leftrightarrow j) = \langle k \rangle/(N-1)$ and $P(i \leftrightarrow j \,|\, k_i) = k_i/(N-1)$, where N is the total number of nodes in the network. It follows that the criterion (1) is equivalent to

$$\kappa \equiv \frac{\langle k^2 \rangle}{\langle k \rangle} = 2\,, \qquad (2)$$

at criticality.

Loops can be ignored below the percolation transition, $\kappa < 2$, because the probability of a bond to form a loop in an s-nodes cluster is proportional to $(s/N)^2$ (i.e., proportional to the probability of choosing two sites in that cluster). The fraction of loops in the system P_{loop} is

$$P_{\text{loop}} \propto \sum_i \frac{s_i^2}{N^2} < \sum_i \frac{s_i S}{N^2} = \frac{S}{N}\,, \qquad (3)$$

where the sum is taken over all clusters, and s_i is the size of the ith cluster. Thus, the overall fraction of loops in the system is smaller than S/N, where S is the size of the largest existing cluster. Below criticality S is smaller than order N (for ER graphs S is of order $\ln N$ [11]), so the fraction of loops becomes negligible in the limit of $N \to \infty$. Similar arguments apply at criticality.

 0031-9007/00/85(21)/4626(3)$15.00

Consider now a random breakdown of a fraction p of the nodes. This would generically alter the connectivity distribution of a node. Consider, indeed, a node with initial connectivity k_0, chosen from an initial distribution $P(k_0)$. After the random breakdown the distribution of the new connectivity of the node becomes $\binom{k_0}{k}(1-p)^k p^{k_0-k}$, and the new distribution is

$$P'(k) = \sum_{k_0=k}^{\infty} P(k_0)\binom{k_0}{k}(1-p)^k p^{k_0-k}. \quad (4)$$

(Quantities after the breakdown are denoted by a prime.) Using this new distribution, one obtains $\langle k\rangle' = \langle k_0\rangle \times (1-p)$ and $\langle k^2\rangle' = \langle k_0^2\rangle(1-p)^2 + \langle k_0\rangle p(1-p)$, so the criterion (2) for criticality may be reexpressed as

$$\frac{\langle k_0^2\rangle}{\langle k_0\rangle}(1-p_c) + p_c = 2 \quad (5)$$

or

$$1 - p_c = \frac{1}{\kappa_0 - 1}, \quad (6)$$

where $\kappa_0 = \langle k_0^2\rangle/\langle k_0\rangle$ is computed from the original distribution, before the random breakdown.

Our discussion up to this point is general and applicable to all randomly connected networks, regardless of the specific form of the connectivity distribution (and provided that loops may be neglected). For example, for random (ER) networks, which possess a Poisson connectivity distribution, the criterion (2) reduces to a known result [11] that the transition takes place at $\langle k\rangle = 1$. In this case, random breakdown does not alter the Poisson character of the distribution, but merely shifts its mean. Thus, the new system is again an ER network, but with new *effective* parameters: $k_{\text{eff}} = k(1-p)$, $N_{\text{eff}} = N(1-p)$. In the case of Cayley trees, the criteria (2) and (6) also yield the known exact results [10].

The case of the Internet is thought to be different. It is widely believed that, to a good approximation, the connectivity distribution of the Internet nodes follows a power law [14]:

$$P(k) = ck^{-\alpha}, \qquad k = m, m+1, \ldots, K, \quad (7)$$

where $\alpha \approx 5/2$, c is an appropriate normalization constant, and m is the smallest possible connectivity. In a finite network, the largest connectivity, K, can be estimated from

$$\int_K^{\infty} P(k)\,dk = \frac{1}{N}, \quad (8)$$

yielding $K \approx mN^{1/(\alpha-1)}$. (For the Internet, $m = 1$ and $K \approx N^{2/3}$.) For the sake of generality, below we consider a range of variables, $\alpha \geq 1$ and $1 \leq m \ll K$. The key parameter, according to (6), is the ratio of second to first moment, κ_0, which we compute by approximating the distribution (7) to a continuum (this approximation becomes exact for $1 \ll m \ll K$, and it preserves the essential features of the transition even for small m):

$$\kappa_0 = \left(\frac{2-\alpha}{3-\alpha}\right)\frac{K^{3-\alpha} - m^{3-\alpha}}{K^{2-\alpha} - m^{2-\alpha}}. \quad (9)$$

When $K \gg m$, this may be approximated as

$$\kappa_0 \to \left|\frac{2-\alpha}{3-\alpha}\right| \times \begin{cases} m, & \text{if } \alpha > 3; \\ m^{\alpha-2}K^{3-\alpha}, & \text{if } 2 < \alpha < 3; \\ K, & \text{if } 1 < \alpha < 2. \end{cases} \quad (10)$$

We see that for $\alpha > 3$ the ratio κ_0 is finite and there is a percolation transition at $1 - p_c = (\frac{\alpha-2}{\alpha-3}m - 1)^{-1}$: for $p > p_c$ the spanning cluster is fragmented and the network is destroyed. However, for $\alpha < 3$ the ratio κ_0 diverges with K and so $p_c \to 1$ when $K \to \infty$ (or $N \to \infty$). The percolation transition does not take place: a spanning cluster exists for arbitrarily large fractions of breakdown, $p < 1$. In *finite* systems a transition is always observed, though for $\alpha < 3$ the transition threshold is exceedingly high. For the case of the Internet ($\alpha \approx 5/2$), we have $\kappa_0 \approx K^{1/2} \approx N^{1/3}$. Considering the enormous size of the Internet, $N > 10^6$, one needs to destroy over 99% of the nodes before the spanning cluster collapses.

The transition is illustrated by the computer simulation results shown in Fig. 1, where we plot the fraction of nodes which remain in the spanning cluster, $P_\infty(p)/P_\infty(0)$, as a function of the fraction of random breakdown, p, for networks with the distribution (7). For $\alpha = 3.5$, the transition is clearly visible: beyond $p_c \approx 0.5$ the spanning

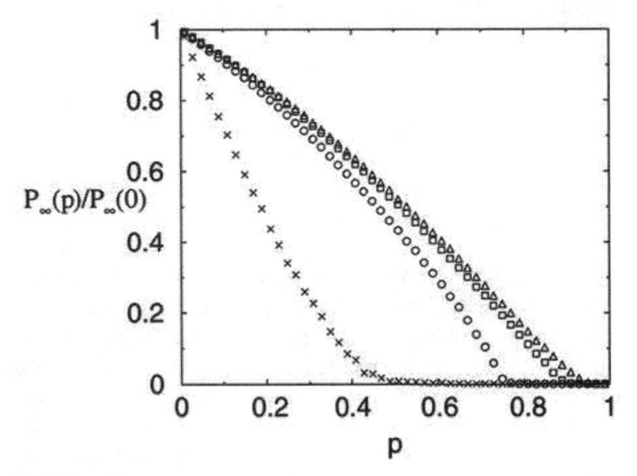

FIG. 1. Percolation transition for networks with power-law connectivity distribution. Plotted is the fraction of nodes that remain in the spanning cluster after breakdown of a fraction p of all nodes, $P_\infty(p)/P_\infty(0)$, as a function of p, for $\alpha = 3.5$ (crosses) and $\alpha = 2.5$ (other symbols), as obtained from computer simulations of up to $N = 10^6$. In the former case, it can be seen that for $p > p_c \approx 0.5$ the spanning cluster disintegrates and the network becomes fragmented. However, for $\alpha = 2.5$ (the case of the Internet), the spanning cluster persists up to nearly 100% breakdown. The different curves for $K = 25$ (circles), 100 (squares), and 400 (triangles) illustrate the finite size effect: the transition exists only for finite networks, while the critical threshold p_c approaches 100% as the networks grow in size.

VOLUME 85, NUMBER 21 PHYSICAL REVIEW LETTERS 20 NOVEMBER 2000

cluster collapses and $P_\infty(p)/P_\infty(0)$ is nearly zero. On the other hand, the plots for $\alpha = 2.5$ (the case of the Internet) show that although the spanning cluster is diluted as p increases [$P_\infty(p)/P_\infty(0)$ becomes smaller], it remains connected even at near 100% breakdown. Data for several system sizes illustrate the finite-size effect: the transition occurs at higher values of p the larger the simulated network. The Internet size is comparable to our largest simulation, making it remarkably resilient to random breakdown.

We have introduced a general criterion for the collapse of randomly connected networks under random removal of their nodes. This criterion, when applied to the Internet, shows that the Internet is resilient to random breakdown of its nodes: a cluster of interconnected sites which spans the whole Internet becomes more dilute with increasing breakdowns, but it remains essentially connected even for nearly 100% breakdown. The same is true for other networks whose connectivity distribution is approximately described by a power law, as in Eq. (7), as long as $\alpha < 3$.

We thank the National Science Foundation for support, under Grant No. PHY-9820569 (D. b.-A.).

Note added.—After completing this manuscript we learned that Eqs. (1) and (2) have been derived earlier using a different approach by Molloy and Reed [15]. We thank Dr. Mark E. J. Newman for bringing this reference to our attention.

*Electronic address: cohenr@shoshi.ph.biu.ac.il

[1] R. Albert, H. Jeong, and A. L. Barabási, Nature (London) **406**, 378 (2000).
[2] V. Paxon, IEEE/ACM Trans. Networking **5**, 601 (1997).
[3] E. W. Zegura, K. L. Calvert, and M. J. Donahoo, IEEE/ACM Trans. Networking **5**, 770 (1997).
[4] A. L. Barabási and R. Albert, Science **286**, 509 (1999).
[5] R. Albert, H. Jeong, and A. L. Barabási, Nature (London) **401**, 130 (1999).
[6] B. A. Huberman and L. A. Adamic, Nature (London) **401**, 131 (1999).
[7] K. Claffy, T. E. Monk, and D. McRobb, *Nature web matters*, http://helix.nature.com/webmatters/tomog/tomog.html, 7 January 1999.
[8] P. L. Krapivsky, S. Redner, and F. Leyvraz, e-print cond-mat/0005139, 2000.
[9] M. E. J. Newman, C. Moore, and D. J. Watts, Phys. Rev. Lett. **84**, 3201 (2000).
[10] J. W. Essam, Rep. Prog. Phys. **43**, 833 (1980).
[11] B. Bollobás, *Random Graphs* (Academic Press, London, 1985), pp. 123–136.
[12] D. J. Watts and S. H. Strogatz, Nature (London) **393**, 440 (1998).
[13] C. Moore and M. E. J. Newman, e-print cond-mat/0001393, 2000.
[14] M. Faloutsos, P. Faloutsos, and C. Faloutsos, Comput. Commun. Rev. **29**, 251 (1999).
[15] M. Molloy and B. Reed, Random Struct. Algorithms **6**, 161 (1995).

VOLUME 85, NUMBER 25 PHYSICAL REVIEW LETTERS 18 DECEMBER 2000

Network Robustness and Fragility: Percolation on Random Graphs

Duncan S. Callaway,[1] M. E. J. Newman,[2,3] Steven H. Strogatz,[1,2] and Duncan J. Watts[3,4]

[1]*Department of Theoretical and Applied Mechanics, Cornell University, Ithaca, New York 14853-1503*
[2]*Center for Applied Mathematics, Cornell University, Ithaca, New York 14853-3801*
[3]*Santa Fe Institute, 1399 Hyde Park Road, Santa Fe, New Mexico 87501*
[4]*Department of Sociology, Columbia University, 1180 Amsterdam Avenue, New York, New York 10027*
(Received 27 July 2000; revised manuscript received 29 September 2000)

Recent work on the Internet, social networks, and the power grid has addressed the resilience of these networks to either random or targeted deletion of network nodes or links. Such deletions include, for example, the failure of Internet routers or power transmission lines. Percolation models on random graphs provide a simple representation of this process but have typically been limited to graphs with Poisson degree distribution at their vertices. Such graphs are quite unlike real-world networks, which often possess power-law or other highly skewed degree distributions. In this paper we study percolation on graphs with completely general degree distribution, giving exact solutions for a variety of cases, including site percolation, bond percolation, and models in which occupation probabilities depend on vertex degree. We discuss the application of our theory to the understanding of network resilience.

PACS numbers: 84.35.+i, 05.50.+q, 64.60.Ak, 87.23.Ge

The Internet, airline routes, and electric power grids are all examples of networks whose function relies crucially on the pattern of interconnection between the components of the system. An important property of such connection patterns is their robustness—or lack thereof—to removal of network nodes [1], which can be modeled as a percolation process on a graph representing the network [2]. Vertices on the graph are considered occupied or not, depending on whether the network nodes they represent (routers, airports, power stations) are functioning normally. Occupation probabilities for different vertices may be uniform or may depend on, for example, the number of connections they have to other vertices, also called the vertex degree. Then we observe the properties of percolation clusters on the graph, particularly their connectivity, as the function determining occupation probability is varied. Previous results on models of this type [1–3] suggest that, if the connection patterns are chosen appropriately, the network can be made highly resilient to random deletion of nodes, although it may be susceptible to an "attack" which specifically targets nodes of high degree. We can also consider bond percolation on graphs as a model of robustness of networks to failure of the links between nodes (e.g., fiber optic lines, power transmission cables, and so forth) or combined site and bond percolation as a model of robustness against failure of either nodes or links.

Percolation models built on networks have also been used to model the spread of disease through communities [4,5]. In such models a node in the network represents a potential host for the disease and is occupied if that host is susceptible to the disease. Links between nodes represent contacts capable of transmitting the disease between individuals and may be occupied with some prescribed probability to represent the fraction of such contacts which actually result in transmission. A percolation transition in such a model represents the onset of an epidemic. Similar models can be used to represent the propagation of computer viruses [6].

The simplest and most widely studied model of undirected networks is the random graph [7], which has been investigated in depth for several decades now. However, random graphs suffer (at least) one serious shortcoming as models of real networks. As pointed out by a number of authors [3,8–11], vertex degrees have a Poisson distribution in a random graph, but real-life degree distributions are strongly non-Poisson, often taking power-law, truncated power-law, or exponential forms. This has prompted researchers to study the properties of generalized random graphs which have non-Poisson degree distributions [12–14].

In this paper we employ the generating function formalism of Newman *et al.* [14] to find exact analytic solutions for site percolation on random graphs with any probability distribution of vertex degree, where occupation probability is an arbitrary function of vertex degree. For the special case of constant occupation probability, we also give solutions for bond and joint site/bond percolation. Our results indicate how robust networks should be to random deletion of vertices or edges, or to the preferential deletion of vertices with particular degree.

We start by examining site percolation for the general case in which occupation probability is an arbitrary function of vertex degree. Let p_k be the probability that a randomly chosen vertex has degree k, and q_k be the probability that a vertex is occupied given that it has degree k. Then $p_k q_k$ is the probability of having degree k and being occupied, and

$$F_0(x) = \sum_{k=0}^{\infty} p_k q_k x^k \tag{1}$$

is the probability generating function for this distribution [15]. (Generating functions of this form have previously

 0031-9007/00/85(25)/5468(4)$15.00

been used by Watts [16] to study cascading failures in networks.) Note that $F_0(1) = q$, where q is the overall fraction of occupied sites. If we wish to study the special case of uniform occupation probability—ordinary site percolation—we simply set $q_k = q$ for all k.

If we follow a randomly chosen edge, the vertex we reach has degree distribution proportional to kp_k rather than just p_k because a randomly chosen edge is more likely to lead to a vertex of higher degree. Hence the equivalent of (1) for such a vertex is [14]

$$F_1(x) = \frac{\sum_k kp_k q_k x^{k-1}}{\sum_k kp_k} = \frac{F_0'(x)}{z}, \tag{2}$$

where z is the average vertex degree.

Now let $H_1(x)$ be the generating function for the probability that one end of a randomly chosen *edge* on the graph leads to a percolation cluster of a given number of occupied vertices. The cluster may contain zero vertices if the vertex at the end of the edge in question is unoccupied, which happens with probability $1 - F_1(1)$, or the edge may lead to an occupied vertex with k other edges leading out of it, distributed according to $F_1(x)$. This means that $H_1(x)$ satisfies a self-consistency condition of the form [14,17,18]

$$H_1(x) = 1 - F_1(1) + xF_1[H_1(x)]. \tag{3}$$

The probability distribution for the size of the cluster to which a randomly chosen *vertex* belongs is similarly generated by $H_0(x)$, where

$$H_0(x) = 1 - F_0(1) + xF_0[H_1(x)]. \tag{4}$$

Together, Eqs. (1)–(4) determine the cluster size distribution for site percolation on a graph of arbitrary degree distribution. From these equations we can determine several quantities of interest such as mean cluster size, position of the percolation threshold, and giant component size, as demonstrated below.

For the special case of uniform (degree-independent) site occupation probability, $q_k = q$ for all k, Eqs. (3) and (4) simplify to

$$H_1(x) = 1 - q + qxG_1[H_1(x)], \tag{5}$$

$$H_0(x) = 1 - q + qxG_0[H_1(x)], \tag{6}$$

where $G_0(x) = \sum_k p_k x^k$ and $G_1(x) = G_0'(x)/z$ are the generating functions for vertex degree alone introduced in Ref. [14]. For bond percolation with uniform occupation probability, we find that

$$H_0(x) = xG_0[H_1(x)], \tag{7}$$

with $H_1(x)$ given by Eq. (5) again, and for joint site/bond percolation with uniform site and bond occupation probabilities q_s and q_b, we have

$$H_1(x) = 1 - q_s q_b + q_s q_b x G_1[H_1(x)], \tag{8}$$

$$H_0(x) = 1 - q_s + q_s x G_0[H_1(x)]. \tag{9}$$

Equations (5)–(7) may be considered special cases of these last two equations when either q_s or q_b is 1.

We now apply these results to the study of network robustness in a variety of cases. First, we consider the case of uniform site occupation probability embodied in Eqs. (5) and (6), which corresponds to random removal of nodes from a network, for example, through failure of routers in a data network or through random vaccination of a population against a disease.

Typically, no closed-form solution exists for Eq. (5), but it is possible to determine the terms of $H_1(x)$ to any finite order n by iterating Eq. (5) $n + 1$ times starting from an initial value of $H_1 = 1$. The probability distribution of cluster sizes can then be calculated exactly by substituting into Eq. (4) and expanding about $x = 0$. To test this method, we have performed simulations [19] of site percolation on random graphs with vertex degrees distributed according to the truncated power law

$$p_k = \begin{cases} 0 & \text{for } k = 0, \\ Ck^{-\tau}e^{-k/\kappa} & \text{for } k \geq 1. \end{cases} \tag{10}$$

Our reasons for choosing this distribution are twofold. First, it is seen in a number of real-world social networks including collaboration networks of movie actors [11] and scientists [20]. The pure power-law distributions seen in Internet data [8–10] are also included in (10) as a special case $\kappa \to \infty$. Second, the distribution has technical advantages over a pure power-law form because the exponential cutoff regularizes the calculations, so that the generating functions and their derivatives are finite. For pure power-law forms on the other hand, the calculations diverge, indicating that real-world networks cannot take a pure power-law form and must have some cutoff (presumably dependent on the system size).

Figure 1 shows the cluster size distribution from our simulations, along with the exact solution from the generating function formalism. The agreement between the two is good.

The sizes of the clusters correspond, for instance, to the sizes of outbreaks of a disease among groups of susceptible individuals. The parameter values used in Fig. 1 are below the percolation threshold for this particular degree distribution, and hence all outbreaks are small and there is no epidemic behavior. The mean cluster size is

$$\langle s \rangle = H_0'(1) = q + qG_0'(1)H_1'(1) = q\left[1 + \frac{qG_0'(1)}{1 - qG_1'(1)}\right], \tag{11}$$

which diverges when $1 - qG_1'(1) = 0$. This point marks the percolation threshold of the system, the point at which a giant component of connected vertices first forms. Thus the critical occupation probability is

$$q_c = \frac{1}{G_1'(1)}. \tag{12}$$

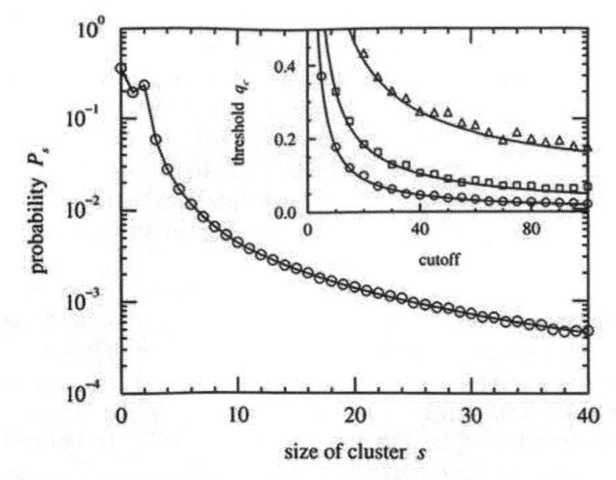

FIG. 1. Probability P_s that a randomly chosen vertex belongs to a cluster of s sites for $\kappa = 10$, $\tau = 2.5$, and $q = 0.65$ from numerical simulation on systems of 10^7 sites (circles) and our exact solution (solid line). Inset: the percolation threshold q_c from Eq. (12) (solid lines) vs computer simulations with $\tau = 1.5$ (circles), 2.0 (squares), and 2.5 (triangles).

A result equivalent to this one has been derived previously by Cohen *et al.* [2] by different means.

In the language of disease propagation q_c is the point at which an epidemic of the disease first occurs. In the language of network robustness, it is the point at which the network achieves large scale connectivity and can therefore function as an effective distribution network. Conversely, if we are approaching the transition from values of q above q_c it is the point at which a sufficient number of individuals are immune to a disease to prevent it from spreading, or the point at which a large enough number of nodes have been deleted from a distribution network to prevent distribution on large scales.

The inset in Fig. 1 shows the behavior of the percolation threshold with the cutoff parameter κ for a variety of values of τ. Note that as the values of κ become large, the percolation threshold becomes small, indicating a high degree of robustness of the network to random deletion of nodes. For $\tau = 2.5$ (roughly the exponent for the Internet data [8]) and $\kappa = 100$, the percolation threshold is $q_c = 0.17$, indicating that one can remove more than 80% of the nodes in the network without destroying the giant component—the network will still possess large-scale connectivity. This result agrees with recent studies of the Internet [1,2] which indicate that network connectivity should be highly robust against the random removal of nodes.

Another issue that has attracted considerable recent attention is the question of robustness of a network to nonrandom deletion targeted specifically at nodes with high degree. Albert *et al.* [1] and Broder *et al.* [3] both looked at the connectivity of a network with power-law distributed vertex degrees as the vertices with highest degree were progressively removed. In the language of our percolation models, this is equivalent to setting

$$q_k = \theta(k_{\max} - k), \tag{13}$$

where θ is the Heaviside step function [21]. This removes (unoccupies) all vertices with degree greater than $k_{\max}$. To investigate the effect of this removal, we calculate the size of the giant component in the network, if there is one. Above the percolation transition the generating function $H_0(x)$ gives the distribution of the sizes of clusters of vertices which are *not* in the giant component [17], which means that $H_0(1)$ is equal to the fraction of the graph which is not occupied by the giant component. The fraction S which *is* occupied by the giant component is therefore given by

$$S = 1 - H_0(1) = F_0(1) - F_0(u), \tag{14}$$

where u is a solution of the self-consistency condition

$$u = 1 - F_1(1) + F_1(u). \tag{15}$$

In cases where this last equation is not exactly solvable we can evaluate u by numerical iteration starting from a suitable initial value. In Fig. 2 we show the results for S from this calculation for graphs with pure power-law degree distributions as a function of $k_{\max}$ for a variety of values of τ. (The removal of vertices with high degree regularizes the calculation in a similar way to the inclusion of the cutoff κ in our earlier calculation, so no other cutoff is needed in this case.) On the same plot we also show simulation results for this problem, and once more agreement of theory and simulation is good.

Opinions appear to differ over whether networks such as this are robust or fragile to this selective removal of vertices. Albert *et al.* [1] point out that only a small fraction of the highest-degree vertices need be removed to destroy the giant component in the network and hence remove all long-range connectivity. Conversely, Broder *et al.* [3] point out that one can remove all vertices with degree greater than $k_{\max}$ and still have a giant component even for surprisingly small values of $k_{\max}$. As we show in Fig. 2, both viewpoints are correct: they are merely different representations of the same data. In the upper frame of the figure, we plot giant component size as a function of the fraction of vertices removed from the network, and it is clear that the giant component disappears when only a small percentage are removed—just 1% for the case $\tau = 2.7$—so that the network appears fragile. In the lower frame we show the same data as a function of $k_{\max}$, the highest remaining vertex degree, and we see that when viewed in this way the network is, in a sense, robust, since $k_{\max}$ must be very small to destroy the giant component completely—just 10 in the case of $\tau = 2.7$.

To conclude, we have used generating function methods to solve exactly for the behavior of a variety of percolation models on random graphs with any distribution of vertex degrees, including uniform site, bond and site/bond

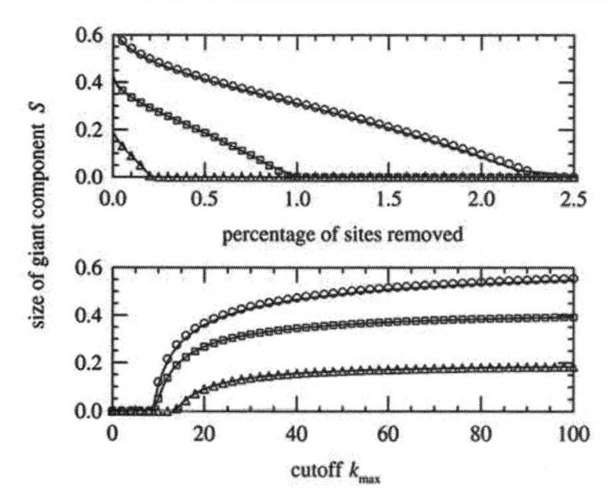

FIG. 2. Size of the giant component S in graphs with power-law degree distribution and all vertices with degree greater than k_{max} unoccupied, for $\tau = 2.4$ (circles), 2.7 (squares), and 3.0 (triangles). Points are simulation results for systems with 10^7 vertices; solid lines are the exact solution. Upper frame: S as a function of fraction of vertices unoccupied. Lower frame: S as a function of the cutoff parameter k_{max}.

percolation, and site percolation in which occupation probability is a function of vertex degree. Percolation systems on graphs such as these have been suggested as models for the robustness of communication or distribution networks to breakdown or sabotage, and for the spread of disease through communities possessing some resistance to infection. Our exact solutions allow us to make predictions about the behavior of such model systems under quite general types of breakdown or interference. Among other results, we find that a distribution network such as the Internet, which has an approximately power-law vertex degree distribution, should be highly robust against random removal of nodes (for example, random failure of routers), but is relatively fragile, at least in terms of fraction of nodes removed, to the specific removal of the most highly connected nodes.

The authors thank Jon Kleinberg for illuminating conversations. This work was funded in part by the National Science Foundation, the Electric Power Research Institute, and the Army Research Office.

[1] R. Albert, H. Jeong, and A.-L. Barabási, Nature (London) **406**, 378 (2000).
[2] R. Cohen, K. Erez, D. ben-Avraham, and S. Havlin, Phys. Rev. Lett. **85**, 4626 (2000).
[3] A. Broder, R. Kumar, F. Maghoul, P. Raghavan, S. Rajagopalan, R. Stata, A. Tomkins, and J. Wiener, Comput. Netw. **33**, 309 (2000).
[4] F. Ball, D. Mollison, and G. Scalia-Tomba, Ann. Appl. Probab. **7**, 46 (1997)
[5] M. E. J. Newman and D. J. Watts, Phys. Rev. E **60**, 7332 (1999).
[6] J. O. Kephart and S. R. White, in *Proceedings of the 1991 IEEE Computer Society Symposium on Research in Security and Privacy, Oakland, CA* (IEEE Computer Society Press, Los Alamitos, CA, 1991).
[7] B. Bollobás, *Random Graphs* (Academic Press, New York, 1985).
[8] M. Faloutsos, P. Faloutsos, and C. Faloutsos, Comput. Commun. Rev. **29**, 251 (1999).
[9] B. A. Huberman and L. A. Adamic, Nature (London) **401**, 131 (1999).
[10] R. Albert, H. Jeong, and A.-L. Barabási, Nature (London) **401**, 130 (1999).
[11] L. A. N. Amaral, A. Scala, M. Barthélémy, and H. E. Stanley, Proc. Natl. Acad. Sci. U.S.A. **97**, 11 149 (2000).
[12] M. Molloy and B. Reed, Random Struct. Algorithms **6**, 161 (1995); Comb. Probab. Comput. **7**, 295 (1998).
[13] W. Aiello, F. Chung, and L. Lu, in *Proceedings of the 32nd Annual ACM Symposium on Theory of Computing, Portland, OR, 2000* (ACM Press, New York, 2000).
[14] M. E. J. Newman, S. H. Strogatz, and D. J. Watts, cond-mat/0007235.
[15] H. S. Wilf, *Generatingfunctionology* (Academic Press, Boston, London, 1994), 2nd ed.
[16] D. J. Watts, Santa Fe Institute Report No. 00-11-062, 2000.
[17] C. Moore and M. E. J. Newman, Phys. Rev. E **62**, 7059 (2000).
[18] The derivation of this equation assumes that all clusters other than the giant component contain no loops. It is straightforward to demonstrate that this assumption is exact in the limit of large system size for degree distributions with finite mean. See Refs. [7], [12], and [13] for discussions of this point.
[19] Simulations were performed using the algorithm of M. E. J. Newman and R. M. Ziff, Phys. Rev. Lett. **85**, 4104 (2000), except for the cluster size distribution, which was calculated using ordinary depth-first search. The value of q_c was assumed equal to the value of q at which the derivative of the size of the larget cluster was greatest.
[20] M. E. J. Newman, Proc. Natl. Acad. Sci. U.S.A. (to be published).
[21] In fact, the approaches of Albert *et al.* [1] and Broder *et al.* [3] differ slightly. Broder *et al.* simply removed the highest degree vertices from the network, whereas Albert *et al.* recalculated vertex degrees after the removal of each vertex and its associated edges, and then removed the next highest degree vertex. Our calculations are equivalent to the method of Broder *et al.*, although in practice there appears to be little difference in qualitative behavior between the two.

Authoritative Sources in a Hyperlinked Environment

JON M. KLEINBERG

Cornell University, Ithaca, New York

Abstract. The network structure of a hyperlinked environment can be a rich source of information about the content of the environment, provided we have effective means for understanding it. We develop a set of algorithmic tools for extracting information from the link structures of such environments, and report on experiments that demonstrate their effectiveness in a variety of contexts on the World Wide Web. The central issue we address within our framework is the distillation of broad search topics, through the discovery of "authoritative" information sources on such topics. We propose and test an algorithmic formulation of the notion of authority, based on the relationship between a set of relevant authoritative pages and the set of "hub pages" that join them together in the link structure. Our formulation has connections to the eigenvectors of certain matrices associated with the link graph; these connections in turn motivate additional heuristics for link-based analysis.

Categories and Subject Descriptors: F.2.2 [**Analysis of Algorithms and Problem Complexity**]: Nonnumerical algorithms and problems—*computations on discrete structures*; H.3.3 [**Information Storage and Retrieval**]: Information Search and Retrieval—*information filtering*; H.5.4 [**Information Interfaces and Presentation (e.g., HCI)**]: Hypertext/Hypermedia

General Terms: Algorithms

Additional Key Words and Phrases: Graph algorithms, hypertext structure, link analysis, World Wide Web

1. *Introduction*

The network structure of a hyperlinked environment can be a rich source of information about the content of the environment, provided we have effective means for understanding it. In this work, we develop a set of algorithmic tools for extracting information from the link structures of such environments, and report on experiments that demonstrate their effectiveness in a variety of contexts on

---

Preliminary versions of this paper appeared in *Proceedings of the Annual ACM-SIAM Symposium on Discrete Algorithms*. ACM, New York, 1998, pp. 668–677, and as IBM Research Report RJ 10076, May 1997.

This work was performed in large part while J. M. Kleinberg was on leave at the IBM Almaden Research Center, San Jose, CA 95120.

J. M. Kleinberg is currently supported by an Alfred P. Sloan Research Fellowship, an ONR Young Investigator Award, and by NSF Faculty Early Career Development Award CLR 97-01399.

Author's address: Department of Computer Science, Cornell University, Ithaca, NY 14853, e-mail: kleinber@cs.cornell.edu.

the World Wide Web (www) [Berners-Lee et al. 1994]. In particular, we focus on the use of links for analyzing the collection of pages relevant to a broad search topic, and for discovering the most "authoritative" pages on such topics.

While our techniques are not specific to the www, we find the problems of search and structural analysis particularly compelling in the context of this domain. The www is a hypertext corpus of enormous complexity, and it continues to expand at a phenomenal rate. Moreover, it can be viewed as an intricate form of populist hypermedia, in which millions of on-line participants, with diverse and often conflicting goals, are continuously creating hyperlinked content. Thus, while individuals can impose order at an extremely local level, its global organization is utterly unplanned—high-level structure can emerge only through a posteriori analysis.

Our work originates in the problem of *searching* on the www, which we could define roughly as the process of discovering pages that are relevant to a given query. The *quality* of a search method necessarily requires human evaluation, due to the subjectivity inherent in notions such as *relevance*. We begin from the observation that improving the quality of search methods on the www is, at the present time, a rich and interesting problem that is in many ways orthogonal to concerns of algorithmic efficiency and storage. In particular, consider that current search engines typically index a sizable portion of the www and respond on the order of seconds. Although there would be considerable utility in a search tool with a longer response time, provided that the results were of significantly greater value to a user, it has typically been very hard to say *what* such a search tool should be computing with this extra time. Clearly, we are lacking objective functions that are both concretely defined *and* correspond to human notions of quality.

1.1. QUERIES AND AUTHORITATIVE SOURCES. We view searching as beginning from a user-supplied *query*. It seems best not to take too unified a view of the notion of a query; there is more than one type of query, and the handling of each may require different techniques. Consider, for example, the following types of queries:

—*Specific queries*. For example, "Does Netscape support the JDK 1.1 code-signing API?"

—*Broad-topic queries*. For example, "Find information about the Java programming language."

—*Similar-page queries*. For example, "Find pages 'similar' to `java.sun.com`."

Concentrating on just the first two types of queries for now, we see that they present very different sorts of obstacles. The difficulty in handling *specific queries* is centered, roughly, around what could be called the *Scarcity Problem*: there are very few pages that contain the required information, and it is often difficult to determine the identity of these pages.

For *broad-topic queries*, on the other hand, one expects to find many thousand relevant pages on the www; such a set of pages might be generated by variants of term-matching (e.g., one enters a string such as "Gates," "search engines," or "censorship" into a search engine such as AltaVista [Digital Equipment Corporation] or by more sophisticated means. Thus, there is not an issue of scarcity here. Instead, the fundamental difficulty lies in what could be called the

Abundance Problem: The number of pages that could reasonably be returned as relevant is far too large for a human user to digest. To provide effective search methods under these conditions, one needs a way to filter, from among a huge collection of relevant pages, a small set of the most "authoritative" or "definitive" ones.

This notion of *authority*, relative to a broad-topic query, serves as a central focus in our work. One of the fundamental obstacles we face in addressing this issue is that of accurately modeling authority in the context of a particular query topic. Given a particular page, how do we tell whether it is authoritative?

It is useful to discuss some of the complications that arise here. First, consider the natural goal of reporting `www.harvard.edu`, the home page of Harvard University, as one of the most authoritative pages for the query `"Harvard"`. Unfortunately, there are over a million pages on the www that use the term "Harvard," and `www.harvard.edu` is not the one that uses the term most often, or most prominently, or in any other way that would favor it under a text-based ranking function. Indeed, one suspects that there is no purely *endogenous* measure of the page that would allow one to properly assess its authority. Second, consider the problem of finding the home pages of the main www search engines. One could begin from the query `"search engines"`, but there is an immediate difficulty in the fact that many of the natural authorities (*Yahoo!*, Excite, AltaVista) do not use the term on their pages. This is a fundamental and recurring phenomenon—as another example, there is no reason to expect the home pages of Honda or Toyota to contain the term "automobile manufacturers."

1.2. ANALYSIS OF THE LINK STRUCTURE. Analyzing the hyperlink structure among www pages gives us a way to address many of the difficulties discussed above. Hyperlinks encode a considerable amount of latent human judgment, and we claim that this type of judgment is precisely what is needed to formulate a notion of authority. Specifically, the creation of a link on the www represents a concrete indication of the following type of judgment: the creator of page p, by including a link to page q, has in some measure *conferred authority* on q. Moreover, links afford us the opportunity to find potential authorities purely through the pages that point to them; this offers a way to circumvent the problem, discussed above, that many prominent pages are not sufficiently self-descriptive.

Of course, there are a number of potential pitfalls in the application of links for such a purpose. First of all, links are created for a wide variety of reasons, many of which have nothing to do with the conferral of authority. For example, a large number of links are created primarily for navigational purposes ("Click here to return to the main menu"); others represent paid advertisements.

Another issue is the difficulty in finding an appropriate balance between the criteria of *relevance* and *popularity*, each of which contributes to our intuitive notion of authority. It is instructive to consider the serious problems inherent in the following simple heuristic for locating authoritative pages: Of all pages containing the query string, return those with the greatest number of in-links. We have already argued that for a great many queries (`"search engines"`, `"automobile manufacturers"`, ...), a number of the most authoritative pages do not contain the associated query string. Conversely, this heuristic would consider a universally popular page such as `www.yahoo.com` or `www.netscape.com` to be highly authoritative with respect to any query string that it contained.

In this work, we propose a link-based model for the conferral of authority, and show how it leads to a method that consistently identifies relevant, authoritative www pages for broad search topics. Our model is based on the relationship that exists between the authorities for a topic and those pages that link to many related authorities—we refer to pages of this latter type as *hubs*. We observe that a certain natural type of equilibrium exists between hubs and authorities in the graph defined by the link structure, and we exploit this to develop an algorithm that identifies both types of pages simultaneously. The algorithm operates on *focused subgraphs* of the www that we construct from the output of a text-based www search engine; our technique for constructing such subgraphs is designed to produce small collections of pages likely to contain the most authoritative pages for a given topic.

1.3. OVERVIEW. Our approach to discovering authoritative www sources is meant to have a *global* nature: We wish to identify the most central pages for broad search topics in the context of the www as a whole. Global approaches involve basic problems of representing and filtering large volumes of information, since the entire set of pages relevant to a broad-topic query can have a size in the millions. This is in contrast to *local* approaches that seek to understand the interconnections among the set of www pages belonging to a single logical site or intranet; in such cases the amount of data is much smaller, and often a different set of considerations dominates.

It is also important to note the sense in which our main concerns are fundamentally different from problems of *clustering*. Clustering addresses the issue of dissecting a heterogeneous population into subpopulations that are in some way more cohesive; in the context of the www, this may involve distinguishing pages related to different meanings or senses of a query term. Thus, clustering is intrinsically different from the issue of distilling broad topics via the discovery of authorities, although a subsequent section will indicate some connections. For even if we were able perfectly to dissect the multiple senses of an ambiguous query term (e.g., "Windows" or "Gates"), we would still be left with the same underlying problem of representing and filtering the vast number of pages that are relevant to *each* of the main senses of the query term.

The paper is organized as follows. Section 2 discusses the method by which we construct a focused subgraph of the www with respect to a broad search topic, producing a set of relevant pages rich in candidate authorities. Sections 3 and 4 discuss our main algorithm for identifying hubs and authorities in such a subgraph, and some of the applications of this algorithm. Section 5 discusses the connections with related work in the areas of www search, bibliometrics, and the study of social networks. Section 6 describes how an extension of our basic algorithm produces multiple collections of hubs and authorities within a common link structure. Finally, Section 7 investigates the question of how "broad" a topic must be in order for our techniques to be effective, and Section 8 surveys some work that has been done on the evaluation of the method presented here.

2. *Constructing a Focused Subgraph of the WWW*

We can view any collection V of hyperlinked pages as a directed graph $G = (V, E)$: the nodes correspond to the pages, and a directed edge $(p, q) \in E$ indicates the presence of a link from p to q. We say that the *out-degree* of a node p is the number of nodes it has links to, and the *in-degree* of p is the number of nodes that have links to it. From a graph G, we can isolate small regions, or *subgraphs*, in the following way. If $W \subseteq V$ is a subset of the pages, we use $G[W]$ to denote the graph *induced* on W: its nodes are the pages in W, and its edges correspond to all the links between pages in W.

Suppose we are given a broad-topic query, specified by a query string σ. We wish to determine authoritative pages by an analysis of the link structure; but first we must determine the subgraph of the www on which our algorithm will operate. Our goal here is to focus the computational effort on relevant pages. Thus, for example, we could restrict the analysis to the set Q_σ of all pages containing the query string; but this has two significant drawbacks. First, this set may contain well over a million pages, and hence entail a considerable computational cost; and second, we have already noted that some or most of the best authorities may not belong to this set.

Ideally, we would like to focus on a collection S_σ of pages with the following properties.

(i) S_σ is relatively small.
(ii) S_σ is rich in relevant pages.
(iii) S_σ contains most (or many) of the strongest authorities.

By keeping S_σ small, we are able to afford the computational cost of applying nontrivial algorithms; by ensuring it is rich in relevant pages we make it easier to find good authorities, as these are likely to be heavily referenced within S_σ.

How can we find such a collection of pages? For a parameter t (typically set to about 200), we first collect the t highest ranked pages for the query σ from a text-based search engine such as AltaVista [Digital Equipment Corporation] or Hotbot [Wired Digital, Inc.]. We will refer to these t pages as the *root set* R_σ. This root set satisfies (i) and (ii) of the desiderata listed above, but it generally is far from satisfying (iii). To see this, note that the top t pages returned by the text-based search engines we use will all contain the query string σ, and hence R_σ is clearly a subset of the collection Q_σ of *all* pages containing σ. Above, we argued that even Q_σ will often not satisfy condition (iii). It is also interesting to observe that there are often extremely few links between pages in R_σ, rendering it essentially "structureless." For example, in our experiments, the root set for the query "`java`" contained 15 links between pages in different domains; the root set for the query "`censorship`" contained 28 links between pages in different domains. These numbers are typical for a variety of the queries tried; they should be compared with the $200 \cdot 199 = 39800$ potential links that could exist between pages in the root set.

We can use the root set R_σ, however, to produce a set of pages S_σ that will satisfy the conditions we are seeking. Consider a strong authority for the query topic—although it may well not be in the set R_σ, it is quite likely to be *pointed to* by at least one page in R_σ. Hence, we can increase the number of strong

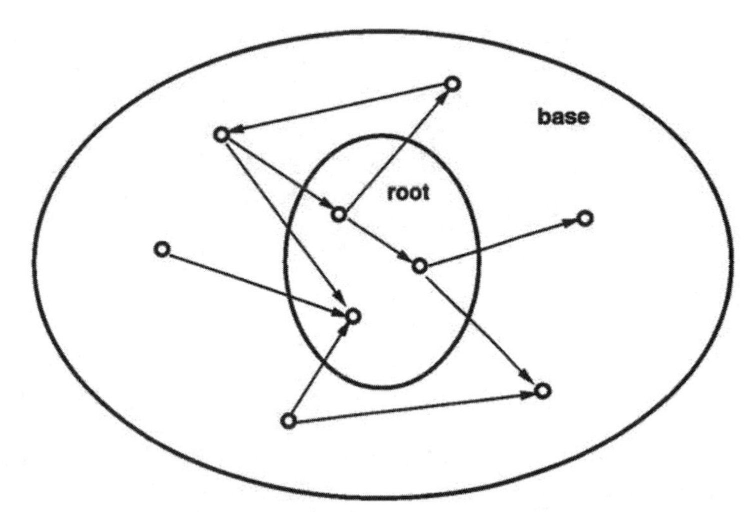

FIG. 1. Expanding the root set into a base set.

authorities in our subgraph by expanding R_σ along the links that enter and leave it. In concrete terms, we define the following procedure:

```
Subgraph(σ,ℰ,t,d)
  σ: a query string.
  ℰ: a text-based search engine.
  t, d: natural numbers.
  Let R_σ denote the top t results of ℰ on σ.
Set S_σ := R_σ
For each page p ∈ R_σ
  Let Γ⁺(p) denote the set of all pages p points to.
  Let Γ⁻(p) denote the set of all pages pointing to p.
  Add all pages in Γ⁺(p) to S_σ.
  If |Γ⁻(p)| ≤ d, then
    Add all pages in Γ⁻(p) to S_σ.
  Else
    Add an arbitrary set of d pages from Γ⁻(p) to S_σ.
End
Return S_σ
```

Thus, we obtain S_σ by growing R_σ to include any page pointed to by a page in R_σ and any page that points to a page in R_σ—with the restriction that we allow a single page in R_σ to bring at most d pages pointing to it into S_σ. This latter point is crucial since a number of www pages are pointed to by several hundred thousand pages, and we can't include all of them in S_σ if we wish to keep it reasonably small.

We refer to S_σ as the *base set* for σ; in our experiments we construct it by invoking the `Subgraph` procedure with the search engine AltaVista, $t = 200$, and $d = 50$. We find that S_σ typically satisfies points (i), (ii), and (iii) above—its size is generally in the range 1000–5000; and, as we discussed above, a strong authority need only be referenced by any one of the 200 pages in the root set R_σ in order to be added to S_σ.

In the next section, we describe our algorithm to compute hubs and authorities in the base set S_σ. Before turning to this, we discuss a heuristic that is very useful for offsetting the effect of links that serve purely a navigational function. First, let $G[S_\sigma]$ denote, as above, the subgraph induced on the pages in S_σ. We distinguish between two types of links in $G[S_\sigma]$. We say that a link is *transverse* if it is between pages with different domain names, and *intrinsic* if it is between pages with the same domain name. By "domain name" here, we mean here the first level in the URL string associated with a page. Since intrinsic links very often exist purely to allow for navigation of the infrastructure of a site, they convey much less information than transverse links about the authority of the pages they point to. Thus, we delete all intrinsic links from the graph $G[S_\sigma]$, keeping only the edges corresponding to transverse links; this results in a graph G_σ.

This is a very simple heuristic, but we find it effective for avoiding many of the pathologies caused by treating navigational links in the same way as other links. There are other simple heuristics that can be valuable for eliminating links that do not seem intuitively to confer authority. One that is worth mentioning is based on the following observation: Suppose a large number of pages from a single domain all point to a single page p. Quite often this corresponds to a mass endorsement, advertisement, or some other type of "collusion" among the referring pages—for example, the phrase "This site designed by ..." and a corresponding link at the bottom of each page in a given domain. To eliminate this phenomenon, we can fix a parameter m (typically $m \approx 4 - 8$) and only allow up to m pages from a single domain to point to any given page p. Again, this can be an effective heuristic in some cases, although we did not employ it when running the experiments that follow.

3. *Computing Hubs and Authorities*

The method of the previous section provides a small subgraph G_σ that is relatively focused on the query topic—it has many relevant pages, and strong authorities. We now turn to the problem of extracting these authorities from the overall collection of pages, purely through an analysis of the link structure of G_σ.

The simplest approach, arguably, would be to order pages by their *in-degree*—the number of links that point to them—in G_σ. We rejected this idea earlier, when it was applied to the collection of *all* pages containing the query term σ; but now we have explicitly constructed a small collection of relevant pages containing most of the authorities we want to find. Thus, these authorities both belong to G_σ and are heavily referenced by pages *within* G_σ.

Indeed, the approach of ranking purely by in-degree does typically work much better in the context of G_σ than in the earlier settings we considered; in some cases, it can produce uniformly high-quality results. However, the approach still retains some significant problems. For example, on the query "`java`", the pages with the largest in-degree consisted of `www.gamelan.com` and `java.sun.com`, together with pages advertising for Caribbean vacations, and the home page of Amazon Books. This mixture is representative of the type of problem that arises with this simple ranking scheme: While the first two of these pages should certainly be viewed as "good" answers, the others are not relevant to the query topic; they have large in-degree but lack any thematic unity. The basic difficulty this exposes is the inherent tension that exists within the subgraph G_σ between strong authorities and pages that are simply "universally popular"; we expect the

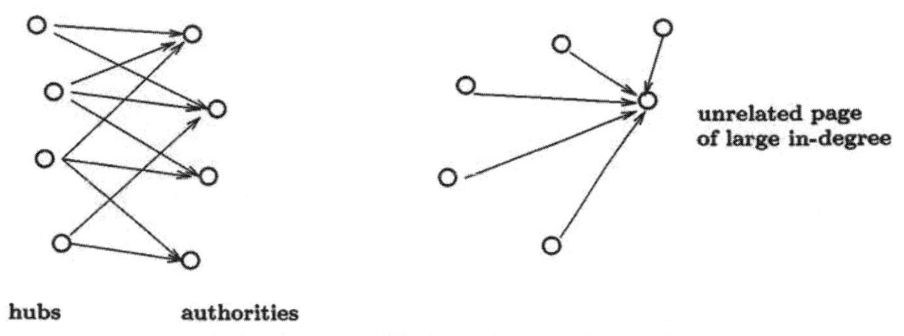

FIG. 2. A densely linked set of hubs and authorities.

latter type of pages to have large in-degree regardless of the underlying query topic.

One could wonder whether circumventing these problems requires making further use of the textual content of pages in the base set, rather than just the link structure of G_σ. We now show that this is not the case—it is in fact possible to extract information more effectively from the links—and we begin from the following observation. Authoritative pages relevant to the initial query should not only have large in-degree; since they are all authorities on a common topic, there should also be considerable overlap in the *sets* of pages that point to them. Thus, in addition to highly authoritative pages, we expect to find what could be called *hub pages*: these are pages that have links to multiple relevant authoritative pages. It is these hub pages that "pull together" authorities on a common topic, and allow us to throw out unrelated pages of large in-degree. (A skeletal example is depicted in Figure 2; in reality, of course, the picture is not nearly this clean.)

Hubs and authorities exhibit what could be called a *mutually reinforcing relationship*: a good *hub* is a page that points to many good authorities; a good *authority* is a page that is pointed to by many good hubs. Clearly, if we wish to identify hubs and authorities within the subgraph G_σ, we need a method for breaking this circularity.

3.1. AN ITERATIVE ALGORITHM. We make use of the relationship between hubs and authorities via an iterative algorithm that maintains and updates numerical weights for each page. Thus, with each page p, we associate a nonnegative *authority weight* $x^{\langle p \rangle}$ and a nonnegative *hub weight* $y^{\langle p \rangle}$. We maintain the invariant that the weights of each type are normalized so their squares sum to 1: $\Sigma_{p \in S_\sigma} (x^{\langle p \rangle})^2 = 1$, and $\Sigma_{p \in S_\sigma} (y^{\langle p \rangle})^2 = 1$. We view the pages with larger x- and y-values as being "better" authorities and hubs, respectively.

Numerically, it is natural to express the mutually reinforcing relationship between hubs and authorities as follows: If p points to many pages with large x-values, then it should receive a large y-value; and if p is pointed to by many pages with large y-values, then it should receive a large x-value. This motivates the definition of two operations on the weights, which we denote by $\mathcal{I}$ and $\mathcal{O}$. Given weights $\{x^{\langle p \rangle}\}$, $\{y^{\langle p \rangle}\}$, the $\mathcal{I}$ operation updates the x-weights as follows:

$$x^{\langle p \rangle} \leftarrow \sum_{q:(q,p) \in E} y^{\langle q \rangle}.$$

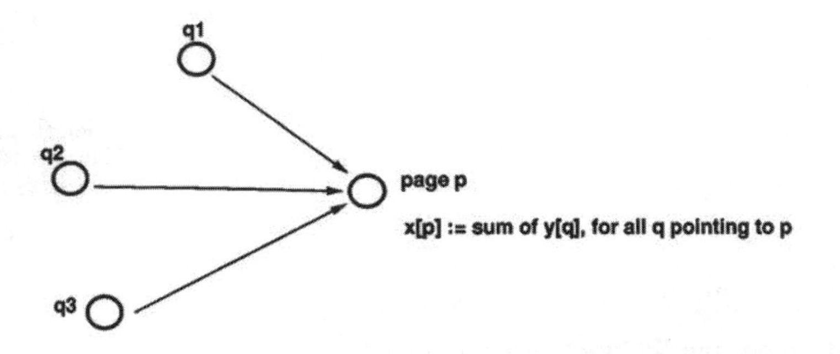

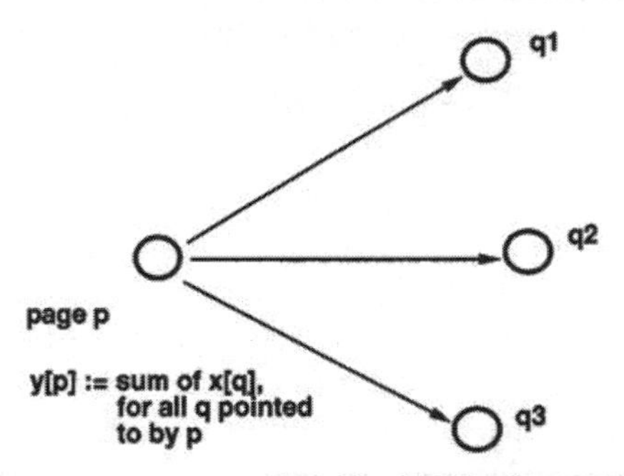

FIG. 3. The basic operations.

The $\mathcal{O}$ operation updates the y-weights as follows:

$$y^{\langle p \rangle} \leftarrow \sum_{q:(p,q) \in E} x^{\langle q \rangle}.$$

Thus, $\mathcal{I}$ and $\mathcal{O}$ are the basic means by which hubs and authorities reinforce one another. (See Figure 3.)

Now, to find the desired "equilibrium" values for the weights, one can apply the $\mathcal{I}$ and $\mathcal{O}$ operations in an alternating fashion, and see whether a fixed point is reached. Indeed, we can now state a version of our basic algorithm. We represent the set of weights $\{x^{\langle p \rangle}\}$ as a vector x with a coordinate for each page in G_σ; analogously, we represent the set of weights $\{y^{\langle p \rangle}\}$ as a vector y.

Iterate(G,k)
  G: a collection of n linked pages
  k: a natural number
  Let z denote the vector $(1, 1, 1, \ldots, 1) \in \mathbf{R}^n$.
  Set $x_0 := z$.
  Set $y_0 := z$.
  For $i = 1, 2, \ldots, k$
    Apply the $\mathcal{I}$ operation to (x_{i-1}, y_{i-1}), obtaining new x-weights x'_i.
    Apply the $\mathcal{O}$ operation to (x'_i, y_{i-1}), obtaining new y-weights y'_i.
    Normalize x'_i, obtaining x_i.
    Normalize y'_i, obtaining y_i.
  End
  Return (x_k, y_k).

This procedure can be applied to filter out the top c authorities and top c hubs in the following simple way:

Filter(G,k,c)
 G: a collection of n linked pages
 k,c: natural numbers
 (x_k, y_k) := `Iterate`(G, k).
 Report the pages with the c largest coordinates in x_k as authorities.
 Report the pages with the c largest coordinates in y_k as hubs.

We will apply the `Filter` procedure with G set equal to G_σ, and typically with $c \approx 5 - 10$. To address the issue of how best to choose k, the number of iterations, we first show that as one applies `Iterate` with arbitrarily large values of k, the sequences of vectors $\{x_k\}$ and $\{y_k\}$ converge to fixed points x^* and y^*.

We require the following notions from linear algebra, and refer the reader to a text such as Golub and Van Loan [1989] for more comprehensive background. Let M be a symmetric $n \times n$ matrix. An *eigenvalue* of M is a number λ with the property that, for some vector ω, we have $M\omega = \lambda\omega$. The set of all such ω is a subspace of $\mathbf{R}^n$, which we refer to as the *eigenspace* associated with λ; the dimension of this space will be referred to as the *multiplicity* of λ. It is a standard fact that M has at most n distinct eigenvalues, each of them a real number, and the sum of their multiplicities is exactly n. We will denote these eigenvalues by $\lambda_1(M), \lambda_2(M), \ldots, \lambda_n(M)$, indexed in order of decreasing absolute value, and with each eigenvalue listed a number of times equal to its multiplicity. For each distinct eigenvalue, we choose an orthonormal basis of its eigenspace; considering the vectors in all these bases, we obtain a set of eigenvectors $\omega_1(M)$, $\omega_2(M), \ldots, \omega_n(M)$ that we can index in such a way that $\omega_i(M)$ belongs to the eigenspace of $\lambda_i(M)$.

For the sake of simplicity, we will make the following technical assumption about all the matrices we deal with:

$$(\dagger)|\lambda_1(M)| > |\lambda_2(M)|.$$

When this assumption holds, we refer to $\omega_1(M)$ as the *principal eigenvector*, and all other $\omega_i(M)$ as *nonprincipal eigenvectors*. When the assumption does not hold, the analysis becomes less clean, but it is not affected in any substantial way.

We now prove that the `Iterate` procedure converges as k increases arbitrarily.

THEOREM 3.1. *The sequences* $x_1, x_2, x_3, \ldots$ *and* $y_1, y_2, y_3, \ldots$ *converge* (*to limits* x^* *and* y^*, *respectively*).

PROOF. Let $G = (V, E)$, with $V = \{p_1, p_2, \ldots, p_n\}$, and let A denote the *adjacency matrix* of the graph G; the (i, j)th entry of A is equal to 1 if (p_i, p_j) is an edge of G, and is equal to 0, otherwise. One easily verifies that the $\mathcal{I}$ and $\mathcal{O}$ operations can be written $x \leftarrow A^T y$ and $y \leftarrow Ax$, respectively. Thus, x_k is the unit vector in the direction of $(A^TA)^{k-1}A^Tz$, and y_k is the unit vector in the direction of $(AA^T)^kz$.

Now, a standard result of linear algebra (e.g., Golub and Van Loan [1989]) states that if M is a symmetric $n \times n$ matrix, and v is a vector not orthogonal to the principal eigenvector $\omega_1(M)$, then the unit vector in the direction of M^kv

converges to $\omega_1(M)$ as k increases without bound. Also (as a corollary), if M has only nonnegative entries, then the principal eigenvector of M has only nonnegative entries.

Consequently, z is not orthogonal to $\omega_1(AA^T)$, and hence the sequence $\{y_k\}$ converges to a limit y^*. Similarly, one can show that if $\lambda_1(A^TA) \neq 0$ (as dictated by Assumption (†)), then A^Tz is not orthogonal to $\omega_1(A^TA)$. It follows that the sequence $\{x_k\}$ converges to a limit x^*. □

The proof of Theorem 3.1 yields the following additional result (in the above notation).

THEOREM 3.2. (SUBJECT TO ASSUMPTION (†)). *x^* is the principal eigenvector of A^TA, and y^* is the principal eigenvector of AA^T.*

In our experiments, we find that the convergence of `Iterate` is quite rapid; one essentially always finds that $k = 20$ is sufficient for the c largest coordinates in each vector to become stable, for values of c in the range that we use. Of course, Theorem 3.2 shows that one can use any eigenvector algorithm to compute the fixed points x^* and y^*; we have stuck to the above exposition in terms of the `Iterate` procedure for two reasons. First, it emphasizes the underlying motivation for our approach in terms of the reinforcing $\mathcal{I}$ and $\mathcal{O}$ operations. Second, one does not have to run the above process of iterated $\mathcal{I}/\mathcal{O}$ operations to convergence; one can compute weights $\{x^{\langle p \rangle}\}$ and $\{y^{\langle p \rangle}\}$ by starting from any initial vectors x_0 and y_0, and performing a fixed bounded number of $\mathcal{I}$ and $\mathcal{O}$ operations.

3.2. BASIC RESULTS. We now give some sample results obtained via the algorithm, using some of the queries discussed in the introduction.

(java) Authorities

.328 http://www.gamelan.com/	*Gamelan*
.251 http://java.sun.com/	*JavaSoft Home Page*
.190 http://www.digitalfocus.com/digitalfocus/faq/ howdoi.html	*The Java Developer: How Do I . . .*
.190 http://lightyear.ncsa.uiuc.edu/~srp/java/javabooks.html	*The Java Book Pages*
.183 http://sunsite.unc.edu/javafaq/javafaq.html	*comp.lang.java FAQ*

(censorship) Authorities

.378 http://www.eff.org/	*EFFweb—The Electronic Frontier Foundation*
.344 http://www.eff.org/blueribbon.html	*The Blue Ribbon Campaign for Online Free Speech*
.238 http://www.cdt.org/	*The Center for Democracy and Technology*
.235 http://www.vtw.org/	*Voters Telecommunications Watch*
.218 http://www.aclu.org/	*ACLU: American Civil Liberties Union*

("search engines") Authorities

.346 http://www.yahoo.com/	*Yahoo!*
.291 http://www.excite.com/	*Excite*
.239 http://www.mckinley.com/	*Welcome to Magellan!*
.231 http://www.lycos.com/	*Lycos Home Page*
.231 http://www.altavista.digital.com/	*AltaVista: Main Page*

(Gates) Authorities

.643 http://www.roadahead.com/	*Bill Gates: The Road Ahead*

.458 http://www.microsoft.com/ *Welcome to Microsoft*
.440 http://www.microsoft.com/corpinfo/bill-g.htm

Among all these pages, the only one that occurred in the corresponding root set R_σ was `www.roadahead.com/`, under the query `"Gates"`; it was ranked 123rd by AltaVista. This is natural in view of the fact that many of these pages do not contain any occurrences of the initial query string.

It is worth reflecting on two additional points here. First, our only use of the textual content of pages was in the initial "black-box" call to a text-based search engine, which produced the root set R_σ. Following this, the analysis ignored the textual content of pages. The point we wish to make here is not that text is best ignored in searching for authoritative pages; there is clearly much that can be accomplished through the integration of textual and link-based analysis, and we will be commenting on this in a subsequent section. However, the results above show that a considerable amount can be accomplished through essentially a "pure" analysis of link structure.

Second, for many broad search topics, our algorithm produces pages that can legitimately be considered authoritative with respect to the www as a whole, despite the fact that it operates without direct access to large-scale index of the www. Rather, its only "global" access to the www is through a text-based search engine such as AltaVista, from which it is very difficult to directly obtain reasonable candidates for authoritative pages on most queries. What the results imply is that it is possible to reliably estimate certain types of global information about the www using only a standard search engine interface; a global analysis of the full www link structure can be replaced by a much more local method of analysis on a small focused subgraph.

4. *Similar-Page Queries*

The algorithm developed in the preceding section can be applied to another type of problem—that of using link structure to infer a notion of "similarity" among pages. Suppose we have found a page p that is of interest—perhaps it is an authoritative page on a topic of interest—and we wish to ask the following type of question: What do users of the www consider to be related to p, when they create pages and hyperlinks?

If p is highly referenced page, we have a version of the Abundance Problem: the surrounding link structure will implicitly represent an enormous number of independent opinions about the relation of p to other pages. Using our notion of hubs and authorities, we can provide an approach to the issue of page similarity, asking: In the local region of the link structure near p, what are the strongest authorities? Such authorities can potentially serve as a broad-topic summary of the pages related to p.

In fact, the method of Sections 2 and 3 can be adapted to this situation with essentially no modification. Previously, we initiated our search with a query string σ; our request to the underlying search engine was "Find t pages containing the string σ." We now begin with a page p and pose the following request to the search engine: "Find t pages pointing to p." Thus, we assemble a *root set* R_p consisting of t pages that point to p; we grow this into a base set S_p as before; and the result is a subgraph G_p in which we can search for hubs and authorities.

Superficially, the set of issues in working with a subgraph G_p are somewhat different from those involved in working with a subgraph defined by a query string; however, we find that most of the basic conclusions we drew in the previous two sections continue to apply. First, we observe that ranking pages of G_p by their in-degrees is still not satisfactory; consider for example the results of this heuristic when the initial page p was `www.honda.com`, the home page of Honda Motor Company.

http://www.honda.com	*Honda*
http://www.ford.com/	*Ford Motor Company*
http://www.eff.org/blueribbon.html	*The Blue Ribbon Campaign for Online Free Speech*
http://www.mckinley.com/	*Welcome to Magellan!*
http://www.netscape.com	*Welcome to Netscape*
http://www.linkexchange.com/	*LinkExchange—Welcome*
http://www.toyota.com/	*Welcome to @Toyota*
http://www.pointcom.com/	*PointCom*
http://home.netscape.com/	*Welcome to Netscape*
http://www.yahoo.com	*Yahoo!*

In many cases, the top hubs and authorities computed by our algorithm on a graph of the form G_p can be quite compelling. We show the top authorities obtained when the initial page p was `www.honda.com` and `www.nyse.com`, the home page of the New York Stock Exchange.

(www.honda.com) Authorities	
.202 http://www.toyota.com/	*Welcome to @Toyota*
.199 http://www.honda.com/	*Honda*
.192 http://www.ford.com/	*Ford Motor Company*
.173 http://www.bmwusa.com/	*BMW of North America, Inc.*
.162 http://www.volvocars.com/	*VOLVO*
.158 http://www.saturncars.com/	*Welcome to the Saturn Web Site*
.155 http://www.nissanmotors.com/	*NISSAN—ENJOY THE RIDE*
.145 http://www.audi.com/	*Audi Homepage*
.139 http://www.4adodge.com/	*1997 Dodge Site*
.136 http://www.chryslercars.com/	*Welcome to Chrysler*

(www.nyse.com) Authorities	
.208 http://www.amex.com/	*The American Stock Exchange—The Smarter Place to Be*
.146 http://www.nyse.com/	*New York Stock Exchange Home Page*
.134 http://www.liffe.com/	*Welcome to LIFFE*
.129 http://www.cme.com/	*Futures and Options at the Chicago Mercantile Exchange*
.120 http://update.wsj.com/	*The Wall Street Journal Interactive Edition*
.118 http://www.nasdaq.com/	*The Nasdaq Stock Market Home Page—Reload Often*
.117 http://www.cboe.com/	*CBOE—The ChicagoBoard Options Exchange*
.116 http://www.quote.com/	*1-Quote.com—Stock Quotes, Business News, Financial Market*
.113 http://networth.galt.com/	*NETworth*
.109 http://www.lombard.com/	*Lombard Home Page*

Note the difficulties inherent in compiling such lists through text-based methods: many of the above pages consist almost entirely of images, with very little text; and the text that they do contain has very little overlap. Our approach, on the other hand, is determining, via the presence of links, what the creators of www pages tend to "classify" together with the given pages www.honda.com and www.nyse.com.

5. *Connections with Related Work*

The analysis of link structures with the goal of understanding their social or informational organization has been an issue in a number of overlapping areas. In this section, we review some of the approaches that have been proposed, divided into three main areas of focus. First, and most closely related to our work here, we discuss research on the use of a link structure for defining notions of *standing*, *impact*, and *influence*—measures with the same motivation as our notion of authority. We then discuss other ways in which links have been integrated into hypertext and www search techniques. Finally, we review some work that has made use of link structures for explicit clustering of data.

5.1. Standing, Impact, and Influence

5.1.1. *Social Networks.* The study of social networks has developed several ways to measure the relative *standing*—roughly, "importance"—of individuals in an implicitly defined network. We can represent the network, as above, by a graph $G = (V, E)$; an edge (i, j) corresponds roughly to an "endorsement" of j by i. This is in keeping with the intuition we have already invoked regarding the role of www hyperlinks as conferrors of authority. Links may have different (nonnegative) *weights*, corresponding to the strength of different endorsements; let A denote the matrix whose (i, j)th entry represents the strength of the endorsement from a node $i \in V$ to a node $j \in V$.

Katz [1953] proposed a measure of standing based on path-counting, a generalization of ranking based on in-degree. For nodes i and j, let $P_{ij}^{(r)}$ denote the number of paths of length exactly r from i to j. Let $b < 1$ be a constant chosen to be small enough that $Q_{ij} = \Sigma_{r=1}^{\infty} b^r P_{ij}^{(r)}$ converges for each pair (i, j). Now Katz defines s_j, the *standing* of node j, to be $\Sigma_i Q_{ij}$—in this model, standing is based on the total number of paths terminating at node j, weighted by an exponentially decreasing damping factor. It is not difficult to obtain a direct matrix formulation of this measure: s_j is proportional to the jth column sum of the matrix $(I - bA)^{-1} - I$, where I denotes the identity matrix and all entries of A are 0 or 1.

Hubbell [1965] proposed a similar model of standing by studying the equilibrium of a certain weight-propagation scheme on nodes of the network. Recall that A_{ij}, the (i, j)th entry of our matrix A, represents the strength of the endorsement from i to j. Let e_j denote an a priori estimate of the standing of node j. Then Hubbell defines the standings $\{s_j\}$ to be a set of values so that the process of endorsement maintains a type of equilibrium—the total "quantity" of endorsement entering a node j, weighted by the standings of the endorsers, is equal to the standing of j. Thus, the standings are the solutions to the system of equations $s_j = e_j + \Sigma_i A_{ij} s_i$, for $j = 1, \ldots, n$. If e denotes the vector of values $\{e_j\}$, then the vector of standings in this model can be shown to be $(I - A^T)^{-1} e$.

Before discussing the relation of these measures to our work, we consider the way in which they were extended by research in the field of bibliometrics.

5.1.2. *Scientific Citations*. *Bibliometrics* [Egghe and Rousseau 1990] is the study of written documents and their citation structure. Research in bibliometrics has long been concerned with the use of citations to produce quantitative estimates of the importance and "impact" of individual scientific papers and journals, analogues of our notion of authority. In this sense, they are concerned with evaluating *standing* in a particular type of social network—that of papers or journals linked by citations.

The most well-known measure in this field is Garfield's *impact factor* [Garfield 1972], used to provide a numerical assessment of journals in Journal Citation Reports of the Institute for Scientific Information. Under the standard definition, the impact factor of a journal j in a given year is the average number of citations received by papers published in the previous two years of journal j [Egghe and Rousseau 1990]. Disregarding for now the question of whether two years is the appropriate period of measurement (see, e.g., Egghe [1988]), we observe that the impact factor is a ranking measure based fundamentally on a pure counting of the in-degrees of nodes in the network.

Pinski and Narin [1976] proposed a more subtle citation-based measure of standing, stemming from the observation that not all citations are equally important. They argued that a journal is "influential" if, recursively, it is heavily cited by other influential journals. One can recognize a natural parallel between this and our self-referential construction of hubs and authorities; we will discuss the connections below. The concrete construction of Pinski and Narin, as modified by Geller [1978], is the following: The measure of standing of journal j will be called its *influence weight* and denoted w_j. The matrix A of connection strengths will have entries specified as follows: A_{ij} denotes the fraction of the citations from journal i that go to journal j. Following the informal definition above, the influence of j should be equal to the sum of the influences of all journals citing j, with the sum weighted by the amount that each cites j. Thus, the set of influence weights $\{w_j\}$ is designed to be a nonzero, nonnegative solution to the system of equations $w_j = \Sigma_i A_{ij} w_i$; and hence, if w is the vector of influence weights, one has $w \geq 0$, $w \neq 0$, and $A^T w = w$. This implies that w is a principal eigenvector of A^T. Geller [1978] observed that the influence weights correspond to the stationary distribution of the following random process: beginning with an arbitrary journal j, one chooses a random reference that has appeared in j and moves to the journal specified in the reference. Doreian [1988; 1994] showed that one can obtain a measure of standing that corresponds very closely to influence weights by repeatedly iterating the computation underlying Hubbell's measure of standing: In the first iteration one computes Hubbell standings $\{s_j\}$ from the a priori weights $\{e_j\}$; the $\{s_j\}$ then become the a priori estimates for the next iteration. Finally, there has been work aimed at the troublesome issue of how to handle journal self-citations (the diagonal elements of the matrix A); see for example, de Solla Price [1981] and Noma [1982].

Let us consider the connections between this previous work and our algorithm to compute hubs and authorities. We also begin by observing that pure in-degree counting, as manifested by the impact factor, is too crude a measure for our purposes, and we seek a type of link-based equilibrium among relative node rankings. But the World Wide Web and the scientific literature are governed by very different principles, and this contrast is nicely captured in the distinction between Pinski–Narin influence weights and the hub/authority weights that we compute. Journals in the scientific literature have, to a first approximation, a common purpose, and traditions such as the peer review process typically ensure that highly authoritative journals on a common topic reference one another extensively. Thus, it makes sense to consider a one-level model in which authorities directly endorse other authorities. The www, on the other hand, is much more heterogeneous, with www pages serving many different functions—individual AOL subscribers have home pages, and multinational corporations have home pages. Moreover, for a wide range of topics, the strongest authorities consciously do not link to one another—consider, for example, the home pages of search engines and automobile manufacturers listed above. Thus, they can only be connected by an intermediate layer of relatively anonymous hub pages, which link in a correlated way to a thematically related set of authorities; and our model for the conferral of authority on the www takes this into account. This two-level pattern of linkage exposes structure among both the set of hubs, who may not know of one another's existence, and the set of authorities, who may not wish to acknowledge one another's existence.

5.1.3. *Hypertext and WWW Rankings*. There have been several approaches to ranking pages in the context of hypertext and the www. In work predating the emergence of the www, Botafogo et al. [1992] worked with focused, stand-alone hypertext environments. They defined the notions of *index nodes* and *reference nodes*—an index node is one whose out-degree is significantly larger than the average out-degree, and a reference node is one whose in-degree is significantly larger than the average in-degree. They also proposed measures of *centrality* based on node-to-node distances in the graph defined by the link structure.

Carrière and Kazman [1997] proposed a ranking measure on www pages, for the goal of re-ordering search results. The rank of a page in their model is equal to the sum of its in-degree and its out-degree; thus, it makes use of a "directionless" version of the www link structure.

Both of these approaches are based principally on counting node degrees, parallel to the structure of Garfield's *impact factor*. In contrast, Brin and Page [1998] have recently proposed a ranking measure based on a node-to-node weight-propagation scheme and its analysis via eigenvectors. Specifically, they begin from a model of a user randomly following hyperlinks: at each page, the user either selects an outgoing link uniformly at random, or (with some probability $p < 1$) jumps to a new page selected uniformly at random from the entire www. The stationary probability of node i in this random process will correspond to the "rank" of i, referred to as its *page-rank*.

Alternately, one can view page-ranks as arising from the equilibrium of a process analogous to that used in the definition of the Pinski–Narin influence weights, with the incorporation of a term that captures the "random jump" to a uniformly selected page. Specifically, assuming the www contains n pages, letting A denote the $n \times n$ *adjacency matrix* of the www, and letting d_i denote the *out-degree* of node i, the probability of a transition from page i to page j in the Brin–Page model is seen to be equal to $A'_{ij} = pn^{-1} + (1 - p)d_i^{-1} A_{ij}$. Let A' denote the matrix whose entries are A'_{ij}. The vector of ranks is then a nonzero, nonnegative solution to $(A')^T r = r$, and hence it corresponds to the principal eigenvector of $(A')^T$.

One of the main contrasts between our approach and the page-rank methodology is that—like Pinski and Narin's formulation of influence weights—the latter is based on a model in which authority is passed directly from authorities to other authorities, without interposing a notion of hub pages. Brin and Page's use of random jumps to uniformly selected pages is a way of dealing with the resulting problem that many authorities are essentially "dead-ends" in their conferral process.

It is also worth noting a basic contrast in the application of these approaches to www search. In Brin and Page [1998], the page-rank algorithm is applied to compute ranks for all the nodes in a 24-million-page index of the www; these ranks are then used to order the results of *subsequent* text-based searches. Our use of hubs and authorities, on the other hand, proceeds without direct access to a www index; in response to a query, our algorithm *first* invokes a text-based search and then computes numerical scores for the pages in a relatively small subgraph constructed from the initial search results.

5.2. OTHER LINK-BASED APPROACHES TO WWW SEARCH. Frisse [1988] considered the problem of document retrieval in singly authored, stand-alone works of hypertext. He proposed basic heuristics by which hyperlinks can enhance notions of *relevance* and hence the performance of retrieval heuristics. Specifically, in his framework, the relevance of a page in hypertext to a particular query is based in part on the relevance of the pages it links to. Marchiori's HyperSearch algorithm [Marchiori 1997] is based on such a methodology applied to www pages: A relevance score for a page p is computed by a method that incorporates the relevance of pages reachable from p, diminished by a damping factor that decays exponentially with distance from p.

In our construction of focused subgraphs from search engine results in Section 2, the underlying motivation ran also in the opposite direction. In addition to looking at where a page p *pointed* to increase our understanding of its contents, we implicitly used the text on pages that *pointed to* p. (For if pages in the root set for "search engines" pointed to `www.yahoo.com`, then we included `www.yahoo.com` in our subgraph.) This notion is related to that of searching based on *anchor text*, in which one treats the text surrounding a hyperlink as a descriptor of the page being pointed to when assessing the relevance of that page. The use of anchor text appeared in one of the oldest www search engines, McBryan's World Wide Web Worm [McBryan 1994]; it is also used in Brin and Page [1998] and Chakrabarti et al. [1998a; 1998b].

Another direction of work on the integration of links into www search is the construction of search formalisms capable of handling queries that involve predicates over both text and links. Arocena et al. [1997] have developed a framework supporting www queries that combines standard keywords with conditions on the surrounding link structure.

5.3. CLUSTERING OF LINK STRUCTURES. Link-based clustering in the context of bibliometrics, hypertext, and the www has focused largely on the problem of decomposing an *explicitly represented* collection of nodes into "cohesive" subsets. As such, it has mainly been applied to moderately sized sets of objects—for example, a focused collection of scientific journals, or the set of pages on a single www site. Earlier, we indicated a sense in which the issues we study here are fundamentally different from those encountered in this type of clustering: Our primary concern is that of representing an enormous collection of pages *implicitly*, through the construction of hubs and authorities for this collection. We now discuss some of the prior work on citation-based and hypertext clustering so as to better elucidate its connections to the techniques we develop here. In particular, this will also be useful in Section 6 when we discuss methods for computing multiple sets of hubs and authorities within a single link structure; this can be viewed as a way of representing multiple, potentially very large clusters implicitly.

At a very high level, clustering requires an underlying *similarity function* among objects, and a method for producing *clusters* from this similarity function. Two basic similarity functions on documents to emerge from the study of bibliometrics are *bibliographic coupling* (due to Kessler [1963]) and *co-citation* (due to Small [1973]). For a pair of documents p and q, the former quantity is equal to the number of documents cited by both p and q, and the latter quantity is the number of documents that cite both p and q. Co-citation has been used as a measure of the similarity of www pages by Larson [1996] and by Pitkow and Pirolli [1997]. Weiss et al. [1996] define linked-based similarity measures for pages in a hypertext environment that generalize *co-citation* and *bibliographic coupling* to allow for arbitrarily long chains of links.

Several methods have been proposed in this context to produce clusters from a set of nodes annotated with such similarity information. Small and Griffith [1974] use breadth-first search to compute the connected components of the undirected graph in which two nodes are joined by an edge if and only if they have a positive co-citation value. Pitkow and Pirolli [1997] apply this algorithm to study the link-based relationships among a collection of www pages.

One can also use principal components analysis [Hotelling 1933; Jolliffe 1986] and related dimension-reduction techniques such as multidimensional scaling to cluster a collection of nodes. In this framework, one begins with a matrix M containing the similarity information between pairs of nodes, and a representation (based on this matrix) of each node i as a high-dimensional vector $\{v_i\}$. One then uses the first few nonprincipal eigenvectors of the similarity matrix M to define a low-dimensional subspace into which the vectors $\{v_i\}$ can be projected; a variety of geometric or visualization-based techniques can be employed to identify dense clusters in this low-dimensional space. Standard theorems of linear algebra (e.g., Golub and Van Loan [1989]) in fact provide a precise sense in which projection onto the first k eigenvectors produces the *minimum* distortion over all k-dimensional projections of the data. Small [1986], McCain [1986], and others have applied this technique to journal and author co-citation data. The application of dimension-reduction techniques to cluster www pages based on co-citation has been employed by Larson [1996] and by Pitkow and Pirolli [1997].

The clustering of documents or hyperlinked pages can of course rely on combinations of textual and link-based information. Combinations of such measures have been studied by Shaw [1991a; 1991b] in the context of bibliometrics. More recently, Pirolli et al. [1996] have used a combination of link topology and textual similarity to group together and categorize pages on the www.

Finally, we discuss two other general eigenvector-based approaches to clustering that have been applied to link structures. The area of *spectral graph partitioning* was initiated by the work of Donath and Hoffman [1973] and Fiedler

[1973]; see the recent book by Chung [1997] for an overview. Spectral graph partitioning methods relate sparsely connected partitions of an *undirected* graph G to the eigenvalues and eigenvectors of its adjacency matrix A. Each eigenvector of A has a single coordinate for each node of G, and thus can be viewed as an assignment of weights to the nodes of G. Each nonprincipal eigenvector has both positive and negative coordinates; one fundamental heuristic to emerge from the study of these spectral methods is that the nodes corresponding to the large positive coordinates of a given eigenvector tend to be very sparsely connected to the nodes corresponding to the large negative coordinates of the same eigenvector.

In a different direction, *centroid scaling* is a clustering method designed for representing two types of objects in a common space [Levine 1979]. Consider, for example, a set of people who have provided answers to the questions of a survey—one may wish to represent both the people and the possible answers in a common space, in a way so that each person is "close" to the answers he or she chose; and each answer is "close" to the people that chose it. Centroid scaling provides an eigenvector-based method for accomplishing this. In its formulation, it thus resembles our definitions of hubs and authorities, which used an eigenvector approach to produce related sets of weights for two distinct types of objects. A fundamental difference, however, is that centroid scaling methods are typically not concerned with interpreting only the largest coordinates in the representations they produce; rather, the goal is to infer a notion of similarity among a set of objects by geometric means. Centroid scaling has been applied to citation data by Noma [1984], for jointly clustering citing and cited documents. In the context of information retrieval, the *Latent Semantic Indexing* methodology of Deerwester et al. [1990] applied a centroid scaling approach to a vector-space model of documents [van Rijsberger 1979; Salton 1989]; this allowed them to represent terms and documents in a common low-dimensional space, in which natural geometrically defined clusters often separate multiple senses of a query term.

6. *Multiple Sets of Hubs and Authorities*

The algorithm in Section 3 is, in a sense, finding the most *densely* linked collection of hubs and authorities in the subgraph G_σ defined by a query string σ. There are a number of settings, however, in which one may be interested in finding several densely linked collections of hubs and authorities among the same set S_σ of pages. Each such collection could potentially be relevant to the query topic, but they could be well-separated from one another in the graph G_σ for a variety of reasons. For example,

(1) The query string σ may have several very different meanings. For example, "`jaguar`" (a useful example we learned from Chekuri et al. [1997]).
(2) The string may arise as a term in the context of multiple technical communities. E.g. "`randomized algorithms`".
(3) The string may refer to a highly polarized issue, involving groups that are not likely to link to one another. For example, "`abortion`".

In each of these examples, the relevant documents can be naturally grouped into several clusters. The issue in the setting of broad-topic queries, however, is not simply how to achieve a dissection into reasonable clusters; one must also deal with this in the presence of the Abundance Problem. Each cluster, in the context of the full www, is enormous, and so we require a way to distill a small set of hubs and authorities out of each one. We can thus view such collections of hubs and authorities as implicitly providing broad-topic summaries of a collection of large clusters that we never explicitly represent. At a very high level, our motivation in this sense is analogous to that of an information retrieval technique such as *Scatter/Gather* [Cutting et al. 1992], which seeks to represent very large document clusters through text-based methods.

In Section 3, we related the hubs and authorities we computed to the principal eigenvectors of the matrices A^TA and AA^T, where A is the adjacency matrix of G_σ. The non-principal eigenvectors of A^TA and AA^T provide us with a natural way to extract additional densely linked collections of hubs and authorities from the base set S_σ. We begin by noting the following basic fact.

PROPOSITION 6.1. *AA^T and A^TA have the same multiset of eigenvalues, and their eigenvectors can be chosen so that $\omega_i(AA^T) = A\omega_i(A^TA)$.*

Thus, each pair of eigenvectors $x_i^* = \omega_i(A^TA)$, $y_i^* = \omega_i(AA^T)$, related as in Proposition 6.1, has the following property: applying an $\mathcal{I}$ operation to (x_i^*, y_i^*) keeps the x-weights parallel to x_i^*, and applying an $\mathcal{O}$ operation to (x_i^*, y_i^*) keeps the y-weights parallel to y_i^*. Hence, each pair of weights (x_i^*, y_i^*) has precisely the *mutually reinforcing relationship* that we are seeking in authority/hub pairs. Moreover, applying $\mathcal{I} \cdot \mathcal{O}$ (respectively, $\mathcal{O} \cdot \mathcal{I}$) multiplies the magnitude of x_i^* (respectively, y_i^*) by a factor of $|\lambda_i|$; thus $|\lambda_i|$ gives precisely the extent to which the hub weights y_i^* and authority weights x_i^* *reinforce* one another.

Now, unlike the principal eigenvector, the nonprincipal eigenvectors have both positive and negative entries. Hence, each pair (x_i^*, y_i^*) provides us with two densely connected sets of hubs and authorities: those pages that correspond to the c coordinates with the most positive values, and those pages that correspond to the c coordinates with the most negative values. These sets of hubs and authorities have the same intuitive meaning as those produced in Section 3, although the algorithm to find them—based on nonprincipal eigenvectors—is less clean conceptually than the method of iterated $\mathcal{I}$ and $\mathcal{O}$ operations. Note also that since the extent to which the weights in x_i^* and y_i^* reinforce each other is determined by the eigenvalue λ_i, the hubs and authorities associated with eigenvectors of larger absolute value will typically be "denser" as subgraphs in the link structure, and hence will often have more intuitive meaning.

In Section 5, we observed that spectral heuristics for partitioning undirected graphs [Chung 1997; Donath and Hoffman 1973; Fielder 1973] have suggested that nodes assigned large positive coordinates in a nonprincipal eigenvector are often well-separated from nodes assigned large negative coordinates in the same eigenvector. Adapted to our context, which deals with directed rather than undirected graphs, one can ask whether there is a natural "separation" between the two collections of authoritative sources associated with the same nonprincipal eigenvector. We will see that in some cases there is a distinction between these two collections, in a sense that has meaning for the query topic. It is worth noting here that the *signs* of the coordinates in any nonprincipal eigenvector represents a purely arbitrary resolution of the following symmetry: if x_i^* and y_i^* are eigenvectors associated with λ_i, then so are $-x_i^*$ and $-y_i^*$.

6.1. Basic Results. We now give some examples of the way in which the application of nonprincipal eigenvectors produces multiple collections of hubs and authorities. One interesting phenomenon that arises is the following: The pages with large coordinates in the first few nonprincipal eigenvectors tend to recur, so that essentially the same collection of hubs and authorities will often be generated by several of the strongest nonprincipal eigenvectors. (Despite being similar in their large coordinates, these eigenvectors remain orthogonal due to differences in the coordinates of smaller absolute value.) As a result, one obtains fewer distinct collections of hubs and authorities than might otherwise be expected from a set of non-principal eigenvectors. This notion is also reflected in the output below, where we have selected (by hand) several distinct collections from among the first few non-principal eigenvectors.

We issue the first query as "`jaguar*`", simply as one way to search for either the word or its plural. For this query, the strongest collections of authoritative sources concerned the Atari Jaguar product, the NFL football team from Jacksonville, and the automobile.

(`jaguar*`) Authorities: principal eigenvector

.370 http://www2.ecst.csuchico.edu/~jschlich/Jaguar/jaguar.html	
.347 http://www-und.ida.liu.se/~t94patsa/jserver.html	
.292 http://tangram.informatik.uni-kl.de:8001/~rgehm/jaguar.html	
.287 http://www.mcc.ac.uk/dlms/Consoles/jaguar.html	*Jaguar Page*

(`jaguar*`) Authorities: 2nd nonprincipal vector, positive end

.255 http://www.jaguarsnfl.com/	*Official Jacksonville Jaguars NFL Website*
.137 http://www.nando.net/SportServer/football/nfl/jax.html	*Jacksonville Jaguars Home Page*
.133 http://www.ao.net/~brett/jaguar/index.html	*Brett's Jaguar Page*
.110 http://www.usatoday.com/sports/football/sfn/sfn30.htm	*Jacksonville Jaguars*

(`jaguar*`) Authorities: 3rd nonprincipal vector, positive end

.227 http://www.jaguarvehicles.com/	*Jaguar Cars Global Home Page*
.227 http://www.collection.co.uk/	*The Jaguar Collection—Official Web site*
.211 http://www.moran.com/sterling/sterling.html	
.211 http://www.coys.co.uk/	

For the query "`randomized algorithms`", none of the strongest collections of hubs and authorities could be said to be precisely on the query topic, though they all consisted of thematically related pages on a closely related topic. They included home pages of theoretical computer scientists, compendia of mathematical software, and pages on wavelets.

("randomized algorithms") Authorities: 1st nonprincipal vector, positive end

.125 http://theory.lcs.mit.edu/~goemans/	*Michel X. Goemans*
.122 http://theory.lcs.mit.edu/~spielman/	*Dan Spielman's Homepage*
.122 http://www.nada.kth.se/~johanh/	*Johan Hastad*
.122 http://theory.lcs.mit.edu/~rivest/	*Ronald L. Rivest: HomePage*

("randomized algorithms") Authorities 1st nonprincipal vector, negative end

−.00116 http://lib.stat.cmu.edu/	*StatLib Index*
−.00115 http://www.geo.fmi.fi/prog/tela.html	*Tela*
−.00107 http://gams.nist.gov/	*GAMS: Guide to Available Mathematical Software*
−.00107 http://www.netlib.org	*Netlib*

("randomized algorithms") Authorities 4th nonprincipal vector, negative end
−.176 http://www.amara.com/current/wavelet.html *Amara's Wavelet Page*
−.172 http://www-ocean.tamu.edu/~baum/wavelets.html *Wavelet sources*
−.161 http://www.mathsoft.com/wavelets.html *Wavelet Resources*
−.143 http://www.mat.sbg.ac.at/~uhl/wav.html *Wavelets*

We also encounter examples where pages from the positive and negative ends of the same nonprincipal eigenvector exhibit a natural separation. One case in which the meaning of this separation is particularly striking is for the query "`abortion`". The natural question is whether one of the nonprincipal eigenvectors produces a division between pro-choice and pro-life authorities. The issue is complicated by the existence of hub pages that link extensively to pages from both sides; but in fact the 2nd nonprincipal eigenvector produces a very clear separation:

(abortion) Authorities: 2nd nonprincipal vector, positive end
.321 http://www.caral.org/abortion.html *Abortion and Reproductive Rights Internet Resources*
.219 http://www.plannedparenthood.org/ *Welcome to Planned Parenthood*
.195 http://www.gynpages.com/ *Abortion Clinics OnLine*
.172 http://www.oneworld.org/ippf/ *IPPF Home Page*
.162 http://www.prochoice.org/naf/ *The National Abortion Federation*
.161 http://www.lm.com/~lmann/feminist/abortion.html

(abortion) Authorities: 2nd nonprincipal vector, negative end
−.197 http://www.awinc.com/partners/bc/commpass/lifenet/lifenet.htm *LifeWEB*
−.169 http://www.worldvillage.com/wv/square/chapel/xwalk/html/peter.htm *Healing after Abortion*
−.164 http://www.nebula.net/~maeve/lifelink.html
−.150 http://members.aol.com/pladvocate/ *Pro-Life Advocate*
−.144 http://www.clark.net/pub/jeffd/factbot.html *The Right Side of the Web*
−.144 http://www.catholic.net/HyperNews/get/abortion.html

7. *Diffusion and Generalization*

Let us return to the method of Section 3, in which we identified a single collection of hubs and authorities in the subgraph G_σ associated with a query string σ. The algorithm computes a densely linked collection of pages without regard to their contents; the fact that these pages are relevant to the query topic in a wide range of cases is based on the way in which we construct the subgraph G_σ, ensuring that it is rich in relevant pages. We can view the issue as follows: Many different topics are represented in G_σ, and each is centered around a competing collection of densely linked hubs and authorities. Our method of producing a focused subgraph G_σ aims at ensuring that the most relevant such collection is also the "densest" one, and hence will be found by the method of iterated $\mathcal{I}$ and $\mathcal{O}$ operations.

When the initial query string σ specifies a topic that is not sufficiently broad, however, there will often not be enough relevant pages in G_σ from which to extract a sufficiently dense subgraph of relevant hubs and authorities. As a result, authoritative pages corresponding to competing, "broader" topics will win out over the pages relevant to σ, and be returned by the algorithm. In such cases, we will say that the process has *diffused* from the initial query.

Although it limits the ability of our algorithm to find authoritative pages for narrow or specific query topics, diffusion can be an interesting process in its own right. In particular, the broader topic that supplants the original, too-specific query σ very often represents a natural generalization of σ. As such, it provides a simple way of abstracting a specific query topic to a broader, related one.

Consider, for example, the query "`WWW conferences`". At the time we tried this query, AltaVista indexed roughly 300 pages containing the string; however, the resulting subgraph G_σ contained pages concerned with a host of more general www-related topics, and the main authorities were in fact very general www resources.

("WWW conferences") Authorities: principal eigenvector

.088 http://www.ncsa.uiuc.edu/SDG/Software/Mosaic/Docs/whats-new.html	*The What's New Archive*
.088 http://www.w3.org/hypertext/DataSources/WWW/Servers.html	*World-Wide Web Servers: Summary*
.087 http://www.w3.org/hypertext/DataSources/bySubject/Overview.html	*The World-Wide Web Virtual Library*

In the context of similar-page queries, a query that is "too specific" corresponds roughly to a page p that does not have sufficiently high in-degree. In such cases, the process of diffusion can also provide a broad-topic summary of more prominent pages related to p. Consider, for example, the results when p was `sigact.acm.org`, the home page of the ACM Special Interest Group on Algorithms and Computation Theory, which focuses on theoretical computer science.

(sigact.acm.org) Authorities: principal eigenvector

.197 http://www.siam.org/	*Society for Industrial and Applied Mathematics*
.166 http://dimacs.rutgers.edu/	*Center for Discrete Mathematics and Theoretical Computer Science*
.150 http://www.computer.org/	*IEEE Computer Society*
.148 http://www.yahoo.com/	*Yahoo!*
.145 http://e-math.ams.org/	*e-MATH Home Page*
.141 http://www.ieee.org/	*IEEE Home Page*
.140 http://glimpse.cs.arizona.edu:1994/bib/	*Computer Science Bibliography Glimpse Server*
.129 http://www.eccc.uni-trier.de/eccc/	*ECCC—The Electronic Colloquium on Computational Complexity*
.129 http://www.cs.indiana.edu/cstr/search	*UCSTRI—Cover Page*
.118 http://euclid.math.fsu.edu/Science/math.html	*The World-Wide Web Virtual Library: Mathematics*

The problem of returning more specific answers in the presence of this phenomenon is the subject of on-going work; in Sections 8 and 9, we briefly discuss current work on the use of textual content for the purpose of focusing our approach to link-based analysis [Bharat and Henzinger 1998; Chakrabarti et al. 1998a; 1998b]. The use of nonprincipal eigenvectors, combined with basic term-matching, can be a simple way to extract collections of authoritative pages that are more relevant to a specific query topic. For example, consider the following fact: Among the sets of hubs and authorities corresponding to the first 20 nonprincipal eigenvectors for the query "WWW conferences", the one in which the pages collectively contained the string "`WWW conferences`" the most was the following.

("WWW conferences") Authorities: 11th nonprincipal vector, negative end

−.097 http://www.igd.fhg.de/www95.html	*Third International World-Wide Web Conference*
−.091 http://www.csu.edu.au/special/conference/WWWWW.html	*AUUG'95 and Asia-Pacific WWW'95 Conference*
−.090 http://www.ncsa.uiuc.edu/SDG/IT94/IT94Info.html	*The Second International WWW Conference '94*
−.083 http://www.w3.org/hypertext/Conferences/WWW4/	*Fourth International World Wide Web Conference*
−.079 http://www.igd.fhg.de/www/www95/papers/	*WWW'95: Papers*

8. *Evaluation*

The evaluation of the methods presented here is a challenging task. First, of course, we are attempting to define and compute a measure, "authority," that is inherently based on human judgment. Moreover, the nature of the www adds complexity to the problem of evaluation—it is a new domain, with a shortage of standard benchmarks; the diversity of authoring styles is much greater than for comparable collections of printed, published documents; and it is highly dynamic, with new material being created rapidly and no complete index of its full contents.

In the earlier sections of the paper, we have presented a number of examples of the output from our algorithm. This was both to show the reader the type of results that are produced, and because we believe that there is, and probably should be, an inevitable component of *res ipsa loquitur* in the overall evaluation—our feeling is that many of the results are quite striking at an obvious level.

However, there are also more principled ways of evaluating the algorithm. Since the appearance of the conference version of this paper, three distinct user studies performed by two different groups [Bharat and Henzinger 1998; Chakrabarti et al. 1998a; 1998b] have helped assess the value of our technique in the context of a tool for locating information on the www. Each of these studies used a system built primarily on top of the basic algorithm described here, for locating hubs and authorities in a subgraph G_σ via the methods discussed in Sections 2 and 3. However, each of these systems also employed additional heuristics to further enhance relevance judgments. Most significantly, they incorporated text-based measures such as anchor text scores to weight the contribution of individual links differentially. As such, the results of these studies should not be interpreted as providing a direct evaluation of the pure link-based method described here; rather, they assess its performance as the core component of a www search tool.

We briefly survey the structure and results of the most recent of these three user studies, involving the CLEVER system of Chakrabarti et al. [1998a] and refer the reader to that work for more details. The basic task in this study was *automatic resource compilation*—the construction of lists of high-quality www pages related to a broad search topic—and the goal was to see how the output of CLEVER compared to that of a manually generated compilation such as the www search service *Yahoo!* [Yahoo! Corp.] for a set of 26 topics.

Thus, for each topic, the output of the CLEVER system was a list of ten pages: its five top hubs and five top authorities. *Yahoo!* was used as the main point of comparison, since its manually compiled resource lists can be viewed as representing judgments of "authority" by the human ontologists who compile them. The top ten pages returned by AltaVista were also selected, so as to provide representative pages produced by a fully automatic text-based search engine. All these pages were collected into a single *topic list* for each topic in the study, without an indication of which method produced which page. A collection of 37 users was assembled; the users were required to be familiar with the use of a Web browser, but were not experts in computer science or in the 26 search topics. The users were then asked to rank the pages they visited from the topic lists as "bad," "fair," "good," or "fantastic," in terms of their utility in learning about the topic. This yielded 1369 responses in all, which were then used to assess the relative quality of CLEVER, *Yahoo!*, and AltaVista on each topic. For approximately 31% of the topics, the evaluations of *Yahoo!* and CLEVER were equivalent to within a threshold of statistical significance; for approximately 50% CLEVER was evaluated higher; and for the remaining 19% *Yahoo!* was evaluated higher.

Of course, it is difficult to draw definitive conclusions from these studies. A service such as *Yahoo!* is indeed providing, by its very nature, a type of human judgment as to which pages are "good" for a particular topic. But even the nature of the quality judgment is not well defined, of course. Moreover, many of the entries in *Yahoo!* are drawn from outside submissions, and hence represent less directly the "authority" judgments of *Yahoo!*'s staff.

Many of the users in these studies reported that they used the lists as starting points from which to explore, but that they visited many pages not on the original topic lists generated by the various techniques. This is, of course, a natural process in the exploration of a broad topic on the www, and the goal of resource lists appears to be generally for the purpose of facilitating this process rather than for replacing it.

9. *Conclusion*

We have discussed a technique for locating high-quality information related to a broad search topic on the www, based on a structural analysis of the link topology surrounding "authoritative" pages on the topic. It is useful to highlight four basic components of our approach.

—For broad topics on the www, the amount of relevant information is growing extremely rapidly, making it continually more difficult for individual users to filter the available resources. To deal with this problem, one needs notions beyond those of relevance and clustering—one needs a way to distill a broad topic, for which there may be millions of relevant pages, down to a representation of very small size. It is for this purpose that we define a notion of "authoritative" sources, based on the link structure of the www.

—We are interested in producing results that are of as a high a quality as possible in the context of what is available on the www *globally*. Our underlying domain is not restricted to a focused set of pages, or those residing on a single Web site.

—At the same time, we infer global notions of structure without directly maintaining an index of the www or its link structure. We require only a basic interface to any of a number of standard www search engines, and use techniques for producing "enriched" samples of www pages to determine notions of structure and quality that make sense globally. This helps to deal with problems of scale in handling topics that have an enormous representation on the www.

—We began with the goal of discovering *authoritative pages*, but our approach in fact identifies a more complex pattern of social organization on the www, in which hub pages link densely to a set of thematically related authorities. This equilibrium between hubs and authorities is a phenomenon that recurs in the context of a wide variety of topics on the www. Measures of impact and influence in bibliometrics have typically lacked, and arguably not required, an analogous formulation of the role that hubs play; the www is very different from the scientific literature, and our framework seems appropriate as a model of the way in which authority is conferred in an environment such as the Web.

This work has been extended in a number of ways since its initial conference appearance. In Section 8, we mentioned systems for compiling high-quality www resource lists that have been built using extensions to the algorithms developed here; see Bharat and Henzinger [1998] and Chakrabarti et al. [1998a; 1998b]. The implementation of the Bharat–Henzinger system made use of the recently developed Connectivity Server [Bharat et al. 1998], which provides very efficient retrieval for linkage information contained in the AltaVista index.

With Gibson et al. [1998a], we have used the algorithms described here to explore the structure of "communities" of hubs and authorities on the www. We find that the notion of topic generalization discussed in Section 7 provides one valuable perspective from which to view the overlapping organization of such communities. In a separate direction, also with Gibson et al. [1998b], we have investigated extensions of the present work to the analysis of relational data, and considered a natural, nonlinear analogue of spectral heuristics in this setting.

There a number of interesting further directions suggested by this research, in addition to the currently on-going work mentioned above. We will restrict ourselves here to three such directions.

First, we have used structural information about the graph defined by the links of the www, but we have not made use of its patterns of traffic, and the paths that users implicitly traverse in this graph as they visit a sequence of pages. There are a number of interesting and fundamental questions that can be asked about www traffic, involving both the modeling of such traffic and the development of algorithms and tools to exploit information gained from traffic patterns (see, e.g., Barrett et al. [1997], Berman et al. [1995], and Huberman et al. [1998]). It would be interesting to ask how the approach developed here might be integrated into a study of user traffic patterns on the www.

Second, the power of eigenvector-based heuristics is not something that is fully understood at an analytical level, and it would be interesting to pursue this question in the context of the algorithms presented here. One direction would be to consider random graph models that contain enough structure to capture certain global properties of the www, and yet are simple enough so that the application of our algorithms to them could be analyzed. More generally, the development of clean yet reasonably accurate random graph models for the www could be extremely valuable for the understanding of a range of link-based algorithms. Some work of this type has been undertaken in the context of the *latent semantic indexing* technique in information retrieval [Deerwester et al. 1990]. Papadimitriou et al. [1998] have provided a theoretical analysis of latent semantic indexing applied to a basic probabilistic model of term use in documents. In another direction, motivated in part by our work here, Frieze et al. [1998] have analyzed sampling methodologies capable of approximating the singular value decomposition of a large matrix very efficiently; understanding the concrete connections between their work and our sampling methodology in Section 2 would be very interesting.

Finally, the further development of link-based methods to handle information needs other than broad-topic queries on the www poses many interesting challenges. As noted above, work has been done on the incorporation of textual content into our framework as a way of "focusing" a broad-topic search [Bharat and Henzinger 1998; Chakrabarti et al. 1998a; 1998b], but one can ask what other basic informational structures one can identify, beyond hubs and authorities, from the link topology of hypermedia such as the www. The means by which interaction with a link structure can facilitate the discovery of information is a general and far-reaching notion, and we feel that it will continue to offer a range of fascinating algorithmic possibilities.

ACKNOWLEDGMENTS. In the early stages of this work, I benefited enormously from discussions with Prabhakar Raghavan and with Robert Kleinberg; I thank Soumen Chakrabarti, Byron Dom, David Gibson, S. Ravi Kumar, Prabhakar Raghavan, Sridhar Rajagopalan, and Andrew Tomkins for on-going collaboration on extensions and evaluations of this work; and I thank Rakesh Agrawal, Tryg Ager, Rob Barrett, Marshall Bern, Tim Berners-Lee, Ashok Chandra, Monika Henzinger, Alan Hoffman, David Karger, Lillian Lee, Nimrod Megiddo, Christos Papadimitriou, Peter Pirolli, Ted Selker, Eli Upfal, and the anonymous referees of this paper, for their valuable comments and suggestions.

REFERENCES

AROCENA, G. O., MENDELZON, A. O., AND MIHAILA, G. A. 1997. Applications of a Web query language. In *Proceedings of the 6th International World Wide Web Conference* (Santa Clara, Calif., Apr. 7–11).

BARRETT, R., MAGLIO, P., AND KELLEM, D. 1997. How to personalize the web. In *Proceedings of the ACM SIGCHI Conference on Human Factors in Computing Systems (CHI '97)* (Atlanta, Ga., Mar. 22–27). ACM, New York, pp. 75–82.

BERMAN, O., HODGSON, M. J., AND KRASS, D. 1995. "Flow-interception problems." In *Facility Location: A Survey of Applications and Methods*, Z. Drezner, ed. Springer-Verlag, New York.

BERNERS-LEE, T., CAILLIAU, R., LUOTONEN, A., NIELSEN, H. F., AND SECRET, A. 1994. The world-wide web. *Commun. ACM 37*, 1 (Jan.), 76–82.

BHARAT, K., BRODER, A., HENZINGER, M. R., KUMAR, P., AND VENKATASUBRAMANIAN, S. 1998. Connectivity server: Fast access to linkage information on the web. In *Proceedings of the 7th International World Wide Web Conference* (Brisbane, Australia, Apr. 14–18).

BHARAT, K., AND HENZINGER, M. R. 1998. Improved algorithms for topic distillation in a hyperlinked environment. In *Proceedings of the 21st Annual International ACM SIGIR Conference on Research and Development in Information Retrieval* (Melbourne, Australia, Aug. 24–28). ACM, New York, pp. 104–111.

BOTAFOGO, R., RIVLIN, E., AND SHNEIDERMAN, B. 1992. Structural analysis of hypertext: Identifying hierarchies and useful metrics. *ACM Trans. Inf. Sys. 10*, 2 (Apr.), 142–180.

BRIN, S., AND PAGE, L. 1998. Anatomy of a large-scale hypertextual web search engine. In *Proceedings of the 7th International World Wide Web Conference* (Brisbane, Australia, Apr. 14–18). pp. 107–117.

CARRIÈRE, J., AND KAZMAN, R. 1997. WebQuery: Searching and visualizing the web through connectivity. In *Proceedings of the 6th International World Wide Web Conference* (Santa Clara, Calif., Apr. 7–11).

CHAKRABARTI, S., DOM, B., GIBSON, D., KUMAR, S. R., RAGHAVAN, P., RAJAGOPALAN, S., AND TOMKINS, A. 1998. Experiments in topic distillation. In *Proceedings of the ACM SIGIR Workshop on Hypertext Information Retrieval on the Web* (Melbourne, Australia). ACM, New York.

CHAKRABARTI, S., DOM, B., GIBSON, D., KLEINBERG, J., RAGHAVAN, P., AND RAJAGOPALAN, S. 1998. Automatic resource compilation by analyzing hyperlink structure and associated text. In *Proceedings of the 7th International World Wide Web Conference* (Brisbane, Australia, Apr. 14–18). pp. 65–74.

CHUNG, F. R. K. 1997. *Spectral Graph Theory*. AMS Press, Providence, R.I.

CHEKURI, C., GOLDWASSER, M., RAGHAVAN, P., AND UPFAL, E. 1997. Web search using automated classification. In *Proceedings of the 6th International World Wide Web Conference* (Santa Clara, Calif., Apr. 7–11).

CUTTING, D. R., PEDERSEN, J., KARGER, D. R., AND TUKEY, J. W. 1992. Scatter/gather: A cluster-based approach to browsing large document collections. In *Proceedings of the 15th Annual International ACM SIGIR Conference on Research and Development in Information Retrieval* (Copenhagen, Denmark, June 21–24). ACM, New York, pp. 330–337.

DE SOLLA PRICE, D. 1981. The analysis of square matrices of scientometric transactions. *Scientometrics 3* 55–63.

DEERWESTER, S., DUMAIS, S., LANDAUER, T., FURNAS, G., AND HARSHMAN, R. 1990. Indexing by latent semantic analysis. *J. Amer. Soc. Info. Sci. 41*, 391–407.

DIGITAL EQUIPMENT CORPORATION. *AltaVista search engine*, http://altavista.digital.com/.

DONATH, W. E., AND HOFFMAN, A. J. 1973. Lower bounds for the partitioning of graphs. *IBM J. Res. Develop. 17.*

DOREIAN, P. 1988. Measuring the relative standing of disciplinary journals, *Inf. Proc. Manage. 24*, 45–56.

DOREIAN, P. 1994. A measure of standing for citation networks within a wider environment. *Inf. Proc. Manage. 30*, 21–31.

EGGHE, L. 1988. Mathematical relations between impact factors and average number of citations. *Inf. Proc. Manage. 24*, 567–576.

EGGHE, L., AND ROUSSEAU, R. 1990. *Introduction to Informetrics*, Elsevier, North-Holland, Amsterdam, The Netherlands.

FIELDER, M. 1973. Algebraic connectivity of graphs. *Czech. Math. J. 23*, 298–305.

FRIEZE, A., KANNAN, R., AND VEMPALA, S. 1998. Fast Monte-Carlo Algorithms for Finding Low-Rank Approximations. In *Proceedings of the 39th IEEE Symposium on Foundations of Computer Science* (Palo Alto, Calif., Nov. 8–11). IEEE Computer Society Press, Los Alamitos, Calif.

FRISSE, M. E. 1988. Searching for information in a hypertext medical handbook. *Commun. ACM 31*, 7 (July), 880–886.

GARFIELD, E. 1972. Citation analysis as a tool in journal evaluation. *Science 178*, 471–479.

GELLER, N. 1978. On the citation influence methodology of Pinski and Narin. *Inf. Proc. Manage. 14*, 93–95.

GIBSON, D., KLEINBERG, J., AND RAGHAVAN, P. 1998. Inferring web communities from link topology. In *Proceedings of the 9th ACM Conference on Hypertext and Hypermedia* (Pittsburgh, Pa., June 20–24). ACM, New York, pp. 225–234.

GIBSON, D., KLEINBERG, J., AND RAGHAVAN, P. 1998. Clustering categorical data: An approach based on dynamical systems. In *Proceedings of the 24th International Conference on Very Large Databases* (New York, N.Y., Aug. 24–27). pp. 311–322.

GOLUB, G., AND VAN LOAN, C. F. 1989. *Matrix Computations*. Johns Hopkins University Press, Baltimore, Md.

HOTELLING, H. 1933. Analysis of a complex statistical variable into principal components. *J. Educ. Psychol. 24*, 417–441.

HUBBELL, C. H. 1965. An input-output approach to clique identification. *Sociometry 28*, 377–399.

HUBERMAN, B., PIROLLI, P., PITKOW, J., AND LUKOSE, R. 1998. Strong regularities in world wide web surfing. *Science*, 280.

JOLLIFFE, I. T. 1986. *Principal Component Analysis*. Springer-Verlag, New York.

KATZ, L. 1953. A new status index derived from sociometric analysis. *Psychometrika 18*, 39–43.

KESSLER, M. M. 1963. Bibliographic coupling between scientific papers. *Amer. Document. 14*, 10–25.
LARSON, R. 1996. Bibliometrics of the world wide web: An exploratory analysis of the intellectual structure of cyberspace. In *Proceedings of the Annual Meeting of the American Society of Information Science* (Baltimore, Md., Oct. 19–24).
LEVINE, J. H. 1979. Joint-space analysis of 'pick-any' data: Analysis of choices from an unconstrained set of alternatives. *Psychometrika, 44*, 85–92.
MARCHIORI, M. 1997. The quest for correct information on the web: Hyper search engines. In *Proceedings of the 6th International World Wide Web Conference* (Santa Clara, Calif., Apr. 7–11).
MCBRYAN, O. 1994. GENVL and WWWW: Tools for taming the web. In *Proceedings of the 1st International World Wide Web Conference* (Geneva, Switzerland, May).
MCCAIN, K. 1986. Co-cited author mapping as a valid representation of intellectual structure. *J. Amer. Soc. Info. Sci. 37*, 111–122.
NOMA, E. 1982. An improved method for analyzing square scientometric transaction matrices. *Scientometrics 4*, 297–316.
NOMA, E. 1984. Co-citation analysis and the invisible college. *J. Amer. Soc. Info. Sci. 35*, 29–33.
PAPADIMITRIOU, C. H., RAGHAVAN, P., TAMAKI, H., AND VEMPALA, S. 1998. Latent semantic indexing: A probabilistic analysis. In *Proceedings of the 17th ACM SIGACT-SIGMOD-SIGART Symposium on Principles of Database Systems* (Seattle, Wash., June 1–3). ACM, New York, pp. 159–168.
PINSKI, G., AND NARIN, F. 1976. Citation influence for journal aggregates of scientific publications: Theory, with application to the literature of physics. *Inf. Proc. Manage. 12*, 297–312.
PIROLLI, P., PITKOW, J., AND RAO, R. 1996. Silk from a sow's ear: Extracting usable structures from the web. In *Proceedings of ACM SIGCHI Conference on Human Factors in Computing Systems (CHI '96)* (Vancouver, B.C., Canada, Apr. 13–18). ACM, New York, pp. 118–125.
PITKOW, J., AND PIROLLI, P. 1997. Life, death, and lawfulness on the electronic frontier. In *Proceedings of ACM SIGCHI Conference on Human Factors in Computing Systems (CHI '97)* (Atlanta, Ga., Mar. 22–27). ACM, New York, pp. 383–390.
SALTON, G. 1989. *Automatic Text Processing*. Addison-Wesley, Reading, Mass.
SHAW, W. M. 1991. Subject and citation indexing. Part I: The clustering structure of composite representations in the cystic fibrosis document collection. *J. Amer. Soc. Info. Sci. 42*, 669–675.
SHAW, W. M. 1991. Subject and citation indexing. Part II: The optimal, cluster-based retrieval performance of composite representations. *J. Amer. Soc. Info. Sci. 42*, 676–684.
SMALL, H. 1973. Co-citation in the scientific literature: A new measure of the relationship between two documents. *J. Amer. Soc. Info. Sci. 24*, 265–269.
SMALL, H. 1986. The synthesis of specialty narratives from co-citation clusters. *J. Amer. Soc. Info. Sci. 37*, 97–110.
SMALL, H., AND GRIFFITH, B. C. 1974. The structure of the scientific literatures I. Identifying and graphing specialties. *Science Studies 4*, 17–40.
SPERTUS, E. 1997. ParaSite: Mining structural information on the web. In *Proceedings of the 6th International World Wide Web Conference* (Santa Clara, Calif., Apr. 7–11).
VAN RIJSBERGEN, C. J. 1979. *Information Retrieval*. Butterworths, London, England.
WEISS, R., VELEZ, B., SHELDON, M. A., NEMPREMPRE, C., SZILAGYI, P., DUDA, A., AND GIFFORD, D. K. 1996. HyPursuit: A hierarchical network search engine that exploits content-link hypertext clustering. In *Proceedings of the 7th ACM Conference on Hypertext* (Washington, D.C., Mar. 16–20). ACM, New York, pp. 180–193.
WIRED DIGITAL, INC. *Hotbot*, `http://www.hotbot.com`.
YAHOO! CORPORATION *Yahoo!*, `http://www.yahoo.com`.

RECEIVED NOVEMBER 1997; REVISED MARCH 1999; ACCEPTED APRIL 1999

PHYSICAL REVIEW E, VOLUME 64, 046135

Search in power-law networks

Lada A. Adamic,[1,*] Rajan M. Lukose,[1,†] Amit R. Puniyani,[2,‡] and Bernardo A. Huberman[1,§]
[1]*HP Labs, Palo Alto, California 94304*
[2]*Department of Physics, Stanford University, 382 Via Pueblo Mall, Stanford, California 94305*
(Received 29 April 2001; revised manuscript received 22 June 2001; published 26 September 2001)

Many communication and social networks have power-law link distributions, containing a few nodes that have a very high degree and many with low degree. The high connectivity nodes play the important role of hubs in communication and networking, a fact that can be exploited when designing efficient search algorithms. We introduce a number of local search strategies that utilize high degree nodes in power-law graphs and that have costs scaling sublinearly with the size of the graph. We also demonstrate the utility of these strategies on the GNUTELLA peer-to-peer network.

DOI: 10.1103/PhysRevE.64.046135 PACS number(s): 89.75.Fb, 05.50.+q, 05.40.−a, 89.75.Da

I. INTRODUCTION

A number of large distributed systems, ranging from social [1] to communication [2] to biological networks [3] display a power-law distribution in their node degree. This distribution reflects the existence of a few nodes with very high degree and many with low degree, a feature not found in standard random graphs [4]. A large-scale illustration of such a network is given by the AT&T call graph. A call graph is a graph representation of telephone traffic on a given day in which nodes represent people and links the phone calls among them. As shown by [1], the out-link degree distribution for a massive graph of telephone calls between individuals has a clean power-law form with an exponent of approximately 2.1. The same distribution is obtained for the case of in links. This power law in the link distribution reflects the presence of central individuals who interact with many others on a daily basis and play a key role in relaying information.

While recent work has concentrated on the properties of these power-law networks and how they are dynamically generated [5–7], there remains the interesting problem of finding efficient algorithms for searching within these particular kinds of graphs. Recently, Kleinberg [8] studied search algorithms in a graph where nodes are placed on a two-dimensional (2D) lattice and each node has a fixed number of links whose placement is correlated with lattice distance to the other nodes. Under a specific form of the correlation, an algorithm with knowledge of the target's location can find the target in polylogarithmic time.

In the most general distributed search context however, one may have very little information about the location of the target. Increasingly a number of pervasive electronic networks, both wired and wireless, make geographic location less relevant. A particularly interesting example is provided by the recent emergence of peer-to-peer networks, which have gained enormous popularity with users wanting to share their computer files. In such networks, the name of the target file may be known, but due to the network's *ad hoc* nature, the node holding the file is not known until a real-time search is performed. In contrast to the scenario considered by Kleinberg, there is no global information about the position of the target, and hence it is not possible to determine whether a step is a move towards or away from the target. One simple way to locate files, implemented by NAPSTER, is to use a central server that contains an index of all the files every node is sharing as they join the network. This is the equivalent of having a giant white pages for the entire United States. Such directories now exist online, and have in a sense reduced the need to find people by passing messages. But for various reasons, including privacy and copyright issues, in a peer-to-peer network it is not always desirable to have a central server.

File-sharing systems that do not have a central server include GNUTELLA and FREENET. Files are found by forwarding queries to one's neighbors until the target is found. Recent measurements of GNUTELLA networks [9] and simulated FREENET networks [10] show that they have power-law degree distributions. In this paper, we propose a number of message-passing algorithms that can be efficiently used to search through power-law networks such as GNUTELLA. Like the networks that they are designed for, these algorithms are completely decentralized and exploit the power-law link distribution in the node degree. The algorithms use local information such as the identities and connectedness of a node's neighbors, and its neighbors' neighbors, but not the target's global position. We demonstrate that our search algorithms work well on real GNUTELLA networks, scale sublinearly with the number of nodes, and may help reduce the network search traffic that tends to cripple such networks.

The paper is organized as follows. In Sec. II, we present analytical results on message passing in power-law graphs, followed by simulation results in Sec. III. Section IV compares the results with Poisson random graphs. In Sec. V we consider the application of our algorithms to GNUTELLA, and Sec. VI concludes.

*Email address: ladamic@hpl.hp.com
†Email address: lukose@hpl.hp.com
‡Email address: amit8@stanford.edu
§Email address: huberman@hpl.hp.com

II. SEARCH IN POWER-LAW RANDOM GRAPHS

In this section we use the generating function formalism introduced by Newman [7] for graphs with arbitrary degree distributions to analytically characterize search-cost scaling in power-law graphs.

A. Random Walk Search

Let $G_0(x)$ be the generating function for the distribution of the vertex degrees k. Then

$$G_0(x)=\sum_0^\infty p_k x^k, \tag{1}$$

where p_k is the probability that a randomly chosen vertex on the graph has degree k.

For a graph with a power-law distribution with exponent τ, minimum degree $k=1$ and an abrupt cutoff at $m=k_{max}$, the generating function is given by

$$G_0(x)=c\sum_1^m k^{-\tau}x^k \tag{2}$$

with c a normalization constant that depends on m and τ to satisfy the normalization requirement

$$G_0(1)=c\sum_1^m k^{-\tau}=1. \tag{3}$$

The average degree of a randomly chosen vertex is given by

$$\langle k\rangle=\sum_1^m kp_k=G_0'(1). \tag{4}$$

Note that the average degree of a vertex chosen at random and one arrived at by following a random edge are different. A random edge arrives at a vertex with probability proportional to the degree of the vertex, i.e., $p'(k)\sim kp_k$. The correctly normalized distribution is given by

$$\frac{\sum_k kp_k x^k}{\sum_k kp_k}=x\frac{G_0'(x)}{G_0'(1)}. \tag{5}$$

If we want to consider the number of outgoing edges from the vertex we arrived at, but not include the edge we just came on, we need to divide by one power of x. Hence the number of new neighbors encountered on each step of a random walk is given by the generating function

$$G_1(x)=\frac{G_0'(x)}{G_0'(1)}, \tag{6}$$

where $G_0'(1)$ is the average degree of a randomly chosen vertex as mentioned previously.

In real social networks, it is reasonable that one one would have at least some knowledge of one's friends' friends. Hence, we now compute the distribution of second neighbors. The probability that any of the second neighbors connect to any of the first neighbors or to one another goes as N^{-1} and can be ignored in the limit of large N. Therefore, the distribution of the second neighbors of the original randomly chosen vertex is determined by

$$\sum_k p_k[G_1(x)]^k=G_0(G_1(x)). \tag{7}$$

It follows that the average number of second neighbors is given by

$$z_{2A}=\left[\frac{\partial}{\partial x}G_0(G_1(x))\right]_{x=1}=G_0'(1)G_1'(1). \tag{8}$$

Similarly, if the original vertex was not chosen at random, but arrived at by following a random edge, then the number of second neighbors would be given by

$$z_{2B}=\left[\frac{\partial}{\partial x}G_1(G_1(x))\right]_{x=1}=[G_1'(1)]^2. \tag{9}$$

In both Eqs. (8) and (9) the fact that $G_1(1)=1$ was used.

Both these expressions depend on the values $G_0'(1)$ and $G_1'(1)$ so we calculate those for given τ and m. For simplicity and relevance to most real-world networks of interest we assume $2<\tau<3$,

$$G_0'(1)=\sum_1^m ck^{1-\tau}\sim\int_1^m x^{\tau-1}dx=\frac{1}{\tau-2}(1-m^{2-\tau}), \tag{10}$$

$$G_1'(1)=\frac{1}{G_0'(1)}\frac{\partial}{\partial x}\sum_1^m ck^{1-\tau}x^{k-1} \tag{11}$$

$$=\frac{1}{G_0'(1)}\sum_2^m ck^{1-\tau}(k-1)x^{k-2} \tag{12}$$

$$\sim\frac{1}{G_0'(1)}\frac{m^{3-\tau}(\tau-2)-2^{2-\tau}(\tau-1)+m^{2-\tau}(3-\tau)}{(\tau-2)(3-\tau)} \tag{13}$$

for large cutoff values m. Now we impose the cutoff of Aiello *et al.* [1] at $m\sim N^{1/\tau}$. Since m scales with the size of the graph N and for $2<\tau<3$ the exponent $2-\tau$ is negative, we can neglect terms constant in m. This leaves

$$G_1'(1)=\frac{1}{G_0'(1)}\frac{m^{3-\tau}}{(3-\tau)}. \tag{14}$$

Substituting into Eq. (8) (the starting node is chosen at random) we obtain

$$z_{2A}=G_0'(1)G_1'(1)\sim m^{3-\tau}. \tag{15}$$

We can also derive z_{2B}, the number of second neighbors encountered as one is doing a random walk on the graph,

$$z_{2B}=[G_1'(1)]^2=\left[\frac{\tau-2}{1-m^{2-\tau}}\frac{m^{3-\tau}}{3-\tau}\right]^2. \quad (16)$$

Letting $m\sim N^{1/\tau}$ as above, we obtain

$$z_{2B}\sim N^{2(3/\tau-1)}. \quad (17)$$

Thus, as the random walk along edges proceeds node to node, each node reveals more of the graph since it has information not only about itself, but also of its neighborhood. The search cost s is defined as the number of steps until approximately the whole graph is revealed so that $s\sim N/z_{2B}$, or

$$s\sim N^{3(1-2/\tau)}. \quad (18)$$

In the limit $\tau\rightarrow 2$, Eq. (16) becomes

$$z_{2B}\sim\frac{N}{\ln^2(N)} \quad (19)$$

and the scaling of the number of steps required is

$$s\sim\ln^2(N). \quad (20)$$

B. Search utilizing high degree nodes

Random walks in power-law networks naturally gravitate towards the high degree nodes, but an even better scaling is achieved by intentionally choosing high degree nodes. For τ sufficiently close to 2 one can walk down the degree sequence, visiting the node with the highest degree, followed by a node of the next highest degree, etc. Let $m-a$ be the degree of the last node we need to visit in order to scan a certain fraction of the graph. We make the self-consistent assumption that $a\ll m$, i.e., the degree of the node has not dropped too much by the time we have scanned a fraction of the graph. Then the number of first neighbors scanned is given by

$$z_{1D}=\int_{m-a}^{m}Nk^{1-\tau}dk\sim Nam^{1-\tau}. \quad (21)$$

The number of nodes having degree between $m-a$ and m, or equivalently, the number of steps taken is given by $\int_{m-a}^{m}k^{-\tau}\sim a$. The number of second neighbors when one follows the degree sequence is given by

$$z_{1D}*G_1'(1)\sim Nam^{2(2-\tau)}, \quad (22)$$

which gives the number of steps required as

$$s\sim m^{2(\tau-2)}\sim N^{2-4/\tau}. \quad (23)$$

We now consider when and why it is possible to go down the degree sequence. We start with the fact that the original degree distribution is a power law

$$p(x)=\left(\sum_1^m x^{-\tau}\right)^{-1}x^{-\tau}, \quad (24)$$

where $m=N^{1/\tau}$ is the maximum degree. A node chosen by following a random link in the graph will have its remaining outgoing edges distributed according to

$$p'(x)=\left[\sum_0^{m-1}(x+1)^{(1-\tau)}\right]^{-1}(x+1)^{(1-\tau)}. \quad (25)$$

At each step one can choose the highest degree node among the n neighbors. The expected number of the outgoing edges of that node can be computed as follows. In general, the cumulative distribution (CDF) $P_{max}(x,n)$ of the maximum of n random variables can be expressed in terms of the CDF $P(x)=\int_0^x p(x')dx'$ of those random variables: $P_{max}(x,n)=P(x)^n$. This yields

$$p'_{max}(x,n)=n(1+x)^{1-\tau}(\tau-2)[1-(x+1)^{2-\tau}]^{n-1}\times(1-N^{2/\tau-1})^{-n} \quad (26)$$

for the distribution of the number of links the richest neighbor among n neighbors has.

Finally, the expected degree of the richest node among n is given by

$$E[x_{max}(n)]=\sum_0^{m-1}xp'_{max}(x,n). \quad (27)$$

We numerically integrated the above equation to derive the ratio between the degree of a node and the expected degree of its richest neighbor. The ratio is plotted in Fig. 1. For a range of exponents and node degrees, the expected degree of the richest neighbor is higher than the degree of the node itself. However, eventually (the precise point depends strongly on the power-law exponent), the probability of finding an even higher degree node in a neighborhood of a very high degree node starts falling.

What this means is that one can approximately follow the degree sequence across the entire graph for a sufficiently small graph or one with a power-law exponent close to 2 ($2.0<\tau<2.3$). At each step one chooses a node with a degree higher than the current node, quickly finding the highest degree node. Once the highest degree node has been visited, it will be avoided, and a node of approximately second highest degree will be chosen. Effectively, after a short initial climb, one goes down the degree sequence. This is the most efficient way to do this kind of sequential search, visiting highest degree nodes in sequence.

III. SIMULATIONS

We used simulations of a random network with a power-law link distribution of $\tau=2.1$ to validate our analytical results. As in the analysis above, a simple cutoff at $m\sim N^{1/\tau}$ was imposed. The expected number of nodes among N having exactly the cutoff degree is 1. No nodes of degree higher than the cutoff are added to the graph. In real-world graphs

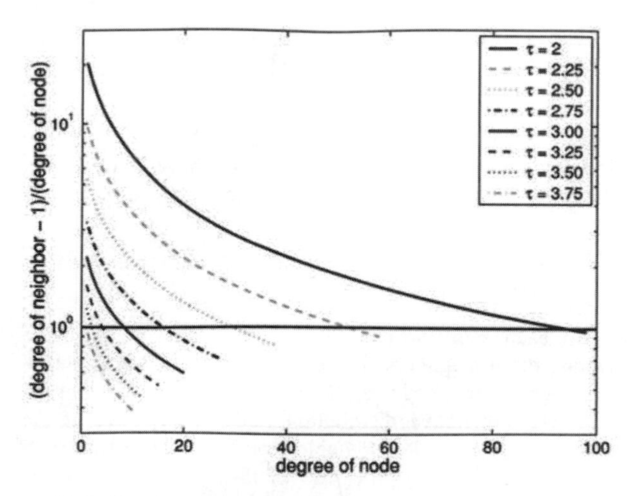

FIG. 1. Ratio r (the expected degree of the richest neighbor of a node whose degree is n divided by n) vs n for τ (top to bottom) =2.0, 2.25, 2.5, 2.75, 3.00, 3.25, 3.50, and 3.75. Each curve extends to the cutoff imposed for a 10 000 node graph with the particular exponent.

one, of course, does observe nodes of degree higher than this imposed cutoff, so that our simulations become a worse case scenario. Once the graph is generated, the largest connected component (LCC) is extracted, that is the largest subset of nodes such that any node can be reached from any other node. For $2<\tau<3.48$ a giant connected component exists [1], and all our measurements are performed on the LCC. We observe that the LCC contains the majority of the nodes of the original graph and most of the links as well. The link distribution of the LCC is nearly identical to that of the original graph with a slightly smaller number of 1 and 2 degree nodes.

Next we apply our message-passing algorithm to the network. Two nodes, the source and the target, are selected at random. At each time step the node that has the message passes it on to one of its neighbors. The process ends when the message is passed on to a neighbor of the target. Since each node knows the identity of all of its neighbors, it can pass the message directly to the target if the target happens to be one of it's neighbors. The process is analogous to performing a random walk on a graph, where each node is "visited" as it receives the message.

There are several variants of the algorithm, depending on the strategy and the amount of local information available.

(1) The node can pass the message onto one of its neighbors at random or it can avoid passing it on to a node that has already seen the message.

(2) If the node knows the degrees of its neighbors, it can choose to pass the message onto the neighbor with the most neighbors.

(3) The node may know only its neighbors or it may know who its neighbors' neighbors are. In the latter case it would pass the message onto a neighbor of the target.

In order to avoid passing the message to a node that has already seen the message, the message itself must be signed by the nodes as they receive the message. Further, if a node has passed the message and finds that all of its neighbors are already on the list, it puts a special mark next to its name, which means that it is unable to pass the message onto any new node. This is equivalent to marking nodes as follows.

White. Node has not been visited.

Gray. Node has been visited, but all its neighbors have not been visited.

Black. Node and all its neighbors have been visited already.

Here we compare two strategies. The first performs a random walk, where only retracing the last step is disallowed. In the message passing scenario, this means that if Bob just received a message from Jane, he would not return the message to Jane if he could pass it to someone else. The second strategy is a self-avoiding walk that prefers high degree nodes to low degree ones. In each case both the first and second neighbors are scanned at each step.

Figure 2(a) shows the scaling of the average search time with the size of the graph for the two strategies. The scaling (exponent 0.79 for the random walk and 0.70 for the high degree strategy) is not as favorable as in the analytic results derived above (0.14 for the random walk and 0.1 for the high degree strategy when $\tau=2.1$) .

Consider, on the other hand, the number of steps it takes to cover half the graph. For this measure we observe a scaling that is much closer to the ideal. As shown in Figure 2(b), the cover time scales as $N^{0.37}$ for the random walk strategy vs $N^{0.15}$ from Eq. (18). Similarly, the high degree strategy cover time scales as $N^{0.24}$ vs $N^{0.1}$ in Eq. (23).

The difference in the value of the scaling exponents of the cover time and average search time implies that a majority of nodes can be found very efficiently, but others demand high search costs. As Figure 2(c) shows, a large portion of the 10 000 node graph is covered within the first few steps, but some nodes take as many steps or more to find as there are nodes in total. For example, the high degree seeking strategy finds about 50% of the nodes within the first 10 steps (meaning that it would take about $10+2=12$ hops to reach 50% of the graph). However, the skewness of the search time distribution brings the average number of steps needed to 217.

Some nodes take a long time to find because the random walk, after a brief initial period of exploring fresh nodes, tends to revisit nodes. It is a well-known result that the stationary distribution of a random walk on an undirected graph is simply proportional to the distribution of links emanating from a node. Thus, nodes with high degree are often revisited in a walk.

A high degree seeking random walk is an improvement over the random walk, but still cannot avoid retracing its steps. Figure 2(d) shows the color of nodes visited on such a walk for a $N=1000$ node power-law graph with exponent 2.1 and an abrupt cutoff at $N^{1/2.1}$. The number of nodes of each color encountered in 50-step segments is recorded in the bar for that time period. We observe that the self-avoiding strategy is somewhat effective, with the total number of steps needed to cover the graph about 13 times smaller than the pure random walk case, and the fraction of visits to gray and black nodes is significantly reduced.

Although the revisiting of nodes modifies the scaling be-

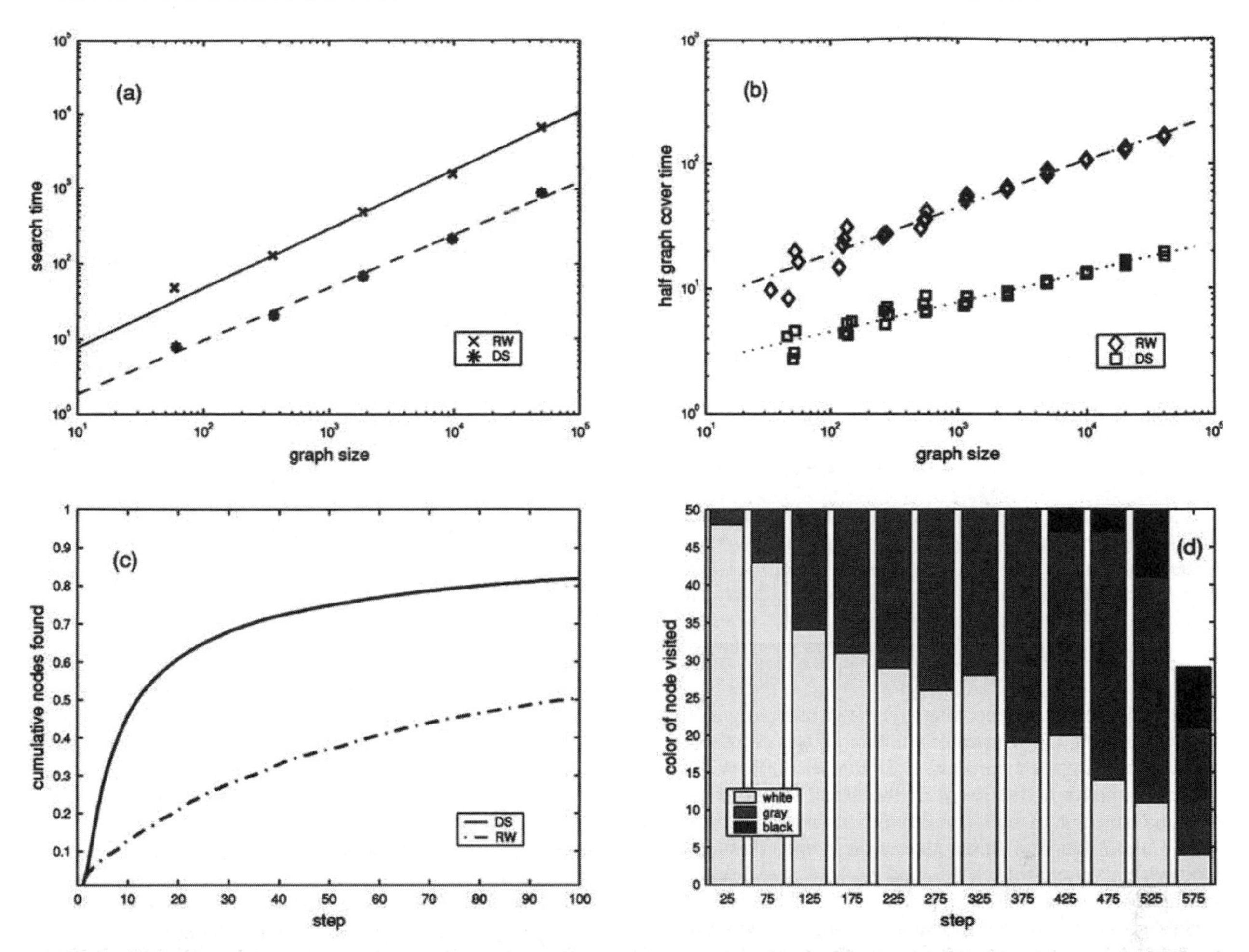

FIG. 2. (a) Scaling of the average node-to-node search cost in a random power-law graph with exponent 2.1, for random walk (RW) and high-degree seeking (DS) strategies. The solid line is a fitted scaling exponent of 0.79 for the RW strategy and the dashed is an exponent of 0.70 for the DS strategy. (b) The observed and fitted scaling for half graph cover times for the RW and DS strategies. The fits are to scaling exponents of 0.37 and 0.24, respectively. (c) Cumulative distribution of nodes seen vs the number of steps taken for the RW and DS strategies on a 10 000 node graph. (d) Bar graph of the color of nodes visited in DS search of a random 1000 node power-law graph with exponent 2.1. White represents a fresh node, gray represents a previously visited node that has some unvisited neighbors, and black represents nodes for which all neighbors have been previously visited.

havior, it is the form of the link distribution that is responsible for changes in the scaling. If nodes were uniformly linked, at every step the number of new nodes seen would be proportional to the number of unexplored nodes in the graph. The factor by which the search is slowed down through revisits would be independent of the size of the graph. Hence, revisiting alone does not account for the difference in scaling.

The reason why the simulated scaling exponents for these search algorithms do not follow the ideal is the same reason why power-law graphs are so well suited to search: the link distribution is extremely uneven. A large number of links point to only a small subset of high degree nodes. When a new node is visited, its links do not let us uniformly sample the graph, they preferentially lead to high degree nodes, which have likely been seen or visited in a previous step. This would not be true of a Poisson graph, where all the links are randomly distributed and hence all nodes have approximately the same degree. We will explore and contrast the search algorithm on a Poisson graph in the following section.

IV. COMPARISON WITH POISSON DISTRIBUTED GRAPHS

In a Poisson random graph with N nodes and z edges, the probability $p=z/N$ of an edge between any two nodes is the same for all nodes. The generating function $G_0(x)$ is given by [7]

$$G_0(x)=e^{z(x-1)}. \tag{28}$$

In this special case $G_0(x)=G_1(x)$, so that the distribution of outgoing edges of a node is the same whether one arrives at the vertex by following a link or picks the node at random.

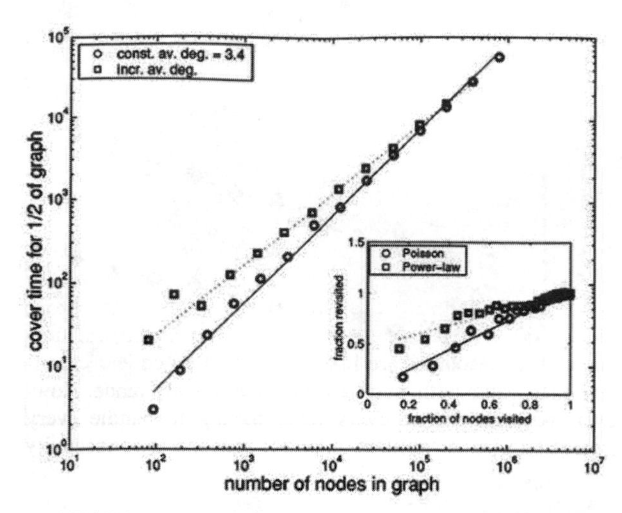

FIG. 3. Squares are scaling of cover time for 1/2 of the graph for a Poisson graph with a constant average degree/node (with fit to a scaling exponent of 1.0). Circles are the scaling for Poisson graphs with the same average degree/node as a power-law graph with exponent 2.1 (with fit to a scaling exponent of 0.85). The inset compares revisitation between search on Poisson vs power-law graphs, as discussed in the text.

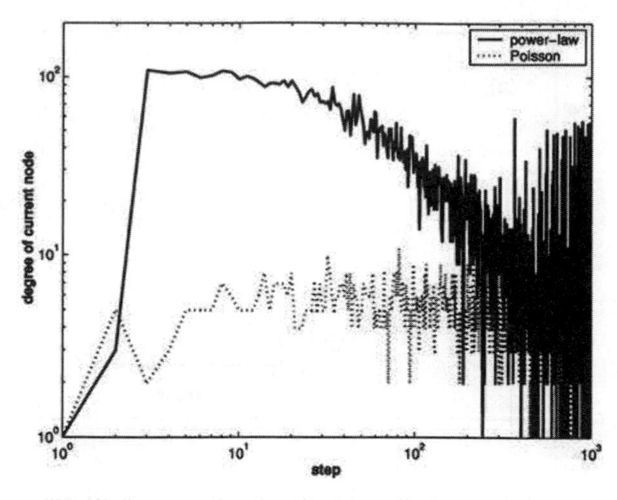

FIG. 4. Degrees of nodes visited in a single search for power law and Poisson graphs of 10 000 nodes.

This makes analysis search in a Poisson random graph particularly simple. The expected number of new links encountered at each step is a constant p. So that the number of steps needed to cover a fraction c of the graph is $s=cN/p$. If p remains constant as the size of the graph increases, the cover time scales linearly with the size of the graph. This has been verified via simulation of the random walk search as shown in Fig. 3.

In our simulations the probability p grows slowly towards its asymptotic value as the size of the graph is increased because of the particular choice of cutoff at $m \sim N^{(1/\tau)}$ for the power-law link distribution. We generated Poisson graphs with the same number of nodes and links for comparison. Within this range of graph sizes, growth in the average number of links per node appears as $N^{0.6}$, making the average number of second neighbors scale as $N^{0.15}$. This means that the scaling of the cover time scales as $N^{0.85}$, as shown in Fig. 3. Note how well the simulation results match the analytical expression. This is because nodes can be approximately sampled in an even fashion by following links.

The reason why the cover time for the Poisson graph matches the analytical prediction and the power-law graph does not is illustrated in Fig. 3 (inset). If links were approximately evenly distributed among the nodes, then if at one point in the search 50% of the graph has already been visited, one would expect to revisit previously seen nodes about 50% of the time. This is indeed the case for the Poisson graph. However, for the power-law graph, when 50% of the graph has been visited, nodes are revisited about 80% of the time, which implies that the same high degree nodes are being revisited before new low degree ones. It is this bias that accounts for the discrepancy between the analytic scaling and the simulated results in the power-law case.

However, even the simulated $N^{0.35}$ scaling for a random, minimally self-avoiding strategy on the power-law graph out performs the ideal $N^{0.85}$ scaling for the Poisson graph. It is also important to note that the the high degree node seeking strategy has a much greater success in the power-law graph because it relies heavily on the fact that the number of links per node varies considerably from node to node. To illustrate this point, we executed the high degree seeking strategy on two graphs, Poisson and power law, with the same number of nodes, and the same exponent $\tau=2$. In the Poisson graph, the variance in the number of links was much smaller, making the high degree node seeking strategy comparatively ineffective as shown in Fig. 4.

In the power-law graph we can start from a randomly chosen node. In this case the starting node has only one link, but two steps later we find ourselves at a node with the highest degree. From there, one approximately follows the degree sequence, that is, the node richest in links, followed by the second richest node, etc. The strategy has allowed us to scan the maximum number of nodes in the minimum number of steps. In comparison, the maximum degree node of the exponential graph is 11, and it is reached only on the 81st step. Even though the two graphs have a comparable number of nodes and edges, the exponential graph does not lend itself to quick search.

V. GNUTELLA

GNUTELLA is a peer-to-peer file-sharing system that treats all client nodes as functionally equivalent and lacks a central server that can store file location information. This is advantageous because it presents no central point of failure. The obvious disadvantage is that the location of files is unknown. When a user wants to download a file, she sends a query to all the nodes within a neighborhood of size ttl, the time to live assigned to the query. Every node passes on the query to all of its neighbors and decrements the ttl by one. In this way, all nodes within a given radius of the requesting node will be queried for the file, and those who have matching files will send back positive answers.

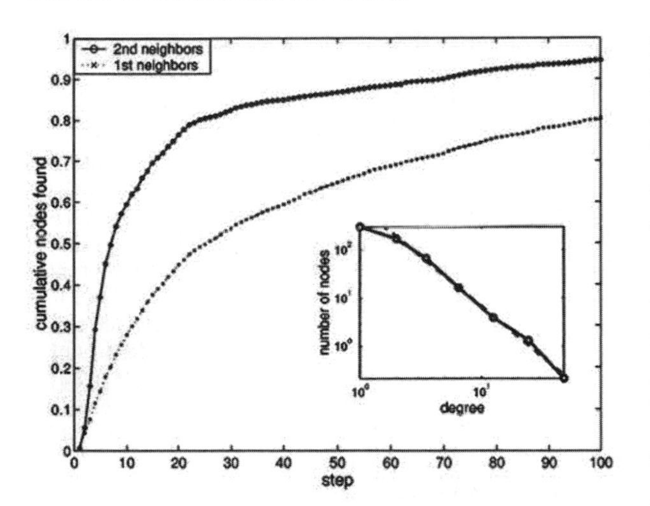

FIG. 5. Cumulative number of nodes found at each step in the GNUTELLA network. The inset shows the measured link distribution of the real GNUTELLA network used in the search simulations and a fit to a power-law link distribution with exponent 2.

This broadcast method will find the target file quickly, given that it is located within a radius of ttl. However, broadcasting is extremely costly in terms of bandwidth. Every node must process queries of all the nodes within a given ttl radius. In essence, if one wants to query a constant fraction of the network, say 50%, as the network grows, each node and network edge will be handling query traffic that is proportional to the total number of nodes in the network.

Such a search strategy does not scale well. As query traffic increases linearly with the size of GNUTELLA graph, nodes become overloaded as was shown in a recent study by the Clip2 company [9]. 56k modems are unable to handle more than 20 queries a second, a threshold easily exceeded by a network of about 1000 nodes. With the 56k nodes failing, the network becomes fragmented, allowing users to query only small section of the network.

The search algorithms described in the previous sections may help ameliorate this problem. Instead of broadcasting a query to a large fraction of the network, a query is only passed onto one node at each step. The search algorithms are likely to be effective because the GNUTELLA network has a power-law connectivity distribution as shown in Fig. 5 (inset).

Typically, a GNUTELLA client wishing to join the network must find the IP address of an initial node to connect to. Currently, *ad hoc* lists of "good" GNUTELLA clients exist [9]. It is reasonable to suppose that this *ad hoc* method of growth would bias new nodes to connect preferentially to nodes that are already fairly well connected, since these nodes are more likely to be "well known." Based on models of graph growth [5,6] where the "rich get richer," the power-law connectivity of *ad hoc* peer-to-peer networks may be a fairly general topological feature.

By passing the query to every single node in the network, the GNUTELLA algorithm fails to take advantage of the connectivity distribution. To implement our algorithm the GNUTELLA clients must be modified to keep lists of the files stored by their first and second neighbors have.[1] This information must be passed at least once when a new node joins the network, and it may be necessary to periodically update the information depending on the typical lifetime of nodes in the network. Instead of passing the query to every node, queries are only passed along to the highest degree nodes. The IP numbers of the nodes already queried are appended to the query, and they are avoided.

The modified algorithm places an additional cost on every node, that of keeping track of the filenames of its neighbors' files. Since network connections saturated by query traffic are a major weakness in GNUTELLA, and since computational and storage resources are likely to remain much less expensive than bandwidth, such a tradeoff is readily made. However, now instead of every node having to handle every query, queries are routed only through high connectivity nodes. Since nodes can select the number of connections that they allow, high degree nodes are presumably high bandwidth nodes that can handle the query traffic. The network has in effect created local directories valid within a two link radius. It is resilient to attack because of the lack of a central server. As for power-law networks in general [11], the network is more resilient than Poisson graphs to random node failure, but less resilient to attacks on the high degree nodes.

Figure 5 shows the success of the high degree seeking algorithm on the GNUTELLA network. We simulated the search algorithm on a crawl by Clip2 company of the actual GNUTELLA network of approximately 700 nodes. Assuming that every file is stored on only one node, 50% of the files can be found in eight steps or less. Furthermore, if the file one is seeking is present on multiple nodes, the search will be even faster.

To summarize, the power-law nature of the GNUTELLA graph means that these search algorithms can be effective. As the number of nodes increases, the (already small) number of nodes that will need to be queried increases sublinearly. As long as the high degree nodes are able to carry the traffic, the GNUTELLA network's performance and scalability may improve by using these search strategies.

We also note that even if a network of clients was not power law, a search strategy that possesses knowledge of its neighbors of a network radius greater than two could still improve search. For example, in the Poisson case, the algorithm could attempt to hold more than the contents of a node's first and second neighbors. How efficient this algorithm is on arbitrary network topologies is the subject of future work. Here we have analyzed the naturally occurring power-law topology.

VI. CONCLUSION

In this paper we have shown that local search strategies in power-law graphs have search costs that scale sublinearly

[1] This idea has already been implemented by Clip2 company in a limited way. 56k modem nodes attach to a high bandwidth Reflector node that stores the filenames of the 56k nodes and handles queries on their behalf.

with the size of the graph, a fact that makes them very appealing when dealing with large networks. The most favorable scaling was obtained by using strategies that preferentially utilize the high connectivity nodes in these power-law networks. We also established the utility of these strategies for searching on the GNUTELLA peer-to-peer network.

It may not be coincidental that several large networks are structured in a way that naturally facilitates search. Rather, we find it likely that these networks could have evolved to facilitate search and information distribution. Networks where locating and distributing information, without perfect global information, plays a vital role tend to be power law with exponents favorable to local search.

For example, large social networks, such as the AT&T call graph and the collaboration graph of film actors, have exponents in the range ($\tau=2.1-2.3$) that according to our analysis makes them especially suitable for searching using our simple, local algorithms. Being able to reach remote nodes by following intermediate links allows communication systems and people to get to the resources they need and distribute information within these informal networks. At the social level, our analysis supports the hypothesis that highly connected individuals do a great deal to improve the effectiveness of social networks in terms of access to relevant resources [12].

Furthermore, it has been shown that the Internet backbone has a power-law distribution with exponent values between 2.15 and 2.2 [2], and web page hyperlinks have an exponent of 2.1 [5]. While in the Internet there are other strategies for finding nodes, such as routing tables and search engines, one observes that our proposed strategy is partially used in these systems as well. Packets are routed through highly connected nodes, and users searching for information on the Web turn to highly connected nodes, such as directories and search engines, which can bring them to their desired destinations.

On the other hand, a system such as the power grid of the western United States, which does not serve as a message passing network, has an exponent $\tau \sim 4$ [5]. It would be fairly difficult to pass messages in such a network without a relatively large amount of global information.

ACKNOWLEDGMENT

We would like to thank the Clip2 company for the use of their GNUTELLA crawl data.

[1] W. Aiello, F. Chung, and L. Lu, in *STOC '00, Proceedings of the Thirty-second Annual ACM Symposium on Theory of Computing*, edited by F. Yao (ACM, New York, 2000), pp. 171–180.

[2] M. Faloutsos, P. Faloutsos, and C. Faloutsos, in *SIGCOMM '99, Proceedings of the Conference on Applications, Technologies, Architectures, and Protocols for Computer Communication*, edited by L. Chapin (ACM, New York, 1999), pp. 251–262.

[3] H. Jeong, B. Tombor, R. Albert, Z.N. Oltvai, and A.-L. Barabasi, Nature (London) **407**, 651 (2000).

[4] P. Erdös and A. Rényi, Publ. Math. Inst. Hung. Acad. Sci., **5**, 17 (1960).

[5] A.-L. Barabasi and R. Albert, Science **286**, 509 (1999).

[6] B.A. Huberman and L.A. Adamic, Nature (London) **401**, 131 (1999).

[7] M.E.J. Newman, S.H. Strogatz, and D.J. Watts, Phys. Rev. E **64**, 026118 (2001).

[8] J.M. Kleinberg, Nature (London) **406**, 845 (2000).

[9] Clip2 Company, Gnutella. http://www.clip2.com/gnutella.html

[10] T. Hong, in *Peer-to-Peer: Harnessing the Benefits of a Disruptive Technology*, edited by Andy Oram (O'Reilley, Sebastopol, CA, 2001), Chap. 14, pp. 203–241.

[11] R. Albert, H. Jeong, and A.-L. Barabasi, Nature (London) **406**, 378 (2000).

[12] M. Gladwell, *The Tipping Point: How Little Things Can Make a Big Difference* (Little Brown, New York, 2000).

Navigation in a small world

It is easier to find short chains between points in some networks than others.

The small-world phenomenon — the principle that most of us are linked by short chains of acquaintances — was first investigated as a question in sociology[1,2] and is a feature of a range of networks arising in nature and technology[3–5]. Experimental study of the phenomenon[1] revealed that it has two fundamental components: first, such short chains are ubiquitous, and second, individuals operating with purely local information are very adept at finding these chains. The first issue has been analysed[2–4], and here I investigate the second by modelling how individuals can find short chains in a large social network.

I have found that the cues needed for discovering short chains emerge in a very simple network model. This model is based on early experiments[1], in which source individuals in Nebraska attempted to transmit a letter to a target in Massachusetts, with the letter being forwarded at each step to someone the holder knew on a first-name basis. The networks underlying the model follow the 'small-world' paradigm[3]: they are rich in structured short-range connections and have a few random long-range connections.

Long-range connections are added to a two-dimensional lattice controlled by a clustering exponent, α, that determines the probability of a connection between two nodes as a function of their lattice distance (Fig. 1a). Decentralized algorithms are studied for transmitting a message: at each step, the holder of the message must pass it across one of its short- or long-range connections; crucially, this current holder does not know the long-range connections of nodes that have not touched the message. The primary figure of merit for such an algorithm is its expected delivery time T, which represents the expected number of steps needed to forward a message between a random source and target in a network generated according to the model. It is crucial to constrain the algorithm to use only local information — with global knowledge of all connections in the network, the shortest chain can be found very simply[6].

A characteristic feature of small-world networks is that their diameter is exponentially smaller than their size, being bounded by a polynomial in $\log N$, where N is the number of nodes. In other words, there is always a very short path between any two nodes. This does not imply, however, that a decentralized algorithm will be able to discover such short paths. My central finding is that there is in fact a unique value of the exponent α at which this is possible.

When $\alpha=2$, so that long-range connections follow an inverse-square distribution, there is a decentralized algorithm that achieves a very rapid delivery time; T is bounded by a function proportional to $(\log N)^2$. The algorithm achieving this bound is a 'greedy' heuristic: each message holder forwards the message across a connection that brings it as close as possible to the target in lattice distance. Moreover, $\alpha=2$ is the only exponent at which any decentralized algorithm can achieve a delivery time bounded by any polynomial in $\log N$: for every other exponent, an asymptotically much larger delivery time is required, regardless of the algorithm employed (Fig. 1b).

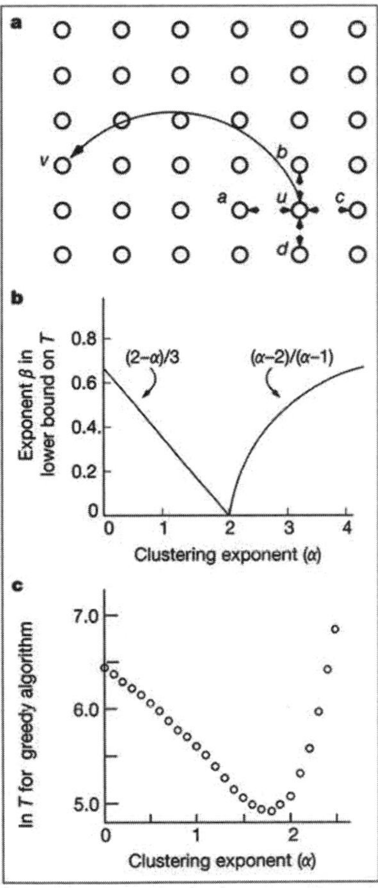

Figure 1 The navigability of small-world networks. **a**, The network model is derived from an $n\times n$ lattice. Each node, u, has a short-range connection to its nearest neighbours (a, b, c and d) and a long-range connection to a randomly chosen node, where node v is selected with probability proportional to $r^{-\alpha}$, where r is the lattice ('Manhattan') distance between u and v, and $\alpha\geq 0$ is a fixed clustering exponent. More generally, for $p,q\geq 1$, each node u has a short-range connection to all nodes within p lattice steps, and q long-range connections generated independently from a distribution with clustering exponent α. **b**, Lower bound from my characterization theorem: when $\alpha\neq 2$, the expected delivery time T of any decentralized algorithm satisfies $T\geq cn^{\beta}$, where $\beta=(2-\alpha)/3$ for $0\leq\alpha<2$ and $\beta=(\alpha-2)/(\alpha-1)$ for $\alpha>2$, and where c depends on α, p and q, but not n. **c**, Simulation of the greedy algorithm on a $20{,}000\times 20{,}000$ toroidal lattice, with random long-range connections as in **a**. Each data point is the average of 1,000 runs.

These results indicate that efficient navigability is a fundamental property of only some small-world structures. The results also generalize to d-dimensional lattices for any value of $d\geq 1$, with the critical value of the clustering exponent becoming $\alpha=d$. Simulations of the greedy algorithm yield results that are qualitatively consistent with the asymptotic analytical bounds (Fig. 1c).

In the areas of communication networks[7] and neuroanatomy[8], the issue of routing without a global network organization has been considered; also in social psychology and information foraging some of the cues that individuals use to construct paths through a social network or hyperlinked environment have been discovered[9,10]. Although I have focused on a very clean model, I believe that a more general conclusion can be drawn for small-world networks — namely that the correlation between local structure and long-range connections provides critical cues for finding paths through the network.

When this correlation is near a critical threshold, the structure of the long-range connections forms a type of gradient that allows individuals to guide a message efficiently towards a target. As the correlation drops below this critical value and the social network becomes more homogeneous, these cues begin to disappear; in the limit, when long-range connections are generated uniformly at random, the result is a world in which short chains exist but individuals, faced with a disorienting array of social contacts, are unable to find them.

Jon M. Kleinberg
Department of Computer Science, Cornell University, Ithaca, New York 14853, USA

1. Milgram, S. *Psychol. Today* **1**, 61–67 (1967).
2. Kochen, M. (ed.) *The Small World* (Ablex, Norwood, NJ, 1989).
3. Watts, D. & Strogatz, S. *Nature* **393**, 440–442 (1998).
4. Albert, R. *et al. Nature* **401**, 130–131 (1999).
5. Adamic, L. in *Proc. 3rd European Conference on Digital Libraries* (eds Abiteboul, S. & Vercoustre, A.-M.) 443–452 (Springer Lecture Notes in Computer Science, Vol. 1696, Berlin, 1999).
6. Cormen, T., Leiserson, C. & Rivest, R. *Introduction to Algorithms* (McGraw-Hill, Boston, 1990).
7. Peleg, D. & Upfal, E. *J. Assoc. Comput. Machinery* **36**, 510–530 (1989).
8. Braitenberg, V. & Schüz, A. *Anatomy of the Cortex* (Springer, Berlin, 1991).
9. Killworth, P. & Bernard, H. *Social Networks* **1**, 159–192 (1978).
10. Pirolli, P. & Card, S. *Psychol. Rev.* **106**, 643–675 (1999).

Chapter Six
Outlook

In this section we close with some speculations about future directions of research on networks. We highlight here a few topics that we find of interest—certainly there are many others, and no doubt many yet to be discovered. We look forward to seeing what the future brings in this exciting field.

Communities within networks

Most of the models and analyses described in the papers reproduced in this volume view networks as essentially homogeneous, with all vertices being roughly equivalent and little large-scale structure. Common experience, however, suggests that this is unlikely to be a very good representation of the real world. In particular, we know from experience that many networks divide into "clusters" or "communities" of vertices, tightly knit groups with dense connections among their members. These groups might represent circles of friends in social networks (Granovetter 1995; Watts *et al.* 2002), functionally related genes or metabolites in a cellular network (Hartwell *et al.* 1999), or Web sites devoted to similar topics on the Web (Gibson *et al.* 1998; Flake *et al.* 2002). One current thread of research deals with the detection of these communities in network data. There are a variety of techniques from computer science and sociology that can be used to perform this task, such as spectral partitioning (Fiedler 1973; Pothen *et al.* 1990) or hierarchical clustering (Wasserman and Faust 1994; Scott 2000), but these are not ideally suited to the types of networks considered here, so a number of other techniques have been developed recently that appear to be more appropriate. The most widely used is probably the algorithm of Girvan and Newman (2002; Newman and Girvan 2004), but a considerable number of others have appeared in recent years (Flake *et al.* 2002; Tyler *et al.* 2003; Radicchi *et al.* 2004; Newman 2004b; Wu and Huberman 2004; Reichardt and Bornholdt 2004; Capocci *et al.* 2004; Fortunato *et al.* 2004). Applications of these and similar algorithms to network data have revealed that indeed many networks are divided into groups, which in many cases have clear structural or functional significance (Girvan and Newman 2002; Guimerà *et al.* 2003; Holme *et al.* 2003; Holme and Huss 2003; Tyler *et al.* 2003; Wilkinson and Huberman 2004; Gleiser and Danon 2003; Newman and Girvan 2004; Boguñá *et al.* 2003a; Radicchi *et al.* 2004; Palla *et al.* 2005).

Hierarchy in networks

The traditional view of a community assumes the existence of a group of nodes that have many links to each other, but only a few to outside nodes. This approach runs into problems, however, when we have nodes of very high degree in the network, as in the scale-free networks discussed in Chapter 4.3. Empirical data indicate that communities and skewed degree distributions are not mutually exclusive; they appear to coexist in a large number of networks, including metabolic networks (**Jeong *et al.* 2000; Wagner and Fell 2001**), protein interaction networks (Jeong *et al.* 2001; Wagner 2001), the World Wide Web (**Albert *et al.* 1999**), and even some social networks (Barabási *et al.* 2002). One way to reconcile the two properties is for a network to have a hierarchical structure (Barabási *et al.* 2001; Ravasz *et al.* 2002; Ravasz and Barabási 2003), whereby numerous small, highly connected modules are joined together into larger modules, which in turn combine into even larger ones, and so forth. Most of the network models described in the literature do not show this sort of hierarchy, including the random graph (**Solomonoff and Rapoport 1951; Erdős and Rényi 1960**), small-world (**Watts and Strogatz 1998**), and preferential attachment models (**Barabási and Albert 1999**), but recently several new models have been proposed that do, and that indeed possess simultaneously a power-law degree distribution and a strong group structure.

Hierarchical structure appears to be common in many real-world networks. Dorogovtsev *et al.* (2002) have observed that in a deterministic hierarchical network the local clustering coefficient C_i (see Eq. (4.8)), which is the fraction of pairs of a vertex's neighbors that are also neighbors of one another, depends on the degree k_i of the vertex in question as $C_i \sim 1/k_i$, and Ravasz *et al.* (2002; Ravasz and Barabási 2003) have argued that this property can be used as a general signature of hierarchies. Behavior of precisely this kind is seen in several large networks, including the World Wide Web, the network of movie actors, language networks, the Internet at the level of autonomous systems, and metabolic and protein networks (Ravasz *et al.* 2002; Ravasz and Barabási 2003; Eckmann and Moses 2002; Vázquez *et al.* 2002; Wuchty *et al.* 2003). By contrast, hierarchy is absent in networks with strong geographical constraints, such as the power grid or the Internet at the router level (Vázquez *et al.* 2002), possibly because of limitations on the lengths of edges. It is worth noting also that, while random graphs, small-world models, and preferential attachment models do not display hierarchical structure, some models with no obvious hierarchical structure still *do* give clustering coefficients going as $C_i \sim 1/k_i$ (**Bianconi and Barabási 2001b**; Kullmann and Kertesz 2001; Dorogovtsev *et al.* 2002; Klemm and Eguiluz 2002; Vázquez 2002; Ravasz and Barabási 2003; Newman 2003c).

Assortative mixing

Vertices in networks typically have various properties associated with them. For instance, individuals in a social network have age, income, race, nationality, and so forth. Pages on the Web have textual and visual content, location, topic, link patterns, and so forth. It seems very likely that these properties would affect where edges in the network fall, and indeed this appears to be the case: ties in social networks are found to depend strongly on race, for instance (Morris 1995; Moody 2001), while Web links depend on content (Gibson *et al.* 1998; Eckmann and Moses 2002). This phenomenon is called *assortative mixing*. Most types of assortative mixing apply only for the corresponding type of network—there is no mixing by page topic in social networks, since people are not pages with topics in the first place. However, there is one property that every vertex in every network has, namely degree. If vertices in a network that have high degree tend to be found most often connected to other high-degree vertices and similarly for low-degree vertices, then we say the network is assortatively mixed by degree. Conversely, if high-degree vertices preferentially associate with low-degree ones, the network is disassortative. Both of these patterns are seen in certain networks, and almost all networks appear to have some level of assortative mixing by degree.

Assortative mixing can be quantified by examining the joint degree distribution $P(k_1, k_2)$ of the degrees of vertices at the ends of edges in a network. In the simplest case, one can simply make a two-dimensional plot of this distribution, as done for instance by Maslov and Sneppen (2002; Maslov *et al.* 2004). A more succinct representation is that introduced by Pastor-Satorras *et al.* (2001; Vázquez *et al.* 2002), who measured the mean degree $\langle k_{\mathrm{nn}} \rangle$ of the neighbors of a vertex as a function of the degree k of that vertex. This quantity will be an increasing function of k for assortative networks and a decreasing one for disassortative networks. Simplifying one step further, one can simply calculate a correlation coefficient for the degrees of adjacent vertices in the network, as done by Newman (2002a, 2003b). This quantity will be positive for assortative networks and negative for disassortative ones.

An interesting result is that most social networks are found to be assortatively mixed by degree, while other types of networks, including technological and biological networks, seem mostly to be disassortative (Newman 2002a). It is not yet fully understood what the explanation for this observation is, although Maslov *et al.* (2004) have recently suggested that it may arise because in most networks there can be only a single edge between any pair of vertices. Another interesting observation is that neither random graphs nor networks created using the preferential attachment model of **Barabási and Albert (1999)** possess any degree correlations (Newman 2002a), suggesting that extensions or modifications of these models will be necessary if they are accurately to mimic the properties of real networks like social networks or the Web.

Identity, perception, and social networks

One of the most exciting findings in the recent flurry of work on networks is that certain features of networks appear to be universal, in the sense that they do not depend on many of the details of the particular system under investigation. For example, many networks, whether social, biological, or technological, display small-world and scale-free properties, and thus can be described, at least in part, by simple models like those described in this volume. But as appealing as they are, universal theories are necessarily limited in their explanatory power. After all, the vast variety of networked systems described in the preceding pages are clearly *not* the same in most respects; any model sufficiently general to model all or most of them must necessarily account only for their most generic features.

An important future direction for network research is therefore the identification not only of similarities between networks, but also their most important differences. Social networks in particular seem to exhibit a number of features that make them very different from other kinds of networks, and these differences are probably central to any scientific understanding of social processes—an outcome that has been a longstanding goal of mathematical sociology in general and social network analysis in particular. **Kleinberg (2000a**, 2000b) has highlighted one such difference in his work on network navigation that appears in Section 5.3—namely, that individuals in social networks can find short paths to others despite having only very limited information about the structure of the network. Subsequent work (Watts *et al.* 2002; Kleinberg 2002) has suggested that this property of network "searchability" depends on the ability of individuals to categorize themselves and each other into socially meaningful groups, that is, to construct notions of social identity. Without a means to estimate what Watts *et al.* call "social distance," individuals are unable to choose which of their network ties is most likely to move a message closer to some remote target. And by making individuals who are similar in one or more "social dimensions" (such as geography, occupation, etc.) more likely to interact, social distance also feeds back into the structure of the network, enabling messages to make successively more precise jumps as they near their target.

But social distance can also be deceiving. As the recent email versions of the small-world experiment (Dodds *et al.* 2003) have shown, individuals are more likely to forward a message when the message's intended recipient appears easier to reach. Thus message chains directed at such "easy" targets experience lower failure rates than those for "difficult" targets, with the result that many more of them reach their target. Differing perceptions of social distance, therefore, can generate large disparities in the *apparent* centrality (as measured by number of completed chains) of individuals in a network—even in the case where all individuals have roughly the same actual centrality. The overall conclusion is that notions of identity and social distance—both real and perceived—not only are unique to social networks (as opposed, say, to biological and technological networks) but are critical to the formation and understanding of these networks' other properties of interest. More generally, progress in understanding network-related processes, such as biological and social contagion, robustness, and collective behavior, will depend on determining which properties of networks are shared universally, and which are specific to certain networks. Remarkably little concrete progress has been made to date in this area, but if the science of networks is to have an impact in policy, business, and technology, then research that explicitly addresses the relationship between network properties and the behavior of networked systems will need to be pursued. We anticipate a great deal of exciting work in this area in the years ahead.

References

Adamic, L. A., 1999. The small world web. In *Lecture Notes in Computer Science*, vol. 1696, pp. 443–454. Springer, New York.

Adamic, L. A. and Huberman, B. A., 2000. Power-law distribution of the world wide web. *Science* **287**, 2115.

Adamic, L. A., Lukose, R. M., Puniyani, A. R., and Huberman, B. A., 2001. Search in power-law networks. *Phys. Rev. E* **64**, 046135.

Adler, J., 1991. Bootstrap percolation. *Physica A* **171**, 453–470.

Agrawal, H., 2002. Extreme self-organization in networks constructed from gene expression data. *Phys. Rev. Lett.* **89**, 268702.

Ahuja, R. K., Magnanti, T. L., and Orlin, J. B., 1993. *Network Flows: Theory, Algorithms, and Applications*. Prentice Hall, Upper Saddle River, NJ.

Aiello, W., Chung, F., and Lu, L., 2000. A random graph model for massive graphs. In *Proceedings of the 32nd Annual ACM Symposium on Theory of Computing*, pp. 171–180. Association of Computing Machinery, New York.

Aiello, W., Chung, F., and Lu, L., 2002. Random evolution of massive graphs. In J. Abello, P. M. Pardalos, and M. G. C. Resende (eds.), *Handbook of Massive Data Sets*, pp. 97–122. Kluwer, Dordrecht.

Alberich, R., Miro-Julia, J., and Rossello, F., 2002. Marvel Universe looks almost like a real social network. Preprint cond-mat/0202174.

Albert, R. and Barabási, A.-L., 2000. Topology of evolving networks: Local events and universality. *Phys. Rev. Lett.* **85**, 5234–5237.

Albert, R. and Barabási, A.-L., 2002. Statistical mechanics of complex networks. *Rev. Mod. Phys.* **74**, 47–97.

Albert, R., Jeong, H., and Barabási, A.-L., 1999. Diameter of the world-wide web. *Nature* **401**, 130–131.

Albert, R., Jeong, H., and Barabási, A.-L., 2000. Attack and error tolerance of complex networks. *Nature* **406**, 378–382.

Alberts, B., Johnson, A., Lewis, J., Raff, M., Roberts, K., and Walter, P., 2002. *Molecular Biology of the Cell*. Garland Publishing, New York, 4th ed.

Amaral, L. A. N., Scala, A., Barthélémy, M., and Stanley, H. E., 2000. Classes of small-world networks. *Proc. Natl. Acad. Sci. USA* **97**, 11149–11152.

Anderson, C., Wasserman, S., and Crouch, B., 1999. A p* primer: Logit models for social networks. *Social Networks* **21**, 37–66.

Anderson, R. M. and May, R. M., 1991. *Infectious Diseases of Humans*. Oxford University Press, Oxford.

ANTHONISSE, J. M., 1971. The rush in a directed graph. Technical Report BN 9/71, Stichting Mathematicsh Centrum, Amsterdam.

APIC, G., GOUGH, J., AND TEICHMANN, S. A., 2001. Domain combinations in archaeal, eubacterial and eukaryotic proteomes. *J. Mol. Biol.* **310**, 311–325.

BAILEY, N. T. J., 1975. *The Mathematical Theory of Infectious Diseases and Its Applications*. Hafner Press, New York.

BAK, P., 1996. *How Nature Works: The Science of Self-Organized Criticality*. Copernicus, New York.

BALL, F., MOLLISON, D., AND SCALIA-TOMBA, G., 1997. Epidemics with two levels of mixing. *Annals of Applied Probability* **7**, 46–89.

BALTHROP, J., FORREST, S., NEWMAN, M. E. J., AND WILLIAMSON, M. M., 2004. Technological networks and the spread of computer viruses. *Science* **304**, 527–529.

BANKS, D. L. AND CARLEY, K. M., 1996. Models for network evolution. *Journal of Mathematical Sociology* **21**, 173–196.

BARABÁSI, A.-L., 2002. *Linked: The New Science of Networks*. Perseus, Cambridge, MA.

BARABÁSI, A.-L. AND ALBERT, R., 1999. Emergence of scaling in random networks. *Science* **286**, 509–512.

BARABÁSI, A.-L., ALBERT, R., AND JEONG, H., 1999. Mean-field theory for scale-free random networks. *Physica A* **272**, 173–187.

BARABÁSI, A.-L., ALBERT, R., AND JEONG, H., 2000a. Scale-free characteristics of random networks: The topology of the World Wide Web. *Physica A* **281**, 69–77.

BARABÁSI, A.-L., ALBERT, R., JEONG, H., AND BIANCONI, G., 2000b. Power-law distribution of the World Wide Web. *Science* **287**, 2115a.

BARABÁSI, A.-L., RAVASZ, E., AND VICSEK, T., 2001. Deterministic scale-free networks. *Physica A* **299**, 559–564.

BARABÁSI, A.-L., JEONG, H., RAVASZ, E., NÉDA, Z., SCHUBERTS, A., AND VICSEK, T., 2002. Evolution of the social network of scientific collaborations. *Physica A* **311**, 590–614.

BARBOUR, A. D. AND REINERT, G., 2001. Small worlds. *Random Structures and Algorithms* **19**, 54–74.

BARRAT, A., 1999. Comment on 'Small-world networks: Evidence for crossover picture'. Preprint cond-mat/9903323.

BARRAT, A. AND WEIGT, M., 2000. On the properties of small-world networks. *Eur. Phys. J. B* **13**, 547–560.

BARTHÉLÉMY, M., 2003. Comment on 'Universal behavior of load distribution in scale-free networks'. *Phys. Rev. Lett.* **91**, 189803.

BARTHÉLÉMY, M. AND AMARAL, L. A. N., 1999a. Small-world networks: Evidence for a crossover picture. *Phys. Rev. Lett.* **82**, 3180–3183.

BARTHÉLÉMY, M. AND AMARAL, L. A. N., 1999b. Erratum: Small-world networks: Evidence for a crossover picture. *Phys. Rev. Lett.* **82**, 5180.

BEARMAN, P. S. AND PARIGI, P., 2004. Cloning headless frogs and other important matters: Conversation topics and network structure. *Social Forces* **83**, 535–557.

BEARMAN, P. S., MOODY, J., AND STOVEL, K., 2004. Chains of affection: The structure of adolescent romantic and sexual networks. *Am. J. Sociol.* **110**, 44–91.

BENDER, E. A. AND CANFIELD, E. R., 1978. The asymptotic number of labeled graphs with given degree sequences. *Journal of Combinatorial Theory A* **24**, 296–307.

BHARAT, K. AND HENZINGER, M. R., 1998. Improved algorithms for topic distillation in a hyperlinked environment. In *Proceedings of the 21st ACM SIGIR Conference on Research and Development in Information Retrieval*, pp. 104–111. Association of Computing Machinery, New York.

BIANCONI, G. AND BARABÁSI, A.-L., 2001a. Bose–Einstein condensation in complex networks. *Phys. Rev. Lett.* **86**, 5632–5635.

BIANCONI, G. AND BARABÁSI, A.-L., 2001b. Competition and multiscaling in evolving networks. *Europhys. Lett.* **54**, 436–442.

BIANCONI, G. AND CAPOCCI, A., 2003. Number of loops of size h in growing scale-free networks. *Phys. Rev. Lett.* **90**, 078701.

BINNEY, J. J., DOWRICK, N. J., FISHER, A. J., AND NEWMAN, M. E. J., 1992. *The Theory of Critical Phenomena*. Oxford University Press, Oxford.

BOGUÑÁ, M., PASTOR-SATORRAS, R., DÍAZ-GUILERA, A., AND ARENAS, A., 2003a. Emergence of clustering, correlations, and communities in a social network model. Preprint cond-mat/0309263.

BOGUÑÁ, M., PASTOR-SATORRAS, R., AND VESPIGNANI, A., 2003b. Epidemic spreading in complex networks with degree correlations. In R. Pastor-Satorras, J. Rubi, and A. Díaz-Guilera (eds.), *Statistical Mechanics of Complex Networks*, no. 625 in Lecture Notes in Physics, pp. 127–147. Springer, Berlin.

BOLLOBÁS, B., 2001. *Random Graphs*. Academic Press, New York, 2nd ed.

BOLLOBÁS, B. AND RIORDAN, O., 2002. The diameter of a scale-free random graph. Preprint, Department of Mathematical Sciences, University of Memphis.

BOLLOBÁS, B., RIORDAN, O., SPENCER, J., AND TUSNÁDY, G., 2001. The degree sequence of a scale-free random graph process. *Random Structures and Algorithms* **18**, 279–290.

BORDENS, M. AND GÓMEZ, I., 2000. Collaboration networks in science. In H. B. Atkins and B. Cronin (eds.), *The Web of Knowledge: A Festschrift in Honor of Eugene Garfield*. Information Today, Medford, NJ.

BORNHOLDT, S. AND EBEL, H., 2001. World Wide Web scaling exponent from Simon's 1955 model. *Phys. Rev. E* **64**, 035104.

BOSS, M., ELSINGER, H., SUMMER, M., AND THURNER, S., 2003. The network topology of the interbank market. Preprint cond-mat/0309582.

BRIN, S. AND PAGE, L., 1998. The anatomy of a large-scale hypertextual Web search engine. *Computer Networks* **30**, 107–117.

BROADBENT, S. R. AND HAMMERSLEY, J. M., 1957. Percolation processes: I. Crystals and mazes. *Proc. Cambridge Philos. Soc.* **53**, 629–641.

BRODER, A., KUMAR, R., MAGHOUL, F., RAGHAVAN, P., RAJAGOPALAN, S., STATA, R., TOMKINS, A., AND WIENER, J., 2000. Graph structure in the web. *Computer Networks* **33**, 309–320.

BUCHANAN, M., 2002. *Nexus: Small Worlds and the Groundbreaking Science of Networks*. Norton, New York.

BUNDE, A. AND HAVLIN, S. (eds.), 1994. *Fractals in Science*. Springer, Berlin.

BUNDE, A. AND HAVLIN, S. (eds.), 1996. *Fractals and Disordered Systems*. Springer, Berlin.

BURDA, Z., JURKIEWICZ, J., AND KRZYWICKI, A., 2004. Network transitivity and matrix models. *Phys. Rev. E* **69**, 026106.

CALDARELLI, G., PASTOR-SATORRAS, R., AND VESPIGNANI, A., 2004. Structure of cycles and local ordering in complex networks. *Eur. Phys. J. B* **38**, 183–186.

CALLAWAY, D. S., NEWMAN, M. E. J., STROGATZ, S. H., AND WATTS, D. J., 2000. Network robustness and fragility: Percolation on random graphs. *Phys. Rev. Lett.* **85**, 5468–5471.

CALLAWAY, D. S., HOPCROFT, J. E., KLEINBERG, J. M., NEWMAN, M. E. J., AND STROGATZ, S. H., 2001. Are randomly grown graphs really random? *Phys. Rev. E* **64**, 041902.

CAMACHO, J., GUIMERÀ, R., AND AMARAL, L. A. N., 2002. Robust patterns in food web structure. *Phys. Rev. Lett.* **88**, 228102.

CAPOCCI, A., SERVEDIO, V. D. P., CALDARELLI, G., AND COLAIORI, F., 2004. Detecting communities in large networks. In S. Leonardi (ed.), *Proceedings of the 3rd Workshop on Algorithms and Models for the Web Graph*, no. 3243 in Lecture Notes in Computer Science. Springer, Berlin.

CARLSON, J. M. AND DOYLE, J., 1999. Highly optimized tolerance: A mechanism for power laws in designed systems. *Phys. Rev. E* **60**, 1412–1427.

CARLSON, J. M. AND DOYLE, J., 2000. Highly optimized tolerance: Robustness and design in complex systems. *Phys. Rev. Lett.* **84**, 2529–2532.

CHAKRABARTI, S., DOM, B., GIBSON, D., KLEINBERG, J. M., RAGHAVAN, P., AND RAJAGOPALAN, S., 1998a. Automatic resource list compilation by analyzing hyperlink structure and associated text. *Computer Networks* **30**, 65–74.

CHAKRABARTI, S., DOM, B., GIBSON, D., KUMAR, S., RAGHAVAN, P., RAJAGOPALAN, S., AND TOMKINS, A., 1998b. Experiments in topic distillation. In *ACM SIGIR Workshop on Hypertext Information Retrieval on the Web*. Association of Computing Machinery, New York.

CHEN, Q., CHANG, H., GOVINDAN, R., JAMIN, S., SHENKER, S. J., AND WILLINGER, W., 2002. The origin of power laws in Internet topologies revisited. In *Proceedings of the 21st Annual Joint Conference of the IEEE Computer and Communications Societies*. IEEE Computer Society.

CHUNG, F. AND LU, L., 2002a. The average distances in random graphs with given expected degrees. *Proc. Natl. Acad. Sci. USA* **99**, 15879–15882.

CHUNG, F. AND LU, L., 2002b. Connected components in random graphs with given degree sequences. *Annals of Combinatorics* **6**, 125–145.

CHUNG, F., LU, L., DEWEY, T. G., AND GALAS, D. J., 2003. Duplication models for biological networks. *Journal of Computational Biology* **10**, 677–688.

COHEN, R. AND HAVLIN, S., 2003. Scale-free networks are ultrasmall. *Phys. Rev. Lett.* **90**, 058701.

COHEN, R., EREZ, K., BEN-AVRAHAM, D., AND HAVLIN, S., 2000. Resilience of the Internet to random breakdowns. *Phys. Rev. Lett.* **85**, 4626–4628.

COHEN, R., EREZ, K., BEN-AVRAHAM, D., AND HAVLIN, S., 2001. Breakdown of the Internet under intentional attack. *Phys. Rev. Lett.* **86**, 3682–3685.

COLLET, P. AND ECKMANN, J.-P., 2002. The number of large graphs with a positive density of triangles. *J. Stat. Phys.* **109**, 923–943.

COOKE, K. L., CALEF, D. F., AND LEVEL, E. V., 1977. Stability or chaos in discrete epidemic models. In *Nonlinear Systems and Applications: An International Conference*, pp. 73–93. Academic Press, New York.

COSTA, L. D., 2004. Reinforcing the resilience of complex networks. *Phys. Rev. E* **69**, 066127.

CRISANTI, A., PALADIN, G., AND VULPIANI, A., 1993. *Products of Random Matrices in Statistical Physics*. Springer, Berlin.

CSÁNYI, G. AND SZENDRŐI, B., 2004. Structure of a large social network. *Phys. Rev. E* **69**, 036131.

DAVIDSEN, J., EBEL, H., AND BORNHOLDT, S., 2002. Emergence of a small world from local interactions: Modeling acquaintance networks. *Phys. Rev. Lett.* **88**, 128701.

DE CASTRO, R. AND GROSSMAN, J. W., 1999. Famous trails to Paul Erdős. *Mathematical Intelligencer* **21**, 51–63.

DE MENEZES, M. A., MOUKARZEL, C., AND PENNA, T. J. P., 2000. First-order transition in small-world networks. *Europhys. Lett.* **50**, 574–579.

DE MOURA, A. P. S., LAI, Y.-C., AND MOTTER, A. E., 2003. Signatures of small-world and scale-free properties in large computer programs. *Phys. Rev. E* **68**, 017102.

DEGENNE, A. AND FORSÉ, M., 1999. *Introducing Social Networks*. Sage, London.

DING, Y., FOO, S., AND CHOWDHURY, G., 1999. A bibliometric analysis of collaboration in the field of information retrieval. *Intl. Inform. and Libr. Rev.* **30**, 367–376.

DODDS, P. S., MUHAMAD, R., AND WATTS, D. J., 2003. An experimental study of search in global social networks. *Science* **301**, 827–829.

DOROGOVTSEV, S. N. AND MENDES, J. F. F., 2000a. Evolution of networks with aging of sites. *Phys. Rev. E* **62**, 1842–1845.

DOROGOVTSEV, S. N. AND MENDES, J. F. F., 2000b. Exactly solvable small-world network. *Europhys. Lett.* **50**, 1–7.

DOROGOVTSEV, S. N. AND MENDES, J. F. F., 2003. *Evolution of Networks: From Biological Nets to the Internet and WWW*. Oxford University Press, Oxford.

DOROGOVTSEV, S. N., MENDES, J. F. F., AND SAMUKHIN, A. N., 2000. Structure of growing networks with preferential linking. *Phys. Rev. Lett.* **85**, 4633–4636.

DOROGOVTSEV, S. N., MENDES, J. F. F., AND SAMUKHIN, A. N., 2001a. Giant strongly connected component of directed networks. *Phys. Rev. E* **64**, 025101.

DOROGOVTSEV, S. N., MENDES, J. F. F., AND SAMUKHIN, A. N., 2001b. Size-dependent degree distribution of a scale-free growing network. *Phys. Rev. E* **63**, 062101.

DOROGOVTSEV, S. N., GOLTSEV, A. V., AND MENDES, J. F. F., 2002. Pseudofractal scale-free web. *Phys. Rev. E* **65**, 066122.

DOYE, J. P. K., 2002. Network topology of a potential energy landscape: A static scale-free network. *Phys. Rev. Lett.* **88**, 238701.

EBEL, H., MIELSCH, L.-I., AND BORNHOLDT, S., 2002. Scale-free topology of e-mail networks. *Phys. Rev. E* **66**, 035103.

ECKMANN, J.-P. AND MOSES, E., 2002. Curvature of co-links uncovers hidden thematic layers in the world wide web. *Proc. Natl. Acad. Sci. USA* **99**, 5825–5829.

EGGHE, L. AND ROUSSEAU, R., 1990. *Introduction to Informetrics*. Elsevier, Amsterdam.

EGUILUZ, V. M., CHIALVO, D. R., CECCHI, G., BALIKI, M., AND APKARIAN, A. V., 2005. Scale-free brain functional networks. *Phys. Rev. Lett.* **94**, 018102.

EISENBERG, E. AND LEVANON, E. Y., 2003. Preferential attachment in the protein network evolution. *Phys. Rev. Lett.* **91**, 138701.

ERDŐS, P. AND RÉNYI, A., 1959. On random graphs. *Publicationes Mathematicae* **6**, 290–297.

ERDŐS, P. AND RÉNYI, A., 1960. On the evolution of random graphs. *Publications of the Mathematical Institute of the Hungarian Academy of Sciences* **5**, 17–61.

ERDŐS, P. AND RÉNYI, A., 1961. On the strength of connectedness of a random graph. *Acta Mathematica Scientia Hungary* **12**, 261–267.

ERGÜN, G. AND RODGERS, G. J., 2002. Growing random networks with fitness. *Physica A* **303**, 261–272.

FALOUTSOS, M., FALOUTSOS, P., AND FALOUTSOS, C., 1999. On power-law relationships of the internet topology. *Computer Communications Review* **29**, 251–262.

FARKAS, I. J., DERÉNYI, I., BARABÁSI, A.-L., AND VICSEK, T., 2001. Spectra of "real-world" graphs: Beyond the semicircle law. *Phys. Rev. E* **64**, 026704.

FARKAS, I. J., JEONG, H., VICSEK, T., BARABÁSI, A.-L., AND OLTVAI, Z. N., 2003. The topology of the transcription regulatory network in the yeast, Saccharomyces cerevisiae. *Physica A* **381**, 601–612.

FEATHERSTONE, D. E. AND BROADIE, K., 2002. Wrestling with pleiotropy: Genomic and topological analysis of the yeast gene expression network. *Bioessays* **24**, 267–274.

FELL, D. A. AND WAGNER, A., 2000. The small world of metabolism. *Nature Biotechnology* **18**, 1121–1122.

FERRER I CANCHO, R. AND SOLÉ, R. V., 2001. The small world of human language. *Proc. R. Soc. London B* **268**, 2261–2265.

FERRER I CANCHO, R. AND SOLÉ, R. V., 2003. Optimization in complex networks. In R. Pastor-Satorras, J. Rubi, and A. Díaz-Guilera (eds.), *Statistical Mechanics of Complex Networks*, no. 625 in Lecture Notes in Physics, pp. 114–125. Springer, Berlin.

FIEDLER, M., 1973. Algebraic connectivity of graphs. *Czech. Math. J.* **23**, 298–305.

FLAKE, G. W., LAWRENCE, S. R., GILES, C. L., AND COETZEE, F. M., 2002. Self-organization and identification of Web communities. *IEEE Computer* **35**, 66–71.

FORD, G. W. AND UHLENBECK, G. E., 1957. Combinatorial problems in the theory of graphs. IV. *Proc. Natl. Acad. Sci. USA* **43**, 163–167.

FORTUNATO, S., LATORA, V., AND MARCHIORI, M., 2004. A method to find community structures based on information centrality. *Phys. Rev. E* **70**, 056104.

FREEMAN, L. C., 1977. A set of measures of centrality based upon betweenness. *Sociometry* **40**, 35–41.

FRONCZAK, A., FRONCZAK, P., AND HOLYST, J. A., 2002a. Exact solution for average path length in random graphs. Preprint cond-mat/0212230.

FRONCZAK, A., HOLYST, J. A., JEDYNAK, M., AND SIENKIEWICZ, J., 2002b. Higher order clustering coefficients in Barabasi-Albert networks. *Physica A* **316**, 688–694.

GIBSON, D., KLEINBERG, J., AND RAGHAVAN, P., 1998. Inferring web communities from link topology. In *Proceedings of the 9th ACM Conference on Hypertext and Hypermedia*. Association of Computing Machinery, New York.

GILBERT, E. N., 1959. Random graphs. *Annals of Mathematical Statistics* **30**, 1191–1141.

GIOT, L., BADER, J. S., BROUWER, C., *et al.*, 2003. A protein interaction map of Drosophila melanogaster. *Science* **302**, 1727–1736.

GIRVAN, M. AND NEWMAN, M. E. J., 2002. Community structure in social and biological networks. *Proc. Natl. Acad. Sci. USA* **99**, 7821–7826.

GIRVAN, M., CALLAWAY, D. S., NEWMAN, M. E. J., AND STROGATZ, S. H., 2002. A simple model of epidemics with pathogen mutation. *Phys. Rev. E* **65**, 031915.

GLADWELL, M., 2000. *The Tipping Point: How Little Things Can Make a Big Difference*. Little Brown, New York.

GLEISER, P. AND DANON, L., 2003. Community structure in jazz. *Advances in Complex Systems* **6**, 565–573.

GLEISS, P. M., STADLER, P. F., WAGNER, A., AND FELL, D. A., 2001. Relevant cycles in chemical reaction networks. *Advances in Complex Systems* **4**, 207–226.

Goh, K.-I., Kahng, B., and Kim, D., 2001a. Spectra and eigenvectors of scale-free networks. *Phys. Rev. E* **64**, 051903.

Goh, K.-I., Kahng, B., and Kim, D., 2001b. Universal behavior of load distribution in scale-free networks. *Phys. Rev. Lett.* **87**, 278701.

Goh, K.-I., Oh, E., Jeong, H., Kahng, B., and Kim, D., 2002. Classification of scale-free networks. *Proc. Natl. Acad. Sci. USA* **99**, 12583–12588.

Goh, K.-I., Ghim, C.-M., Kahng, B., and Kim, D., 2003. Comment on 'Universal behavior of load distribution in scale-free networks'. *Phys. Rev. Lett.* **91**, 189804.

Govindan, R. and Tangmunarunkit, H., 2000. Heuristics for Internet map discovery. In *Proceedings of the 19th Annual Joint Conference of the IEEE Computer and Communications Societies*, pp. 1371–1380. Institute of Electrical and Electronics Engineers, New York.

Granovetter, M., 1995. *Getting a Job: A Study in Contacts and Careers*. University of Chicago Press, Chicago.

Grossman, J. W., 2002. The evolution of the mathematical research collaboration graph. *Congressus Numerantium* **158**, 202–212.

Grossman, J. W. and Ion, P. D. F., 1995. On a portion of the well-known collaboration graph. *Congressus Numerantium* **108**, 129–131.

Guardiola, X., Guimerà, R., Arenas, A., Diaz-Guilera, A., Streib, D., and Amaral, L. A. N., 2002. Macro- and micro-structure of trust networks. Preprint cond-mat/0206240.

Guhr, T., Müller-Groeling, A., and Weidenmüller, H. A., 1998. Random-matrix theories in quantum physics: Common concepts. *Phys. Rep.* **299**, 189–425.

Guimerà, R. and Amaral, L. A. N., 2003. Modelling the world-wide airport network. *Eur. Phys. J. B* **38**, 381–385.

Guimerà, R., Danon, L., Díaz-Guilera, A., Giralt, F., and Arenas, A., 2003. Self-similar community structure in organisations. *Phys. Rev. E* **68**, 065103.

Hammersley, J. M., 1957. Percolation processes: II. The connective constant. *Proc. Cambridge Philos. Soc.* **53**, 642–645.

Harary, F., 1995. *Graph Theory*. Perseus, Cambridge, MA.

Hartwell, L. H., Hopfield, J. J., Leibler, S., and Murray, A. W., 1999. From molecular to modular cell biology. *Nature* **402**, C47–52.

Herrero, C. P., 2002. Ising model in small-world networks. *Phys. Rev. E* **65**, 066110.

Hethcote, H. W., 2000. The mathematics of infectious diseases. *SIAM Review* **42**, 599–653.

Holland, P. W. and Leinhardt, S., 1981. An exponential family of probability distributions for directed graphs. *J. Amer. Stat. Assoc.* **76**, 33–50.

Holme, P. and Huss, M., 2003. Discovery and analysis of biochemical subnetwork hierarchies. In R. Gauges, U. Kummer, J. Pahle, and U. Rost (eds.), *Proceedings of the 3rd Workshop on Computation of Biochemical Pathways and Genetic Networks*, pp. 3–9. Logos, Berlin.

Holme, P., Huss, M., and Jeong, H., 2003. Subnetwork hierarchies of biochemical pathways. *Bioinformatics* **19**, 532–538.

Holme, P., Edling, C. R., and Liljeros, F., 2004. Structure and time-evolution of an Internet dating community. *Social Networks* **26**, 155–174.

Hong, H., Kim, B. J., and Choi, M. Y., 2002. Comment on 'Ising model on a small world network'. *Phys. Rev. E* **66**, 018101.

HUBERMAN, B. A. AND ADAMIC, L. A., 1999. Growth dynamics of the World-Wide Web. *Nature* **401**, 131.

HUGHES, D. AND PACZUSKI, M., 2003. Scale-free magnetic networks: Comparing observational data with a self-organizing model of the coronal field. Preprint astro-ph/0309230.

ITO, T., TASHIRO, K., MUTA, S., OZAWA, R., CHIBA, T., NISHIZAWA, M., YAMAMOTO, K., KUHARA, S., AND SAKAKI, Y., 2000. Toward a protein–protein interaction map of the budding yeast: A comprehensive system to examine two-hybrid interactions in all possible combinations between the yeast proteins. *Proc. Natl. Acad. Sci. USA* **97**, 1143–1147.

ITO, T., CHIBA, T., OZAWA, R., YOSHIDA, M., HATTORI, M., AND SAKAKI, Y., 2001. A comprehensive two-hybrid analysis to explore the yeast protein interactome. *Proc. Natl. Acad. Sci. USA* **98**, 4569–4574.

JACOBS, J., 1961. *The Death and Life of Great American Cities*. Random House, New York.

JAIN, S. AND KRISHNA, S., 1998. Autocatalytic sets and the growth of complexity in an evolutionary model. *Phys. Rev. Lett.* **81**, 5684–5687.

JANSON, S., ŁUCZAK, T., AND RUCINSKI, A., 1999. *Random Graphs*. John Wiley, New York.

JEONG, H., TOMBOR, B., ALBERT, R., OLTVAI, Z. N., AND BARABÁSI, A.-L., 2000. The large-scale organization of metabolic networks. *Nature* **407**, 651–654.

JEONG, H., MASON, S., BARABÁSI, A.-L., AND OLTVAI, Z. N., 2001. Lethality and centrality in protein networks. *Nature* **411**, 41–42.

JEONG, H., NÉDA, Z., AND BARABÁSI, A.-L., 2003. Measuring preferential attachment in evolving networks. *Europhys. Lett.* **61**, 567–572.

JIN, E. M., GIRVAN, M., AND NEWMAN, M. E. J., 2001. The structure of growing social networks. *Phys. Rev. E* **64**, 046132.

JONES, J. H. AND HANDCOCK, M. S., 2003a. An assessment of preferential attachment as a mechanism for human sexual network formation. *Proc. R. Soc. London B* **270**, 1123–1128.

JONES, J. H. AND HANDCOCK, M. S., 2003b. Sexual contacts and epidemic thresholds. *Nature* **423**, 605–606.

KARINTHY, F., 1929. Chains. In *Everything is Different*. Budapest.

KAUTZ, H., SELMAN, B., AND SHAH, M., 1997. ReferralWeb: Combining social networks and collaborative filtering. *Comm. ACM* **40**, 63–65.

KEELING, M. J., 1999. The effects of local spatial structure on epidemiological invasion. *Proc. R. Soc. London B* **266**, 859–867.

KIM, J., KRAPIVSKY, P. L., KAHNG, B., AND REDNER, S., 2002. Infinite-order percolation and giant fluctuations in a protein interaction network. *Phys. Rev. E* **66**, 055101.

KLEINBERG, J. M., 1999. Authoritative sources in a hyperlinked environment. *J. ACM* **46**, 604–632.

KLEINBERG, J. M., 2000a. Navigation in a small world. *Nature* **406**, 845.

KLEINBERG, J. M., 2000b. The small-world phenomenon: An algorithmic perspective. In *Proceedings of the 32nd Annual ACM Symposium on Theory of Computing*, pp. 163–170. Association of Computing Machinery, New York.

KLEINBERG, J. M., 2002. Small world phenomena and the dynamics of information. In T. G. Dietterich, S. Becker, and Z. Ghahramani (eds.), *Proceedings of the 2001 Neural Information Processing Systems Conference*. MIT Press, Cambridge, MA.

KLEINBERG, J. M., KUMAR, S. R., RAGHAVAN, P., RAJAGOPALAN, S., AND TOMKINS, A., 1999. The Web as a graph: Measurements, models and methods. In *Proceedings of the International Conference on Combinatorics and Computing*, no. 1627 in Lecture Notes in Computer Science, pp. 1–18. Springer, Berlin.

KLEMM, K. AND EGUILUZ, V. M., 2002. Highly clustered scale-free networks. *Phys. Rev. E* **65**, 036123.

KRAPIVSKY, P. L. AND REDNER, S., 2001. Organization of growing random networks. *Phys. Rev. E* **63**, 066123.

KRAPIVSKY, P. L., REDNER, S., AND LEYVRAZ, F., 2000. Connectivity of growing random networks. *Phys. Rev. Lett.* **85**, 4629–4632.

KREBS, V. E., 2002. Mapping networks of terrorist cells. *Connections* **24**, 43–52.

KRETSCHMER, H., 1994. Coauthorship networks of invisible college and institutionalized communities. *Scientometrics* **30**, 363–369.

KRETZSCHMAR, M. AND MORRIS, M., 1996. Measures of concurrency in networks and the spread of infectious disease. *Math. Biosci.* **133**, 165–195.

KULKARNI, R. V., ALMAAS, E., AND STROUD, D., 2000. Exact results and scaling properties of small-world networks. *Phys. Rev. E* **61**, 4268–4271.

KULLMANN, L. AND KERTESZ, J., 2001. Preferential growth: Exact solution of the time-dependent distributions. *Phys. Rev. E* **63**, 051112.

KUMAR, R., RAGHAVAN, P., RAJAGOPALAN, S., SIVAKUMAR, D., TOMKINS, A. S., AND UPFAL, E., 2000a. Stochastic models for the Web graph. In *Proceedings of the 42st Annual IEEE Symposium on the Foundations of Computer Science*, pp. 57–65. Institute of Electrical and Electronics Engineers, New York.

KUMAR, R., RAGHAVAN, P., RAJALOPAGAN, S., SUVAKUMAR, D., TOMKINS, A. S., AND UPFAL, E., 2000b. The web as a graph. In *Proceeedings of the 19th ACM Symposium on Principles of Database Systems*, pp. 1–10. Association of Computing Machinery, New York.

KUPERMAN, M. AND ABRAMSON, G., 2001. Small world effect in an epidemiological model. *Phys. Rev. Lett.* **86**, 2909–2912.

KUPERMAN, M. AND ZANETTE, D. H., 2002. Stochastic resonance in a model of opinion formation on small world networks. *Eur. Phys. J. B* **26**, 387–391.

LAUMANN, E. O., GAGNON, J. H., MICHAEL, R. T., AND MICHAELS, S., 1994. *The Social Organization of Sexuality: Sexual Practices in the United States*. University of Chicago Press, Chicago.

LAWRENCE, S. AND GILES, C. L., 1998. Searching the world wide web. *Science* **280**, 98–100.

LAWRENCE, S. AND GILES, C. L., 1999. Accessibility of information on the web. *Nature* **400**, 107–109.

LEHMANN, S., LAUTRUP, B., AND JACKSON, A. D., 2003. Citation networks in high energy physics. *Phys. Rev. E* **68**, 026113.

LI, L., ALDERSON, D., WILLINGER, W., AND DOYLE, J., 2004a. A first principles approach to understanding the internet's router-level topology. *Computer Communications Review* **34**, 3–14.

LI, S., ARMSTRONG, C. M., BERTIN, N., *et al.*, 2004b. A map of the interactome network of the metazoan C. elegans. *Science* **303**, 540–543.

LILJEROS, F., EDLING, C. R., AMARAL, L. A. N., STANLEY, H. E., AND ÅBERG, Y., 2001. The web of human sexual contacts. *Nature* **411**, 907–908.

LLOYD, A. L. AND MAY, R. M., 2001. How viruses spread among computers and people. *Science* **292**, 1316–1317.

LONGINI, I. M., 1988. A mathematical model for predicting the geographic spread of new infectious agents. *Math. Biosci.* **90**, 367–383.

ŁUCZAK, T., 1992. Sparse random graphs with a given degree sequence. In A. M. Frieze and T. Łuczak (eds.), *Proceedings of the Symposium on Random Graphs, Poznań 1989*, pp. 165–182. John Wiley, New York.

LYNCH, N. A., 1996. *Distributed Algorithms*. Morgan Kauffman, San Francisco, CA.

MACCLUER, C. R., 2000. The many proofs and applications of Perron's theorem. *SIAM Rev.* **42**, 487–498.

MASLOV, S. AND SNEPPEN, K., 2002. Specificity and stability in topology of protein networks. *Science* **296**, 910–913.

MASLOV, S., SNEPPEN, K., AND ZALIZNYAK, A., 2004. Detection of topological patterns in complex networks: Correlation profile of the internet. *Physica A* **333**, 529–540.

MEDINA, A., MATTA, I., AND BYERS, J., 2000. On the origin of power laws in Internet topologies. *Computer Communications Review* **30**, 18–28.

MEHTA, M. L., 1991. *Random Matrices*. Academic Press, New York, 2nd ed.

MELIN, G. AND PERSSON, O., 1996. Studying research collaboration using co-authorships. *Scientometrics* **36**, 363–377.

MENGER, K., 1927. Zur allgemeinen Kurventheorie. *Fundamenta Mathematicae* **10**, 96–115.

MERTON, R. K., 1968. The Matthew effect in science. *Science* **159**, 56–63.

MILGRAM, S., 1967. The small world problem. *Psychology Today* **2**, 60–67.

MILO, R., SHEN-ORR, S., ITZKOVITZ, S., KASHTAN, N., CHKLOVSKII, D., AND ALON, U., 2002. Network motifs: Simple building blocks of complex networks. *Science* **298**, 824–827.

MOLLISON, D., 1977. Spatial contact models for ecological and epidemic spread. *Journal of the Royal Statistical Society B* **39**, 283–326.

MOLLOY, M. AND REED, B., 1995. A critical point for random graphs with a given degree sequence. *Random Structures and Algorithms* **6**, 161–179.

MOLLOY, M. AND REED, B., 1998. The size of the giant component of a random graph with a given degree sequence. *Combinatorics, Probability and Computing* **7**, 295–305.

MONASSON, R., 1999. Diffusion, localization and dispersion relations on 'small-world' lattices. *Eur. Phys. J. B* **12**, 555–567.

MONTOYA, J. M. AND SOLÉ, R. V., 2002. Small world patterns in food webs. *J. Theor. Bio.* **214**, 405–412.

MONTOYA, J. M. AND SOLÉ, R. V., 2003. Topological properties of food webs: From real data to community assembly models. *Oikos* **102**, 614–622.

MOODY, J., 2001. Race, school integration, and friendship segregation in America. *Am. J. Sociol.* **107**, 679–716.

MOORE, C. AND NEWMAN, M. E. J., 2000a. Epidemics and percolation in small-world networks. *Phys. Rev. E* **61**, 5678–5682.

MOORE, C. AND NEWMAN, M. E. J., 2000b. Exact solution of site and bond percolation on small-world networks. *Phys. Rev. E* **62**, 7059–7064.

MOREIRA, A. A., ANDRADE, JR., J. S., AND AMARAL, L. A. N., 2002. Extremum statistics in scale-free network models. *Phys. Rev. Lett.* **89**, 268703.

MORRIS, M., 1993. Telling tails explain the discrepancy in sexual partner reports. *Nature* **365**, 437–440.

MORRIS, M., 1995. Data driven network models for the spread of infectious disease. In D. Mollison (ed.), *Epidemic Models: Their Structure and Relation to Data*, pp. 302–322. Cambridge University Press, Cambridge.

MOUKARZEL, C. F., 1999. Spreading and shortest paths in systems with sparse long-range connections. *Phys. Rev. E* **60**, 6263–6266.

NAGURNEY, A., 1993. *Network Economics: A Variational Inequality Approach*. Kluwer Academic Publishers, Dordrecht.

NEWMAN, M. E. J., 2001a. The structure of scientific collaboration networks. *Proc. Natl. Acad. Sci. USA* **98**, 404–409.

NEWMAN, M. E. J., 2001b. Scientific collaboration networks: I. Network construction and fundamental results. *Phys. Rev. E* **64**, 016131.

NEWMAN, M. E. J., 2001c. Scientific collaboration networks: II. Shortest paths, weighted networks, and centrality. *Phys. Rev. E* **64**, 016132.

NEWMAN, M. E. J., 2001d. Clustering and preferential attachment in growing networks. *Phys. Rev. E* **64**, 025102.

NEWMAN, M. E. J., 2002a. Assortative mixing in networks. *Phys. Rev. Lett.* **89**, 208701.

NEWMAN, M. E. J., 2002b. Spread of epidemic disease on networks. *Phys. Rev. E* **66**, 016128.

NEWMAN, M. E. J., 2002c. The structure and function of networks. *Computer Physics Communications* **147**, 40–45.

NEWMAN, M. E. J., 2003a. Ego-centered networks and the ripple effect. *Social Networks* **25**, 83–95.

NEWMAN, M. E. J., 2003b. Mixing patterns in networks. *Phys. Rev. E* **67**, 026126.

NEWMAN, M. E. J., 2003c. Properties of highly clustered networks. *Phys. Rev. E* **68**, 026121.

NEWMAN, M. E. J., 2004a. Coauthorship networks and patterns of scientific collaboration. *Proc. Natl. Acad. Sci. USA* **101**, 5200–5205.

NEWMAN, M. E. J., 2004b. Fast algorithm for detecting community structure in networks. *Phys. Rev. E* **69**, 066133.

NEWMAN, M. E. J. AND GIRVAN, M., 2004. Finding and evaluating community structure in networks. *Phys. Rev. E* **69**, 026113.

NEWMAN, M. E. J. AND PARK, J., 2003. Why social networks are different from other types of networks. *Phys. Rev. E* **68**, 036122.

NEWMAN, M. E. J. AND WATTS, D. J., 1999a. Renormalization group analysis of the small-world network model. *Phys. Lett. A* **263**, 341–346.

NEWMAN, M. E. J. AND WATTS, D. J., 1999b. Scaling and percolation in the small-world network model. *Phys. Rev. E* **60**, 7332–7342.

NEWMAN, M. E. J., MOORE, C., AND WATTS, D. J., 2000. Mean-field solution of the small-world network model. *Phys. Rev. Lett.* **84**, 3201–3204.

NEWMAN, M. E. J., STROGATZ, S. H., AND WATTS, D. J., 2001. Random graphs with arbitrary degree distributions and their applications. *Phys. Rev. E* **64**, 026118.

NEWMAN, M. E. J., FORREST, S., AND BALTHROP, J., 2002a. Email networks and the spread of computer viruses. *Phys. Rev. E* **66**, 035101.

NEWMAN, M. E. J., JENSEN, I., AND ZIFF, R. M., 2002b. Percolation and epidemics in a two-dimensional small world. *Phys. Rev. E* **65**, 021904.

NEWMAN, M. E. J., WATTS, D. J., AND STROGATZ, S. H., 2002c. Random graph models of social networks. *Proc. Natl. Acad. Sci. USA* **99**, 2566–2572.

OHNO, S., 1970. *Evolution by Gene Duplication*. Springer, New York.

OVERBEEK, R., LARSEN, N., PUSCH, G. D., D'SOUZA, M., SELKOV, JR., E., KYRPIDES, N., FONSTEIN, M., MALTSEV, N., AND SELKOV, E., 2000. Wit: Integrated system for high-throughput genome sequence analysis and metabolic reconstruction. *Nucleic Acids Res.* **28**, 123–125.

OZANA, M., 2001. Incipient spanning cluster on small-world networks. *Europhys. Lett.* **55**, 762–766.

PALLA, G., DERENYI, I., FARKAS, I., AND VICSEK, T., 2005. Uncovering the overlapping community structure of complex networks in nature and science. *Nature* **435**, 814–818.

PANDIT, S. A. AND AMRITKAR, R. E., 2001. Random spread on the family of small-world networks. *Phys. Rev. E* **63**, 041104.

PARK, J. AND NEWMAN, M. E. J., 2003. The origin of degree correlations in the Internet and other networks. *Phys. Rev. E* **68**, 026112.

PARK, J., LAPPE, M., AND TEICHMANN, S. A., 2001. Mapping protein family interactions: Intramolecular and intermolecular protein family interaction repertoires in the PDB and yeast. *J. Mol. Biol.* **307**, 929–938.

PASTOR-SATORRAS, R. AND VESPIGNANI, A., 2001. Epidemic spreading in scale-free networks. *Phys. Rev. Lett.* **86**, 3200–3203.

PASTOR-SATORRAS, R. AND VESPIGNANI, A., 2004. *Evolution and Structure of the Internet*. Cambridge University Press, Cambridge.

PASTOR-SATORRAS, R., VÁZQUEZ, A., AND VESPIGNANI, A., 2001. Dynamical and correlation properties of the Internet. *Phys. Rev. Lett.* **87**, 258701.

PĘKALSKI, A., 2001. Ising model on a small world network. *Phys. Rev. E* **64**, 057104.

PENNOCK, D. M., FLAKE, G. W., LAWRENCE, S., GLOVER, E. J., AND GILES, C. L., 2002. Winners don't take all: Characterizing the competition for links on the web. *Proc. Natl. Acad. Sci. USA* **99**, 5207–5211.

PERSSON, O. AND BECKMANN, M., 1995. Locating the network of interacting authors in scientific specialties. *Scientometrics* **33**, 351–366.

PLISCHKE, M. AND BERGERSEN, B., 1994. *Equilibrium Statistical Physics*. World Scientific, Singapore, 2nd ed.

POOL, I. DE S. AND KOCHEN, M., 1978. Contacts and influence. *Social Networks* **1**, 1–48.

POTHEN, A., SIMON, H., AND LIOU, K.-P., 1990. Partitioning sparse matrices with eigenvectors of graphs. *SIAM J. Matrix Anal. Appl.* **11**, 430–452.

POTTERAT, J. J., PHILLIPS-PLUMMER, L., MUTH, S. Q., ROTHENBERG, R. B., WOODHOUSE, D. E., MALDONADO-LONG, T. S., ZIMMERMAN, H. P., AND MUTH, J. B., 2002. Risk network structure in the early epidemic phase of HIV transmission in Colorado Springs. *Sexually Transmitted Infections* **78**, i159–i163.

PRICE, D. J. DE S., 1965. Networks of scientific papers. *Science* **149**, 510–515.

PRICE, D. J. DE S., 1976. A general theory of bibliometric and other cumulative advantage processes. *J. Amer. Soc. Inform. Sci.* **27**, 292–306.

PROVERO, P., 2002. Gene networks from DNA microarray data: Centrality and lethality. Preprint cond-mat/0207345.

RADICCHI, F., CASTELLANO, C., CECCONI, F., LORETO, V., AND PARISI, D., 2004. Defining and identifying communities in networks. *Proc. Natl. Acad. Sci. USA* **101**, 2658–2663.

RAPOPORT, A., 1957. Contribution to the theory of random and biased nets. *Bulletin of Mathematical Biophysics* **19**, 257–277.

RAPOPORT, A., 1963. Mathematical models of social interaction. In R. D. Luce, R. R. Bush, and E. Galanter (eds.), *Handbook of Mathematical Psychology*, vol. 2, pp. 493–579. Wiley, New York.

RAPOPORT, A. AND HORVATH, W. J., 1961. A study of a large sociogram. *Behavioral Science* **6**, 279–291.

RAVASZ, E. AND BARABÁSI, A.-L., 2003. Hierarchical organization in complex networks. *Phys. Rev. E* **67**, 026112.

RAVASZ, E., SOMERA, A. L., MONGRU, D. A., OLTVAI, Z., AND BARABÁSI, A.-L., 2002. Hierarchical organization of modularity in metabolic networks. *Science* **297**, 1551–1555.

REDNER, S., 1998. How popular is your paper? An empirical study of the citation distribution. *Eur. Phys. J. B* **4**, 131–134.

REICHARDT, J. AND BORNHOLDT, S., 2004. Detecting fuzzy community structures in complex networks with a q-state Potts model. Preprint cond-mat/0402349.

RITTER, J. P., 2000. Why Gnutella can't scale. No, really. URL `http://www.darkridge.com/~jpr5/doc/gnutella.html`.

ROZENFELD, H. D. AND BEN-AVRAHAM, D., 2004. Designer nets from local strategies. *Phys. Rev. E* **70**, 056107.

SÁNCHEZ, A. D., LÓPEZ, J. M., AND RODRÍGUEZ, M. A., 2002. Nonequilibrium phase transitions in directed small-world networks. *Phys. Rev. Lett.* **88**, 048701.

SANDER, L. M., WARREN, C. P., SOKOLOV, I., SIMON, C., AND KOOPMAN, J., 2002. Percolation on disordered networks as a model for epidemics. *Math. Biosci.* **180**, 293–305.

SATTENSPIEL, L. AND SIMON, C. P., 1988. The spread and persistence of infectious diseases in structured populations. *Math. Biosci.* **90**, 341–366.

SCHNEEBERGER, A., MERCER, C. H., GREGSON, S. A. J., FERGUSON, N. M., NYAMUKAPA, C. A., ANDERSON, R. M., JOHNSON, A. M., AND GARNETT, G. P., 2004. Scale-free networks and sexually transmitted diseases—A description of observed patterns of sexual contacts in Britain and Zimbabwe. *Sexually Transmitted Diseases* **31**, 380–387.

SCHWARTZ, N., COHEN, R., BEN-AVRAHAM, D., BARABÁSI, A.-L., AND HAVLIN, S., 2002. Percolation in directed scale-free networks. *Phys. Rev. E* **66**, 015104.

SCOTT, J., 2000. *Social Network Analysis: A Handbook.* Sage, London, 2nd ed.

SEARLS, D. B., 2003. Data integration—connecting the dots. *Nature Biotechnology* **21**, 844–845.

SEGLEN, P. O., 1992. The skewness of science. *J. Amer. Soc. Inform. Sci.* **43**, 628–638.

SERRANO, M. A. AND BOGUÑÁ, M., 2003. Topology of the world trade web. *Phys. Rev. E* **68**, 015101.

SETHNA, J. P., DAHMEN, K., KARTHA, S., KRUMHANSL, J. A., ROBERTS, B. W., AND SHORE, J. D., 1993. Hysteresis and hierarchies: Dynamics of disorder-driven first-order phase transformations. *Phys. Rev. Lett.* **70**, 3347–3350.

SHAW, S., 2003. Evidence of scale-free topology and dynamics in gene regulatory networks. In *Proceedings of the ISCA 12th International Conference on Intelligent and Adaptive Systems and Software Engineering*, pp. 37–40.

SIMMEL, G., 1950. *The Sociology of Georg Simmel.* Free Press, New York.

SIMON, H. A., 1955. On a class of skew distribution functions. *Biometrika* **42**, 425–440.

SMITH, R. D., 2002. Instant messaging as a scale-free network. Preprint cond-mat/0206378.

SOLÉ, R. V. AND MONTOYA, J. M., 2001. Complexity and fragility in ecological networks. *Proc. R. Soc. London B* **268**, 2039–2045.

SOLÉ, R. V., PASTOR-SATORRAS, R., SMITH, E., AND KEPLER, T. B., 2002. A model of large-scale proteome evolution. *Advances in Complex Systems* **5**, 43–54.

SOLOMONOFF, R. AND RAPOPORT, A., 1951. Connectivity of random nets. *Bulletin of Mathematical Biophysics* **13**, 107–117.

SPENCER, M., 2002. Rapoport at ninety. *Connections* **24**, 104–107.

STANLEY, H. E., 1971. *Introduction to Phase Transitions and Critical Phenomena*. Oxford University Press, Oxford.

STAUFFER, D. AND AHARONY, A., 1992. *Introduction to Percolation Theory*. Taylor and Francis, London, 2nd ed.

STRAUSS, D., 1986. On a general class of models for interaction. *SIAM Review* **28**, 513–527.

STROGATZ, S. H., 1994. *Nonlinear Dynamics and Chaos*. Addison-Wesley, Reading, MA.

TANIZAWA, T., PAUL, G., COHEN, R., HAVLIN, S. AND STANLEY, H. E., 2005. Optimization of network robustness to waves of targeted and random attacks. *Phys. Rev. E* **71**, 047101.

TRAVERS, J. AND MILGRAM, S., 1969. An experimental study of the small world problem. *Sociometry* **32**, 425–443.

TSALLIS, C. AND DE ALBUQUERQUE, M. P., 1999. Are citations of scientific papers a case of nonextensivity? *Eur. Phys. J. B* **13**, 777–780.

TYLER, J. R., WILKINSON, D. M., AND HUBERMAN, B. A., 2003. Email as spectroscopy: Automated discovery of community structure within organizations. In M. Huysman, E. Wenger, and V. Wulf (eds.), *Proceedings of the First International Conference on Communities and Technologies*. Kluwer, Dordrecht.

UETZ, P., GIOT, L., CAGNEY, G., MANSFIELD, T. A., JUDSON, R. S., KNIGHT, J. R., LOCKSHON, D., NARAYAN, V., SRINIVASAN, M., POCHART, P., QURESHI-EMILI, A., LI, Y., GODWIN, B., CONOVER, D., KALBFLEISCH, T., VIJAYADAMODAR, G., YANG, M., JOHNSTON, M., FIELDS, S., AND ROTHBERG, J. M., 2000. A comprehensive analysis of protein–protein interactions in saccharomyces cerevisiae. *Nature* **403**, 623–627.

VALVERDE, S. AND SOLÉ, R. V., 2002. Self-organized critical traffic in parallel computer networks. *Physica A* **312**, 636–648.

VÁZQUEZ, A., 2001. Statistics of citation networks. Preprint cond-mat/0105031.

VÁZQUEZ, A., 2002. Growing networks with local rules: Preferential attachment, clustering hierarchy and degree correlations. *Phys. Rev. E* **67**, 056104.

VÁZQUEZ, A., PASTOR-SATORRAS, R., AND VESPIGNANI, A., 2002. Large-scale topological and dynamical properties of the Internet. *Phys. Rev. E* **65**, 066130.

VÁZQUEZ, A., FLAMMINI, A., MARITAN, A., AND VESPIGNANI, A., 2003. Modeling of protein interaction networks. *Complexus* **1**, 38–44.

VIANA LOPES, J., POGORELOV, Y. G., LOPES DOS SANTOS, J. M. B., AND TORAL, R., 2004. Exact solution of Ising model on a small-world network. *Phys. Rev. E* **70**, 026112.

WAGNER, A., 2001. The yeast protein interaction network evolves rapidly and contains few redundant duplicate genes. *Mol. Biol. Evol.* **18**, 1283–1292.

WAGNER, A. AND FELL, D., 2001. The small world inside large metabolic networks. *Proc. R. Soc. London B* **268**, 1803–1810.

WALSH, T., 1999. Search in a small world. In T. Dean (ed.), *Proceedings of the 16th International Joint Conference on Artificial Intelligence*. Morgan Kaufmann, San Francisco, CA.

WASSERMAN, S. AND FAUST, K., 1994. *Social Network Analysis*. Cambridge University Press, Cambridge.

WATTS, D. J., 1999. *Small Worlds*. Princeton University Press, Princeton.

WATTS, D. J., 2002. A simple model of global cascades on random networks. *Proc. Natl. Acad. Sci. USA* **99**, 5766–5771.

WATTS, D. J., 2003. *Six Degrees: The Science of a Connected Age*. Norton, New York.

WATTS, D. J. AND STROGATZ, S. H., 1998. Collective dynamics of 'small-world' networks. *Nature* **393**, 440–442.

WATTS, D. J., DODDS, P. S., AND NEWMAN, M. E. J., 2002. Identity and search in social networks. *Science* **296**, 1302–1305.

WEST, D. B., 1996. *Introduction to Graph Theory*. Prentice Hall, Upper Saddle River, NJ.

WHITE, H. C., 1970. Search parameters for the small world problem. *Social Forces* **49**, 259–264.

WHITE, J. G., SOUTHGATE, E., THOMPSON, J. N., AND BRENNER, S., 1986. The structure of the nervous system of the nematode C. Elegans. *Phil. Trans. R. Soc. London* **314**, 1–340.

WIGNER, E. P., 1955. Characteristic vectors of bordered matrices with infinite dimensions. *Ann. Math.* **62**, 548–564.

WIGNER, E. P., 1957. Characteristic vectors of bordered matrices with infinite dimensions II. *Ann. Math.* **65**, 203–207.

WIGNER, E. P., 1958. On the distributions of the roots of certain symmetric matrices. *Ann. Math.* **67**, 325–327.

WILF, H., 1994. *Generatingfunctionology*. Academic Press, London, 2nd ed.

WILKINSON, D. M. AND HUBERMAN, B. A., 2004. A method for finding communities of related genes. *Proc. Natl. Acad. Sci. USA* **101**, 5241–5248.

WINFREE, A. T., 1974. Rotating chemical reactions. *Sci. Am.* **230**(6), 82–95.

WINFREE, A. T., 2000. *The Geometry of Biological Time*. Springer, New York, 2nd ed.

WU, F. AND HUBERMAN, B. A., 2004. Finding communities in linear time: A physics approach. *Eur. Phys. J. B* **38**, 331–338.

WUCHTY, S., 2001. Scale-free behavior in protein domain networks. *Mol. Biol. Evol.* **18**, 1694–1702.

WUCHTY, S., 2002. Interaction and domain networks of yeast. *Proteomics* **2**, 1715–1723.

WUCHTY, S., OLTVAI, Z., AND BARABÁSI, A.-L., 2003. Evolutionary conservation of motif constituents within the yeast protein interaction network. *Nature Genetics* **35**, 176–179.

YOOK, S. H., JEONG, H., AND BARABÁSI, A.-L., 2001. Modeling the Internet's large-scale topology. *Proc. Natl. Acad. Sci. USA* **99**, 13382–13386.

ZHABOTINSKY, A. M., 1991. A history of chemical oscillations and waves. *Chaos* **1**, 379–386.

ZHU, J.-Y. AND ZHU, H., 2002. Introducing small-world network effects to critical dynamics. *Phys. Rev. E* **67**, 026125.

Index